CONVERSION FACTORS AND SELECTED CONSTANTS

$\ln x = 2.303 \log_{10} x$
$\ln 10 = 2.303 \ldots$
$e = 2.718 \ldots$
$\pi = 3.1416 \ldots$
1 in. $= 2.54$ cm $= 0.0254$ m
1 lb mass $= 454$ g $= 0.454$ kg
1 lb force $= 4.44$ newtons
1 Å $= 10^{-8}$ cm $= 0.1$ nm
$°C = (°F - 32)/1.8$
$K = °C + 273$
$°R = °F + 460$
1 poise $= 0.1$ Pa-sec
1 lb force/in.^{2} $= 6.9 \times 10^3$ N/m^2 $= 6.9 \times 10^{-3}$ MPa $= 7.03 \times 10^{-4}$ kg/mm^2
1 cal $= 4.186$ J
1 eV $= 1.6 \times 10^{-19}$ J
1 erg $= 1 \times 10^{-7}$ J
R = gas constant $= 1.987$ cal/mole·K
k = Boltzmann's constant $= 1.3 \times 10^{-16}$ erg/K $= 1.3 \times 10^{-23}$ J/K
h = Planck's constant $= 6.62 \times 10^{-27}$ erg-sec $= 6.62 \times 10^{-34}$ J·sec
Avogadro's number $= 6.02 \times 10^{23}$ atoms/at. wt.
$\qquad\qquad\qquad = 6.02 \times 10^{23}$ molecules/mol. wt.
Density of water $= 1$ g/cm^3 $= 62.4$ lb/ft^3 $= 0.0361$ lb/in.^{3}
1 gal water weighs 8.33 lb
Electron charge = electron hole charge $= 1.6 \times 10^{-19}$ coul
1 Bohr magneton $= 9.27 \times 10^{-24}$ amp-m^2 $= 0.927$ erg/gauss
c = velocity of light $= 3 \times 10^{10}$ cm/sec

Prefixes

giga	G	10^9
mega	M	10^6
kilo	k	10^3
milli	m	10^{-3}
micro	μ	10^{-6}
nano	n	10^{-9}

Engineering Materials and Their Applications

FOURTH EDITION

Richard A. Flinn
University of Michigan, Ann Arbor

Paul K. Trojan
University of Michigan, Dearborn

JOHN WILEY & SONS, INC.
New York • Chichester • Brisbane • Toronto • Singapore

Credits for chapter-opening figures follow:

Chapter 1—Kirk Schlea/Allsport Photography   Chapter 2—E. Bischoff and O. Sbaizero, University of California, Santa Barbara   Chapter 3—Wide World Photos   Chapter 4—John Mardinly   Chapter 5—Bob Heitzenrater, Senior Scientist, Corning Inc.   Chapter 6—Courtesy of Pratt and Whitney Aircraft Group   Chapter 7—Leroy F. Grannis   Chapter 8—Richard Flinn   Chapter 9—International Nickel Company   Chapter 10—Corning Glass Works, Corning, NY   Chapter 11—Garrett Auxiliary Power Division, Allied-Signal Aerospace Co.   Chapter 12—© 1988 Marvin Lewiton, shot at Ceramics Process Systems Co.   Chapter 13—Stephen Krause, Department of Mechanical Engineering, Arizona State University at Tempe   Chapter 14—Mad River Canoe Co., Waitsfield, VT   Chapter 15—Ford Motor Company, Dearborn, MI   Chapter 16—M. Ruhle, University of California, Santa Barbara   Chapter 17—R. P. Kingston/Stock Boston   Chapter 18—Richard Flinn   Chapter 19—Wide World Photos   Chapter 20—Gregory Head   Chapter 21—C. C. Wu and B. Roessler   Chapter 22—Courtesy Electrical Power Research Institute   Chapter 23—Courtesy Textron Specialty Materials.

Cover photo by Gabe Palmer, Palmer/Kane Inc./The Stock Market

To Edwina and Barbara

CONTENTS OVERVIEW

CONTENTS

TO THE INSTRUCTOR

Change is rampant among the world's engineering campuses, especially in departments in which Materials Science and Engineering are taught. As new, rewarding prospects develop for engineers with backgrounds in ceramics, plastics, and composites, courses must change to educate engineers in these disciplines as well as in traditional metallurgical topics. It is equally important for other engineers—mechanical, civil, chemical, and electrical—to understand these new topics so that they can apply their professional backgrounds profitably in these new fields.

In writing this Fourth Edition, we have made substantial changes to provide balanced coverage of metals, ceramics, polymers, and composites. The intensive coverage of metallurgical topics in previous editions has actually been strengthened by providing a wider range of examples and concepts from the other materials for comparison.

At first glance the text seems large for a general one-semester course. But let us explain. We toured the country interviewing our colleagues and submitted several widely circulated questionnaires. From the opinions gathered, it appeared that we had reached an impasse. Professor A said "We use your book because of the excellent coverage of concrete and wood." Professor B said "Leave out that material on concrete and add more on modern composites." Professor C asked for emphasis on electronic materials. We found that all three professors were correct: in case A, the students were predominantly Civil Engineers; in Case B, students were Mechanical Engineers interested in aircraft industry positions; and in case C, students were Electrical Engineers.

To accommodate these groups and others, we developed the text as follows:

- We begin with a five-chapter core of basic concepts needed to understand, develop, and use any material. After covering these five chapters, professors are free to take up any combination of other chapters suited to their backgrounds and their students' needs. The table of Possible Choices of Topics for Different Engineering Programs illustrates how the textbook might be used for different areas of interest.
- In the first five core chapters we analyze the different structures of metals, ceramics, and polymers and how stress and temperature affect them. Then we illustrate how to optimize the structure of a material by using equilibrium data (phase diagrams) and nonequilibrium conditions, especially precipitation hardening.
- After this general background, we discuss the structures, properties, and applications of the important materials in each field: metals (Chapters 6, 7, 8, 9); ceramics (Chapters 10, 11, 12); polymers (Chapters 13, 14, 15); and composites (Chapters 16 and 17).

Possible Choices of Topics for Different Engineering Programs
(15-Week Semester)

			Department				
Chapter Topic	*Gen.*	*M. E.*	*Comp. & E. E.*	*C. E.*	*Aero.*	*Ch. E.*	*I. E.*
1 Bonding and Structure	o	o	o	o	o	o	o
2 Mechanical Properties	o	o	o	o	o	o	o
3 Thermal Effects	o	o	o	o	o	o	o
4 Phase Diagrams	o	o	o	o	o	o	o
5 Nonequilibrium Structures	o	o	o	o	o	o	o
6 Metals — Processing		o					o
7 Nonferrous Alloys	o	o		o	o	o	o
8 Ferrous Alloys	o	o	o	o	o	o	o
9 Highly Alloyed Steels and Irons	o			o	o		
10 Traditional Ceramics	o			o	o	o	o
11 Structural Ceramics		o	o		o		
12 Processing Ceramics							o
13 High-Volume Polymers	o	o	o	o	o	o	
14 Special Polymers		o	o			o	
15 Processing Polymers						o	o
16 Synthetic Composites	o	o	o	o	o	o	o
17 Natural Composites	o			o			o
18 Corrosion	o	o	o	o	o	o	
19 Failure Analysis		o		o	o	o	o
20 Electrical and Optical Materials	o		o				
21 Magnetic Materials			o				
22 Processing Electrical Optical Magnetic Materials			o				
23 Selection of Materials	o	o	o	o	o	o	o
	15	15	15	15	15	15	15

- The final chapters are devoted to topics common to all materials: corrosion and oxidation (Chapter 18), failure analysis (Chapter 19), electrical and optical properties (Chapter 20), magnetic properties (Chapter 21), processing of electrical and magnetic materials (Chapter 22), materials selection and specification (Chapter 23).
- To highlight the new materials, special chapters on advanced ceramics, specialized polymers, and synthetic composites are included. For those interested in large volume engineering materials such as aluminum alloys, steel, polymers, cast iron, concrete, and wood, there are detailed chapters on these topics.
- Although most instructors will not use the entire text, we have found that many electrical engineers have entered the construction industry and were

glad to find some "user friendly" sections on conventional materials in the text. Conversely, many traditional mechanical engineers have later needed to study some of the basics in electronic devices or composites.

- A new feature of this fourth edition is the separate discussion of processing in each field so that the instructor can use as much of this material as desired. It is not our intent in this text to give a detailed analysis of processing operations. We want to provide enough background so that the student will understand which processes can be used for a given material and what effects a process may have on the structure of a material. We shudder at the thought, for example, of one of our students specifying rolled .060 in. cast iron sheet for auto fenders!

- A number of requests were made to expand the treatment of aluminum and titanium alloys, and we have complied. Also, to facilitate choice and assignment of the material on ferrous alloys, the material has been divided into two chapters. The material has not been expanded, but revised to reflect new developments. A new concluding chapter on selection of materials and specification has been added.

- As any subject new to most readers, Engineering Materials has its own unique vocabulary. To understand the concepts, one must be able to speak the language. Therefore, generous use of italics in the text and definitions at chapter endings provide the necessary vocabulary. Highlighted page numbers in the index provide easy access to definition location.

- The use of computers is certainly appropriate in Engineering Materials. In several places in the textbook we refer to how computers can be used as support tools, although no programs are included with this textbook. The examples range from direct manipulation of databases to how computers can play an active role in the production of a reliable and saleable engineered component.

- Answers are not provided for the problems at the end of each chapter. However, the answers to the examples that appear every five to ten pages are very complete. The lack of problem answers is not intended to penalize students, but rather to help them develop a proper thought process leading to the answer. The objective is to learn the logic associated with materials selection as it is forever changing. Singular answers suggest a degree of "absoluteness" that is often inappropriate. However, the solutions to the problems are discussed in detail in the Instructors Manual, and it is a simple matter to copy and post these solutions if desired.

TO THE STUDENT

If you have read the preceding section, "To the Instructor," you will understand why this textbook is so extensive and that in general you won't be asked to study the entire textbook in detail.

We do hope, however, that you will read the unassigned sections, because you may wish to refer to the material later. After mastering the first five chapters, you will be able to progress to almost any other section of the book.

Let's take a simple example. Almost every engineer at some time becomes involved in building a home or cabin. Ninety percent of concrete is poured under nonoptimum conditions. Many contractors do not understand that for maximum strength a rather low level of water should be used while mixing the concrete but, after pouring, the concrete should be kept wet to develop the gel structure (Chapters 10 and 17).

We move rather rapidly to different topics, but we use two devices to make the transition easier between chapters. First, chapter-opening photographs with accompanying text provide an immediate example of chapter content. For example, in failure analysis we show the problem of a catastropic brittle fracture of a ship in which the steel was quite ductile at room temperature.

Second, each chapter overview focuses on a few important concepts. At the end of each chapter the discussion of these critical points is summarized.

Now for a word about the problems. In pure science or mathematics courses the usual trick is to find which formula applies and then plug and chug. However, in this course, we have two general types of problems: those with a specific or unique answer and the more professional type of problem that requires qualification of the answer and engineering thinking.

For ease of identification, we have divided the study and example problems into two categories called Engineering Science and Engineering Judgment. Problems are thus designated [ES] or [EJ].

Engineering Science problems are usually completely defined, with problem constraints and data necessary for solution provided. Numerical solutions are often required, and even the solution method may be unique. The objective is to introduce the scientific principles and vocabulary associated with engineering materials.

In Engineering Judgment problems single answers may not be appropriate, as the data may be incomplete or be of questionable reliability or assumptions may be required before a solution can be reached. Logic and application of the principles are necessary to reach an optimal solution.

Both varieties of problems are important to the understanding of engineering materials. The scientific base of engineering is necessary to comprehension of the subject matter and precision in the use of language. However, the judgment base reflects the day-to-day decision making, from an everchanging set of data, that results in creation of a product useful to society. A successful engineer therefore applies sound judgment through the application of scientific principles.

A simple example will show that a mere change in problem wording and emphasis can modify the categorization as an ES or an EJ problem.

ES Wording (True or False): A single ethylene molecule, C_2H_4, will have a higher melting point than $(C_2H_4)_n$, n chemically joined ethylene molecules known as polyethylene.

Answer: False. For a given symmetry of molecule, an increase in molecular weight is accompanied by an increase in the melting point. Polyethylene, with $n > 1$, would have a higher molecular weight, and thus a higher melting point, than a single ethylene molecule.

EJ Wording (True or False): The maximum temperature to which polyethylene can be heated before melting is fixed, and it is consistent from one manufacturer to another.

Answer: False. The number of ethylene molecules tied together is not given. The value of n is variable, as is the molecular weight and hence the melting point. Handbook values indicate the melting range to be 90° to 135°C. Chapters 13 and 14 discuss variables other than n that influence the melting point of polyethylene.

Some problems are marked [ES/EJ], which indicates that a unique solution may be required, but judgment, assumptions, and approximate data are required to arrive at the solution.

Throughout the problems, metric, English, and SI units will be used. Although this may be confusing at first, it is necessary to become proficient in changing from one set of units to another, for we live in an age in which unit standardization has not been accepted. As reference books are not consistent with units, it is important to gain experience presented through the examples and problems. Also, it will become apparent that a unit analysis is an important tool for problem solution.

Finally, the concept of the percentage change, as in a material property or characteristic, will be used in the text. By definition,

$$\% \text{ Change} = \frac{\text{final condition} - \text{initial condition}}{\text{initial condition}} \times 100$$

Therefore, the percentage change can be either a positive or a negative number, depending on the final and initial values. It should also be noted that two of the three variables are necessary to completely understand the phenomenon.

Acknowledgments

First, we would like to thank all who offered criticisms of the previous editions. We have made every effort to satisfy these criticisms. In addition, a group of reviewers were especially helpful in following the development of this edition. Their efforts were so detailed, constructive, and enthusiastic that we feel the resulting textbook is a team effort, for which we are very grateful. These substantial reviews have been received from

Prof. D. Bruce Masson, Washington State University
Prof. Robert N. Pangborn, Pennsylvania State University
Prof. William F. Hosford, University of Michigan
Prof. F. L. Riley, University of Leeds (England)
Prof. Jack L. Tomlinson, California State Polytechnic University
Prof. M. T. Simnad, University of California, San Diego
Prof. H. Thomas McClelland, Montana State University
Prof. Aaron D. Krawitz, University of Missouri
Prof. R. Edward Barker, Jr., University of Virginia

Second, we received data and illustrations that are at the cutting edge of materials science and engineering and point out the fields of research and development effort ranging from electronic circuits to sports equipment. We are happy to acknowledge the contributions of

Dr. Ben P. Winter, Ford Motor Company
Drs. A. W. Urquhart and John Beal, Lanxide Corporation
Mr. Glen Lazar, Griffin Wheel Company
Mr. Alan Petro, Siemens Corporation
Mr. Rob Center, Mad River Canoe Company
Mr. Ron van Oostendorp, Fanatic Corporation
Mr. Robert Heitzenrater, Corning Glass Works
Mr. John Flinn, NEC
Mr. Melvin Mittnick, Textron Corporation
Ms. Tannyjha Goodfellow, Siecor Corporation

The respondents to our questionnaire provided valuable information and insight. We are grateful to those who participated:

Prof. Richard Marriott, University of Maryland
Prof. M. S. Devgun, Iowa State University
Prof. M. F. Berard, Iowa State University
Prof. Dave McDowell, Georgia Institute of Technology
Prof. A. Sutko, Midwestern State University
Prof. George N. Starr, Memphis State University
Prof. Theodore D. Taylor, Clemson University
Prof. Allan Rustad, Olympic College
Prof. John R. Collier, Ohio University
Prof. Roy E. Westcott, Indiana University/Purdue University at Indianapolis
Prof. John W. Shue, Cuyamaca College
Prof. Stephen Fogle, County College of Morris
Prof. William Rosenstein, County College of Morris
Prof. Himanshu Jain, Lehigh University
Prof. H. E. Lindner, Norwich University (Vermont)
Dr. Ali Ogut, Rochester Institute of Technology
Prof. H. B. Kendall, Ohio University
Prof. Aaron Krawitz, University of Missouri, Columbia

Prof. Neil B. Johnson, Tri-State University
Dr. Carl Wissemann, University of Texas, Arlington
Prof. Tseng Huang, University of Texas, Arlington
Prof. John Morral, University of Connecticut
Prof. Charles E. Lyman, Lehigh University
Prof. Robert N. Bruce, Jr., Tulane University
Prof. Richard W. Hanks, Brigham Young University
Prof. Max Deibert, Montana State University
Prof. Trish Vergos, University of California, Davis
Prof. William E. Johns, Washington State University
Prof. Thomas McClelland, Montana State University
Prof. Peter Thrower, Penn State University
Prof. Mahbub Uddin, Trinity University
Prof. Isa Bar-on, Worcester Polytechnic Institute
Prof. David Young, Miami University
Prof. J. A. Dantzig, University of Illinois
Prof. Jack Petersen, Iowa State University
Prof. Patricia Shamamy, Lawrence Technological University
Prof. Robert S. Bray, California State Polytech University
Prof. Carl D. Spear, Utah State University
Prof. Martin J. Siegel, University of Southern California
Prof. Richard D. Schile, University of Bridgeport
Prof. Phil Rogers, Saddleback College
Prof. M. T. Simnad, University of California, San Diego
Prof. Benki Narayan, Ohlone College
Prof. D. Marriott, University of Illinois
Prof. Vec Wan Cho, University of Missouri
Prof. F. Xavier Spiegel, Loyola College
Dr. J. R. Battenburg, California State University, Fresno
Dr. Jerry Bergman, N.W.T. College
Prof. Thomas W. Butler, US Naval Academy and Johns Hopkins University
Prof. Richard A. Hultin, Rochester Institute of Technology
Prof. John E. Ritter, University of Massachusetts
Prof. Suresh B. Patel, Morris County College
Prof. Arnold R. Marder, Lehigh University
Prof. R. Newcomb, Embry Riddle Aeronautical University
Prof. Robert E. Haag, Waterbury State Technical College
Calvin H. Baloun, Ohio University
R. J. Arsenault, University of Maryland
Jeffery C. Gibeling, University of California, Davis
A. A. Hendrickson, Michigan Technological University
Thomas J. Rockett, University of Rhode Island

We are particularly grateful for the micrographs and supporting data for ceramics and ceramic composites provided by Profs. M. Rühle, A. G. Evans, and Fred Lange—University of California, Santa Barbara.

Finally, the great contribution of Ms. Joy Warrick and Mr. Geoffrey Hosker in manuscript preparation is especially appreciated.

Engineering Materials
and Their Applications

PART I

Fundamentals

Before entering the deep woods, a good hiker will study the map to plan the trail through the mountains and bogs ahead. In the same way, before plunging into Chapters 1–5, let us view the outstanding features along the path we will take to investigate engineering materials.

In Chapter 1 we begin with a close look at the 105 atoms that are our elementary building blocks. We then examine the different ways in which these can bond to each other and review all the other concepts from chemistry that we need for later sections of the text. Then, while the bonding concepts are in mind, we go on to the formation of crystals and glassy structures. As an example of the importance of bonding type we consider diamond and graphite. Both are pure carbon, but the covalent bonding in diamond leads to great hardness while the weaker bonding in graphite produces a soft lubricant.

In Chapter 2 we apply stress to these structures and examine on an atomic scale and on an "engineering" scale the movements that occur. These movements range from dislocations in a few groups of atoms to gross fracture. For clarity in this discussion we take up the simple metal structures first, then the ceramics, and finally the polymers.

In Chapter 3 we consider the effects of temperature: expansion and contraction, change in structure, heat conduction, and atomic diffusion. Then we explore the effects of superimposing stress that can lead to recrystallization, creep at high temperatures, and cases of brittle fracture at low temperatures.

In Chapter 4 we add the variable of several phases and discuss their formation and control under equilibrium conditions. Adding a second phase, for example, may increase strength and hardness.

In Chapter 5 we investigate how optimum properties may be obtained by controlling the precipitation of additional phases under nonequilibrium conditions, such as heat treatment and quenching.

After studying these chapters you will have a broad view of Materials Engineering. You will know the structures of the principal materials, how they react to stress and temperature, and how polyphase structures form and are controlled to produce optimum properties. You can then advance quickly to examine the structure and properties of available metals, ceramics, polymers, and composites. Next you can explore the effects of other environments, such as corrosion, different modes of failure, electrical, optical, and magnetic properties and their relation to structure, and the steps needed to select a material for a given engineering component.

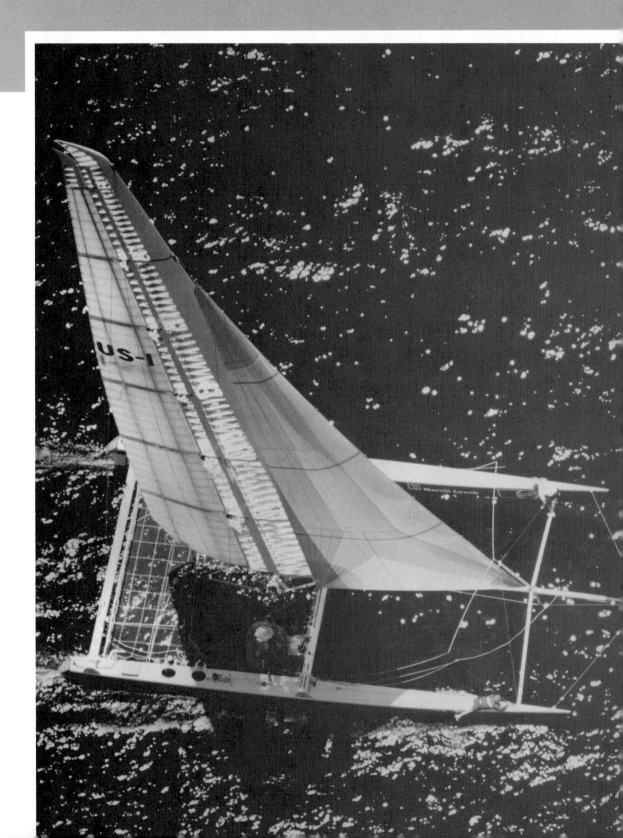

CHAPTER 1

The Structure of Engineering Materials

Whether we analyze the materials in a boat such as the *Stars and Stripes*, an automobile or a computer chip, we find that there are only three classes of materials: metals, ceramics and polymers. Each class has distinct characteristics.

In a yacht, for example, the highly stressed fittings subject to wear and corrosion are metal (stainless steel); the sails and lines are polymers; and the hull is a ceramic polymer composite.

In this chapter we will investigate the differences in bonding among atoms in metals, ceramics and polymers and how these lead to different crystalline and noncrystalline structures with widely different properties.

1.1 Overview

Unlike many introductory chapters, gentle, historical descriptions, this chapter is probably the most important in the whole book for you to understand thoroughly. After mastering these concepts, you will be able to sit in the same spot you are now, look around at the same materials and objects, and see beneath their surface. You will have a new depth of understanding of the atoms of which metals, ceramics, and plastics are composed and of how these atoms are bonded to form the *structures* that determine all their properties.

Every engineer is aware that the chemical analysis of a material is important because it tells us *which* atoms are present. On the other hand, very few engineers understand thoroughly the importance of how these atoms are *arranged* and *bonded* to form the structure.

To illustrate this, let us run a simple experiment. We take two identical steel wire coat hangers and use one in the standard condition as it comes from the factory (Figure 1.1*a*). We take the second and heat it to redness (about 900°C) for several minutes and allow it to cool. Then we use it, with the disastrous results shown in Figure 1.1(*b*). What has happened? As a first step in finding out, we cut, polish, and etch a small sample of each hanger to look at under the microscope. We see that the two differ considerably in *structure* and hardness (Figure 1.1*c* and *d*).

The important lesson to be learned from this simple experiment is that it is not enough for us to know which atoms are present (the chemical composition of the hanger we heated did not change). We must also find out how these atoms are arranged. At this point you can see from the photos taken through the microscope (photomicrographs) that there is a difference between the two structures. After studying the first three chapters of this book, you will be able to describe, in terms of the motions and positions of the iron atoms, the processing (heat treatment and cold working) that produced these differences and why they lead to vastly different properties. This basic knowledge will equip you to specify the processing needed to attain the combination of properties you require in a material for a given application. And should failures occur, you will be able to understand why.

In this chapter you will take the first important step: learning the structures encountered in the three basic groups of engineering materials — metals, ceramics, and polymers. (The composites and special electrical materials will be discussed later.) We have organized this material under the following four topics:

1. *Atomic structure.* All materials are made from the 109 known elements.*
   We need to know the differences and similarities among these elements.

---

*We do not limit ourselves to the naturally occurring elements because the artificially made ones are also valuable. Technetium is being made in kilogram quantities for medical use, for example. We will not discuss some recently discovered elements because of their short half-lives; only 105 elements will be considered.

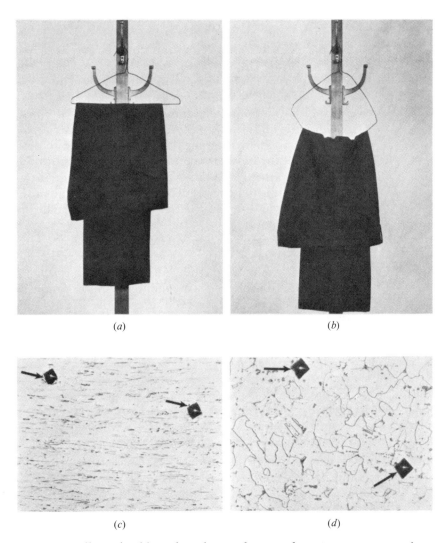

(a)   (b)

(c)   (d)

FIGURE 1.1   Effect of cold work and annealing on the microstructure and service performance of a clothes hanger. (a) Cold-worked hanger, normal form. Grains are elongated in the direction in which the wire was drawn through a die. (b) Annealed clothes hanger. The larger equiaxed grains make the hanger soft and weak. (c) Normal hanger, Vickers hardness number (HV) 173 to 183; 500×, 2% nital etch. (d) Hanger annealed at 1700°F (910°C) $\frac{1}{2}$hr, HV 108–118; 500×, 2% nital etch.

Grain structures shown are from samples that were longitudinally cut, then polished and etched with 2% nitric acid in alcohol (nital). The "normal" hanger was worked-hardened by drawing through a die. When it was heated, new, softer grains were formed. The HV was obtained by pressing a light (25-g) load into individual grains with a pyramid-shaped diamond indenter (arrows).

For our purposes, the most important characteristics of any element are the number of orbital electrons and how they are arranged around the nucleus, because this arrangement determines the nature and strength of bonding between atoms.

2. *Bonding.* Here we study how the number and arrangement of the electrons determine the type of primary bonding (ionic, covalent, metallic) that occurs in pure elements or in combinations of elements. We also find that the secondary bonding forces (van der Waals, hydrogen bonding) can be related to the orbital electrons.

3. *Crystalline/Noncrystalline Structures.* Now that we have an understanding of bonding in small groups of atoms, we need to determine how these are organized into *larger structures*. These structures fall into one of two groups: crystalline (this does not necessarily mean with shiny crystal faces, but exhibiting the regular positioning of the atoms on a space lattice) or noncrystalline (an example is the glassy condition found widely in ceramics and polymers). More important, we can control the degree of crystallinity in many engineering materials.

4. *Structure Summary.* With this background, we will examine the *structure* of important examples of the *three groups of basic materials* and predict some of the differences in mechanical, thermal, optical, and electrical properties we can expect to encounter in the chapters ahead.

ATOMIC STRUCTURE

1.2 Electronic Structure of the Atom and the Periodic Table

Let us start with a fresh, atomic-scale approach to engineering materials by examining the 105 elements that are the basic building blocks of all substances. Every atom of all the elements is made up of a positively charged nucleus that is balanced in charge by the negatively charged electrons outside the nucleus. The number of electrons around the nucleus of the neutral atom determines the *atomic number* from 1 to 105. Thus hydrogen (at. no. 1) has 1 electron, iron (at. no. 26) has 26 electrons, and so on.

Now instead of dealing with 105 unrelated building blocks, we can avail ourselves of many useful similarities and relations by referring to the *periodic table* (Table 1.1). The elements are simply written down in horizontal rows in order of atomic number; a new row is begun after each *noble* gas (helium, neon, argon, krypton, xenon, and radon) is encountered.

From this table we find first of all that the elements in each *vertical column* have similar properties. Such groups include the noble gases; the very active metals lithium, sodium and potassium; and the noble metals copper,

TABLE 1.1 Periodic Table of the Elements*

Main-Group Elements

Transition Metals

Inner-Transition Metals

Key:

1	
H	Atomic number / Symbol
1.00794	Atomic weight

Period	IA	IIA	IIIB	IVB	VB	VIB	VIIB	VIIIB			IB	IIB	IIIA	IVA	VA	VIA	VIIA	VIIIA
1	1 **H** 1.00794																	2 **He** 4.00260
2	3 **Li** 6.941	4 **Be** 9.01218											5 **B** 10.81	6 **C** 12.011	7 **N** 14.0067	8 **O** 15.9994	9 **F** 18.998403	10 **Ne** 20.179
3	11 **Na** 22.98977	12 **Mg** 24.305											13 **Al** 26.98154	14 **Si** 28.0855	15 **P** 30.97376	16 **S** 32.06	17 **Cl** 35.453	18 **Ar** 39.948
4	19 **K** 39.0983	20 **Ca** 40.08	21 **Sc** 44.9559	22 **Ti** 47.88	23 **V** 50.9415	24 **Cr** 51.996	25 **Mn** 54.9380	26 **Fe** 55.847	27 **Co** 58.9332	28 **Ni** 58.69	29 **Cu** 63.546	30 **Zn** 65.38	31 **Ga** 69.72	32 **Ge** 72.59	33 **As** 74.9216	34 **Se** 78.96	35 **Br** 79.904	36 **Kr** 83.80
5	37 **Rb** 85.4678	38 **Sr** 87.62	39 **Y** 88.9059	40 **Zr** 91.22	41 **Nb** 92.9064	42 **Mo** 95.94	43 **Tc** (98)	44 **Ru** 101.07	45 **Rh** 102.9055	46 **Pd** 106.42	47 **Ag** 107.8682	48 **Cd** 112.41	49 **In** 114.82	50 **Sn** 118.69	51 **Sb** 121.75	52 **Te** 127.60	53 **I** 126.9045	54 **Xe** 131.29
6	55 **Cs** 132.9054	56 **Ba** 137.33	57 **La*** 138.9055	72 **Hf** 178.49	73 **Ta** 180.9479	74 **W** 183.85	75 **Re** 186.207	76 **Os** 190.2	77 **Ir** 192.22	78 **Pt** 195.08	79 **Au** 196.9665	80 **Hg** 200.59	81 **Tl** 204.383	82 **Pb** 207.2	83 **Bi** 208.9804	84 **Po** (209)	85 **At** (210)	86 **Rn** (222)
7	87 **Fr** (223)	88 **Ra** 226.0254	89 **Ac*** 227.0278	104 † (261)	105 † (262)	106 † (263)	107 †	108 †	109 †									

* Lanthanides

57 **La** 138.9055	58 **Ce** 140.12	59 **Pr** 140.9077	60 **Nd** 144.24	61 **Pm** (145)	62 **Sm** 150.36	63 **Eu** 151.96	64 **Gd** 157.25	65 **Tb** 158.9254	66 **Dy** 162.50	67 **Ho** 164.9304	68 **Er** 167.26	69 **Tm** 168.9342	70 **Yb** 173.04	71 **Lu** 174.967

** Actinides

89 **Ac** 227.0278	90 **Th** 232.0381	91 **Pa** 231.0359	92 **U** 238.0289	93 **Np** 237.0482	94 **Pu** (244)	95 **Am** (243)	96 **Cm** (247)	97 **Bk** (247)	98 **Cf** (251)	99 **Es** (252)	100 **Fm** (257)	101 **Md** (258)	102 **No** (259)	103 **Lr** (260)

Legend:
- Metal
- Metalloid
- Nonmetal

†Element synthesized, but no official name assigned

Note: Additional elements with short half-lives have been synthesized.

*The upper atomic number in each box also gives us the number of orbital electrons. The atomic weight is based on the mass of the atom relative to the nitrogen isotope of 14.00. The atomic weight values are used extensively in calculations.

[Reprinted with permission from *CRC Handbook of Chemistry and Physics*. Copyright CRC Press, Inc., Boca Raton, FL]

silver and gold. We will examine the reasons for these periodic relations in a moment, but first let us consider a few engineering applications.

During a national emergency in World War II, tungsten, which came principally from China, became scarce. The production of 18% tungsten tool steel, which was vital for many machine tools, was threatened. Because molybdenum is found directly above tungsten in the periodic table, a number of steels in which molybdenum was substituted for tungsten were made and tested. It was soon established that an 8% Mo, 2% W steel is equivalent to the 18% W steel, and the new formula is still used today. Another shortage that developed had an amusing personal aftermath. Tellurium became scarce, and we needed a substitute element for controlling the structure of iron castings. We were glad to find that an equivalent amount of selenium (just above Te) was a satisfactory substitute, but were astonished to be informed by our friends that we were coming home with the same garlic breath that our previous exposure to tellurium produced!

Another striking feature of the periodic table is that the elements on the left-hand side of the diagonal dividing band (see Table 1.1) are metals, whereas those on the right-hand side are nonmetals. The elements within the band are metalloids or semi-metals. It is to our great advantage to understand the reasons for these phenomena. To this end, we need to review a few basic applications of quantum theory to atomic structure. Although these quantum relations may seem arbitrary and even a little fanciful at first, abundant experimental proof exists and the results are quite useful.

1.3 Quantum Numbers

The first point is that the 1 to 105 electrons revolving about the nucleus do not have a random assortment of energies but occupy only definite levels. Accordingly, the early concepts of atomic structure had the electrons rotating at a fixed distance from the nucleus in shells, (Figure 1.2). These shells were labeled *K, L, M, N*, and so on, starting with the innermost shell at the lowest energy level. This model fitted a number of experimental facts. For example, every time the outer shell contained eight electrons, as in the case of neon, this was found to be a *stable* configuration and the element was unreactive. On the other hand, the intense reaction of sodium and fluorine to form sodium fluoride was explained by the strong tendency of fluorine (with an outer shell containing seven electrons) to attract an electron with which to form a stable group of eight. At the same time, sodium readily sheds its single outer electron to expose a group of eight beneath (Figure 1.3). As a result of this electron transfer, we have Na^+ and F^- *ions*, which attract each other and form a sodium fluoride crystal.

This simple model was useful, but it did not explain major differences such as the properties of diamond and graphite, which are made up of atoms of the same element (carbon). A subtler but important additional fact was that many spectroscopic data could not be explained adequately. Remember

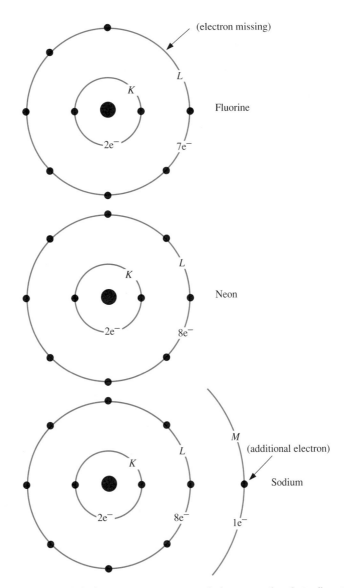

FIGURE 1.2   Simple model of atomic structure and electron orbitals in fluorine, neon, and sodium

that the spectral lines (such as the yellow line of sodium) are due to energy emitted as photons of light when an electron falls from an outer shell to an inner one. These spectroscopic data showed, for example, that the spectra produced by electrons falling from the same outer shell to the same inner shell were different. Therefore, although we can retain the general concepts of groups in a shell, we need to define the energy levels more distinctly.

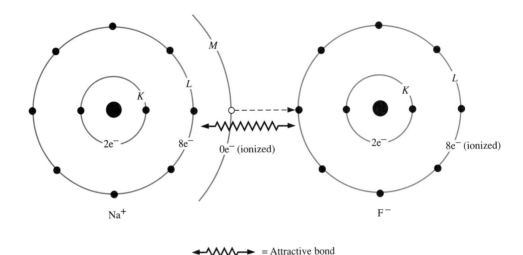

FIGURE 1.3   Formation of sodium fluoride to produce an ionic bond

An improved model was developed by applying quantum theory, and we will use the results without derivation. Instead of the simple model showing two similar electrons in the first shell, eight in the second, and so on, we assign to each electron a unique set of four *quantum numbers*. These are calculated via a simple set of rules. We shall see that, once assigned to the orbital electrons, these numbers have great significance in the interpretation and prediction of atomic behavior.

The first quantum number, n, has values of a positive integer starting with 1. Thus the value for the single electron of hydrogen is $n = 1$. As we follow the rules for assignment of other quantum numbers, we find that we can have only two electrons with $n = 1$ and eight electrons with $n = 2$. These and certain other restrictions applied at higher values of n lead to the general rule that the maximum number of electrons of a given shell is $2n^2$.

Because n, the principal quantum number, is by far the most significant in fixing the energy of an electron, it is customary to use the old concept of shells of electrons, at different energy levels, to match the values of n (see Figure 1.2). Thus the first shell, which has the lowest energy and $n = 1$, is called the K shell. The other shells and corresponding quantum numbers are as follows:

$$n = 1 \quad 2 \quad 3 \quad 4 \quad 5 \quad 6$$

$$\text{shell} \quad K \quad L \quad M \quad N \quad O \quad P$$

Although this concept of electrons revolving at a fixed distance from the nucleus to form a shell is useful for cataloging, not only are there variations in *distance* from the nucleus in a given shell, but some electrons also show considerable *deviation from a spherical orbit*. The significance of the difference in *shape* and *orientation* of these orbits is that different types of bonding are associated with them.

The second quantum number, *l*, indicates the value of the angular momentum. Just as *n* has quantized values, *l* is also restricted. The regions of high probability for finding an electron with a given value of *l* can be calculated (Figure 1.4). Because we are dealing with an extremely small quantity of matter and charge, we cannot fix its location and velocity exactly. However, we can calculate the shape of an *envelope* in space in which there is 99% probability of finding a given electron. It turns out that having this knowledge is even more helpful in understanding bonding than being able to specify the exact position of the electron.

The value of *l* must be a positive integer not greater than *n* − 1. However, in this case 0 is considered a positive integer. It is a little confusing that in scientific discussion, instead of referring to these different numbers, scientists prefer to use lowercase letters, which are assigned as follows:

$$l = 0 \quad 1 \quad 2 \quad 3 \quad 4 \quad 5$$

$$\text{letter}^* = s \quad p \quad d \quad f \quad g \quad h$$

Thus, for example, if *n* = 2, *l* can have values of 0 or +1 and we call these 2*s* and 2*p electrons*, respectively.

We can use this designation in shorthand form to catalog the electrons of an element by giving the first two quantum numbers. For example, for boron (at. no. 5), we have two 1*s* electrons, two 2*s* electrons, and one 2*p* electron. This is written B (at. no. 5): $1s^2, 2s^2, 2p^1$.

We mentioned earlier that the second quantum number, *l* specifies the shape of the envelope in which the electron is likely to be found. In Figure 1.4 we show the calculated (from quantum theory) envelopes for *s*, *p*, and *d* electrons. In the case of the *s* electrons, the envelope is spherical; for *p* electrons, it is dumbbell-shaped; and for *d* electrons, it is clover-shaped in four cases and dumbbell-shaped in one.

The third quantum number, m_l, indicates that in addition to angular momentum, there is a quantized orbital *magnetic* moment that is very important in understanding magnetic properties of the elements.

To obtain an axis of reference, we apply a weak magnetic field and the atom, because of its electrical charges, aligns itself relative to the field. Then by definition we designate the *z* axis of the atom as parallel to the field. The *x* and *y* axes are at right angles to the *z* axis and to each other but are not

---

*The sequence of letters *s*, *p*, *d f* is not capricious but arises from older spectrographic language wherein the lines produced by these electrons were called "sharp," "principal," "diffuse," and "fundamental."

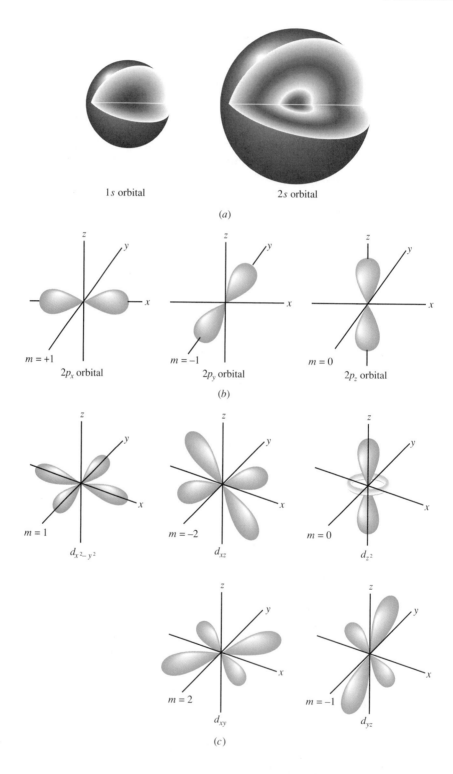

1*s* orbital

2*s* orbital

(*a*)

$m = +1$
2p_x orbital

$m = -1$
2p_y orbital

$m = 0$
2p_z orbital

(*b*)

$m = 1$
$d_{x^2-y^2}$

$m = -2$
d_{xz}

$m = 0$
d_{z^2}

$m = 2$
d_{xy}

$m = -1$
d_{yz}

(*c*)

FIGURE 1.4 (*a*) Cutaway diagrams showing the spherical shape of *s* orbitals. In both diagrams, the upper part of each orbital is cut away to reveal the electron distribution of the orbital. (*b*) The three 2*p* orbitals. Each orbital consists of two lobes, with the lobes oriented along a given axis. The $2p_x$ orbital, for example, has its lobes along the *x* axis. (*c*) The five 3*d* orbitals. These are labeled by subscripts, as in d_{xy}, that describe their mathematical characteristics. *Note:* It is important to begin with these pictures of the orbits in the free atom because in later discussions, the combination of orbits (hybridization) is very significant.

[D. Ebbing, *General Chemistry*, 2nd ed., pp. 206–207. Copyright © 1987 Houghton Mifflin Company. Reprinted by permission.]

located more specifically. The integral values of m_l must be equal to or less than *l* but may be positive or negative.

$$\text{Thus for } n = 2 \quad \text{and} \quad l = 1, 0, \qquad m_l = 1, 0, -1$$
$$n = 3 \quad \text{and} \quad l = 2, 1, 0, \qquad m_l = 2, 1, 0, -1, -2$$

EXAMPLE 1.1 *[ES]*

According to the Pauli exclusion principle, no more than two electrons can occupy the same orbit, and these must have opposite spins. Do the *p* electrons in neon meet this criterion?

Answer The structure of neon is $1s^2, 2s^2, 2p^6$. There are three pairs of 2*p* electrons with $m = +1, -1,$ and 0, and the envelope of each pair has a different orientation, as shown in Figure 1.4(*b*).

The fourth quantum number, m_s, electron spin, represents a real triumph of theory. It was predicted that the electron, as a spinning charge, could spin in one of two directions. If we took a stream of hydrogen atoms we would expect that the single electron in each might be spinning in one of two directions and that the atoms would thus be deflected differently by a magnetic field. Experiments confirm that the atoms are indeed affected in two ways, so we have two values of a fourth quantum number m_s. By convention, these are labeled $+1/2$ and $-1/2$.

We should discuss the assignment of m_s a little further. This is best displayed by using an *orbital diagram* and taking iron—Fe (at. no. 26)—as an example. The first two shells are the same as for sodium, as we have noted, so let us focus on the values of m_s for the 3 and 4 quantum shells: $[3s^2 3p^6 3d^6][4s^2]$.

$3s^2$	$3p^6$	$3d^6$	$4s^2$
(↑↓)	(↑↓)(↑↓)(↑↓)	(↑↓)(↑)(↑)(↑)(↑)	(↑↓)

The up arrows are used as symbols of $m_s = +1/2$, and the down arrows represent $m_s = -1/2$. In the assignment of m_s, all the $+1/2$ levels in a given shell are filled before the $-1/2$ shell levels start to fill (Hund's rule). We will use these orbital diagrams in discussing bonding and also in explaining electrical and magnetic properties. The $4s$ shell is filled before the $3d$ shell, so both $+$ and $-$ spins are present. The presence of the four unbalanced values of m_s in the $3d$ shell leads to ferromagnetism, as discussed in Chapter 21.

Now let us put these relations to work in formulating atomic structures. Note, in the following example, that we can use an abbreviated formula giving only the first two quantum numbers. Thus for the lithium atom (at. no. 3), we have $1s^2, 2s^1$ which is read two $1s$ electrons (not $1s$ squared!), one $2s$ electron.

EXAMPLE 1.2 *[ES]*

Assign all four quantum numbers to the electrons for elements of at. no. 9, 10, 11, and give the abbreviated formulas.

Answer We begin with at. no. 9 and then add the electrons for at. no. 10 and at. no. 11.

Electron	$n =$	$l =$	$m_l =$	$m_s =$	*Abbreviated Designation*
1	1	0	0	$+1/2$	$1s^2$
2	1	0	0	$-1/2$	
3	2	0	0	$+1/2$	$2s^2$
4	2	0	0	$-1/2$	
5	2	1	-1	$+1/2$	
6	2	1	0	$+1/2$	
7	2	1	$+1$	$+1/2$	$2p^5$
8	2	1	-1	$-1/2$	
9	2	1	0	$-1/2$	
10	2	1	$+1$	$-1/2$	$2p^6$
11	3	0	0	$+1/2$	$3s^1$

Note that in assigning m_s at $l = 1$, we use the full number of possible $+$ spins first and then the $-$ spins. Also by convention, m_l at minus values precedes zero and positive values.

1.4 Electron Configurations of Common Elements

Let us look at the abbreviated designations for some common materials.

Atomic Number	Element	Electron Configuration
12	magnesium	$1s^2, 2s^2, 2p^6, 3s^2$
13	aluminum	$1s^2, 2s^2, 2p^6, 3s^2, 3p^1$
26	iron	$1s^2, 2s^2, 2p^6, 3s^2, 3p^6, 3d^6, 4s^2$
29	copper	$1s^2, 2s^2, 2p^6, 3s^2, 3p^6, 3d^{10}, 4s^1$

(As a review, check to be sure the sum of the electrons in each case equals the atomic number.)

We note that in forming the structure of iron, we have filled the $4s$ level before the $3d$ level, which can eventually contain ten electrons. The other transition elements are shown in the complete table of elements, Table 1.2. An exception is copper, wherein the tendency to form a complete d shell of ten is stronger than the tendency to have two $4s$ electrons. Copper is still a transition element. An analogous effect is seen for the elements with atomic numbers 40–57, where the $5s$ quantum shell builds up before the $4f$.

An important point is that the $3s^2$ electrons of Mg and the $3s^2$, $3p^1$ electrons of Al have an $n = 2$ shell of eight beneath and that, therefore, Mg and Al are very active. In the case of iron and copper, the $3d$ and $4s$ electrons are close in energy and the elements are less reactive.

TABLE 1.2   Electron Configurations of the Elements*

Atomic Number	Element	K	L		M			N				O				P				Q
		1	2		3			4				5				6				7
		s	s	p	s	p	d	s	p	d	f	s	p	d	f	s	p	d	f	s
1	H	1																		
2	He	2																		
3	Li	2	1																	
4	Be	2	2																	
5	B	2	2	1																
6	C	2	2	2																
7	N	2	2	3																
8	O	2	2	4																
9	F	2	2	5																
10	Ne	2	2	6																

TABLE 1.2   *(Continued)*

		K	L		M			N				O				P				Q
		1	2		3			4				5				6				7
Atomic Number	Ele-ment	s	s	p	s	p	d	s	p	d	f	s	p	d	f	s	p	d	f	s
11	Na	2	2	6	1															
12	Mg	2	2	6	2															
13	Al	2	2	6	2	1														
14	Si	2	2	6	2	2														
15	P	2	2	6	2	3														
16	S	2	2	6	2	4														
17	Cl	2	2	6	2	5														
18	Ar	2	2	6	2	6														
19	K	2	2	6	2	6		1												
20	Ca	2	2	6	2	6		2												
21	Sc	2	2	6	2	6	1	2												
22	Ti	2	2	6	2	6	2	2												
23	V	2	2	6	2	6	3	2												
24	Cr	2	2	6	2	6	$5^\dagger$	1												
25	Mn	2	2	6	2	6	5	2												
26	Fe	2	2	6	2	6	6	2												
27	Co	2	2	6	2	6	7	2												
28	Ni	2	2	6	2	6	8	2												
29	Cu	2	2	6	2	6	$10^\dagger$	1												
30	Zn	2	2	6	2	6	10	2												
31	Ga	2	2	6	2	6	10	2	1											
32	Ge	2	2	6	2	6	10	2	2											
33	As	2	2	6	2	6	10	2	3											
34	Se	2	2	6	2	6	10	2	4											
35	Br	2	2	6	2	6	10	2	5											
36	Kr	2	2	6	2	6	10	2	6											
37	Rb	2	2	6	2	6	10	2	6	··		1								
38	Sr	2	2	6	2	6	10	2	6	··		2								
39	Y	2	2	6	2	6	10	2	6	1		2								
40	Zr	2	2	6	2	6	10	2	6	2	··	2								
41	Nb	2	2	6	2	6	10	2	6	$4^\dagger$	··	1								
42	Mo	2	2	6	2	6	10	2	6	5	··	1								
43	Tc	2	2	6	2	6	10	2	6	6	··	1								
44	Ru	2	2	6	2	6	10	2	6	7	··	1								
45	Rh	2	2	6	2	6	10	2	6	8	··	1								
46	Pd	2	2	6	2	6	10	2	6	$10^\dagger$	··	··								
47	Ag	2	2	6	2	6	10	2	6	10	··	1								
48	Cd	2	2	6	2	6	10	2	6	10	··	2								
49	In	2	2	6	2	6	10	2	6	10	··	2	1							
50	Sn	2	2	6	2	6	10	2	6	10	··	2	2							
51	Sb	2	2	6	2	6	10	2	6	10	··	2	3							
52	Te	2	2	6	2	6	10	2	6	10	··	2	4							
53	I	2	2	6	2	6	10	2	6	10	··	2	5							

TABLE 1.2   *(Continued)*

		K	L		M			N				O				P				Q
		1	2		3			4				5				6				7
Atomic Number	Element	s	s	p	s	p	d	s	p	d	f	s	p	d	f	s	p	d	f	s
54	Xe	2	2	6	2	6	10	2	6	10		2	6							
55	Cs	2	2	6	2	6	10	2	6	10	··	2	6	··	··	1				
56	Ba	2	2	6	2	6	10	2	6	10	··	2	6	··	··	2				
57	La	2	2	6	2	6	10	2	6	10	··	2	6	1	··	2				
58	Ce	2	2	6	2	6	10	2	6	10	2^+	2	6	··	··	2				
59	Pr	2	2	6	2	6	10	2	6	10	3	2	6	··	··	2				
60	Nd	2	2	6	2	6	10	2	6	10	4	2	6	··	··	2				
61	Pm	2	2	6	2	6	10	2	6	10	5	2	6	··	··	2				
62	Sm	2	2	6	2	6	10	2	6	10	6	2	6	··	··	2				
63	Eu	2	2	6	2	6	10	2	6	10	7	2	6	··	··	2				
64	Gd	2	2	6	2	6	10	2	6	10	7	2	6	1	··	2				
65	Tb	2	2	6	2	6	10	2	6	10	9^+	2	6	··	··	2				
66	Dy	2	2	6	2	6	10	2	6	10	10	2	6	··	··	2				
67	Ho	2	2	6	2	6	10	2	6	10	11	2	6	··	··	2				
68	Er	2	2	6	2	6	10	2	6	10	12	2	6	··	··	2				
69	Tm	2	2	6	2	6	10	2	6	10	13	2	6	··	··	2				
70	Yb	2	2	6	2	6	10	2	6	10	14	2	6	··	··	2				
71	Lu	2	2	6	2	6	10	2	6	10	14	2	6	1	··	2				
72	Hf	2	2	6	2	6	10	2	6	10	14	2	6	2	··	2				
73	Ta	2	2	6	2	6	10	2	6	10	14	2	6	3	··	2				
74	W	2	2	6	2	6	10	2	6	10	14	2	6	4	··	2				
75	Re	2	2	6	2	6	10	2	6	10	14	2	6	5	··	2				
76	Os	2	2	6	2	6	10	2	6	10	14	2	6	6	··	2				
77	Ir	2	2	6	2	6	10	2	6	10	14	2	6	9^+	··	0				
78	Pt	2	2	6	2	6	10	2	6	10	14	2	6	9	··	1				
79	Au	2	2	6	2	6	10	2	6	10	14	2	6	10	··	1				
80	Hg	2	2	6	2	6	10	2	6	10	14	2	6	10	··	2				
81	Tl	2	2	6	2	6	10	2	6	10	14	2	6	10	··	2	1			
82	Pb	2	2	6	2	6	10	2	6	10	14	2	6	10	··	2	2			
83	Bi	2	2	6	2	6	10	2	6	10	14	2	6	10	··	2	3			
84	Po	2	2	6	2	6	10	2	6	10	14	2	6	10	··	2	4			
85	At	2	2	6	2	6	10	2	6	10	14	2	6	10	··	2	5			
86	Rn	2	2	6	2	6	10	2	6	10	14	2	6	10	··	2	6			
87	Fr	2	2	6	2	6	10	2	6	10	14	2	6	10	··	2	6	··	··	1
88	Ra	2	2	6	2	6	10	2	6	10	14	2	6	10	··	2	6	··	··	2
89	Ac	2	2	6	2	6	10	2	6	10	14	2	6	10	··	2	6	1	··	2
90	Th	2	2	6	2	6	10	2	6	10	14	2	6	10	··	2	6	2	··	2
91	Pa	2	2	6	2	6	10	2	6	10	14	2	6	10	2^+	2	6	1	··	2
92	U	2	2	6	2	6	10	2	6	10	14	2	6	10	3	2	6	1	··	2
93	Np	2	2	6	2	6	10	2	6	10	14	2	6	10	4	2	6	1	··	2
94	Pu	2	2	6	2	6	10	2	6	10	14	2	6	10	6	2	6	··	··	2
95	Am	2	2	6	2	6	10	2	6	10	14	2	6	10	7	2	6	··	··	2

TABLE 1.2   *(Continued)*

Atomic Number	Element	K	L		M			N				O				P				Q
		1	2		3			4				5				6				7
		s	s	p	s	p	d	s	p	d	f	s	p	d	f	s	p	d	f	s
96	Cm	2	2	6	2	6	10	2	6	10	14	2	6	10	7	2	6	1	··	2
97	Bk	2	2	6	2	6	10	2	6	10	14	2	6	10	9	2	6	··	··	2
98	Cf	2	2	6	2	6	10	2	6	10	14	2	6	10	10	2	6	··	··	2
99	Es	2	2	6	2	6	10	2	6	10	14	2	6	10	11	2	6	··	··	2
100	Fm	2	2	6	2	6	10	2	6	10	14	2	6	10	12	2	6	··	··	2
101	Md	2	2	6	2	6	10	2	6	10	14	2	6	10	13	2	6	··	··	2
102	No	2	2	6	2	6	10	2	6	10	14	2	6	10	14	2	6	··	··	2
103	Lr	2	2	6	2	6	10	2	6	10	14	2	6	10	14	2	6	1	··	2
104	Rf	2	2	6	2	6	10	2	6	10	14	2	6	10	14	2	6	2	··	2
105	Ha	2	2	6	2	6	10	2	6	10	14	2	6	10	14	2	6	3	··	2

*Devised by Laurence S. Foster.
†Note irregularity.
References: Therald Moeller, *Inorganic Chemistry*, John Wiley & Sons, New York, 1952, pp. 98–101. Joseph J. Katz and Glenn T. Seaborg, *The Chemistry of the Actinide Elements*, Methuen & Co., Ltd., London, 1957; John Wiley & Sons, New York, 1957, p. 464.

BONDING

1.5 Types of Bonding

We have gone into detail in studying the atomic structures because they are the key to understanding bonding. In turn, the properties and applications of materials depend on the bond strengths.

You should understand at the outset that there are two radically different types of bonds. In *primary bonds* there are strong atom-to-atom attractions produced by changes in electron position of the outer (valence) electrons. These changes can range from sharing to the almost complete transfer of the outer-shell electron(s) from the field of one atom to that of the other. In *secondary bonds*, which are much weaker than primary bonds, atoms or molecules are attracted by *overall electric fields*, which often result from the transfer of electrons in the primary bonding.

We show the whole picture in Figure 1.5 (page 19). In the *ionic bond*, electrons are *transferred* from the metal to the nonmetal, creating ions that attract each other throughout the mass. In the *covalent bond*, electrons are *shared* between atoms to produce a stable group of eight. In the *metallic bond*, the electrons are *delocalized*, or given up to form a common sea of electrons surrounding the positive ion cores.

By contrast, in secondary bonding we see that in the van der Waals forces known as permanent dipole or dipole–dipole forces, polar attraction occurs when dipolar molecules are formed first by primary bonding, which results in

Primary bonds

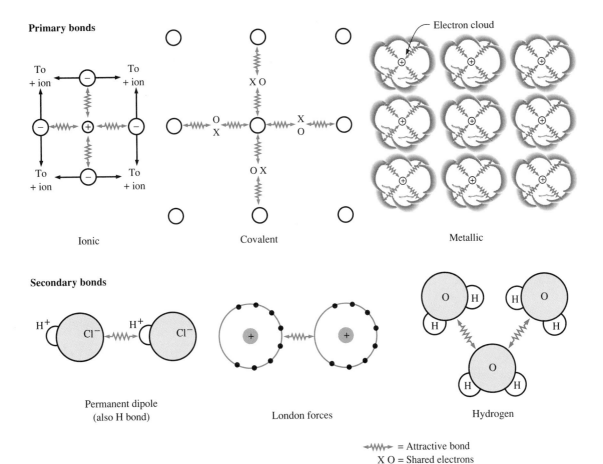

| Ionic | Covalent | Metallic |

Secondary bonds

| Permanent dipole (also H bond) | London forces | Hydrogen |

-WWW- = Attractive bond
X O = Shared electrons

FIGURE 1.5   Summary of primary and secondary bonds in materials

a molecule with a permanent charge difference between its ends. Even when a dipole is absent, *London forces* caused by a temporary charge difference cause attraction. Finally, a hydrogen bond is associated with the residual positive field near a hydrogen atom.

Let us now go into detail about the individual types of bonds.

1.6 Ionic Bonding

In the ionic bond, one element sheds or gives up its outer-shell electron(s) to uncover a stable inner shell of eight. The electron(s) are attracted to a second element in which they can serve to complete its outer shell of eight. Sodium, for instance, gives up its $3s^1$ electron, leaving an L shell of eight; and chlorine, with an outer shell of seven $(3s^2 3p^5)$, attracts that electron to form an outer shell of eight. In this way two ions, Na^+ and Cl^-, are formed.

These ions are attracted by an attractive force that varies inversely as the square of the distance between ions. However, when the ions are close to one another, a repulsive force develops between the electron fields. From these observations, a graph of potential energy versus interatomic distance can be derived (Figure 1.6). The minimum in the curve corresponds to the distance at which the attractive force equals the repulsive force.

The balance between these two forces is arrived at as follows. The attraction between charges is the well-known coulombic equation:

$$F_A = \frac{-k_0(Z_1q)(Z_2q)}{r_0^2} \tag{1-1}$$

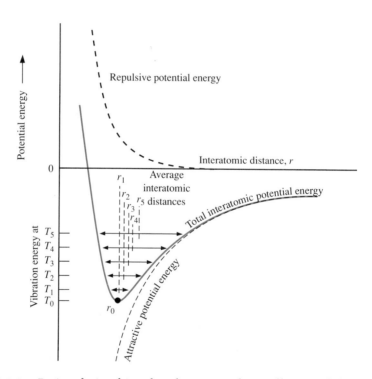

FIGURE 1.6   Basic relationships that determine the coefficient of thermal expansion. The change in interatomic distance with temperature follows. At 0 K, as the electron clouds overlap, the balance between the attractive forces of oppositely charged ions and the repulsive forces due to the overlap leads to the interatomic distance r_0. As the material is heated, the average distance between atoms increases, becoming r_1, r_2, and so forth. The change in distance per kelvin is the coefficient of expansion.

(Adapted from R. M. Rose, L. A. Shepard, and J. Wulff, *The Structure and Properties of Materials*, Vol. 4: Electronic Properties, John Wiley & Sons, Inc., New York, 1966).

where Z_1 and Z_2 are the valences of the charged ions (+1 for sodium and −1 for chlorine); q = the charge of a single electron (0.16×10^{-18} coul); k_0 is aproportionality constant (9×10^9 V · m/coul); and r_0 is the center-to-center distance between Na^+ and Cl^-, or the sum of their ionic radii.

The repulsive force is

$$F_R = \lambda_e^{-r/p} \tag{1-2}$$

where λ and p are constants for a given ion pair.

EXAMPLE 1.3 *[ES]*

Calculate the attractive bonding force between Na^+ and Cl^- at the equilibrium distance $r_0 = 0.278$ nm at 20°C. Note that 1 V · coul = 1 J and 1 J/m = 1 N.

Answer

$$F_A = -\frac{k_0(Z_1 q)(Z_2 q)}{r_0^2}$$

$$= -\frac{(9 \times 10^9 \text{ V} \cdot \text{m/coul})[(+1)(0.16 \times 10^{-18} \text{ coul})][(-1)(0.16 \times 10^{-18} \text{ coul})]}{(0.278 \times 10^{-9} \text{ m})^2}$$

$$= 2.98 \times 10^{-9} \text{ N}$$

Although this number seems small, there are many ion pairs in a block of sodium chloride. Its strength is seen in the salt columns that are used when mining underground for structural support where material is removed from between columns.

It is interesting to go a little further and investigate the effect of temperature on the curve in Figure 1.6. At a higher temperature, the thermally induced vibration changes the curves unevenly and leads to a larger average value of r. This effect is the basic reason for a very important engineering property: the thermal coefficient of expansion, which is the change in length per degree increase in temperature. For example, some ceramics (such as pyroceram and fused silica) have a low coefficient and therefore resist thermal shock very well. Another property, the bulk modulus of elasticity, is an index of how difficult it is to compress a material. Ceramics have a high modulus, polymers a low value. This is due to the fact that the ceramics are very tightly bonded in the basic structure, whereas the bonds between polymer molecules are looser and more easily compressed.

1.7 Covalent Bonding

In the covalent bond, atoms attain a stable group of eight by *sharing* outer-shell electrons. In the case of fluorine gas, one electron from each atom is used to form a common pair so that each atom is surrounded by a stable group of eight (Figure 1.7). This is called a molecular bond because the stable molecule F_2 is formed. Another way of representing this is to show the merging of two of the $2p$ electron envelopes (Figure 1.8a). Instead of drawing the entire ring, it is customary to show only the pair of shared electrons as two dots or as a single line (Figure 1.8b).

1.8 Hybridization of Electrons

Other cases of covalent bonding are more intricate. For example, we must pay careful attention to understand the structure of hydrocarbons used in polymers and the difference in structure between diamond and graphite.

So far we have discussed the s and p electrons as occupying definite energy states. Thus the $2p$ electrons are slightly higher in energy than the $2s$. We have also pointed out that the orbits of the s electrons are spherical, whereas the three $2p$ orbits are mutually perpendicular and dumbbell-shaped.

In order to explain the tetrahedral structure of a CH_4 (methane) molecule and also of diamond, we must introduce the concept of *hybridization* of the s

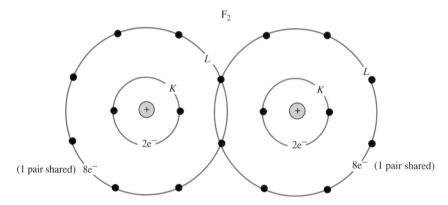

FIGURE 1.7   Simple model of covalent bond in F_2

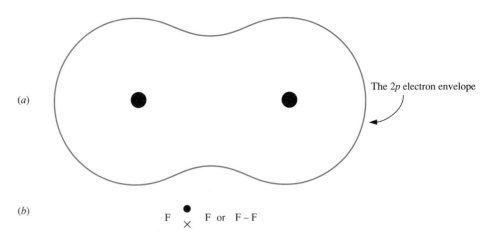

(a)

The 2p electron envelope

(b)

F •
   × F or F—F

FIGURE 1.8 The F_2 molecule: (a) Envelope of electron positions in F_2; (b) Schematic (dot) designation of covalent bond

and *p* electrons. We recall from biology that in forming a *hybrid*, we take two different strains to produce a single new strain. So in this case, we take one *s* electron and three *p* electrons of carbon to form a hybridized group of four electrons with orbits along four evenly spaced (tetrahedral) axes (Figure 1.9).

Here are the details: If we look only at the *p* electrons in the atomic structure of carbon, $1s^2$, $2s^2$, $2p^2$, we might predict (incorrectly) that the $2p$ electrons might bond to hydrogen, giving CH_2. However, this is not a stable compound. Instead a two-step process occurs. First, one of the $2s$ electrons is

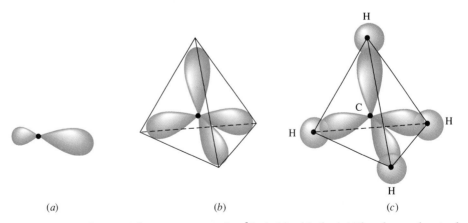

(a)   (b)   (c)

FIGURE 1.9 The spatial arrangement of sp^3 hybrid orbitals. (a) The shape of a single sp^3 hybrid orbital. (b) The four hybrid orbitals are arranged tetrahedrally in space (small lobes are omitted for clarity). (c) Bonding in CH_4. Each C—H bond is formed by the overlap of a 1s orbital from hydrogen with an sp^3 hybrid orbital of the carbon atom.

promoted to the 2p state (this takes energy, but it is more than recovered in formation of the new C—H bonds). Next the 2s electron and three 2p electrons hybridize. The up arrow signifies $m_s = +1/2$, the down arrow, $m_s = -1/2$.

(initial) ground state in an isolated carbon atom	$(\uparrow \downarrow) (\uparrow \downarrow) \ (\uparrow \uparrow)$ $1s^2 \quad 2s^2 \quad 2p^2$
s electron *promoted* to *p* (transition state)	$(\uparrow \downarrow) (\uparrow \ )(\uparrow \uparrow \uparrow)$ $1s^2 \ 2s^1 \ 2p^3$
2s and 2p electrons *hybridize*	$(\uparrow \downarrow) \ (\uparrow \uparrow \uparrow \uparrow)$ $1s^2 \qquad 2sp^3$ \qquad (all four electrons at same energy level)

Four *equal* C—H bonds are then formed to produce the tetrahedral structure of the CH_4 molecule (Figure 1.9). The tetrahedral bonding of carbon atoms to each other to form the hard, covalently bonded structure of diamond occurs in the same way (Figure 1.11*a*).

It is worthwhile to illustrate the structure of graphite, which is also pure carbon. Before doing this, we need to describe σ (sigma) and π (pi) bonding. If, in building up a molecule of fluorine, the electron envelopes of two bonding electrons from adjacent atoms are lined up *end to end*, this is called a sigma bond (Figure 1.10). If the electron fields align parallel, this is called a π (pi) bond; it is weaker.

The structure of graphite (Figure 1.11) is radically different from that of diamond. Although the structural formula is written with double bonds, we often read that the second double bond is not fixed but "resonates." This is not a very satisfactory explanation, because we have to account for the excellent electrical conductivity in the plane of the hexagonal structure. If we apply molecular bond theory, however, we can say that the electrons indeed hybridize but in a different way than in diamond. After promotion of one 2s electron to the 2p state, a group of three electrons $(2sp^2)$ hybridizes, giving bonds oriented at 120° and one electron remains unhybridized.

$$(\uparrow \downarrow) \quad (\uparrow) \quad (\uparrow \uparrow \uparrow)$$
$$1s^2 \qquad 2s^1 \qquad 2sp^2$$

Thus in the case of graphite, we use only three of the electrons to form the *permanent bonds* to each carbon, as σ bonds, making the hexagonal structure in one plane. The remaining *s* electron forms weak π bonds and is said to be *delocalized*; that is, it is not attached to any particular carbon atom. This explains the high electrical and thermal conductivity in the hexagonal plane — over 100 times that normal to the plane. The sheets of carbon atoms are attracted to each other only by van der Waals forces (London forces; see Section 1.10). They cleave (separate) easily, which accounts for the typical "greasy" feel of graphite. To take advantage of the high strength in the plane

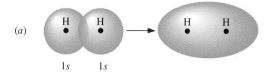

(a)

1s   1s

(b)

2p   2p

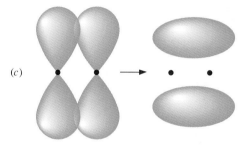

(c)

FIGURE 1.10 (a) The formation of a σ bond by the overlap of two s orbitals as found in hydrogen (H_2). (b) A σ bond can also be formed by the overlap of two $2p$ orbitals along their axes as found in the fluorine model (F_2). (c) When two $2p$ orbitals overlap sidewise, a π bond is formed.

[D. Ebbing, *General Chemistry*, 2nd ed., fig. 9.12. Copyright © 1987 Houghton Mifflin Company. Reprinted by permission.]

of the hexagon, graphite fibers are deliberately grown in a hexagonal direction and used in such composites as tennis racquet frames.

1.9 Metallic Bonding

In the metallic bond encountered in pure metals and metallic alloys, the atoms contribute their outer-shell electrons to a generally shared electron cloud for the whole block of metal (Figure 1.12).* This model suggests that good electrical conductivity would be characteristic of metallic bonding due to the high electron mobility.* Indeed, when an electrical potential is applied across a block of metal, the electrons are easily driven to provide the desired current.

---

*These electrons can be considered as related to the pi electrons discussed earlier. However, while in metals the s, p, and d electrons may become delocalized like pi electrons, the s electrons in polymers form sigma bonds.

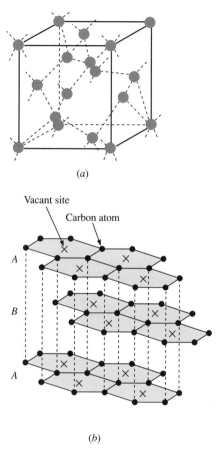

(a)

(b)

FIGURE 1.11 (*a*) Structure of diamond, shown as a three-dimensional array of carbon atoms. (*b*) The layer structure of graphite (carbon).

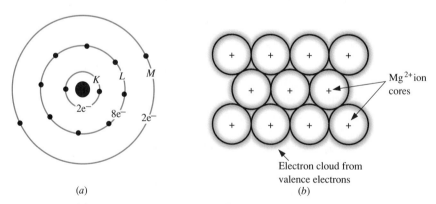

(a)   (b)

FIGURE 1.12 (*a*) A magnesium atom with its electron shells. (*b*) Atoms (ions) in solid magnesium surrounded by "electron gas."

1.10 Secondary Bonds — Intermolecular Forces

The forces between molecules are much weaker than the primary binding forces *between* atoms because there is no electron exchange. However, such forces are extremely important; the strength of a plastic made up of large molecules is governed by these *inter*molecular attractions, not by the bond strength within the molecule. There are two types of secondary bonds: van der Waals forces and hydrogen bonding (see Figure 1.5). In turn, van der Waals forces themselves fall into two groups: dipole–dipole and London.

In *dipole–dipole forces*, the arrangement of electrons and positive nuclei results in a positively charged field at one end of the molecule and a negative field at the other. For example, hydrogen chloride, HCl (Figure 1.13), is a polar molecule because the region near the hydrogen nucleus is highly positive and the opposite region near the chlorine nucleus is negative. Therefore, the negative region of an adjacent molecule is attracted to the positive region of the first molecule.

London forces, which are a little more difficult to understand (Figure 1.14a), explain why there is attraction between nonpolar molecules or the single atoms of the inert gases. The attraction depends on the probability that because the electrons are in motion, *temporary dipoles* will develop and lead to attractions between molecules.

Hydrogen bonding is really a special case of intermolecular attraction produced between certain covalently bonded hydrogen atoms and lone pairs of electrons of another atom. There is a positive field adjacent to the hydrogen and a negative field around the electron pair. This force is very important in plastics and in biological molecules such as DNA.

Let us take the simple case of water. The electrons in the O—H bond are strongly attracted to the oxygen, leaving the *partially exposed protons* to attract lone pairs of electrons on adjacent oxygen atoms (Figure 1.14b). This gives a strong dipole.

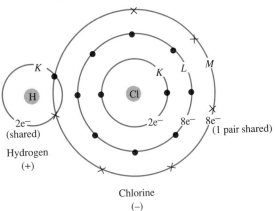

FIGURE 1.13 Formation of a polar molecule (HCl)

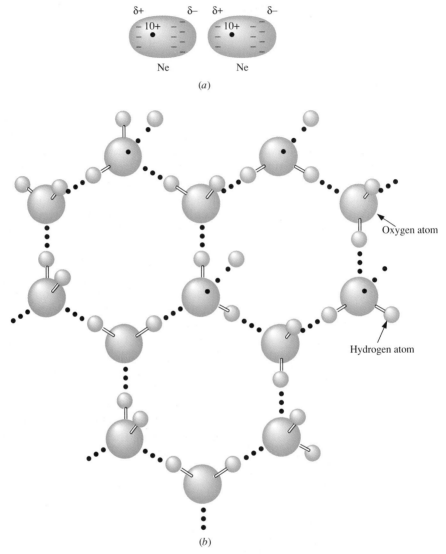

FIGURE 1.14 *(a) Origin of the London force.* At some instant, there are more electrons on one side of a neon atom than on the other. If this atom is near another neon atom, the electrons on that atom are repelled. The result is two instantaneous dipoles, shown by δ^+ and δ^-, which give an attractive force. Later the electrons on both atoms may move, but they move together, so there is always an attractive force between the atoms. *(b) The structure of a portion of ice.* Oxygen atoms are represented by large spheres, hydrogen atoms by small spheres. Each oxygen atom is tetrahedrally surrounded by four hydrogen atoms. Two are close, giving the H_2O molecule. Two are further away, held by hydrogen bonding (represented by dots). The distribution of hydrogen atoms in these two positions is random.

[D. Ebbing, *General Chemistry*, 2nd ed., p. 493. Copyright © 1987 Houghton Mifflin Company. Reprinted by permission.]

1.11　Quantitative Discussion of Bond Strength

To emphasize the difference between primary and secondary bonds and the importance of the hydrogen bond, the general ranges of bond strength can be given as follows:

	Energy kJ/mol
Intermolecular Forces	
Van der Waals (London, dipole–dipole) forces	0.1–10
Hydrogen bonding	10–40
Chemical Bonding	
Ionic	50–1000
Covalent	200–1000
Metallic Bonding	50–1000

These are rather wide ranges, so it is worthwhile to examine the concept of electronegativity, which influences the values.

Electronegativity may be defined as the tendency of the atom in a molecule to attract electrons. For example, we would expect chlorine to have a high electronegativity and sodium to have a low electronegativity. Pauling devised a method, based on the bond strength, to obtain the relative values shown in Table 1.3.

We can measure (by thermochemical methods) the thermal energy ΔH required to break the bonds in a gaseous material such as methane.

$$CH_4 \rightarrow C + 4H \qquad \Delta H = 1662 \text{ kJ/mol} \tag{1-3}$$

This gives a value of 416 kJ per bond. Because the energy is approximately the same as that required to break the C—H bonds in other hydrocarbons, we arrive at an average value of 411 kilojoules per mol per C—H bond. A summary of bond energies is given in Table 1.4

Pauling derived the following expression for *relative* electronegativity.

$$X_A - X_B = \left[\frac{BE(A \text{ to } B) - 1/2[BE(A \text{ to } A) + BE(B \text{ to } B)]}{R} \right]^{1/2} \tag{1-4}$$

where BE is bonding energy, R is 96.5 kJ (a constant to convert kJ to electronegativity units), and X_A and X_B are the electronegativities of elements A and B.

To obtain the values of Table 1.3, the electronegativity of fluorine was arbitrarily assigned as 4.0. The reasoning behind the formula is that because there is a large difference between the value for the two elements that are bonded together, $BE(A \text{ to } B)$, and the value for the average bond strength of the individual molecules, $1/2[BE(A \text{ to } A) + BE(B \text{ to } B)]$, a large difference in electronegativity exists.

TABLE 1.3   Electronegativities of the Elements*

1a	2a	3b	4b	5b	6b	7b	8	8	8	1b	2b	3a	4a	5a	6a	7a	0
1 H 2.1																	2 He –
3 Li 1.0	4 Be 1.5											5 B 2.0	6 C 2.5	7 N 3.0	8 O 3.5	9 F 4.0	10 Ne –
11 Na 0.9	12 Mg 1.2											13 Al 1.5	14 Si 1.8	15 P 2.1	16 S 2.5	17 Cl 3.0	18 Ar –
19 K 0.8	20 Ca 1.0	21 Sc 1.3	22 Ti 1.5	23 V 1.6	24 Cr 1.6	25 Mn 1.5	26 Fe 1.8	27 Co 1.8	28 Ni 1.8	29 Cu 1.9	30 Zn 1.6	31 Ga 1.6	32 Ge 1.8	33 As 2.0	34 Se 2.4	35 Br 2.8	36 Kr –
37 Rb 0.8	38 Sr 1.0	39 Y 1.2	40 Zr 1.4	41 Nb 1.6	42 Mo 1.8	43 Tc 1.9	44 Ru 2.2	45 Rh 2.2	46 Pd 2.2	47 Ag 1.9	48 Cd 1.7	49 In 1.7	50 Sn 1.8	51 Sb 1.9	52 Te 2.1	53 I 2.5	54 Xe –
55 Cs 0.7	57 Ba 0.9	57–71 La–Lu 1.1 1.2	72 Hf 1.3	73 Ta 1.5	74 W 1.7	75 Re 1.9	76 Os 2.2	77 Ir 2.2	78 Pt 2.2	79 Au 2.4	80 Hg 1.9	81 Tl 1.8	82 Pb 1.8	83 Bi 1.9	84 Po 2.0	85 At 2.2	86 Rn –
87 Fr 0.7	88 Ra 0.9	89–103 Ac–Lr 1.1 1.7	104 (Rf)	105 (Ha)													

*Adapted from Linus Pauling: *The Nature of the Chemical Bond, Third Edition*. Copyright © 1960 by Cornell University. Used by permission of the publisher Cornell University Press.

TABLE 1.4 Bond Energies (in kJ/mol)*

| | *Single Bonds* | | | | | | | | |
	H	C	N	O	S	F	Cl	Br	I
H	432								
C	411	346							
N	386	305	167						
O	459	358	201	142					
S	363	272	—	—	226				
F	565	485	283	190	284	155			
Cl	428	327	313	218	255	249	240		
Br	362	285	—	201	217	249	216	190	
I	295	213	—	201	—	278	208	175	149

	Multiple Bonds					
C=C	602	C=N	615	C=O		799
C≡C	835	C≡N	887	C≡O		1072
N=C	418	N=O	607	S=O (in SO_2)		532
N≡N	942	O=O	494	S=O (in SO_3)		469

*Data are taken from J. E. Huheey, *Inorganic Chemistry*, 2nd ed. (New York: Harper & Row, 1978), pp. 842–850. Used by permission.

EXAMPLE 1.4 *[ES]*

Calculate the electronegativity of carbon, using fluorine (4.0) as a reference and the data of Table 1.4.

Answer
$$X_A - X_B = \left[\frac{BE(A \text{ to } B) - 1/2[BE(A \text{ to } A) + BE(B \text{ to } B)]}{96.5 \text{ kJ}} \right]^{1/2}$$

$$X_F - X_C = \left[\frac{485 \text{ kJ} - 1/2(346 + 155) \text{ kJ}}{96.5 \text{ kJ}} \right]^{1/2}$$

$$= 1.56$$

Therefore, carbon is 1.56 below fluorine, or 2.44, which is close to the value of 2.5 given in Table 1.3.

Mixed Ionic and Covalent Bonding

We rarely encounter cases of perfect ionic or covalent bonding—that is, cases in which the electrons are completely donated or equally shared. Another Pauling formula can be used to calculate the percent ionic bonding (the rest of the bonding is covalent).

$$\text{Fraction ionic} = 1 - e^{-1/4(X_A - X_B)^2} \tag{1-5}$$

where X_A and X_B are electronegativities of the two species.

CRYSTALLINE AND NONCRYSTALLINE GLASS STRUCTURES

1.12 Comparison of Crystalline and Noncrystalline Structures — General

Now that we have studied the electronic structures of the atoms and how they lead to bonding into interatomic groups, we are ready to see how these groups are assembled in engineering materials as larger groups that make up *crystals, grains*, the *noncrystalline* structures of glasses, including glassy metals, the mixed glass–ceramics, and the glassy and crystalline polymers. Before getting into the details of these two groups, let us compare their principal features.

In the typical crystalline material, the atoms occupy regular positions that we can describe by using a space lattice. Metals, for example, are crystalline under normal conditions. We developed a rather striking example by accident (Figure 1.15). We were attempting to dissolve magnesium in liquid cast iron using a pressure chamber with an argon atmosphere. After cooling we found fine, brilliant crystals of magnesium on the inside of the cover. (At these temperatures magnesium vaporized from the melt and condensed on the cover.) The hexagonal faces were quite evident; they are manifestations of the hexagonal arrangement of magnesium atoms within. In the usual case, as in a bar of magnesium, we would expect to find not crystal faces but a number of grains (Figure 1.16, page 34). However, if we examine the structures of the individual grains, we find the atoms arranged in an independent space lattice in each grain, with the same hexagonal structure as in the single crystals. This is called *long-range order*.

As an example of a ceramic crystal, we may consider the structure of quartz. We find a regular, hexagonal network of Si and O atoms (Figure 1.17*a*, page 34).

By contrast, in the *noncrystalline* materials there is no *regular space* lattice. If we take a crystal of quartz, melt it, and cool the melt in air, we obtain a glassy mass called fused silica (Figure 1.17*b*). We still find that each silicon atom is surrounded by four oxygen atoms, forming a tetrahedron, but there is no *regular repetition* of this structure on a space lattice. Another major difference from the original quartz, which exhibits a definite melting point, is that the noncrystalline mass *gradually* softens upon heating and finally turns to a viscous liquid.

In addition to the ceramic glasses, the polymers are entirely or partly glassy[*]

---

[*]Sometimes the word *amorphous* is used to describe the lack of crystallinity. However, we will require that a material have no long- *or* short-range order for it to be called amorphous.

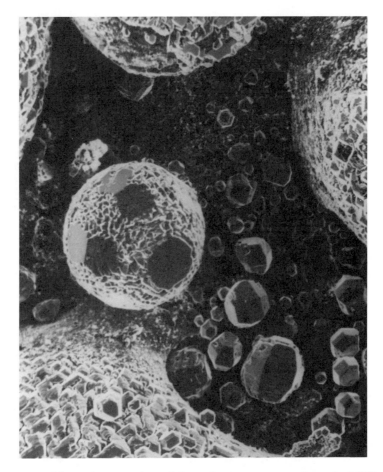

FIGURE 1.15 Magnesium condensed from the vapor: approximately 1,000×
[Stephen Krause, Department of Mechanical Engineering, Arizona State University]

(Figure 1.18, page 35). For example, in the case of the phenol formaldehyde resin (often called "bakelite") encountered in a bowling ball, we have a single, three-dimensional network that is essentially a giant molecule but lacks the long-range order of a crystal (Figure 1.18a). In another type of polymer in which the giant molecules are linear, and which is made up of a backbone of a chain of carbon atoms, some crystallization may occur when the chain is bent back and forth regularly (Figures 1.18b, c).

With those distinctions in mind, let us review the crystalline structures typically encountered in metals and some ceramics and then consider the noncrystalline structures encountered in other ceramics and polymers.

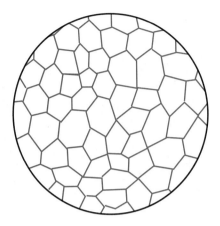

FIGURE 1.16   Grain structure in bar of magnesium

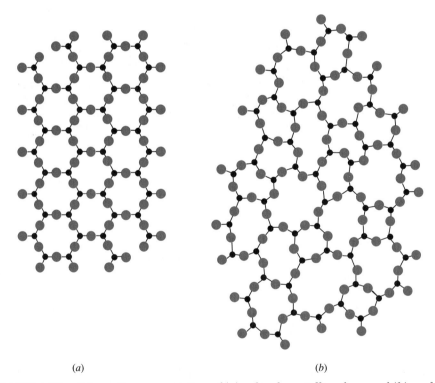

(a)   (b)

FIGURE 1.17   Schematic representation of (a) ordered crystalline form and (b) random-network glassy form of the same composition

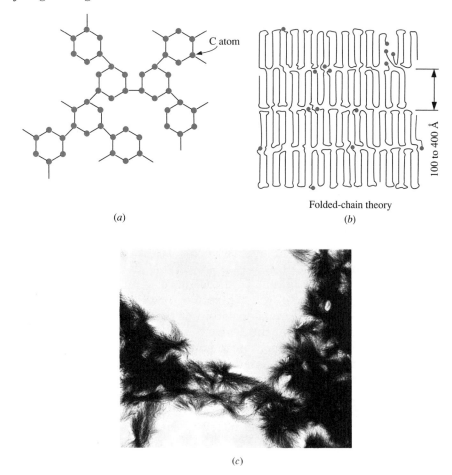

Folded-chain theory
(b)

(a)

(c)

FIGURE 1.18 (a) Giant noncrystalline molecule of phenol formaldehyde. (b) Crystallinity created by the folding back of chains. The growth mechanism involves the folding over of the planar zigzag chain on itself at intervals of about every 100 chain atoms. A single crystal may contain many individual molecules. (c) Spherulites (spherelike areas of crystallinity) in polycarbonate polymer; transmission electron micrograph, 66,000×.

[Parts (a) and (b) from L. E. Nielsen, *Mechanical Properties of Polymers*, Litton Educational Publishing, Inc., New York, 1962. Reprinted by permission of Van Nostrand Reinhold Company. Part (c) courtesy Jim de Rudder, University of Michigan.]

1.13 Crystalline Structures — Metals

We have already agreed that the hallmark of a crystal structure is the regular repeating arrangement of atoms. It will be very useful to us in the next chapter, when we investigate the effects of stress on crystal structure, to have a clear, simple mathematical model of the atoms' positions. To create one, we begin with a space lattice to suit our particular metal. For example, if the crystal is cubic, we choose a lattice made up of cubes. By describing a single cube, called a *unit cell*, we can specify the entire lattice. This is analogous to describing just one office module in a skyscraper made up of identical modules.

1.14 The Unit Cell

To describe the unit cell and later the movement of an atom in the cell, we need a system for specifying (1) atom positions or *coordinates*, (2) directions in the cell, and (3) planes in the cell.

Position

The position of an atom is described with reference to the axes of the unit cell and the unit dimensions of the cell. Suppose that a grain or crystal is built of unit cells of dimensions a_0,* b_0, and c_0 angstroms,† as shown in Figure 1.19. In this case the axes are at right angles to each other. To construct the cell, we merely lay out a_0 (the lattice parameter) in the x direction, b_0 in the y direction, and c_0 in the z direction. The figure shows the coordinates of several atoms in the cell. An atom at the center would have coordinates $\frac{1}{2}, \frac{1}{2}, \frac{1}{2}$, whereas an atom in the center of the face in the xy plane would have coordinates $\frac{1}{2}, \frac{1}{2}, 0$. It is important to note that commas after coordinates in space are a signal that we are referring to *points* in that space. These coordinates are not enclosed in parentheses; we do not want you to confuse them with planes, which we shall discuss in a later section.

Up to this point we have used a relatively simple cell as an example. In nature, 14 different types of crystal lattices are found (Figure 1.20). These lattices cover the variations in the length of a, b, and c and in the angles between the axes.

Fortunately, in metals we find mostly the three simple types of cells shown in Figure 1.21 (page 39): body-centered cubic (BCC), face-centered cubic (FCC), and hexagonal close-packed (HCP). Some of the other types are encountered occasionally in a few metals, ceramics and polymers.

Direction

To specify a direction in the unit cell, we merely place the base of the arrow of the direction ray at the origin and follow the shaft until we encounter integral coordinates (Figure 1.22, page 40). Instead of constructing other cells,

---

*The subscript 0 specifies that this dimension is measured at a standard temperature, usually 68°F (20°C). When the unit cell expands with temperature, a, b, and c change.

†The angstrom (Å) = 10^{-8} cm. In the SI system of units, the linear unit of measure is the meter; 1 Å = 10^{-10} meter (m) = 0.1 nanometer (nm), where 1 nm = 10^{-9} m. Both Å and nm will be used in the following discussion of unit cell dimensions. It should also be noted that some experimenters prefer picometers (10^{-12} m) in order to have whole numbers rather than decimals associated with the larger nanometer.

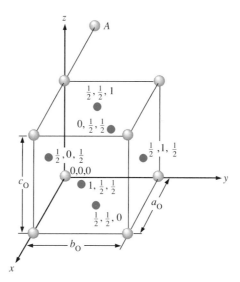

FIGURE 1.19 Coordinates of atoms in the face-centered positions in a unit cell. The atom A shown in the next unit cell would have coordinates $\bar{1}, 0, 1$. Note: $\bar{1}$ is equivalent to -1.

we can use a point that has fractional intercepts in the unit cell and multiply by the least common denominator. Thus direction A is obviously [111], but B has coordinates $1, \frac{1}{2}, 0$ at the edge of the cell. These become [210]. Note that in specifying a direction, we put square brackets around the numbers to distinguish direction from the notation for coordinates and parentheses for a plane, described later. Note also that we can have negative-direction indices, as shown by C. We indicate these with an overbar.

If we wish to find the *indices* of a direction that does not pass through the origin, we merely pass a parallel direction through the origin and proceed as before.

In the cubic system the dimensions of the unit cell are the same: $a_0 = b_0 = c_0$. The order in a given set of indices, such as [110], depends on which directions we happened to choose for x, y, and z in the crystal, because in the cubic system the indices are identical. Often we wish to signify a set of directions that are related, such as [110], [101], and [011]. These are all directions across the face diagonal of the cube. To specify these similar directions, we use pointed brackets and the indices of only one direction. Thus $\langle 110 \rangle = [110]$, [101], and [011]. Directions with negative indices, such as [$\bar{1}$10], are also considered part of the same set.

It is important to recognize that unit distances in the directions x, y, and z are a_0, b_0, and c_0, respectively. Therefore the [111] direction in a noncubic system [for example, the orthorhombic (three sides, $a_0 \neq b_0 \neq c_0$) in Figure 1.20] refers to a direction starting at coordinates $0, 0, 0$, with the tip at a_0 distance in the x direction, b_0 distance in the y direction, and c_0 distance in the z direction.

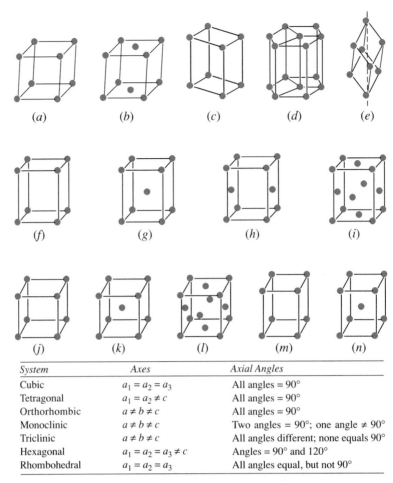

System	Axes	Axial Angles
Cubic	$a_1 = a_2 = a_3$	All angles = 90°
Tetragonal	$a_1 = a_2 \neq c$	All angles = 90°
Orthorhombic	$a \neq b \neq c$	All angles = 90°
Monoclinic	$a \neq b \neq c$	Two angles = 90°; one angle ≠ 90°
Triclinic	$a \neq b \neq c$	All angles different; none equals 90°
Hexagonal	$a_1 = a_2 = a_3 \neq c$	Angles = 90° and 120°
Rhombohedral	$a_1 = a_2 = a_3$	All angles equal, but not 90°

FIGURE 1.20   The 14 crystal lattices and their geometric relationships. The lattices continue in three dimensions. (*a*) simple monoclinic. (*b*) End-centered monoclinic. (*c*) Triclinic. (*d*) Hexagonal. (*e*) Rhombohedral. (*f*) Simple orthorhombic. (*g*) Body-centered orthorhombic. (*h*) End-centered orthorhombic. (*i*) Face-centered orthorhombic. (*j*) Simple cubic. (*k*) Body-centered cubic. (*l*) Face-centered cubic. (*m*) Simple tetragonal. (*n*) Body-centered tetragonal.

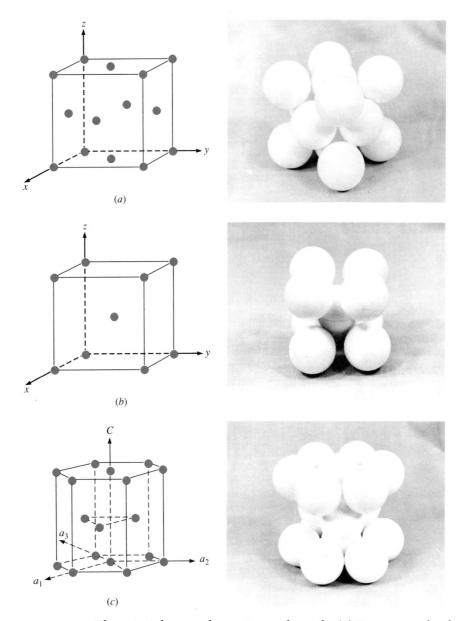

FIGURE 1.21 The principle crystal structures of metals. (*a*) Face-centered cubic. (*b*) Body-centered cubic. (*c*) Close-packed hexagonal. For each case, a sketch and a photograph of the hard-sphere model are shown.

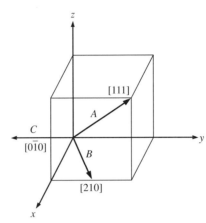

FIGURE 1.22   Specification of directions

Plane

The method for finding the Miller indices of a plane differs from that for describing coordinates and directions (Figure 1.23). Follow this sequence:

1. Select a plane in the unit cell that does not pass through the origin. For example, to describe the plane at face A that passes through $0, 0, 0$, we use plane B, which is in the same set of planes because it is a unit distance away from A. (That is, B would have the same number of atoms and would be parallel to the original plane.)
2. Record the intercepts of the plane as multiples of a_0, b_0, and c_0 on the x, y, and z axes, in that order.
3. Take reciprocals and clear fractions. This gives (001) for plane B.* Note that we use parentheses in this case to show that we are describing a plane rather than a direction. Commas do not separate the indices, as they do in atom positions. The plane should be read as the "zero-zero-one" plane.

Following the same rules, we find the indices for plane C of intercepts $1, \frac{2}{3}, \frac{1}{3}$. We take reciprocals: $1, \frac{3}{2}, 3$. Then we clear fractions: $2, 3, 6$, or (236).

As in the case of directions, we often want to specify a family of planes, such as the cube faces. Here we use braces; thus

$$\{100\} = (100), (010), (001)$$

We can also have the corresponding negative values, such as $(\bar{1}00)$, depending on our choice of position for the origin $0, 0, 0$. With positive and negative indices we have six planes that describe the six faces of the cube.

---

* $\frac{1}{\infty}, \frac{1}{\infty}, \frac{1}{1}$

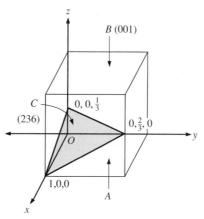

FIGURE 1.23 Calculation of Miller indices of a plane

EXAMPLE 1.5 *[ES]*

Determine the Miller indices of the plane shown in the figure.

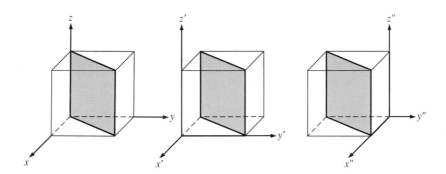

Answer Because the plane passes through the origin, we must go through a unit translation of the axes to obtain the intersection of the plane with the axes. This unit translation does not influence the outcome of our problem because of the symmetry of the unit cells. There are two possible translations:

In x', y', z', the intercepts are $-1, 1, \infty$, which give the Miller indices $(\bar{1}10)$. In x'', y'', z'', the intercepts are $1, -1, \infty$, or $(1\bar{1}0)$. These indices are equivalent, so we can see the advantage of referring to families of planes, such as $\{110\}$.

1.15 Correlation of Data on Unit Cells with Measurements of Density

Let us see whether we can check the dimensions of the unit cell with engineering data, such as density. After all, we are not going to use the metals in unit-cell-size pieces. We should satisfy ourselves that the characteristics of the unit cell relate to engineering-size components.

For example, the density of copper is 8.96 g/cm³ (8.96 × 10³ kg/m³ = 8.96 Mg/m³) at 20°C (68°F). If our data are correct, we should find the same value when we use the mass of copper in the volume of a unit cell.

We can find the mass (M) and volume (V) from previous definitions. We know the weight of an atom of copper, because each gram atomic weight of copper (63.5 g/at. wt) has 6.02 × 10²³ atoms (Avogadro's number). The volume of a unit cell is merely the lattice dimension cubed. Before we can finish calculating it, however, we must determine the number of atoms per unit cell.

Figure 1.24 shows how to determine the number of atoms in a single cell. If we were to set up thin glass walls as faces of the FCC, these would pass through the centers of the atoms in the faces and extend only to the centers of the corner atoms. Counting the portions of atoms within the cell, we have

$$6 \text{ face atoms of which } \tfrac{1}{2} \text{ of each is inside the glass} = 3 \text{ atoms}$$
$$8 \text{ corner atoms of which } \tfrac{1}{8} \text{ of each is inside the glass} = \underline{1 \text{ atom}}$$
$$\text{Total atoms per unit cell} = 4 \text{ atoms}$$

In a similar manner we find that the number of atoms per unit cell in a BCC is two, whereas that in an HCP is six.

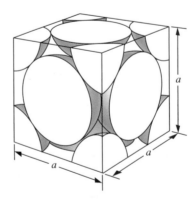

FIGURE 1.24   Illustration of the number of atoms (four) in an FCC unit cell

---

EXAMPLE 1.6 *[ES]*

We know from x-ray data discussed later in this chapter that $a_0 = 3.615$ Å (0.3615 nm) at 20°C (68°F) for copper. Calculate the theoretical density of copper, using the knowledge that the cell of this element is an FCC.

Answer

$$\text{Density} = \frac{M}{V} = \frac{4 \text{ atoms/unit cell} \dfrac{63.546 \text{ g/at. wt}}{6.02 \times 10^{23} \text{ atoms/at. wt}}}{(3.615 \times 10^{-8})^3 \text{ cm}^3/\text{unit cell}}$$

$$= 8.938 \frac{\text{g}}{\text{cm}^3} \left(8.937 \times 10^3 \frac{\text{kg}}{\text{m}^3} = 8.937 \text{ Mg/m}^3 \right)$$

This is quite close to the density of a block of copper (8.930 g/cm³). In Chapter 3 we shall see that there are missing atoms that explain the difference.

1.16 Other Unit Cell Calculations

Atomic Radius

When we discuss the design of alloys, we shall see that when two metals are combined, atoms of the second element can substitute for some of the atoms of the major element. The extent to which this substitution can take place is governed by the similarity of the two metals. One important measure of similarity is the *atomic radius*, which we can easily calculate once we know the dimensions of the unit cell. First we consider the atoms as spheres and find some dimension of the unit cell along which the spheres are in contact. We can calculate any dimension of the cube, such as a body or face diagonal, by geometry. Then we merely divide the figure by the number of atomic radii present (Figure 1.25).

Planar Density

We will see that when slip occurs under stress (plastic deformation), it takes place on the planes on which the atoms are most densely packed. To calculate *planar density*, we use the following convention.

If an atom belongs entirely to a given area, such as the atom in the center of a face in an FCC structure, we note that the trace of the atom on the plane is a circle (Figure 1.26). Therefore, in the area a_0^2 we count one atom for the center but one-quarter atom for each of the corners, because each has a trace of only one-quarter circle on the area a_0^2. The planar density is $2/a_0^2$ (atoms/Å² or atoms/nm²). It should be added that in all of these density calculations, one of the ground rules is that a plane or a line must pass through the center of an atom or else the atom is not counted in the calculations.

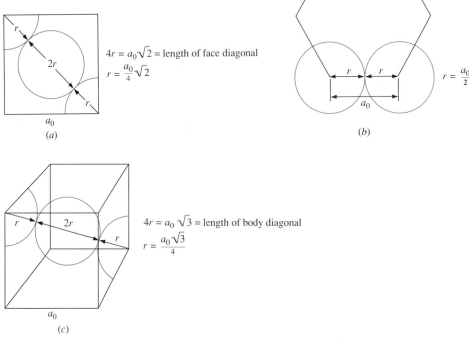

FIGURE 1.25 Calculation of atomic radius. (*a*) FCC unit cell: spheres (atoms) in contact along the face diagonal. (*b*) HCP unit cell: spheres (atoms) in contact on the edge. (*c*) BCC unit cell: spheres (atoms) in contact along the body diagonal.

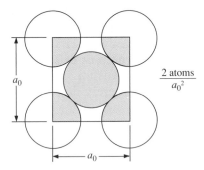

FIGURE 1.26 Calculation of planar density on the face plane of an FCC unit cell

Linear Density

Linear density is an important concept because when planes slip over each other, the slip takes place in the direction of the closest packing of atoms on the planes. We calculate it by using the following convention.

If a line passes completely through an atom, the trace of the atom on the line is one diameter (Figure 1.27). In an FCC face, the center atom counts for

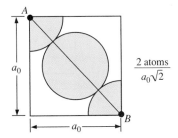

FIGURE 1.27   Calculation of linear density on the face diagonal of an FCC unit cell

one atom on the face diagonal. The corner atoms make traces equal to only one-half diameter each on the line of length AB. Therefore, the linear density in the AB direction [110] in an FCC structure is

$$\frac{1 + \frac{1}{2} + \frac{1}{2} \text{ atoms}}{a_0\sqrt{2}} \frac{}{\text{Å}} = \frac{2}{a_0\sqrt{2}} \frac{\text{atoms}}{\text{Å}} \quad \text{or} \quad \frac{\text{atoms}}{\text{nm}} \qquad (1\text{-}6)$$

By the same reasoning, the linear density in a BCC in the [111] direction is

$$\frac{2}{a_0\sqrt{3}} \frac{\text{atoms}}{\text{Å}} \quad \text{or} \quad \frac{\text{atoms}}{\text{nm}} \qquad (1\text{-}7)$$

(Recall that the atoms touch along the body diagonal of a BCC.)

EXAMPLE 1.7 *[ES]*

Calculate linear and planar atomic density for a face-centered cubic structure in the [112] direction and the (111) plane, respectively.

Answer   When determining linear atomic density, we must remain in the reference cell. Therefore the centers of the atoms cut in an FCC are at $0, 0, 0$ and $\frac{1}{2}, \frac{1}{2}, 1$. There are 2 radii, or 1 atom. We can determine the length of the ray in several ways. One is to use the geometry of the large right triangle.

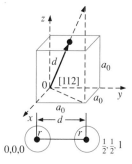

$$(2d)^2 = (a_0\sqrt{2})^2 + (2a_0)^2 \text{ from which } d = \frac{a_0\sqrt{6}}{2}$$

Therefore,

$$\text{Linear atomic density} = \frac{1 \text{ atom}}{a_0\sqrt{6}/2} = \frac{2}{a_0\sqrt{6}} \frac{\text{atoms}}{\text{Å}} \quad \text{or} \quad \frac{\text{atoms}}{\text{nm}}$$

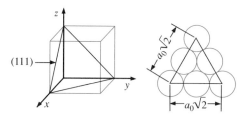

$$\text{Number of atoms} = 3 \times \tfrac{1}{6} + 3 \times \tfrac{1}{2} = 2 \text{ atoms}$$

$$\text{Area of equilateral triangle} = \frac{\sqrt{3}a_0^{\,2}}{2} \quad \text{(see Example 1.8 for derivation)}$$

Therefore,

$$\text{Planar atomic density} = \frac{2}{\sqrt{3}\,a_0^{\,2}/2} = \frac{4}{a_0^{\,2}\sqrt{3}}\ \frac{\text{atoms}}{\text{Å}^2} \quad \text{or} \quad \frac{\text{atoms}}{\text{nm}^2}$$

1.17 Close-Packed Hexagonal Metals

To describe hexagonal structures, we need a few simple modifications of the Miller indices of directions and planes. These are called *Miller–Bravais indices* (Figure 1.28). Instead of three axes x, y, z, we use four axes—three in the horizontal xy plane at 120° to each other, called a_1, a_2, a_3, and the fourth, c, in the z direction. Use of the extra axis makes it easier to see the relations be-

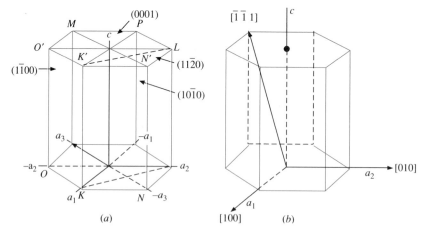

FIGURE 1.28   (*a*) Indices of some planes in a hexagonal crystal. (*b*) Indices of some directions in a hexagonal crystal.

[Part (*a*) from C. S. Barrett and T. B. Massalski, *The Structure of Metals*, 3d ed., McGraw-Hill, New York, 1973, Fig. 1.7, p. 12. Used with permission of T. B. Massalski.]

tween similar planes in the hexagonal structure. Let us locate some planes using this reference frame (see Figure 1.28a).

The most important planes in an HCP cell are the basal planes. Finding the intercepts of one of these ($K'N'LPMO'$), we obtain $a_1 = \infty, a_2 = \infty, a_3 = \infty$, $c = 1$. Taking reciprocals as before, we find that this is a (0001) plane. Now taking the intercepts of $KK'N'N$, we get $1, \infty, \bar{1}, \infty$, or ($10\bar{1}0$). Similarly, for $KOO'K'$ we find intercepts $1, \bar{1}, \infty, \infty$, or ($1\bar{1}00$). For the other faces we find other combinations of $1\bar{1}00$. Thus these planes are of the family $\{10\bar{1}0\}$. We call these Miller–Bravais indices $h, k, i,$ and l. If we select $h, k,$ and l, then i cannot be independent and by vector geometry always equals $-(h + k)$.

The notation for a direction is developed by using either three axes (a_1, a_2, and c) or four axes. Note again that a_1 and a_2 are at 120°. We shall use the simple three-axes notation. Figure 1.28b shows that the method is similar to that explained for the cubic system.

In a hexagonal close-packed system, additional atoms are present in the unit cell (Figure 1.21). The packing on the additional plane is the same as on the planes above and below. However, the stacking sequence changes.

Note: An HCP structure and an FCC structure exhibit similar packing. The figures for the (111) plane of an FCC and the (0001) plane of an HCP show the same atomic arrangement. Using the hard, perfect-sphere model for atoms, we find that the ratio c/a_0 for an HCP is 1.63, which is derivable from the geometry of a regular tetrahedron. (The height of a regular tetrahedron with side length a is $a\sqrt{\frac{2}{3}}$, or $0.816a$. So this is the distance from the basal plane to the center of the close-packed atoms. The total HCP height is twice this value, or $1.63a$.) The *atomic packing factor*, defined as the volume of atoms divided by the volume of the unit cell, is 0.74 for both the FCC and an ideal HCP. This is further proof of their similarity. The similarity is important to our later discussion of why FCC and HCP structures form extensive solid solutions (Section 1.21).

Most HCP metals do *not* exhibit perfect packing with a c/a_0 of 1.63. The atoms in this arrangement do not behave as perfect spheres. Examples of experimental c/a_0 ratios are Be (1.57), Ti (1.58), Mg (1.62), Zn (1.86), and Cd (1.89).

The basic difference between FCC and HCP structures lies in the way the close-packed planes are stacked above one another. In the HCP the stacking is ABAB..., whereas in the FCC the stacking is ABCABC..., where A, B, and C refer to relative atom positions in adjacent layers.

Here is a summary of the various crystal structures that are important in metals.

Structure	Atoms/Unit Cell	Unit Cell–Radius Relationships
Body-center cubic	2	$4r = a_0\sqrt{3}$
Face-centered cubic	4	$4r = a_0\sqrt{2}$
Hexagonal close-packed	6	$2r = a_0; c/a_0 = 1.63^*$

*See text and Example 1.8 for experimental ratios that are not this ideal value.

1.18 Definitions of a Phase; Solid Phase Transformations in Metals

At this point let us explain what we mean by a *phase* and a *phase transformation*. Everyone knows that H_2O can exist as a gas, a liquid, and a solid. These are three different phases. *The general definition of a phase is a homogeneous aggregation of matter.*

Under different conditions of temperature and pressure, different crystal structures (that is, different unit cells) for ice are formed, called ice I, II, III, etc. These are different solid phases with the same chemical composition.

Similarly, we encounter different crystal structures in the same metal. A bar of iron at room temperature has a BCC structure. However, if we heat the bar above 912°C (1674°F), the structure changes to FCC. Instead of calling these forms iron I and iron II, as we did with ice, we call the BCC iron α (alpha) and the FCC iron γ (gamma). In general, Greek letters are used to designate the different phases in solid metals.

Table 1.5 summarizes the crystal structures of the elements. It shows that many metallic elements can have different crystal structures. Note that most of the metallic structures are BCC, FCC, or HCP. When a material has more than one crystal structure, we say it exhibits *allotropy* and has different *allotropic* forms.

Now let us use some lattice parameter data to estimate volume changes resulting from transformation of a metal from one solid phase to another.

EXAMPLE 1.8 *[ES]*

On cooling through 880°C (1615°F), titanium goes through a phase change analogous to that of iron, except that in this case the crystal structure changes from BCC to HCP.

> BCC: $a = 3.32$ Å (0.332 nm)
>
> to HCP: $a = 2.956$ Å (0.2956 nm) $c = 4.683$ Å (0.4683 nm)

What is the volume change?

Answer   The volume of the BCC is $(3.32 \text{ Å})^3 = 36.6 \text{ Å}^3$ for the two atoms per unit cell.

To determine the volume of the HCP, observe that there are six equilateral triangles in the basal plane.

$$\text{Area} = \tfrac{1}{2}ah$$

$$\sin 60° = \frac{\sqrt{3}}{2} = \frac{h}{a}$$

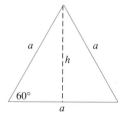

TABLE 1.5 Periodic Table of the Elements and Their Crystal Structures*

Period	1a	2a	3b	4b	5b	6b	7b	8	8	8	1b	2b	3a	4a	5a	6a	7a	0
1st							**1 H** A3, A1											**2 He** (A3), (A2)
2nd	**3 Li** A2, A1, A3	**4 Be** (A2), A3											**5 B** H, T, R	**6 C** H, H, A4	**7 N** H, C	**8 O** C, (R)	**9 F**	**10 Ne** A1
3rd	**11 Na** A2, A3	**12 Mg** A3											**13 Al** A1	**14 Si** A4	**15 P** C, O, C	**16 S** O, M, R	**17 Cl** A1, A3, A2	**18 A** A1
4th	**19 K** A2	**20 Ca** A2, A1	**21 Sc** (A2), A3	**22 Ti** A2, A3	**23 V** A2	**24 Cr** (A1), A2	**25 Mn** A2, A1, C, C	**26 Fe** A2, A1, A2	**27 Co** A1, A3	**28 Ni** A1	**29 Cu** A1	**30 Zn** A3	**31 Ga** O	**32 Ge** A4	**33 As** A7	**34 Se** A8, M	**35 Br** O	**36 Kr** A1
5th	**37 Rb** A2	**38 Sr** A2, A3, A1	**39 Y** A2, A3	**40 Zr** A2, A3	**41 Nb** A2	**42 Mo** A2	**43 Tc** A3	**44 Ru** A3	**45 Rh** A1	**46 Pd** A1	**47 Ag** A1	**48 Cd** A3	**49 In** A6	**50 Sn** A5, A4	**51 Sb** A7	**52 Te** A8	**53 I** O	**54 Xe** A1

*These other designations are used to denote the structures of the allotropic forms, which are given in the order of their appearance. A1 = FCC; A2 = BCC; A3 = HCP; A4 = diamond cubic; A5 = body-centered tetragonal; A6 = face-centered tetragonal; A7 = rhombohedral; A8 = trigonal; H = hexagonal (usually *ABAC*...close-packed); R = rhombohedral; O = orthorhombic; C = complex cubic; T = tetragonal; M = monoclinic; ( ) = uncertain. (Adapted from C. S. Barrett and T. B. Massalski, *The Structure of Metals,* 3d ed., McGraw-Hill Book Company, New York, 1983, p. 227. Used with permission of T. B. Massalski.)

TABLE 1.5 *(Continued)*

55 Cs	56 Ba	57 La	72 Hf	73 Ta	74 W	75 Re	76 Os	77 Ir	78 Pt	79 Au	80 Hg	81 Tl	82 Pb	83 Bi	84 Po	85 At	86 Rn
A2	A2	A2	A2	A2	A2	A3	A3	A1	A1	A1	R	A2	A1	A7	R		
	(T)	A1	A3								T	A3			C		
	(H)	H															

87 Fr	88 Ra	89 Ac												104 (Rf)	105 (Ha)	106
		A1														

6th (continued)

58 Ce	59 Pr	60 Nd	61 Pm	62 Sm	63 Eu	64 Gd	65 Tb	66 Dy	67 Ho	68 Er	69 Tm	70 Yb	71 Lu
A2	A2	A2		(A2)	A2	(A2)	(A2)	(?)	(?)	A3	A3	A2	(?)
A1	H	H		R		A3	A3	A3	A3			A1	A3
H													

7th (continued)

90 Th	91 Pa	92 U	93 Np	94 Pu	95 Am	96 Cm	97 Bk	98 Cf	99 Es	100 Fm	101 Md	102 No	103 Lr
A2	T	A2	A2	A2	H								
A1		T	T	T									
		O	O	A1									
				O									
				M									
				M									

or

$$\text{Area of the triangle} = \frac{\sqrt{3}}{4}a^2$$

Therefore the area of the hexagonal base is $(6\sqrt{3}/4)a^2$, and the volume of the HCP equals

$$\left(6\frac{\sqrt{3}}{4}a^2\right)(c) = \left(\frac{3}{2}\sqrt{3}\right)(2.956)^2(4.683) = 106.3 \text{ Å}^3 \ (0.1063 \text{ nm}^3)$$

However, this HCP structure has six atoms per unit cell. To compare the volumes, we must consider the same number of atoms in each structure. We therefore compare one-third of the HCP volume (two atoms) with the two atoms in the BCC, or

$$36.6 \text{ Å}^3(0.0366 \text{ nm}^3) \rightarrow \frac{106.3}{3} = 35.4 \text{ Å}^3(0.0354 \text{ nm}^3)$$

Therefore there is a 3.3% contraction when BCC changes to HCP. Note that the c/a ratio of the HCP is not 1.63. We cannot calculate the volume on the basis of an assumed *hard-sphere* model.

1.19 Effects of the Addition of Other Elements on the Structure of Pure Metals

Only a few elements are widely used commercially in their pure form; pure copper in electrical conductors is one example. Generally, as in the case of iron, other elements are added to produce greater strength or to improve corrosion resistance or are simply left in as impurities because of the cost of refining. Whatever the reason, we need to know the effect these elements have on structure, because structure determines properties.

When a second element is added, two basically different structural changes are possible.

1. The atoms of the new element form a solid solution with the original element, but there is still only one phase, such as an FCC or a BCC.
2. The atoms of the new element form a new second phase, which usually contains some atoms of the original element. The entire microstructure may change to this new phase, or two phases may both be present.

Let us examine each possibility separately.

1.20 Solid Solutions

If we take a crucible containing 70 g of liquid copper and add 30 g of nickel while heating to maintain a completely liquid melt, we obtain a liquid solution of copper and nickel. If we cool the solution slowly and examine the grains, we find that we have an FCC structure and that a_0, the edge of the unit cell, is between the values for copper and nickel. This is called a *substitutional solid solution*, because nickel atoms have substituted for copper atoms in the FCC structure (Figure 1.29). This structure is stronger than either pure copper or pure nickel because the interaction between the atoms gives greater resistance to slip.

Another type of *solid solution* is called an *interstitial solid solution*. The new atoms are located in the interstices, or spaces, between the atoms in the matrix. Naturally, the atoms that form interstitial solid solutions are those with small radii, such as hydrogen, carbon, boron, and nitrogen. Carbon, for example, forms an interstitial solid solution in iron, giving steel. Position C $\frac{1}{2}, \frac{1}{2}, \frac{1}{2}$ in Figure 1.30 is an interstitial position in the FCC allotropic form of iron that can be filled with carbon.

When a substitutional or interstitial solid solution is formed, the Greek letter that was used for the pure metal is retained.

1.21 New Phases, Intermediate Phases

If instead of nickel we added 30 wt % lead to the copper melt, we would again obtain a liquid solution. However, when we examined the frozen melt, we would find two different phases. There would be grains of copper with a very small amount of lead in solid solution, and the balance of the lead would have the form of small spherical grains of practically pure lead. Both struc-

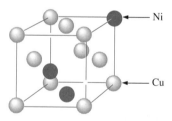

FIGURE 1.29   Substitutional solid solution of nickel in the FCC copper unit cell

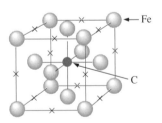

FIGURE 1.30   Interstitial solid solution of carbon in the FCC form of iron. Positions marked x are positions in the FCC equivalent to $\frac{1}{2}, \frac{1}{2}, \frac{1}{2}$ position.

tures are FCC, but the unit cell of lead is much larger, which means that the atomic radius is greater. Therefore we find two phases, α and β. Although the added lead lowers the strength, a material of this type is quite useful in bearings. When a load is applied, the lead from the spherical grains tends to flow out and coat the entire surface of the bearing with a film that has a low coefficient of friction.

In this case, the new phase has the same structure as the element being introduced. Frequently the new phase has a structure different from either element. This is called an *intermediate phase*. For example, if we add 15 wt % tin to copper, some of the tin will go into solution in the FCC copper phase, but we will find another phase with the approximate formula $Cu_{31}Sn_8$. This is a hard phase completely different from copper or tin.

We may ask if there are any general principles that determine whether a given element B will form a separate phase or a solid solution when added to another element A. This is important in alloy development because, in general, a solid solution is more ductile and more easily shaped, whereas a two-phase material usually exhibits greater hardness and strength.

The most important considerations in forming solid solutions are related to the atomic radii and chemical properties. To form an extensive solid solution (greater than 10 at. % soluble*), we should be aware of the following general rules of W. Hume-Rothery.

1. The difference in atomic radii should be less than 15%.
2. Proximity within the periodic table is important (similar electronegativities).
3. For a complete series of solid solutions, the metals must have the same crystal structure.
4. When two elements have different valency, the higher valency is more likely to dissolve in the lower valency element. For example, more Al is more likely to dissolve in Cu than Cu in Al.

Elements close to one another in the periodic table have similar electronegativities (Table 1.3) and are more compatible than elements that are further apart.

As an example, in Figure 1.31 we have plotted the atomic diameters ($2r$) of a number of elements. The atomic diameter of copper $\pm 15\%$ is shown by the dashed heavy lines, that of silver $\pm 15\%$ by the dashed light lines. If we include elements with an FCC or HCP structure, we find that zinc, which is in the favorable zone for both copper and silver, forms extensive solid solutions with both elements. Cadmium, however, forms extensive solid solutions with silver but not with copper, because it is inside the favorable zone of silver but outside that of copper.

---

*To find atomic percent, calculate the percentage atoms of element B for the total number of atoms present.

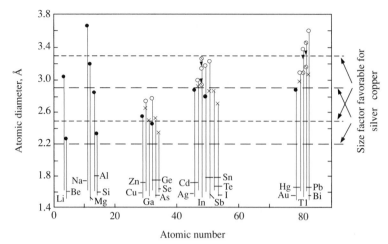

FIGURE 1.31   Favorable atomic diameter for elements soluble in copper and silver. When the size is within ±15%, there is extensive solid solution.

(Courtesy of W. Hume-Rothery)

EXAMPLE 1.9   *[ES/EJ]*

Predict whether extensive substitutional solid solubility will occur between copper as the solvent and aluminum, nickel, and chromium as the solute elements.

Element	Atom Radius (Å)	Crystal Structure	Electronegativity
Cu	1.28	FCC	1.9
Al	1.43	FCC	1.5
Ni	1.25	FCC	1.8
Cr	1.25	BCC	1.6

Answer   The radius difference is as follows:

$$\left(\frac{final - initial}{initial} \times 100\right) \quad \text{or} \quad \frac{r_x - r_{Cu}}{r_{Cu}} \times 100$$

$$Cu\text{-}Al \quad +11.7\%$$

$$Cu\text{-}Ni \quad -\ 2.3\%$$

$$Cu\text{-}Cr \quad -\ 2.3\%$$

All the elements fall within the required ±15% radius range. The electronegativity values are not vastly different. This indicates somewhat similar chemical activity for the metals, and possibly more than 10 at. % solubility.

However, the atom packing within the BCC Cr is not the same as in the FCC Cu. Therefore, we would not expect extensive solid solubility between the two elements. The order of expected solid solubility in copper, on the basis of radius difference and crystal structure, along with the observed experimental values, is as follows:

Element	Observed At. % Soluble
Ni (best)	100
Al	17
Cr (poorest)	<1

The importance of considering atomic percentage rather than weight percentage can be seen when 17 at. % Al is converted to wt %.

Basis: 100 atoms (17 atoms Al, 83 atoms Cu)

$$\frac{17 \text{ atoms} \times 26.98 \text{ g/at. wt}}{6.02 \times 10^{23} \text{ atoms/at. wt}} = \text{weight of 17 Al atoms}$$

$$\frac{83 \text{ atoms} \times 63.54 \text{ g/at. wt}}{6.02 \times 10^{23} \text{ atoms/at. wt}} = \text{weight of 83 Cu atoms}$$

Therefore

$$\text{Wt \% Al} = \frac{\dfrac{17 \times 26.98}{6.02 \times 10^{23}}}{\dfrac{17 \times 26.98}{6.02 \times 10^{23}} + \dfrac{83 \times 63.54}{6.02 \times 10^{23}}} \times 100 = 8.0$$

Because aluminum has a low atomic weight compared to copper, the weight percentage is much less than the atomic percentage. Nickel, however, is close to copper in atomic weight, so values for the atomic and weight percentages are very close.

1.22 Ceramic Crystals

We discussed metal crystals first because they involve the simplest crystal structures. Why is this true? In the case of metals, we were dealing with atoms of the same size (in a pure metal) or close to the same size (in a solid solution). We should point out that if we shake a box of tennis balls, or any hard spheres for that matter, we find the closest packed arrangements: FCC, HCP, or a mixture of both. For this reason many metals show these structures.

In the case of a ceramic, the ions are of different sizes and may have different charges, as in alumina, Al_2O_3. In laying out a structure on a space lattice, we need to have the formula ratio of atoms of different types, and we need overall charge neutrality. The atoms need not occupy lattice points — for example, we may have a cluster of four atoms, as in a carbonate ion, $(CO_3)^{2-}$, around a lattice point or at a constant distance from such a point.

Because the ions are of different sizes, we reach a limit in the number of ions of one type that can be placed around another and maintain contact. This is called the *coordination number* (CN) or the number of adjacent ions touching an adjacent ion. In the case of some metals, an examination of the basal plane in the hexagonal structure shows CN = 12. There are 6 nearest neighbors in the plane and groups of three above and below that touch the center atom in the base. This is the highest possible CN.

By contrast, in the sodium chloride structure (Figure 1.32, page 60) that is typical of many ceramics, we find a coordination number of 6 for sodium (6 chlorines touching each sodium); and in quartz (Figure 1.17), we find a coordination number of 4 for silicon (4 oxygens touching each silicon).

These different coordination numbers arise from the fact that, for a stable ionic configuration, the ions must touch. When we look up the ionic radii of sodium and chlorine in Table 1.6 (page 57), we see that the sodium ions will contact the surrounding chlorine ions (Figure 1.32). By contrast, a silicon ion could not be substituted; it would "rattle around" because of its small size.

Let us see if this model checks for the *density* of salt.

EXAMPLE 1.10 *[ES]*

Calculate the density of sodium chloride from the crystal structure (Figure 1.32) and the atomic weights of Na^+ and Cl^-.

Answer We know that

$$\text{Density} = \frac{\text{mass}}{\text{volume}} = \frac{\text{mass of a unit cell}}{\text{volume of a unit cell}} \tag{1-8}$$

In Figure 1.32 the larger ions (Cl^-) form an FCC-like arrangement. From our earlier discussion, we know that the FCC structure has four atoms per unit cell. Therefore, sodium chloride has four Cl^- ions, and to maintain the 1:1 stoichiometry (ratio of chemical species) it must have four Na^+ ions.

$$\frac{\text{Mass}}{\text{Unit cell}} = \frac{4 \; Na^+ \text{ ions} \times 22.99 \text{ g/at. wt} + 4 \; Cl^- \text{ ions} \times 35.45 \text{ g/at. wt}}{6.02 \times 10^{23} \text{ atoms/at. wt}}$$

$$= 3.88 \times 10^{-22} \text{ g}$$

Even though the structure in Figure 1.32 resembles an FCC structure, the ions touch along an edge as in a simple cubic, *not* along a face diagonal. (For ionic radii see Table 1.6.)

$$\text{Volume/unit cell} = (2 \times r_{Cl^-} + 2 \times r_{Na^+})^3$$
$$= (2 \times 1.81 \times 10^{-8} \text{ cm} + 2 \times 0.98 \times 10^{-8} \text{ cm})^3$$
$$= 1.737 \times 10^{-22} \text{ cm}^3$$
$$\text{Density} = \frac{3.88 \times 10^{-22} \text{ g}}{1.737 \times 10^{-22} \text{ cm}^3} = 2.24 \text{ g/cm}^3 \ (2.24 \times 10^3 \text{ kg/m}^3)$$
$$= 2.24 \text{ Mg/m}^3$$

TABLE 1.6 Atomic and Ionic Radii of the Elements

Atomic Number	Symbol	As Element Type of Structure*	As Element Coordination Number	As Element Interatomic Distances, Å†	As Element Atomic Radii, Å	As Ion State of Ionization	As Ion Goldschmidt Ionic Radii, Å
1	H	HCP	6, 6	—	0.46	H$^-$	1.54
2	He	—	—	—	—	—	—
3	Li	BCC	8	3.03	1.52	Li$^+$	0.78
4	Be	HCP	6, 6	2.22; 2.28	1.14	Be^{2+}	0.54
5	B	—	—	—	0.97	B^{3+}	0.2
6	C	Dia.	4	1.54	0.77	C^{4+}	<0.2
		Hex.	3	1.42	0.71		
7	N	Cubic	—	—		N^{5+}	0.1 to 0.2
8	O	Orthorh.	—	—	0.60	O^{2-}	1.32
9	F	—	—	—	—	F$^-$	1.33
10	Ne	FCC	12	3.20	1.60	—	—
11	Na	BCC	8	3.71	1.86	Na$^+$	0.98
12	Mg	HCP	6, 6	3.19; 3.20	1.60	Mg^{2+}	0.78
13	Al	FCC	12	2.86	1.43	Al^{3+}	0.57
14	Si	Dia.	4	2.35	1.17	Si^{4-}	1.98
						Si^{4+}	0.39
15	P	Orthorh.	3	2.18	1.09	P^{5+}	0.3 to 0.4
16	S	FC Orthorh.	—	2.12	1.06	S^{2-}	1.74
						S^{6+}	0.34
17	Cl	Orthorh.	—	2.14	1.07	Cl$^-$	1.81
18	Ar	FCC	12	3.84	1.92	—	—
19	K	BCC	8	4.62	2.31	K$^+$	1.33
20	Ca	FCC	12	3.93	1.97	Ca^{2+}	1.06
		HCP	6, 6	3.98; 3.99	2.00		
21	Sc	FCC	12	3.20	1.60	Sc^{2+}	0.83
		HCP	6, 6	3.23; 3.30	1.64		
22	Ti	HCP	6, 6	2.91; 2.95	1.47	Ti^{2+}	0.76
						Ti^{3+}	0.69
						Ti^{4+}	0.64
23	V	BCC	8	2.63	1.32	V^{3+}	0.65
						V^{4+}	0.61
						V^{5+}	~0.4
24	Cr	BCC (α)	8	2.49	1.25	Cr^{3+}	0.64
		HCP (β)	6, 6	2.71; 2.72	1.36	Cr^{6+}	0.3 to 0.4
25	Mn	Cubic (α)	—	2.24 to 2.96	1.12	Mn^{2+}	0.91
		Cubic (β)	—	2.36 to 2.68	1.18	Mn^{3+}	0.70
		FCT (γ)	8, 4	2.58; 2.67	~1.37	Mn^{4+}	0.52

TABLE 1.6 *(Continued)*

Atomic Number	Symbol	As Element				As Ion	
		*Type of Structure**	*Coordination Number*	*Interatomic Distances, Å†*	*Atomic Radii, Å*	*State of Ionization*	*Goldschmidt Ionic Radii, Å*
26	Fe	BCC (α)	8	2.48	1.24	Fe^{2+}	0.87
		FCC (γ)	12	2.52	1.26	Fe^{3+}	0.67
27	Co	HCP (α)	6, 6	2.49, 2.51	1.25	Co^{2+}	0.82
		FCC (β)	12	2.51	1.26	Co^{3+}	0.65
28	Ni	HCP (α)	6, 6	2.49; 2.49	1.25	Ni^{2+}	0.78
		FCC (β)	12	2.49	1.25		
29	Cu	FCC	12	2.55	1.28	Cu^+	0.96
30	Zn	HCP	6, 6	2.66; 2.91	1.33	Zn^{2+}	0.83
31	Ga	Orthorh.	—	2.43 to 2.79	1.35	Ga^{3+}	0.62
32	Ge	Dia.	4	2.44	1.22	Ge^{4+}	0.44
33	As	Rhomb.	3, 3	2.51; 3.15	1.25	As^{3+}	0.69
						As^{5+}	~0.4
34	Se	Hex.	2, 4	2.32; 3.46	1.16	Se^{2-}	1.91
						Se^{6+}	0.3 to 0.4
35	Br	Orthorh.	—	2.38	1.19	Br^-	1.96
36	Kr	FCC	12	3.94	1.97	—	—
37	Rb	BCC	8	4.87	2.51	Rb^+	1.49
38	Sr	FCC	12	4.30	2.15	Sr^{2+}	1.27
39	Y	HCP	6, 6	3.59; 3.66	1.81	Y^{3+}	1.06
40	Zr	HCP	6, 6	3.16; 3.22	1.58	Zr^{4+}	0.87
		BCC	8	3.12	1.61		
41	Nb	BCC	8	2.85	1.43	Nb^{4+}	0.74
						Nb^{5+}	0.69
42	Mo	BCC	8	2.72	1.36	Mo^{4+}	0.68
						Mo^{6+}	0.65
43	Tc	—	—	—	—	—	—
44	Ru	HCP	6, 6	2.64; 2.70	1.34	Ru^{4+}	0.65
45	Rh	FCC	12	2.68	1.34	Rh^{3+}	0.68
						Rh^{4+}	0.65
46	Pd	FCC	12	2.75	1.37	Pd^{2+}	0.50
47	Ag	FCC	12	2.88	1.44	Ag^+	1.13
48	Cd	HCP	6, 6	2.97; 3.29	1.50	Cd^{2+}	1.03
49	In	FCT	4, 8	3.24; 3.37	1.57	In^{3+}	0.92
50	Sn	Dia.	4	2.80	1.58	Sn^{4-}	2.15
		Tetra.	4, 2	3.02; 3.18	—	Sn^{4+}	0.74
51	Sb	Rhomb.	3, 3	2.90; 3.36	1.61	Sb^{3+}	0.90
52	Te	Hex.	2, 4	2.86; 3.46	1.43	Te^{2-}	2.11
						Te^{4+}	0.89
53	I	Orthorh.	—	2.70	1.36	I^-	2.20
						I^{5+}	0.94
54	Xe	FCC	12	4.36	2.18	—	—
55	Cs	BCC	8	5.24	2.65	Cs^+	1.65
56	Ba	BCC	8	4.34	2.17	Ba^{2+}	1.43
57	La	HCP	6, 6	3.75; 3.75	1.87	La^{3+}	1.22
		FCC	12	3.75	1.87		
58	Ce	HCP	6, 6	3.63; 3.65	1.82	Ce^{3+}	1.18
		FCC	12	3.63	1.82	Ce^{4+}	1.02
59	Pr	Hex.	6, 6	3.63; 3.66	1.83	Pr^{3+}	1.16
		FCC	12	3.64	1.82	Pr^{4+}	1.00
60	Nd	Hex.	6, 6	3.62; 3.65	1.82	Nd^{3+}	1.15

TABLE 1.6 *(Continued)*

Atomic Number	Symbol	As Element				As Ion	
		Type of Structure*	Coordination Number	Interatomic Distances, Å†	Atomic Radii, Å	State of Ionization	Goldschmidt Ionic Radii, Å
61	Pm	Hex.	—	—	—	Pm^{3+}	1.06
62	Sm	Rhomb.	—	—	1.81	Sm^{3+}	1.13
63	Eu	BCC	8	3.96	2.04	Eu^{3+}	1.13
64	Gd	HCP	6, 6	3.55; 3.62	1.80	Gd^{3+}	1.11
65	Tb	HCP	6, 6	3.51; 3.59	1.77	$\begin{cases} Tb^{3+} \\ Tb^{4+} \end{cases}$	1.09 0.89
66	Dy	HCP	6, 6	3.50; 3.58	1.77	Dy^{3+}	1.07
67	Ho	HCP	6, 6	3.48; 3.56	1.76	Ho^{3+}	1.05
68	Er	HCP	6, 6	3.46; 3.53	1.75	Er^{3+}	1.04
69	Tm	HCP	6, 6	3.45; 3.52	1.74	Tm^{3+}	1.04
70	Yb	FCC	12	3.87	1.93	Yb^{3+}	1.00
71	Lu	HCP	6, 6	3.44; 3.51	1.73	Lu^{3+}	0.99
72	Hf	HCP	6, 6	3.13; 3.20	1.59	Hf^{4+}	0.84
73	Ta	BCC	8	2.85	1.47	Ta^{5+}	0.68
74	W	$\begin{cases} BCC\ (\alpha) \\ Cubic\ (\beta) \end{cases}$	8 12; 2, 4	2.74 2.82; 2.52, 2.82	1.37 1.41	$\begin{cases} W^{4+} \\ W^{6+} \end{cases}$	0.68 0.65
75	Re	HCP	6, 6	2.73; 2.76	1.38	Re^{4+}	0.72
76	Os	HCP	6, 6	2.67; 2.73	1.35	Os^{4+}	0.67
77	Ir	FCC	12	2.71	1.35	Ir^{4+}	0.66
78	Pt	FCC	12	2.77	1.38	$\begin{cases} Pt^{2+} \\ Pt^{4+} \end{cases}$	0.52 0.55
79	Au	FCC	12	2.88	1.44	Au^{+}	1.37
80	Hg	Rhomb.	6	3.00	1.50	Hg^{2+}	1.12
81	Tl	$\begin{cases} HCP \\ BCC \end{cases}$	6, 6 8	3.40; 3.45 3.36	1.71 1.73	$\begin{cases} Tl^{+} \\ Tl^{3+} \end{cases}$	1.49 1.06
82	Pb	FCC	12	3.49	1.75	$\begin{cases} Pb^{4-} \\ Pb^{2+} \\ Pb^{4+} \end{cases}$	2.15 1.32 0.84
83	Bi	Rhomb.	3, 3	3.11; 3.47	1.82	Bi^{3+}	1.20
84	Po	—	—	2.81	1.40	Po^{6+}	0.67
85	At	—	—	—	—	At^{7+}	0.62
86	Rn	—	—	—	—	—	—
87	Fr	—	—	—	—	Fr^{+}	1.80
88	Ra	—	—	—	—	Ra^{+}	1.52
89	Ac	—	—	—	—	Ac^{3+}	1.18
90	Th	FCC	12	3.60	1.80	Th^{4+}	1.10
91	Pa	—	—	—	—	—	—
92	U	Orthorh.	—	2.76	1.38	U^{4+}	1.05
93	Np	—	—	—	—	—	—
94	Pu	—	—	—	—	—	—
95	Am	—	—	—	—	—	—
96	Cm	—	—	—	—	—	—

The structure of calcium fluoride is somewhat more complex. The simplest way to visualize it is to note that the calcium ions form an FCC array and the eight fluoride ions form a simple cubic structure within the FCC (Figure 1.33). This satisfies the formula CaF_2 because we have four Ca^{2+} (see

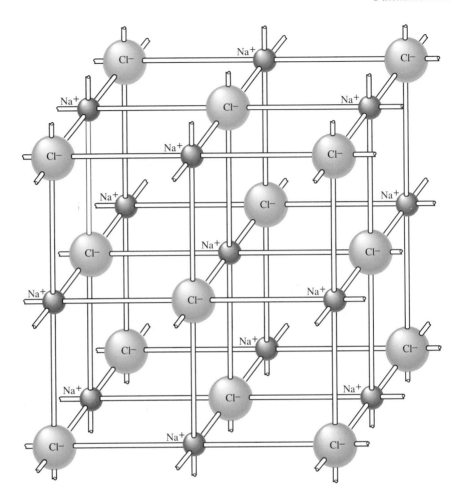

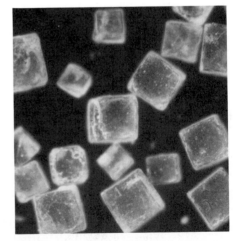

FIGURE 1.32 The sodium chloride crystal. (*a*) A model of a portion of the crystal, detailing the regular arrangement of sodium ions and chloride ions. Each sodium ion is surrounded by six chloride ions, and each chloride ion by six sodium ions. (*b*) A photograph showing the cubic shape of sodium chloride crystals.

[Part (*a*), D. Ebbing, *General Chemistry*, 2nd. ed., p. 41. Copyright © 1987 Houghton Mifflin Company. Reprinted by permission. Part (*b*), Martin Rotker/Taurus Photos]

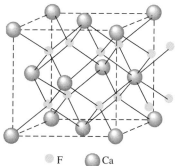

FIGURE 1.33   Fluorite structure (calcium fluoride)

[*Introduction to Ceramics*, 2nd ed., by W. D. Kingery, H. K. Bowen and D. R. Uhlmann, fig. 2.27. Copyright © 1976 John Wiley & Sons, Inc. Reprinted by permission of John Wiley & Sons, Inc.]

the density calculations for copper in Example 1.6) and the eight F^- ions are clearly within the unit cell, giving CaF_2.

The Perovskite Structure

So far we have only considered structures with two ions, but those with three or more are very important. The $20 billion electroceramics industry is based on such structures as perovskite, spinel, and garnet. We will consider only the perovskite structure here, leaving other details for the chapters on electrical, optical, and magnetic properties.

Until a few years ago, the mineral perovskite was only seen in a few dusty, rare-mineral collections. It was first identified in the 1830s by G. Rose and named after the Russian mineralogist Count Perovski. The chemical formula is $CaTiO_3$, and the charges are balanced: Ca^{2+}, Ti^{4+}, $3(O^{2-})$. The unit cell is shown in Figure 1.34. We find the calcium and oxygen ions making up a FCC box with a titanium ion at the center. Calcium titanate itself is of no industrial importance, but if we substitute other ions for the calcium and titanium while *retaining the same unit cell still called the perovskite structure*, we have very important electrical materials. Barium titanate is used in capacitors and piezoelectric or electro-mechanical devices. The complex perovskites, with more than three ions, are the basis of the new ceramic superconductors such as $YBa_2Cu_3O_{6.5}$ (called 1, 2, 3 type because of the ratio of the metal ions)

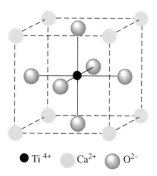

FIGURE 1.34   Perovskite structure (idealized)

[*Introduction to Ceramics*, 2nd ed., by W. D. Kingery, H. K. Bowen and D. R. Uhlmann, fig. 2.28. Copyright © 1976 John Wiley & Sons, Inc. Reprinted by permission of John Wiley & Sons, Inc.]

● Ti^{4+} Ca^{2+} O^{2-}

(Chapter 20). We shall see that the source of the unusual electrical properties is the small deviation in ion positions from the cubic lattice, which leads to an unsymmetrical charge distribution.

We can form solid solutions in ceramic crystals by following guidelines similar to the Hume-Rothery rules we noted when we talked about metallic alloys. For example, FeO can be dissolved in MgO in all proportions, and the crystal structure NaCl type remains the same. The dimensions a_0, the edge of the unit cell, changes uniformly from the value for MgO to that for FeO. Because in this case Fe^{2+} is involved, there is no problem with charge balance. The ionic radii of Mg^{2+} and Fe^{2+} are close, so a continuous series of solid solutions occurs. The atomic size difference of $\pm 15\%$ in the Hume-Rothery rules for metals may not hold for these ionic materials. It is more important that we maintain the same coordination number, a point that will be discussed further in Chapter 10.

By contrast, consider the very important silica–alumina system. The components are soluble in all proportions in the liquid state, but upon freezing, a silica structure, an alumina structure (corundum), and a compound of Si, Al, and O (mullite) are encountered (Chapter 10).

1.23 Polymer Crystals

We have already mentioned that the giant molecules of the polymers are either long snake-like structures or three-dimensional random networks. How can we expect to find crystals in these materials? It develops that crystals are encountered only in the linear polymers and that 100% crystallinity is not attained. However, the development of even partial crystallinity is important because of its effects on mechanical, thermal, and optical properties. This crystallinity is achieved in a unique fashion. Very long molecules such as chains of 10,000 carbon atoms are produced first by chemical reaction (polymerization, as discussed in Chapter 13). Upon cooling, from the liquid these molecules bend back and forth to form individual crystals. The lattice spacings can be determined and density calculations performed, just as for metals and ceramics.

EXAMPLE 1.11 *[ES]*

The polyethylene structure (Figure 1.35) is an example of a crystalline space lattice in a polymer. The giant molecules fold back and forth to generate a regular lattice.

Calculate the density of 100% crystalline polyethylene. Although in this case the unit cell is not drawn through atom centers, assume two molecules (mers) of C_2H_4 per cell.

Answer

$$\text{Mass/unit cell} = \frac{4 \text{ C atoms} \times 12 \text{ g/at. wt} + 8 \text{ H atoms} \times 1 \text{ g/at. wt}}{6.02 \times 10^{23} \text{ atoms/at. wt}}$$

$$= 9.30 \times 10^{-23} \text{ g}$$

$$\text{Volume/unit cell} = [(2.53 \text{ Å} \times 4.93 \text{ Å} \times 7.40 \text{ Å})(10^{-8} \text{ cm/Å})]^3$$

$$= 9.23 \times 10^{-23} \text{ cm}^3$$

$$\text{Density} = \frac{9.30 \times 10^{-23} \text{ g}}{9.23 \times 10^{-23} \text{ cm}^3}$$

$$= 1.01 \text{ g/cm}^3 \ (1.01 \times 10^3 \text{ kg/m}^3 = 1.01 \text{ Mg/m}^3)$$

The experimental density is 0.92–0.96, depending on the degree of crystallinity. The difference is due to poorer packing within and between molecules called "free space."

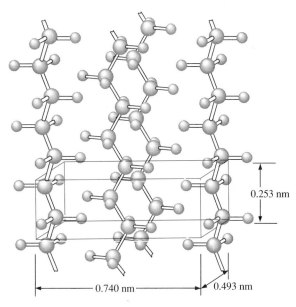

FIGURE 1.35 Molecular crystal (polyethylene). The chains are aligned longitudinally. The unit cell is orthorhombic (90° angles between axes).

[Gordon, *High Polymers*, © 1964, Addison-Wesley Publishing Co., Inc., Reading, Massachusetts. Reprinted with permission of the publisher. After C. W. Bunn, *Chemical Crystallography*, Oxford.]

1.24 Defect Structures in Crystals

So far we have described perfect lattice structures, but there are many reasons why atoms or groups of atoms may be missing. These defects may either raise or lower strength, or they may be advantageous in promoting electrical conductivity or diffusion. It is important, therefore, to catalog these defects and trace their origins. Defects may be classified as point defects, line defects (dislocations), or space defects, as in cavities. Here we will cover only *point defects* (Figure 1.36).

1. *Missing atom* (vacancy). This defect can arise from three sources: thermal agitation, accidents in growth, and charge balance requirements. In the case of thermal agitation, which is covered in detail in Chapter 3, a small percentage of atoms or molecules have a high enough energy to move away from the normal lattice positions, leaving vacancies. The accidents in growth are particularly marked at grain boundaries, where mismatches of the space lattices develop. The necessity for charge balance (electric neutrality) leads to regular vacancies. For example, it is important to stabilize (that is, prevent the transformation of) zirconia, ZrO_2, by incorporating calcium ions in the lattice. However, when we replace Zr^{4+} ions with Ca^{2+} ions, we reduce the overall positive charge. To reduce the negative charge proportionately, it is necessary to remove some O^{2-} ions, leaving cation (negative ion) vacancies. The opposite case, an anion vacancy, develops when Fe^{3+} ions are added to ferrous (Fe^{2+}) oxide.

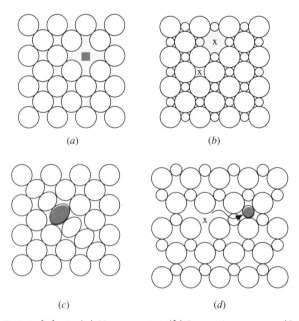

(a)   (b)

(c)   (d)

FIGURE 1.36   Point defects (a) Vacancy, □. (b) Ion-pair vacancy (Schottky defect). (c) Interstitialcy. (d) Displaced ion (Frenkel defect).

2. *Schottky defect.* In this case, in a balanced ionic lattice we have a missing anion (+) and a cation (−).

3. *Frenkel defect.* Here the ion leaves a normal site and occupies an interstitial site, leaving a vacancy.

1.25 Determination of Crystal Structure with Electrons and X-ray Methods

When the typical engineer is asked about applications of electrons and x rays in studying materials, the answer that comes to mind first is the uses of the scanning electron microscope and of radiographs such as a chest x ray. Two other uses of great significance are related to the concepts discussed in this chapter.

1. The use of *characteristic* x rays to identify the location of specific atoms in the microstructure. For example, in Figure 1.37 we see a specimen that was cut from an aluminum alloy engine block and is about 17% Si, 2% Cu, and 81% Al. The microstructure consists of several phases, as shown in the upper left-hand corner. We want to find how the elements are distributed among these phases. We do this by having the atoms of each element in turn emit x rays, which appear as bright regions. For example, the dark script-like phase in the upper left has little aluminum but is practically pure silicon (lower left).

2. The use of x rays of known wavelength to determine crystal structure and unit cell dimensions. In this case we determine the distances between definite planes in the lattice and the density of packing of atoms on the planes. From these data we can calculate unit cell dimensions and atom locations.

Let us investigate the first case, the generation of characteristic x rays. We define *characteristic x rays* as those that have a definite wavelength, unlike the *white x rays* of many wavelengths used in radiography. To generate characteristic or monochromatic x rays, we accelerate electrons from a hot filament to a metal target by a selected potential of 20–50 kilovolts. When the energy of the electrons is raised to a threshold value by the potential, they can knock out *K*-shell electrons from the target atoms (Figure 1.38). The entering electron and the *K*-shell electron depart, and then an *L*-shell electron falls to the empty state in the *K* shell. This is a precise energy change and a photon is emitted of a definite wavelength, as shown by the equation

$$\lambda = \frac{hc}{E} \tag{1-9}$$

where

λ = wavelength

h = Planck's constant

E = energy difference between L and K shells

c = velocity of light

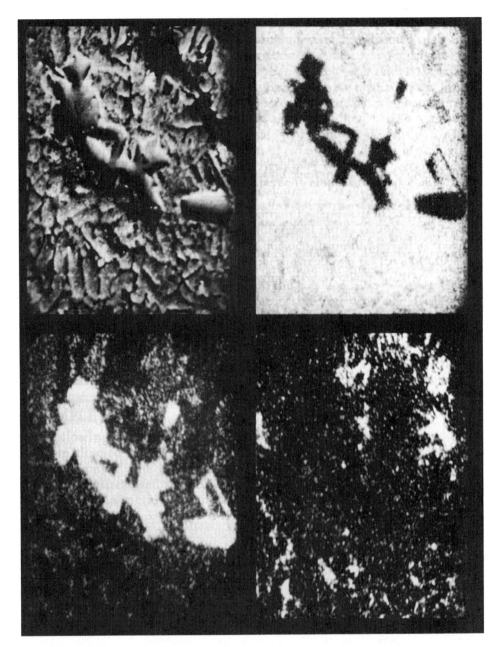

FIGURE 1.37   A microprobe analysis of an aluminum alloy: Upper left — electron micrograph; Lower left — bright areas show silicon-rich phase; Upper right — bright areas show aluminum-rich matrix; Lower right — copper-rich portions appear bright.

[John Mardinly, Department of Materials Science, University of Michigan]

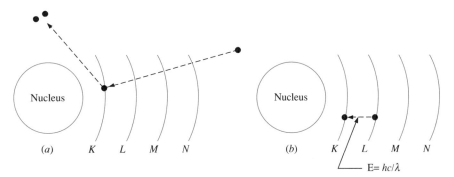

FIGURE 1.38 (*a*) In the production of a characteristic x ray, an incident electron strikes a *K*-shell electron and both leave the atom. (*b*) The movement of an *L*-shell electron into the *K* shell produces the characteristic x ray known as K_α.

The important point is that *E*, the difference in energy between the *K* shell and the *L* shell is characteristically different for each element. The following short table shows this effect when *L* shell electrons fall to the *K* shell, giving photons which have the K_α wavelength shown below and in Figure 1.38.

Element	Atomic Number	K_α Wavelength, Å (nm)
Cr	24	(0.229)
Fe	26	(0.194)
Co	27	(0.179)
Cu	29	(0.154)
Mo	42	(0.0709)
W	74	(0.0209)

In actual operation, to obtain the photos of Figure 1.37 an electron beam was moved across the sample. Characteristic x rays of the aluminum, copper, and silicon atoms were emitted from regions rich in these elements. Then x-ray maps were obtained by using the different emitted x rays in turn to show the location of different elements in a given field. The value of these x-ray maps is that they reveal the effect of different elements on the structure. For example, because the script-like phase is a hard, silicon-rich material, we can either increase the amount by raising the silicon content to make the engine block more wear-resistant or soften it for better machinability by reducing the silicon content.

Now let us examine the x-ray diffraction method. It does not measure the positions of the individual atoms directly but instead the distances between planes of atoms. To visualize this in two dimensions, consider a simple unit cell *ABCD* (Figure 1.39). If we determine the distance between parallel planes containing *AB* and *CD* and between parallel planes containing *AC* and *BD*, we have determined the cell.

We measure the interplanar distances by x-ray diffraction in the manner shown in Figure 1.40, where the planes are similar to *AB* and *CD* in Figure 1.39. The differences between diffraction and reflection are important. In reflection (as with a mirror), we can shine a light beam at a surface at any angle and it will be reflected at the same angle. In diffraction, the x-ray beam penetrates beneath the surface. When it strikes the atoms, they re-emit the radiation at the same wavelength. However, if the radiations being emitted are not in phase, there will be destructive interference and no diffracted beam.

If the radiations are in phase, there will be a beam, as shown in Fig. 1.40. To obtain this beam we must have a certain relationship among the angle of incidence, the wavelength of the radiation λ, and the interplanar distance d. For example, the path of the x-ray beam (1–1') that goes one planar distance beneath the surface is longer than that of the beam diffracted by the surface atoms (2–2') by the distance $2d \sin \theta$. For the radiations to be in phase,

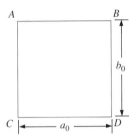

FIGURE 1.39   Calculation of $a_0 b_0$. By finding distances between planes *AC* and *BD*, one can determine a_0.

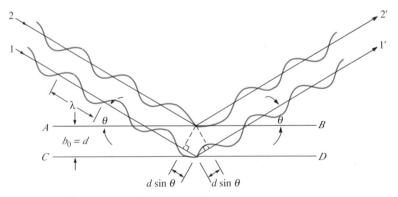

FIGURE 1.40   Diffraction of x rays from planes of atoms

$2d \sin \theta$ must equal either the wavelength λ or some multiple $n\lambda$. This principle is known as *Bragg's law* and is written

$$n\lambda = 2d \sin \theta \qquad (1\text{-}10)$$

Therefore if we know θ and λ, we can find d. We obtain a beam of x rays of constant wavelength λ from an x-ray tube and observe the angle θ at which diffraction occurs.

EXAMPLE 1.12 *[ES]*

A sample of BCC chromium was placed in an x-ray beam of $\lambda = 1.54$ Å. Diffraction from 110 planes was obtained at $\theta = 22.2°$. Calculate a_0. Assume n (the order of diffraction) = 1.

Answer

$$\lambda = 2d \sin \theta \qquad d = \frac{1.54 \text{ Å}}{2(0.378)} = 2.04 \text{ Å (0.204 nm)}$$

If we inspect the BCC unit cell, we see that the distance between 110 planes is one-half the face diagonal, or

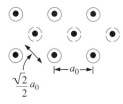

Therefore,

$$a_0 = 2.04\sqrt{2} = 2.88 \text{ Å (0.288 nm)}$$

(This is a very much simplified case. Determining crystal structure by diffraction methods is an interesting and important field covered in advanced courses. Note also that parentheses are not used for the experimentally determined diffraction indices to distinguish these planes from the idealized Miller index planes.)

Generalizing the procedure in this example to the determination of distances between planes in cubic structures, we have the relationship

$$d_{hkl} = \frac{a_0}{\sqrt{h^2 + k^2 + l^2}} \qquad \text{(for cubics only)} \qquad (1\text{-}11)$$

In practice, only specific planes diffract, as Bragg's law is derived for the mutual support of the x radiation. The existence or extinction of diffraction from specific sets of planes lets us determine what type of cubic structure is diffracting the x rays. For example, for BCC structures, $h + k + l$ must be even if diffraction is to occur, whereas for FCC, h, k, and l must be either all odd or all even if diffraction is to occur.

1.26 Noncrystalline Structures

Ceramics

Glass and ceramics with partly glassy structures have been made since ancient times. They should be discussed first because much of the terminology used for polymers, such as *glass transition temperature*, is derived from glass technology. We need to understand the structure of glass to explain its behavior on cooling from the liquid, as well as its mechanical and other properties. (Metallic glasses will be covered separately later.)

Every student who has ever worked with glass tubing from a chemistry set knows that as the glass is heated, it gradually softens and finally, if overheated, becomes a syrupy liquid. This is in sharp contrast to a pure crystalline substance such as an ice cube, which melts sharply at 0°C.

Another way of looking at this phenomenon is to measure the volume of a pure material that crystallizes when cooled slowly or forms a glass when cooled at a faster rate (Figure 1.41). The melt crystallizes to a solid with a sharp change in volume at constant temperature (T_m, melting point) when cooled slowly. When cooled more rapidly, the liquid becomes more and more viscous as the temperature falls. At temperature T_g, the glass transition temperature, the change in volume with falling temperature becomes less and the glass will support modest loads.

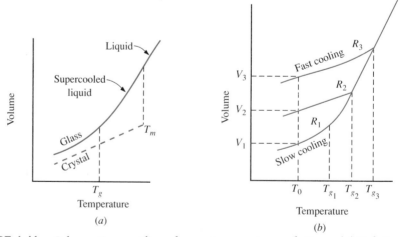

FIGURE 1.41 Schematic specific volume–temperature relations. (*a*) Relations for liquid, glass, and crystal. (*b*) Glasses formed at different cooling rates $R_1 < R_2 < R_3$.

[*Introduction to Ceramics*, 2nd ed., by W. D. Kingery, H. K. Bowen and D. R. Uhlmann. Copyright © 1976 John Wiley & Sons, Inc. Reprinted by permission of John Wiley & Sons, Inc.]

X-ray diffraction data for the crystalline material indicate a regular lattice, whereas the data for the glass show no long-range order. Both structures in the example shown for silica are made up of the same basic $(SiO_4)^{4-}$ tetrahedra, but in the glass these are linked in random fashion. An indication of the random arrangement of the tetrahedra in the glass compared to the arrangement of those in the crystal is shown in the broader distribution of bond angles in the glass (Figure 1.42).

Polymers

At this point we will leave the inorganic glasses for later discussion and go on to the organic (polymer) glass structures. If we cool a melt of a linear polymer

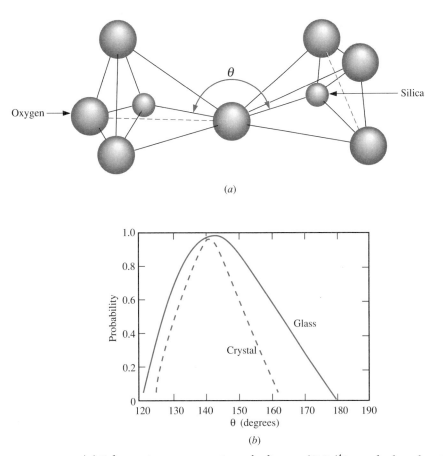

FIGURE 1.42   (a) Schematic representation of adjacent $(SiO_4)^{4-}$ tetrahedra, showing Si—O—Si bond angle. (b) Distribution of Si—O—Si bond angles in fused silica and crystalline crystobalite, a specific crystal structure in silica.

[From R. L. Mozzi, Sc.D. thesis, MIT, 1967. (*R. L. Mozzi and B. E. Warren, *J. Appl. Cryst.*, **2**, 164 (1969)].

from the liquid, we may obtain a volume–temperature curve showing glassy behavior or crystalline behavior as with the inorganic glass. There is, however, an important difference in the details of the development of the glassy condition. In the linear polymer we have many large molecules. Above the glass transition temperature T_g, the bonding force is weak — a few cross links between long chains and some mechanical linkage. Below T_g the molecules are closer and the thermal agitation is less, so van der Waals forces and hydrogen bonding between molecules become important. The material becomes rigid or elastic.

If the material crystallizes, the more regular and closer packing increases the attractive forces, and in general the strength and stiffness of the crystalline material are higher than for the glassy material.

The structure of the polymer molecules is very important in determining their degree of crystallinity. To begin with, the typical thermosetting three-dimensional polymer structures do not crystallize. The linear polymers, such as polyethylene, crystallize by folding individual molecules back and forth into a regular arrangement. Furthermore, when the linear polymer contains bulky side groups, as in the case of polystyrene, crystallization is rare.

We will examine other features of polymer structure in Chapter 13.

Metals — Metallic Glasses

There is a strong driving force for metals and alloys to assume a crystal structure upon solidification, because the equilibrium structures are so simple and therefore easily formed compared to the complex ionic and covalently bonded compounds of the ceramics and polymers. Accordingly, it is necessary to cool at a very fast rate to develop amorphous structures. As an analogy, consider two ways of delivering tennis balls to a bin. If we perform the transfer slowly while vibrating the bin, we will get the close-packed, regular arrangement discussed earlier. On the other hand, if we dump a truckload of balls suddenly into the bin, the arrangement will be random, as in an amorphous material. To attain this random arrangement in metals, cooling rates of the order of one million degrees per second are needed. This is accomplished by splat cooling — projecting liquid droplets against the interior of a rotating cold metal drum or pouring liquid between closely spaced rolls. These *metallic glasses* have superior corrosion resistance (no grain boundaries for preferential attack) and superior strength (no regular slip planes or well-defined dislocations that lower strength — See Chapter 2). However, vacancies and irregular dislocations are present.

Summary

At the beginning of the chapter we promised that after mastering this material, you would be able to look at the same group of objects with deeper insight born of a knowledge of their basic structure. We have tried to give you this insight by progressing in four closely linked steps.

1. *Atomic structure.* We began with the 105 atoms as building blocks for all materials and pointed out that for our purposes, the most important characteristic was the number of orbital electrons, the atomic number.

   From these numbers we laid out the periodic table, which enabled us to understand the properties of the elements in groups rather than as 105 unrelated entities.

   In order to examine the individual properties of the elements and their tendency to bond, we described the energy and location of the electrons around the nucleus, using a set of four quantum numbers for each electron.

2. *Bonding.* With this background we were able to understand primary and secondary bonds. We found that ionic bonding occurred between a nonmetal such as chlorine (seeking an electron to build its outer shell from seven to a stable group of eight) and a metal such as sodium (ready to release its single outer-shell electron to expose a stable group of eight beneath).

   In covalent bonds we found that two atoms that are the same or close in electronegativity, such as carbon or silicon and oxygen, share electrons. Many cases (such as a silicon and oxygen combination) show bonding that is partly covalent and partly ionic.

   Hybridization is an important aspect of covalent bonding in which, in the case of diamond, $2s$ and $2p$ electrons form identical envelopes around the nucleus, leading to perfect tetragonal bonding and the diamond crystal.

   In metallic bonding the atoms shed their outer-shell electrons to form a negative common cloud that bonds the positive ion cores. Because of mutual repulsion, the positive ions assume positions on a regular space lattice.

   Secondary bonds caused by permanent dipoles, London forces, and hydrogen bonding establish the mechanical properties of many polymers* and of some ceramics, such as mica and graphite.

3. *Crystalline and noncrystalline materials.* As a result of bonding forces and methods of processing, materials are found in crystalline or noncrystalline forms or in mixtures of the two. We learned that crystalline structures consist of grains that are made up of unit cells of atoms on a regular space lattice. The atom positions can be established by x-ray methods or high-resolution electron microscopy. Various types of defects were described, which will later be shown to have an important influence on mechanical, thermal, and electrical properties. Long-range order and short-range order are present in crystal structures.

   In glassy structures only short-range order, at most, is present. These materials do not exhibit a sharp melting point, as a pure crystalline material does, but gradually soften above the glass temperature.

---

*However, by producing fibers of crystalline covalently bonded polymers, it is possible to attain very high strengths in the longitudinal direction (Kevlar).

4. *Groups of materials.* The following table summarizes the atoms of which the three groups of materials are composed, the types of bonding that each exhibits, and whether each is crystalline or glassy in structure.

Some Characteristics of Metals, Ceramics, Polymers

	Atoms Present	Bonding	Crystalline or Noncrystalline
Metals and metallic alloys	Principally the elements on the left side of the periodic table with outer shell of 1–3 electrons. Other elements may be present in minor amounts.	Principally metallic. Some precipitates may form with covalent or ionic bonding.	Crystalline, except under very fast cooling.
	Characteristics: High electrical and thermal conductivity because of free electrons; good strength because of metallic bond; good ductility because of packing and metal bonding.		
Ceramics	Metal plus O, N, C, H; Si also very important.	Covalent, ionic or both.	Crystalline or glassy or *mixed.*
	Characteristics: Insulating because electrons tightly bonded; high strength and low ductility because of ionic and covalent bonds; high stiffness, low coefficient of expansion.		
Polymers	C + H, N, O, S, Cl, etc.	Covalent within molecules and secondary forces *between* molecules.	Glassy or mixed crystalline–glassy.
	Characteristics: Insulating because electrons tightly bonded: lower strength* and softening point because of only secondary bonds between molecules; low stiffness, high coefficient of expansion.		

*See previous footnote about Kevlar.

Definitions

Allotropy   The occurence of two or more crystal structures at the same chemical composition.

Amorphous   Said of an atomic structure (or of a molecular arrangement viewed as an atomic structure) that is completely lacking in order.

Atomic radius   A measure calculated from the unit cell dimensions by finding a direction in which atoms are in contact (for example, the diagonal of the cube face in an FCC). Because the diagonal length can be calculated from a_0, the edge of the unit cell, the atomic radius can be found.

Ceramic materials   Materials exhibiting ionic or covalent bonds or both.

Coordinates   Location of an atom in the unit cell, found by moving specified distances from the origin in the *x*, *y*, and *z* directions.

Coordination number   The number of equidistant nearest neighbors to a given atom. The sodium ion, for instance, has six equidistant chloride ions in sodium chloride, so $CN = 6$.

Crystalline polymer   Substance in which a molecule or molecules are aligned in a regular array for which a unit cell can be described. Crystallinity is generally not complete.

Defects in atomic structures

    Point defect   An atom is missing in an atomic arrangement.

    Frenkel defect   An ion leaves a normal lattice site and occupies an interstitial site.

    Schottky Defect   In an ionic lattice both a positive and negative ion are missing, such that electroneutrality is maintained.

Density   Mass divided by volume, usually expressed in grams per cubic centimeter, kilograms per cubic meter or megagrams per cubic meter.

Electronegativity   The relative tendency of an atom to attract electrons.

Fluorite   The crystal structure of the compound CaF_2.

Glass   A noncrystalline solid that softens (rather than melting) at a distinct temperature because of its random three-dimensional network of atoms.

Glass transition temperature   The temperature at which the rate of contraction of a cooling noncrystalline material changes to a lower value.

Hybridization   The combination of electrons from two separate energy levels to produce a new, single electron energy level.

Indices of direction   Descriptions of direction in a unit cell found by translating the direction passing through the origin, finding the coordinates of the point at which the arrow leaves the unit cell, and clearing fractions. These indices are enclosed in brackets.

Interstitial solid solution   A solution formed when a dissolved element fills holes (interstices) in the lattice of the solvent element.

Linear density   The number of atoms that have their centers located on a given line of direction for a given length.

Metallic glasses   Metals that have been rapidly cooled from the liquid state (10^5 to 10^8 °C/s) to preserve the amorphous state.

Metallic materials   Materials characterized by metallic bonds.

Miller–Bravais indices   Descriptions of a plane found by taking reciprocals of the intercepts of the plane on the *x*, *y*, and *z* axes and clearing fractions. Miller indices are enclosed in parentheses.

Noncrystalline   Characterized by the lack of a regular arrangement or order in a collection of atoms in space.

Number of atoms per unit cell   A number found by connecting the corners of the cell to form an imaginary box, determining the fraction of each atom contained within the box (for example, an atom at a corner position of a cube is counted as one-eighth, because it is shared equally by eight unit cells meeting at the center of the atom), and adding the fractions.

Periodic table   A tabular grouping of the elements in order of atomic number, with atomic number increasing as one moves across the table from left to right. The metallic elements are on the left-hand side, because in general these elements contain only one, two, or three electrons in the outer shell.

Perovskite   The crystal structure of the compound $BaTiO_3$ or $CaTiO_3$.

Phase   A homogeneous aggregation of matter. In solid metals, grains composed of atoms of the same unit cells are the same phase. Alloying atoms may also be present in a one-phase structure.

Planar density   The number of atoms with centers located within a given area of the plane. The planar area selected should be representative of the repeating groups of atoms in the plane.

Plastics (high polymers)   Materials exhibiting principally covalent bonding. The residual bonding forces, van der Waals bonds, are also important.

Polyethylene   A plastic or polymer material composed of many ethylene (C_2H_4) molecules covalently bonded to one another.

Primary bonds

    Covalent bond   Bond in which electrons are shared by atoms of the same element (such as carbon) or different elements, so a strong bond is produced.

    Ionic bond   Bond between metal and nonmetal ions in which the metal gives up an electron or electrons, which are taken up by the outer shell of the nonmetal, producing positive ions of the metal and negative ions of the nonmetal, which attract each other.

    Metallic bond   Bond in which metal atoms give up their electrons to an electron gas and take up a regular arrangement.

Quantum number   One of a set of four numbers that describes the characteristics of each electron in an atom.

Secondary bonds

    Dipole   The positive and negative centers of a molecule are not coincident, which results in polarity that attracts other species.

    Hydrogen bonding   Covalently bonded hydrogen acts positively and is attracted to another negatively acting atom. The attractive force depends on the species involved.

    London forces   Electron orbits around the positive nucleus give temporary dipoles because of the nonsymmetry of the orbit.

    Van der Waals forces   Hydrogen and/or London force bonding.

Shorthand notation   A system for describing the number of electrons in each shell of an atom and their subgrouping.

Silica   The compound SiO_2 which is composed of tetrahedra of silicon and oxygen.

Solid solution   A solution formed when the addition of one or more new elements still results in a single-phase structure.

Structure   The organization of a material, defined by specifying which atoms are present and in what amounts, how they are arranged in unit cells, and what grains result from the unit cells. The presence of several different unit cells leads to grains or phases with different properties.

Substitutional solid solution   A solution formed when the dissolved element substitutes for (replaces) an atom or atoms of the solvent element in its unit cell.

Thermal expansion   The change in unit length per unit of temperature change, $\Delta l/\Delta T$. Weaker materials generally show greater thermal expansion.

Transition elements   A group of elements having partially filled $3d$ electron orbitals with one or two electrons at the $4s$ level.

Unit cell   A geometric figure illustrating the grouping of atoms in a solid. This group, or module, is repeated many times in space within a grain or crystal.

X-ray diffraction   Diffraction of short-wavelength radiation by atoms on specific planes of a crystal structure.

     Bragg's law   An x-ray diffraction equation relating the wavelength to the geometry of the crystal structure.

     Characteristic x rays   Fixed-wavelength x radiation resulting from displacement of a K-shell electron followed by insertion of an electron from an adjacent shell.

     White x rays   The x rays of many wavelengths produced when electron displacement and insertion occur from all available shells.

Problems

1.1   *[ES/EJ]* For the following elements, write out the shorthand electron notation and explain whether the elements are active, inactive, or intermediate metals or nonmetals: nitrogen, calcium, chromium, silicon, and sulfur. (Sections 1.1 through 1.4)

1.2   *[EJ]* We know that both calcium and magnesium remove dissolved oxygen from molten iron by forming the insoluble compounds CaO or MgO. Why might both elements perform the same task, and why is the melting point of pure calcium higher than that of pure magnesium? (Sections 1.1 through 1.4)

1.3   *[ES]* Give the un-ionized and ionized shorthand electron notation for the following elements. (Sections 1.1 through 1.4)

$$Cu \longrightarrow Cu^+ \text{ and } Cu^{2+} \qquad \text{atomic number} = 29$$
$$Fe \longrightarrow Fe^{2+} \text{ and } Fe^{3+} \qquad \text{atomic number} = 26$$
$$K \longrightarrow K^+ \qquad\qquad\qquad \text{atomic number} = 19$$
$$V \longrightarrow V^{3+} \text{ and } V^{5+} \qquad \text{atomic number} = 23$$
$$Ga \longrightarrow Ga^{3+} \qquad\qquad\quad \text{atomic number} = 31$$

1.4   *[ES/EJ]* Write out the shorthand electron notation for the following elements and justify their designation as active or inactive. (Sections 1.1 through 1.4)

   a. Helium — inactive

   b. Fluorine — active

   c. Krypton — inactive

   d. Cerium — active

1.5   *[ES/EJ]* An unbalance in spin in an atom can give rise to a net magnetic moment. Write out the quantum numbers for samarium in the fourth electron shell, and indicate which electrons could contribute to an unbalance in spin and hence ferromagnetism. (Sections 1.1 through 1.4)

1.6 *[ES]* Write out all four quantum numbers of the following common elements for each electron. (Sections 1.1 through 1.4)

K   (19) $1s^2, 2s^2, 2p^6, 3s^2, 3p^6, 4s^1$
Fe (26) $1s^2, 2s^2, 2p^6, 3s^2, 3p^6, 3d^6, 4s^2$
Cu (29) $1s^2, 2s^2, 2p^6, 3s^2, 3p^6, 3d^{10}, 4s^1$
Mo (42) $1s^2, 2s^2, 2p^6, 3s^2, 3p^6, 3d^{10}, 4s^2, 4p^6, 4d^5, 5s^1$

1.7 *[ES]* Write out the four quantum numbers for sodium and chlorine in the un-ionized and ionized states. (The ionized states would be present in the compound NaCl.) (Sections 1.1 through 1.4)

1.8 *[ES/EJ]* Why does it require four quantum numbers, rather than three or five numbers, to represent the electrons of an atom? (Sections 1.1 through 1.4)

1.9 *[EJ]* Despite the strength of the ionic bond, many materials with ionic bonds are not considered "good engineering materials." Why not? [*Hint:* Consider how ionics react to different environments.] (Sections 1.5 through 1.11)

1.10 *[ES]* Sketch the valence-electron configuration for an NO_3^- ion and for a single carbon dioxide molecule, CO_2. (Sections 1.5 through 1.11)

1.11 *[ES]* Sketch the valence-electron configuration for water (H_2O) and formaldehyde (CH_2O). (Sections 1.5 through 1.11)

1.12 *[ES]* Silica (SiO_2) can be treated as either a covalently or an ionically bonded material. (The silicon-to-oxygen distance is essentially the same whether the radii are treated as atomic or ionic.) Sketch the valence-electron configuration for silica both ways. (Sections 1.5 through 1.11)

1.13 *[EJ]* Producing finished engineering components normally requires processing of the raw materials. Considering the bond type in ceramic materials, indicate the limitations on the following types of processing. (Sections 1.5 through 1.11)
   a. Casting a ceramic material—that is, heating it to the molten state and pouring it into a mold.
   b. Forming a ceramic material by both a hot and a cold rolling technique.

1.14 *[EJ]* The bonds in polymers range from completely covalent to a combination of covalent and van der Waals. Indicate the limitations on the following types of processing. (Sections 1.5 through 1.11)
   a. Casting a polymer—that is, heating it to the molten state and pouring it into a mold.
   b. Forming a polymer by both a hot and a cold rolling process.

1.15 *[EJ]* Metallic elements do not all exhibit the same strength, yet we say the bond *type* is the same for all metals. What must we consider besides the valence electron interaction? What indices of metallic bond strength do we have in addition to mechanical strength? (Sections 1.5 through 1.11)

1.16 *[ES/EJ]* MgO has the same crystal structure as NaCl. The melting point of MgO is 2800°C, and that of NaCl is 800°C. (Sections 1.5 through 1.11)
   a. Calculate the attractive bonding force between magnesium and oxygen in MgO if the interionic distance is 0.210 nm.
   b. Indicate what features besides the attractive bonding force might contribute to the difference in melting point between MgO and NaCl.

1.17 *[ES]* In the text, carbon is used as an example of hybridization. In effect, it produces two different crystal forms, graphite and diamond. Indicate two other elements that might hybridize their *s* and *p* electrons. (Sections 1.5 through 1.11)

1.18 *[ES/EJ]* The table in Section 1.11 suggests that hydrogen bonding forces are stronger secondary bonding mechanisms than dipole–dipole or London forces. Why might this be the case, and in what class of engineering materials is hydrogen bonding important? (Sections 1.5 through 1.11)

1.19 *[ES]* Calculate the electronegativity of sulfur, using a basis of 4.0 for the electronegativity of fluorine. (See Example 1.4.) Determine the percentage difference between your calculated value and that given in Table 1.3. (Sections 1.5 through 1.11)

1.20 *[ES/EJ]* Calculate the fraction ionic character for sodium chloride. (Sections 1.5 through 1.11)
   a. Because we have used NaCl as our example of ionic bonding, what is the significance of a calculated ionic fraction not equal to 1.0?
   b. Repeat the calculation for a carbon–hydrogen bond and indicate the significance of the fraction ionic character.

1.21 *[ES/EJ]* Why is it easier to produce an amorphous material in polymers and in some ceramics than in metals? (*Hint:* See Figures 1.17 and 1.18.) (Sections 1.12 through 1.17)

1.22 *[ES]* Indicate whether the following statements are correct or incorrect, and justify your answer. (Sections 1.12 through 1.17)
   a. The number of atoms intersected by the [110] direction in an FCC is the same as the number intersected by the [111] direction in a BCC.
   b. The linear density in the [110] direction in an FCC is the same as the linear density in the [111] direction in a BCC.
   c. The number of atoms intersected by the (100) plane and the number intersected by the (110) plane of an FCC are the same.
   d. The planar density on the (100) plane and on the (110) plane of an FCC is the same.

1.23 *[ES]* Calculate the planar density of atoms (atoms/mm^2) in FCC gold on the (100), (110), and (111) planes. (Sections 1.12 through 1.17)

1.24 *[ES]* Calculate the linear density (atoms/Å) of atoms in the [100], [110], and [111] directions in BCC iron ($a_0 = 2.86$ Å). (Sections 1.12 through 1.17)

1.25 *[ES]* In FCC lead, calculate the planar density of atoms (atoms/m^2) on the (100) and (110) planes and the linear density of atoms (atoms/m) in the [110] and [111] directions. (Sections 1.12 through 1.17)

1.26 *[ES]* In BCC barium, calculate the planar density of atoms (atoms/m^2) on the (111) and (110) planes and the linear density of atoms (atoms/m) in the [100] and [111] directions. (Sections 1.12 through 1.17)

1.27 *[EJ]* A student says that the ratio of the experimental density to the calculated density is the same as the ratio of atoms actually present to available atom positions in the unit cell. Under what circumstances is this conclusion not correct? (Sections 1.12 through 1.17)

1.28 *[ES]* One of your associates has calculated the density of body-centered cubic iron (atomic radius = 1.24 Å and atomic weight = 55.85) as follows:

$$\text{Density} = \frac{\dfrac{4 \text{ atoms/u.c.} \times 55.85 \text{ g/at. wt}}{6.02 \times 10^{23} \text{ atoms/at. wt}}}{\left[\dfrac{2(1.24 \times 10^{-8} \text{ cm})}{\sqrt{2}}\right]^3} = 68.8 \text{ g/cm}^3$$

Materials do not have densities this high, so errors must have been made. What are these errors? (Assume that the errors are not a result of having pushed the wrong buttons on the calculator.) (Sections 1.12 through 1.17)

1.29 *[ES/EJ]* We have a hypothetical FCC element Q with an atomic radius of 0.131 nm and an experimental density of 8.41×10^3 kg/m³. (Sections 1.12 through 1.17)
   a. Calculate the approximate atomic weight of element Q.
   b. Why is the calculation only approximate?

1.30 *[ES]* Calculate the planar density of atoms in the (0001) plane of zinc. Why is the answer the same as for the (111) plane of copper? Will this be true for any HCP and FCC structures? (Sections 1.12 through 1.17)

1.31 *[ES/EJ]* Show that an HCP structure can be treated as having either six atoms per unit cell or two atoms per unit cell. (Sections 1.12 through 1.17)

1.32 *[EJ]* As noted in the text, the c/a ratio of an idealized HCP structure should be 1.63. The measured c/a ratios for HCP metals are never quite equal to this value. Using as an example, zinc, with a c/a ratio of 1.86, describe how one would arrive at a value for the atomic radius. (Sections 1.12 through 1.17)

1.33 *[EJ]* Several years ago a national emergency occurred as a result of a shortage of tungsten, used in producing high-speed steel. A typical analysis of this steel was 18% W, 4% Cr, 1% V, balance Fe. It was found that molybdenum could be substituted for more than half the tungsten. What basic relationships would lead you to believe that these elements can be substituted for one another? Why might substituting titanium for tungsten not work so well? (Sections 1.18 through 1.21)

1.34 *[ES/EJ]* Refer to Problem 1.33. (Sections 1.18 through 1.21)
   a. Calculate the atomic percentage of each of the elements present in the alloy if it is assumed that the composition is given as a weight percentage.
   b. If one-half of the tungsten is to be replaced by molybdenum, as stated, is this likely to be a weight or atomic percentage replacement? Justify your answer.

1.35 *[ES]* During the recent accelerated development of titanium alloys for aircraft and underwater research, one group concentrated on alloys that would be single-phase. Decide which of the following metals would be expected to form extensive solid solutions with titanium.

Ti is HCP	$a_0 = 2.95$ Å	V is BCC	$a_0 = 3.04$ Å
Be is HCP	$a_0 = 2.28$ Å	Cr is BCC	$a_0 = 2.88$ Å
Al is FCC	$a_0 = 4.04$ Å		

For those elements that form extensive solid solutions, calculate the value in weight percent that would correspond to 10 at. % in titanium. (Use a definition of extensive solid solutions as 10 at. %.) (Sections 1.18 through 1.21)

1.36 *[ES]* As we mentioned, pure iron undergoes an allotropic transformation at 912°C (1674°F). The BCC form is stable at temperatures below 912°C, whereas the FCC form is stable above 912°C. Calculate the volume change for the transformation BCC → FCC, if, at 912°C, $a = 3.63$ Å for FCC and $a = 2.93$ Å for BCC. (Sections 1.18 through 1.21)

1.37 *[EJ]* A number of metals undergo an allotropic transformation from one crystal structure to another at a specific temperature. There is generally a volume change that accompanies the transformation (see Problem 1.36). What is the practical significance of such data? (Sections 1.18 through 1.21)

1.38 *[ES/EJ]* Solid solutions occur to some extent in all alloys. Given the FCC unit cell of iron, $a_0 = 3.60$ Å, calculate the radius of an atom for which you would expect a high probability of *interstitial* solid solution. Why would carbon not fill all of the available interstitial positions? Calculate the range in radius for an atom for which you would expect extensive *substitutional* solid solution. (Sections 1.18 through 1.21)

1.39 *[EJ]* Criteria for extensive substitutional solid solubility are given in the text. However, there is no mention of the temperature at which the solubility is measured.

What would be the effect of temperature on the solid solubility, and why might it not be uniform between element pairs? (Sections 1.18 through 1.21)

1.40 *[ES]* Titanium nitride (TiN) has the same crystal structure as NaCl. The side of the unit cell of TiN is 4.235 Å (0.4235 nm). What is its density? (Sections 1.22 through 1.24)

1.41 *[ES]* The strontium oxide (SrO) structure can be described as similar to a BCC with an Sr^{2+} ion at the center and O^{2-} ions at the cube corners. (Sections 1.22 through 1.24)
a. Using the ionic radii given in Table 1.6, calculate the unit cell size. (The ions touch along a body diagonal.)
b. Calculate the theoretical density of the structure in units of g/cm^3.

1.42 *[ES/EJ]* Review the comments in Example 1.11, and explain the difference between high- and low-density polyethylene. Which variety would be stronger and which would be more pliable? (*Note:* Both types find uses as commercial products.) (Sections 1.22 through 1.24)

1.43 *[ES/EJ]* Indicate whether the following statements are correct or incorrect, and justify your answer. (Sections 1.22 through 1.24)
a. Substitutional solid solutions in ceramic materials follow the same general rules as those developed for metals.
b. Alloying in polymers is impossible.

1.44 *[ES/EJ]* Assuming that the materials with point defects as shown in Figure 1.36 are ceramics, explain how their strength might be modified by the defect. (Sections 1.22 through 1.24)

1.45 *[ES]* X rays with a wavelength λ of 2.29 Å are diffracted from BCC iron (lattice constant = 2.866 Å) at a Bragg angle θ of 34.5° (sin 34.5° = 0.566). Is this diffraction from 110 planes? (Sections 1.25 through 1.26)

1.46 *[ES]* We are given an x-ray tube with anode material unlabeled. We obtain an x-ray diffraction pattern for FCC aluminum and find the diffraction angle to be 24.5° for the 111 planes. Determine the anode material from the wavelengths of elements given in Section 1.25 (Sections 1.25 through 1.26)

1.47 *[ES]* A student conducting an x-ray diffraction experiment indicates a rather poor reproducibility of the diffraction angle, at ±0.50°. Using the data of Problem 1.46, calculate the range in atomic radius for pure aluminum for this reproducibility. (Sections 1.25 through 1.26)

1.48 *[ES]* A specimen of silver (FCC, r_0 = 1.44 Å) is placed in an x-ray camera and irradiated with molybdenum-characteristic radiation (0.709 Å). It is observed that θ for 111 planes decreases by 0.11° as the silver is heated from room temperature to 800°C (1470°F). Given that the crystal structure remains the same, find the change in *a* due to the heating. [*Hint:* $d_{111} = a/\sqrt{3}$.] (Sections 1.25 through 1.26)

1.49 *[EJ]* Indicate whether the following statements are correct or incorrect, and justify your answer. (Sections 1.25 through 1.26)
a. The *d* values obtained from an x-ray diffraction pattern are the same whether the examined material is a pure metal or a solid solution alloy of the same metal.
b. A 111 plane from any FCC structure exhibits the same measured *d* value.
c. The distance between 110 planes of an FCC cannot be directly determined from x-ray diffraction data.

1.50 *[EJ]* Describe the type of experiment that would produce an amorphous metal and an amorphous polymer. Indicate how you would treat the two materials differently to make them amorphous and how you would test to be sure that crystallinity was not present. (Sections 1.25 through 1.26)

1.51 *[EJ]* Why has window glass, which seems to be a perfectly good solid material, been called a supercooled liquid? (Sections 1.25 through 1.26)

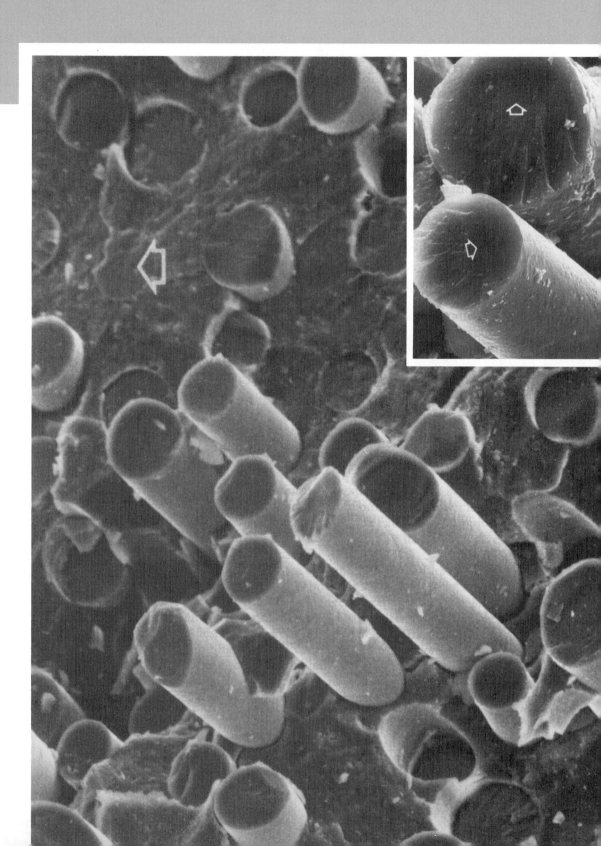

CHAPTER 2

Effects of Stress on Structure

One of the most important characteristics of a material is its response to stress. In modern engineering it is essential to understand the reactions of all three types of materials because the rapidly growing field of composites utilizes all types. In this illustration two types of ceramics are used to produce a composite with fracture toughness rivaling many metals.

2.1 Overview

Now that we have a picture of atomic arrangement and bonding, as well as of crystal, grain, and amorphous structures, we are prepared to develop a basic understanding of the effects of stress on these structures. To do so, we will examine seven key topics:

1. The elastic and plastic deformation that occurs in all materials under stress*
2. Test methods and stress–strain relations in metals
3. Test methods and stress–strain relations in ceramics
4. Test methods and stress–strain relations in polymers
5. Fracture toughness of all three groups of materials
6. Fatigue of materials
7. Wear and abrasion of materials

The most important requirement of most components—from aircraft wings through auto axles to our own bones—is that they endure stress without breaking and usually without undergoing "damaging" deformation. Let us see how the different families of materials respond to stress (Figure 2.1). We perform a very simple tensile test on each type of material and, in each case, measure first the stress needed to produce a small elastic elongation (only 1 ten-thousandth of an inch in a 1-inch gage length). We select a stress sufficient to deform the material only a small amount, so that it will spring back to very close to its original dimensions. Of course different materials can endure different levels of stress before being deformed permanently, but what we want to compare are the amounts of stress required to produce a given elastic deformation. This is important because most components work in the elastic range. After observing the elastic deformations at low stress, we will compare the materials at higher stresses that produce permanent deformation and fracture.

When stress is applied to a material, two things occur: The structure deforms first elastically and then permanently (plastically). There is, of course, an additional amount of elastic deformation during the second stage. We will also make simple observations of how the deformation changes with the duration of the test. This is very important at room temperature for the polymers and at "high" temperatures for the other materials.

Now to our observations of the tests. Everyone thinks of copper as very ductile—that is, easily hammered or bent to shape. But it is important to realize that at "low" stresses, copper is deformed elastically and springs back to its original shape. To produce an elastic deformation of 0.0001 in. (0.01%) in a 1-in. gage length, a stress of 1600 psi (11.04 MPa) is needed. In the case of a specimen of hardened steel in the elastic range, a stress of 3000 psi (20.7 MPa)

Elastic deformation is the change in dimensions under stress that *disappears* after the stress is removed. *Plastic deformation* is the change under stress that *persists* after the stress is removed.

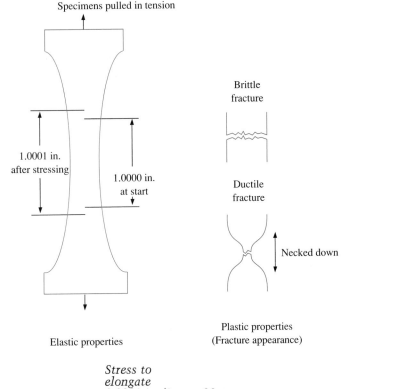

Elastic properties   Plastic properties
                                  (Fracture appearance)

Material		Stress to elongate 0.0001 in./in. (0.01 %) lbf/in² (MPa)	Max Stress lbf/in² (MPa)	% Plastic elongation	Fracture Appearance
Metals	Copper (annealed)	1600 (11.04)	30,000 (207)	45	Ductile
	Steel 1%C (hardened)	3000 (20.7)	250,000 (1725)	3	Low ductility
Ceramics	Glass (Pyrex)	1000 (6.9)	10,000 (69)	0	Brittle
	Alumina	5500 (37.95)	60,000 (414)	0	Brittle
Polymers	Thermo plastic	10 (0.069)	3000 (20.7)	100	Ductile
	Thermo set	100 (0.69)	10,000 (69)	1	Brittle
Elastomer		1 (0.0069)	3000 (20.7)	1	Brittle

FIGURE 2.1   An overview of elastic and plastic properties

is required to produce a 0.01% deformation. This is not because the steel is hardened but is, rather, a basic characteristic of iron alloys. It is due to the stronger bonding between iron atoms than between copper atoms.

As the stress is increased in the copper to 10,000 psi (69 MPa), considerable plastic deformation occurs. At 30,000 psi (207 MPa) and about 45% elongation (that is, an extension of 0.45 in. of a 1-in. gage length), ductile fracture occurs. By contrast, the hardened steel endures a much higher stress—about 250,000 psi (1725 MPa)—but ruptures with 4% elongation and a less ductile fracture.

In the case of the ceramic, a polycrystalline alumina specimen, the stress required for a 0.01% elongation is higher than for any of the metals—5500 psi (37.95 MPa). This high value of stiffness is characteristic of many strong ceramics and glass as a result of covalent and ionic bonds. On the other hand, we find that after rupture at 60,000 psi (414 MPa), there is no evidence of plastic deformation. Except for a few cases of testing in selective directions of single crystals or at elevated temperatures, brittle fracture is characteristic of ceramics. The surface of the fracture is irregular or may show *cleavage*. For this reason, bend and compression tests are commonly used, and tensile stresses are avoided in service where possible. Pyrex glass behaves in similar fashion.

The two polymers behave quite differently. Polyethylene (PE), made up of linear molecules, deforms elastically more easily than the phenol formaldehyde (PF), which is one giant molecule. Elastic stress values for 0.01% elongation are 10 psi (0.069 MPa) and 100 psi (0.69 MPa), respectively. If higher loads are used and kept constant, the specimen continues to elongate (creep). The values of rupture stress are in favor of the stiffer PF—3000 psi (20.7 MPa) for PE vs. 10,000 psi (69 MPa) for PF—but there is a great difference in plastic elongation—over 100% for PE compared to about 1% for PF.

In summary, we see that the different families of materials respond in various ways to stress. As a result, different test procedures have been developed for quality control and specifications. For example, although tensile tests are used widely for metallic materials, they are rarely employed for ceramics because a single flaw anywhere in the test section would give a low value. By contrast, in the bend test only a small region is under maximum stress and the probability of encountering a flaw is low. Also, the expense of machining a tensile specimen of hard ceramic is high.

To accommodate the differences in test procedures, we will discuss the tests for each family of materials separately. This will be followed by an overall discussion of crack initiation and growth and of fatigue testing. We shall see that in many cases, the sensitivity to crack propagation (fracture toughness) is more important than the conventional measures of strength developed from the tensile test. Finally, the resistance of materials to wear and abrasive conditions will be discussed.

2.2 Testing of Metallic Materials

Before going into the details of conventional test methods, let's spend a few moments watching atom movements under stress.

As a first step, we observe the side face of a bent copper bar at the regions of highest stress (Figure 2.2). We see by inspection that the upper layers of the bar are under tension and that the lower ones (near the support) are under compression, so there must be a "neutral axis" near the center. If we have

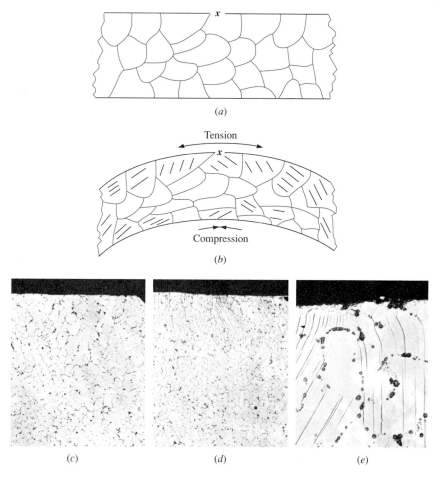

FIGURE 2.2   (*a*) Grains in a polished bar of copper. (*b*) Slip occurring during bending. (*c*) Photomicrograph of region *x* in (*a*), 100×. (*d*) Photomicrograph of region *x* in (*b*), 100×. (*e*) Photomicrograph of region *x* in (*b*), 500×.

taken care to polish and etch the side face to provide a good surface for microscopic examination (called metallographic examination), we find typical grains or crystals of copper before the test. As the bending progresses, we notice that fine lines appear in the grains in *similar* fashion on *both* the tension and compression sides but not at the neutral axis. From the change in shape of the grains, we can see that the motion on both sides of the neutral axis is in *shear* — in other words, planes of atoms are sliding over each other, *not* pulling apart. Furthermore, we find that this motion, called *slip*, starts earlier in some grains than in others, even when these grains are not at the region of maximum tensile or compressive stress. Another important effect is that slip may start in one location, stop, and then start in a new location. These are all interesting phenomena, but we need to return to our atomic-scale models of structure to explain them.

It is a good deal simpler to start with a *single crystal* as a test specimen and then progress to such polycrystalline materials as our copper bend-test specimen. (Years ago it was considered academic to discuss single crystals at all, but today they are an important commercial product. For example, silicon single crystals are the starting point for the manufacture of semiconductor chips!) We will discuss the growth of single crystals later, so for the present let us assume we have a single crystal of iron grown in the shape of a cylinder. Before we apply a load to the sample, let us define stress, strain, and modulus of elasticity (Young's modulus). We will call these terms *engineering stress, S,* and *engineering strain, e,* because later on we will define *true stress, σ,* and *true strain, ε.*

$$\text{Engineering stress, } S = \frac{\text{load}}{\text{original area of cross section}}$$

$$= \frac{P}{A_0} \tag{2-1}$$

The units for S are pounds force per square inch (psi) or newtons per square meter (N/m^2), which can also be expressed as pascals ($1 \ N/m^2 = 1$ Pa).

$$\text{Engineering strain, } e = \frac{\text{change in length}}{\text{original length}}$$

$$= \frac{\Delta l}{l_0} \tag{2-2}$$

The units for e are dimensionless: in./in. or m/m.

To indicate the elastic strain produced by a given stress, we define the modulus of elasticity, E, also called Young's modulus.

$$E = \frac{S}{e} \tag{2-3}$$

The units for E are psi or N/m^2 (Pa).

As long as we are careful not to reach too high a stress, the deformation is principally elastic and E is a constant. For each group of materials E has a characteristic value. For example, $E = 30 \times 10^6$ psi (2.07×10^5 MPa)* for all steels and 10×10^6 psi (0.69×10^5 MPa) for aluminum alloys. The modulus is related to the bonding between atoms.

This is the macroscopic picture of elastic strain, and it can be deceivingly simple. Let us go to the other extreme and test a single crystal of iron rather than a wire, which contains thousands of grains or crystals.

---

*1 psi = $6.9 \times 10^3 \ N/m^2 = 6.9 \times 10^{-3} \ MN/m^2 = 7.03 \times 10^{-4} \ kg/mm^2$ (MN = meganewton). 1MPa = 1 MN/m^2 (MPa = megapascal). We have chosen to use psi as the primary units in this chapter for the sake of the student. It is easier to visualize stress as force per unit area with units of pounds force per square inch compared to the single unit of Pascals. However, in all cases conversion between units is given for completeness.

If we stress the single crystal along different crystal directions, we get values quite different from 30 million psi (2.07×10^5 MPa).

Crystal direction:	[111]	[100]
$E(10^6 \text{ psi})$	41	18
$E(10^5 \text{ MPa})$	2.83	1.24

Although this seems astonishing at first, recall that in BCC structures (such as iron at room temperature), the atomic packing is densest in [111], the direction of highest E. We would expect that the interatomic forces would be greatest along this direction and therefore that the stress required to produce a given strain would be highest.

How do we explain the practically constant value of 30×10^6 psi (2.07×10^5 MPa) for steel? We get this value when there are many crystals of different orientations, because our measurement gives the average value. There are occasional important exceptions for which we must remember the properties of the single crystal. Let us consider some examples.

A number of identical dentures for teeth were cast of Vitallium, a cobalt alloy, and tested for deflection under constant stress. The deflections differed, although the cross sections were the same. It was found that because of the thin cross section, only one or two grains were present at the highly stressed region. The modulus E varies with direction in a grain, so the orientation of these grains determined the amount of deflection.

In some cases we want to produce a part with the crystals oriented in one direction. This is called development of *preferred orientation*. An outstanding example is the production of transformer steel sheet with what is termed a cubic texture. By special processing, the ⟨100⟩ directions of BCC iron are aligned in the plane of the sheet. In this case the ease of magnetization, like the modulus, varies with direction in the crystal. Thus when transformer laminations are stamped from sheet with this preferred orientation, the hysteresis losses (magnetic energy losses manifested as heat) in the finished part are lower.

A third example is a process that casts blades with controlled directional properties for aircraft gas turbines so that the best value of strength is in the direction in which the operating stress is highest (see page 259).

2.3 Plastic Strain, Permanent Deformation, Slip

There is a well-known physics experiment in which a wire is suspended from the ceiling and loaded with weights at the bottom. Occasionally an inventive student adds a really massive weight from some other equipment and finds that (1) the extension in length is more than expected from the equation

$e = S/E$, and (2) when the load is removed, the wire does not return to its original length. The portion of the total deformation under load that does not disappear when the load is removed is called *plastic* or *permanent* deformation. Let us now discuss what occurs within the structure during plastic deformation.

2.4 Critical Resolved Shear Stress for Plastic Deformation

To illustrate plastic deformation, let us assume that we grow a number of single crystals of zinc in the form of rods. To do this, we melt the zinc in a test tube in a vertical furnace and then lower the tube very slowly out of the bottom of the furnace. If we are careful, a single crystal will form and the rest of the metal will freeze with this crystal as a nucleus, giving us a single crystal in the form of a rod. After breaking away the glass, suppose that we grip the ends of the rods in a tensile machine and pull until there is appreciable permanent deformation. The specimens will differ greatly in the axial stress required to cause this permanent deformation, or strain. In all cases, however, *the flow occurs by shearlike movement (called "slip") on the {0001} planes.* The specimens that require the least stress for slip (plastic flow) are those with both the normal to the {0001} planes and the ⟨110⟩ directions at 45° to the axis of the specimen.

To explain this we need an important but simple derivation.

Consider the rod in Figure 2.3, which is a single crystal of zinc. We locate the orientation of the {0001} planes, such as A_2, via x rays. The (0001) planes are the basal planes of the HCP structure in the rod crystal. Slip takes place on these planes and in the [110] direction.

We first find the component of the *force* in the slip plane. This is $F \cos \lambda$. The shear stress in the slip plane is this force divided by the area A_2. Next A_2

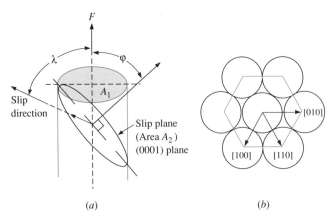

(a)   (b)

FIGURE 2.3   (a) Critical resolved shear stress model. (b) Slip directions in HCP.

may be related to the known area A_1 by trigonometry: $A_2 = A_1/\cos \varphi$. Therefore the shear stress τ is

$$\tau = \frac{F}{A_1} \cos \lambda \cos \varphi = S \cos \lambda \cos \varphi \qquad (2\text{-}4)$$

If we take the different axial stresses required to cause slip and the angular measurements and calculate the resolved shear stress in each case, we get the same value (within experimental error) for all. This is the *critical resolved shear stress* τ_c. When $\varphi = \lambda = 45°$, the maximum shear stress is obtained from a given axial stress.

We finally have an explanation for the unexpected behavior of the grains of copper in the bar in Figure 2.2. Although slip occurred in some regions away from the region of maximum tensile stress, these were regions of maximum resolved shear stress. That is, because of the favorable *orientation* of these grains, τ_c was reached sooner than in the other grains.

In hexagonal close-packed metals the slip occurred in the {0001} planes and in the ⟨110⟩ directions. As illustrated in Figure 2.3(b), three directions and one plane are involved. We define a plane and a slip direction in the plane as a *slip system*; therefore at room temperature, magnesium, an HCP, has three slip systems. The slip system is usually made up of the planes with highest atomic density (closest packing) and the directions with highest linear density. In FCC metals the {111} planes and the ⟨110⟩ directions reach maximum density. There are no exceptions to this rule. In BCC metals the ⟨111⟩ directions are most densely packed, and slip always occurs in these directions. There is no plane of maximum packing density, so a number of planes such as (110) are involved. In HCP metals the c/a ratio (see Chapter 1) is never the ideal value of 1.633. When this value is other than 1.633, planes other than the {0001} planes may slip.

EXAMPLE 2.1 *[ES]*

An investigator prepares four single crystals of magnesium in cylinders of the same cross-sectional area and finds that different axial stresses are required for permanent deformation (0.2% plastic strain). Furthermore, by using x rays the investigator finds the orientation of the slip plane and of the slip direction relative to the axis.

Are the differences in yield strength due to imperfections in the crystals? What are the critical resolved shear stresses? The data are as follows:

Crystal	Yield Stress g/mm^2	φ	λ
1	200	45°	54°
2	230	30°	66°
3	400	60°	66°
4	1,000	70°	76°

Answer

Crystal	F/A	$\times$	$(\cos \varphi)$	$\times$	$(\cos \lambda)$	$=$	$\tau_c, g/mm^2$
1	200		0.707		0.587		83
2	230		0.866		0.407		81
3	400		0.500		0.407		81
4	1,000		0.342		0.242		83

The critical resolved shear stresses are the same (within experimental error). Difference in orientation, not imperfections, causes variation in yield strength.

2.5 Twinning

Another type of plastic deformation is called *twinning*. It is particularly important in hexagonal crystals, because normally slip can occur on only one plane, (0001). If this plane is normal to the specimen axis, there is no shearing stress, and brittle fracture would occur if twinning could not take place. The essential difference between the two types of plastic deformations is that in slip each atom on one side of the slip plane moves a constant distance, whereas in twinning the movement is proportional to the distance from the twin boundary (Figure 2.4). Twinning is most common in BCC and HCP metals. It can

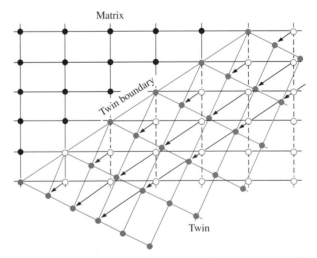

FIGURE 2.4   Formation of a twin in a tetragonal lattice by a uniform shearing of atoms parallel to the twin boundary. The dashed lines represent the lattice before twinning; the solid lines represent the lattice after twinning.

(H. W. Hayden, W. G. Moffatt, and John Wulff, *The Structure and Properties of Materials*, Vol. 3: Mechanical Behavior, John Wiley, New York, 1965, Fig. 5.10, p. 111. By permission of John Wiley & Sons, Inc.)

take place much more rapidly than slip. Therefore we often find mechanically formed twins rather than slip bands after shock loading. In FCC structures, twins are usually formed only on heating (annealing) of cold-worked structures (Chapter 3).

The differences between slip and mechanical twinning are sometimes hard to understand. Figure 2.5 shows two single crystals after plastic deformation. In Figure 2.5(a) the deformation is by slip. Each plane, approximately 1000 atoms thick, moves an integral amount relative to an adjacent plane. The movement is analogous to that brought about by putting your hands on the top and bottom of a squared deck of cards and then moving them in opposite directions.

Figure 2.5(b) shows deformation by twinning. The atoms move a distance proportional to their distance from the twin boundaries. The layer of atoms between the boundaries is called the mechanical twin.

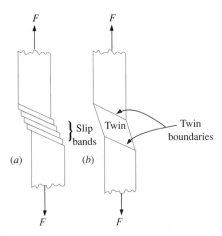

FIGURE 2.5   (a) Single crystal that exhibits slip. (b) A mechanical twin shown in a single crystal.

2.6 Engineering Stress–Strain Curves

Now let us look at the way in which the strength of commercial materials is tested and specified. In the specifications for materials of the principal engineering groups, such as the American Society for Testing and Materials (ASTM), the American Iron and Steel Institute (AISI), and the Society of Automotive Engineers (SAE), the most common basis for testing is the 0.505-in.-(12.8-mm-) diameter tensile specimen (Figure 2.6). In other words, whether we are buying material for a bridge or a crankshaft, the same tensile test is used to evaluate the material's mechanical properties. Usually we are interested in determining the modulus of elasticity, tensile strength, yield strength, percent elongation, and percent reduction of area of a given specimen. Let us discuss how these are determined.

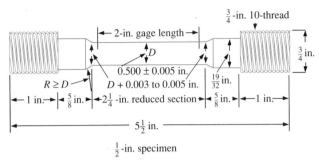

FIGURE 2.6   Design of a tensile test specimen. (To convert to millimeters, multiply by 25.4.)

The "0.505 bar," as it is called from its diameter, is machined from the stock being bought or from separately cast specimens from the same ladle of metal. The bar is screwed into a pair of grips, which are part of a tensile testing machine (Figure 2.7).

Next the grips are pulled apart by mechanical means. The load on the specimen is recorded continuously. The load can be converted to engineering stress by multiplying by 5, because the area of the 0.505-in.-diameter specimen is 0.2 in.^{2} and engineering stress is load divided by the original area.

It is important to measure the extension or strain of the bar at the same time. A variety of devices, from mechanically operated extensometers to

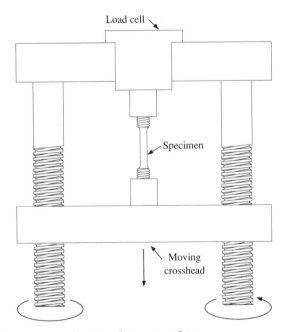

FIGURE 2.7   Cross section of a tensile test machine

(H. W. Hayden, W. G. Moffatt, and John Wulff, *The Structure and Properties of Materials*, Vol 3: Mechanical Behavior, John Wiley, New York, 1965, Fig. 1.1, p. 2. By permission of John Wiley & Sons, Inc.)

electric strain gages, are used. In the usual mechanical extensometer, the gage is anchored to a 2-in. gage length at the start of the test. Note that the raw data are in terms of *load* and *extension* (change in length). These are converted as follows:

$$\text{Engineering stress} = S = \frac{\text{load}}{\text{original area}}$$

$$\text{Engineering strain} = e = \frac{\text{change in length}}{\text{original length}}$$

We then plot the stress vs. strain and obtain *engineering stress–strain curves*, as shown in Figure 2.8(*a* and *b*).

The following data, obtained from a tensile test, are used in specifications.

Modulus of elasticity (psi or MPa)

= stress/strain) in elastic range (slope of stress–strain curve)

Tensile strength or ultimate tensile strength (psi or MPa)

= maximum stress on the stress–strain curve

Yield strength at 0.2% offset (psi or MPa)

= stress at which 0.2% permanent or plastic strain is present

Percent elongation at fracture

= $(l_f - l_o)/l_o \times 100$, where l_f = final length and l_o = original length

Percent reduction of area at fracture

= $(A_o - A_f)/A_o \times 100$, where A_o = original area and A_f = final area

Breaking stress, the engineering stress at fracture, is also noted but is not included in specifications for ductile materials. For brittle materials the breaking stress is hard to distinguish from the tensile strength.

The *modulus of elasticity* is used to calculate the deflection of a given part under load. For example, a crankshaft will bend a certain amount between bearings, and clearance must be allowed for this. Also, we may not substitute one material for another of the same strength without considering the modulus. For instance, an aluminum bolt will stretch three times as much as a similar steel one because the modulus is 10×10^6 psi (0.69×10^5 MPa) rather than 30×10^6 (2.07×10^5 MPa) psi. Note that the modulus does not change with strength; that is, it does not vary significantly within an alloy family. (See Figure 2.8*a* and Table 2.1.)

The *tensile strength* is an index of the quality of the material. It is not used much in design for ductile materials, because there has been a great deal of plastic strain by the time the tensile strength has been reached. However, it is a good indication of defects. If the bar has flaws or harmful inclusions, it will not reach the same maximum stress as a bar of higher-quality material.

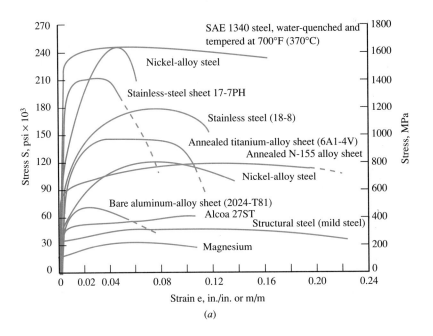

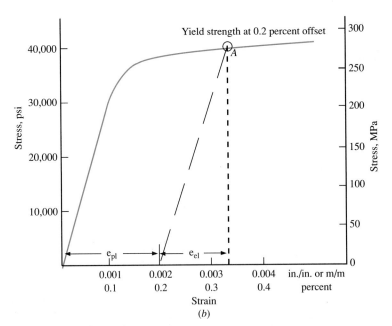

FIGURE 2.8 (*a*) Stress–strain curves for various alloys. (*b*) Details of an engineering stress–strain curve for mild steel (0.3% carbon).

[Part (*a*) from Joseph Marin, *Mechanical Behavior of Engineering Materials*, © 1962, Fig. 1.10, p. 24. Reprinted by permission of Prentice-Hall, Inc., Englewood Cliffs, N.J.]

TABLE 2.1 Moduli of Elasticity for Several Metallic Materials

Material	Modulus of Elasticity, psi $\times 10^6$ (MPa $\times 10^5$)
All steels, alloyed and unalloyed	30 (2.07)
Nickel alloys	26 to 30 (1.79 to 2.07)
Copper alloys	15 to 18 (1.04 to 1.24)
Aluminum alloys	10 to 11 (0.69 to 0.76)
Magnesium alloys	6.5 (0.45)
Cast iron, depending on amount and type of graphite	15 to 22 (1.04 to 1.52)
Ductile iron, depending on amount of graphite	22 to 25 (1.52 to 1.73)
Malleable iron, depending on amount of graphite	26 to 27 (1.79 to 1.86)
Molybdenum	47 (3.24)

The *yield strength* is the most important value for design. The significance of the phrase "0.2% offset" needs to be explained. As the tensile specimen is loaded, two types of strain develop: elastic and plastic. As we mentioned earlier, the elastic strain disappears upon unloading, whereas the plastic strain resulting from permanent deformation by slip will remain. It was once thought that the bar behaved elastically up to a certain point, called the *yield point*, after which plastic strain began. However, we know now from precise strain measurements and from microscopic observation of slip that minor plastic strain begins at very low stresses. The question then becomes "How much set (plastic elongation) can the designer tolerate?" The percentage of set that can be tolerated in the spring of a delicate balance is different from the percentage that can be tolerated in the boom of a steam shovel. However, for most engineering uses, a plastic strain of 0.2% can be tolerated. The stress at which this plastic strain occurs is called the 0.2% *offset yield strength*.

We calculate this value using the following procedure (Figure 2.8*b*). First, 0.002 strain (0.2% strain) is marked off from the origin on the *x* axis. Next, a line is drawn from this point parallel to the straight-line portion of the stress–strain curve until it intersects the curve. This point of intersection gives the value of the yield strength, which, for the material graphed in Figure 2.8(*b*), is 40,000 psi (276 MPa). If we follow normal testing procedures with a tensile specimen to this point and then remove the load, we will have 0.2% set, which is tolerable in most cases. In general, to find the amount of permanent and elastic strain at any point on a stress–strain curve, we draw a line parallel to the straight-line portion from the point to the strain axis, as shown in Figure 2.8(*b*). For the mild steel shown in that figure, the total strain at *A* is approximately 0.0033 in./in. (m/m), the elastic component e_{el} is 0.0013 in./in. (m/m), and the plastic component e_{pl} is 0.002 in./in. (m/m). If less plastic strain can be tolerated, the yield strength is specified at 0.1% offset or lower.

In other alloys, such as copper, aluminum, and magnesium, the stress–strain graph begins to curve at low stresses. In this case the yield strength is usually specified as the stress at 0.5% *total* strain. This strain is read directly from the graph as the sum of the elastic and plastic components.

The calculation of the *percent elongation at fracture* serves several purposes. It can be a better index of quality than the tensile strength, because if porosity or inclusions are present, the elongation is drastically lowered. Also, the elongation multiplied by the tensile strength is an index of toughness at low rates of strain. The toughest steel available, Hadfield's 12% manganese steel, is used for railroad crossings, for safe parts, and in ore crushers, as discussed in Chapter 9. It has an elongation of over 40% and tensile strength of over 100,000 psi (690 MPa). The gage length over which the percent elongation is calculated must be noted since percent elongation will change with gage length.

The *percent reduction in area* is important to deformation processing such as rolling. Note that percent *reduction* (decrease) rather than percent *change* is specified. If we were consistent with our definition of percentage change, the final value would be a negative number instead of a positive number.

As an example, suppose that the original area is equal to 0.20 in.^{2} and the final area is 0.10 in.^{2}

$$\% \text{ Change in area} = \frac{(0.10 - 0.20)}{(0.20)} \times 100 = -50\%$$

$$\% \text{ Reduction (or decrease in area)} = \frac{-(0.10 - 0.20)}{(0.20)} \times 100 = +50\%$$

EXAMPLE 2.2   *[ES]*

The following data were obtained for a high-strength aluminum alloy. A 0.505-in.-diameter (12.8 mm) tensile specimen with a 2-in. (50.8 mm) gage length was used. The load and gage length were obtained experimentally, whereas the stress and strain were calculated. Plot the engineering stress–strain curve.

Load, lb	Stress, psi (MPa)	Gage Length, in.	Strain
0	0	2.0000	0
4,000	20,000 (138)	2.0041	0.002
8,000	40,000 (276)	2.0079	0.004
10,000	50,000 (345)	2.0103	0.005
12,000	60,000 (414)	2.0114	0.006
13,000	65,000 (449)	2.0142	0.007
14,000	70,000 (483)	2.0202	0.010
16,000	80,000 (552)	2.0503	0.025
16,000 (maximum)	80,000 (552)	2.0990	0.050
15,600 (fracture)	78,000 (538)	2.1340	0.067

Calculate the modulus of elasticity, yield strength at 0.2% offset, percent elongation, and percent reduction of area. The diameters at maximum load and fracture are 0.485 in. (12.3 mm) and 0.468 in. (11.9 mm), respectively.

Answer The data for load are converted to stress by multiplying by 5, because the area of a 0.505-in.-diameter tensile test bar is 0.2 in.^{2}. The data are then plotted, as shown in Figure 2.9. Note that to find the modulus of elasticity and yield strength, we use a magnified scale on the *x* axis.

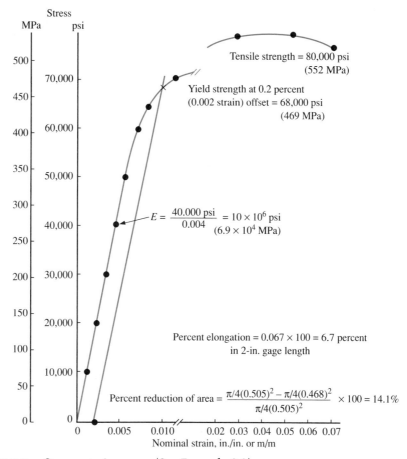

FIGURE 2.9   Stress–strain curve. (See Example 2.2)

EXAMPLE 2.3 *[ES]*

Assume that we have a rectangular bar 1 in. × 0.5 in. × 10 in. (25.4 mm × 12.7 mm × 254 mm) long of the aluminum alloy in Example 2.2. The bar is hung from one end and a 35,000-lb (156,250-N) load is placed on the other end. The bar undergoes an increase in length. Determine the stress, the elastic and plastic components of strain, and the bar length when under load.

Answer

$$\text{Stress} = \frac{\text{force}}{\text{area}} = \frac{35{,}000 \text{ lb}}{1 \text{ in.} \times 0.5 \text{ in.}} = 70{,}000 \text{ psi } (483 \text{ MPa})$$

At this stress the total strain is 0.010 in./in. (Example 2.2). Therefore the strained bar length is

$$10.0 \text{ in.} + (10.0 \text{ in.} \times 0.010 \text{ in./in.}) = 10.10 \text{ in } (256.5 \text{ mm})$$

When the bar is unloaded, the stress follows a straight line that depends on the modulus of elasticity E. Therefore we can calculate the elastic component of strain.

$$e_{el} = \frac{\text{stress}}{E} = \frac{70{,}000 \text{ psi}}{10 \times 10^6 \text{ psi}} = 0.007$$

Because $e_{total} = e_{el} + e_{pl}$

$$e_{pl} = 0.010 - 0.007 = 0.003$$

(Note that the elastic component of strain is not fixed. It is dependent on the stress value even after plastic deformation begins. This is an important consideration in plastic deformation processes such as forming a fender, where we must know the "spring back" if we are to control the final dimensions.)

2.7 True Stress–True Strain Relations

True stress–true strain curves are used in development and research more than in routine testing, because data on true stress and strain are more difficult to obtain and plot than data on engineering stress and strain. We shall discuss the basic features here and add further material in the problem section. *True stress* is simple to understand because it follows the real definition of stress,

$$\sigma = \frac{\text{load}}{\text{true area at the time}} \tag{2-5}$$

In calculating engineering stress, we used the original area throughout. During testing the area decreases only a small amount in the elastic range, but it decreases significantly in the plastic range. To obtain true stress, we must measure the diameter at several intervals during testing.

True strain, ε, is a slightly more difficult concept. Suppose we stretched the gage length of a bar of material from 2 in. to 3 in. Now let us stretch the bar another 0.1 in. The added engineering strain is $\Delta l/l_o$, or 0.1/2. However, if we considered strain as the change in length divided by the length *at the time,* the added strain would be 0.1/3.05. In this case 3.05 is the average length during the additional straining.

We obtain true strain, therefore, by summing a succession of Δl's divided by the length at the time the Δl is produced. The solution is found by calculus to be

$$\varepsilon = \ln \frac{l}{l_o} \tag{2-6}$$

Because of volume conservation, this expression is equivalent to

$$\varepsilon = \ln \frac{A_o}{A} = 2.3 \log_{10} \frac{A_o}{A} \tag{2-7}$$

Note that we use the symbols S = engineering stress, σ = true stress, e = engineering strain, and ε = true strain. Figure 2.10 shows a typical curve.

The true stress–true strain data are most helpful in planning forming operations that involve high plastic deformation, such as the production of auto bodies.

When we plot the true stress–true strain curve using log–log scales, we obtain a straight line for the plastic portion (Figure 2.11). Therefore, from the graph we have the equation of a straight line:

$$y = mx + b$$

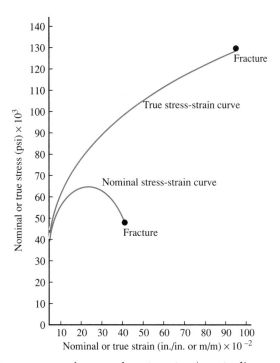

FIGURE 2.10 Comparison of true and engineering (nominal) stress–strain curves on the basis of measurements of diameter. The material is a low-carbon steel (0.20%C).

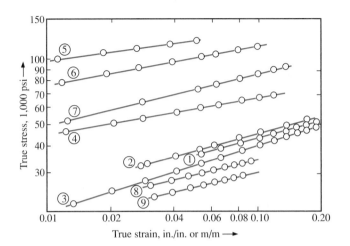

Material	Treatment
1. 0.05 percent carbon rimmed steel	Annealed
2. 0.05 percent carbon killed steel	Annealed and temper-rolled
3. Same as 2, completely decarburized	Annealed in wet hydrogen
4. 0.05 to 0.07 percent phosphorus low-carbon steel	Annealed
5. SAE 4130 steel	Annealed
6. SAE 4130 steel	Normalized and temper-rolled
7. Type 430 stainless steel (17 percent chromium)	Annealed
8. 2024 aluminum alloy	Annealed
9. 2014 aluminum alloy	Annealed

FIGURE 2.11 Logarithmic true stress–true strain $(\sigma - \varepsilon)$ tensile relations for various materials. Tests are by J. R. Low and F. Garafalo. (To obtain MPa, multiply psi by 6.9×10^{-3}. To obtain kg/mm^2, multiply psi by 7.03×10^{-4}.)

or

$$\ln \sigma = n \ln \varepsilon + K$$

Taking antilogs, we can write this as

$$\sigma = K\varepsilon^n \qquad (2\text{-}8)$$

In this form, n is called the *strain hardening exponent.* For annealed pure aluminum the equation is

$$\sigma = 25,000\varepsilon^{0.21}$$

The values of K and n depend on the material and its structure. In Figure 2.11 we show true stress–true strain relations for a variety of materials. Note that the aluminum alloys and low-carbon steels require the lowest stresses for deformation. Because the added cost of alloys does not seem excessive, the general public often asks why auto bodies are not made of stainless steel. The answer lies in the more expensive forming operations required for stainless steel because of strain hardening.

It can also be shown that engineering stress and true stress are related by the equation

$$\sigma = S(1 + e) \tag{2-9}$$

and engineering and true strain by

$$\varepsilon = \ln(1 + e) \tag{2-10}$$

EXAMPLE 2.4 *[ES]*

Using the data in Example 2.2, determine the true stress–true strain curve for the aluminum alloy.

Answer   We can calculate the true areas from the conservation-of-volume relationship up to necking (when uniform elongation no longer occurs, but rather a local decrease in cross-sectional area takes place).

$$l_0 A_0 = lA \quad \text{or} \quad A = \frac{2.000 \text{ in.} \times \pi/4(0.505 \text{ in.})^2}{l} = \frac{0.400 \text{ in.}^3}{l}$$

However, we must use the actual diameters at the tensile strength and at higher strains, because the material is necking and the gage length is well beyond this localized area. Therefore actual areas are necessary. We must also use the area relationship for the true strain calculation—that is, $\varepsilon = \ln l/l_0$ for strains up to necking (tensile strength) and $\varepsilon = \ln A_0/A$ for higher strains or deformation. Figure 2.12 shows the required curve.

	Load, lb	Gage Length, in.	Area, in.²	True Stress, psi	True Strain
Plastic	0	2.0000	0.2000	0	0
	4,000	2.0041	0.1996	20,040	0.0020
	8,000	2.0079	0.1992	40,160	0.0039
	10,000	2.0103	0.1990	50,250	0.0051
	12,000	2.0114	0.1989	60,330	0.0057
Elastic	13,000	2.0142	0.1986	65,460	0.0071
	14,000	2.0202	0.1980	70,710	0.0100
	16,000	2.0503	0.1951	82,010	0.0248
	16,000 (max)	2.0990	0.1847*	86,630*	0.0796*
	15,600 (break)	2.1340	0.1720*	90,700*	0.1508*

*Values based on measured diameters

Note: If we assumed conservation of volume and uniform deformation along the gage length, the area at fracture would be 0.400 in³/2.1340 = 0.1874 in² and the breaking stress would be 15,600/0.1874 = 83,240 psi. This value is impossible; the true breaking strength cannot be less than the true stress at the engineering tensile strength.

For uniaxial loading in the *elastic region*, strength of materials texts show that $\Delta v/v_0 = (Al - A_0l_0)/A_0l_0 = (1 - 2\nu)\varepsilon$. For values of ν (Poisson's ratio) other than 0.5, Al is not exactly equal to A_0l_0. At a 12,000-lb load with $\nu = 0.35$ for the aluminum alloy in this example and with the values for A_0, l_0, and l the same as above, A would equal 0.1992 in², which is not significantly different from the 0.1989 in² that we obtained assuming volume conservation.

Supplementary data useful in design can be derived by tensile testing or by using tests that impose stress states other than tension. For example, the fact that a tensile sample necks when the engineering tensile strength is exceeded suggests that a strain perpendicular to the tension strain must exist. The elastic component ratio of this lateral to the longitudinal strain is called *Poisson's ratio*, ν, and in metals has a range of 0.25–0.50.

If we apply a shearing force to a block of material as shown in Figure 2.13, we have a new stress state called the *shear stress*, τ, which is still defined, however, as the force per unit area.

Similarly, the *shear strain*, γ, is still a displacement divided by the distance over which it occurs, as shown in Figure 2.13. The *shear modulus*, G, is again the elastic stress divided by the elastic strain; that is, $G = \tau/\gamma$.

The relationship between tensile and shear elastic moduli can be expressed as $E = 2G(1 + \nu)$. Because of the possible range in Poisson's ratio, the tensile modulus is 2.5 to 3 times the shear modulus.

Tensile testing is much more common than shear testing, so it is often necessary to approximate the shear values. For example, yield strength in shear generally falls between 0.5 and 0.6 of the yield strength in tension.

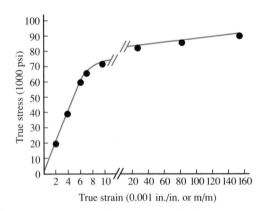

FIGURE 2.12   True stress–true strain curve for the high-strength aluminum alloy whose engineering, or nominal, stress–strain curve is shown in Figure 2.9.

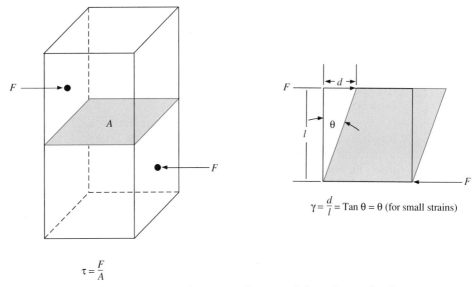

$$\tau = \frac{F}{A}$$

FIGURE 2.13 Shear stresses and strains. Shear modulus, G, equals τ/γ.

Another useful elastic constant is the compressibility. When we apply hydrostatic pressure to a material, we observe a change in volume with an increase in pressure:

$$K = -\frac{P}{\Delta V} \qquad (2\text{-}11)$$

where K is the *bulk modulus*. The minus sign is required because the volume decreases as the pressure is increased. We can again relate the constant K to tensile characteristics by the relationship

$$K = \frac{E}{3(1 - 2\nu)} \qquad (2\text{-}12)$$

The three elastic moduli E, G, and K therefore characterize a material's ability to react to several different stress/strain fields.

2.8 Dislocations

There used to be an afternoon colloquium in Cambridge, Massachusetts, in which metallurgists and physicists of leading institutions met after tea to present new data. Often a budding metallurgist, beaming with success, would present data for a new steel with greatly improved tensile strength, such as 300,000 psi (2.07×10^3 MPa). Invariably, at the end of the presentation a pipe-puffing theoretician would puncture the young metallurgist with a statement such as "Interesting—but when are you going to reach at least the order of a million psi called for by calculations using simple ball models of perfect crystals?"

Finally one day some scientists at the Bell Telephone Laboratories received samples of tiny tin crystals called "whiskers," which were shorting out capacitors. Out of curiosity they built a microtesting device and found that these whiskers did indeed have the theoretical strength level of over 1 million psi $(6.9 \times 10^3$ MPa).

For a long time it was suspected that actual and theoretical values differed because of the presence of microdefects. Indeed, Griffith had shown that by testing glass fibers of fine diameters (where the probability of a defect was low), he could obtain strengths of 500,000 psi $(3.45 \times 10^3$ MPa).

The line defect, or *dislocation*, is of primary importance in understanding the reason for the gap between commercial and theoretical strengths. To illustrate this, let us suppose a block of metal has an extra plane of atoms extending halfway through it (Figure 2.14a). This will result in a core of unstable material going back into the block, as shown by the five circled atoms in the face section. This core is a line defect called an *edge dislocation*. An important characteristic of this core is that the atoms in the upper side are in compression, those in the lower side in tension. Clearly the bonding forces are not so strong as in a perfect lattice.

Now let us apply a shearing stress (Figure 2.14b). In the perfect lattice we need to attain a uniformly high stress level to displace the A atoms; this is the theoretical yield strength obtained in whiskers. However, where we have a dislocation, the atoms nearby are not so firmly held, and the dislocation moves easily to the right. In other words, the yield strength value is lower.

By now it should be clear that the high strength of the tin whiskers tested at Bell Labs was due to the absence of dislocations in the direction in which they were tested. Whiskers of many materials have now been tested and found to be exceptionally strong. As a matter of fact, sapphire whiskers (Al_2O_3), glass fibers, and graphite fibers are being used as strengtheners in composite materials. Such strengtheners will be discussed in Chapter 16.

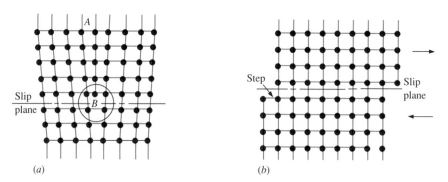

FIGURE 2.14   (a) Atomic packing near edge dislocation. The dislocation B is circled. The atoms at A are in a normal configuration. (b) Correlation of slip with a line dislocation.

To apply a simple analogy for dislocation motion, suppose we wish to move a large rug 6 in. in one direction. We tug mightily on one edge without results. Then we decide to push one edge inward, creating a 6-in. wrinkle. We find it is relatively easy to work this across the floor, and thus we can move the whole rug (Figure 2.15). The atomic model is shown in Figure 2.16 (a).

With this analogy we can introduce another concept, the *dislocation pin*. Suppose that as we made our move across the room, we encountered a large, heavy lamp on the rug. Our progress would stop because of this pin. With enough force to slide or shear the rug under the lamp, we could break the pin. In a material, foreign atoms or phases can act as dislocation pins.

For the sake of thoroughness, we should also describe a *screw dislocation*, which is equally important. There is another way in which we could move the rug gradually. Suppose that instead of pushing inward, we tugged parallel to the floor (Figure 2.15), gradually moving across the rug, and slid element by element by a shearing force relative to the floor. We would again have a core of unstable, easy-to-move material. On an atomic model this is shown in Figure 2.16(b).

To distinguish between the edge and screw types of dislocations, we use the *Burgers vector* (Figure 2.17). This is similar to the error of closure that surveyors encounter when they make a survey around an area. In the case of the edge dislocation, we find that when we make a circuit around the dislocation, we wind up one atomic spacing away from our starting point. The direction in which we would have to add this spacing is perpendicular to the core of the dislocation. In the case of an edge dislocation, we say that the Burgers vector (the error of closure) is perpendicular to the line of the dislocation, whereas in the screw dislocation it is parallel.

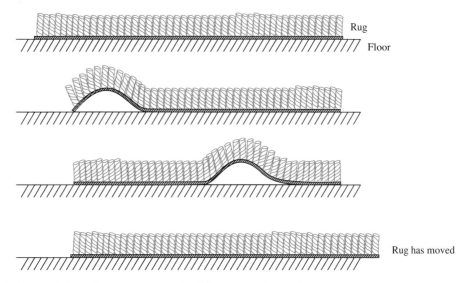

FIGURE 2.15   Rug movement by "dislocation" travel

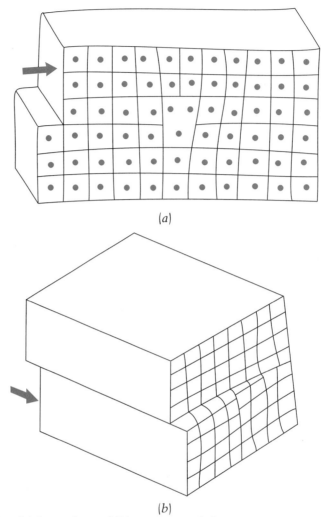

FIGURE 2.16   (a) Pure edge and (b) pure screw dislocations occurring during plastic deformation

[*Introduction to Ceramics*, 2nd ed., by W. D. Kingery, H. K. Bowen and D. R. Uhlmann, fig. 4.14. Copyright © 1976 John Wiley & Sons, Inc. Reprinted by permission of John Wiley & Sons, Inc.]

The Burgers vector is important in another way involving the strain energy of a dislocation. Although an equilibrium number of isolated point vacancies is to be expected, under equilibrium conditions it takes energy to produce a dislocation. As a rough indication, consider the force required to insert a foreign plane of atoms into the lattice shown in Figure 2.14(a). After insertion there is strong elastic tension in the lattice beneath the core and elastic compression above. It can be shown that the strain energy of a dislocation is Gb^2l, where G is the shear modulus, b the Burgers vector, and l the length of the dislocation.

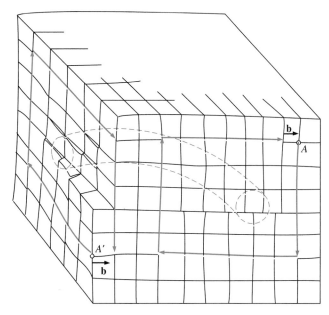

FIGURE 2.17   Combination edge and screw dislocation. Burgers vector **b** shown for pure screw and for pure edge. Dislocation line connecting these is shown.

[*Introduction to Ceramics*, 2nd ed., by W. D. Kingery, H. K. Bowen and D. R. Uhlmann, fig. 4.16. Copyright © 1976 John Wiley & Sons, Inc. Reprinted by permission of John Wiley & Sons, Inc.]

Comparing the same dislocation length *l* in a metal to a complex ceramic would lead to a higher strain energy in the ceramic. We have both a large Burgers vector because of large ion clusters around lattice points and a large shear modulus due to the strong bonding inherent in the ceramic.

A question often occurs when we observe the movement of an edge dislocation to the side of a crystal: "This explains only the movement of one atom–atom spacing. Where do additional dislocations come from?" One mechanism called a Frank–Read source is shown in Figure 2.18. We have a dislocation that is pinned at two points by solute atoms or by interaction with other dislocations. Then with the applied shear force we can develop additional dislocation loops, which carry on the movement on the slip plane. Note the unit steps. (We will use these concepts of dislocation motion later in further discussions of deformation.)

EXAMPLE 2.5 *[ES]*

Show that the stress for slip that is to move a layer of atoms *B* over layer *A* for pure magnesium is over 600,000 psi (4140 MPa), assuming no defects are present. The shear modulus of magnesium = *G* is 2.5×10^6 psi (0.17×10^5 MPa).

Answer   First let us review from physics the definition of the shear modulus. Just as we defined Young's modulus in tension as the ratio of tensile stress to tensile strain, we define G, the shear modulus, as the ratio of shear stress to shear strain. In the model shown, the shear strain is a/b.

Let us examine the stress needed to move one plane over another (Figure 2.19). If we start with the equilibrium position 1 and move to an intermediate position 2, from here on the stress cycle is repetitive; that is, it has reached the highest stress. At the intermediate position the elastic strain $\gamma = a/b$ is close to 1/4, because $a = r/2$ and $b = d$ (the atomic diameter $= 2r$). Note that the stress is always highest when $a = \tau/2$.

We can then say

$$G = 2.5 \times 10^6 \text{ psi} = \frac{\tau}{\gamma} = \frac{\tau}{1/4}$$

$$\tau = 625{,}000 \text{ psi } (4313 \text{ MPa})$$

The actual value of shear stress for pure magnesium is only about 100 psi (0.69 MPa) in a single crystal, giving a ratio of theoretical to actual of 6250 to 1. This illustrates the importance of the concept of dislocations.

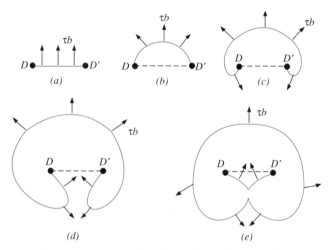

FIGURE 2.18   Frank–Read source. The plane of the figure is the slip plane of dislocation DD'; the dislocation leaves the plane of the figure at the fixed points D and D'. An applied stress produces a glide force τb on the dislocation and makes it bulge. The initially straight dislocation (a) acquires a curvature proportional to τ. If τ is increased beyond a critical value corresponding to position (b), where the curvature is a maximum, the dislocation becomes unstable and expands indefinitely. The expanding loop doubles back on itself, (c) and (d). Unit slip occurs in the area swept out by the bulging loop. In (e) the two parts of the slipped area have joined; now there is a closed loop of dislocation, and the section DD' is ready to bulge again and give off another loop.

[From W. T. Read, *Dislocation in Crystals*, McGraw-Hill Book Company, Inc., New York (1953).] Figure from A. L. Ruoff "Introduction to Materials Science" Prentice-Hall, 1972, p. 535. Used by permission.

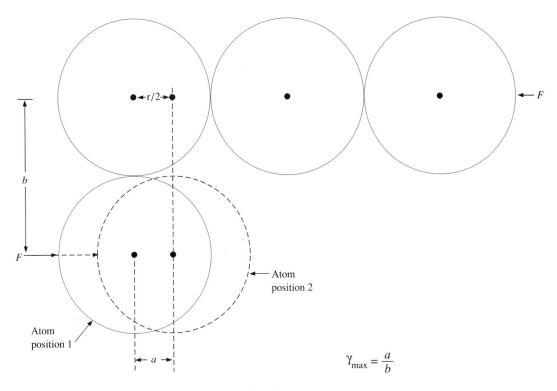

$$\gamma_{max} = \frac{a}{b}$$

FIGURE 2.19 Shear strain in simple atom displacement

2.9 Cold Work, or Work Hardening

We have already seen the phenomenon of work hardening in the stress–strain curve. To illustrate, let us take a bar of metal and apply stress until we reach point *A* on the curve, far above the yield strength of 50,000 psi (345 MPa), but below the tensile strength of 100,000 psi (690 MPa) (Figure 2.20*a*).

Now let us unload the sample, take it from the machine, and give it to another technician for testing. This technician will obtain a yield strength of over 80,000 psi (552 MPa) (point *B* in Figure 2.20*a*) instead of 50,000 psi (345 MPa). We can verify this effect by recalling that on unloading (Figure 2.20*a*) and on restressing, the points would follow the line *AA'*.[*]

Now visualize what is happening. As we stress into the region of plastic strain, slip takes place on the favorably oriented planes, producing dislocations and their movement. However, as more and more slip occurs, dislocations interact and pile up, and dislocation tangles form. This makes it more

---

[*]There are small "aftereffects" of a lesser order of magnitude that we need not consider at this time.

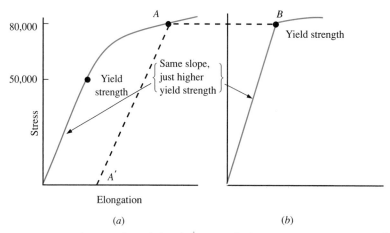

FIGURE 2.20 Correlation of work hardening with the stress–strain curve. The units of stress are psi.

and more difficult for further slip to take place. The rising stress–strain curve shows that to produce more strain, more stress is needed.

It is important to remember that slip requires both movement of atoms and dislocations. Because dislocations also have a stress field and more dislocations are created by slip, they get in one another's way, forming the dislocation tangles. The net effect is to make further slip more difficult.

We come now to the key point of the argument. When we reached point *A* in Figure 2.20(*a*), all the planes and dislocation sites fo easy slip had been used up. When the load was removed, this situation did not change. Therefore, when the load was reapplied, no plastic strain could take place until the stress level of point *A* was reached. Thus we encountered only elastic strain to a much higher level than in Figure 2.20(*a*), and the yield strength was correspondingly higher.

It is important to realize that a few isolated dislocations decrease the strength of an ideal material that might not contain dislocations. Dislocation collections and their entanglement increase the strength, however, by blocking atom motion necessary for slip to occur.

This phenomenon is called *work hardening, strain hardening,* or *cold working.* The term *cold* is relative. It means working at a temperature that does not alter the structural changes produced by the work. In other words, cold work causes atoms to move and dislocation tangles to form. As we shall see in Chapter 3, this effect can be removed by working at higher temperatures. We define percentage cold work as the change in cross-sectional area per unit area times 100.

2.10 Methods of Work Hardening

Naturally, if we wish to raise the yield strength by work hardening, we do not need to put the part in a tensile machine. We need only to produce slip. From our experiences with the copper bar, we know that slip can be produced by shearing under compressive stress. Various methods of treating a material so as to produce slip are discussed in Chapter 6. Note that at the same time we can be producing both slip and the final shape we desire.

2.11 Hardness Testing

Hardness is usually defined as resistance to penetration. However, other tests are also used, such as amount of rebound of a weight and scratch tests, especially in polymers and ceramics. Let us review a few of the most common tests and see how closely they fit this definition.

Brinell Hardness Number (HB)

The Brinell hardness test is one of the oldest tests. It is still the most common standard (Figure 2.21).

A specimen with a flat upper surface is placed on an anvil. A steel or tungsten carbide ball is pressed into the sample with a load of either 500 or 3000 kg. The lighter load is used for the softer, nonferrous metals, such as copper and aluminum, and the heavier load is used for iron, steel, and hard alloys. The load is left in place for 30 s and then removed. Some Brinell test machines give a reading of the *Brinell hardness number* (HB), but most require that the observer read the diameter of the impression in millimeters with a low-power microscope with a filar (measuring) eyepiece and then read the Brinell hardness number (HB) that corresponds to the impression diameter from a table of values for the load used. We will not analyze the method used to derive the numbers. We merely note that the more difficult the penetration, the higher the HB. The table is developed so that the HB is about the same whether the 500- or the 3000-kg load is used, though obviously the impression diameter is different. The lighter load is used with very soft materials because the 3000-kg load would continue to penetrate them until the ball was deeply sunken.

Vickers Hardness Number (HV)

The Vickers hardness test is an improvement on the Brinell test. Here a diamond pyramid is pressed into the sample under loads much lighter than those used in the Brinell test. The diagonal of the square impression is read, and the *Vickers hardness number* (HV) is read from a chart. As shown in Figure 2.22, the HV is close to the HB from 250 to 600. The figure does not show that the HV climbs steadily with strength at higher values of both strength and HV, whereas the HB is not used above 750. The Vickers test is better for obtaining hardness measurements at high levels and for measuring the hardness of a small region. On the other hand, the HB gives a better averaging effect because of the larger impression. Knoop hardness is used as an alternate to Vickers for hard materials.

Test	Indenter	Shape of Indentation — Side View	Shape of Indentation — Top View	Load	Formula for Hardness Number
Brinell	10-mm sphere of steel or tungsten carbide			P	$HB = \dfrac{2P}{\pi D(D-\sqrt{D^2-d^2})}$
Vickers	Diamond pyramid	136°		P	$HV = 1.72\,P/d_1^2$
Knoop microhardness	Diamond pyramid	$l/b = 7.11$ $b/t = 4.00$		P	$HKN = 14.2P/l^2$
Rockwell A C D	Diamond cone	120°		60 kg 150 kg 100 kg	HRA = HRC = 100–500t HRD =
B F G	$\frac{1}{16}$-in. diameter steel sphere			100 kg 60 kg 150 kg	HRB = HRF = 130–500t HRG =
E	$\frac{1}{8}$-in. diameter steel sphere			100 kg	HRE =

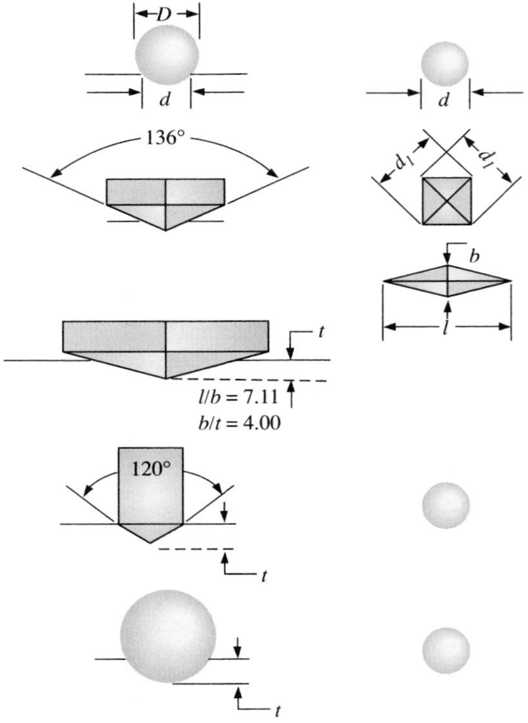

FIGURE 2.21 Hardness testing methods

(Adapted from H. W. Hayden, W. G. Moffatt, and John Wulff, *The Structure and Properties of Materials*, Vol. 3: *Mechanical Behavior*, John Wiley, New York, 1965. By permission of John Wiley & Sons, Inc.)

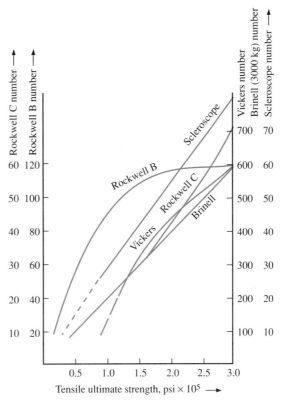

FIGURE 2.22   Conversion values for Brinell, Vickers, and Rockwell tests

(From Joseph Marin, *Mechanical Behavior of Engineering Materials*, © 1962, Table 10.2, p. 450. Reprinted by permission of Prentice-Hall, Inc., Englewood Cliffs, N.J.)

Rockwell Hardness Numbers (HRA, HRB, etc.)

The chief advantage of the Rockwell test is that the hardness is read directly from a dial. The indenter for the HRC test is a suitably supported diamond cone, or "brale." The observer first turns a handle that presses the diamond cone a slight standard amount into the sample. This is called the "preload." Next the standard HRC load of 150 kg (Figure 2.21) is released. This forces the diamond farther into the sample. The same lever is used to remove the load. At this point the observer reads the *Rockwell hardness number* (here, HRC) from the dial and then unloads the specimen. The principle of this test is that the dial, through a lever system, records the depth of penetration between the preload and the 150-kg load and reads directly in HRC. The HRC is approximately $\frac{1}{10}$ HB (Figure 2.22). The HRB scale is used for softer materials. It employs a $\frac{1}{16}$-in.-diameter ball and a 100 kg load. It is also direct-reading.

Scleroscope Hardness Number

The scleroscope hardness test is used chiefly for checking large rolls on which it is difficult to use the other tests. The value is obtained by measuring the height of rebound of a small weight under standard conditions.

2.12 Correlation of Hardness, Tensile Strength, and Cold Work

All the hardness tests depend on resistance to plastic deformation. It is no surprise, therefore, to find that the hardness of a bar is higher after cold working, because all the sites for easy slip have been used up. Thus there is good correlation between hardness and strength. For steel, a simple relation to remember is that the tensile strength in pounds per square inch is 500 times the HB (derived from Figure 2.22).

2.13 Solid-Solution Strengthening

Up to this point we have emphasized the effect of cold working on increasing strength and hardness. However, the strength may be further improved by alloying the material with one or more additional elements to provide solid-solution strengthening. The combined effects of solid-solution strengthening and cold working can be seen in two common aluminum alloys.

	Yield Strength, psi	Elongation, %	HB
Commercially pure aluminum annealed	4,000 (27.6 MPa)	43	19
Cold-worked aluminum, 50% reduction	18,000 (124.2 MPa)	6	35
Aluminum with 2.5% Mg (5052 alloy)	12,300 (84.9 MPa)	23	46
Cold-worked 2.5% Mg alloy, 50% reduction	24,700 (170.4 MPa)	5	75

These data show that adding magnesium to aluminum raised the yield strength from 4000 to 12,300 psi (27.6 to 84.9 MPa) in the annealed condition and from 18,000 to 24,700 psi (124.2 to 170.4 MPa) in the cold-worked condition. We shall encounter many single-phase alloys of this type in which solid-solution strengthening and cold working are combined to obtain high strength.

In general, the greater the difference in atomic radius, the greater the strengthening. Unfortunately, the Hume–Rothery rules limit the amount of the alloying element that we can add: The greater the difference in radius, the lower the solubility.

We have discussed conventional tensile and hardness testing for metals. Next we will cover standard tests for ceramics and polymers. Then we will take up the new concepts of fracture toughness and its design implications for all materials.

2.14 Ceramics

In our introductory comparison of different materials, the brittle behavior of a typical ceramic was evident. This feature can lead to irregular tensile test results because a small amount of misalignment can reduce the test values substantially. Also, the presence of flaws can lead to results well below the strength of sound material. The theoretical and measured strengths of two important ceramic fibers, alumina and silicon carbide, are shown in Table 2.2. Although the strength of fibers is close to the theoretical, polycrystalline material exhibits only about 1/100 of this value. (Here we will discuss only tensile testing, but the compressive strength is higher, as we will discover in Chapters 10 and 11.)

A variety of test specimens have been used for ceramics, but the three-point and four-point *bending tests*, which are compared with the tensile strength type in Figure 2.23 are the most common. The average value of tensile strength increases as we go from the tensile test to the four-point test, because of a smaller region under tension that reduces the role of flaws. Similarly, in the three-point specimen only a narrow region opposite the fulcrum

TABLE 2.2 Comparison of Theoretical Strength and Actual Strength

Material	E, GPa (psi)	Estimated theoretical strength, GPa (psi)	Measured strength of fibers, GPa (psi)	Measured strength of polycrystalline specimen, GPa (psi)
Al_2O_3*	380 (55×10^6)	38 (5.5×10^6)	16 (2.3×10^6)	0.4 (60×10^3)
SiC	440 (64×10^6)	44 (6.4×10^6)	21 (3.0×10^6)	0.7 (100×10^3)

*From R. J. Stokes, in *The Science of Ceramic Machining and Surface Finishing*, NBS Special Publication 348, 1972, U.S. Government Printing Office, Washington, D. C., p. 347.

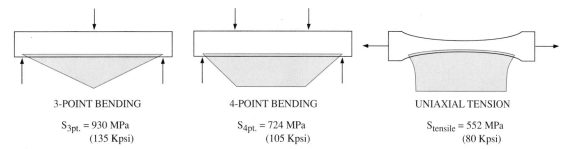

3-POINT BENDING

$S_{3pt.} = 930$ MPa
(135 Kpsi)

4-POINT BENDING

$S_{4pt.} = 724$ MPa
(105 Kpsi)

UNIAXIAL TENSION

$S_{tensile} = 552$ MPa
(80 Kpsi)

FIGURE 2.23 Comparison of the tensile stress distributions for three-point, four-point, and uniaxial tensile test specimens, along with typical average strengths as measured by each technique for Norton NC-132 hot-pressed Si_3N_4. Shaded area represents the tensile stress, which ranges from zero at the supports of the bend specimens to maximum at midspan and is uniformly maximum along the whole gage length of the tensile specimen. Note that in the three-point specimen, there is only a limited region of maximum stress compared to the stress over the entire gage length region in the tensile sample.

(Richerson, *Modern Ceramic Engineering*, Dekker, p. 88)

is under maximum tension, so the role of flaws is further reduced. It is still possible to have failure at a flaw away from the region of maximum tension in the three-point specimen, but the calculated value of tensile stress for failure will be higher than the stress at the flaw. In other words, the flaw leads to failure at a lower stress than that which the sound material beneath the fulcrum is enduring. You may find it a little disconcerting to have the results from the bend test reported as *modulus of rupture.* This is an old convention, and the word *modulus* is misleading because it is not a stress-to-strain ratio but simply the maximum tensile stress in the outer fibers at rupture. In the three-point test it is calculated from the standard formula for stresses in a beam, as described in Figure 2.24. The formula is accurate in representing tensile stress at rupture only for brittle materials, because if the material deforms plastically in the tensile region, the load is transferred disproportionately to inner portions of the bar. For ceramic materials the calculated tensile stress is accurate but is labeled modulus of rupture (MOR).

It is interesting to examine on an atomic basis why the ceramics are brittle compared to the metals, because many have the same unit cells. In the case of a metal lattice, only positively charged ions are present, and these reach positions determined by the balance of the repulsive positive-ion forces and the attraction of the electron cloud. By contrast, there is strong interatomic attraction with the ionic bond. When we shear two planes (Figure 2.25),

FIGURE 2.24   Bend test for specimens of ceramics and powder metals. The centers of the two bottom supports are 1 in. (2.54 cm) apart. Applying a load to the top member places the bottom of the beam in tension. The tensile stress is given approximately by the formula $S = 3PL/(2bh^2)$, where S = transverse rupture strength, psi (kg/mm^2); P = load required to fracture, lb (kg); L = length of span, in. (mm); b = width of specimen, in. (mm); and h = thickness of specimen, in. (mm).

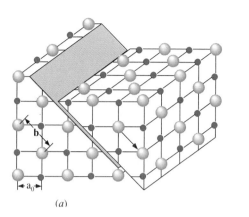
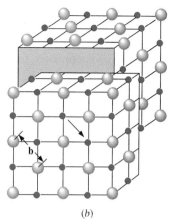

(a)   (b)

FIGURE 2.25 Translation gliding in the ⟨110⟩ direction and on (a) the {110} plane and (b) the {100} plane for crystals with the rock salt structure. {110}⟨110⟩ glide is preferred. Note that ⟨100⟩ glide would place similar atoms opposite developing high repulsion.

[*Introduction to Ceramics*, 2nd ed., by W. D. Kingery, H. K. Bowen and D. R. Uhlmann, fig. 14.10. Copyright © 1976 John Wiley & Sons, Inc. Reprinted by permission of John Wiley & Sons, Inc.]

bonds are broken and fracture takes place. However, slip is possible on some planes where the Na^+ ions move under a uniform attractive energy environment of Cl^- ions. Plastic deformation is actually obtained in NaCl single crystals, but in polycrystalline material the need to accommodate different slip orientations at the grain boundary is too demanding,* and cleavage occurs.

The phenomenon of cleavage, which is the fracture along definite planes of atoms such as {111}, is a very important mode of fracture in ceramics (and also in metals at low temperatures). It follows certain basic "rules" that are analogous to those for slip:

1. With layer lattices such as graphite, the cleavage is parallel to the layers. We know already that there are strong primary bonds within the layer and only secondary bonds between layers, so this effect is predictable.
2. Cleavage planes are the most widely spaced. Because the widely spaced planes are the densest packed, this is related to rule 1. Also, the further apart, the easier it is to displace two planes in shear.
3. Cleavage does not cut through radicals or ionic complexes. This would require more energy than other paths.
4. Cleavage acts to expose planes of anions (negative ions) if its tendency to do so does not violate rule 3. This reasoning is related to that for slip in ceramics.
5. In AX crystals such as NaCl, cleavage occurs on {100} planes. These are densely packed.

---

*Taylor has shown for ductile behavior in polycrystalline material that five slip systems are required.

2.15 Polymers

In contrast to the universally brittle ceramics, many polymers show greater ductility than metals whereas others are relatively brittle. The difference among polymers can be traced directly to the two great groups of structures discussed in Chapter 1. The thermoplastics (linear molecules) usually show great ductility because the adjacent giant molecules are held together only by van der Waals forces and can slide past each other without fracturing. As shown in Figure 2.1, elongation of 100% is attained. In the other case, we have a giant molecule held together by covalent bonds, and a stiffer, less ductile material is encountered.

Unlike metals and ceramics, polymers exhibit a change in strain as a function of *time* as the molecules uncoil. This is shown by a change in modulus (Figure 2.26). This problem, which is called *viscoelasticity*, will be discussed further as a function of temperature in the section on creep (Chapter 3). Upon removal of the stress, the specimen gradually contracts.

Conventional mechanical testing is usually performed on round, injection-molded, tensile specimens for thermoplastics and on compression-molded

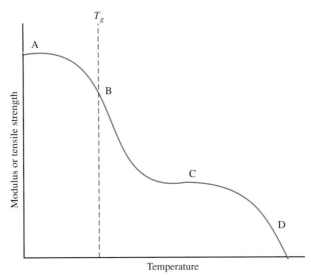

Period
A   Hooke's law is obeyed: material is glassy and has low ductility.
B   Side chains loosen; some main chain movement occurs, material shows mixed elastic and viscous behavior.
C   Main chain loosen; material is rubbery.
D   Thermal excitation overcomes bonds; material flows.

FIGURE 2.26   Variations in modulus and tensile strength as a function of temperature for a thermoplastic (schematic). T_g is the glass transition temperature. *Note:* A change in strain rate gives a similar relationship. At high strain rates the thermoplastic may be brittle, whereas at low strain rates it may be more rubbery.

specimens for thermosets. Machined flat specimens from large premolded sheets may also be used. As discussed later there is usually a difference between specimens cut from components and test bar specimens because of variation in molding flow patterns.

2.16 Fracture Toughness

For many years engineers have worked intensively to develop materials with higher yield and tensile strength, because lighter components (especially in aircraft) would make possible savings in both materials and energy. Many of the new designs failed, however, even though the design stress was below the yield strength. The components were especially vulnerable under conditions of plane strain.*

The mechanism of failure was growth of cracks originating from small flaws or inclusions. Finally it was concluded that a new criterion, fracture toughness, must be imposed in designing with these new materials. *Fracture toughness* is a *material* property that indicates its susceptibility to the spread of a crack. Because we can detect cracks and other flaws only above 0.020 in. (0.51 mm), we accept the fact that smaller flaws exist and use stress levels at which the cracks will not propagate. The relationship among stress, fracture toughness, and crack length for plane strain (covered in detail in Chapter 19) is

$$\sigma_f = \frac{Y K_{IC}}{\sqrt{\pi a}} \tag{2-13}$$

σ_f = stress to cause failure

K_{IC} = fracture toughness

a = crack length for edge crack (1/2 length for center crack)

Y = a shape factor calculated from the geometry of the part

The explanation of fracture toughness is rather lengthy, so we will defer it until Chapter 19. However, because of the increasing use of this criterion, it is important to include this short description in our discussion of mechanical properties. We will also indicate briefly its importance in fatigue.

EXAMPLE 2.6 *[ES]*

Two rectangular supports carry equal tensile loads and were originally quenched to martensite (a hardened steel microstructure). Support A is fabri-

---

*This is produced by a stress system that causes strain in two directions at right angles (which determine a plane). The strain at right angles to the plane is zero. We will apply this concept in Chapter 19 in further discussion of failure.

cated from 4340V (Ni, Cr, Mo + Vanadium) steel given a 427°C (800°F) temper, and support B is fabricated from 4340V steel given a 260°C (500°F) temper. (See Chapter 8 for a complete discussion of microstructure control in steels.) The thickness of each member has been adjusted so that each supports a stress σ_D equal to 60% of the respective yield strength. For each support, what is the longest edge crack (see figure) that can be tolerated without causing catastrophic failure?

Answer

Step 1. * Determine the appropriate expression for the stress-intensity factor for the geometry in question.

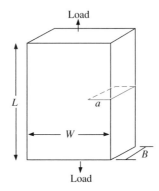

The stress intensity for the configuration of the support is given by[†]

$$K_I = \sigma\sqrt{\pi a}\, Y(a/w)$$

where $Y(a/w)$ is the geometric correction factor that takes into account the finite width of the plate. It is calculated as

$$Y(a/w) = 0.265\left(1 - \frac{a}{w}\right)^4 + \frac{0.857 + 0.265\left(\dfrac{a}{w}\right)}{\left(1 - \dfrac{a}{w}\right)^{3/2}}$$

[*Note:* In this discussion, $Y(a/w)$ by convention is not intended to represent an algebraic form, but rather Y is a function of a/w.]

Step 2. Determine the relevant dimensions of the supports and the appropriate material properties for use in the calculations.

---

*This step may be omitted by assuming $Y = 1$ to give approximate values.
[†]Taken from *The Stress Analysis of Cracks Handbook*, Hiroshi Tada, Del Research Corporation, Hellertown, Pa., 1973.

Support Dimensions	Support A	Support B
L	16 in. (0.406 m)	16 in. (0.406 m)
W	4 in. (0.102 m)	4 in. (0.102 m)
B	0.7 in. (0.018 m)	0.59 in. (0.015 m)
Material Properties	4340V 427°C (800°F)	4340V 260°C (500°F)
$\sigma_{Y.S.}$	191 ksi	228 ksi
σ_D	115 ksi	137 ksi
K_{IC}	97 ksi $\sqrt{\text{in.}}$	51 ksi $\sqrt{\text{in.}}$
(ksi × 6.9 = MPa; ksi $\sqrt{\text{in.}}$ × 1.1 = MPa$\sqrt{\text{m}}$)		

Step 3. Calculate the critical flaw sizes. At fracture the governing equation is

$$\frac{K_{IC}}{\sigma_D} = \sqrt{\pi a_c}\left[0.265\left(1 - \frac{a_c}{w}\right)^4 + \frac{0.857 + 0.265\dfrac{a_c}{w}}{\left(1 - \dfrac{a_c}{w}\right)^{3/2}}\right]$$

where a_c is the critical flaw size and K_{IC} is the critical stress intensity. Substitute the appropriate values of K_{IC} and σ_D for each support and solve for a_c by iteration (trial and error). Calculated critical flaw sizes are

Support A:* $a_c = 0.171$ in. (4.34 mm)

Support B: $a_c = 0.035$ in. (0.89 mm)

Note: Even though both members are designed with the same safety factor in terms of the yield stress, the size of the critical flaw is less than one-fifth as large, and hence much harder to detect, in support B, which is fabricated from the higher-strength steel. Of course, support B is lighter in section.

2.17 Fatigue

A survey of the broken parts in any automobile scrap yard reveals that the majority failed at stresses below the yield strength. This is a result not of imperfections in the material but of the phenomenon called *fatigue*. If a bar of steel is loaded a number of times to a stress—80% of the yield strength, for example—it will ultimately fail if it is stressed through enough cycles.

*1 + Y is taken as 1. The approximate values are A = 0.23 and B = 0.044. The agreement is within control of crack depth.

Furthermore, even though the steel would show 30% elongation in a normal tensile test, no elongation is evident in the appearance of the fatigue fracture. A typical crankshaft failure is shown in Figure 2.27.

Fatigue testing in its simplest form involves preparing test specimens with carefully polished surfaces and testing them at different stresses to obtain an *S-N* curve relating *S* (stress required for failure) to *N* (number of cycles) (Figure 2.28). As we would expect, the lower the stress, the greater the number of cycles to failure.

The curve for ferrous materials exhibits what is called the *fatigue strength*, or *endurance limit* (Figure 2.29). In other words, beyond 10^6 cycles, further cycling does not cause failure. On the other hand, for nonferrous materials such as aluminum, the curve continues to decline. The material must be tested for the number of cycles it will encounter in service.

Fatigue tests are time-consuming, and the fatigue data require statistical treatment because of reproducibility difficulties, so attempts have been made to relate the tensile test to fatigue data. The *fatigue ratio*, or *endurance ratio*,

FIGURE 2.27 Typical fatigue failure of a crankshaft. The crack began at the stress concentration at the hole (front of picture). Steps in the fracture caused by successive cycles of loading led to a clamshell effect. The final failure took place suddenly, as shown by the small fracture surface at the rear.

(Courtesy of H. Mindlin, Battelle)

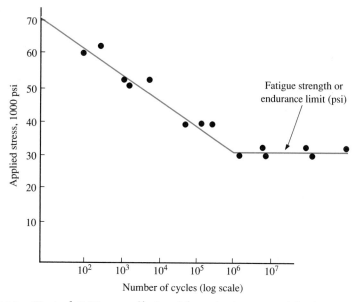

FIGURE 2.28 Typical S-N curve (fatigue) for a ferrous material, showing an endurance limit (schematic)

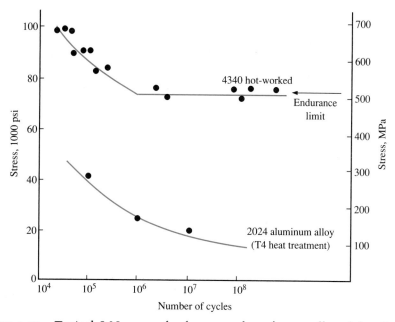

FIGURE 2.29 Typical S-N curves for ferrous and nonferrous alloys (after Garwood and Alcoa). 2024 aluminum: 4 Cu, 0.6 Mn, 1.5 Mg; 4340 steel: 0.4 C, 0.7 Mn, 0.25 Si, 1.85 Ni, 0.3 Cr, 0.25 Mo.

is defined as the ratio of fatigue strength to tensile strength. It has values of 0.45 to 0.25, depending on the material.

The fatigue strength is also greatly affected by the following variables:

1. Stress concentrations, such as radii at fillets and possible notches
2. Surface roughness, which indicates that results depend on the type of machining used
3. Surface residual stress
4. Environment, such as corrosion

Ceramics, particularly certain glasses and oxides, exhibit a phenomenon known as *static fatigue*. These materials and even some ultra-high-strength alloys will withstand a high static load for a period of time and then fail suddenly. This type of failure does not take place in dry air or *in vacuo*, so it is related to a chemical reaction between the water in the atmosphere and the highly stressed surface. (See Chapter 18.)

2.18 Relationship Between Fatigue and Fracture Toughness

Just as we encounter crack growth in tensile testing, the same effects involving fracture toughness are found in fatigue. In this case the crack grows if we exceed a threshold stress ΔK_{th} (Figure 2.30). This crack growth is shown by the expression da/dN, where a is edge crack length and N is number of cycles. On the x axis we plot ΔK_I, which is the stress intensity in the tensile range at the root of the crack. This value is calculated from the range of stress (maximum stress minus minimum stress) during a fatigue cycle and the instantaneous crack length. The calculation of safe values of stress for a given number

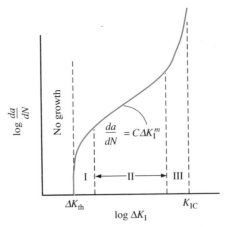

FIGURE 2.30   Growth rate of fatigue cracks da/dN as a function of the stress-intensity range, ΔK_I

of cycles is given in Chapter 19. The advantage of this method, compared to simply consulting an S-N curve, is that we can factor into the solution our knowledge of maximum starting crack length based on inspection methods.

2.19   Wear

The most important characteristic of wear is its unpredictability. For example, a gray cast iron is an excellent engine-block material because it retains lubricant and thus prevents seizing of the piston. However, the graphitic areas in gray cast iron crumble under high stresses and cannot be present in a component such as a railroad car wheel or a grinding ball. Although a general "rule" is to avoid contact between similar metals, gray iron piston rings perform well against a gray iron engine block, and gray iron valve lifters* do well against a gray iron camshaft.*

In spite of this apparent confusion, problems of wear can be approached reasonably if the mechanism of wear is known. Often a careful observation of worn parts discloses which structures failed and what conditions developed at the interface between parts. Let us consider these fundamentals of wear:

1. Nature of wearing surfaces
2. Effects of pressing surfaces together (static contact)
3. Interaction between sliding surfaces
4. Effects of lubrication
5. Resistance of wear of different material combinations

The Nature of Wearing Surfaces

Even the best polished surface has two types of inhomogeneity. By using an accurate diamond stylus we can find pits and scratches of the order of 1000 Å (100 nm), or hundreds of times greater than the size of the unit cell. In addition to these physical differences, we know that variations in grain orientation and in the nature of phases and inclusions can occur at the wearing surface. In other words, even with the best machining, we have two relatively rough inhomogeneous mating surfaces. Furthermore, we know that it is possible to have two very different microstructures of the same overall macrohardness. For example, one structure might contain martensite at HB 600 microhardness plus graphite flakes that would lower the overall hardness (macrohardness) to HB 500. Another structure might be homogeneous martensite at HB 500. Under one type of wear, involving high loads and good lubrication, the second structure would probably perform better, particularly if the load crushed the iron at the edges of the graphite. However, with moderate loads

---

*These contain controlled amounts of massive iron carbide, which is hard and soft graphite flakes in a steel matrix.

and intermittent lubrication, the first structure would resist galling (sticking) better than the second.

Because wear is a surface phenomenon, a number of wear-resistant surfaces have been developed. *Carburizing* of steel is the addition of carbon to a surface to produce a higher carbon content and hence harder martensite. In *nitriding*, iron–nitrogen compounds are formed that resist thermal decomposition, which may occur in carburized surfaces. *Ion implantation* is a more recent technique in which foreign atoms are implanted in a metal surface from a beam of ionized particles. Although the implanted species may be only 0.1–0.2 μm (1000–2000 Å) deep, the wear resistance of a surface can be greatly enhanced by this process. For example, nitrogen implants in steel have lengthened the tool life of dies and rolls by 3–10 times. Chapter 8 will provide a more complete discussion of microstructure control in steels.

Static Contact

Consider the simple case of a ball bearing on a flat plate. This is incorrectly called *point contact*; actually the ball and plate both deform, giving a circular area of contact. It can be shown that the shear stress is maximum not at the surface, but beneath it. If a is the radius of the circle of contact, the maximum shear stress is at $0.6a$ beneath the surface. We can calculate the maximum stress at this level as a function of the radius of the ball or other "point" contact. From this value we can calculate the load required to produce instantaneous plastic deformation for different radii of the ball and different plate materials (Table 2.3).

In conventional terms, a high load is required to deform tool steel with a ball. However, when we consider that a large ball has tiny rough points of radii as small as 10^{-4} cm, then we must expect some deformation even with very light loading (1.4×10^{-3} g).

Interaction Between Sliding Surfaces

When surfaces slide over each other, the projections produce very high stresses and flow in the areas of contact. If we accept this, we can better understand

TABLE 2.3   Load Required to Produce Flow in Different Materials Using Indenters (Steel Balls) of Different Radii

Plate Material	Load, g, for Ball Radius Indicated		
	10^{-4} cm	10^{-2} cm	1 cm
Copper	2.5×10^{-6}	2.5×10^{-2}	250
Mild steel	4.7×10^{-5}	0.47	4,700
Tool steel	1.4×10^{-3}	14	140,000

the breaking-in period necessary for machines. In many cases the high stresses result in high friction, and welding followed by shearing takes place between the parts.

1. If the weld is weaker than either metal, little wear takes place. This is the case, for example, when a tin-base-alloy bearing wears against steel.
2. If the junction is stronger than one of the metals, then shear takes place in the weaker metal. For example, when steel wears on lead, the fracture takes place in the lead.
3. If the junction if stronger than both metals (for example, when martensite forms in the junction), tearing occurs in both metals and wear is rapid.

The effects of surface temperature are also important. Under many simple conditions of sliding wear, lead actually melts and covers the surface, whereas steel is hardened by local heating above the austenite transformation temperature. The presence of freshly formed martensite is often noted in steel or iron parts subjected to wear, such as brake drums (see Chapters 8 and 9).

Effects of Lubrication

When a lubricant is present, there is either hydrodynamic lubrication or boundary lubrication. In hydrodynamic lubrication, we try to preserve a fluid film between surfaces so that they do not touch. Many factors operate against this, including intermittent operation, breakdown of the lubricant molecule, and heavy loads with slow speeds. Under these conditions welding or galling can take place.

With boundary lubrication, we try to coat the surface with lubricant molecules. The metal reacts with the lubricant to form a metal soap. In other cases, extreme-pressure lubricants containing sulfur, phosphorus, or chlorine react with the metals to reduce seizing. Molybdenum disulfide and graphite are also used in the bearing area, where extreme pressure exists, to serve as solid fragments with easy basal cleavage.

2.20 Wear-Resistant Combinations

It is possible to dissolve lead in liquid copper alloys. Upon solidification, the lead precipitates in fine globules. These alloys are often used against steel in slow-speed bearings with intermittent lubrication. The copper is hardened by the addition of tin and zinc, as in the alloy 85% Cu, 5% Sn, 5% Zn, 5% Pb. For some applications a higher-strength alloy is needed, and aluminum bronze (89% Cu, 11% Al) is used. For gears a favorite combination is hardened steel wearing against a nickel-tin bronze (87% Cu, 11% Sn, 2% Ni). The tin provides a hard phase called δ, and the nickel gives solid-solution hardening of the copper.

Many applications involve ceramic and plastic bearing materials. Sapphire bearings have long been used in precision equipment and watches, and it is possible to obtain sapphire balls of 1-in. (2.54-cm) diameter. Nylon composites are used in applications in which metal was required previously, such as fishing reel gears.

2.21 Resistance to Abrasion

In the usual wear application we avoid the problem of abrasion by not introducing gritty foreign material. However, in some cases, such as grinding ore and conveying sand, the abrasive is part of the system. The chief variables in abrasion are the hardness of the abrasive and the type of loading of the equipment (whether impact or steady compression).

Unfortunately, the hardest and most abrasion-resistant alloys, such as tungsten carbide, can be used in only a few installations because the carbides crack out. The most useful abrasion-resistant alloys in mining and other grinding are martensitic white cast iron, hardened steel, and austenitic manganese steel. The white irons are the most wear-resistant, and the manganese steel is the toughest. In some instances, in which the part is not heated or cut by the abrasive, rubber-coated parts are very successful. Conveyor belts and gloves for sandblast operators are examples of abrasion-resistant items made of rubber. (See Chapters 7–9 for a complete discussion of the metallic materials.)

Summary

As increasing stress is applied to an engineering material, there are two important effects. Up to the stress called the yield strength, the strain in the sample is principally elastic. That is, when the stress is removed, the specimen returns to close to its original dimensions (within 0.2%). The important ratio S/e is the modulus of elasticity for the material. Above the yield strength, permanent (plastic) deformation takes place.

In common metals the modulus is from 6×10^6 psi to 50×10^6 psi (0.414×10^5 MPa to 3.45×10^5 MPa), and the plastic elongation from 2% to 50%. The high values of plastic deformation are caused by dislocation movement and slip on closely packed planes of atoms. The tensile test is the most common test for strength and elongation.

In ceramics slip occurs only in special single crystals, and practically no plastic deformation occurs at room temperature in polycrystalline material. Cleavage or irregular brittle fracture occurs. Because of the strong, primary ionic and covalent bonding, the hardness, strength, and modulus are high. Because of the effects of flaws, bend tests are preferred to the tensile test. The modulus is from 10×10^6 psi to 70×10^6 psi (0.69×10^5 MPa to 4.83×10^5 MPa).

In polymers the modulus is much lower (0.025×10^6 psi to 1.5×10^6 psi or 0.17×10^3 MPa to 10.4×10^3 MPa) than for other materials, because only the secondary bonding forces are active. Furthermore, the modulus changes with the duration of loading, because the molecules move apart and uncoil as a function of time. The strength and hardness are also lower because of the weak secondary bonds. However, in thermoplastic material, elongations of over 100% can be encountered, because the molecules can slide past each other as well as uncoil. The thermosets are relatively brittle (less than 2% elongation).

The fracture toughness of all three types of materials is important in highly stressed components, particularly in cases of plane strain. In these instances the design stress should be calculated on the basis of the largest expected flaw and the fracture toughness K_{IC}, rather than on the basis of yield strength.

Service performance in wear and abrasion cannot be predicted from simple measurements such as hardness and strength because of the surface interactions. Often the strength of locally bonded irregularities is more important than the individual hardness of the materials. The most satisfactory procedure is to test a variety of materials under controlled conditions and observe the mechanism of wear of the structure.

Fatigue is important because more failures take place by this mechanism than by static tensile, shear, or compressive loading. The relationships among flaw size, fracture toughness, and number of stress cycles are highly significant in the design of parts.

Finally, it should be emphasized that the behavior under stress we have just discussed is that which occurs at ambient temperature, about 20°C. In the next chapter we shall see that brittle ceramics become plastic at elevated temperatures and that BCC metals and many ductile plastics become brittle at low temperatures.

Definitions

Brinell hardness number, *HB*   The value obtained from a Brinell indentation hardness test.

Bulk modulus   The elastic change in the volume of a material while it is under hydrostatic pressure.

Burgers vector   An error of closure around a dislocation.

Cleavage   Fracture along specific crystallographic planes, as characterized by failure in ceramic materials.

Cold working   The deformation of material at a temperature at which dislocations are moved by mechanical rather than thermal energy.

Critical resolved shear stress, τ_c   The applied stress resolved in the slip direction in the slip plane, $(F/A) \cos \lambda \cos \varphi$.

Dislocation   A collection of point defects that results in a line defect.

Edge dislocation   A line defect due to a plane of missing atoms around which a strain field is developed.

Elastic deformation   The temporary displacement of atoms from their normal positions, which can be the result of applying a tensile or a compressive stress. When the stress is removed, the atoms return to the normal spacing.

Endurance limit A stress below which a material that is cyclically loaded does not fail regardless of the number of cycles.

Engineering strain, *e* Change in length divided by original gage length, $\Delta l/l_o$.

Engineering stress, S Load divided by original area, P/A_o.

Engineering stress–strain curve The results, usually of a tensile test, plotted with S as the y axis and e as the x axis.

Fatigue strength A measure of the stress required to produce failure in a specimen subjected to a specified number of cycles, usually 10^6, of loading and unloading in tension, compression, or bending.

Fracture toughness A measure of the stress intensity required to cause catastrophic failure. A ductile material generally has high fracture toughness, and a brittle material exhibits low fracture toughness.

Modulus of elasticity, *E* Stress divided by strain, S/e, in the tensile elastic range.

Modulus of rupture (MOR) A measure of the tensile stress required to cause failure in the bending of a beam of a brittle material such as a ceramic.

Percent cold work $\dfrac{\text{Change in cross-sectional area}}{\text{Original area}} \times 100$

Percent elongation at fracture Engineering (plastic) strain times 100,

$$\frac{l_f - l_o}{l_o} \times 100$$

Percent reduction of area $\dfrac{\text{Change in area}}{\text{Original area}} \times 100 = \dfrac{A_o - A_f}{A_o} \times 100$

Plastic deformation The permanent displacement of atoms from a given starting position, as by slip or twinning.

Rockwell hardness number, *HRA*, *HRB*, *HRC*, etc. The number obtained from a Rockwell hardness test.

Screw dislocation A line defect arranged in a spiral or screw-like configuration leading to a strain field.

Shear modulus, *G* Shear stress divided by shear strain in the elastic region. $G = \tau/\gamma$.

Shear strain, γ The angular displacement θ due to a shear stress. $\gamma = \theta$ when the strains are small.

Shear stress, τ A force couple divided by the area over which the force acts. $\tau = F/A$.

Slip system A combination of a slip direction and a slip plane containing the direction.

Strain hardening exponent The exponent n in the equation $\sigma = K\varepsilon^n$, which relates the true stress to the true strain in the plastic deformation region of a tensile test.

Tensile strength The maximum engineering stress encountered during a tensile test.

True strain, ε $\ln(l/l_o) = 2.3 \log_{10}(l/l_o)$.

True stress, σ Load divided by area at the given load, P/A.

Twinning The shifting of the atoms on one side of a twinning plane by an amount proportional to the distance from the plane. One part of the grain becomes a mirror image of the other.

Vickers hardness number, *HV* The value obtained from a Vickers test.

Viscoelasticity The relationship among time, strain, and temperature that can lead to permanent deformation in materials such as polymers.

Work hardening An increase in hardness and strength that is due to plastic deformation.

Yield strength The stress at which a specified amount of plastic strain, usually 0.2%, is produced.

Problems

2.1 *[EJ]* In Figure 2.2(b), some grains on the compression surface show no slip, whereas others closer to the center exhibit slip. Explain how this can be a normal occurrence and why it is likely also to happen on the tension side of the small copper bar. (Sections 2.1 through 2.7)

2.2 *[ES]* Slip planes and directions for three common metal structures are shown below.

Structure	Directions	Planes
FCC	$\langle 110 \rangle$	$\{111\}$
BCC	$\langle 111 \rangle$	$\{110\}$
HCP	$\langle 110 \rangle$	$\{0001\}$

Using the FCC as an example, compare the linear atomic density in the [100], [110], and [111] directions and the planar atomic density on the (100), (110), and (111) planes to show that the given slip directions and planes have the densest packing. (Sections 2.1 through 2.7)

2.3 *[EJ]* Two rods in the form of single crystals of pure zinc show widely different yield strengths at 0.2% offset. What index of yield strength should be the same? (Sections 2.1 through 2.7)

2.4 *[ES]* Given that the critical resolved shear stress of a material is 82 psi (0.566 MPa), plot axial stress for slip as a function of $\cos \varphi \cos \lambda$. See Example 2.1. (Sections 2.1 through 2.7)

2.5 *[EJ]* Explain how we might use data from a tensile test to determine whether a pure metal is iron or nickel. (*Hint:* Consider *all* the values that might be obtained in a tensile test.) (Sections 2.1 through 2.7)

2.6 *[ES]* From the following data, plot the engineering stress–strain curve of the material and calculate the modulus of elasticity, yield strength at 0.2% offset, percent elongation, percent reduction of area, and tensile strength. (Sections 2.1 through 2.7)

Load, lb	Gage Length, in.
0	2.0000
3000	2.00098
4000	2.00134
5000	2.00164
6000	2.00202
7000	2.0026
6900	2.0034
8000	2.0056
9000	2.0086
10,000	2.0124
11,000	2.0200
12,500 (maximum)	2.157
11,500 (fracture)	2.560 (after fracture)
Original diameter: 0.505 in.	
Diameter at maximum load: 0.479 in.	
Diameter at fracture region: 0.350 in.	

To convert stress to MPa, multiply psi by 6.9×10^{-3}.

2.7 *[ES]* Rather than machining a round tensile specimen out of a piece of sheet metal, we cut a flat tensile specimen out of an 0.100-in.- (2.54 mm-) thick sheet, as shown on the accompanying graph. The specimen is 0.500 in. (12.7 mm) wide in the gage length. From the data given (load versus change in 1-in. (25.4 mm) gage length), calculate the tensile strength, yield strength (0.2% offset), modulus of elasticity, and percent plastic elongation (fracture). (Sections 2.1 through 2.7)

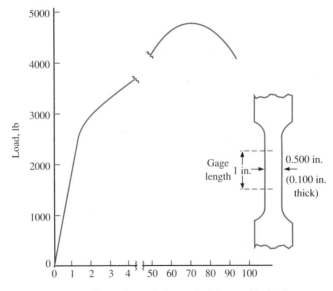

Change in one-inch gage length caused by load.
Scale in *thousanths* of an inch.

2.8 *[ES]* A rectangular bar originally 0.75 in. × 0.25 in. × 12 in. long (19.05 mm × 6.35 mm × 304.8 mm long) is loaded in tension so that it elongates 0.015 in./in. (mm/mm) while under load. (Sections 2.1 through 2.7)
a. What is the length of the bar while loaded?
b. If the new unloaded length of the bar is 12.120 in. (307.85 mm), determine the elastic and plastic components of strain when the bar is loaded.
c. If the modulus of elasticity is 20×10^6 psi (1.4×10^5 MPa), determine the load necessary to produce the strain of 0.015 in./in. (mm/mm).

2.9 *[ES]* A complete stress–strain curve is often not determined in the daily gathering of data. From the information in the accompanying table, determine the yield strength, tensile strength, modulus of elasticity, percent reduction of area, and percent elongation. The initial gage length is 2.000 in.; the initial diameter is 0.505 in. (Sections 2.1 through 2.7)

Load (lb force)	Gage Length
2000	2.001 in. (all elastic deformation)
6000	2.004 in. (all plastic deformation)
8500 (max)	2.300 in. (all plastic deformation)
7800 (failed)	2.450 in. after failure
	0.423 in. diameter after failure

2.10 *[EJ]* Not all metals use a 0.2% offset for a definition of the yield strength. Using your knowledge of how the offset yield strength is experimentally determined, describe some limitations on the precision of the determination. What might the term *proof stress* mean? (Sections 2.1 through 2.7)

2.11 *[EJ]* It is noted in the text that the gage length should be specified in a calculation of the percent elongation. What might be expected for the percent elongation value if the gage length is increased from 2.000 in. (50.80 mm) to 4.000 in. (101.60 mm)? Why are two values, elongation and reduction of area, used as measures of ductility? (Sections 2.1 through 2.7)

2.12 *[ES/EJ]* The following data are for a circular cross-section metal rod that is stressed: length = 24 in., radius = 0.50 in., force = 40,000 lb, and elongation = 1/16 in. Answer the following questions and *list the assumptions* necessary to arrive at your answer. Definitions of terms will dictate the assumptions necessary to arrive at a numerical solution. (Sections 2.1 through 2.7)
a. What is the stress on the rod?
b. What is the percentage elongation of the rod?
c. What is the strain on the rod?
d. What is the elastic modulus of the rod material?

2.13 *[ES]* Calculate the true stress–true strain curve using the data of Prob. 2.6. Use the relation

$$\text{True strain} = \ln \frac{A_o}{A_f} = 2.3 \log \frac{A_o}{A_f}$$

for the strain at maximum load and fracture. (Sections 2.1 through 2.7)

2.14   *[ES]* The accompanying graph is a recorder plot from a tensile test of a 0.505-in.-
(12.83-mm-) diameter specimen with a 2-in. (50.8-mm) gage length, along with other
data. Calculate E (modulus of elasticity), yield strength at 0.2% offset, ultimate tensile
strength, percent reduction in area, and true stress at fracture. (Sections 2.1 through 2.7)

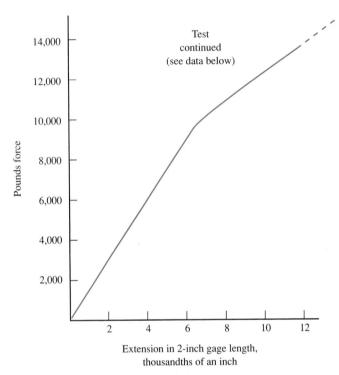

Additional data: Maximum load,   16,000 pounds force
                  Load at fracture,   12,000 pounds force
                  Diameter of fracture,   0.400 inch
                  Gage length at fracture, 2.50 inch

2.15   *[EJ]* The text discussion of true stress and true strain indicates that area ratios should
be used rather than length ratios in calculating the true strain beyond the engineering
tensile strength. Why is volume not conserved beyond the engineering tensile strength?
Under what conditions is $\ln(l/l_o) = \ln(A_o/A)$? (Sections 2.1 through 2.7)

2.16   *[ES/EJ]* The curves in Fig. 2.11 are for plastic strain only and are governed by the equation
$\sigma_t = K\varepsilon^n$, where K is a constant and n is known as the strain hardening exponent.
   a. Determine the general equation (that is, determine K and n) for type 430 stainless
      steel (Curve 7) and 2024 aluminum (Curve 8).
   b. If the area under the true stress–true strain curve is an index of the work required
      to deform the metal, which of the alloys in Fig. 2.11 would require the most work?
      (Sections 2.1 through 2.7)

2.17 *[ES]* Derive these two relationships. (Sections 2.1 through 2.7)

    a. $\sigma = S(1 + e)$ (*Hint:* Define σ, S, and e and conserve volume.)

    b. $\varepsilon = \ln(1 + e)$ (*Hint:* Define ε and e.)

2.18 *[ES/EJ]* Indicate whether the following statements are correct or incorrect, and justify your answer. (Sections 2.8 through 2.13)

    a. The presence of more dislocations increases the yield strength of an annealed metal.

    b. Dislocations move more readily in a solid-solution alloy than in a pure metal.

    c. A fine-grained alloy will have a higher yield strength than a coarse-grained alloy of the same chemical composition.

2.19 *[ES/EJ]* The ASTM (American Society for Testing and Materials) grain size number of a metal is defined as $N = 2^{n-1}$, where

    N = number of grains per in^2 at 100 magnifications (actual area = 1×10^{-4} in^2 or 0.01 in. $\times$ 0.01 in.)

    n = ASTM grain size number (2 to 4 would be coarse and 6 to 10 would be fine-grained)

    a. In Figure 2.2(*a*), approximate the ASTM grain size number if the magnification is 250×.

    b. Why is this number n of interest? (Sections 2.8 through 2.13)

2.20 *[ES/EJ]* Using dislocations in your argument, explain what is happening in the following portions of a stress–strain diagram for a metal. (Sections 2.8 through 2.13)

    a. In the elastic region where $E = S/e$.

    b. Between the yield strength and the tensile strength.

    c. Between the tensile strength and the breaking strength.

2.21 *[ES/EJ]* In Example 2.5, the experimental shear stress (τ) for magnesium is much less than the theoretical value. G is always equal to τ/γ. Explain how the presence of dislocations modify τ (shear stress) and γ (shear strain). (Sections 2.8 through 2.13)

2.22 *[ES/EJ]* Figure 2.18 gives a method for dislocation multiplication. Explain why the dislocation generation does not take place forever. (Sections 2.8 through 2.13)

2.23 *[EJ]* Explain why cold-forming operations might produce a material with directional properties. This phenomenon is called anisotropy. (Sections 2.8 through 2.13)

2.24 *[EJ]* Explain why the effects of cold work may not be uniform through a cross section. (*Hint:* Consider how cold work such as rolling is accomplished.) (Sections 2.8 through 2.13)

2.25 *[EJ]* Explain how work hardening influences the following metal components. (Sections 2.8 through 2.13)

    a. Nail

    b. Automotive fender

    c. Electrical wire

    d. Clothes hanger

2.26 *[ES/EJ]* Figure 2.22 shows a linear relationship between HB and the ultimate tensile strength for steels. (Sections 2.8 through 2.13)

    a. Using the data in the figure, determine a general equation relating the tensile strength to HB.

    b. Why is the tensile strength generally not related to HRC in equation form?

2.27 *[EJ]* Indicate why hardness testing is sometimes done to 100% of engineered components that are to be used in critical applications. (Sections 2.8 through 2.13)

2.28 *[ES]* Using the rules for cleavage of ceramics given in Section 2.14, explain the following observations. Reference to various figures in the text will be required. (Sections 2.14 through 2.15)

a. The cleavage of NaCl is cubic.

b. The cleavage of CaF_2 is octahedral.

c. The cleavage of graphite is parallel to the basal plane.

d. The cleavage of zinc is on the (0001) planes.

2.29 *[ES]* A bend test used for ceramic materials is shown in Figure 2.23*b*. If the tensile strength of a ceramic is 25,000 psi (172.5 MPa), what load will a standard small test beam support at fracture? (Sections 2.14 through 2.15)

2.30 *[ES/EJ]* Indicate whether the following statements are correct or incorrect, and justify your answer. (Sections 2.14 through 2.15)

a. The tensile strength of a ceramic material in a table of values requires a notation of the method of testing by the author.

b. There are likely to be fewer flaws in a ceramic fiber than in a bulk polycrystalline ceramic.

c. All polymer materials have a low tensile strength.

d. The tensile strength of a thermoplastic polymer is higher if loaded over a shorter period of time.

2.31 *[ES]* A measuring tape is made out of a thermoplastic material and has the following characteristics: tensile strength, 11,000 psi; elastic modulus, 4.0×10^5 psi; cross section 0.020 in. $\times$ 0.500 in., and length, 100 ft. We measure something 100 ft long and pull on the tape with a 15-lb force during the measurement. Calculate the true measured length. (Sections 2.1 through 2.7 and 2.14 through 2.15)

2.32 *[ES/EJ]* For what types of engineering materials are fracture toughness values more important than strength values? Explain (Sections 2.16 through 2.21)

2.33 *[ES/EJ]* Explain why it is necessary to quantitatively define flaws in a material that is to be exposed to cyclic loading. (Sections 2.16 through 2.21)

2.34 *[EJ]* A designer is replacing a steel that has an endurance limit of 75,000 psi (518 MPa) (10^6 cycles) with an aluminum alloy that has an endurance strength of 24,000 psi (166 MPa). She calculates that the cross section should increase by a factor of 75/24. Do you agree? (Sections 2.16 through 2.21)

2.35 *[ES/EJ]* The following table gives the tensile strength and endurance limit, or endurance strength, for two steels, two copper-based alloys, and an aluminum alloy. (Sections 2.16 through 2.21)

Alloy	Tensile Strength	Endurance Limit/Strength
Cold-worked 1040 steel	118,000 psi	59,000 psi
Cast 0.25% C Steel	54,000 psi	25,000 psi
Aluminum 2014-T4	62,500 psi	20,000 psi (5×10^8)
Cold-worked brass	70,000 psi	22,000 psi (5×10^7)
Free-cutting brass	56,000 psi	14,000 psi (3×10^8)

a. Why do the nonferrous alloys show the number of cycles to failure in parentheses, whereas the steels do not?

b. Determine the endurance ratios for the alloys.

c. If you conducted fatigue tests on the same alloys, why might your values not be exactly the same as those given in the table?

2.36 *[EJ]* Why are there some exceptions to the widely published generality "Wear is approximately inversely proportional to hardness"? (Sections 2.16 through 2.21)

2.37 *[EJ]* Wear in metals is sometimes rapid at first but then appears to lessen as the time in service increases. From your limited present knowledge about metal structures, explain how this nonuniform wear rate might occur. (Sections 2.16 through 2.21)

2.38 *[ES/EJ]* Explain how, when a surface wears, small chunks pull out and cause an appearance like pits, even though corrosion has not taken place. (Sections 2.16 through 2.21)

CHAPTER 3

Effects of Temperature on Structure and Mechanical Properties

We have now explored the effects of stress on materials at room temperature. Many challenging applications such as the stealth bomber shown, however, require good mechanical properties at elevated temperatures. For example, the evolution of aircraft has required continual improvement in materials, because increased speed raises the heating of the skin from friction and increased power raises the temperature of the engine. Skin materials have progressed from wood and fabric to advanced alloys of aluminum, nickel, and titanium and composite materials containing graphite and polymers. In engines a similar progression has occurred, leading to ceramic coatings on high-alloy metal blades. The stealth bomber has the added material requirement of minimum radar detection.

3.1   Overview

We ended the Summary in Chapter 2 with a warning that the mechanical properties we observe in an engineering material at room temperature may be very different from those at low and elevated temperatures. Many catastrophic failures have resulted from designs based only on properties observed at room temperature. These effects vary drastically with the material. For example, little change occurs in the properties of steel and ceramics when they are heated from room temperature to 250°C (482°F), but some plastics lose shape and even a high-strength aluminum alloy loses strength. At the other end of the temperature scale, many plastics and some steels become brittle at low temperatures.

To reach a basic understanding of these and other important concepts involving temperature, we need to discuss first the thermal properties of materials (Figure 3.1): how they expand and contract, change structure, and conduct heat, and how the atoms move (diffuse). Then we will explore how temperature affects the atoms' movements under stress, the materials' low-temperature properties, the heating (annealing) of cold-worked structures, and finally the creep of ceramics, metals, and polymers.

3.2   Thermal Expansion

Data on *thermal expansion* must be the most abused handbook material in all engineering, because the average engineer is unaware of the profound effects of solid-state phase transformations, particularly in iron alloys and in ceramics. Let us consider the graph shown in Figure 3.1(*a*). As the specimen is heated, its length increases uniformly while the structure is solid 1. Then a sudden contraction takes place while solid 2 forms at T_1. After the transformation, the new solid expands at a greater rate than solid 1. Next solid 2 reverts to the crystal structure of solid 1 at T_2. Finally the material melts (an expansion) at T_3, and the liquid expands at a still greater rate than solid 1 or 2. This graph is not a bizarre abstraction—it is an approximation of the behavior of ordinary iron!

The *coefficient of thermal expansion* alpha, $\alpha^\dagger$, is defined as the change in length per °C or °F. Most handbooks give a value for iron of 14.0×10^{-6}/°C (7.8×10^{-6}/°F), and this is typical for solid 1. However, many designers have extrapolated this value into the range of solid 2 with disastrous results. And, as we shall see in Chapter 8, many steels, such as certain stainless steels, exhibit solid 2 structure (FCC) at room temperature. If they are attached to other components that are solid 1 in structure, stresses develop when the assembly is heated or cooled.

---

†This is not to be confused with the use of alpha for BCC iron.

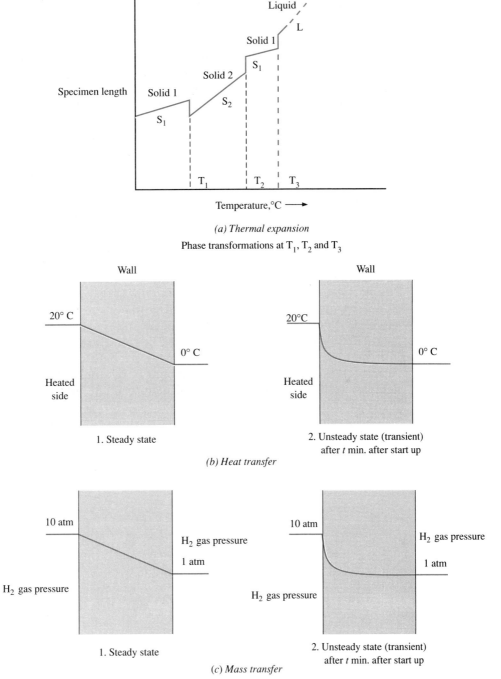

(a) Thermal expansion

Phase transformations at T_1, T_2 and T_3

(b) Heat transfer

(c) Mass transfer

FIGURE 3.1 Examples of (a) thermal expansion, (b) heat transfer, and (c) mass transfer in engineering materials

The coefficient of expansion is even more important to ceramists because the components are brittle — that is, incapable of plastic deformation to accommodate differential movement. Some of the length changes that ceramics undergo upon transformation are more severe than those in metals. Those of silica, SiO_2, are particularly important, as shown in Figure 3.2. This figure shows change in *volume* among the *silica phases*, rather than change in length (for small changes, the change in volume is approximately three times the change in length).

To interpret the data, let us begin with the graph for α-quartz, which is the stable form of silica at room temperature, as in beach sand. When we heat α-quartz, we encounter a pronounced expansion at about 600°C as it is transformed into β-quartz. Even more spectacular is the severe change in length of the high-temperature equilibrium structure, cristobalite, as it cools below 225°C. In the use of silica bricks for furnace roofs, in order to avoid this contraction, the roof is kept above 275°C once it has been heated and transformed to cristobalite. By contrast with these crystalline structures and their transformations, we find no sharp changes in vitreous silica, which has a noncrystalline structure.

In the polymers, the outstanding feature is that the coefficients of expansion are much higher than those in either metals or ceramics (Table 3.1). For example, low-density polyethylene has a value of 180×10^{-6}/°C, compared to 14.0×10^{-6}/°C for iron and 6.7×10^{-6}/°C for alumina ceramic. The reason for the high value is that the molecules are held together only by van der Waals forces; the input of a small amount of thermal energy causes more molecular oscillation than in a material held together with one of the three primary bonds.

Before discussing thermal expansion in greater depth, we should emphasize here that heat and mass transfer are also important characteristics of engineering materials. These phenomena are shown schematically in Figures 3.1*b* and *c* and will be treated more fully later in the chapter.

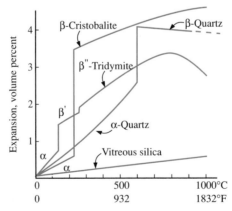

FIGURE 3.2   Changes in volume with temperature in silica structures

TABLE 3.1 Thermal Properties of Selected Materials

Material	Specific Heat C_p (at 300 K) cal/g-K*	Atomic or Molecular Weight, g/mol	Linear Coefficient of Expansion [at 300 $(K)^{-1}$] $\times 10^{-6}$	Lorenz Number[†] $L = \sigma_T/\sigma_e T$, $(volt/K)^2 \times 10^{-8}$	Melting Point T_m, K
Aluminum	0.22	27	24.1	2.2	933
Carbon		12			
Graphite	0.18		2.3 to 2.8		3760
Diamond	0.12		1.2		
Copper	0.092	63.5	17.6	2.23	1356
Gold	0.031	197	13.8	2.35	1336
Iron	0.11	55.9	14.0	2.47	1810
Lead	0.32	207.2	28	2.47	600
Molybdenum	0.065	26	5.55	2.61	2880
Nickel	0.105	58.7	13.3	2.2	1725
Niobium	0.074	92.9	7.4		2740
Platinum	0.031	195.1	8.8	2.51	2040
Silver	0.056	107.9	19.5	2.31	1333
Tantalum	0.036	181	6.7		3270
Tin	0.54	118.7	23.5	2.52	505
Tungsten	0.034	183.9	3.95	3.04	3680
Type 304 stainless steel (austenitic)	0.14		17.3		~1690
Invar	0.12	101.9	1.26		~1700
Alumina ceramic	0.18		6.7		2470
Boron nitride		24.8	7.72		1950
Magnesia (MgO)	0.21	40.3	14		3020
Fused silica (or fused quartz)	0.19	60.1	0.05		1950
Glass	0.2		7.2		
Mica	0.12 to 0.25		40		
Phenolic resins	~0.4		15 to 45		
Polymethylmethacrylate	0.35		81		
Polytetrafluoroethylene	0.25		100		
Low-density polyethylene	0.5		180		
Natural rubber			670		
Silicone rubber			1,200		

*Multiply by 4.186 to obtain J/g-K.
[†]At 20°C (293 K), defined as thermal conductivity/electrical conductivity × absolute temperature.
From R. M. Rose, L. A. Shepard, and J. Wulff, *The Structure and Properties of Materials*, Vol. 4: *Electronic Properties*, John Wiley, New York 1966. By permission of John Wiley & Sons, Inc.

3.3 Structural View of Thermal Expansion

Now that we have defined thermal expansion, let us take an atomic-scale view to explain the wide range of the values that appear in Table 3.1.

We begin with a question: If atoms (and ions) merely vibrate over a wider amplitude upon heating, shouldn't their average positions stay the same, and shouldn't there therefore be no overall expansion? To answer this question, let us return to the graph showing the attractive and repulsive forces in sodium chloride, Figure 1.6. At absolute zero there is very little vibration and the ion centers are at a distance r_0. As we heat, the ions must oscillate and the limiting approach is determined on one side by repulsive forces and on the other by attractive forces. As the temperature is raised, the average position has a larger value for interatomic distance and hence expansion. Modifications in the coefficient of thermal expansion variation with temperature are discussed in ceramics, Chapter 10.

It is a stunning confirmation of the x-ray diffraction technique that when we measure the effect of temperature on a_0 (the edge of the unit cell) and calculate $\Delta a/\Delta T$, the resulting coefficient of expansion is very close to that of a large specimen. We should hasten to add, however, that when the unit cell is anisotropic and exhibits poor crystalline symmetry, the coefficient of expansion depends on direction. For example, when we use high-strength graphite fibers, we must realize that in the high-strength plane of the hexagonal structure, the coefficient of expansion is $1 \times 10^{-6}/°C$, whereas it is $27 \times 10^{-6}/°C$ normal to the plane. Remember that *in* the plane we have strong covalent bonding, whereas perpendicular to the planes there are only van der Waals forces. Therefore, thermal agitation produces greater movement between more weakly bonded atoms (see Figure 1.11*b*).

We may use this atomistic picture to explain both the steps and the differences in slope of the expansion curve for a pure solid shown in Figure 3.1*a*. At low temperature we have solid 1 (S_1), and the expansion is approximately a linear function of temperature. This is why many handbooks give a single value for the coefficient of expansion.

As the structure changes to S_2, a denser arrangement develops (such as occurs when BCC iron is transformed into FCC iron) and a decrease in length (and volume) is encountered at constant temperature. The new structure has a higher coefficient of expansion in this case. Finally, the solid melts as the thermal agitation destroys the crystal structure, and the length (and volume) increase.* The liquid has a higher coefficient of expansion than does S_2.

The magnitude and direction of length changes may be different for different materials, but it is important to note the strong effect of transformation. Let us suppose we are cooling a bar of brittle material that has a tensile

---

*There is also a narrow temperature region in iron wherein the FCC iron becomes BCC iron before melting, as discussed more fully in Chapter 8. This reversion is an exception to typical behavior.

strength of 30,000 psi (207 MPa) at 100°C and transforms at 100°C on cooling with a contraction strain due to transformation of 1200×10^{-6} in./in. Note that this is only 0.12%. If the surface of the bar is only 1°C colder than the interior, it will undergo this contraction upon reaching 100°C while the interior remains over 100°C. As an approximation, if we assume the surface layer is restrained completely and the modulus of elasticity is 30×10^6 psi, the resulting stress will be $S = Ee = 30 \times 10^6$ psi $\times 1200 \times 10^{-6}$ in./in. $= 36,000$ psi (248.4 MPa).

The bar may crack at the surface because of the transformation contraction, even though the temperature gradient is very small. Let us consider another interesting application of differences in thermal expansion in the operation of a bimetal strip. If we bond two strips of different metals together by hot rolling, the strip will bend in proportion to the temperature and can be used to actuate an electrical relay at a desired temperature.

EXAMPLE 3.1 *[ES/EJ]*

(This example will help you understand the action of a thermostat.) Given strips of iron and copper 0.010 in. wide, 0.005 in. thick, and 1 in. long (0.254 × 0.127 × 25.4 mm), bond them together lengthwise along the flat faces at 70°F (21°C). Then heat the bimetallic strip to 170°F (77°C).

A. Which way will the strip curl?
B. If the bimetal is restrained from curling (as in a die), what will be the stress developed in the iron portion and in the copper portion?
   Assume the following data:

	Linear Coefficient of Expansion in./in.-°F (m/m-K)	Modulus of Elasticity, psi (MPa)
Iron	7.8×10^{-6} (14×10^{-6})	30×10^6 (2.07×10^5)
Copper	10.0×10^{-6} (18×10^{-6})	16×10^6 (1.10×10^5)

Answers

A. If it were unbonded, the copper strip would expand more than the iron strip. The bonded strip, therefore, curls into an arc with the copper on the outside and the iron on the inside.
B. If the bonded strip is restrained between flat plates, the copper is prevented from attaining the length it would reach if it were not bonded to the iron. It is under compressive stress. The iron, on the other hand, is stretched by the tendency of the copper to expand a greater amount, and so the iron is under tension. Because the cross-sectional areas are equal, the compressive stress in the copper is the same magnitude as the tensile stress in the iron (otherwise movement due to the unbalanced force would occur):

$$S_{Cu} = S_{Fe}$$

Now consider for a moment the expansion that would have occurred in each metal if it had not been bonded to the other during heating from 70°F to 170°F. The change in length in the iron would be:

$$\Delta l = (170 - 70)°\text{F} \times 7.8 \times 10^{-6} \text{ in./in.-°F}$$

$$= 780 \ \mu\text{in. (in an inch length) or 780 micron/m}$$

For the copper the change in length would be:

$$\Delta l = (170 - 70)°\text{F} \times 10.0 \times 10^{-6} \text{ in./in.-°F}$$

$$= 1000 \ \mu\text{in. (in an inch length) or 1000 micron/m}$$

Therefore the free iron specimen would be 1.000780 in. long and the free copper would be 1.001000 in. long.

Returning to the bonded case, we find that the specimen will have a length between these values. Furthermore, the difference in elastic strain between the copper and the iron will equal $1000 - 780 \ \mu\text{in./in.}$, or $220 \ \mu\text{in./in.}$ In other words, the copper attempts to expand $1000 \ \mu\text{in./in.}$ and the iron only $780 \ \mu\text{in./in.}$ An accommodation is reached in which the copper side is under compression and the iron side in tension. The important point is that the elastic strain in the copper plus that in the iron equal $220 \ \mu\text{in./in.}$ or 220 micron/m:

$$e_{\text{Cu}} + e_{\text{Fe}} = 220 \times 10^{-6}$$

Substituting the relation $e = s/E$ in each case:

$$\frac{S_{\text{Cu}}}{E_{\text{Cu}}} + \frac{S_{\text{Fe}}}{E_{\text{Fe}}} = 220 \times 10^{-6}$$

But we know that

$$S_{\text{Cu}} = S_{\text{Fe}} \quad \text{or} \quad E_{\text{Cu}}e_{\text{Cu}} = E_{\text{Fe}}(220 \times 10^{-6} - e_{\text{Cu}})$$

Substituting the numerical values given for E_{Cu} and E_{Fe}, we obtain:

$$e_{\text{Cu}} = 143 \times 10^{-6} \text{ in./in.} \quad \text{or} \quad 143 \text{ micron/m}$$
$$S_{\text{Cu}} = 2300 \text{ psi or 15.9 MPa compression}$$

$$e_{\text{Fe}} = 77 \times 10^{-6} \text{ in./in.} \quad \text{or} \quad 77 \text{ micron/m}$$
$$S_{\text{Fe}} = 2300 \text{ psi or 15.9 MPa tension}$$

3.4 Heat Capacity, Specific Heat

The specific heat is measured in units of cal/g-K: the heat we have to put into a gram of a material to raise its temperature 1 K (or 1°C). If we want to heat a load of steel parts to 870°C (1600°F), we can calculate the energy required (calories) by using the average* value of 0.11 for iron obtained from Table 3.1. Of course, the efficiency of the furnace or other heating medium must also be considered. Another important application of specific heat is in calculating the change in temperature with time at different locations as a wall is heated. With higher specific heat, the thermal gradient is steeper and the time to reach a desired internal temperature is longer.

Effect of Phase Transformation

When a solid transforms to another solid, to a liquid, or to a gas, we observe a change in the heat content of the material called the heat of transformation, the heat of fusion, or the heat of vaporization respectively. In other words, to heat from T_1 to T_3 for a pure solid that undergoes a phase transformation, we have to provide:

$$\text{Heat for (solid 1)}_{T_1 \text{ to } T_2} + (\text{solid 1} \rightarrow \text{solid 2})_{\text{at } T_2} + (\text{solid 2})_{T_2 \rightarrow T_3}$$

Similarly, if we placed a thermocouple in a block of solid 2 during cooling, we would find a steady decrease in temperature to the transformation temperature; then an arrest, or period of constant temperature, as the heat of transformation was released; then continued cooling. This is a useful method for determining the temperature of phase changes.

3.5 Thermal Conductivity

The whole world became aware of the importance of thermal conductivity when missing tiles on the first manned Space Shuttle compromised the safety of the crew. In a more mundane field, the building legislation (designed to conserve energy) that requires minimum "R values" for walls has sparked new interest in the insulating properties of different materials. For instance, the advocates of log homes and conventional construction engineers endlessly debate the relative expense of heating up a house compared with the cost of maintaining a given temperature. Another way of saying this is that we must consider both transient and steady-state heat transfer. And later on, we can deal with mass transfer (the diffusion of atoms, which is important in changing the surface composition of a component) by using the same mathematics that we apply in analyzing heat transfer. These two cases were illustrated in Figure 3.1*b* and *c* and are also shown in Figure 3.3.

Let us begin with steady-state heat transfer, which is simpler. If we reach equilibrium in a heated building and maintain the inner wall surface at 20°C (68°F) via warm air, a constant temperature distribution develops through the wall. See Figure 3.1*b*, wherein we assume an outdoor temperature of 0°C

---

*The heat capacity is a function of temperature and structure, so this is an approximation.

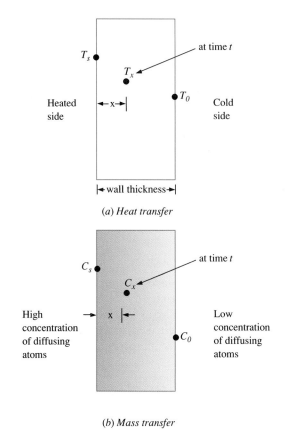

(a) *Heat transfer*

(b) *Mass transfer*

FIGURE 3.3   Heat transfer and mass transfer under transient conditions

(32°F). The heat flow through the wall per unit of area is called the thermal flux J:

$$J = K\frac{\Delta T}{\Delta x} \tag{3-1}$$

K = thermal conductivity, $\dfrac{\text{BTU-ft}}{°\text{F-hr-ft}^2}$ or $\dfrac{\text{cal-m}}{\text{K-min-m}^2}$

$\dfrac{\Delta T}{\Delta x}$ = thermal gradient — that is, temperature change per unit wall thickness

Let us state the equation in words: The heat flow J in steady state is proportional to the thermal gradient (change in temperature with thickness). The proportionality constant is K, which is called the *thermal conductivity*.

EXAMPLE 3.2 *[ES]*

Calculate the energy loss per hour through a brick wall 12 ft × 8 ft × 6 in. (3.66 m × 2.44 m × 0.15 m) thick when the interior is at 68°F (20°C) and the exterior is 32°F (0°C). Assume

$$K = 0.7 \frac{\text{BTU-ft}}{\text{°F-hr-ft}^2}$$

Answer

$$J = \frac{0.7 \text{ BTU-ft}}{\text{°F-hr-ft}^2} \frac{(68 - 32\text{°F})}{0.5 \text{ ft}}$$

$$J = 50.4 \frac{\text{BTU}}{\text{hr-ft}^2} \left(2278 \frac{\text{cal}}{\text{min.-m}^2} \right)$$

The total heat loss, then, is

$$50.4 \frac{\text{BTU}}{\text{hr-ft}^2} \times (12 \text{ ft} \times 8 \text{ ft}) = 4838 \text{ BTU/hr} \quad (20{,}320 \text{ cal/min.})$$

Now let us consider transient conditions, such as the case when we return to a cold house and the wall is uniformly cold. We want to know how fast the wall will heat as a function of time. The equation governing this case is derived in standard heat-transfer texts. It is

$$\frac{T_s - T_x}{T_s - T_0} = \text{erf} \frac{x}{2\sqrt{ht}} \tag{3-2}$$

where
T_s = heated surface temperature

T_0 = original temperature throughout

T_x = temperature at a distance x from the heated surface

h = thermal diffusivity

$\quad = \dfrac{K \text{ (thermal conductivity)}}{C_p \text{ (heat capacity)} \times \rho \text{ (density)}}$

t = time after heating started

erf = error function

This equation arises from the solution of the differential equation on which it is based. Table 3.2, page 158, gives values for the error function showing values ranging from 0.0 to 1.0.

If we compare this equation with that for steady-state heat transfer, we see that here, the temperature profile we obtain by solving for different values of x varies with time. Also, instead of thermal conductivity we have thermal diffusivity, which takes into account the effect of heat capacity and density. As expected, when the thermal capacity is high, T_x increases slowly.

EXAMPLE 3.3 *[ES]*

Develop graphs showing the temperature profile in a wall 20 cm thick at 1 min and 1 hr after the heat is turned on in a house, providing a constant internal-surface temperature of 20°C (68°F). Assume the original temperature was 2°C (36°F).

$$K = 0.0014 \frac{\text{cal-cm}}{\text{cm}^2\text{-K-sec}} \qquad \rho = 2.3 \text{ g/cm}^3 \qquad C_p = \frac{0.25 \text{ cal}}{\text{g-K}}$$

Answer

$$\frac{T_s - T_x}{T_s - T_0} = \text{erf} \frac{x}{2\sqrt{ht}} \tag{3-2}$$

$$h = \frac{0.0014 \dfrac{\text{cal-cm}}{\text{cm}^2\text{-K-sec}}}{\dfrac{0.25 \text{ cal}}{\text{g-K}} \cdot \dfrac{2.3 \text{ g}}{\text{cm}^3}} = 0.0024 \text{ cm}^2/\text{sec}$$

$$\frac{20 - T_x}{20 - 2} = \text{erf} \frac{x}{2\sqrt{ht}}$$

$$T_x = 20 - 18 \text{ erf} \frac{x}{2\sqrt{ht}}$$

For 1 min, $T_x = 20 - 18 \text{ erf} \dfrac{x}{2\sqrt{0.0024 \, (60)}} = 20 - (18 \text{ erf } 1.32x)$
(60 sec)

For 1 hr, $T_x = 20 - 18 \text{ erf} \dfrac{x}{2\sqrt{0.0024 \, (3600)}} = 20 - (18 \text{ erf } 0.17x)$
(3600 sec)

At x =	*0.1 cm*	*0.5 cm*	*1 cm*	*2 cm*	*5 cm*
For 1 minute					
1.32x	0.132	0.66	1.32	2.64	6.6
erf(1.32x)	0.15	0.65	0.94	1.0	1.0
18(erf 1.32x)	2.7	11.7	16.9	18	18
20 − (18 erf 1.32x)	17.3°C	8.3°C	3.1°C	2.0°C	2.0°C

At x =	0.1 cm	0.5 cm	1 cm	2 cm	5 cm	10 cm
For 1 hour						
0.17x	0.017	0.085	0.17	0.34	0.85	1.70
erf(0.17x)*	0.018	0.096	0.19	0.37	0.77	0.98
18[erf(0.17x)]	0.32	1.73	3.42	6.66	13.9	17.6
20 − [18 erf(0.17x)]	19.7°C	18.3°C	16.6°C	13.3°C	6.1°C	2.4°C

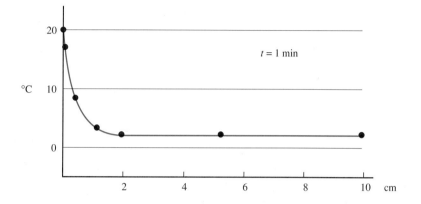

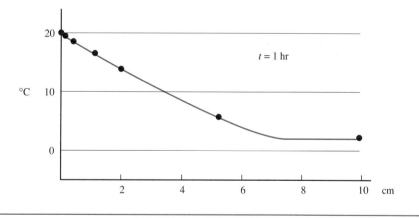

3.6 Mass Diffusion, Mass Transfer

In materials engineering, the transfer of mass—that is, atom movement—is analogous to heat transfer. This can be appreciated readily by comparing Figure 3.1*b* and *c*. The thermal and mass profiles are similar for the steady state and transient conditions. There are many processes in which the surface

---

*See Table 3.2, page 158, for error function values.

layers are improved by diffusing new atoms inward. Among the oldest is raising the carbon content in the surface of steel gears by heating in a carbonaceous atmosphere (*carburizing*), but similar treatments using nitrogen (nitriding) are now common. Some of the most spectacular developments have occurred in the processing of semiconductors, in which small amounts of second elements (dopants) are diffused into the semiconductor. We also encounter some harmful effects, such as carbon loss from the surface layers by heating in an oxidizing atmosphere (decarburization). Another important effect is *diffusion bonding*, in which several layers of material are pressed together and heated. The U.S. quarter is an example. Atom movement across the interface obliterates the mechanical boundary between the layers and raises the strength.

Diffusion Mechanisms—Atom Vacancies

Before going into mass-transfer calculations, we should reflect a moment on the mechanism of mass transfer. It is simple for the diffusing atoms to make use of vacancies. Of course, when line defects exist, such as dislocations or the channels of vacancies that occur at grain boundaries, diffusion is rapid along these paths. However, the number of vacancies per dislocation is variable, but there is always an equilibrium number of point vacancies, which we may calculate as follows.

In order to leave a lattice site and create a vacancy, an atom needs a certain amount of energy ΔE, which is called the *activation energy*. Although the energy of an atom depends on temperature, at any given temperature there is a broad spectrum of energies called the Boltzman distribution. At a given temperature a certain fraction n of the total number of atoms n_t have enough energy to leave the regular lattice points, providing vacancies. From chemistry we find that the activation energy ΔE needed to create the vacancy is given by the relationship

$$\frac{n}{n_t} = Ce^{-\Delta E/kT} \tag{3-3}$$

C = a material constant

where

k = Boltzman's constant, 1.38×10^{-23} J/K   (or 1.38×10^{-16} erg/K)

T = absolute temperature

ΔE = activation energy

e = 2.718

We can simplify this equation and solve it to provide a plot by taking natural logarithms of both sides.

$$\ln \frac{n}{n_t} = \ln C - \frac{\Delta E}{k} \frac{1}{T} \tag{3-4}$$

This is the equation of a straight line.

$$y = mx + b, \quad \text{where } y = \ln \frac{n}{n_t},$$

$$m = \frac{-\Delta E}{k} \quad \text{and} \quad x = \frac{1}{T} \quad \text{with } b = \ln C$$

Therefore, if we have data for two temperatures, we can solve for the general equation as shown in the following example. We can also find the number of vacancies by using the dimensions of the unit cell to calculate the density of vacancy-free material and compare this with the actual density measured. We are assuming no microporosity in our experimental sample.

EXAMPLE 3.4* *[ES]*

The comparison of density values obtained using the conventional method and using x-ray diffraction data reveals that the percentages of vacant lattice sites in aluminum are 0.08% at 923 K and 0.01% at 757 K. Calculate the number of vacant lattice sites per cubic centimeter at 527°C, using Equation 3-4,

$$\ln \frac{n}{n_t} = \ln C - \frac{\Delta E}{kT}.$$

Answer

A. At 923 K,

$$\ln \frac{0.08}{100} = \ln (0.0008) = \ln C - \Delta E/(1.38 \times 10^{-16} \text{ erg/vacancy-K})(923 \text{ K})$$

B. At 757 K,

$$\ln \frac{0.01}{100} = \ln (0.0001) = \ln C - \Delta E/(1.38 \times 10^{-16} \text{ erg/vacancy-K})(757 \text{ K})$$

Solving simultaneously, we find that $\Delta E = 1.21 \times 10^{-12}$ erg/vacancy. Substituting yields $\ln C = 2.37$.

C. The general equation for aluminum is then

$$\ln \frac{n}{n_t} = 2.37 - \frac{1.21 \times 10^{-12}}{kT}$$

where

$$n_t \approx \frac{2.70 \text{ g/cm}^3 \times 6.02 \times 10^{23} \dfrac{\text{atoms}}{\text{at. wt}}}{26.99 \text{ g/at. wt}}$$

*After L. H. Van Vlack, *Materials Science for Engineers*, 2d ed., Addison-Wesley, Reading, Mass., 1970.

At 527°C (800 K),

$$\ln \frac{n}{n_t} = 2.37 - [1.21 \times 10^{-12} \text{ erg/vacancy}$$
$$/(1.38 \times 10^{-16} \text{ erg/vacancy-K}) (800 \text{ K})]$$
$$= -8.59$$

or

$$\frac{n}{n_t} = 1.86 \times 10^{-4}$$

Then

$$\left(1.86 \times 10^{-4} \frac{\text{vacancies}}{\text{atom}}\right) \left(\frac{2.70 \text{ g/cm}^3}{26.99 \text{ g/at.wt.}}\right) \left(6.02 \times 10^{23} \frac{\text{atoms}}{\text{at.wt.}}\right)$$

$$= 1.12 \times 10^{19} \frac{\text{vacancies}}{\text{cm}^3}$$

Now that we have discussed diffusion by vacancies, let us take up mass-transfer calculations. We can calculate mass diffusion or transfer by using the same concepts that we applied in calculating heat transfer as shown in Figure 3.3b. The unsteady state is nearly always more important, and the equation we use in such cases is known as Fick's second law.* It is

$$\frac{C_s - C_x}{C_s - C_0} = \text{erf}\left(\frac{x}{2\sqrt{Dt}}\right) \tag{3-6}$$

where

C_s = concentration at surface after start of diffusion

C_0 = original concentration

C_x = concentration at point x

x = distance from interface

D = diffusivity

The symbols for concentration of the diffusing species correspond to those for temperature in heat transfer. The mass diffusivity corresponds to the thermal diffusivity h and has the same units.

Let's use these concepts in a sample problem.

---

*The simpler case, Fick's first law of diffusion, corresponds to steady-state heat transfer. It is

$$J = -D\frac{\Delta c}{\Delta x} \tag{3-5}$$

In other words, the net atom flow is proportional to the concentration gradient. The proportionality constant is the *diffusivity D*, and the minus sign indicates that atom flow is from a higher to a lower concentration. Note the similarity to steady state-heat transfer.

EXAMPLE 3.5 *[ES]*

Many sliding and rotating parts, such as gears, call for a hard structure in the surface layers backed by a tough structure in the interior. The first step in producing such parts is to diffuse carbon into the surface of a steel, raising the level from the original level of about 0.2% carbon to 0.5–0.9% carbon for 0.005–0.050 in. (0.0127–0.127 cm).

If we place the gear in a furnace at 1000°C with an atmosphere rich in hydrocarbon gas, the surface reaches a carbon content of about 0.9% very rapidly. The carbon content beneath the surface then rises gradually as a function of time. Calculate the carbon content C_x at 0.010 in. (0.0254 cm) beneath the surface after 10 hr (36,000 sec) at 1000°C.

Answer

$$\frac{C_s - C_x}{C_s - C_0} = \text{erf}\left(\frac{x}{2\sqrt{Dt}}\right) \tag{3-6}$$

$$C_s = 0.9 \ (C_x \text{ is the desired value})$$

$$C_0 = 0.2 \ (\text{assume an SAE 1020 steel is used. See Chapter 8.})$$

$$D = 0.298 \times 10^{-6} \text{ cm}^2/\text{sec (from Example 3.7)}$$

Let $x = 0.01$ in. $= 0.0254$ cm. because D is in cm^2/sec. Then

$$\frac{0.9 - C_x}{0.9 - 0.2} = \text{erf}\left(\frac{0.0254 \text{ cm}}{2\sqrt{(0.298 \times 10^{-6} \text{ cm}^2/\text{sec})(3.6 \times 10^4 \text{ sec})}}\right)$$

$$= \text{erf } 0.123$$

To find the error function of the number 0.123, we interpolate from Table 3.2.

z	$\text{erf } z$
0.150	0.1680
0.123	x
0.100	0.1125

$$\frac{0.150 - 0.123}{0.150 - 0.100} = \frac{0.1680 - x}{0.1680 - 0.1125} \quad \text{or} \quad x = 0.1380$$

Therefore,

$$\frac{0.9 - C_x}{0.7} = 0.138$$

$$C_x = 0.803$$

TABLE 3.2　Table of the Error Function

z	erf(z)	z	erf(z)	z	erf(z)	z	erf(z)
0	0	0.40	0.4284	0.85	0.7707	1.6	0.9763
0.025	0.0282	0.45	0.4755	0.90	0.7970	1.7	0.9838
0.05	0.0564	0.50	0.5205	0.95	0.8209	1.8	0.9891
0.10	0.1125	0.55	0.5633	1.0	0.8427	1.9	0.9928
0.15	0.1680	0.60	0.6039	1.1	0.8802	2.0	0.9953
0.20	0.2227	0.65	0.6420	1.2	0.9103	2.2	0.9981
0.25	0.2763	0.70	0.6778	1.3	0.9340	2.4	0.9993
0.30	0.3286	0.75	0.7112	1.4	0.9523	2.6	0.9998
0.35	0.3794	0.80	0.7421	1.5	0.9661	2.8	0.9999

EXAMPLE 3.6　*[ES]*

Using the information given in Example 3.5, calculate the time necessary to raise the carbon level to 0.60% at 0.010 in. (0.0254 cm) beneath the surface. The diffusion coefficient is again 0.298×10^{-6} cm^2/sec at 1000°C.

Answer

$$\frac{C_s - C_x}{C_s - C_0} = \mathrm{erf}\left(\frac{x}{2\sqrt{Dt}}\right) \tag{3-6}$$

where $C_s = 0.9$, $C_0 = 0.2$, and $C_x = 0.6$. Thus

$$\frac{0.9 - 0.6}{0.9 - 0.2} = \mathrm{erf}\left(\frac{0.0254 \text{ cm}}{2\sqrt{(0.298 \times 10^{-6} \text{ cm}^2/\text{sec}) \times t}}\right)$$

$$0.4286 = \mathrm{erf}\left(\frac{23.26}{\sqrt{t}}\right)$$

We need a number Z whose error function is 0.4286. We refer to Table 3.2 and find that this number is 0.40. Therefore,

$$0.40 = \frac{23.26}{\sqrt{t}}$$

$$t = 3{,}381 \text{ sec} = 56.35 \text{ min (approximately 1 hr)}$$

Lowering the carbon requirement from 0.80 to 0.60 at 0.010 in. (0.0254 cm) below the surface decreases the furnace time from 10 hr to 1 hr, a considerable saving in time. Chapter 8 will treat the carbon requirements for this diffusion process.

3.7 Effects of Temperature

In Fick's first and second laws, the movement of the diffusing material is proportional to the diffusivity D. We would expect D to increase with temperature, since the atomic motion and the number of vacancies both increase. The relationship is found to obey the equation

$$D = Ae^{-Q/(RT)} \qquad (3\text{-}7)$$

where A = constant, Q = the constant for the diffusing substance and solvent involved, R = the gas constant (1.987 cal/mole-K), and T = the absolute temperature (K). Therefore, taking $\log_e$ of both sides, we have

$$\ln D = \ln A - \frac{Q}{RT} \qquad (3\text{-}8)$$

which is the equation of a straight line if we plot $\ln D$ vs. $1/T$.

This approach is important because if we determine D for only two temperatures, we can solve for A and Q and get a general relation for any temperature. In a practical case, if we wish to estimate the effect of changing a furnace temperature for carburizing a gear, we can calculate the time needed at the new temperature to obtain equivalent results.

EXAMPLE 3.7 *[ES]*

In Example 3.5 the value of D was given. This time calculate D at 1000°C, given A = 0.25 cm^2/sec and Q = 34,500 cal/mole for the diffusion of carbon in γ iron.

Answer

$$D = (0.25 \ \text{cm}^2/\text{sec})e^{-\dfrac{34{,}500 \ \text{cal/mole}}{(1.987 \ \text{cal/mole-K})(1273 \ \text{K})}} = \frac{0.25}{e^{13.64}}$$

Since

$$\ln e^{13.64} = 13.64,$$

or

$$\log_{10} e^{13.64} = \frac{13.64}{2.3} = 5.92 \quad \text{antilog } 5.92 = 8.38 \times 10^5$$

$$e^{13.64} = 8.38 \times 10^5$$

Thus

$$D = \frac{0.25}{8.38 \times 10^5} = 0.298 \times 10^{-6} \ \text{cm}^2/\text{sec}$$

3.8 Other Diffusion Phenomena

It is interesting to analyze the value of D for different combinations of materials (Figure 3.4). In particular, elements with small atomic diameters that form interstitial solid solutions diffuse very rapidly. The fact that carbon diffuses faster in BCC iron than in FCC iron seems paradoxical (much more carbon will dissolve in FCC iron than in BCC iron) until we recall that the BCC cell is a less dense structure. Therefore, although the FCC cell centers have more room for the carbon atoms, the passageways through the unit cell are tighter for carbon movement.

The role of dislocations and grain boundaries is also of interest. The looser packing caused by dislocations leads to a more rapid movement of the diffusing substance, which can lead to ten times greater penetration along grain boundaries.

Diffusion in Ceramics and Polymers

In general, diffusion is slower and more complex in ceramics. For example, if we are to examine the diffusion of Al_2O_3 through a barrier layer of another material, we must consider the independent motion of Al^{3+} and O^{2-} ions. The Al^{3+} ion is smaller and usually penetrates the barrier faster. This is shown by the data for Al and O in Al_2O_3 (Figure 3.5). If the ions were diffusing through

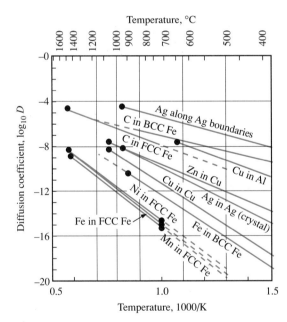

FIGURE 3.4 Variation in diffusivity D as a function of temperature for various materials (units of D are cm²/sec.)

[Lawrence Van Vlack, *Elements of Materials Science*, © 1964, Addison-Wesley Publishing Co., Inc., Reading, Massachusetts. Reprinted with permission of the publisher.]

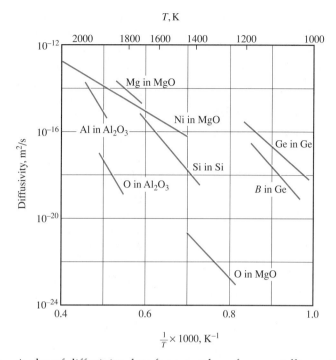

FIGURE 3.5   A plot of diffusivity data for a number of nonmetallic systems

[Reprinted with permission of Macmillan Publishing Company from *Introduction to Materials Science for Engineers* by James E. Shakelford. Copyright © 1984 by Macmillan Publishing Company.]

another material, the number of cation and anion vacancies, defect structures, and the like affect the result.*

In polymers we are dealing with different phenomena, but the diffusion loss of small molecules used as "plasticizers" can result in embrittlement with time. Similarly, the diffusion of oxygen into rubber may cause loss of elasticity.

EFFECTS OF TEMPERATURE ON MECHANICAL PROPERTIES

3.9 General Effects

Many of the most critical engineering applications of materials such as aircraft engines, cutting tools for machining, power generators, and rocket components depend on a combination of mechanical properties exhibited at elevated temperatures or at low temperatures. The purpose of this discussion

---

*For further discussion of the silver-sulfur system, refer to Chapter 18.

is to illustrate these effects; we will leave the detailed discussion of properties and composition to later chapters. In this section we will begin by investigating the combined effects of stress and low temperature and then take up high-temperature effects. In both cases, the typical 0.505 in. (12.8 mm) diameter specimen is used. Other tensile test samples are inadequate to disclose the information that engineers need to design components for these temperature ranges.

3.10 Low-Temperature Effects

For a long time, engineers noticed that steel parts often failed with a brittle fracture when exposed to low temperatures. Tensile tests of the failed material showed normal ductility, so a search was made for a test that would produce a brittle fracture at low temperatures. It was believed that a brittle fracture could be induced by combining two factors: impact and the presence of a notch. Two *impact tests*, the Charpy and the Izod tests, were developed; they use the same equipment with different specimen designs. The Charpy test is shown in Figure 3.6a.

A specimen of the material to be tested, in the shape of a square bar with a V notch, is struck by a calibrated swinging arm, and the energy absorbed is measured. It is relatively simple to examine the effects of temperature by immersing several specimens in advance in liquids at different temperatures and then transferring them quickly to the test fixture.

The types of data obtained are shown schematically in Figure 3.6b. FCC metals show high impact values and no important change with temperature. However, BCC metals, polymers, and ceramics show a transition temperature below which brittle behavior is found. It should be emphasized that the actual transition temperature varies greatly for different materials. For metals and polymers it is between −200 and 200°F (−129 and 93°C); for ceramics it is above 1000°F (538°C).

There is a distinct difference in the appearance of the fractures of low-carbon steels, depending on whether the specimen was tested and broken below or above the transition temperature. As indicated in Figure 3.7b, the appearance of Charpy V notch specimens varies from ductile to brittle as the specimen temperature is reduced from 200 to −321°F (93 to −196°C). Careful observation shows that a shear-type fracture, indicated by the presence of a shear lip, is characteristic of the specimens tested at higher temperatures, whereas shear is absent in the specimens tested at the lowest temperatures. Photomicrographs taken with the scanning electron microscope provide additional evidence (Figure 3.7 c, d, and e). The specimen with highest impact strength shows a dimpled surface with myriads of cuplike projections of deformed metal. In contrast, the brittle sample shows mainly cleavage fractures, almost like split platelets of mica. The specimen at 77°F (25°C) shows a mixture of both types of fractures, as expected.

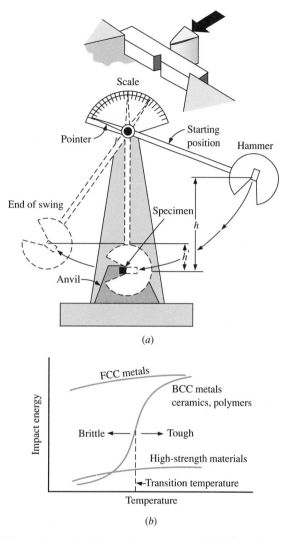

FIGURE 3.6   (*a*) Operation of a Charpy impact test. (*b*) Effect of temperature on the impact strength of various materials (schematic).

[Part (a) from H. W. Hayden, W. G. Moffatt, and John Wulff, *The Structure and Properties of Materials*, Vol. 3: *Mechanical Behavior*, John Wiley, New York, 1965. By permission of John Wiley & Sons, Inc.]

One of the most striking illustrations of the importance of transition temperature is the failure of the Liberty ships produced during World War II. These ships were made of low-carbon steel, which showed good ductility in the usual tensile test. However, 25% of the ships developed severe cracks, often while anchored in the harbor, and many broke in two. The failures were analyzed as being the result of constraint (stress concentration) caused by square hatches in the deck, coupled with the use of steel with a transition temperature (Charpy) near the operating temperature. However, the Charpy

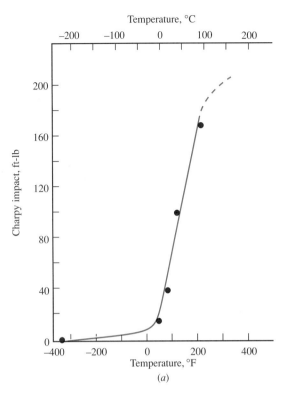

(a)

FIGURE 3.7 (*a*) Charpy impact vs. temperature for SAE 1020 steel, hot-rolled. (*b*) Charpy impact fractures after testing at (right to left) $-196, 0, 25, 50, 93°C$. Parts (*c*), (*d*), and (*e*) are scanning electron micrographs of Charpy specimens, showing shear failure above transition temperature and cleavage at intermediate temperature; 1020 steel, hot-rolled, 1500×. (*c*) Tested at 93°C, ductile (shear) fracture, 170 ft-lb. (*d*) Tested at 25°C, mixed fracture, 40 ft-lb. (*e*) Tested at $-196°C$, cleavage fracture, <1 ft-lb.

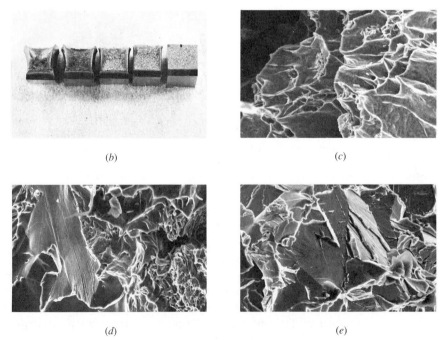

(b)

(c)

(d)

(e)

test does not give a wholly satisfactory answer, as the recent failures of oil tankers show, and the Naval Research Laboratory has developed a new test called the *dynamic tear test*. With this test, notched samples of varying sections can be tested with a drop weight. The transition temperatures determined by this test are more reliable indicators of steel quality.

3.11 Effects of Elevated Temperature on Work-Hardened Structures

We do not always want maximum strength and hardness in a part, because as the hardness increases the ductility decreases, as shown by the percent elongation (Figure 3.8). Suppose we clamp one end of a metal alloy rod in a vise and try to bend the free end through an angle of 120°. At approximately 90° the rod begins to crack. We have "used up" the plastic ductility. We encounter similar failures in other cold-forming operations. What can we do to permit completion of the rod-bending operation?

Basically we wish to restore the original structure by eliminating the extensive slip and dislocation tangles. Remember that atoms are not rigidly fixed but can diffuse from their positions. Diffusion increases rapidly with rising temperature. Thus, if we heat the part, the atoms redistribute between strained and unstrained regions. This is adjustment to strain on a microscopic scale. The partly formed shape does not change dimensions.

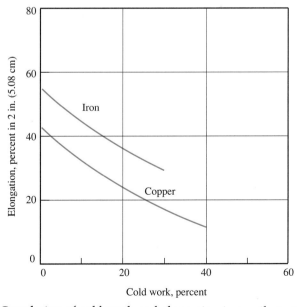

FIGURE 3.8   Correlation of cold work and elongation in tensile testing

[Lawrence Van Vlack, *Elements of Materials Science*, © 1964, Addison-Wesley Publishing Co., Inc., Reading, Massachusetts. Reprinted with permission of the publisher.]

Therefore, we take the cold-worked part and heat-treat it in a furnace in the process called *annealing.* The metal softens as a function of heating temperature (Figure 3.9) and time at temperature. We can see that profound changes take place in the microstructure (Figure 3.10). At elevated temperatures and in the absence of foreign nuclei, new small grains of equal dimensions in all directions (*equiaxed* grains) grow inside the old distorted grains and at the old grain boundaries.* At higher temperatures the grain size is larger. Hardness and strength decrease with increasing grain size, but elongation increases. This can be explained from a microscopic perspective. Slip on a given plane stops when a grain boundary is reached, a region of high dislocation density and entanglements. Thus the more grain boundaries, the greater the limitation to slip and the higher the strength. Ductility (elongation) is lower because slip is limited by the larger grain boundary area found in fine grained materials.

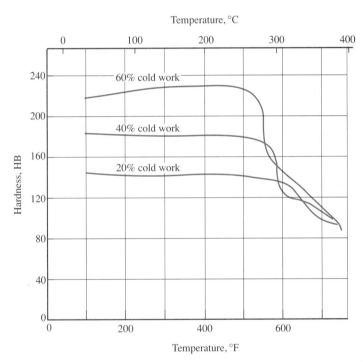

FIGURE 3.9   Effect of heating on hardness of cold-worked 65% Cu, 35% Zn brass, 1 hr

[Lawrence Van Vlack, *Elements of Materials Science,* © 1964, Addison-Wesley Publishing Co., Inc., Reading, Massachusetts. Reprinted with permission of the publisher.]

---

*In steels to which small amounts of aluminum have been added, called aluminum-killed steels, the presence of aluminum nitride inclusions at the grain boundaries inhibits the formation of equiaxed grains. Similar grain boundary inclusions can prevent formation of truly equiaxed grains in alloy systems other than steels.

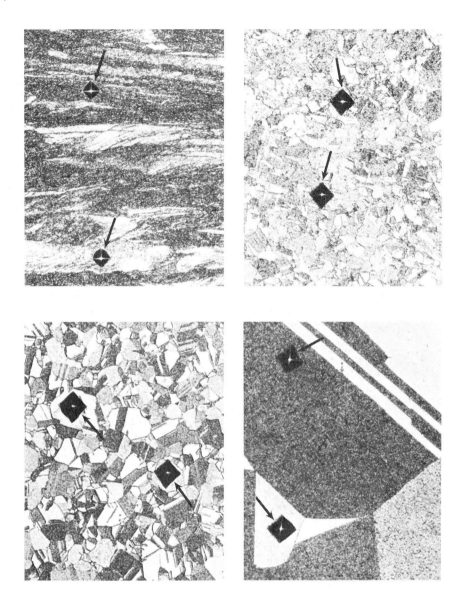

FIGURE 3.10 Strip of single-phase 30% Zn brass, 50% reduction. Magnification 500×, peroxide etch. Upper left: cold-worked, microhardness 140 to 170 HV, 25-g load. Upper right: partly recrystallized after 1 hr at 600°F (316°C), microhardness (arrows) 93 to 109 HV. Lower left: recrystallized at 880°F (471°C) for 1 hr, small grains, microhardness 90 to 100 HV. Lower right: recrystallized at 1400°F (760°C) for 1 hr, large grains, microhardness 65 to 87 HV.

(Paul W. Flinn, Adjunct Professor, Kansas State University)

3.12 Recovery, Recrystallization, and Grain Growth

It is useful to divide the effects of temperature and time on cold-worked material into three regions, in order of increasing temperature (Figure 3.11).

Recovery

Recovery takes place in the temperature range just below recrystallization. With the electron microscope we can see that stresses are relieved in the most severely slipped regions. Dislocations move to lower-energy positions, giving rise to subgrain boundaries in the old grains. This process is called *polygonization*. Hardness and strength do not change greatly, but corrosion resistance is improved. Electrical resistivity decreases slightly.

Recrystallization

When the temperature rises above the recovery range, *recrystallization* takes place. In this higher temperature range the formation of new stress-free and equiaxed grains leads to lower strength and higher ductility. There is an interesting relation between the amount of previous cold work and the grain size of the recrystallized material. With less cold work there are fewer nuclei for the new grains, and the resulting grain size is larger. Even though inclusions

	Copper, OFHC			70% Cu, 30% Zn Brass				
	Prior Cold Work			*Prior Cold Work*		*Tensile Strength, psi × 10³*	*Percent Elongation in 2 in.*	*Grain Size, mm*
	30%	*50%*	*80%*	*50% F.G.*	*50% C.G.*			
Initial	86 HRH	91 HRH	95 HRH	99 HRX	97 HRX	80	8	
30 min @								
150°C	85	90	94	101	98	81	8	
200°C	80	88	93	102	100	82	8	
250°C	74	75	65	103	101	82	8	
300°C	61	54	42	82	98	76	12	
350°C	46	40	34	66	80	60	28	0.02
450°C	24	22	27	50	58	46	51	0.03
600°C	15	17	22	38	34	44	66	0.06
750°C				20	14	42	70	0.12
Final grain size (mm)	0.15	0.12	0.10	0.08	0.12			

F.G. = originally fine-grained; C.G. = originally coarse-grained; HRH = Rockwell scale, $\frac{1}{8}$-in. ball, 60-kg load; HRX = $\frac{1}{16}$-in. ball, 75-kg load.

FIGURE 3.11   Recovery, recrystallization, and grain growth produced by heating cold-worked material. Dislocation redistribution upon heating may cause an initial increase in hardness in some solid solutions, such as 70% Cu, 30% Zn.

[After Brick and Phillips, *Structure and Properties of Alloys*, McGraw-Hill, New York, 1949. Used by permission.]

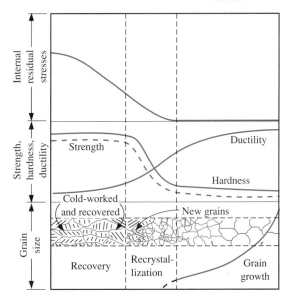

FIGURE 3.12   Summary of annealing effects, time- and temperature-dependent (schematic representation)

(After Sachs)

at the grain boundaries may prevent completely equiaxed grains, the strength and ductility are still modified and are normally lowered.

Grain Growth

As the temperature is raised further, the grains continue to grow because the large grains have less surface area per unit of volume. There are fewer atoms leaving a unit grain boundary area in a large grain than in a small grain, because the grains minimize their surface area. Therefore, the larger grains grow at the expense of the smaller.

It should be added that all three effects are time-dependent as well. These effects are summarized in Figure 3.12.

3.13 Selection of Annealing Temperature

The temperature required for recrystallization varies with the metal. It is approximately one-third to one-half the melting temperature of a pure metal, expressed on an absolute scale (Figure 3.13). Therefore steel is annealed at red heat (1600–1800°F, or 870–980°C), whereas aluminum would be liquid at this temperature. On the other hand, lead can hardly be cold-worked at room temperature because its recrystallization temperature is so low. Furthermore, the recrystallization temperature is really a range, not a sharp point. Also, the

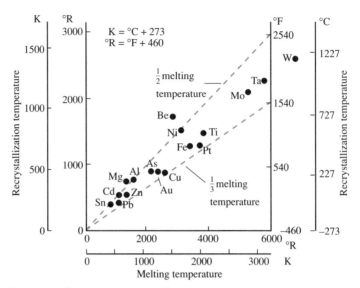

FIGURE 3.13   Correlation of recrystallization temperature with melting point of pure metals. See text for reasons why the recrystallization temperature is not fixed.

[Lawrence Van Vlack, *Elements of Materials Science*, © 1964, Addison-Wesley Publishing Co., Inc., Reading, Massachusetts. Reprinted with permission of the publisher.]

greater the amount of cold work, the lower the recrystallization temperature, because of the stored energy. The sudden drop in hardness for the 65% Cu, 35% Zn brass in Figure 3.9 shows the order of magnitude of the drop in recrystallization temperature that is to be expected with increased cold work.

We may now summarize the variables that influence the recrystallization temperature. In effect it is not a fixed temperature. Rather, it depends on the alloy, the percent cold work, the original grain size, and the time held at temperature. The recrystallization temperature increases with increased alloying. However, it decreases with increased cold work, finer grain size, and longer holding times.

The effects of cold work and grain size can be explained by the fact that the new grains can be more readily nucleated at sites of high dislocation density (grain boundaries and slip planes). The alloy and time effects depend on diffusion. However, it is reasonable to expect that atom motion is required in recrystallization and that the presence of foreign atoms (alloying elements) makes atom motion more difficult. Similarly, longer holding times give atoms more time to move, so recrystallization can occur at a lower temperature. It is common practice to use 1 hr at different temperatures to determine the actual recrystallization temperature.

3.14   Effect of Grain Size on Properties

The preceding discussion showed that we can control the grain size of the final part by a combination of cold work and annealing. With large amounts

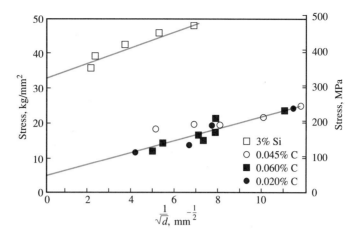

FIGURE 3.14   Relation between yield stress and grain diameter of iron alloys

[Reprinted with permission of Macmillan Publishing Company from *Elements of Mechanical Metallurgy* by W. J. M. Tegart. Copyright © 1966 by William John McGregor Tegart.]

of cold work and rapid annealing, the grain size will be small, because many nuclei for growth of new grains are present. Furthermore, grain boundaries are an impediment to slip. We would expect, then, that a fine-grained material would have higher strength than a coarse-grained one. This is expressed in the Petch relation: $S = S_0 + kd^{-1/2}$, where S is the yield strength, d is the grain size, and S_0 and k are constants for a particular material. Added advantages of fine-grained material are that the plastic flow under stress is more even and the surface of stamped or formed parts is smoother than with coarse-grained material. Figure 3.14 gives an example of the effect of grain size on various iron alloys.

3.15 Engineering Application of Cold Work and Annealing

The cold-work–annealing sequence may be repeated as desired. We can start with a simple sheet of metal and roll or form it extensively. Each time the material becomes difficult to work or is in danger of cracking, we can anneal it. Furthermore, we can develop the desired combination of strength and hardness on the one hand and ductility on the other by the final sequence of operations. There are two possibilities:

1. We can finish the part with higher hardness and lower ductility than desired and then anneal it to obtain the desired combination.
2. We can use the proper amount of cold work for the final operation to reach the required hardness level.

3.16 Hot Working

At this time we may well ask, "Instead of cold working plus annealing, what would happen if we worked the part *at* the annealing temperature?" Would not recrystallization take place, so that we could continue to work the part without interruption? The answer is yes, and therefore we can conduct practically all the desired working at the annealing temperature or above.* Steel, for example, is passed through a set of rolls at a high temperature. In some cases the heat produced by the mechanical work actually increases the temperature! The disadvantages of *hot working* are oxidation of the metal surface and the shorter die life caused by heating from the part. Cold working can produce a better surface because the part can be annealed in a furnace with a protective atmosphere after working. Also, the strength and hardness of parts finished by hot working are lower.

In many cases a combination of hot and cold working is used. A large shape such as an ingot is cast, hot-rolled to bar stock, and then cleaned. The surface is then preserved by cold rolling and annealing in protective atmospheres.

EXAMPLE 3.8 *[EJ]*

In the manufacture of practically all stamped and cold-formed parts, it is necessary to balance the properties desired in the final part with the economic and engineering problems of forming the part. Let us recall the following aspects of the problem.

Treatment	Strength	Hardness	Percent Elongation	Percent Reduction of Area
Increased work before testing	Increased	Increased	Decreased	Decreased
Increased annealing temperature	Decreased	Decreased	Increased	Increased

For maximum workability we want a well-annealed metal, but for highest strength and hardness we want maximum cold work. Suppose we have some annealed 70% Cu, 30% Zn brass bar stock (cylindrical rod) 0.35 in. (8.89 mm) in diameter. Also assume that this particular alloy cannot be cold-worked over 60%.

Problem Produce bar stock 0.21 in. (5.33 mm) in diameter with a tensile strength of over 60,000 psi (414 MPa) and an elongation of over 20%.

*Recrystallization is not the only reason for greater formability. Increased atomic mobility provides greater ductility.

Answer   One way to solve the problem is first to determine what the final step in cold working should be. From Figure 3.15 we see that to obtain 60,000 psi (414 MPa) tensile strength we need more than 15% cold work. Also, to obtain the required elongation we should cold-work less than 23%. Let us choose cold working 20% as the final working step. This will meet the tensile strength and elongation specifications. Then

$$\text{Percent cold work} = \frac{\Delta A}{A} = \frac{\frac{1}{4}\pi d^2 - \frac{1}{4}\pi(0.21)^2}{\frac{1}{4}\pi d^2} \times 100 = 20$$

where d is the diameter at which we wish to start the final deformation (starting, of course, with an annealed structure). Solving, we have

$$d = 0.235 \text{ in. (5.96 mm)}$$

Therefore, we obtain 20% cold work going from 0.235- to 0.21-in. (5.96- to 5.33-mm) diameter.

The 20% cold work required to meet the final specification is not to be confused with the 20% elongation specification. The cold work is a forming

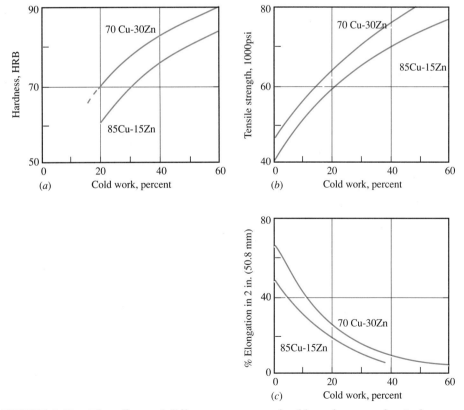

FIGURE 3.15   The effects of different amounts of cold work on mechanical properties. (For the tensile strength, multiply psi by 6.9×10^{-3} to obtain MPa.)

operation, $\Delta A/A$, whereas the elongation specification is derived from a tensile test of a specimen that may have been previously cold-worked. It is defined as $\Delta l/l$.

Now we have to reduce the diameter from 0.35 to 0.235 in. (8.89 to 5.96 mm). The cold work required in this step is

$$\frac{\frac{1}{4}\pi(0.35)^2 - \frac{1}{4}\pi(0.235)^2}{\frac{1}{4}\pi(0.35)^2} \times 100 = 54.9\%$$

Because we received the material in the annealed state, we could cold-work from 0.35- to 0.235-in. (8.89- to 5.96-mm) diameter, then anneal, and finally cold-work the required 20% to achieve our tensile strength and elongation specifications, along with a final diameter of 0.21 in. (5.33 mm). Had the original step been more than 60% cold work, we would run the risk of fracture during cold work. We would have to use several cold-work and annealing steps. Remember, the effects of cold work are cumulative, but they can be removed by annealing.

An alternative would be to hot-work (at or above the recrystallization temperature) to a 0.235-in. (5.96-mm) diameter. After hot working, the material would have a recrystallized structure that could then be cold-worked. The most economical solution would depend on the surface quality desired and the equipment available.

3.17 Properties in Single-Phase Alloys

Let us review the methods of controlling properties in single-phase alloys. (In Chapters 4 and 5 we shall discuss methods of controlling properties when more than one phase is present.)

Alloying

In general, with element substitution, or alloying, the strength and hardness of a metal increase, whereas the ductility may increase or decrease depending on the substitutional element. Figure 3.16 shows the effect of two substitutional solid-solution elements (Zn and Ni) on the mechanical properties of copper. Note that zinc in copper is one of the few cases where ductility increases with increasing solute element.

Cold Work

As shown in Figure 3.15, cold work also increases the tensile strength and hardness, while significantly decreasing the ductility.

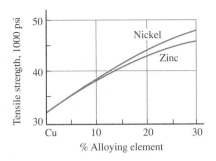

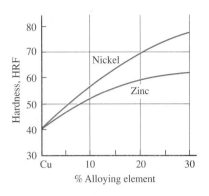

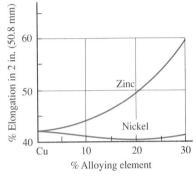

FIGURE 3.16 The influence of the solid-solution elements zinc and nickel on annealed mechanical properties of copper-base alloys. Nickel forms a complete solid solution with copper, whereas the maximum solubility of zinc in copper is approximately 40 wt %.

Annealing

Annealing removes the influence of cold work, leading to a gain in ductility but a loss in strength and hardness.

As we noted earlier, a combination of alloying and cold working may produce the most desirable combination of mechanical properties.

EXAMPLE 3.9 *[EJ]*

We require a bar of copper-base alloy with a minimum tensile strength of 35,000 psi (241.5 MPa) and a minimum elongation of 30%. We have in stock an annealed 85% Cu, 15% Ni alloy, an annealed 85% Cu, 15% Zn alloy, and a 20% cold-worked 85% Cu, 15% Zn alloy. Which alloy would you select to meet the requirement?

Answer   Using Figures 3.15 and 3.16, we find the following properties.

Alloy	Tensile Strength, psi	Percent Elongation
Annealed 85% Cu, 15% Ni	42,000 (290 MPa)	41
Annealed 85% Cu, 15% Zn	41,000 (283 MPa)	46
20% cold-worked 85% Cu, 15% Zn	59,000 (407 MPa)	15

The cold-worked alloy does not meet the minimum elongation specification of 30%. Therefore either annealed alloy would be a proper selection. Nickel is much more expensive than zinc, so the annealed 85% Cu, 15% Zn alloy would have the lowest initial cost. But if salt-water corrosion were involved, as in desalinization equipment, the copper–nickel alloy would give longer service and the lowest cost per hour of service.

3.18   Creep

In Chapter 2 and our discussion of Figure 2.1, we found that when a thermoplastic polymer is stressed a constant amount, the deformation may continue to increase with time. This same phenomenon, which is called *creep*, occurs in metals at higher temperatures and in ceramics at higher temperatures still. Let us look at the overall engineering consequences in design and then at the atom movements involved. These data are obtained via *high-temperature testing*.

Figure 3.17 shows the types of data obtained in creep testing. Bars are equalized at temperature in a test furnace, and then a fixed load is applied

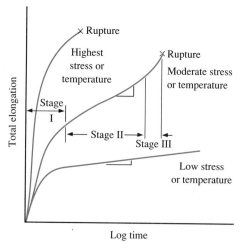

FIGURE 3.17   Typical creep curves

through a lever system. The extension is plotted as a function of time. Note that we obtain increased *creep rates* (slopes) and shorter rupture times by increasing either the load or the temperature.

Three stages of creep are recognized: stage I, in which initial yielding is fairly rapid; stage II, in which a straight line is obtained; and stage III, in which yielding is again rapid and failure occurs. From the kinds of data illustrated in Figure 3.17 it is possible to plot the stress required to produce various amounts of deformation and to calculate rupture as a function of time.

The test temperature has a pronounced effect on the creep and rupture properties of all materials. This will be discussed again in Chapter 9.

Stress relief is an important concept related to creep. High internal or residual stresses may develop in components as a result of uneven temperature distribution during processing. The residual stress is actually present as an elastic strain, just as though the section were held in a tensile test machine under load. If we heat the entire part to a temperature at which creep takes place, the elastic strain is converted to plastic strain. If the part has been cold-worked, recrystallization may also occur. The part is then cooled slowly to avoid setting up new elastic strains resulting from temperature gradients in the material during cooling.

EXAMPLE 3.10 *[EJ]*

It is common practice to give strength, ductility, and hardness specifications for engineering components. For the following applications, indicate whether creep, impact, or fatigue specifications should also be included: (*a*) automotive exhaust manifold, (*b*) fan blade, (*c*) automotive disc brake rotor, (*d*) paneling nail.

Answers

A. *Automotive exhaust manifold.* This type of part is in service primarily at high temperatures, so creep and rupture are important. After dark it is possible to see the manifold of a car glowing red after expressway travel, evidence of the elevated temperatures that the metal has attained.

B. *Fan blade.* Because the fan rotates at high speed, it is subjected to cyclic stress. Thus fatigue resistance is important. If the fan blade operates at high temperature, as in a turbine, then creep must be considered.

C. *Automotive disc brake rotor.* The rotor heats during braking, so we might suspect that creep resistance should be specified. However, failure is often due to thermal fatigue, because the cyclic heating and cooling of the disc brake rotor cause thermally induced cyclic stresses.

D. *Paneling nail.* Impact resistance is an obvious specification. The existence of notches in a serrated nail tends to lower the impact strength.

3.19   Atom Movements in Creep

Let us examine the effects that are added as a result of increasing temperature. At lower temperatures (less than 0.3 times the melting point, T_m), plastic deformation proceeds in crystals at stresses above the yield strength via dislocation glide on slip planes. As increased numbers of dislocations are formed, tangles develop and dislocations are pinned. Foreign particles also block movement.

At a higher temperature (0.3–0.5 times the melting point, T_m), dislocation climb becomes important (Figure 3.18). Consider a dislocation blocked by a foreign particle. (For simplicity we have shown only a very small particle.) In this case vacancies, which are relatively numerous at higher temperatures, diffuse into the region. The dislocation climbs and, when above the block, is able to resume slip in the typical direction. This is called dislocation creep and increases with temperature according to the equation

$$\dot{\varepsilon} = A\sigma^n e^{-Q/RT}\text{*}$$ (3-9)

where
$\dot{\varepsilon}$ = creep (strain) rate

A = constant

σ = stress

Q = activation energy

R = 1.987 cal/mol $\cdot$ K

T = absolute temperature

n = constant (stress exponent)

This effect, which is called power law creep, falls within the steady-state

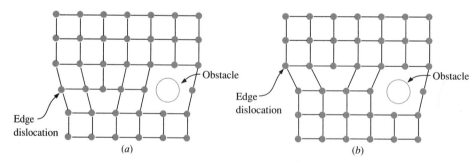

(a)   (b)

FIGURE 3.18   By absorbing vacancies, a dislocation can climb out of its slip plane to where its glide is not hindered by an obstacle.

[*Introduction to Ceramics*, 2nd ed., by W. D. Kingery, H. K. Bowen and D. R. Uhlmann, fig. 14.31. Copyright © 1976 John Wiley & Sons, Inc. Reprinted by permission of John Wiley & Sons, Inc.]

---

*Note the similarity to the equation for the temperature dependence of diffusivity.

region of Figure 3.17. The common values of n are 3–8, and the constant A depends on the material.

As we lower the stress, the driving force for dislocation climb is decreased. Another mechanism takes over; it involves atom movements along grain boundaries (see the grain boundary diffusion segment of the curve in Figure 3.19). By this method, grains elongate in the direction of the tensile stress. It is interesting that the diffusion driving force is mechanical (stress) rather than a chemical concentration difference.

Figure 3.19 also shows that plastic strain occurs at a still higher rate at higher stresses as a result of gross atom movements called bulk diffusion. Note that the schematic curve holds only for a specific temperature; new curves would be necessary at other temperatures. For example, at a constant stress level we would expect a higher creep strain rate at a higher temperature than that shown in the figure.

Ashby has developed a diagram showing all of these effects. It is reproduced in schematic form in Figure 3.20. Let us review these stress–temperature–creep relationships.

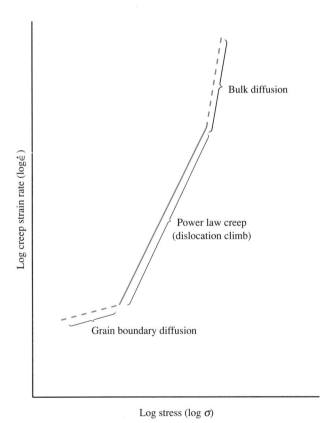

FIGURE 3.19   Schematic representation of creep strain rate as related to stress

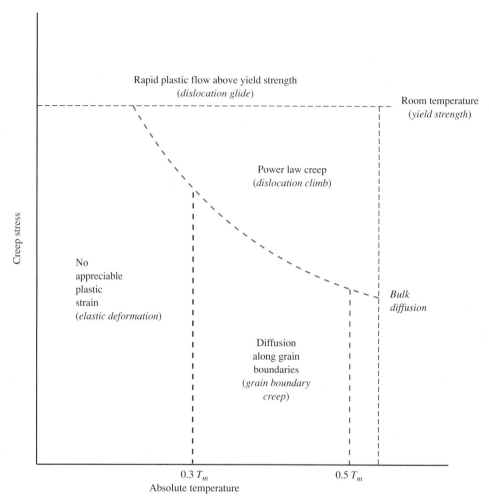

FIGURE 3.20 Simplified deformation map after Ashby. (In the actual maps, stress is plotted as normalized shear stress, γ/G, and temperature as homologous temperature T/T_m for comparison of different materials.)

First, at lower temperatures (less than $0.3T_m$), only *elastic deformation* is encountered. At high temperatures and above the yield strength at room temperature, conventional plastic flow by *dislocation glide* is obtained, as expected, giving the upper horizontal boundary.

If we select a stress and temperature just below the yield strength, power law creep or *dislocation climb* takes place. At higher temperatures, *bulk diffusion* dominates. It should be noted, however, that these regions are a function of both stress and temperature, as shown in Figure 3.20.

Finally, as the temperature is increased at a low stress level, we progress from *elastic strain* to *grain boundary creep* and then to *bulk diffusion* again.

In actual deformation maps, we would superimpose lines of constant creep strain rates so that all creep parameters were represented. As noted in the legend to Figure 3.20, a normalized shear stress and a homologous temperature are generally used.

The homologous temperature is a particularly interesting concept. As a first approximation, we can determine equivalent temperatures of materials at which the creep strain would be the same. For example, comparing lead with a melting point of 600 K and nickel with a melting point of 1726 K, we find that an equivalent creep strain of 0.5 would result in the following temperatures:

$$\text{Lead} \quad T/600 \text{ K} = 0.5 \quad \text{or} \quad T = 300 \text{ K } (27°C)$$

$$\text{Nickel} \quad T/1726 \text{ K} = 0.5 \quad \text{or} \quad T = 863 \text{ K } (590°C)$$

Taking this approximation is merely recognizing that the bond energy is related to the absolute melting temperature and that the creep strain is also related to the bond energy. (This assumes that the creep mechanism at the indicated temperatures is the same for both materials.)

Creep in Ceramics

After our discussion about the homologous temperature, it should not be surprising that the creep rate in ceramic materials is low because of relatively high melting temperatures. On the other hand, the schematic curves shown in Figures 3.19 and 3.20 have to be modified: Because of the ionic and covalent bonding within the ceramic materials, the precise creep mechanisms are not the same. Furthermore, the brittleness of ceramics can inhibit their effectiveness, especially when strength at low temperature is also required in a component.

Many of these low- and high-temperature specifications are closely related to the physical processing characteristics of the ceramics, which we have not yet discussed. We will examine them in Chapters 10 and 11.

Creep in Polymers

The reason why we limited our discussion of the effect of stress on polymers in Chapter 2 was that it is important to discuss creep first because it is present at room temperature. As we mentioned in discussing Figure 2.1, the extension may continue with time when we stress a polymer at room temperature.

A basic summary of this effect is given in Figure 3.21, which shows the change in modulus and strength with temperature. Below the glass temperature, *which is close to room temperature* for many polymers, the material obeys Hooke's law — that is, a fixed amount of instantaneous elastic strain is developed proportional to the stress. Above T_g there is movement in the side chains and between molecules as a function of time, and creep occurs. We are therefore faced with designing for creep at room temperature. One conservative procedure is to obtain a creep curve for the material and temperature of

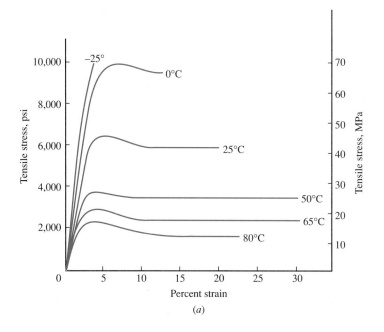

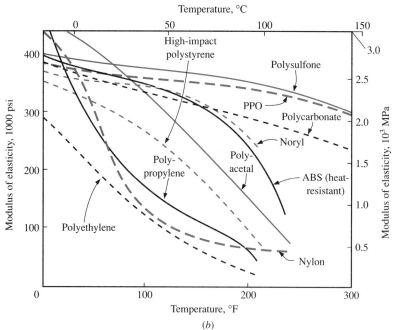

FIGURE 3.21　(a) Effect of temperature on the stress–strain curve of cellulose acetate. (b) Effects of temperature on properties of various resins.

[Part (a) after T. S. Carswell and H. K. Nason. Part (b) from *Modern Plastics Encyclopedia.* Copyright © 1968, reprinted by permission of McGraw-Hill Publishing Co.]

operation, read off the strain for the time of service expected, and then use this strain with the expected stresses to develop a conservative value of "modulus" for design purposes. A further discussion of these effects is closely related to individual polymers, so we will go into further detail in Chapter 13.

Summary

The principal thrust of this chapter is to help you avoid two-dimensional, "room temperature" thinking about structure and properties and progress to the three-dimensional world of structure, properties, and *temperature*.

The first part of the chapter was devoted to defining the important thermal properties—expansion, heat capacity, and thermal conductivity—in an engineering fashion and then, in each case, to discussing the relationship of the property to structure. We saw that it was essential to use thermal data that applied to the structure at the temperature of design, and we used as an example the great changes that occur in the properties of iron as the structure changes at elevated temperature. The thermal expansion of ceramics is the lowest because of the strong ionic and covalent bonds; that of the metals is somewhat higher as a result of metallic bonding; and that of the polymers is several orders of magnitude higher because secondary bonding is dominant. The metals exhibit the highest thermal conductivity, because free electrons act as carriers. The covalently and ionically bonded electrons in other materials are not conductors, and insulating properties are encountered.

The phenomenon of mass diffusion follows the same mathematical principles as thermal diffusion, but here the flux is in terms of atom motion rather than heat flow.

The effect of temperature on mechanical properties is dramatically linked to the material's structure. Most polymers and BCC metals are brittle at sub-zero temperatures. In the case of polymers, no molecular motion occurs to accommodate the stress, and in BCC metals the mode of fracture changes to a cleavage type. At elevated temperatures in crystalline materials, creep is encountered. The motion of dislocations is freer, and dislocation climb occurs when pins or tangles are encountered. The influence of creep is especially important in polymers, because it is encountered near room temperature and must be considered in the design of parts that are to operate at normal temperatures.

The effect of temperature in producing recovery, recrystallization, and grain growth in metals is especially helpful in attaining the desired balance between strength and ductility for a given application.

Definitions

Activation energy The energy required to initiate a process or chemical reaction that is usually self-sustaining after initiation.

Annealing   In general, a heat treatment in which a part is heated to soften the material; the treatment leading to the recrystallization of cold-worked material.

Bimetal   A two-layer structure made by rolling or plating a layer of one metal or alloy on another.

Carburizing   A process for adding carbon to the surface layer of an iron part.

Creep   Permanent strain that increases as a function of time under stress.

Creep rate   The slope of a curve of elongation versus log time at a constant temperature. The normal slope is during the stage II or constant region and has units such as %/1000 hr.

Diffusion   Migration of atoms in a solid, liquid, or gas. Even in a pure material, atoms change position; this process is called *self-diffusion*.

Diffusivity, D   The rate of diffusion of a substance at a given temperature. D varies exponentially with temperature: $D = Ae^{-Q/(RT)}$.

Equiaxed grains   Grains that are of equivalent dimensions in all directions (that is, they are not longer in one direction than in another, as they commonly are after cold-working).

Grain growth   The development of larger grain size by preferential growth of larger crystals (grains) after recrystallization.

Heat capacity, C_p   Also called the *specific heat*. The heat required to raise the temperature of a specific quantity of a system by one degree. Normal units are cal/g-K.

High-temperature testing   Testing of properties such as creep and stress required for rupture in order to obtain test data of stress vs. time. For practically every material, there is a range of temperature in which it softens and deforms in increasing amounts as a function of time.

Homologous temperature   The ratio of ambient or test temperature to the melting temperature (Kelvin) of a material.

Hot working   The deformation of material at or above its recrystallization temperature.

Impact test   A test in which a bar specimen of a material, usually notched, is struck by a calibrated pendulum and the energy required to cause fracture is measured. Although the term *impact strength* is used, the test results are expressed in terms of energy. The principal reason for the fracture is not the impact but the presence of the notch, which causes a brittle stress condition known as a triaxial stress condition. This is discussed in texts on strength of materials.

Interstitial diffusion   Migration of atoms through a space lattice using interstitial positions.

Lorenz number L   The ratio

$$\frac{\text{thermal conductivity}}{\text{electrical conductivity} \times \text{absolute temperature}}$$

Normalized shear stress   Ratio of the shear stress to the shear modulus. It varies with temperature because of the thermal change in shear modulus.

Recovery The relief of elastic strain in the early stages of annealing.

Recrystallization The growth of new stress-free, equiaxed crystals in cold-worked material.

Silica phases Crystalline phases: *quartz, tridymite, cristobalite;* amorphous phase: *vitreous silica.*

Single-phase material A material in which only one phase is present, such as BCC iron. Any number of elements may be present, but these must be in solid solution.

Steady-state condition A system in which diffusion or heat transfer is taking place and there is no change in composition or temperature at different points in the system with the passage of time.

Stress relief The disappearance of elastic strain, especially during annealing.

Substitutional diffusion Migration using substitutional positions.

Thermal conductivity The quantity of heat transmitted through a unit volume in a unit time. Normal units are cal/cm-sec-K.

Thermal expansion The change in unit length per unit of temperature change: $\Delta l/\Delta T = \text{m/m-K} = (\text{K})^{-1}$.

Transition temperature A low temperature, such as 0°C, at which a material (such as steel) shows a rapid fall in impact strength. The fracture changes from ductile to brittle at this point. (This brittleness with temperature usually does not exist in FCC metals and occurs at higher temperatures in ceramics.

Unsteady-state condition A system in which temperature or composition is changing as a function of time. Also called *transient condition.*

Problems

3.1 *[ES/EJ]* Figure 1.16 shows schematically the energy trough (minimum energy) and equilibrium spacing of atoms for sodium chloride. (Sections 3.1 through 3.5)
a. Schematically show the trough for a more strongly bonded material.
b. Repeat part *a* for a more weakly bonded material.
c. Indicate the qualitative relationships among bond strength, coefficient of expansion, melting point, and modulus of elasticity.

3.2 *[ES]* Calculate the stresses for a temperature change of 100°F (55.6°C) in a bimetallic strip made of austenitic stainless steel and SAE 1020 steel. The coefficient of expansion of austenitic stainless steel is 9.6×10^{-6} in./in.-°F (17.3×10^{-6} m/m-°C), and that of SAE 1020 steel is 7.8×10^{-6} in./in.-°F (14.0×10^{-6} m/m-°C). Assume that the individual metals have equal cross-sectional areas. (Sections 3.1 through 3.5)

3.3 *[ES]* A piece of polymethylmethacrylate (a polymer) measures 0.015 in. thick × 0.25 in. wide × 5 in. long (0.38 mm × 6.35 mm × 127 mm). It is firmly attached to a piece of pure iron that has the same width and length but is 5 times thicker than the piece of polymer. The modulus of the polymer is 0.42×10^6 psi (0.29×10^4 MPa), and that of the steel is 30×10^6 psi (20.7×10^4 MPa). (Sections 3.1 through 3.5)
a. Calculate the stress in the longitudinal direction in each material if both have zero strain at 25°C and are cooled to −25°C.
b. Calculate whether the stresses would be significantly different if the ratio of the thicknesses were increased to 20.

3.4 *[ES/EJ]* As a first approximation, the isothermal expansion during a phase transformation is proportional to the amount of the phase present. Using this relationship and the data given in Figure 3.2, plot the change in volume versus temperature for a silica brick composed of equal weight percentages of quartz, tridymite, cristobalite, and vitreous silica. Indicate why this brick might be better than an all-quartz brick. (Sections 3.1 through 3.5)

3.5 *[ES/EJ]* Refer to Example 3.3 and answer the following questions. (Sections 3.1 through 3.5)

a. Why are the curves not carried out to the full 20-cm wall thickness?

b. Show by calculation the time when the temperature would be 10°C halfway through the wall.

c. What does your answer in part *b* suggest about the relative thermal insulating quality of the wall material? Relate your discussion to each of the parameters that make up the thermal diffusivity.

3.6 *[ES]* The heat capacity is a function of temperature and can be represented by the equation

$$C_p = a + bT - c/T^2 \qquad (3\text{-}10)$$

where C_p is in units of cal/mol-K, T is in K, and a, b, and c are material constants dependent on whether the material is a liquid, solid, or gas. Data for pure aluminum are given below:

	a	b	c	
Solid	4.80	3.22×10^{-3}	0	Heat input $= \int_{T_1}^{T_2} C_p \, dT$
Liquid	7.00	0	0	

Heat absorbed at 931.7 K = 2550 cal/mol. (Sections 3.1 through 3.5)

a. What is the heat absorbed at 931.7 K called?

b. Calculate the heat (cal) required to heat 1 kg of pure aluminum from room temperature (25°C) to 100°C above its melting point.

3.7 *[ES/EJ]* The Lorenz number given in Table 3.1 is defined as follows

$$L = \frac{\text{thermal conductivity}}{\text{electrical conductivity} \times T}$$

where T is the absolute temperature (20°C = 293 K). Why are values noted for pure metals and not alloys? (*Hint:* Note that alloys can be either single-phase or multiphase when solubility is exceeded.) (Sections 3.1 through 3.5)

3.8 *[EJ]* The steady-state heat flow in a material is directly proportional to the thermal conductivity. Why are bricks for furnaces deliberately made with voids? What are the characteristics of home insulation? (Sections 3.1 through 3.5)

3.9 *[ES]* Assume that the heat-treatment supervisor in charge of processing the gears described in Example 3.5 is on an economy program directed toward getting better life from the electric furnaces. The supervisor recommends reducing the carburizing temperature from 1000 to 900°C, saying that the furnace life is much longer at the lower temperature and that the carburizing time needed to get the same results will be only

1000/900 times, or about 10%, longer. Show that the supervisor is sadly in error by calculating D for 900°C and calculating the time required to reach the same percentage of carbon as in Example 3.5. (Sections 3.6 through 3.8)

3.10 *[ES]* When segregation or nonuniform chemistry within a phase exists as a result of solidification conditions, we can use diffusion calculations to estimate the time required to homogenize the material. The simplest calculation gives the time needed to reduce the difference in concentration to one-half. For example, suppose that we have a carbon concentration C_s at one point and zero carbon at another. We calculate the time required for C_x to equal 0.5 C_s. (Sections 3.6 through 3.8)

    a. Show that the homogenization time is approximately $t = x^2/D$.

    b. Refer to Figure 3.4. What would be the relative homogenization times for nickel in FCC iron compared with carbon in FCC iron at 1000°C?

3.11 *[ES]* Given an activation energy Q of 40,000 cal/mol and an initial temperature of 1000 K, find the temperature that will increase the diffusion coefficient D by a factor of 10. (Sections 3.6 through 3.8)

3.12 *[EJ]* A 15% Cr, 85% Fe alloy, a variety of stainless steel, is brazed with a 70% Cu, 30% Zn alloy at approximately 1000°C. The stainless steel becomes embrittled and cracks. Explain whether copper or zinc atoms might diffuse more rapidly and why the crack might appear to be intergranular (between the grains). (Sections 3.6 through 3.8)

3.13 *[ES/EJ]* Pure zinc is to be diffused into pure copper by dipping copper into molten zinc at 500°C. (Sections 3.6 through 3.8)

    a. Calculate how long it would take to obtain 30 wt% Zn at 1.0 mm beneath the copper surface.

    b. Diffusion is more rapid at higher temperatures. However, what practical problems might be involved in using a calculated value at 1000°C?

3.14 *[ES/EJ]* Refer to Example 3.4. (Sections 3.6 through 3.8)

    a. Calculate the number of vacancies in pure aluminum at 20°C.

    b. If atom movement is primarily through lattice vacancies, what does this suggest about the temperature dependency of diffusion?

    c. What is the upper temperature limit of the general equation?

    d. Why is the value of n_t only approximate in the example?

3.15 *[ES/EJ]* Give several reasons why the equation for the temperature dependency of the diffusivity D is similar to that for the vacancy relationship (n/n_t). (Sections 3.6 through 3.8)

3.16 *[ES/EJ]* Refer to Figure 3.5. (Sections 3.6 through 3.8)

    a. At a given temperature, why would you expect the diffusivity of oxygen in MgO to be less than that of magnesium in MgO?

    b. Similarly, at a given temperature, why does the magnesium diffusivity in MgO lie close to, but higher than, that of nickel in MgO? (*Hint:* Consideration should also be given to the thermal vibrational energy of the species.)

3.17 *[EJ]* Mark the following statements true or false *and* illustrate your answers with schematic graphs of impact energy as a function of temperature. Be sure to mark the units on the axes. (*Hint:* Steels may be obtained with both the BCC and FCC crystal structure.) (Sections 3.9 through 3.19)

    a. Two steels can exhibit about the same percent elongation in tension and the same Brinell hardness, but have impact properties that vary by 2:1 at 0°F (-18°C).

    b. It is possible for one steel to have about the same percent elongation at 0°F (-18°C) and 70°F (21°C) but to have impact properties that vary by 3:1 at these temperatures.

3.18 *[E]* Would it be important to specify the impact strength and transition temperature for the following applications? Explain. (Sections 3.9 through 3.19)
   a. Steel axle in an automobile
   b. Copper pipe
   c. Polymer garbage can

3.19 *[E]* You are going to bend a piece of steel rod at 90°. Why might a torch be required to complete the bend? What are the disadvantages of using a torch? (Sections 3.9 through 3.19)

3.20 *[ES/E]* Explain why the phenomenon of creep is so closely related to diffusion (Sections 3.9 through 3.19)

3.21 *[ES/E]* A rolled 70% Cu, 30% Zn brass plate 0.500 in. (12.8 mm) thick has 2% elongation as received from the supplier. Desired specifications for the final sheet are a thickness of 0.125 in. (3.17 mm), a tensile strength of 70,000 psi (483 MPa) minimum, and an elongation of 7% minimum. Assume that the rolling is conducted so that the width of the sheet is unchanged. (This means that in calculating the area, $A = wt$, with width w constant.) Specify all steps in the procedure, including heat treatments. (Sections 3.9 through 3.19)

3.22 *[E]* Sketch the microstructures (longitudinal and transverse to the rolling direction) that you would expect to see in the material of Problem 3.21 as received, after intermediate annealing, and after final working. Show the differences in grain shape, slip bands, and annealing twins. (Sections 3.9 through 3.19)

3.23 *[ES/E]*
   a. For power law creep, show why the equation should result in the straight-line portion described in Figure 3.19.
   b. Relate power law creep to Figure 3.17. Show how the creep strain rate increases when the temperature increases at constant creep stress. (Sections 3.9 through 3.19)

3.24 *[ES/E]* We have a piece of brass that we know to be cold-worked, but we are uncertain whether the metal is a 70-30 or a 85-15 alloy. However, someone has run a tensile test for us and found the tensile strength to be 60,000 psi (414 MPa) with an elongation of 15%. Determine the alloy composition, amount of cold work, and tensile strength in the annealed condition. (Sections 3.9 through 3.19)

3.25 *[ES/E]* Because polymers can also creep, the modulus of elasticity is a function of time as well as of temperature. As we will discuss in Chapter 17, asphalt is really a polymer. Compare the effects of creep on the characteristics of an asphalt-bonded aggregate road bed in hot summer temperatures and cold winter temperatures. (Sections 3.9 through 3.19)

3.26 *[E]* The following properties are obtained for a cold-worked 90% Cu, 10% Ni (cupronickel) alloy.

Percent Cold Work	Tensile Strength, psi	Percent Elongation	Hardness HRF	Hardness HRB
10	50,000 (345 MPa)	25	80	30
20	55,000 (380 MPa)	16	95	63
30	59,000 (407 MPa)	12	100	70
40	63,000 (435 MPa)	11	102	74
50	65,000 (449 MPa)	10	105	79

Graph these data. Select a degree of cold work for both a 90% Cu, 10% Ni alloy and a 85% Cu, 15% Zn alloy to achieve the following final properties:

Minimum tensile strength:	54,000 psi (373 MPa)
Maximum hardness:	65 HRB
Minimum elongation:	12%

When annealed properties are compared, do copper–zinc alloys or copper–nickel alloys appear to show more work hardening? (Comparison can be made only for the low solute contents.) (Sections 3.9 through 3.19)

3.27 *[ES]* Stress relaxation can be defined as the adjustment of a material to its stress, and it is related to stress relief or creep. It is governed by the equation

$$\sigma = \sigma_0 e^{-t/\lambda} \tag{3-11}$$

where

σ_0 = stress at time zero

σ = stress at time t

λ = relaxation time (function of material and temperature)

If a cold-worked alloy has a relaxation time of 0.5 hr at 500°C, how long would it take to remove 50% of the residual stress if the material were annealed at 500°C? How much time would be necessary to remove 90% of the residual stress at 500°C? (Sections 3.9 through 3.19)

3.28 *[ES/EJ]* Figure 3.14 is a graphical interpretation of the Petch relation. (Sections 3.9 through 3.19)
a. For the plain carbon steels (lower curve in Figure 3.14), determine the constants S_0 and k in the Petch relation.
b. Why do the experimental data points fall within a rather narrow band of grain diameters?

3.29 *[EJ]* Two students are discussing how to cold-work an initially annealed round rod of 70% Cu, 30% Zn from 0.500 in. (12.7 mm) to 0.275 in. (7.0 mm) in diameter. One student suggests cold-working it in one operation. The other student says that this cold work would exceed the ductility of the alloy and that an intermediate cold-worked diameter followed by annealing will be necessary. Determine which student is correct and give reasons for your decision. (Sections 3.9 through 3.19)

3.30 *[EJ]* The text says that the effects of cold work are additive. Therefore, 30% cold work followed by 25% cold work gives a total of 55%. The properties and the physical size of the component are dictated by the total percentage of cold work. Show with an appropriate example that the addition of percentages of cold work must be handled with care. (Sections 3.9 through 3.19)

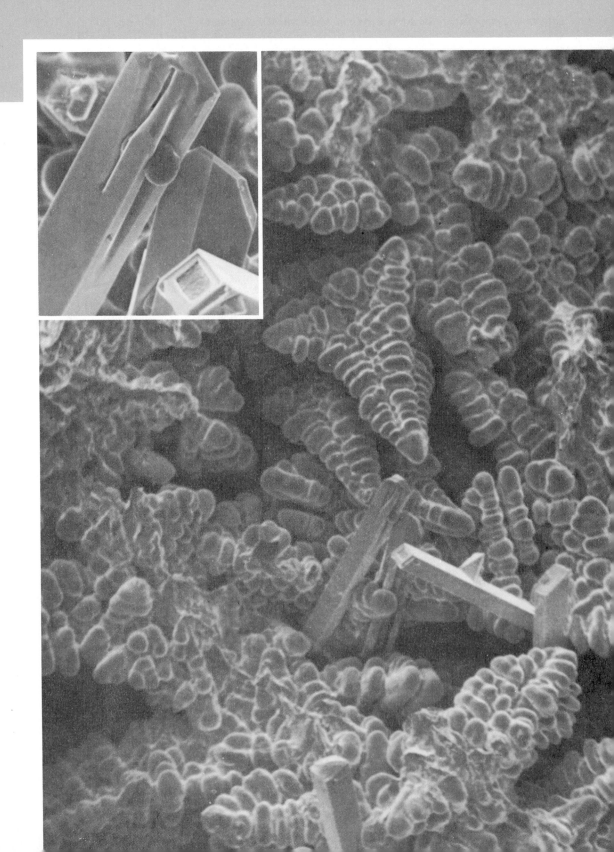

CHAPTER 4

The Control of Structure Under Equilibrium Conditions (Phase Diagrams)

The micrographs (scanning electron micrographs at 200× and 2000×) show the growth of two different structures in a copper tin alloy. The rounded dendritic (treelike) phase is rich in copper and crystallized first. The smooth-faced columnar crystals are rich in tin, and they crystallized later. In general, the presence of a second phase leads to higher strength.

The reason that these crystals are shown clearly in a cavity is that the solid alloy is denser than the liquid. Therefore, as freezing progressed toward the center of the casting, no liquid was available to fill the void and crystallization was arrested.

4.1 Overview

After studying three chapters on the structures of materials and their properties, we might naturally ask, "How, then, do we get the structure we want? As a starter, how about making some diamonds?"

Fine, we will. And doing so, we will illustrate the key concept of this chapter: how to control structure by using phase diagrams.

A *phase diagram* is a map showing the temperature, pressure, and composition we need to obtain a given structure under equilibrium conditions. True to our promise, in Figure 4.1 we offer an equilibrium diagram showing the pressure–temperature range over which the diamond structure is stable. We would also like to know how fast it forms, which is the subject of Chapter 5. Nonetheless, understanding equilibrium diagrams is valuable as a base for kinetic experiments.

Before taking up phase diagrams, however, consider that so far we have concentrated on *single-phase* materials, such as pure metals, solid solutions, and intermetallic compounds. These are useful materials, but the metallic structures with highest strength, hardness, and wear resistance are *polyphase materials* — that is, they are composed of two or more phases in a well-controlled dispersion. This microstructure is usually not obtained in the original casting or ingot. It is formed by carefully controlled processing that involves hot working and heat treatment. For example, the properties of a common aluminum alloy, 2014, used for structural members in aircraft, are as follows:

	Yield Strength at 0.2% Offset, psi $\times$ 10^3*	Tensile Strength, psi $\times$ 10^3*	Percent Elongation
Alloy 2014,[†] annealed	14	27	18
Alloy strengthened by heat treatment	60	70	13

*To obtain MN/m² (MPa), multiply by 6.9 $\times$ 10^{-3}.
[†]4% Cu, 0.8% Si, 0.8% Mn, 0.6% Mg, balance Al.

If an engineer had to design on the basis of yield strength of the annealed material, an aircraft would have to be four times heavier. It might never get off the ground!

Here we introduce a general statement of the greatest importance for understanding the relationship between structure and properties, not only for metals but for all materials. *The properties of a material depend on the nature, amount, size, shape, distribution, and orientation of the phases.* This point is illustrated in large scale fashion in Figure 4.2a, which shows how the properties of a reinforced concrete slab may vary. Let us review this figure and consider the equivalent effects in other microstructures.

1. *Nature* of the phases. In the concrete slab, the individual properties of the concrete and the steel affect the overall strength of the slab. In the aluminum alloy 2014, two phases are involved: a solid solution of copper in

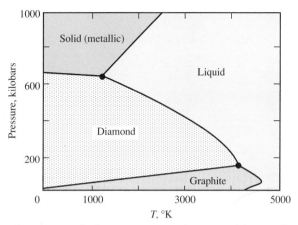

FIGURE 4.1   High-pressure, high-temperature phase equilibrium diagram for carbon
[From C. G. Suits, *Am. Sci.*, 52, 395 (1964). Used by permission.]

aluminum and the intermetallic compound $CuAl_2$. The individual properties of each phase determine the characteristics of the alloy.

2. *Amount* of the phases. Just as the relative amounts of steel and concrete are important, so are the amounts of the two phases in the aluminum alloy.

3. *Size* of the phases. If there are only a few large steel reinforcing rods in the concrete slab, the structure is not so strong as it is when the same amount of steel is used in thinner bars. Similarly, a fine dispersion of $CuAl_2$ is preferred.

4. *Shape* of the phases. If the steel is in square bars, stress concentration occurs at the corners, causing cracking of the concrete; hence round bars are preferred. In the case of a precipitate in a microstructure, the properties obtained depend on whether the shape of the precipitate is spheroidal or plate-like.

5. *Distribution* of the phases. If the slab is to encounter bending stresses, the steel should be placed close to the surface rather than at the center. Similarly, in a microstructure, properties differ when a precipitate phase is localized at grain boundaries instead of being uniformly distributed.

6. *Orientation* of the phases. If the slab is to be used to resist typical traffic, a horizontal orientation of the steel is preferred. An aligned precipitate will also show directional mechanical properties. We will see later that for optimal magnetic properties, a certain alignment of particles or precipitates in a magnetic tape is very important.

These examples show that it is essential to understand the methods of controlling the development of polyphase structures.

We shall begin with a discussion of the structural changes that take place under equilibrium (slow cooling), because these treatments are useful in softening a material for machining or forming, and because the nonequilibrium

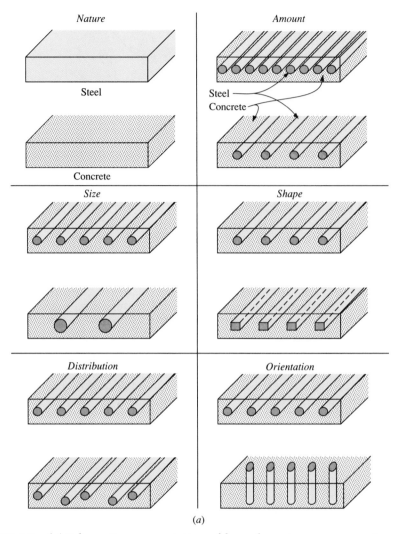

(a)

FIGURE 4.2   (a) Schematic representation of how the nature, amount, size, shape, distribution, and orientation of phases control physical properties (as applied to steel and concrete). Photographs show the microstructure of an aluminum–silicon alloy used in an aluminum automobile engine block. The composition is 17% Si, 4% Cu, 0.5% Mg, balance Al. The large gray crystals are β (almost pure silicon) and are precipitated from the liquid. The balance of the structure is a mixture of fine α, β, and a trace of copper-rich phase θ (light gray). (b) Silicon crystals 1070 HV; matrix 130 HV, 500×, etched. (c) Silicon gray, θ light gray, matrix white, 1500×, unetched. (Note: See the description of the microhardness test in Section 4.2.)

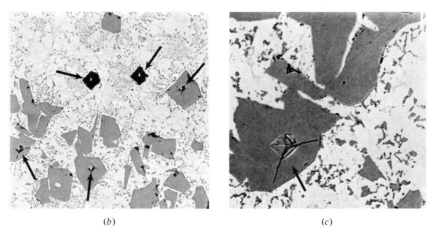

(b)   (c)

FIGURE 4.2   (continued)

treatments discussed later are usually based on the control or suppression of the equilibrium reactions.

4.2 Phase Diagrams for Polyphase Materials

Up to this point we have discussed only the simple but important case of single-phase alloys. In these, all the grains in the metal have the same crystal structure, the same analysis, and the same properties. Alloys in which two or more phases are present are called the polyphase alloys. Most metallic materials are of this type.

As an example of a *two-phase material,* let us look at the microstructure of an aluminum–silicon alloy used in many engine blocks (Figure 4.2b and c). It is apparent that two different major phases are present. (Recall that a phase is a homogeneous and physically distinct region of matter.)

We can show that the properties of the phases are different through a *microhardness test.* In this procedure we position a very small diamond indenter (carefully ground to a pyramid-shaped point) above the phase we wish to test. A load—of 25 g, for example—is applied to the indenter to press it into the sample, as in the Vickers test discussed earlier (except that a much lower load is used). The diagonal of the impression is measured and related to the hardness; the smaller the value, the harder the phase.

In the sample shown in Figure 4.2, the original melt or liquid contained 17% silicon, and the balance was essentially aluminum. On cooling, it separated into an aluminum-rich phase, α, with some silicon in solid solution, plus a silicon-rich phase, β, with very little aluminum. (Greek letters are used to distinguish the different solid phases in a given alloy. These solid phases are seldom pure, but rather are solid solutions or compounds. The symbol L is used for the liquid phase; L_1 and L_2 are used if there are two different liquid phases.) It is apparent from the hardness tests that the alloy contains a hard structure in the softer background or continuous phase, usually called the *matrix*. Furthermore, some of the silicon phase is in fine needles, whereas other particles are large. The wear resistance of the cylinder walls of an engine block depends on the combination of these hard particles and the soft matrix.

4.3 Equilibrium Phase Diagrams — the Aluminum–Silicon Diagram

The most useful tool in understanding and controlling polyphase structures is a knowledge of *phase diagrams*. The phase diagram is simply a map showing the structures or phases present as the temperature and overall composition of the alloy are varied. We shall discuss one very important phase diagram in detail at this point, the aluminum–silicon diagram. Let us construct a phase diagram of the aluminum–silicon alloy by using direct observations from several experiments. If we heat pure aluminum in a crucible, we find that it melts sharply at 660°C. Similarly, on cooling, crystallization from the liquid takes place at the same constant temperature.

Now let us remelt the aluminum and add 5% silicon to the bath. The silicon dissolves just like sugar in water. It is important to avoid saying that the silicon "melts," for the melting point of silicon is far above the temperature of the liquid aluminum. The silicon *diffuses* from the surface of the solid into the liquid. After the silicon is dissolved, we cool the melt. We find two important differences between the freezing of the alloy and the freezing of the pure aluminum.

1. The alloy starts to freeze or crystallize at a lower temperature, 626°C, than pure aluminum (660°C).
2. The alloy exists in a mushy condition (liquid plus solid) over a range of temperatures, rather than freezing at a constant temperature. More and more solid precipitates from the liquid until the alloy is finally solid at 577°C. We find the same type of behavior at 10% silicon, but the start of freezing is delayed until 593°C.

Now we have enough data to begin constructing the aluminum–silicon phase diagram (Figure 4.3). We mark off percentage silicon by weight on the *x* axis and temperature (°C) on the *y* axis and plot the points we have obtained.

Above line *AB* the melt is entirely liquid. *AB* is called the *liquidus*. Below *DE* the melt is entirely solid. *DE* is called the *solidus*. Between *AB* and *DE* we label the region liquid plus solid (L + α).

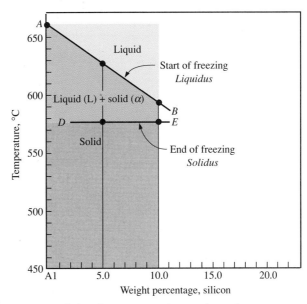

FIGURE 4.3   A portion of the aluminum–silicon phase diagram

If we continue our experiments, we can complete the diagram, as shown in Figure 4.4. This diagram contains many new lines, which we will now discuss. First, as the percentage of silicon increases, the liquidus temperature falls until 12.6% silicon is reached; then the temperature rises. This pattern is quite common in alloys. The composition at the minimum freezing point has a special name, *eutectic* (*E*), which means "of low melting point." (In practice, the word is used to mean both the composition at the minimum freezing point and the temperature (the *eutectic temperature*) at which a liquid of *eutectic composition* freezes.) Note that although the melting point of silicon is twice that of aluminum, adding silicon *lowers* the melting range of aluminum until the alloy has 12.6% silicon. Note also that the phases present to the right of the eutectic and above the solidus are β + L instead of α + L. In other words, when we cool an alloy with 16% silicon, the first phase to crystallize is β (crystals of almost pure silicon). Only after the solidus at 577°C is crossed do α crystals begin to form.

Let us interrupt this discussion to point out the practical significance of what we have learned. First, the aluminum–silicon alloys are important engine-block materials. They have been substituted for cast iron in some cases. It is essential that these castings be poured at the proper temperature. If the liquid is too cold, it will not fill the mold cavity completely; if it is too hot, it can react with the mold wall, producing a poor surface. Usually the pouring temperature is set at about 110°C above the liquidus. We see, therefore, that the upper line of the diagram gives us a guide for establishing pour-

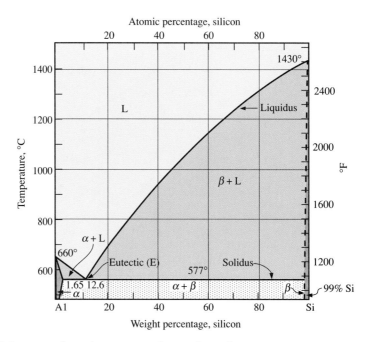

Atomic percentage, silicon

FIGURE 4.4 Complete aluminum–silicon phase diagram

[Reprinted with permission from *Metals Handbook*, 8th ed., Vol. 8, *Metallography, Structures, and Phase Diagrams*, American Society for Metals, Metals Park, OH, 1973.]

ing temperature. At any silicon level we merely add 110°C to the liquidus to obtain the pouring temperature. Second, many alloys are heat-treated. The *solidus* gives us a guide for determining the maximum heating temperature, because above this line the parts begin to melt. (As a matter of fact, it is advisable to stay 20°C below the solidus because melting may start at lower temperatures when impurities, which lower the solidus, are present.)

We can get a good deal of additional information from the phase diagram. To illustrate, let us discuss the determination of line *AD* in Figure 4.5.

If we make a melt with 0.2% silicon, we find that it starts to freeze at point 1 in the enlarged diagram and is solid at point 2. Similarly, using 0.4% silicon, we obtain the liquidus and solidus points 3 and 4, respectively. By connecting the solidus points, we determine *AD*, the line between the *single-phase field*, marked α, and the *two-phase field*, marked α + L.

Now the question is how to determine the solubility line *DG*. The line *DG* is sometimes referred to as a *solvus*, or the locus of temperatures where the compositions of one solid are in equilibrium with the compositions of another solid—in our example, α and β. If we follow the cooling of our sample containing 0.2% silicon, we see that it is in a single-phase α field from 648 to 470°C. However, when we cross *DG* at 470°C, from points 5 to 6 in Figure 4.5, the specimen enters a two-phase field, which means that β begins to precipitate from solid α. This is the key to our problem.

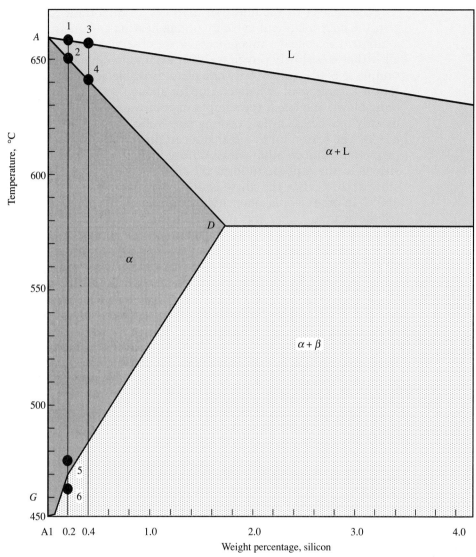

FIGURE 4.5 Enlarged region of the aluminum–silicon phase diagram (the high aluminum end)

The answer to the question involves an important tool of the metallurgist, the heat treatment cycle involving heating and quenching. Let us digress a moment to consider this process. If we observe a sample of our 0.4% silicon alloy that was slowly cooled to 20°C, we find two phases. Now let us heat the sample to 550°C, where only one phase is stable according to the diagram. We let the sample soak at temperature for about 1 hr to let the $\alpha + \beta$ change to α only. Now we quench the sample in cold water. The rapid cooling rate gives us only α, whereas a slow cooling rate would result in the equilibrium phases $\alpha + \beta$. The reason for this is that a certain amount of time is required for β to precipitate. Silicon atoms must diffuse from their dispersed solid-solution positions in the α phase to form the new 99% silicon β phase. Diffusion is so slow at 20°C that the alloy contains the same structure (α) it showed at 550°C. In other words, quenching a sample from a given temperature tends to fix the structure that was present at that temperature. There are important exceptions that we shall discuss later, but the technique can be used successfully in many cases.

Now we are ready to consider the location of line *DG*. This time let us take five samples of the alloy, heat them all to 550°C, and soak them for 1 hr. We then slow-cool a sample to 525°C and quench it. We slow-cool the other samples to 500, 475, 450, and 425°C, respectively, and quench them. We find the structures shown in Figure 4.6.

Since no β was present under slow cooling (equilibrium) conditions followed by quenching at 500°C but β appeared in the 475°C sample, the line *DG* is between 475 and 500°C. By further testing we could locate the line with greater accuracy.

EXAMPLE 4.1 *[EJ]*

We have used the expressions "soak for 1 hr" and "slow-cool" in our examples. Why are these expressions necessary to the discussion of phase diagrams?

Answer   Phase diagrams could also be treated as graphical representations of phase equilibria. Here we will emphasize equilibrium, recognizing that infinite time may be necessary to achieve true equilibrium. More realistically, approximate equilibrium is achieved in a finite time interval. Therefore "soaking for 1 hr" approximates equilibrium at the temperature in question, and "slow cooling" allows the equilibrium phases to form upon cooling.

The kinetics of these phase transformations are very important to engineering alloys. In fact, heat treatment is used both to promote and to inhibit equilibrium phases, as best suits the engineering application. Several important examples will be given in the latter parts of this chapter.

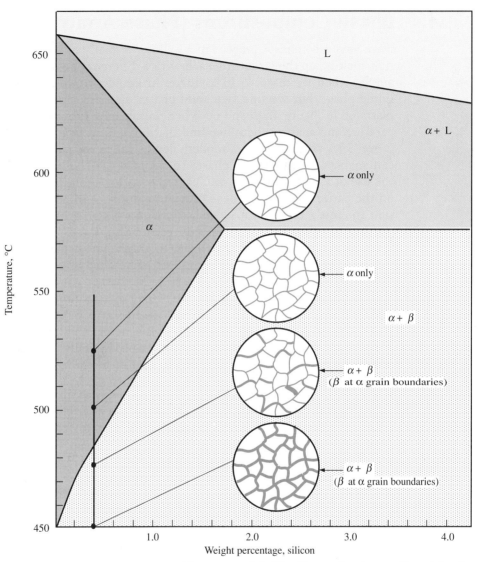

FIGURE 4.6 Microstructure of heat-treated 0.4% silicon–aluminum alloys (slowly cooled from 550°C to the indicated temperature and then quenched to room temperature)

4.4   Phase Compositions (Phase Analyses)

If we have two phases present in an alloy, we really have two different materials making up the structure, each with a different chemical analysis. For example, when we dissolve 15% silicon in liquid aluminum, we have a single liquid phase, but we find two solid phases after freezing, similar to those illustrated in Figure 4.2(*b* and *c*). When we separate these phases at room temperature and analyze them, we find that α contains over 98% aluminum, and β over 99% silicon. (With a modern device called the *electron microprobe* we can analyze the individual phases in the sample. The procedure is related to the use of x rays, discussed in Chapter 1. A tiny beam of electrons is focused on the particular phase. These cause the atoms of the elements that are present to emit x rays. Because each element gives off x rays of its own characteristic wavelength, we need only record this x-ray spectrum, just as the visible spectrum is recorded in a spectrograph. The percentage of a given element in the phase is related to the intensity of radiation of its characteristic wavelength. See also Figure 1.37.)

Let us suppose we have an alloy of 97% aluminum and 3% silicon at 600°C. This is definitely in the two-phase field $\alpha + L$. To find the phase compositions, we simply draw a horizontal line across the two-phase field on the phase diagram until it touches the single-phase fields (Figure 4.7). Then we drop a perpendicular from each contact point to the x axis and read off the particular phase composition. Thus we see that the phase composition of α in the 3% silicon alloy at 600°C is 1.2% silicon. Similarly, using the same horizontal, we see that the composition of the liquid is 9.0% silicon. Using the same construction, we find that even if the overall composition of the alloy changes from 1.2 to 9.0% silicon, the α phase at this temperature still contains 1.2% silicon, and the liquid phase at this temperature still contains 9% silicon. There is nothing artificial about this procedure; it is merely a way of recovering the information that we found experimentally. This method of determining the phase compositions in a two-phase field is so important that we will state it as a simple rule.

RULE 1. FOR DETERMINING PHASE COMPOSITIONS (CHEMICAL ANALYSIS OF PHASES) FROM A PHASE DIAGRAM. In a two-phase field draw a horizontal line (called a "tie line") at the temperature desired, touching the single-phase fields. Drop perpendiculars from the points where the tie line meets these fields, and read the phase compositions on the x axis. *Tie lines are present in two-phase fields and not in one-phase fields.*

Note that in a phase diagram, single-phase fields are always at the ends of the tie line in a two-phase field. This is a consequence of the phase rule, which we will discuss later.

Note also that we have used several terms in this section that have exactly the same meaning and are used interchangeably. These terms are *phase composition*, *phase analysis*, and *chemical analysis of the phases*.

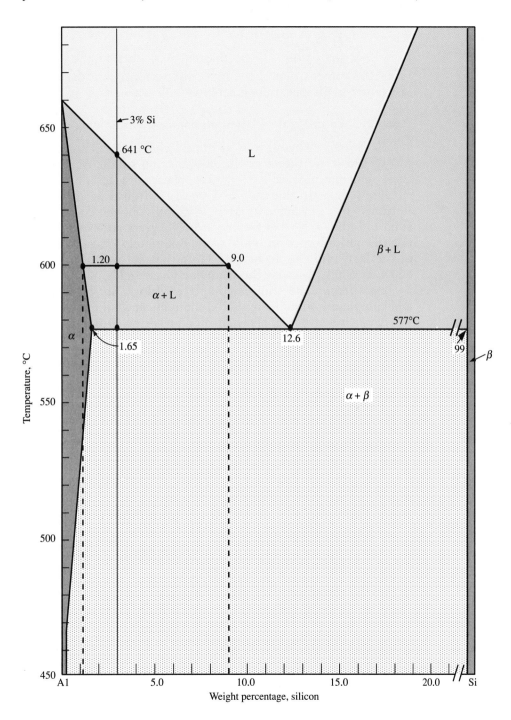

FIGURE 4.7 Determination of phase composition from a phase diagram

EXAMPLE 4.2 *[ES]*

For a 1.0% and a 2.0% silicon alloy, what are the liquidus and solidus temperatures, the composition of the first solid to form, and the composition of the
last liquid to disappear upon cooling?

Answer   From Figure 4.7 we can obtain the following information.

1% silicon: The alloy is all liquid above 655°C and all solid (α) below
610°C. These are therefore the liquidus and solidus temperatures. Between
these temperatures, we have the two phases α + L.

Drawing a tie line at 655°C, we find liquid with 1% silicon in equilibrium with α of approximately 0.1% silicon (the composition of the first
solid to form). A second tie line at 610°C gives an α composition of 1% silicon in equilibrium with liquid of approximately 7% silicon (the composition
of the last liquid to disappear). Note that we would not draw tie lines at
660°C or 600°C, because these are one-phase fields.

2% silicon: The alloy is completely liquid above approximately 645°C
and completely solid below 577°C (liquidus and solidus, respectively).

A tie line at 645°C gives an α composition of 0.3% silicon and a liquid
composition of 2.0% silicon. A tie line at 578°C (just within the α + L) region
gives an α composition of 1.65% silicon and a liquid composition of 12.6%
silicon. Note that any total composition (alloy composition) between 1.65
and 12.6% silicon will give the same liquid-phase and solid-phase compositions at 578°C.

4.5 Amounts of Phases

It is important to know the *amounts* of phases, because the properties of a
two-phase mixture depend on these percentages. These amounts can be found
through calculation or through a simple graphical treatment of the phase diagram. Although the graphical method is faster, the calculation will be presented first because it deepens our understanding of what is meant by *overall*
composition compared with *phase* composition.

Considering again the two-phase region α + L of the aluminum–silicon
diagram (Figure 4.7), let us ask the following question: What are the relative
amounts of solid and liquid in the alloy containing 97% aluminum and 3%
silicon overall composition when it is at equilibrium at 600°C? In other
words, if we pour off the liquid and weigh the liquid and solid separately,
what percentage of the total weight will each be? (It is evident from the diagram that both solid and liquid are present.)

Let us start by making a homogeneous melt at 700°C, using 97 g of aluminum and 3 g of silicon. Now let us cool the all-liquid alloy to 600°C, where

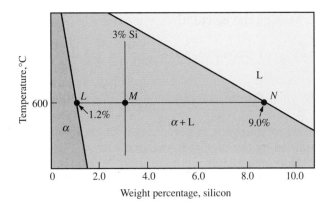

FIGURE 4.8   Calculation of phase amounts from a phase diagram

two phases are present. Even though there are two phases, a total of 3 g of silicon must still be present in these phases, so we know that the amount of silicon in α plus the amount of silicon in the liquid equals 3 g. We add to this the information obtained from a horizontal at 600°C (Figure 4.8) — that from rule 1 the compositions of the phases are

$$\alpha:\ \ 98.8\%\ \text{Al},\ 1.2\%\ \text{Si} \qquad \text{L:}\ \ 91.0\%\ \text{Al},\ 9.0\%\ \text{Si}$$

Now let the grams of the α phase equal x. Then the grams of the liquid phase equal $100 - x$. The total amount of silicon is

(grams of α) (fraction of silicon in α) + (grams of L)

$$\times\ (\text{fraction of silicon in L}) = \text{total grams of silicon}$$

or $\qquad\qquad (x)(0.012) + (100 - x)(0.090) = 3$

$$-0.078x = -6.0$$

$$x = 76.9\ (\text{grams of } \alpha)$$

Therefore we have 76.9% by weight α and 23.1% by weight liquid.

Now let us use a simpler graphical method called the *inverse lever rule.*

RULE 2. LEVER RULE TO DETERMINE AMOUNTS (PERCENTAGES) OF PHASES PRESENT. We again draw a tie line on the phase diagram, crossing the two-phase field at the desired temperature (600°C) and touching the one-phase fields (Figure 4.8). Next we draw a vertical line at the overall composition of the melt (3% silicon). The inverse lever rule (which can be proved by geometry) states that the percentage of α is $(MN/LN) \times 100$ and the percentage of L is $(LM/LN) \times 100$. Note that we find the amount of α by taking the length of the tie line on the *far* side of the overall composition; hence the expression *inverse* lever rule.

Making the calculation, we have

$$\text{Percentage of } \alpha = \frac{9.0 - 3}{9.0 - 1.2} \times 100 = 76.9, \text{ or } 76.9 \text{ g of } \alpha$$

$$\text{Percentage of L} = \frac{3 - 1.2}{9.0 - 1.2} \times 100 = 23.1, \text{ or } 23.1 \text{ g of L}$$

These are the weight percentages of the phases. They are the same as those obtained by the algebraic method.

Note again that we have used several terms interchangeably in this section. The terms are *phase amounts, quantity of phases,* and *percentage of phases present.* Some confusion usually arises when "% silicon" is used in the inverse lever rule to determine the "% phases." It is simply for convenience that we use the percent silicon scale on the x-axis as a convenient ruler to determine the fractional lengths of the tie line, as demonstrated in the following example.

EXAMPLE 4.3 *[ES]*

Calculate the phase amounts present at 600°C for a 2% and a 6% silicon alloy.

Answer   First we must determine *what* phases are present. They are α (a solid solution of silicon in the aluminum lattice) and liquid (Figure 4.8).

Next the compositions of the phases in equilibrium are needed. These are found to be the same as discussed previously.

$$\alpha: \quad 98.8\% \text{ Al, } 1.2\% \text{ Si} \qquad \text{L: } \quad 91.0\% \text{ Al, } 9.0\% \text{ Si}$$

In other words, the same two phases with the same phase compositions exist at 600°C for both the 3% and the 6% alloy. Only the quantities change. These are determined by the inverse lever rule.

We could use a ruler to determine the lengths between the total composition and the ends of the tie lines. However, there is a "built in" ruler on each phase diagram: the composition scale. Therefore,

$$\text{2\% silicon alloy:} \qquad \text{Percentage of } \alpha = \frac{9.0 - 2.0}{9.0 - 1.2} \times 100 = 89.7$$

$$\text{Percentage of L} = \frac{2 - 1.2}{9.0 - 1.2} \times 100 = 10.3$$

$$\text{6\% silicon alloy:} \qquad \text{Percentage of } \alpha = \frac{9.0 - 6.0}{9.0 - 1.2} \times 100 = 38.5$$

$$\text{Percentage of L} = \frac{6.0 - 1.2}{9.0 - 1.2} \times 100 = 61.5$$

Thus we see that increasing the silicon of the alloy (total composition) increases the amount of liquid while decreasing the amount of solid. This is not totally unexpected, because increasing the silicon of the alloy at 600°C pushes us closer to the liquidus. A materials balance for the 6% silicon alloy serves as a check on our previous calculation.

Basis: 100 g of alloy; silicon balance

Total silicon in alloy = silicon in α + silicon in liquid

$$= (38.5 \text{ g of } \alpha)(0.012 \text{ Si in } \alpha)$$
$$+ (61.5 \text{ g of L})(0.090 \text{ Si in L})$$
$$= 0.46 + 5.54 = 6 \text{ g Si in 100 g of alloy,}$$
$$\text{or 6\% Si}$$

4.6 Phase Fraction Chart of Amounts of Phases Present as a Function of Temperature

A phase fraction chart (Figure 4.9) is quite useful for summarizing our interpretation of a phase diagram. This chart shows how the percentages of phases present in an alloy of fixed overall composition change as the temperature changes. Let us use the 3% silicon alloy just discussed as an example. In Figure 4.9 the vertical axis represents the percentage of phases present, and the horizontal axis the temperature. A separate line graph is used for each phase. Taking the data for α as an example, if we follow the vertical overall composition line, we can calculate and show the following:

1. α appears at 641°C (liquidus).
2. At 600°C we have 76.9% α and 23.1% liquid, as calculated before.
3. At just above the eutectic, say at 578°C, we have, by the inverse lever rule,

$$\text{Percentage of } \alpha = \left(\frac{12.6 - 3}{12.6 - 1.65}\right) 100 = 87.7$$

Percentage of L = 12.3

4. At 576°C (just below the eutectic), we have $\alpha + \beta$ present and no liquid.

$$\text{Percentage of } \alpha = \left(\frac{99 - 3}{99 - 1.65}\right) 100 = 98.6$$

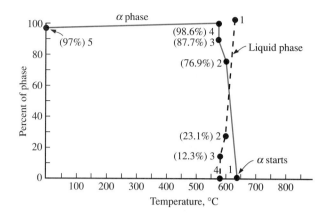

FIGURE 4.9 Phase fraction chart calculated from a phase diagram for 3% silicon in aluminum. Only α and liquid are shown. The β phase is 1.4% at 576°C and increases to 3% at 20°C. The numbers after the percentages refer to the enumerated steps in the text.

$$\text{Percentage of } \beta = \left(\frac{3 - 1.65}{99 - 1.65} \right) 100 = 1.4$$

The change in α from 87.7% to 98.6% is abrupt, because the 12.3% of liquid of eutectic composition that was present at 578°C formed $\alpha + \beta$ at the eutectic temperature.

5. At 450°C there is more β.

$$\text{Percentage of } \alpha = \left(\frac{99 - 3}{99 - 0} \right) 100 = 97$$

(Note that Figure 4.7 shows approximately 0 wt % Si soluble in aluminum from 450°C to room temperature, 20°C.)

$$\text{Percentage of } \beta = \left(\frac{3 - 0}{99 - 0} \right) 100 = 3$$

It is important to our later discussions dealing with control of the shape of phases to note that the α precipitated under two different temperature conditions. In the temperature interval 641 to 578°C, the α precipitated alone from a liquid. This is called the *primary* α. It forms large crystals, or dendrites. The second important temperature point was at the eutectic. Here α and β precipitated together, giving a mixture of fine α and β crystals. Therefore, 87.7% of the α will be in the form of large primary grains, and 10.9% will be small grains in the eutectic mixture. [Total α (98.6%) = primary α (87.7%) + eutectic α (10.9%) at 576°C.]

This concludes our simplified discussion of phase diagrams.

EXAMPLE 4.4 *[ES]*

Calculate the volume percentages of phases present in an alloy of 16% by weight silicon and 84% by weight aluminum. [Density of silicon = 2.35 g/cm^3 (2.35 × 10^3 kg/m^3); density of aluminum = 2.70 g/cm^3 (2.70 × 10^3 kg/m^3).]

Answer   From the phase diagram (Figure 4.7) we find that we will have approximately 16 g of β and 84 g of α, because at room temperature these phases are essentially pure silicon and aluminum. That is, very little silicon dissolves in aluminum or aluminum in silicon, even though measurable solubility is shown at higher temperatures.

$$\text{Volume of } \alpha = \frac{84 \text{ g}}{2.70 \text{ g/cm}^3} = 31.2 \text{ cm}^3$$

$$\text{Volume of } \beta = \frac{16 \text{ g}}{2.35 \text{ g/cm}^3} = 6.8 \text{ cm}^3$$

$$\text{Total volume} = 31.2 + 6.8 = 38.0 \text{ cm}^3$$

$$\text{Volume percentage of } \alpha = \frac{31.2 \text{ cm}^3}{38 \text{ cm}^3} \times 100 = 82$$

$$\text{Volume percentage of } \beta = \frac{6.8 \text{ cm}^3}{38 \text{ cm}^3} \times 100 = 18$$

Note: Physical properties are more easily related to the volume percentages of phases than to the weight percentages.

4.7 The Phase Rule

So far we have used phase diagrams as useful maps to tell us which phases to expect at equilibrium. However, if you have a healthy scientific curiosity, you may have been wondering whether some rules of nature govern these diagrams. Isn't it curious that when we draw a horizontal line across a diagram, we go from a one-phase to a two-phase field, then back to one phase, and so on? And why do we always show a horizontal line at the eutectic temperature?

A professor at Yale, J. Willard Gibbs, developed the *phase rule* to address questions of this type over a hundred years ago. He published it in a poorly circulated Connecticut journal, but it achieved prominence only after a German reviewer found it and applied it to the exploitation of the great European salt beds. The derivation is really quite simple and serves two purposes. It gives us a test for validity of a phase diagram and secondly a better understanding of the concepts of phase composition and phase equilibrium.

The crux of the derivation depends on the simple mathematical principle that if we have three variables and two equations, we need another piece of information to fix the values of the variables.* In phase rule terminology, we have one degree of freedom. We can express this situation as an equation:

No. of degrees of = total number − relations among
freedom   of variables   variables

or

$$F = U - J \tag{4-1}$$

Now let us consider a system in equilibrium. Let us take a test tube of water and dissolve in it a few crystals of iodine, which gives a brownish solution. Now let us add a few milliliters of carbon tetrachloride and shake. The CCl_4 turns to a dark blue layer, and the water layer grows paler. In other words, a good portion of the iodine enters the CCl_4. Also, some CCl_4 dissolves in the water, and vice versa. We call H_2O, I, and CCl_4 *components*, so we have three components.

If we wanted to describe to a colleague in California the characteristics of the *water phase*, we would need to communicate

1. The concentration of iodine.
2. The concentration of CCl_4. (We would *not* need to specify the concentration of water, because it is already fixed as $100 - (\%I + \%CCl_4)$.
3. The pressure and temperature.

The total number of variables in the system are as follows: in the water phase, conc. I, conc. CCl_4; in the CCl_4 phase, conc. I, conc. CCl_4; and the overall pressure and temperature. Therefore the total number of variables is $U = 6$.

In general then $U = P(C - 1) +$ pressure, temperature. The total number of variables $U =$ no. of phases (no. of components − 1) + 2 (for pressure and temperature)

$$U = 2(3 - 1) + 2$$

$$= 6$$

Now let us find J, the number of relations. If we experiment with different amounts of iodine, we find that the percentage of iodine in the CCl_4 is greater than that in the water. However, there is a fairly constant ratio[†] of the percentages; it can be expressed as the distribution coefficient k (k_I when we are referring to iodine).

---

*Although an additional equation is suggested, it is just as reasonable to remove one of the variables.

[†]When the ratio varies with concentration, we use the activity instead of percent as discussed in chemistry, but the general derivation is the same.

$$k_I = \frac{\% \text{ I in } H_2O}{\% \text{I in } CCl_4}$$

We have additional equations for the ratio of CCl_4 in the water layer to CCl_4 in the CCl_4 layer and for water in the water layer to water in the CCl_4 layer. Therefore, for three components and two phases, we have three relations. (If we included another phase, for example, the gas in the space above the liquid, we could write three more equations for distribution coefficients between the upper water layer and the space.)

In general, we find the number of relations fixed by nature

$$J = (\text{no. of phases} - 1)(\text{no. of components})$$

$$= (P - 1)C$$

Substituting yields the number of *degrees of freedom, F*:

$$F = U - J$$

$$= P(C - 1) + 2 - (P - 1)C$$

$$F = C - P + 2 \tag{4-2}$$

In words,

$$\begin{matrix} \text{number of} \\ \text{degrees of freedom} \end{matrix} = \begin{matrix} \text{number of} \\ \text{components} \end{matrix} - \text{number of phases} + 2$$

This is the celebrated Gibbs phase rule.

In the usual case we are investigating a phase diagram at a fixed pressure of 1 atm. This means that we have reduced the number of variables F that we can change. We must rewrite the equation as

$$F = C - P + 1 \qquad (\text{pressure} = 1 \text{ atm}) \tag{4-3}$$

Now let us apply the phase rule to the aluminum–silicon diagram. First, consider the case of a one-phase field (liquid, for example). From the phase rule we have

$$F = C - P + 1 = 2 - 1 + 1 = 2$$

This means that to describe fully to a colleague the inherent characteristics of a liquid in the one-phase liquid field, we would need to fix two things. For example, if we specified the composition of the liquid and the temperature, our colleague could obtain a liquid of the same density, conductivity, or any other measurable property. If we specified only the composition, this would not be true, because if the temperature of the liquid is not fixed, the other properties vary.

Now let us look at the two-phase field $\alpha + L$. The phase rule states that

$$F = C - P + 1 = 2 - 2 + 1 = 1$$

Therefore, in a two-phase field, we have only one degree of freedom. Let us test this result by using up our one degree of freedom.

1. If we fix temperature to "use up" our one degree of freedom, we also fix the *phase* compositions. These are given by the intersections of the horizontal tie line (temperature) with the single-phase fields at this temperature. (We want to specify the *nature* of the phases, *not* their amounts. The overall composition can be any value in the two-phase field.)
2. Instead of temperature, let us fix the phase composition of α in equilibrium with L. This fixes temperature and the composition of the L phase, again by the tie line relationship.

Suppose an investigator reports a *field* where α, β, and L are present in a two-component system at constant pressure.

$$F = C - P + 1 = 2 - 3 + 1 = 0$$

The phase rule states that if two components and three phases are present, there is no degree of freedom. This is illustrated by the eutectic. At 577°C, the temperature fixed by nature, not by us, we have three phases of fixed phase composition in equilibrium (α: 1.65% silicon, L: 12.6% silicon, and β: 99% silicon). Thus if we say $P = 3$, then $F = 0$, and all conditions are already fixed. We may change the *overall* composition horizontally along the line, but the composition of each individual phase (the phase rule variables) does not change. Therefore $\alpha + \beta + L$ does not occur in a field, as reported by the investigator, but rather along a unique three-phase tie line — the eutectic in our example.

4.8 Formation of Microstructures

Just as we developed rules for finding analyses and amounts of phases in any two-phase field, we can develop an understanding of how structures form in any phase diagram. Let us begin with the aluminum–silicon system with which we are already familiar and then investigate more advanced diagrams. We call the alloys with less than the eutectic percentage of silicon *hypoeutectic* and those with more than the eutectic percentage *hypereutectic* (Figure 4.10). This nomenclature appears in many specifications and is easy to remember by recalling that someone who is "hyper" is above a given level.

Let us see how we obtain the final microstructures shown in Figure 4.10a. In the case of the hypoeutectic alloy containing 5% Si, we read from the diagram, using rule 1, that practically pure (only 1% Si) aluminum precipitates. These crystals are called *primary phase* because they precipitate before the

FIGURE 4.10   Schematic representation of microstructures in (*a*) hypoeutectic and (*b*) hypereutectic aluminum–silicon alloys

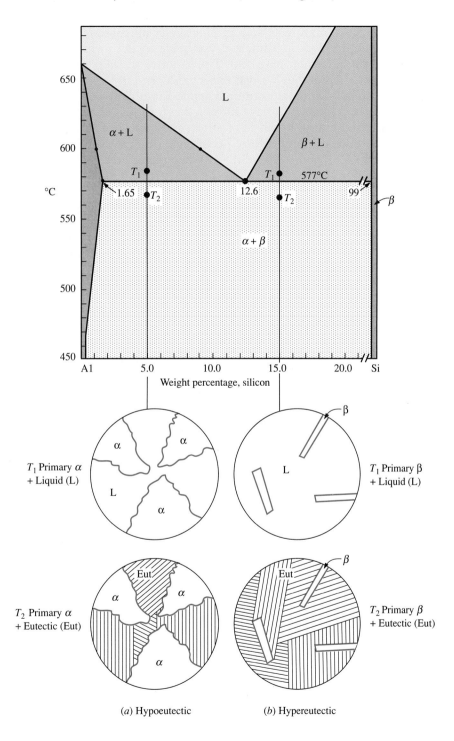

(a) Hypoeutectic   (b) Hypereutectic

eutectic mixture.* Now in practically all cases of crystallization of metallic structures, these primary crystals are in the form of dendrites, as illustrated in the photo on page 190. The kinetics of their growth are discussed in the next chapter, but we can expect the primary crystals in our important metal systems—iron, aluminum, copper, and nickel—to show this form. After dendritic crystals of α have formed down to just above the eutectic temperature, the remaining liquid is of eutectic composition (rule 1). As cooling through the eutectic temperature occurs, we observe simultaneous precipitation of α (1.65% Si) and β (99% Si). This two-phase precipitation gives a sandwich-like structure of α and β called the eutectic structure.

Because the eutectic reaction L $\rightarrow$ solid 1 + solid 2 is encountered in many alloy systems, we should point out that different microstructures occur. For example, rods of one phase surrounded by the other, alternating plates, and independent growth of the two phases are all encountered. Sometimes, because nuclei of one phase are already present, this phase forms alone at the eutectic temperatures and below, until the composition is close to that of the second phase. The second phase then forms alone, and the result is called a *divorced eutectic*.

K. Jackson, of the Bell Laboratories, has shown that some basic factors govern the shape of the phases formed. The rounded dendritic shapes of the metals occur with small entropy changes—that is, small changes in order between liquid and solid. On the other hand, when complex organic crystals form from the liquid, there is a large entropy change and faceted crystals are formed. The complex tetrahedral hybridized bonding in silicon leads to the faceted diamond cubic crystals shown in Figure 4.2b.

In general, we would expect spherical phases to form from the liquid, and indeed these are encountered in some cases, such as lead in copper. However, because the growth rate is usually different in different directions, spherical shapes are not as common as expected.

On the hypereutectic side of the eutectic at 15% Si, the primary crystals are β (99% Si), and in the typical condition long, faceted shapes are obtained. When the eutectic temperature is reached, the eutectic liquid is of the same composition as in the 5% Si alloy, and the same eutectic structure is obtained (Figure 4.10b).

4.9 Complex Phase Diagrams

The most complex binary diagram is made up of just two types of lines, horizontal and nonhorizontal. Let us take up the nonhorizontal lines first. All of them are very simple, because they merely give the boundary between a one-phase field and a two-phase field. Let us check the copper–zinc diagram (Figure 4.11) to convince ourselves of this. Two two-phase fields never touch

---

*These crystals may also be called proeutectic.

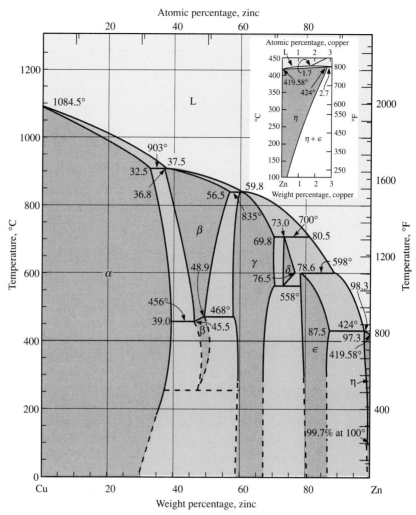

FIGURE 4.11 Copper–zinc phase diagram. (See the text for an explanation of the significance of the dashed solubility lines.)

[Reprinted with permission from *Metals Handbook*, 8th ed., Vol. 8, *Metallography, Structures, and Phase Diagrams*, American Society for Metals, Metals Park, OH, 1973, p. 301.]

along a vertical line, say $\alpha + \beta$ and $\beta + \gamma$, because this would mean that three phases—α, β, and γ—could exist over a range of temperatures at the composition of the dividing line. Varying the temperature would leave one degree of freedom to be specified, which is contrary to our previous phase rule analysis for two components and three phases in equilibrium at 1 atm ($F = 0$). Note therefore that as we go across the phase diagram at a given temperature, we go alternately through one- and two-phase fields.

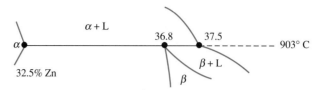

FIGURE 4.12   Peritectic transformation in the copper–zinc diagram

Now let us consider the horizontal lines in the copper–zinc diagram. At 903°C we have the situation shown in Figure 4.12. Let us consider an alloy with 36.8% zinc at 902°C containing β. If this alloy is heated to 904°C, we find $\alpha + L$. Thus one solid phase has been transformed to a new solid phase plus liquid. When we cool the same material, we have a solid α (S_1) that reacts with liquid to form one new solid S_2:

$$S_1 + L \xrightarrow{\text{cooling}} S_2$$
$$(\alpha) \qquad\qquad (\beta)$$

This reaction is different from the *eutectic* reaction in the 12.6% silicon alloy in the aluminum–silicon system, which reacts upon cooling:

$$L \xrightarrow{\text{cooling}} S_1 + S_2$$
$$(\alpha) \quad (\beta)$$

We therefore give the $S_1 + L \xrightarrow{\text{cooling}} S_2$ reaction a different name: *peritectic reaction.*

At 835°C we encounter the same type of reaction, $\beta + L \rightarrow \gamma$, at 59.8% zinc, and at 598°C we have $\delta + L \rightarrow \varepsilon$ at 78.6% zinc.

Another type of reaction, $\delta \rightarrow \gamma + \varepsilon$, is encountered upon cooling an alloy of 74% zinc at 558°C. This reaction is analogous to the eutectic reaction, but it involves an all-solid transformation, so it has the special name *eutectoid*.

There are a few other types of horizontal lines involving three phases, which may be encountered in other binary phase diagrams. Briefly, these are

$$S_1 + S_2 \xrightarrow{\text{cooling}} S_3 \ (\textit{peritectoid})$$

and

$$L_1 \xrightarrow{\text{cooling}} L_2 + S_1 \ = (\textit{monotectic})$$

These three-phase equilibria will be pointed out when they occur in later discussion.

Let us make a few final comments about the copper–zinc phase diagram. First, there are several dotted lines that may be present on phase diagrams for two reasons. One possibility is that the exact position of the solubility lines has not been firmly established. A second consideration, especially at lower temperatures, is that because in theory equilibrium can take an infinite amount of time, equilibrium may require extended times to achieve, and so the phase relationships may not be observed in reasonable lengths of time. For example, the horizontal lines at 250°C in the copper–zinc system are dotted because a 40% zinc alloy at room temperature would be found to consist of α and a small amount of β' rather than α and γ as indicated by the equilibrium diagram.

A further source of confusion in the copper–zinc system is the occurrence of β and β'. The primed notation β' refers to an ordered solid solution or preferred positions for the solute atom in the solvent lattice. The unprimed β, on the other hand, occurs at high temperatures, where we might expect to find a random or more disordered structure.

EXAMPLE 4.5   *[ES/EJ]*

In a two-component system, why are there tie lines for three-phase equilibria but not for two-phase equilibria?

Answer   From the phase rule,

$$F = C - P + 1 \text{ (constant pressure)} = 2 - P + 1 = 3 - P$$

Thus

$$F = 0 \quad \text{for three phases in equilibrium}$$

and

$$F = 1 \quad \text{for two phases in equilibrium}$$

When two phases are present, the one degree of freedom allows us to fix another variable. We might do this by fixing the temperature—in other words, by providing a tie line. However, because two phases exist over a temperature range, the number of tie lines is infinite. It is unreasonable to include all of these on a phase diagram.

When three phases are in equilibrium, the temperature is fixed because there are zero degrees of freedom. The tie line tying together the three phase compositions is no longer variable and therefore appears on the phase diagram.

4.10 Ceramic Phase Diagrams

These diagrams follow the same principles as metallic diagrams, but the components are usually pure compounds, particularly oxides. An important diagram for structural ceramics is the ZrO_2–CaO system (Figure 4.13).

Pure zirconia can exist in either the tetragonal or the monoclinic crystal

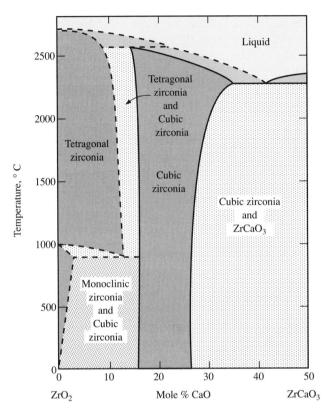

FIGURE 4.13   The binary system ZrO_2–$ZrCaO_3$

[P. Duwez, F. Odell and F. H. Brown, Jr., "Stabilization of Zirconia with Calcia and Magnesia," *Journal of the American Ceramic Society*, Vol. 35, p. 109, 1952. Reprinted by permission of the American Ceramic Society.]

structure. Upon cooling, the tetragonal structure undergoes a great volume change in transforming which can shatter a component. However, we see from the diagram that a cubic form of zirconia is formed if 20% CaO is added; this is a stable and a very useful refractory. An excess of CaO results in $ZrCaO_3$. Note that the diagram ends at $ZrCaO_3$ because this is the second component. The scale on the calibrated *x* axis is calibrated in mole % CaO for convenience, not because CaO is a component. Further diagrams for ceramics will be presented in Chapter 10.

4.11 Polymer Phase Diagrams

Phase diagrams are also used to show relations in polymers, but not as widely as in metals and ceramics. These are *not equilibrium diagrams* but may be called *behavior diagrams*; they will be described more fully in Chapter 14. In Figure 4.14, we see the structures in a copolymer of components A and B. The diagram shows that the typical copolymer in region 5 is glassy and brittle.

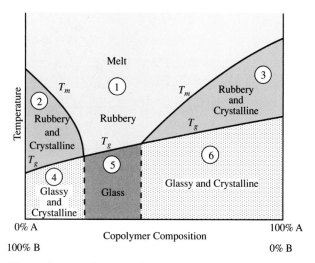

FIGURE 4.14   Phase diagram for a random copolymer system

[From *Fundamental Principles of Polymeric Materials* by Stephen L. Rosen, Copyright © 1971 by Barnes & Noble, Inc. Reprinted by permission of Harper & Row, Publishers, Inc.]

Above the T_g temperature on both sides of the diagram, the material is rubbery and crystalline. Obviously, we cannot apply the phase rule to this diagram, but it is a useful map of temperature and composition effects.

4.12 Ternary Diagrams

So far we have discussed only systems with two components which lead to binary diagrams. In many alloys we have three principal elements present, as in 18/8 stainless steel, which contains 18% chromium, 8% nickel, and about 74% iron. Let us compare the degrees of freedom of such a *ternary system* and of a binary system:

$$\text{Binary:} \quad F = C - P + 2 = 4 - P$$

$$\text{Ternary:} \quad F = C - P + 2 = 5 - P$$

If we fix pressure at 1 atm in the binary system, $F = 3 - P$. Thus if we specify that three phases are in equilibrium, we have nothing else to specify; the temperature and phase compositions are fixed by nature.

However, in a ternary system under the same conditions (fixed pressure, $F = 4 - P$), we can encounter three phases over a range of temperatures and four phases at a given temperature.

Obviously, we need some sort of three-dimensional map to represent this added variable. We use a triangular graph (Figure 4.15).

Let us consider first the representation of composition at constant temperature, then the added dimension for varying temperature. Perhaps the easi-

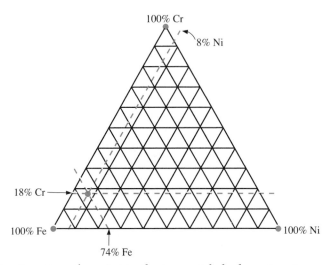

FIGURE 4.15   Location of a point in the iron–nickel–chromium ternary diagram

est way to visualize the net shown is an equilateral triangle. The point at the iron corner is 100% iron. Any point on the 90% iron line contains that amount of iron. Next let us locate the 18% chromium, 8% nickel, and 74% iron stainless steel called 18/8. The 74% iron line is shown, and the composition must lie somewhere on this line.

Recognizing that the chromium corner represents 100% chromium, we move away to find the 18% chromium line. The point where this line intersects the 74% iron line represents 18% chromium and 74% iron. We do not need to draw the 8% nickel line except as a check, because the percentage of nickel is determined by the difference from 100% (that is, % Ni = 100 − 74% Fe − 18% Cr).

RULE. To find the composition of a three-element alloy using a triangular graph, locate the proper isocomposition lines for each element, starting from that element's corner of the triangle.

4.13 Representation of Temperature in Ternaries

A complete three-component, or ternary, diagram at constant pressure but varying temperature is shown in Figure 4.16.

In some cases, determination of the liquidus can be a million-dollar research problem. For example, during pig iron production in a blast furnace, the slag must be kept liquid or a very expensive shutdown occurs. The principal components of the ternary diagram for the slag are three oxides—CaO, SiO_2, and Al_2O_3—instead of three metals. The million-dollar liquidus surface is shown in Figure 4.17. Note that the liquidus is no longer a line as in a two-

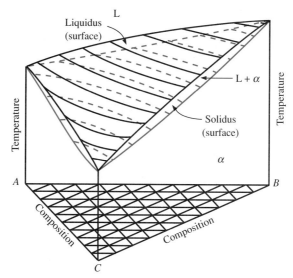

FIGURE 4.16 Temperature–composition space diagram of a ternary isomorphous system (complete solubility in the solid state)

(F. Rhines, *Phase Diagrams in Metallurgy*, McGraw-Hill, New York, 1956. Used with permission of McGraw-Hill Book Company)

FIGURE 4.17 Liquidus surface of the ternary system CaO–SiO_2–Al_2O_3. The horizontal lines or planes represent constant temperature contours. The area above the top surface would be all liquid. For a given composition, when the temperature falls to that of the top surface, solid begins to form.

component system; it becomes a surface when three components are present. This occurs because the addition of another component increases the number of degrees of freedom by 1. The significant fact is that a low melting combination occurs in the range 49% CaO, 39% SiO_2, 12% Al_2O_3 at 1315°C. Knowing this, the metallurgist adds the right amounts of oxides to the ore to avoid a freeze-up and, equally important, to create a fluid slag that is active in removing impurities such as sulfur from the metal.

The ternary diagram is equally important in dealing with metals. A third element can be added to lower the melting point of an alloy. For example, an alloy of 51% bismuth, 40% lead, and 9% cadmium melts in boiling water (212°F, 100°C), whereas the individual melting points of the elements are higher—bismuth, 520°F (271°C); lead, 621°F (327°C); cadmium, 609°F (320°C). This type of alloy is used in automatic sprinkler heads for plugs that must melt when a fire occurs.

Summary

In this chapter we adopted a new approach. Instead of merely observing structures and their properties, we took a dynamic tack, the first step in learning to control them, getting acquainted with equilibrium diagrams. The amount of information stored in such a diagram is really tremendous. For example, we shall see that the principal equilibrium structures encountered in thousands of compositions of steel are all given in one simple diagram, the iron–carbon diagram introduced in Chapter 8. Similarly, the melting points of many refractory brick compositions are given in the silica–alumina diagram that appears in Chapter 10.

In studying this chapter, you have learned (1) to read which phases are present, their phase analyses, and their amounts from an equilibrium diagram, and (2) to anticipate the microconstituents that form upon cooling from a melt and upon heating. This knowledge will prove invaluable in discussing heat treatment, so be sure you understand the simple rules we have formulated.

Definitions

Degrees of freedom, *F*   The number of variables that the experimenter can specify or control and still have the number of phases and components originally substituted into the phase-rule equation. For example, in the aluminum–silicon system, if two phases (α + L) are present, $F = 2 - 2 + 1 = 1$, and the temperature can be varied over a wide range without leaving the two-phase field (constant pressure).

Eutectic composition   The composition of a liquid that reacts to form two solids at the eutectic temperature. In the aluminum–silicon system, the eutectic composition is 12.6% silicon and 87.4% aluminum. Note, however, that some liquid of eutectic composition is obtained during the freezing of any alloy with an overall composition of between 1.65 and 99% silicon.

Eutectic temperature   The temperature at which a liquid of eutectic composition freezes to form two solids simultaneously under equilibrium conditions; also, the temperature at which a liquid and two solids are in equilibrium.

Eutectoid   Transforming from a solid phase to two other solid phases upon cooling.

Hypereutectic   Exhibiting greater than the eutectic composition. In the aluminum–silicon system (eutectic at 12.6% silicon), for example, 5% silicon is hypoeutectic, whereas 16% silicon is hypereutectic. See also *hypoeutectic*.

Hypoeutectic   Exhibiting less than the eutectic composition. See also *hypereutectic*.

Isomorphous   Of the same structure. Note that there is only one solid phase α in Fig. 4.16; hence only one structure is present. However, the lattice parameter changes continuously with composition.

Liquidus   The temperature at which a liquid begins to freeze upon cooling under equilibrium conditions (solid first forms).

Microhardness test   A test for measuring the hardness of a grain of a particular phase by using a very light load on a small indenter.

Monotectic   Transforming from a liquid phase to a solid phase plus another liquid phase upon cooling.

Number of components, *C*   The number of materials, elements, or compounds for which the phase diagram applies. For example, the iron–nickel and iron–iron-carbide systems are both binary, or two-component, systems.

Number of phases, *P*   The sum of all the solid, liquid, and gas phases. There can only be one gas phase, because all gases are intersoluble in all proportions. (See Chapter 1 for a detailed discussion of phases.)

Peritectic reaction   A reaction in which a solid goes to a new solid plus a liquid on heating, and the reverse occurs upon cooling:

$$S_1 \underset{\text{cooling}}{\overset{\text{heating}}{\rightleftharpoons}} S_2 + L$$

Peritectoid   Transforming from two solid phases to a third solid phase upon cooling.

Phase diagram   A graph showing the phase or phases present for a given composition as a function of temperature (a collection of solubility lines).

Phase rule $F = C - P + 2$, where F = degree of freedom, C = number of components, and P = number of phases in equilibrium. This formula is usually written +1 instead of +2, because we use up one degree of freedom in fixing the pressure at 1 atm.

Polyphase material   A material in which two or more phases are present.

Primary phase   The phase that appears first upon cooling. For example, a 5% silicon–aluminum alloy shows primary α before α and β appear together as a eutectic. Primary α is sometimes referred to as *proeutectic* α, because it occurs before the eutectic (simultaneous formation of α and β).

Solidus   The temperature at which the liquid phase disappears on cooling or at which melting begins on heating.

Solvus   The locus of temperatures for compositions of solid phases that are in equilibrium with one another; that is, solubility lines between solid phases.

Ternary system   A three-component system. A ternary system has one more degree of freedom than a binary system.

Two-phase material   A material in which two different phases are present and grains or crystals of two different materials can be found.

Problems

4.1 *[ES]* Consider an alloy of 5% Si, 95% Al. (Sections 4.1 through 4.4)
  a. What is the percentage of silicon in the α phase at 640, 600, 577, and 550°C?
  b. What is the percentage of silicon in the liquid phase at 640, 600, and 577°C?
  c. What is the percentage of silicon in the β phase at 550°C? (To answer this last part, draw the horizontal tie line until it touches the β field.)

4.2 *[ES]* Consider the typical engine-block alloy, which is 16% Si, 84% Al. (Sections 4.1 though 4.4)
  a. At what temperature will the first crystals of solid appear upon slow cooling of the melt?
  b. At what temperature will the alloy be completely solid?
  c. Just before the alloy is all solid — at, say, 578°C — what will be the analyses of the β and the liquid, respectively?
  d. At this temperature will the analysis of the liquid be greatly different from that of the liquid in the alloy of Problem 4.1?
  e. What will be the phase analyses of the α and β in this alloy at 550°C?

4.3 *[ES/EJ]* The copper–zinc equilibrium diagram is shown in Fig. 4.11. Although complex, the high-copper portion of the diagram is similar to that of the aluminum–silicon system in that a liquid and a solid occur over a temperature range. (Sections 4.1 through 4.4)
  a. What are the liquidus and solidus temperatures for a 70% Cu, 30% Zn alloy?
  b. If casting is easier with a narrower liquidus-to-solidus temperature range, which of the following alloys has better castability: 80% Cu, 20% Zn; 70% Cu, 30% Zn; or 60% Cu, 40% Zn?
  c. What is the maximum annealing temperature for a 70% Cu, 30% Zn alloy? [*Hint:* A 25°F (14°C) temperature variation within a furnace is not unlikely.]

4.4 *[EJ]* The complete aluminum–silicon equilibrium diagram is shown in Fig. 4.4. There are two scales for percentage silicon. Why are we interested in both scales? (Sections 4.1 through 4.4)

4.5 *[ES/EJ]* Now that we know how to make diamonds from the equilibrium diagram in Figure 4.1 why do we not flood the market with synthetic diamonds? (Sections 4.1 through 4.4)

4.6 *[ES]* Under what circumstances is the total composition the same as the phase composition or compositions? Under what circumstances is it different? (Sections 4.1 through 4.4)

4.7 *[ES]* Why do we label only the one-phase regions in the Cu–Zn phase diagram in Figure 4.11? (Sections 4.1 through 4.4)

4.8 *[ES]* At the eutectic in the Al–Si phase diagram, what phase(s) is (are) present? Give the chemical analysis of the phase(s). (Sections 4.1 through 4.4)

4.9 *[ES]* Prepare a fraction chart for the hypereutectic 16% Si, 84% Al alloy. (Sections 4.5 through 4.7)

4.10 *[ES]* What are the percentages of α and liquid in a 5% Si, 95% Al alloy at 620, 600, and 578°C? What are the percentages of α and β in this alloy at 576 and 550°C? (Sections 4.5 through 4.7)

4.11 *[ES]* What are the percentages of α and liquid in a 1% Si, 99% Al alloy at 630, 600, and 578°C? (Sections 4.5 through 4.7)

4.12 *[ES]* In a *hyper*eutectic (more than eutectic) analysis, 16% Si, 84% Al, calculate the amounts of liquid and β at 578°C and of α and β at 576°C. (Sections 4.5 through 4.7)

4.13 *[ES/EJ]* The lead–antimony equilibrium diagram is given in Problem 4.21. Suppose you view an equilibrium structure under the microscope at room temperature and estimate the volume percentage of α as 20%. (The densities of antimony and lead are 6.62 and 11.36 g/cm^3, respectively.) (Sections 4.5 through 4.7)
a. Calculate the weight percentage antimony in the alloy.
b. Why is your answer to part *a* only an estimate?
c. Determine the liquidus and the solidus for the alloy.
d. How much of each phase is present at 250°C?

4.14 *[EJ]* A student attempts to apply the phase rule to a 5% Si, 95% Al alloy in the α + L field. He sees that he is in a two-phase field, and at a pressure of 1 atm the equation is $F = C - P + 1$ or $F = 2 - 2 + 1 = 1$. He says that because this is an area where there are two variables, composition and temperature, the phase rule is incorrect in giving $F = 1$. Do you agree? (Sections 4.5 through 4.7)

4.15 *[EJ]* Another student applies the phase rule to the eutectic temperature for the aluminum–silicon system. She says, quite correctly, that there are three phases and two components; hence $F = 0$. She then complains that three phases will be in equilibrium from 1.6 to 99% silicon at 577°C, so there is still a variable to be fixed if she is to describe the system to another student. Can you help her with this problem? (Sections 4.5 through 4.7)

4.16 *[ES/EJ]* Sketch Figure 4.1 and label all areas, lines, and points in terms of the number of degrees of freedom. Indicate the practical significance of knowing the degrees of freedom. (Sections 4.5 through 4.7)

4.17 *[ES]* Using the diagram at the top of page 226, plot the weight fraction of the α phase as a function of the temperature that would be encountered under equilibrium conditions in an alloy containing 10% R, 90% K and in one containing 60% R, 40% K. (Sections 4.8 through 4.13)

4.18 *[ES/EJ]* In a 60% Cu, 40% Zn alloy called "Muntz metal," α begins to form at approximately 750°C. At 600°C the zinc content *in both* the α phase and the β phase is greater than at 750°C. Where does the added zinc in the two phases come from? (Sections 4.8 through 4.13)

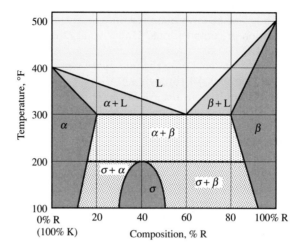

4.19 *[ES/EJ]* Given the following data, sketch a phase diagram. Be as accurate and as thorough as your knowledge permits. Cooling curves show horizontal temperature arrests at 715°F for pure A and 655°F for pure B. (From the phase rule we know that for the pure metals this indicates two phases in equilibrium.) A horizontal arrest in the two-component system at 500°F for 25% A, 75% B indicates that three phases are in equilibrium. (Sections 4.8 through 4.13)

Melting point of *A* = 715°F; melting point of B = 655°F

501°F, 25% A and 75% B; all liquid
499°F, 25% A and 75% B; 25% of a BCC metal, 75% of an FCC metal
499°F, 35% A and 65% B; 50% of a BCC metal, 50% of an FCC metal
400°F, 40% A and 60% B; 60% of a BCC metal, 40% of an FCC metal
400°F, 20% A and 80% B; 20% of a BCC metal, 80% of an FCC metal

4.20 *[ES]* In the copper–zinc diagram (Figure 4.11), locate the temperatures at which three phases can exist at equilibrium and name the reactions (eutectic, peritectic, etc.). (Sections 4.8 through 4.13)

4.21 *[ES/EJ]* A contractor has a specification for soldering sections of copper water pipe. The recommendation is for 50% Pb, 50% Sb, and the contractor suspects that 50% Pb, 50% Sn is intended. Both phase diagrams are given in the accompanying diagrams. (Sections 4.8 through 4.13)
   a. What are the liquidus and solidus temperatures for each of the alloys?
   b. What is the primary phase for each alloy?
   c. Calculate the percentage primary phase present in part *b* for each alloy at 1°C below the eutectic temperature.
   d. Why would a 50% lead, 50% tin alloy be easier to use as a solder than a 50% lead, 50% antimony alloy?

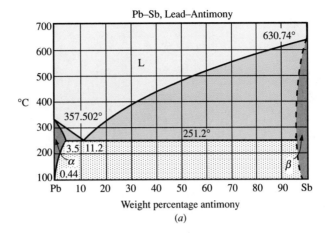

(a)

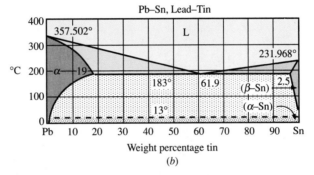

(b)

4.22 *[ES/EJ]* In Section 4.8 we define the term *primary phase*. Why would we be interested in the percentage of primary phase in the aluminum–silicon system? (Sections 4.8 through 4.13)

4.23 *[ES]* Indicate by name and temperature the three-phase equilibria in the ZrO_2–$ZrCaO_3$ phase diagram in Figure 4.13. Why are some of the lines dashed? (Sections 4.8 through 4.13)

4.24 *[ES/EJ]* Indicate whether the following statements are correct or incorrect, and justify your answer. (Sections 4.8 through 4.13)

a. β-brass is likely to have the same mechanical properties as α-brass.

b. In Figure 4.14, the crystalline polymers in areas 4 and 6 are distinctly different from one another.

c. In a ternary isomorphous system (Figure 4.16), the three binary systems are also isomorphous.

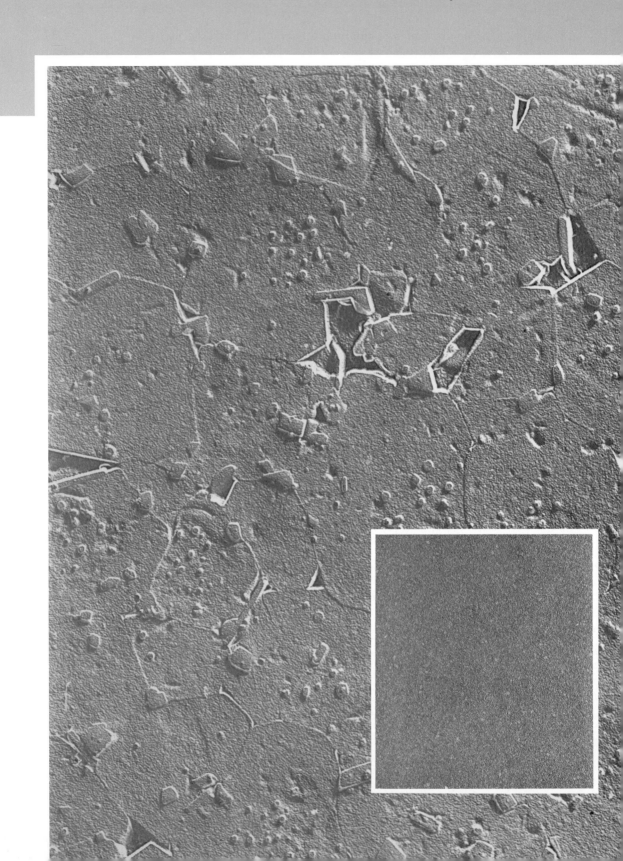

CHAPTER 5

The Control of Nonequilibrium Structures

In this chapter we will study nonequilibrium structures, not only because they can be processed more easily but also because they can provide final structures that are better than those obtained under equilibrium conditions.

For example, every cook is familiar with Pyroceram plates and baking dishes because they are strong and heat resistant. These properties are obtained by manipulating nonequilibrium structures. First the material is shaped into the easy-to-form glassy structure shown in the insert, and then it is heat-treated to precipitate the tough, low-expansion microstructure shown in the large photo. The grains in the ceramic are of a low-expansion material that is nucleated by the blade-like and angular particles.

5.1 Overview

The control of nonequilibrium structures is one of the most exciting branches of processing—yet it is also the subtlest, because no macroscopic changes in shape occur. Suppose, for example, we assembled an automobile with gears and bearings in the *un*treated condition. The dimensions of these shiny parts cannot be distinguished visually from heat-treated parts. However, after traveling only a few miles, our car becomes a rattling, smoking mess because of friction and wear of the soft surfaces.

Let us take a few minutes to explain this on a qualitative basis as an example of the importance of thermal treatment to control nonequilibrium structure. (Chapter 8 covers the quantitative details for steel.) The process used for hardening these parts, and practically all steels, consists of three basic steps:

1. The untreated steel parts at room temperature consist of BCC iron grains accompanied by grains of iron carbide (Figure 5.1). We heat the steel to a

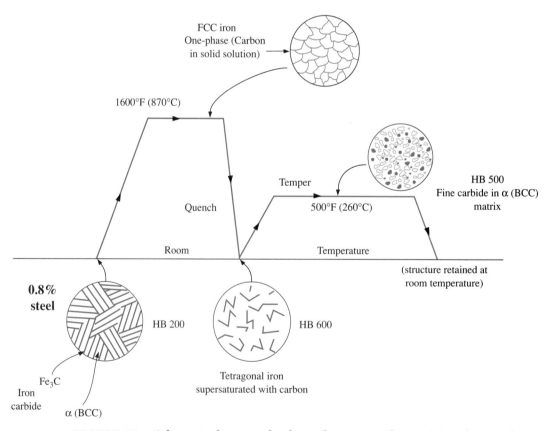

FIGURE 5.1   Schematic diagram of a thermal treatment for a 0.8% carbon steel

bright red color—about 870°C (1600°F)—and the iron structure changes to FCC. There is now more empty space at the center region of this unit cell than there was in the BCC structure. Accordingly, the iron carbide "dissolves" and the carbon atoms take up interstitial positions at the centers of the unit cells.

2. Next the hot steel is quenched in water or oil. The FCC iron changes to a distorted BCC structure (really a tetragonal structure) upon cooling, but the carbon atoms are still in an atomic dispersion (a supersaturated solution).

3. Finally, the steel is "tempered" by heating to a low temperature 150–425°C (300–800°F) where the carbon atoms migrate and form *fine,* more or less spherical carbides. This "hardened and tempered" structure is much harder than the initial structure because the carbides are finely dispersed and prevent easy slip. It is this structure that is required for hardened gears and bearings and for many other components.

As illustrated by the preceding example the purpose of this chapter is to examine the nonequilibrium methods that can be used to provide optimal structures. We will consider both liquid–solid reactions and solid–solid reactions, drawing examples from metals, ceramics, and polymers.

To understand the structures formed by nonequilibrium crystallization from the liquid (and practically all freezing of this type), we need to lay a foundation in three areas: (1) nucleation and growth, (2) segregation, and (3) dendrite formation.

5.2 Nucleation and Growth

It is important to re-emphasize that we frequently find phases that are not indicated by the phase diagram—phases that are out of equilibrium. In discussing the aluminum–silicon diagram, we mentioned that we could quench an all-α alloy from a high temperature and retain the same structure at room temperature, even though the equilibrium diagram called for $\alpha + \beta$ at approximately 0.1% silicon or more at 20°C. The reason is that silicon atoms must diffuse from their random solid solution position in the FCC lattice to form the silicon crystals with a new structure. This diffusion occurs slowly at 20°C; the quenching operation halts the diffusion process before it is complete. For a complete structural change to occur, the material needs sufficient time for diffusion. In addition, the silicon phase must form a nucleus.

The phenomenon of *nucleation* has wide application. For example, clouds can be seeded to produce rain by supplying silver iodide crystals as nuclei. Pure water can be cooled to −30°C without freezing in a special container, and common metals such as nickel can be cooled to 200°C below the equilibrium freezing point without solidifying. The usual impression is that supercooling or *undercooling* is accomplished by fast cooling, but actually liquids can be maintained for a period of time at these low temperatures. Once a liquid is cooled below its equilibrium freezing temperature, there is a driving force for solid to precipitate. This force is the difference in bulk free energy, ΔG_v, between the liquid and the solid. For instance, for a *spherical* volume the difference is [(free energy of solid/mm³) − (free energy of liquid/mm³)] × [volume of sphere, mm³] or $\Delta G_v \frac{4}{3}\pi r^3$. ΔG_v is always negative below the equilibrium freezing temperature.

Bulk free energy is a quantity that is precisely defined in chemistry. For those unfamiliar with the concept, a mechanical analogy may be useful. Consider an old-fashioned screen door that slams shut when released because of an overhead spring. As we open the door, tension builds up in the spring. The further the door is opened, the greater the energy in the spring. When a liquid is cooled below the equilibrium freezing temperature, the greater the undercooling, the greater the driving force for solidification (or the greater the change in the bulk free energy).

However, there is a force to be overcome in growing a solid sphere of radius r in place of the liquid. A new surface has to be formed. We need something like the energy required to blow a soap bubble against surface-tension effects. The energy required to create a new surface for the sphere is equal to the surface area times the surface tension, or $4\pi r^2 \sigma$, where σ is the surface tension in ergs per square millimeter of surface area.

If we plot these two energies and their sum, we obtain the curve shown in Figure 5.2. Because the difference in free energy or bulk free energy is a cubic function, whereas the surface energy is a quadratic equation, their sum results in a maximum, which we call r^*. In nature a system tends to minimize its energy, so for r values less than r^*, the easiest way to go to lower energy is to have r values become still smaller, or to have the nucleus dissolve. On the other hand, for values of r^* or greater, minimum energy is obtained by an increase in r, so the nucleus will grow. These are two important cases.

CASE 1. If no nuclei are present in the melt, it will have to be supercooled greatly for nucleation to occur. The lower the temperature, the greater the difference in bulk free energy and the greater the driving force to transform. (The surface tension does not change appreciably with temperature.) With

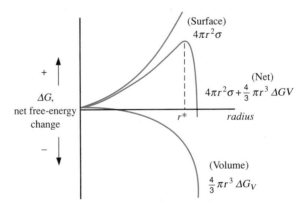

FIGURE 5.2   Change in free energy as a function of the nucleus radius

great undercooling, r^* becomes very small and *homogeneous nucleation*, or nucleation at many points within the liquid, takes place.

CASE 2. If we introduce a solid nucleus of radius greater than r^*, it will grow. This nucleus may be another material that is "wet" by the liquid and hence acts like a nucleus of the metal. This is called *heterogeneous nucleation*. Because this nucleus is usually relatively large, it produces freezing close to the equilibrium temperature.

Once a nucleus or nuclei are present, growth occurs. The typical growth curve is shown in Figure 5.3. At first the rate of growth is slow because there is limited solid surface. Then, as the surface increases, the growth curve rises rapidly. Finally the rate of growth slows down because solid surfaces come in contact, reducing the area of the solid–liquid interface.

These nucleation and growth phenomena are also present in solid–solid reactions, wherein an additional nucleation effect may affect the structure of the solid precipitate. The solid often comes out in the form of needles or plates rather than spheres or cubes, because an additional factor operates against the growth of the nucleus. This factor is related to the volume change in going from one solid to another. If the second solid is more voluminous than the first, the nucleus builds up compressive stress. This effect is diminished if the nucleus is needle-like in shape instead of spherical. Therefore, in transformations that occur at low temperatures, where the matrix has high strength and strains are not easily redistributed, the precipitate is usually in the form of needles or plates. The needle shape may still occur at high temperatures when the interfacial energy prevents a spherical shape formation.

A practical application is found in the control of the size and shape of the silicon crystals in the aluminum–silicon engine block alloy. For many years hypereutectic (greater than 12.6% silicon) alloys could not be used because large, hard, blade-like silicon crystals formed from the liquid. These were difficult to machine and gave the block a rough surface. Finally it was found that if phosphorus was added to the melt, many small aluminum phosphide

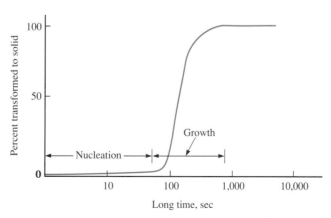

FIGURE 5.3 Representation of a liquid transforming to a solid as a function of time

crystals formed, which are similar to silicon. These in turn nucleated many small silicon crystals, giving the desired refinement.

5.3 Segregation

Segregation may be defined as variation in composition in a phase and, as we shall see, can lead in some cases to the formation of a second phase not predicted by the phase diagram. Consider Figure 5.4.

Assume we have a material of composition (1), Figure 5.4*b*. Under equilibrium conditions, below T_2 the material will be a one-phase solid of uniform composition (1). However, consider what is necessary to attain this structure. The first solid to crystallize at T_1 is of composition A, with less than half the % B contained in the liquid. As more solid separates at lower temperatures, the composition grows richer in B but still falls to the left of the starting composition. In order for *all* the solid to attain the equilibrium concentration at temperature T_2, diffusion of B must take place throughout the solid.

By contrast, to illustrate extreme segregation, let us assume that no diffusion takes place in the solid and plot the average composition as the temperature falls, as shown by the dashed line. We find that at the eutectic temperature T_3, the average composition of the solid is still to the left of the overall composition. As a consequence we must still have liquid present. Furthermore, the analysis will reveal this liquid to be *eutectic* in composition. Upon further cooling, we have the eutectic reaction in this liquid: $L \rightarrow \alpha + \beta$. This microstructure is shown schematically in Figure 5.4*a*.

There are two nonequilibrium structures of great importance. First, the crystals (dendrites) of solid will show *coring*—that is, a difference in concentration from center to edge. Second, there is the presence of a second phase that could lead to problems in rolling, to corrosion, and possibly to reduced strength in the case of a brittle grain-boundary precipitate.

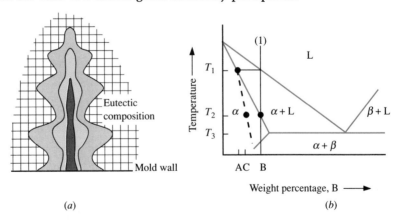

(a)   (b)

FIGURE 5.4   (*a*) Segregation in dendrite growth. (*b*) Phase diagram illustrating segregation. The dashed line shows the new solidus temperature.

In many cases, reheating the structure (below the eutectic temperature!), which is called *"homogenization,"* permits enough diffusion to occur. The second phase dissolves in the solid state and a uniform single phase results. Let us perform a simple calculation involving homogenization, using the principles developed in Chapter 3.

EXAMPLE 5.1 *[ES/E]]*

It can be determined that $t \simeq x^2/D$ for reducing the concentration difference to 50% when a material is homogenized.

A. Recalculate the time (in terms of x and D) if the concentration difference is to be only 5% ($C_x = 0.95\ C_s$; $C_0 = 0$).
B. What is the physical significance of these homogenization values?
C. What effect would this have on efforts to remove the cored microstructure shown in Fig. 5.4?

Answer

A.

$$\frac{C_s - 0.95C_s}{C_s - 0} = 0.05 = \text{erf}\,\frac{x}{2\sqrt{Dt}}$$

(Table 3.2)

z	erf(z)
0.025	0.0282
x	0.0500
0.050	0.0564

interpolation gives

$$\frac{0.050 - x}{0.050 - 0.025} = \frac{0.0564 - 0.0500}{0.0564 - 0.0282}$$

$$x = 0.044$$

therefore

$$0.044 = \frac{x}{2\sqrt{Dt}}$$

$$0.088 = \frac{x}{\sqrt{Dt}}$$

$$11.4x = \sqrt{Dt}$$

$$129x^2 = Dt$$

$$t = \frac{129x^2}{D}$$

B. 50% homogenization $\approx x^2/Dt$
   95% homogenization $\approx 129x^2/Dt$
   Increasing the degree of homogenization requires 129 times longer. The time may be prohibitive. For example, if 1 hr is required for 50% homogenization, we would not heat-treat for 129 hours to gain the added homogeneity.

C. The increased time is like the argument for a frog jumping halfway to a line with each hop. When are we there for all practical purposes? We may never completely remove the cored microstructure in a finite time, but we can certainly reduce it significantly with several hours of heat treatment. (*Note:* Hotworking uses both mechanical and thermal effects to increase atom motion and hence to homogenize a cored structure efficiently.)

5.4 Dendrite Growth

Let us begin with an experiment to illustrate the different types of structure we can form during crystallization. We mold a rectangular cavity in sand to measure 12 in. × 2 in. × 2 in. (305 mm × 51 mm × 51 mm). At one end we have a "chill" made of graphite to provide a fast cooling rate, and at the other end we have a large cylindrical reservoir of liquid metal. We pour pure liquid copper into the mold and, after about a minute, invert the mold to allow any liquid to drain out. We then section the specimen longitudinally and find that the metal was freezing with a smooth liquid–solid interface (Figure 5.5*a*).

Now we repeat the experiment with scrap copper, which has small amounts of impurities. In this case we do not find a smooth interface, but rather we see that the metal has crystallized by forming a network of *dendrites* (Figure 5.5*b*). This type of structure often contains shrinkage voids that

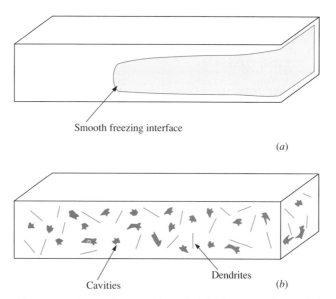

FIGURE 5.5   Cross sections of partially solidified copper bars. (*a*) Pure copper. (*b*) Copper containing impurities.

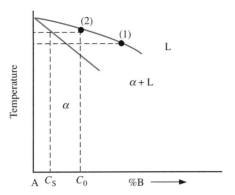

FIGURE 5.6   Phase diagram for a pure metal with soluble impurities, leading to a binary diagram

lower strength, and it often exhibits coring. Let us examine the reasons for the difference between the two copper bars.

Assume that the impure alloy has the phase diagram shown in Figure 5.6. The solid that forms first has a lower concentration, C_s, of B (the impurity) than the starting liquid, which has concentration C_0. This will sweep B from the solid zone into the liquid just ahead of the freezing front. When we plot the concentration of B along the bar, as shown in Figure 5.7a, we see that the concentration of B just ahead of the solid is greater than the starting concentration in the liquid (assuming there is no diffusion in the liquid).

If we chemically analyze a sample of metal at the interface and another a short distance away, we can find the liquidus temperatures (1) and (2) on the phase diagram (Figure 5.6). We find that the temperature for start of freezing is higher at point (2), away from the interface (Figure 5.7b). The actual temperatures at points (1) and (2) in the bar are shown as (1) and (2′) in Figure 5.7(b). We see that the actual temperature at (2′) is significantly below the equilibrium liquidus temperature at point (2). Solid is already nucleated at point (1), and there is a strong tendency to grow at point (2), which provides a spike of solid penetrating out into the liquid. Another way of putting this is to say that the interface is unstable and that rapid branching growth on a dendrite, occurs (Figure 5.7c). This phenomenon is called *constitutional supercooling* (Figure 5.7b), because we develop supercooled metal not by temperature control but by changes in constitution—that is, in the chemical composition. The liquid whose actual temperature is less than the equilibrium liquidus is therefore "constitutionally supercooled" (Figure 5.7c).

EXAMPLE 5.2 *[ES]*

Which alloy, (1) or (2), would be expected to show more dendrite formation under the same circumstances?

Answer

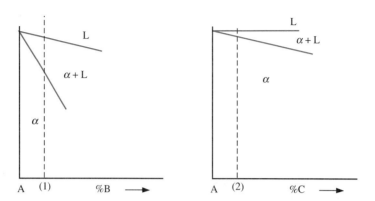

The concentration profile between the liquid and the solid for alloy (1) would be steeper than for alloy (2) in front of the freezing interface. Therefore, the constitutional supercooling would be greater, as would be the dendrite formation.

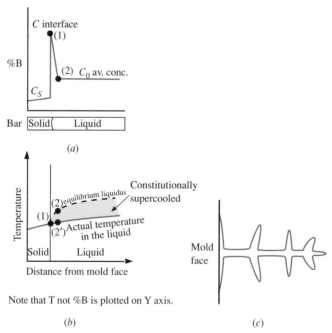

Note that T not %B is plotted on Y axis.

(b)　　　　　　　　　　　　　　　　(c)

FIGURE 5.7 (a) The concentration gradient of solute element B during solidification of a bar. (b) Development of an actual temperature below the equilibrium liquidus is called constitutional supercooling. (c) A result of the supercooling is a fern-like spike, called a dendrite, extending into the liquid. The driving force (amount of constitutional supercooling) is greater *away* from the L–S interface.

EXAMPLE 5.3 *[ES/EJ]*

Three methods for growing silicon single crystals for memory chips are shown in Figure 5.8. In *a*, a solid crystal of the desired orientation is introduced at the cooler end of a boat of liquid silicon. The boat is gradually moved out of the heating coils, and the liquid crystallizes as a single crystal. In *b*, the most common method, called Czochralski crystal pulling, a small crystal on the end of a rod is lowered into the melt and withdrawn while being rotated. The diameter of the crystal is governed by the speed of withdrawal. In *c*, a floating zone of liquid starts at one end of the sample. As it moves up through the solid, crystallization begins and continues behind the floating zone.

In each case, how does the procedure avoid dendrite growth, lead to a single crystal, and produce some refining?

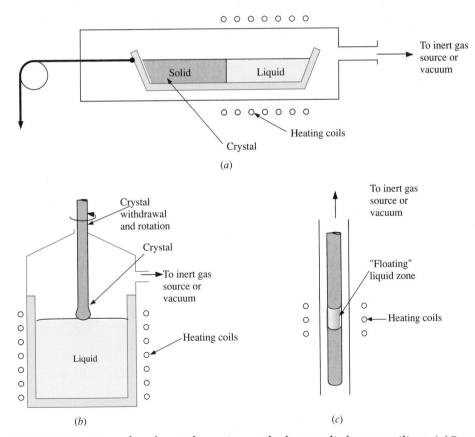

FIGURE 5.8   Examples of crystal-growing methods as applied to pure silicon. (*a*) Boat method. (*b*) Czochralski crystal pulling. (*c*) Floating zone method.

[Fleming, *Physics*, © 1978, Addison-Wesley Publishing Co., Inc., Reading, Massachusetts. Fig. 1.1. Reprinted with permission of the publisher.]

Answer   There is a steep, constant thermal gradient at the interface that prevents constitutional supercooling. The nucleation by a single crystal of the same material provides a larger radius than any other available nucleus and the greatest driving force for crystallization. In each case the solid is either withdrawn from the liquid or directionally solidified, leaving impurities in the remaining liquid or in the last solid to form. These impurities can then be discarded. The starting silicon melt is also quite pure.

5.5   Control of Liquid-to-Solid Reactions

We have already mentioned the nucleation of primary silicon crystals during solidification of aluminum–silicon alloys. Another example of a change of great engineering importance is the change in the shape of graphite particles in iron alloys. In normal *gray cast iron*, the graphite precipitates in flake-like particles (Figure 5.9a). The matrix of the material is similar to that of steel with good ductility, but the flakes act as notches, so that the plastic elongation of the material as a whole is below 1%. Adding a small amount of magnesium (0.05%) causes the graphite to crystallize in the form of spheres

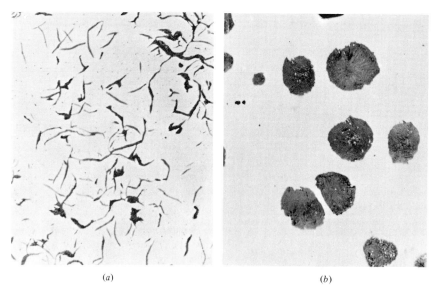

(a)   (b)

FIGURE 5.9   (a) Flake graphite in gray cast iron; 100 × , unetched. (b) Spheroidal graphite in ductile cast iron; 100 × , unetched. Both irons have approximately the same analysis (3.5% carbon, 2.5% silicon), but the graphite shape is spheroidal in (b) because of the presence of 0.05% magnesium. As a result, the strength and ductility of this iron are more than twice as great as the strength and ductility of the iron shown in (a).

(Figure 5.9*b*). The elongation is raised to as high as 20%, and the strength is increased several times. This material, which is discussed in detail in Chapter 9, is known as *ductile cast iron* or as nodular iron or spheroidal graphitic iron.

EXAMPLE 5.4 *[EJ]*

Figure 5.3 is a representation of a phase transformation as a function of time. What other variable should be considered in the discussion of nucleation and growth?

Answer Temperature. We have already pointed out that nucleation increases as the temperature is decreased (undercooling). On the other hand, growth depends on diffusion, and we know that diffusion rates decrease as the temperature is decreased. This is shown schematically below.

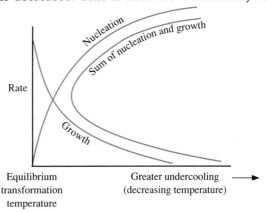

The C-shape of the curve that represents the sum of the individual nucleation and growth curves suggests that there exists a temperature (where the nucleation and growth rate curves cross) that gives the new phase in the shortest period of time. The concept becomes more important if we *do not* want the equilibrium low-temperature phase but rather wish to retain the high-temperature phase at low temperatures. In Section 5.6 on age-hardening reactions, we will discuss nucleation and growth control in the solid state.

5.6 Control of Solid-State Precipitation Reactions

As examples of the effect of solid-state precipitation reactions on the properties of a material, let us discuss the very important control of precipitation of iron carbide in steel first and then follow with the age hardening of aluminum alloys by the $CuAl_2$ phase.

There are two heat-treatment cycles for the hardening of steel. Both refine the dispersion of iron carbide, but the mechanisms are quite different.
The first is a time-dependent nucleation and growth reaction; the second is a
rapid shear-like change in structure. Let us consider both these reactions in a
typical steel containing 0.77% carbon.

Pearlite Formation

Figure 5.10 shows the pertinent section of the iron–iron-carbide phase diagram. Note that at 0.77% carbon a single phase, γ, is present above 1341°F
(727°C). This consists of FCC iron with all the carbon in interstitial solid solution. The diagram shows that under equilibrium conditions, on cooling below 1341°F (727°C) γ changes to two phases — α, which is essentially pure,
soft BCC iron, and iron carbide, which is hard. This reaction is called a *eutectoid reaction*. The two-phase mixture is called *pearlite*. Under slow-cooling
conditions the dispersion of plates of Fe_3C in α is coarse (Figure 5.11a), and
the hardness low, HV 230. If we heat a piece of the same steel into the γ range
and then quench it in a bath of liquid lead or salt at 1000°F (538°C), the reaction will take place over a period of less than 1 min, and we will get a much
finer dispersion (Figure 5.11b) with a higher hardness, HV 300. We can obtain

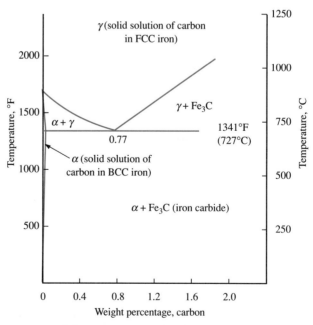

FIGURE 5.10   Section of the iron–iron carbide phase diagram

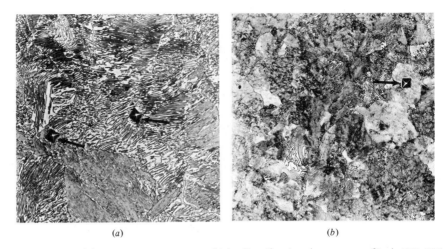

(a)   (b)

FIGURE 5.11 (a) Coarse iron–iron carbide distribution (coarse pearlite), HV 196 to 266; 500 × , 2% nital etch. (b) Fine iron–iron carbide distribution (fine pearlite), HV 270 to 320; 500 × , 2% nital etch. Distribution of the finer carbide results in greater hardness.

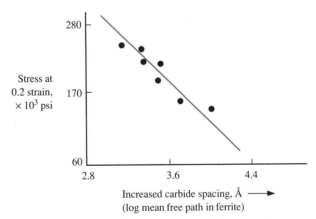

FIGURE 5.12 The effect of carbide spacing on yield strength

(Gensamer et al.)

still finer dispersion and further increases in hardness and strength by transforming the γ to α plus carbide at still lower temperatures. Figure 5.12 shows the effect of these changes on the properties of the steel. The strength increases as the spacing between carbide particles decreases.*

---

*The average spacing is called the *mean free path*. We get a straight-line relationship when we plot the $\log_{10}$ of the mean free path against yield strength.

Martensite Reaction

The second heat-treatment cycle is more complex. If the 0.77% carbon steel, originally all γ, is quenched quickly to below 400°F (205°C), a very rapid shear-like change in structure occurs. Instead of α plus carbide, a single new phase, called *martensite*, is formed. The structure is BCT (body-centered tetragonal), which can be considered an elongated BCC (Figure 5.13). The carbon is still distributed in atomic form as in γ, but in a supersaturated solid solution. When the martensite is reheated at a relatively low temperature [400°F (205°C), for example], the carbon precipitates as very fine iron carbide particles, and the tetragonal iron structure changes to the normal BCC structure of iron. By using this processing we can obtain the finest carbide dispersions. This reaction involves two steps: quenching to low temperature to form martensite and reheating (called "tempering") to form fine iron carbide inferrite.

In both the pearlite and the martensite reactions, the C-shaped curve discussed in Example 5.4 becomes important. Obviously, we must quench fast to avoid pearlite if martensite is to be formed. This point will be discussed in depth in Chapter 8.

5.7 Dispersion Hardening and Age Hardening

We have already seen several examples of *dispersion hardening*. In the aluminum–silicon alloy used for engine blocks, we have a dispersion of hard silicon particles that give added strength and wear resistance. Actually we can call this a micro-composite of silicon and aluminum. In another case (just de-

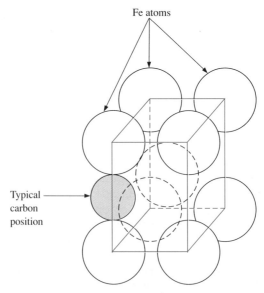

FIGURE 5.13   Body-centered tetragonal structure of martensite

scribed), we saw plates of carbide dispersed in iron as pearlite and, in the case of tempered martensite (reheated martensite), fine spheres of carbide. In all these cases, small, hard particles block the movement of dislocations and thus raise the yield strength. For a given volume of precipitate, the amount of blocking increases with fineness. This is why refining treatments are used in the liquid–solid reactions and in solid–solid precipitation.

Age hardening, though it also involves the formation of fine precipitates, deserves separate attention because the precipitate is *coherent* with the matrix. By "coherent," we mean that certain matrix planes match specific planes of the precipitate quite closely in atom spacing and are continuous throughout the precipitate. This small difference in spacing produces a strain field around the precipitate particle, which causes blocking of dislocations for some distance from the actual precipitate. In this way the effect of the precipitate is greatly enhanced. In general, coherency develops when there is a specific relation between the spacing of certain planes and directions in the precipitate and the matrix. Usually the strain field can exist only while the precipitate is small. When the precipitate grows larger and shears to form its equilibrium space lattice, only dispersion hardening remains. In this case, the precipitate particles themselves inhibit dislocation movement, but there are no large elastic strain fields around them.

Age-Hardening Treatment

The objective of this treatment is to begin with a coarse, two-phase structure and end with a finely dispersed second phase that provides higher strength and hardness. This mechanism is the most important heat treatment for strengthening aluminum, magnesium, copper, and other nonferrous alloys.

The treatment consists of three steps, as shown in Figure 5.14 for an aluminum–copper alloy.

1. Heat to dissolve the second phase. Note that a solvus line showing increased solubility of the second component in the α phase is needed. We are using an example in which all of the second phase is dissolved, but the process can be applied to an alloy in which some second phase remains. This is called the *solution heat treatment*.
2. Quench to room temperature. (Sometimes hot water is used as a quenchant to reduce stresses due to uneven contraction during quenching.) The objective of this step is to prevent precipitation of the second phase during quenching.
3. Age the part at room temperature (*natural aging*) to develop a precipitate or at a slightly elevated temperature for a controlled time (*artificial aging*). The aging step is critical in developing maximum strength.

In the case of aluminum–copper alloys, several distinct features of precipitation in step 3 have been observed, and interesting graphs of hardness versus time have resulted (Figure 5.15).

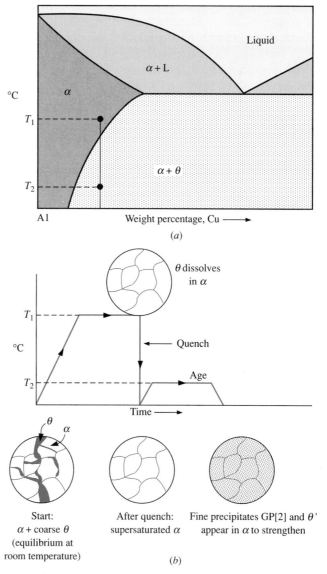

FIGURE 5.14   Age-hardening heat treatment. (*a*) Phase diagram. (*b*) Chart of heat treatment and microstructures.

In the sample aged at 130°C (266°F), the first increase in hardness — marked GP[1] — is due to the formation of plate-like regions higher in copper content than the average and only 1 nm thick and 10 nm in diameter. This results from *homogeneous* nucleation, which distorts the lattice and makes dislocation movement difficult. A coherency stress or strain develops because the platelets are coherent with the matrix. The stress regions are called Guinier–Preston (GP) zones in honor of their discoverers.

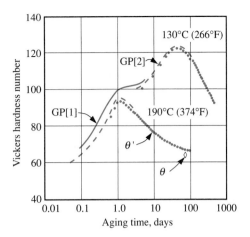

FIGURE 5.15   Correlation of structure and hardness of Al-4% Cu alloy aged at two temperatures

[J. M. Silcock, T. J. Heal, and H. K. Hardy, *Journal of the Institute of Metals*, 1953–1954. Used by permission.]

With further heating, the GP zones increase in size to about 10 nm thick and 100 nm in diameter, at which point they are called GP[2] zones. This precipitate results in maximum strengthening because of the close spacing. For movement to occur, a dislocation must cut these particles or pass with difficulty through the adjacent strain field.

At higher temperatures and longer times, a new precipitate is formed. It is called θ' because it has the composition of θ (CuAl$_2$) but has different unit cell dimensions as a result of its coherency with the matrix. This precipitate forms by *heterogeneous* nucleation at matrix dislocations. These particles grow, and the smaller GP[2] particles disappear. Appreciable strengthening is still produced by the θ' because it is still coherent with the matrix. Finally, θ' particles grow to a size at which coherency with the matrix is lost and only dispersion hardening is left. This is called *overaging*. The higher hardness encountered with GP[2] and θ' is due to the cutting through of the particles and their strain fields by dislocations, which requires more energy than passing through the strain-free region around θ particles. The lower hardness accompanying the θ dispersion is also caused by the lower number of larger θ particles compared to the larger number of smaller θ' in the preceding precipitate.

Several age-hardening systems operate in typical commercial alloys, and it is necessary to develop time–temperature curves for aging, as shown in Figure 5.16.

It should be added that in commercial casting alloys, the content of copper is beyond the end of the α field (Figure 5.14), and the θ that is undissolved acts as a dispersion to provide additional hardness. However, there are still advantages of a coherent precipitate originating from the dissolved θ due to the large strain field compared to that of dispersion hardening. These effects are summarized in Figure 5.17.

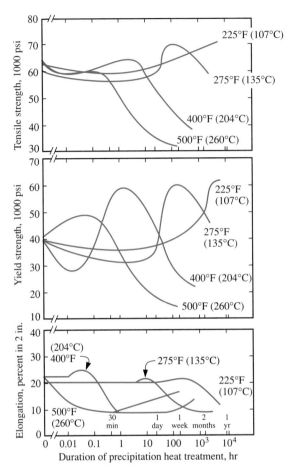

FIGURE 5.16   Effect of age hardening on the mechanical properties of a 4.5% copper–aluminum alloy (2014)

[Reprinted with permission from *Metals Handbook*, 8th ed., Vol. 2, *Heat Treating, Cleaning and Finishing*, American Society for Metals, Metals Park, OH, 1964.]

5.8 Summary of Strengthening Mechanisms in Multiphase Metals

As we pointed out at the beginning of the chapter, the way to obtain desired mechanical properties in a material is to control the nature, size, shape, amount, distribution, and orientation of the phases. This is accomplished via liquid-to-solid reactions or solid-to-solid reactions, which may be characterized as follows:

A. Liquid-to-solid reactions
   1. Variations in grain size
   2. Eutectic
   3. Change in phase shape (for example, graphite in iron alloys)

(a) Coherent precipitate

(b) Incoherent precipitate

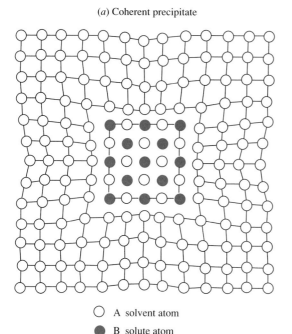

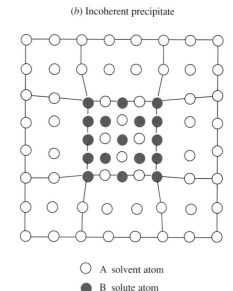

○ A solvent atom

● B solute atom

1. Low elastic strains
2. Dislocations at the interface

○ A solvent atom

● B solute atom

1. High elastic strain
2. Strain extends for large distances

FIGURE 5.17 (*a*) Large strain fields around a coherent precipitate. Note distortion of 4 atom layers. (*b*) A small strain field when a precipitate is incoherent (as in dispersion hardening or overaging). Note distortion of only one atom layer around precipitate. Less hardening than in (*a*).

B. Solid-to-solid reactions
1. Eutectoid
2. Order–disorder solid solutions*
3. Martensite
4. Age hardening
(Phase shape and size may be superimposed on the foregoing variables.)

The differences between age-hardening and martensite reactions may be confusing. The reactions are shown schematically below. The important difference is that a new crystal structure is formed in martensite. Both processes use quenching and reheating to give a more desirable distribution of the two phases. Chapter 8 treats the martensite reaction in more detail.

---

*An ordered solid solution has preferred positions for the solute atom in the solvent lattice.

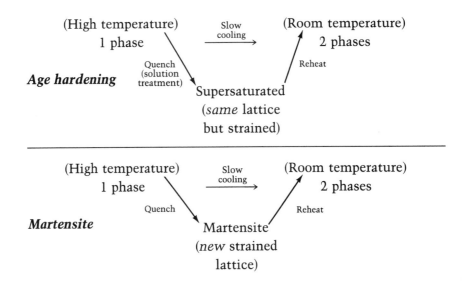

(High temperature)   Slow   (Room temperature)
  1 phase   cooling →   2 phases

Age hardening   Quench
                      (solution
                      treatment)   Supersaturated
                                      (*same* lattice
                                      but strained)

(High temperature)   Slow   (Room temperature)
  1 phase   cooling →   2 phases

Martensite   Quench   Reheat
                              Martensite
                              (*new* strained
                              lattice)

EXAMPLE 5.5 *[EJ]*

Four alloys have been identified on the accompanying phase diagram. Indicate the most probable strengthening mechanism(s) for each.

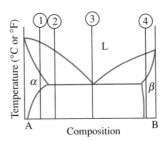

Answer   Alloy ①: Age hardening is likely to be the best strengthening method because of decreasing solid solubility. If β is soft and ductile, a slow-cooled alloy may be work-hardened.

Alloy ②: At room temperature the microstructure would be composed of α plus a small amount of eutectic. The α may still be age-hardened after solution treatment just below the eutectic temperature.

Alloy ③: Strengthening is by the eutectic reaction. In general the ductility of eutectics is low. Therefore work hardening is not used. The α-phase portion of the eutectic does not generally respond to age hardening, because after solution treatment and aging, the β phase grows larger rather than forming new β within the α of the eutectic. In other words, diffusion distances between the α and β are very short, and growth takes less energy than nucleation. In instances where the eutectics are very coarse with larger diffusion distances, there may be some response to age hardening.

Alloy ④: Because the alloy is single-phase and already solid-solution–strengthened, we may only cold-work the alloy to increase its strength.

5.9 Multiphase Reactions in Ceramics and Polymers

A number of reactions in ceramics are similar to those in metals. In liquid–solid reactions we have the controlled precipitation of solids, the formation of glass ceramics, and the reconstructive and displacive reactions, which are analogous to pearlite and martensite reactions.

The liquid–solid reactions in ceramics display more variety than those in metals, because it is possible to retain any desired portion of the glassy phase. We will investigate the crystallization of low-expansion phases such as β-spodumene from a glassy matrix to produce high-strength materials resistant to thermal shock in Chapter 11.

To illustrate the difference between reconstructive and displacive transformations, it is helpful to recall that pearlite grows from the parent austenite by a process of nucleation and growth, forming a new structure; this is similar to a reconstructive transformation. By contrast, martensite is formed by a small shift of atom positions, as in a displacive transformation. A schematic sketch of the two types is shown in Figure 5.18. Note that in *displacive transformation* the new structure is obtained by cocking the unit cells of the parent lattice, whereas in *reconstructive transformation* we have to start all over to make the new structure. The reason why it is important to distinguish between the two types is that the displacive (shear) type occurs rapidly and the reconstructive very slowly—far slower than the austenite-to-pearlite reaction in steel even though two new structures are formed.

Another reaction that is quite important in ceramics is transformation toughening. Earlier, as an example of a phase diagram, we used the zirconia–calcia system (Figure 4.13). If enough CaO is added, we get stabilized—that is, cubic—zirconia. This material has low fracture toughness, but if we add a smaller amount of CaO, we obtain a mixture of the stabilized cubic phase and the unstable monoclinic phase. When a crack develops, the energy added to the material ahead of the crack causes transformation to a more voluminous phase, which stops the crack by introducing compressive stress.

In the case of polymers, we have two principal methods of controlling multiphase dispersions: the degree of crystallinity and the formation of copolymers. A copolymer may be thought of as analogous to an alloy in metals where relative miscibility becomes important.

We can control the relative amounts of glassy (amorphous) phase and crystalline phase by cooling rate or by thermal treatment after a component

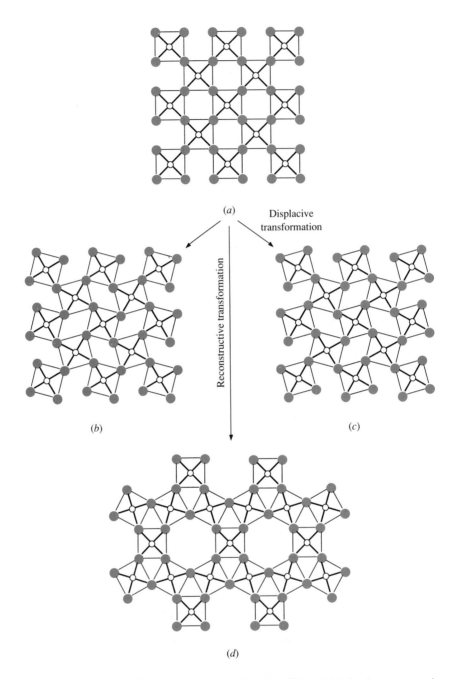

FIGURE 5.18 (*a*) Open form of structure, showing (*b*) and (*c*) displacive transformations into collapsed forms and (*d*) reconstructive transformation into a basically different form

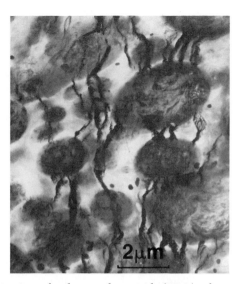

FIGURE 5.19 A dispersion of polypropyleneoxide (PPO) spheres in 6/6 nylon (white). Under stress, crazing-like slip begins in the PPO, which leads to debonding at the interface with the nylon matrix and ultimate cracking of the nylon.

Photo courtesy of Professor Alfred F. Yee, University of Michigan

is formed. The crystalline polymer is generally stronger. We would also find that crystallinity is suppressed when we produce a copolymer due to the irregular molecular structure characteristic of the copolymers.

However, copolymers can also be quite strong. Figure 5.19 shows that when two immiscible polymers are mixed mechanically, an emulsion of the two polymers can be produced. The ultimate strength is dependent upon the interfacial bonding between the two polymers. An important variation is to produce an emulsion of A and B polymers first and then use the emulsion to bond together separate regions of A and B. Without the emulsion there would be little interfacial strength between pure A and pure B regions.

Summary

The subject of this chapter is controlling the shape and size of the second phase in the matrix under nonequilibrium conditions in order to obtain enhanced properties. Addressing metals in the liquid state, we investigated the use of inoculating agents to control the precipitation of the second phase in aluminum alloys and ductile iron. Then we discussed dispersion, martensite, and age hardening in the solid state. We found that similar structural displacement takes place in ceramics and that an additional phenomenon, transformation toughening, may be exploited. We will soon expand our study to interface bonding in composites and to the effects of processing variables on the properties of polymers.

Definitions

Age hardening   A three-step process consisting of heating an alloy to dissolve all or part of a second phase in the matrix phase, quenching to retain the solute in a supersaturated solid solution, and then allowing the second phase to precipitate in fine particles that are coherent with the matrix.

Coherent precipitate   The stage intermediate between a supersaturated solvent phase and a distinct precipitating second phase. The precipitate has no distinct phase boundary but has a region of high elastic energy at the interface.

Constitutional supercooling   Phenomenon whereby a portion of a material is liquid even though it is below its equilibrium liquidus temperature. This is due to a severe chemical gradient at the solid–liquid interface.

Coring   A difference in chemical concentration between the center and the edge of a microscopic grain.

Dendrite   A fern-like crystal growing from a solid interface into a liquid during solidification.

Dispersion hardening   Precipitation process in which the phase coming out of solution is not coherent with the matrix.

Displacive transformation   Small, angular displacement between ions that produces new dimensions within the same crystal structure.

Ductile cast iron   Spheroidal graphite in an alloyed iron matrix.

Gray cast iron   Flake graphite in an alloyed iron matrix.

Heterogeneous nucleation   The development of a new phase by the addition of foreign material (seeding).

Homogeneous nucleation   The development of a new phase by the formation of nuclei of the new phase from the parent phase.

Martensite   A metastable, body-centered tetragonal phase formed by quenching γ iron containing carbon. γ iron is called *austenite*.

Matrix   The continuous phase in a two-phase material. The matrix usually forms the "background" of the microstructure.

Nucleation   The development of nuclei that act as centers of crystallization for a new phase.

Pearlite   A two-phase mixture of α iron and iron carbide produced when γ iron is transformed by a eutectoid reaction. α iron is called *ferrite*.

Reconstructive transformation   Formation of a new crystal structure via the complete rearrangement of ions.

Segregation   The development of a concentration gradient or nonequilibrium structure as a result of freezing under nonequilibrium conditions.

Shrinkage cavity   A void produced within a casting or ingot because the solid, which is denser, cannot fill the mold volume originally occupied by the liquid.

Undercooling   Condition wherein a high-temperature phase temporarily exists below its equilibrium transformation temperature.

Problems

5.1 *[EJ]* A 60% Cu, 40% Zn alloy is melted and cast in a sand mold, one face of which is graphite (heat is rapidly removed by the graphite). Tensile samples are cut parallel to, but at different distances from, the chilled face, and it is found that the mechanical properties are not the same throughout the casting. Explain how the properties change as you move away from the chilled surface.(Sections 5.1 through 5.4)

5.2 *[EJ]* It is possible to purify a metal by using the knowledge that the composition of the first solid to form is not the same as the composition of the liquid from which it first forms. The method is called "zone refining." Use a phase diagram to explain how this might work. (Sections 5.1 through 5.4)

5.3 *[ES]* A critical radius size of r^* is given for a spherical nucleus in Figure 5.2. Say we are to obtain a fine-grained material. Show by a sketch what must happen to the net free-energy curve in Figure 5.2. What primary factor can cause the net free-energy curve to shift? (Sections 5.1 through 5.4)

5.4 *[ES/EJ]* What is the role of *seed crystals*, as in cloud seeding? (Sections 5.1 through 5.4)

5.5 *[ES/EJ]* Figure 5.4 refers to segregation that can occur on a microscopic level. Explain how *macrosegregation* can occur in a large ingot, and speculate about the possible meaning of the terms *negative segregation* and *positive segregation*. (Sections 5.1 through 5.4)

5.6 *[ES/EJ]* Explain why *overheating* in a hot-working operation might be undesirable. (Sections 5.1 through 5.4)

5.7 *[ES/EJ]* Why might the ease of producing a sound casting of a copper–zinc alloy be dependent on the zinc content? (Sections 5.1 through 5.4)

5.8 *[ES/EJ]* Example 5.3 discusses several methods for producing pure silicon single crystals. Why would these techniques generally begin with silicon that has been purified by other methods? (Sections 5.1 through 5.4)

5.9 *[ES/EJ]* Indicate why computer models that use phase diagrams as a data base for solidification studies may incorporate errors. (Sections 5.1 through 5.4)

5.10 *[ES/EJ]* In this portion of a simple phase diagram, shaded areas indicate those alloys that are considered "commercial." For each group (1, 2, and 3), indicate the principal hardening or strengthening mechanism. (Sections 5.5 through 5.9)

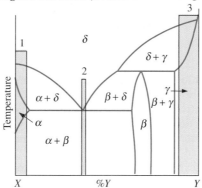

5.11 *[ES/EJ]* An example of phase shape control in a liquid–solid reaction is given for graphite in cast irons in Figure 5.9. What might be the role of *grain refiners* as added to other alloy systems? (Sections 5.5 through 5.9)

5.12 *[ES]* In chemical kinetics, the equation $1/t = Ae^{-B/T}$ can be used to represent the initiation of a reaction (A and B are constants, t is time, and T is absolute temperature). Indicate how two such equations can be used to represent the graphical interpretation of kinetics shown in Example 5.4. (Sections 5.5 through 5.9)

5.13 *[ES]* From the data given in Figure 5.12 in the text, develop an equation that relates yield stress to iron carbide spacing in steel. (Sections 5.5 through 5.9)

5.14 *[ES/EJ]* Why are some alloy systems that exhibit decreasing solid solubility, and are hence age-hardening candidates, nevertheless not used commercially? (Section 5.5 through 5.9)

5.15 *[EJ]* Why is a longer time required to reach the maximum hardness at lower aging temperatures? (See Figure 5.16.) Why is the maximum hardness higher for lower aging temperatures? (Sections 5.5 through 5.9)

5.16 *[ES/EJ]*

   a. Which of the alloys (1, 2, 3, and/or 4) in the accompanying figure could be strengthened by age hardening? [Assume that β forms a coherent precipitate in α.] (Sections 5.5 through 5.9)

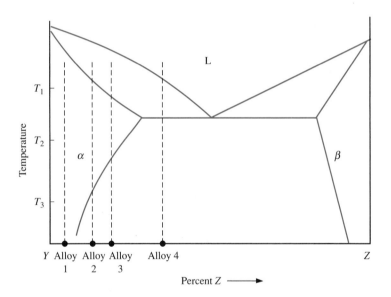

   b. In which *one* of these alloys (give the number) could the greatest amount of element Z be dissolved in solid solution α?

   c. Select a treatment from the following list that would represent the $T5$ condition, which is defined as "artificially aged only," for alloy 3: (1) heat to T_1, hold 3 hr, water quench, reheat to T_3, hold 4 hr, air cool; (2) heat to T_3 for 4 hr, air cool; (3) heat to T_2, hold 3 hr, water quench, reheat to T_3, hold 4 hr, air cool.

5.17 *[ES/EJ]* Indicate whether the following statements are correct or incorrect and justify your answer. (Sections 5.5 through 5.9)

   a. In age hardening a 2014 aluminum–copper alloy (see the data given in Figure 5.16), an aging temperature of 400°F is more likely to be used than one of 275°F.

   b. When a 1.5% silicon in aluminum alloy is cast, it is possible to have some eutectic ($\alpha + \beta$) present in the final room-temperature microstructure.

5.18 *[E]* How might we obtain a *glass* ceramic or polymer via modification of its thermal treatment? [*Hint:* It will be necessary to use Example 5.4 in conjunction with phase diagrams. (Sections 5.5 through 5.9 and Sections 4.10 through 4.11)

5.19 *[ES/EJ]* Complete each sentence by writing the correct number, word, or abbreviation in the blank at the right. The abbreviations I, N, and D stand for the following: I = increase, N = not change significantly, D = decrease. (Sections 5.5 through 5.9 and review of earlier chapters)

Example: Water has a melting point of _____. When it is heated, its temperature will (I/N/D).

0°

I

1. Cold working a metal causes:
   a. its yield strength to (I/N/D)
   b. its hardness to (I/N/D)
   c. its ductility to (I/N/D)
   d. the concentration of dislocations in it to (I/N/D)
   e. the mobility of dislocation in it to (I/N/D)
   f. its elastic modulus to (I/N/D)

2. The curve shows the hardness of a cold-rolled copper alloy after annealing for 1 hr at various temperatures.

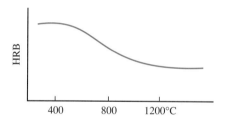

   a. To achieve stress relief without appreciable recrystallization, this alloy should be heated for 1 hr at _____ °C.
   b. To achieve complete recrystallization without appreciable grain growth, this alloy should be annealed for 1 hr at _____ °C.

3. The addition of up to 30% zinc as an alloying element in copper causes:
   a. the melting point to (I/N/D)
   b. the cost to (I/N/D)
   c. the yield strength to (I/N/D)
   d. the critical resolved shear stress to (I/N/D)

4. Three elements that form interstitial solid solutions in iron are carbon, _____ , and _____ .

PART II

Metals and Alloys

Now that we have a general view of the role of metals and alloys in the broad field of materials, it is time to focus on their structures, properties, and applications. For example, from what has been said so far, you could not be expected to specify a steel for an I beam in an office building or to distinguish among specifications for metal fasteners.

So far we have emphasized that the properties depend on the structure, but we haven't said anything about how we get the structure. This is the province of metal processing—casting, rolling, forging, forming, powder metallurgy, welding, and machining. The structure, and therefore the properties, are greatly affected by those operations. An intergral part of writing a specification for an alloy is specifying the processing. We will therefore begin with a short investigation of the processes. Then we will take up the nonferrous alloys, and finally the ferrous alloys. You should realize that a good background in metallurgy is essential for understanding not only the many new alloys but the metal matrix composites (MMC) as well.

Now as to the problem of surveying the thousands of existing metal compositions. The first simplification is to divide the alloys into two groups: nonferrous and ferrous. The first group will have little ferrous iron in the analysis. Here we will find large families, such as aluminum-base, copper-base, and so on. By base we mean that the element named is the principal constituent in the analysis. For example, if we have a 96% Al 4% Cu alloy, this is called aluminum base, alloyed with copper. On the other hand, if we find a 95% Cu 5% Al alloy, we consider it copper base, alloyed with aluminum. The principal nonferrous alloy systems are aluminum, magnesium, copper, nickel, titanium, and zinc. These are covered in Chapter 7.

Turning to the ferrous, or iron-base, alloys we can make two divisions: the plain carbon and low-alloy steels (less than 5% alloy); and the high-alloy steels and cast irons. These are covered in Chapters 8 and 9 respectively.

It is interesting to note from the tables of specifications in these chapters which alloys are available in the cast vs. wrought conditions. This can greatly affect the shapes in which they can be obtained, and is discussed further in Chapter 23.

259

CHAPTER 6

Processing Metals into Components

These three production techniques for turbine blades illustrated here show the important effect of processing method on structure and properties.

In the first case, the conventional production technique, liquid metal is poured into a preheated ceramic mold (alumina) and the metal solidifies in a random crystallization pattern.

In the second case, the mold is chilled at the base and insulated at the top. As a result, columnar crystals grow from the base. Because there are no grain boundaries across the blade, resistance to rupture by centrifugal force at high temperatures is improved.

In the third case, the crystallization in the constricted region is only wide enough for one crystal to grow. This orientation is encountered throughout the blade. The elimination of grain boundaries results in a further increase in strength at elevated temperatures.

6.1 Overview

At this point we have a general understanding of how atoms are bonded to form structures, how these structures respond to stress and temperature, and how they may be changed by working and heat treatment to give enhanced properties in test bars. Before we go deeper into the test-bar properties of metals, it is time for a story to explain the need for this chapter.

In our research laboratory at ABEX, we had just developed a new alloy called austenitic ductile cast iron that offered a unique combination of strength, elongation, and machinability. We requested a conference with the Engineering Department of one of the major aircraft companies to discuss the use of this material. We gave a rather thorough presentation with elegant slides of microstructures and tables of test-bar data. After all of this, the vice president of engineering turned to a dour Scot in charge of aircraft design seated in the corner of the room and said, "Angus, what do you think?" "Weel," Angus said, "I canna fly test bars, kin ye gie me c-a-a-stings?"

Angus's concern was well-founded. Many new alloys with outstanding properties in test bars literally tear themselves apart when cast into complex shapes or subjected to forging or rolling.

In order to be able to bridge the gap between a desirable structure and producing an engineered component, you must understand the different processing operations. In this chapter we will take up the methods that are used for metals. Later chapters will address the processing of ceramics and polymers.

The unit operations common to all engineering materials are as follows:

1. *Casting* (transformation from liquid to solid)
2. *Deformation* (shape control via forming)
3. *Particulate processing* (the processing of powders)
4. *Joining* (welding or the use of adhesive)
5. *Machining* (removal of material for geometry control)
6. *Thermal treatment* (heat treatment or thermal processing)
7. *Finishing — quality control** (surface treatment)

These seven operations are techniques that may be applied to any component to generate a salable product. However, some of the processes may be inappropriate, such as the cold forming of ceramic products. In the case of metals, it is also important to realize that over 98% of components begin as castings; the rest are made by powder metallurgy and other processes. It is usually necessary to employ more than one process on the same component, yet each process has its own unique technology.

We will now cover these unit operations separately. We will begin the description of each technique with a simple example of the process and intro-

---

*Quality control should be a state of mind that permeates *all* of the unit operations. For ease of discussion, however, we have included it with the finishing operations to show how a final check on quality may reveal defects that originated early in a processing sequence.

duce any special vocabulary. This will be followed by identification of principles introduced in the first five chapters that are relevant to the process; then any new principles will be developed as required. Finally, we will explain potential defects that may arise as a consequence of the processing.

The emphasis here will be on the processing of metals and the relationship of processing to structure and product integrity. The use of these concepts for other classes of materials will be addressed in subsequent chapters.

6.2 Casting

We begin with *casting* because most metals are initially melted and cast into shapes, even though they may finally end up in a wrought form (Figure 6.1). In these cases, the cast shape may be an ingot or slab that receives further processing.

The following "hands-on" description of casting a frying pan will provide a feel for how castings can be produced to a more or less final shape that requires only minimal further processing, such as machining.

In this elementary example, the casting itself is used as a pattern, although we will not obtain a true duplicate because of the phenomenon of metal shrinkage. (This method is crude, but in an emergency it can be used to replace a failed part that can be glued together and used as a pattern.)

We begin by positioning the pan on a plate Figure 6.2*a* and placing an open box called a *drag flask* over the pattern. Next we prepare a green* sand mixture of approximately 93% silica sand, 4% clay (bentonite), and 3% water. Because of the clay–water–silica bond, this mixture has adequate strength to form a relatively firm surface when rammed. After the flask is rammed full (Figure 6.2*b*), the surface is leveled with a simple bar called a *strike*; then a *bottom plate* is positioned on the struck surface and the mold is rolled over. The sand does not fall out of the mold box because of its strength and because of the lip of the flask, which retains the sand.

Next the *parting surface* is cut with a trowel and finished with a smaller tool called a *slick* (Figure 6.2*c*). It is important to design the parting surface such that the pattern can be *drawn* later — that is, pulled cleanly from both halves of the mold without tearing the sand surface.

A second flask called the *cope flask* is placed in position over the first one (Figure 6.2*d*). Pins that project from the upper half fit into bushed holes to fix the cope in position over the drag. A channel is needed to introduce liquid metal into the mold. This channel is called the *gating*. Loose pieces of wood were already positioned in the drag to form this channel, as shown in Figure 6.2*a*. However, the vertical portion of the gating (*downsprue*) is in the cope (Figure 6.2*d*). Once the cope is in position, the surface of the drag is

---

*The sand is called green not because of its color but because of its water content!

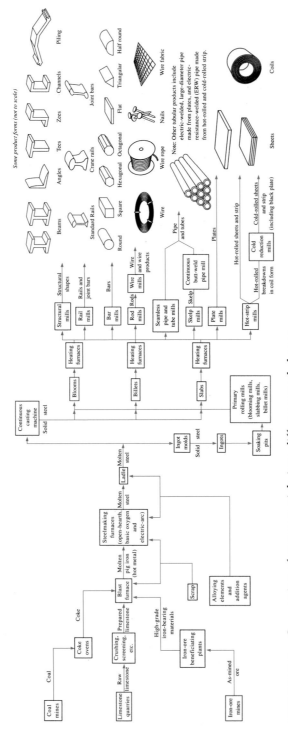

FIGURE 6.1   Conversion of raw materials into different steel shapes

[Courtesy Association of Iron and Steel Engineers, *The Making, Shaping and Treating of Steel*, 10th Edition, p. 2, Fig. 1-1.

dusted lightly with a ceramic powder to prevent the sand of the drag surface from bonding to the sand of the cope.

The cope is then rammed, and the upper surface is leveled with a strike. Next vents are made to the inner cope surface by piercing the mold with a 0.1 in. (2.5 mm) diameter wire. This prevents mold gases from accumulating in the top of the mold and retarding filling. Finally, a *pouring basin* is cut around the vertical portion of the gating (Figure 6.2*e*). The pattern for the downsprue has been drawn upward because its draft is designed to taper downward for better metal flow.

The cope is then carefully separated from the pattern and the drag. Next the pattern is removed from the drag with the aid of a magnet* (Figure 6.2*f*). The gating is removed with a wood screw.

After the casting cavity has been inspected, the flask sections are re-assembled accurately with the aid of the pins and bushings and clamped together. The mold is poured with liquid cast iron at approximately 1370°C (2500°F) (Figure 6.2*g*).

After 30 min the mold is shaken out (Figure 6.2*h*). Then the casting is sandblasted, the gating is cut off, and the casting is ground smooth (Figure 6.2*i*). The casting will be approximately 1% smaller than the pattern because of the shrinkage of the metal from the solidification temperature, about 1150°C (2100°F) to room temperature.

6.3 Review of Principles Related to Casting

The following concepts were introduced in earlier chapters. Their relationship to the casting process requires a brief review.

Phase Diagrams

We commented in our treatment of the aluminum–silicon system that the presence of β phase gives a material of lower ductility than results when α aluminum only is present. This suggests a cast product because processing by deformation may result in failure. Therefore, a phase diagram can predict the occurrence of less ductile phases that may make the casting process the only reasonable technique. A similar conclusion is reached for the gray cast iron frying pan shown in Figure 6.2.

Further review of the aluminum–silicon diagram (Figure 6.3) reveals other features that are important. The pouring temperature is dependent on the alloy composition and is selected as the *superheat* above the liquidus necessary to fill the mold cavity. A common *fluidity* requirement may be 50–100°C for large sections and possibly 300°C for thin castings.

Low pouring temperatures result in the mold not filling, whereas high temperatures may cause the liquid metal to react with mold surfaces. The reaction product can then result in *nonmetallic inclusions* in the casting. This

---

*Generally a lifting screw is used.

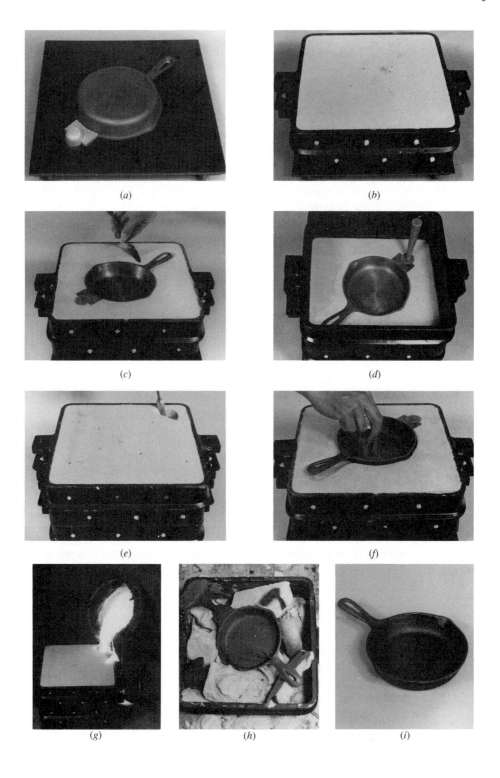

(a)

(b)

(c)

(d)

(e)

(f)

(g)

(h)

(i)

FIGURE 6.2   (*a*) Frying pan pattern positioned on a bottom plate with a wood gating system attached. (*b*) Drag flask filled with sand and leveled (struck), ready for roll over. (*c*) Parting surface being cut with a slick in order to make possible easy removal of the pattern. (*d*) Cope flask in place over the pattern. The vertical portion of the gating (downsprue) is centered over the gating well. (*e*) Cutting a pouring basin in the cope. The sprue has been removed. Loose sand is blown out of the sprue after removal of the cope. (*f*) Cast iron pattern being removed from the drag with the aid of a magnet. (*g*) Casting being poured into clamped flasks. (*h*) Casting shaken out of mold. Note that some metal ran up into one of the vent holes at the handle. (*i*) Casting after removal of the gate, grinding, and sandblasting.

suggests that phase diagrams for the oxides commonly used as mold materials are also important.

Finally, note that between the liquidus and the solidus there is a mushy structure. As a general rule, the wider the temperature range at which both liquid and solid exist, the more difficult it is to produce a casting devoid of shrinkage. We will discuss this point further in Section 6.4.

Diffusion/Segregation

The unsteady-state diffusion model necessary to attain a homogeneous structure in Chapter 3 predicts that long times are necessary. Castings seldom cool slowly, and a homogenizing heat treatment may then be required. Remember

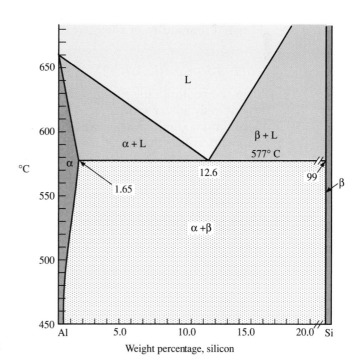

FIGURE 6.3
The aluminum–
silicon phase diagram

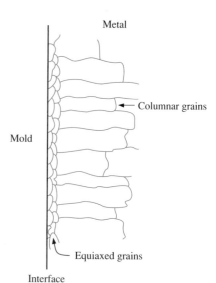

FIGURE 6.4   Fine grains occur at the
metal surface in contact with the mold.

that the mechanical properties that depend on alloy chemistry cannot always
be treated as uniform throughout the casting if equilibrium has not been
achieved.

The nonequilibrium structures and the effects of alloy segregation were
discussed more fully in Chapter 5.

Nucleation and Growth

When the cooling rates are higher, there is a higher nucleation rate but a correspondingly lower growth rate for the solidifying grains because of the reduced temperature. Heat is extracted more rapidly at the mold interface, and
hence the surface grain size is smaller, as shown schematically in Figure 6.4.
The advantage of smaller grains is somewhat higher strength. Fortunately,
many machine members are loaded in service in such a way that the max-
mum stress is at the surface; an example is bending in a simple beam.

When a casting is machined, however, the stronger, fine-grained surface
may be removed and the strength advantage lost. Casting to closer tolerances,
which minimizes machining, (near net shape) then becomes an important design consideration.

6.4 Further Principles Involved in Casting

We will now introduce several new concepts that have not been covered previously or that require an expanded discussion.

Shrinkage

We have already discussed thermal contraction in the solid state, but the contraction when liquid transforms to solid may lead to serious defects called *shrinkage cavities*. While in most cases shrinkage takes place in metals during solidification, exceptions occur when primary silicon grains form in the aluminum–silicon system (β in Figure 6.3) and when graphite forms in gray cast iron (Section 9.11). In each of these latter examples, the average density of the precipitating solid is less than that of the liquid, which gives a volumetric expansion.

When specific volume (represented as 1/density, with units of volume/mass) is plotted versus the temperature, we obtain a decrease in volume. That is, *shrinkage* takes place during solidification, as shown in Figure 6.5. In a pure metal the shrinkage occurs at a constant temperature, the melting point. In an alloy, shrinkage occurs over the temperature range between the liquidus and the solidus.

The amount of shrinkage is from 2 to 6 vol. % and depends on the composition of the alloy. Thus a solidifying sphere freezes last at its center and may then contain a shrinkage pore of up to 5 vol. % if we do not provide a source of molten metal to fill the centrally located void.

In order to produce a sound casting, a *riser*, or liquid metal reservoir, is attached to the casting to provide a source of liquid metal. An example is shown in Figure 6.6, a cross sectional slice from a bar casting. Only the riser

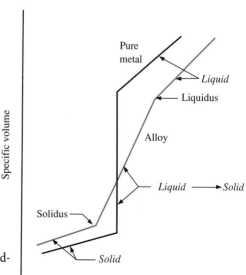

FIGURE 6.5 During the liquid-to-solid transformation, a decrease in volume occurs.

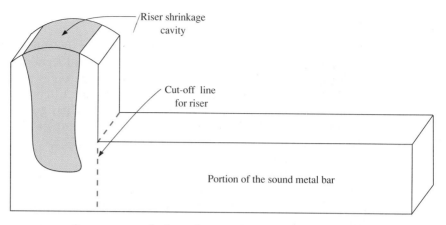

FIGURE 6.6 Cross sectional slice of a cast bar wherein all of the shrinkage is located in the riser.

has the shrinkage, and it is cut off for remelting at the time the gates are removed. To be effective, the riser must freeze after the casting and must be large enough to meet the volumetric shrinkage requirements of the casting.

Note in Figure 6.5 that the liquid metal also shrinks because of the coefficient of thermal expansion in the liquid state. We pour with superheated liquid metal, so our riser must also accommodate the liquid shrinkage before the onset of freezing. Higher pouring temperatures, (more superheat) means that larger risers may be required.

Dendrites

The foregoing discussion addresses only riser size; it does not tell you where to place a riser or whether more than one riser might be advantageous. The key to answering these questions is the liquidus-to-solidus temperature difference, which has been identified as one index of alloy castability.

A phase diagram may show the coexistence of liquid and solid over a temperature range, but this does not mean that there is a random mixture of the two phases. The completely solid interface has fern-like solid spikes that grow out into the liquid and are called *dendrites*. If insufficient liquid is present to feed the voids created by volumetric shrinkage of the remaining liquid when it solidifies, shrinkage porosity results. The dendritic shrinkage cavities are shown in Figure 6.7.

If we examine a two-dimensional representation of a solidifying bar and view the cross section during solidification, we can see the effects of a dendrite network. With a wide range between liquidus and solidus, the mushy region is large and the dendrite network is also extensive, making liquid

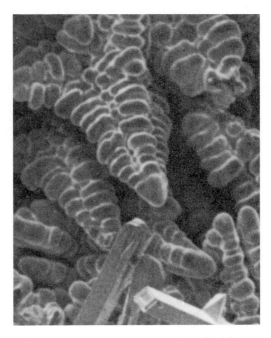

FIGURE 6.7   A shrinkage cavity in copper, as viewed with a scanning electron microscope at 100×. The terminations of the dendrites (fern-like grains) is apparent.

John Mardinly, Department of Materials Science, University of Michigan

penetration from the riser difficult (Figure 6.8). Shrinkage porosity can result when the dendrite arms join to block the formation of liquid-metal feeding paths. An alloy composition with a wide liquidus-to-solidus temperature range is then more prone to shrinkage defects. The key to casting such alloys is to reduce the extensive dendrite network and to approximate the smoother interfaces like alloys with a narrow freezing range. This is accomplished by increasing the solidification rate, which gives a narrower region of liquid plus solid within the casting and provides stubbier dendrites. This effect is shown schematically in the cross sectional bar slice in Figure 6.9.

Here we use a *chill* (an inserted material to remove heat more rapidly than the mold) that provides a narrower mushy region and hence more open paths for liquid-metal flow from the riser.

Several risers with chills between them may be necessary when a casting is large. However, the use of chills to provide open feeding channels is more extensive when the alloy liquidus-to-solidus temperature range is large. Although it is difficult to generalize because different thermal constants occur in the many commercial alloys, a solidification range of 25°C or less can be considered narrow, whereas 100°C or greater is a wide freezing range.

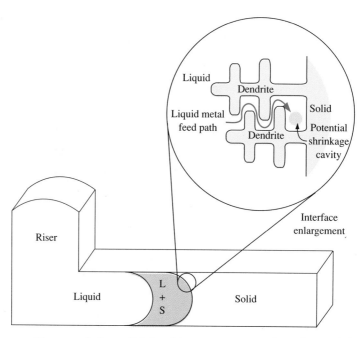

FIGURE 6.8 Closure of the solid dendrites can prevent liquid-metal feeding, and shrinkage results (cross sectional slice in a solidifying bar).

Residual Stresses and Hot Tears

When restraint is imposed on the normal thermal contraction of the casting, a strain and hence a stress can develop. During solidification with some liquid present, such strain can result in fracture or form what is called a *hot tear.* When solidification is complete, adjustment to the strain is possible via creep as long as the temperature is still in the plastic strain range of the alloy.

At temperatures below the creep range, elastic strains can build up as a result of unequal cooling rates in castings of variable section size. Solid-state phase transformations that occur in an alloy system may also contribute to these elastic strains, which lead to what are called residual stresses.

To illustrate the importance of such stresses, consider the following example. A large casting (an experimental railroad car wheel) was subjected to a new heat treatment in which the entire part was heated to 1600°F (871°C). Next the rim was subjected to a water spray to form a hard martensitic surface. After this, the wheel was laid aside to cool. In the middle of the night, the wheel exploded into several pieces. (One traveled several hundred feet!) Examination of the pieces showed that the metal was perfectly sound, exhibited the desired structure, and had a tensile strength of 60,000 psi (414 MPa) in the fracture area.

The reason for this catastrophic failure was the presence of residual stresses resulting from high elastic strain that developed in the part. We shall

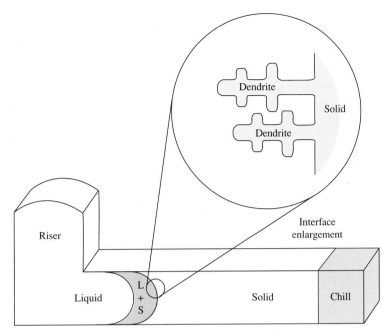

FIGURE 6.9 When a chill is used, the dendrite network is less extensive, providing more open liquid-metal feeding channels.

discuss first the determination of residual stress, then how it develops in a simple wheel, and finally some methods of controlling it.

Let us recall the simple tensile test of Chapter 2. Suppose a student stressed a tensile specimen of steel to 30,000 psi (207 MPa), causing an elastic strain of 0.001 in. (0.025 mm). If the student locked the machine with the load intact, we would have a system in which one member (the specimen) has a residual stress of 30,000 psi (207 MPa). If the dials on the machine were covered, how could we determine the residual stress? We could inscribe two marks very accurately, 1 in. apart on the gage length, and then make a saw cut above the gage length across the bar. The saw cut would unload the specimen and relieve the elastic strain. When we remeasured the distance between the scribe marks, we would find that it had contracted to 0.999 in., indicating 0.001 in./in. elastic strain. We could then calculate residual stress (from $S = eE$ as $0.001 \times 30 \times 10^6$ psi = 30,000 psi (207 MPa).

In practice, instead of using scribe marks (which are relatively inaccurate), we use electric strain gages. These are essentially fine wires held in a paper matrix and cemented to the specimen in the stressed condition. We take an initial reading of the electrical resistance of the gage with a precision of 1 μohm. We then relieve the elastic strain by sectioning and take a second resistance reading. If the gage length shortens, the resistance decreases, indicating residual *tension*; the resistance increases with residual compression. The precision of the method is excellent, because 1 μohm change is equivalent to about 0.000001 in./in. strain, or a residual stress of 30 psi (0.21 MPa) for steel.

When residual stresses are present in several directions, we use strain gage rosettes with three strain gages. A Mohr circle analysis is used to interpret the results. This is discussed in standard texts on mechanics.

Although the fracture usually relieves the residual stresses in a failed part, it is often possible to obtain a similar unfractured part for stress analysis. In many cases a project engineer has felt secure with a steel of 60,000 psi (414 MPa) tensile strength operating in a given service, only to find that the processing operations led to a residual stress of 50,000 psi (345 MPa) and that, consequently, an additional service stress of only 10,000 psi (69 MPa) was required for failure.

Now let us consider a case of residual stress and how its location and sign (compressive or tensile) may be predicted. We will investigate metallic materials, but the same principles apply to ceramics and polymers.

Only two simple rules are involved:

1. Metals expand when heated and contract when cooled, except in temperature ranges in which a phase transformation occurs. For example, most low-alloy steels and cast irons have a coefficient of expansion of approximately $14 \times 10^{-6}/°C$ ($7.8 \times 10^{-6}/°F$) up to the ferrite–pearlite–austenite transformation.* When ferrite transforms to austenite, a substantial contraction occurs, and the resulting austenite has a coefficient of expansion of about $23 \times 10^{-6}/°C$ ($12.8 \times 10^{-6}/°F$).
2. Metals are plastic (that is, they do not store elastic strain) at "elevated" temperatures. In the case of ferrous alloys, the plastic range is above 1000°F. For copper and aluminum alloys, the range is lower (at approximately the recrystallization temperature band). For thermoplastics the range is much lower, and it is generally higher for ceramics.

Now let us apply these rules in an example that investigates the development of residual stress in a simple wheel casting.

EXAMPLE 6.1 *[ES/EJ]*

Estimate the direction and sign of residual stresses in the wheel casting shown. Assume the material is medium carbon (0.3 C) steel.

---

*See Chapter 8 for a fuller discussion of the phases present in steel. Ferrite—body-centered cubic (a low-temperature phase); austenite—face-centered cubic (a high-temperature phase); pearlite—a mixture of ferrite and iron carbide.

Answer First we note that the spokes and hub are thinner than the rim. If the mold is poured in green sand and the part mold-cooled, the entire light-section region will cool more rapidly than the rim. As an example, suppose the average temperature of the light sections is 1050°F when that of the rim is 1200°F. The spokes will have contracted $(150°F)(7.8 \times 10^{-6}\text{in./in.°F})$ or $1.17 \times 10^{-3}\text{in./in.}$ relative to the rim. At room temperature this strain, if elastic, would result in a residual stress of $(1.17 \times 10^{-3}\text{in./in.})(30 \times 10^6 \text{ psi}) = 35,100$ psi (242 MPa). However, in this temperature range we must consider rule 1 and realize that it is impossible to develop a stress of this magnitude in a hot, plastic material. Instead, plastic strain occurs.

We now arrive at the elastic range (~1000°F) with practically no elastic strain but with a temperature difference between the spokes and the rim. Suppose the spokes are at 800°F and the rim is at 1000°F. The structure in

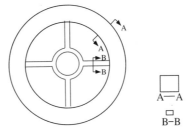

both cases is ferritic (pearlite has almost the same coefficient of expansion). If free to act alone, the spokes would only contract $(800°F - 70°F)(7.8 \times 10^{-6}\text{in./in.} \cdot °F)$ in cooling to room temperature, whereas the rim would contract $(1000°F - 70°F)(7.8 \times 10^{-6}\text{in./in.} \cdot °F)$. The result is that the rim "attempts" to reach a smaller diameter than the spokes will allow. This places the spokes in a state of residual radial compressive stress and the rim in circumferential tension. If the rim were relatively massive and all the strain were assigned to the spokes, a radial compressive stress of $(200°F)(7.8 \times 10^{-6}\text{in./in.} \cdot °F)(30 \times 10^6 \text{ psi})$, or 46,800 psi (323 MPa), would develop.

6.5 Other Casting Processes

The process described for making the frying pan is used to produce castings of from 30 g to 150 metric tons (1 metric ton = 2205 lb). There are a number of other casting methods, a few of which are described below.

In *permanent-mold casting* the mold is made of metal or graphite. The cooling rate is faster than in sand, which results in finer grain size to greater depth and decreases the extensive dendrite network.

Die Casting also uses a permanent mold, but the metal is injected under pressure. This procedure is particularly useful for metals with low melting points—such as zinc, aluminum, and magnesium—which do not attack the metal die rapidly.

Investment casting is a method for producing small precision castings. A wax or polymer replica of the desired part is made, and a liquid, cementlike slurry is poured around it in a mold. After the cement (which is called an "investment") sets, the mold is heated. The heating eliminates the pattern by melting and preheats the mold. Then liquid metal is poured into the cavity.

One variation on the expendable pattern used in investment casting is the use of a polymer pattern that is burned out ahead of the liquid metal during the pouring operation. This *evaporative pattern* technique employs loose sand as the molding material and is used for complex shapes (such as aluminum automotive intake manifolds) wherein machining is to be minimized.

Centrifugal casting utilizes a spinning mold to facilitate mold filling by centrifugal force. Cylindrical shapes such as cast water pipes are one application. When used in conjunction with an investment mold, centrifugal castings of small and detailed complex shapes can be produced. Examples include jewelry and dental castings.

Finally, Figure 6.1 shows that a cast product is really the starting point for a product that is to be formed. The large ingot step may be eliminated by *continuous casting*. A continuous strand of metal issues from the bottom of a water-cooled die. Cross sections can be rectangular or circular, depending on which shape is easier to process.

6.6 Casting Defects

The major defects that occur in casting are caused by gas porosity, shrinkage porosity, nonmetallic inclusions, and hot tears. Figure 6.10 displays these in some detail and keys them, by number, to the following list.

1. *Cut and wash.* The metal erodes the sharp corners of the ingate and the protruding core. Letters A and B denote the sequence of steps.
2. *Rat tail, buckle, scab.* Heat from the stream of metal—either radiant heat or heat caused by conduction—locally and nonuniformly expands the sand, producing either a minor extension (rat tail), a buckle, or spalling away (a scab). The material from the mold forms a sand inclusion at another spot.
3. *Fusion, penetration.* Sand can fuse by action of the hot metal, producing a mass that may adhere to the casting. Metal can penetrate the sand grains after the surface of the grains has fused.
4. *Crush.* When the mating halves of a mold do not fit, the protruding portion of the casting is crushed.

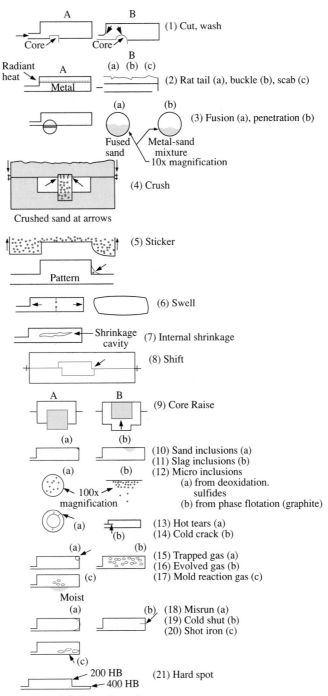

FIGURE 6.10 Typical defects found in casting

5. *Sticker.* Sand sticks to the pattern, producing a projection of ragged edges on the metal after casting.
6. *Swell.* Hydrostatic pressure of the metal (plus expansion upon solidification in some cases) pushes the mold walls outward.
7. *Shrinkage.* When the density of the solid metal is greater than that of the liquid metal (the usual case), porosity is produced.
8. *Shift.* The halves of the mold are mismatched.
9. *Core raise.* Flotation of the core is caused by the buoyancy of the liquid metal.
10. *Sand inclusions.* Sand is entrapped in the mold as a result of crushing or spalling, or as a result of loose sand being in the mold.
11. *Slag inclusions.* These result from the presence of slag in the metal.
12. *Microinclusions.* Sulfide and oxide inclusions occur because of separation from the melt. Graphite flotation may occur in hypereutectic ductile iron. (See Chapters 8 and 9)
13. *Hot tears.* Interdendritic ruptures are produced by tensile stresses on the casting when it is near the solidus temperature.
14. *Cold cracks.* These are produced by residual stresses or by improper removal of gate and riser.
15. *Trapped gas.* Gas can be mechanically entrapped in the mold.
16. *Evolved gas.* This is gas that evolves from precipitation of gas dissolved from the liquid metal.
17. *Mold reaction gas.* Gas can be evolved when the liquid metal reacts with sand that is damp or that contains volatile hydrocarbons in the mold. *Note:* Pores due to gas are often spherical in shape, whereas pores due to internal shrinkage show an irregular dendritic surface.
18. *Misrun.* A misrun is the incomplete filling of the mold that results when the metal is of low fluidity (cold metal).
19. *Cold shut.* A section of the casting is imperfectly bonded because the metal is poured at too low a temperature, and different metal streams do not bond.
20. *Shot iron.* Particles of frozen metal can be caught in the stream of molten metal and improperly bonded; this is often caused by interrupted flow.
21. *Hard spot.* A hard spot occurs, especially in gray and ductile iron. More rapid rate of solidification of light sections and corners may result in a very hard carbidic structure.

Nonmetallic inclusions are of two types and require further explanation. One is a gross type made up of foreign materials such as sand, slag, or oxide dross. These inclusions can be controlled by being careful during the pouring operation and by designing an appropriate gating system for the casting. The second type of inclusion, made up of such materials as sulfides or complex silicates, forms upon solidification.

Controlling these inclusions is more difficult. We control them either by reducing the offending element to very low levels or by making ladle additions that change the shape of the inclusions to a less harmful geometry—for example, from a continuous-grain boundary shape to a spherical shape.

6.7 The Deformation Process

We are already familiar with metal deformation from our discussion of hot and cold working in Chapter 3. Here we will expand on, and explore more carefully, the relationship among structure, process, and performance.

It is convenient to divide the processes into two categories:

1. *In bulk deformation,* a marked change is observed in the cross-sectional thickness or area of the component. Examples are rolling, drawing, extrusion, and forging (Figure 6.11). A number of very important engineering shapes are produced by these methods.
2. *In sheet forming,* little change is found in the cross section, yet there is a great modification in shape. A simple example is the formation of a cup shown in Figure 6.12. Here a "blankholder" can be incorporated to control more carefully excessive "thinning" of the sidewalls and "wrinkling" of the upper edges, points that will be discussed more fully later.

First, we should review the concepts introduced earlier that are important to deformation.

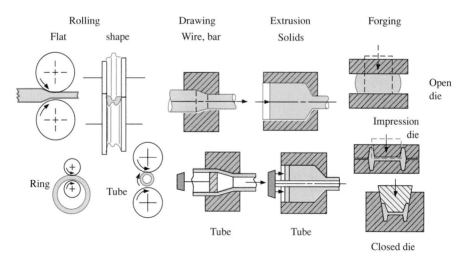

FIGURE 6.11   Typical metal bulk-deformation processes

[Adapted from J. A. Schey, *Introduction to Manufacturing Processes*, McGraw-Hill, New York, © 1977. Reprinted by permission of McGraw-Hill Publishing Company.]

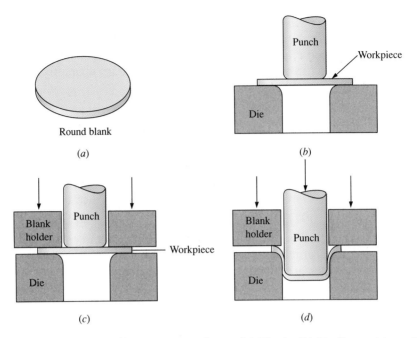

FIGURE 6.12   Steps in drawing a cup shape. (*a*) Blank. (*b*) Tooling without hold-down. (*c*) Tooling with a blankholder. (*d*) Forming the cup.

6.8 Review of Principles Related to Deformation

General Principles

All of the phenomena associated with plastic deformation, such as crystal slip systems and the relationship of bond energy to temperature, are important to the deformation process. Furthermore, we must consider diffusion to take into account the effect of time. For example, atoms can be moved by both a strain gradient and a concentration gradient. But these are not instantaneous processes, and we need the concept of time to explain some of the practical observations.

In casting, segregation and coring (Figure 5.4) are difficult to break up by heat treatment alone. However, by combining the high-temperature diffusion with a bulk-deformation process such as rolling, we can achieve much more thorough homogenization in a shorter time period.

Similarly, the definitions of *cold work* and *hot work* incorporate the recrystallization temperature, which is also time-dependent. When annealing, we might allow an hour for recrystallization, yet hot-working temperatures are commonly well above this recrystallization temperature where the effect is very rapid.

The lower energy requirement for deformation and more or less instantaneous recrystallization at higher temperatures suggests that we should work as hot as possible. On the other hand, there are limitations (such as the possible formation of a liquid phase) determined by a phase diagram. Often the phase diagram also dictates a high-temperature deformation to dissolve brittle phases in the solid state, as for example, in high-carbon steels (see Chapter 8).

It should also be noted that higher temperatures result in more oxidation and scale formation, on the deformed product. Although a surface scale may not be harmful to a construction beam, it is not desirable for an automotive bumper. Furthermore, hot working removes the strength increase associated with cold working. Finally, the dies used for hot working must also be able to withstand the high temperatures it entails. Generally, the higher the alloy content of the dies, the lower the diffusion rates and the greater the resistance to a microstructural change at the elevated temperatures. However, these high-alloy dies are more costly.

Stress and Strain

Because the design of any deformation process is dependent on the strength of the materials (both the material to be deformed and the die), the stress–strain curves are important considerations in that design.

Recall that we defined two methods for measuring stress and strain: engineering (nominal) values and true values. The true stress and strain are more important to our discussion here, because we are concerned with actual area reductions during a given deformation process.

An idealized true stress–true strain curve is shown in Figure 6.13. Considering a sample that is loaded to point A, we note the following:

1. When the sample is loaded, the total strain at C is made up of an elastic and a plastic component. When the sample is unloaded, the elastic strain

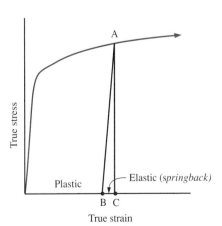

FIGURE 6.13   Portion of a schematic true stress–true strain curve

is recovered and the resultant permanent strain is at point B. The elastic recovery is called *springback*; we can observe it by winding a metal wire around a wooden dowel and then releasing the wire end. We have no difficulty removing the wire from the dowel, because the coil springs to a larger diameter.

2. Data such as those reflected in Figure 6.13 would normally be developed at room temperature. At high temperatures, such as those we encounter when hot-working, the elastic-strain component is converted to plastic strain *if* sufficient time is allowed for this "self-annealing". The implication is that the stress–strain curve is sensitive to both temperature and strain rate.

6.9 Further Principles Involved in Deformation

Three areas demand further discussion if we are to gain a better understanding of the practical considerations involved in deformation processing. These are energy requirements, anisotropy, and complex strains.

Energy Requirements

As a first approximation, the total work required in deformation is equivalent to the area under the true stress–true strain curve, as shown in Figure 6.14. The larger the area, the more work is required. However, the use of experimental stress–strain diagrams is somewhat cumbersome when we are calculating energy.

A simplification is possible when we recall the linearity of log true stress versus log true strain, as described in Figure 2.11. This relationship can be expressed mathematically

$$\sigma = K\epsilon^n \qquad \text{(from } \ln \sigma = \ln K + n \ln \epsilon \tag{6-1}$$

$$\text{or } \log \sigma = \log K + n \log \epsilon)$$

where n is the slope in Figure 2.11 and is called the *strain-hardening exponent*. Hence knowing the plastic deformation (true strain) enables us to calculate the true stress and hence the force (stress = force divided by the true area). The actual force requirement would be higher because of friction inherent in the process and within the equipment itself.

EXAMPLE 6.2 *[ES/EJ]*

A particular component is normally cold-formed from a low-carbon steel. Criticize the following statement: "We can use the same dies for making the component out of an aluminum alloy."

Answer Because the two aluminum alloys given as examples in Figure 2.11 are beneath those of the low-carbon steel, we might expect a lower energy requirement for the aluminum. If it worked for steel, the die should be able to withstand the forces in aluminum deformation. We may have to modify the lubricant used to prevent sticking, however.

A more subtle problem is one of dimensional control. Aluminum has a lower modulus of elasticity, and the nature of its stress–strain curve is certainly not the same as for the steel. Hence, the amount of springback will also be different, and die modification may be necessary if equivalent dimensional control is required for the same component made from the two families of alloys.

Anisotropy

Anisotropy implies properties that are directional in nature. Isotropy, on the other hand, means that properties are uniform and equal in all directions—a condition that is often assumed for convenience in mathematical modeling but is seldom achieved completely. Although anisotropy in castings is to be expected because of different cooling rates and hence variable grain size, anisotropy in deformed products can be more severe and can arise from two sources.

One form of anisotropy is due to nonmetallic inclusions in the original ingot, billet, or slab that become elongated in the forming direction during primary processing such as hot rolling. These inclusions constitute areas of weakness that are orientation-sensitive. A second form of anisotropy is *crystallographic texture*, or preferred orientation of the grains that results from the crystallographic nature of the slip process.

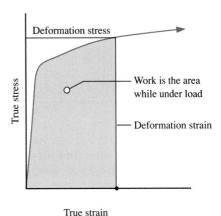

FIGURE 6.14 The "work" required during deformation is represented by the area under the true stress–true strain curve.

Whatever the source of the anisotropy, further processing and product design require recognition of the dependence of properties on direction. As an example of product design, consider cutting an open-end wrench shape from hot-rolled plate stock, as shown in Figure 6.15. Parts *a* and *b* illustrate that it is possible to make a wrench that will break in service because of a faulty relationship between the rolling direction and the forces (depicted by the arrows) that the tool will undergo.

On the other hand, a wrench that has an opening formed as in Figure 6.15*c* — it is upset-forged — has a final texture such that the forces dur-

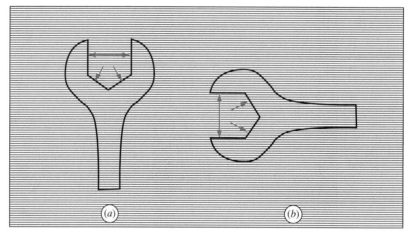

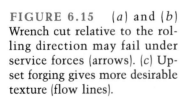

Rolling direction

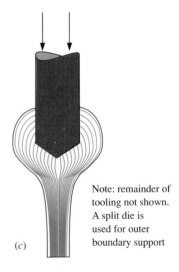

FIGURE 6.15   (*a*) and (*b*) Wrench cut relative to the rolling direction may fail under service forces (arrows). (*c*) Upset forging gives more desirable texture (flow lines).

(*c*)

Note: remainder of tooling not shown. A split die is used for outer boundary support

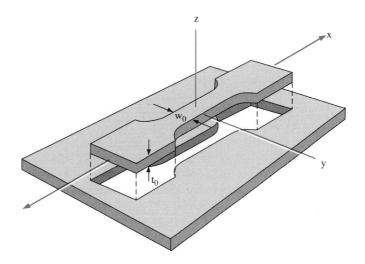

FIGURE 6.16   Tensile strip specimen cut from sheet to determine the anistropy

[Walter A. Backofen (ed.) *et al.*, *Fundamentals of Deformation Processing* (Syracuse, N.Y.: Syracuse University Press, 1964), Figure 10 by permission of the publisher.]

ing service are roughly perpendicular to the texture. This wrench may be expected to have a longer life.

In thin sheet metals it is possible to quantitatively evaluate the anisotropy by cutting tensile samples at angles of 0°, 45°, and 90° to the rolling direction. Three strains are measured (Figure 6.16) in a tensile test:

$$\epsilon_w = \ln(w/w_0) \quad \text{width strain} \tag{6-2}$$

$$\epsilon_t = \ln(t/t_0) \quad \text{thickness strain} \tag{6-3}$$

$$\epsilon_l = \ln(l/l_0) \quad \text{length strain, normally 15\% elongation} \tag{6-4}$$

An R value is then defined as (ϵ_w/ϵ_t); it varies with orientation of the sample. Refer to Figure 6.12 as you consider the following two consequences of the anisotropy.

1. When R is equal to 1.0, the material is isotropic; when $R > 1.0$, it is resistant to uneven thinning; when $R < 1.0$, local thinning of the side walls can occur. Fracture may even result.
2. R varies with the direction, or orientation, of the sample, and

$$\Delta R = (R_{0°} + R_{90°} - 2R_{45°})/2 \tag{6-5}$$

When ΔR varies from the isotropic value of zero, *earing* can occur (Figure 6.17). The ears must be trimmed, which adds to the cost.

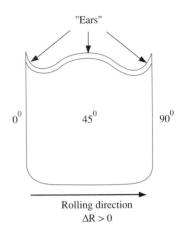

FIGURE 6.17   Earing in a drawn cup that occurs due to anisotropy. (Angles refer to the relationship to the rolling direction.)

Complex Stresses

We have made frequent reference to the tensile stress–strain diagram, and it would appear that the reduction in area during the test is the limit of the deformation process. Although this conclusion is correct for simple tension, processes such as drawing include a compressive-strain component and less tension in one direction than in another.

These strain combinations are not indexed by the tensile test, so in sheet metal deformation, a *forming-limit diagram* is developed. The limits of major strains (highest tension) to minor strains (lowest tension or most compression) that will cause failure are then plotted in the diagram. Die designs and processing variables such as the blankholder pressure shown in Figure 6.12 can then be determined to be such that failure does not occur.

6.10   Other Deformation Processes

The methods employed to deform metals range from a blacksmith forge to ultrasonic deformation. All such techniques incorporate the following design considerations, which become part of the name used to describe the particular deformation process.

1. *Die design.* Dies may be *open* or *closed*, as in forging (Figure 6.11). The formation of relatively simple shapes such as nails or rivets may require special intermediate steps. In *upsetting*, or *heading* (Figure 6.18), a *coning* operation may have to precede the actual heading when the unsupported length is large in order to prevent buckling.
2. *Means of applying force.* The forces are certainly applied in different ways in rolling, forging, bending, extrusion, and so on. Furthermore, the rate of deformation may also become part of the description of the process. For

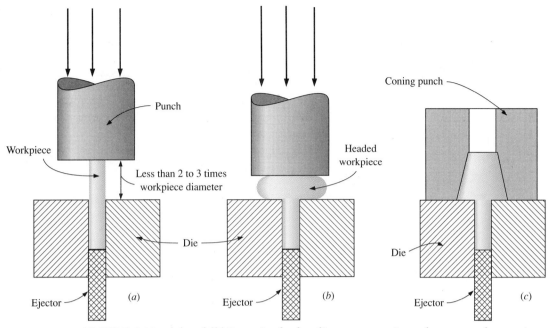

FIGURE 6.18 (*a*) and (*b*) Steps in the heading, or upsetting, of a part such as a rivet. (*c*) Coning as a first step to minimize buckling.

example, *impact extrusion* suggests a higher strain rate than normal extrusion. In *explosive forming*, a shock wave supplies a sudden force.

3. *Material characteristics.* Here we refer to the material being deformed, although the die material is equally important. *Cold forming, warm forming,* and *hot forming* are relative terms that can be applied to any process. Sensitivity to strain rate may also become part of the description. Some metals (such as titanium alloys) are *superplastic* at high temperatures and *low* strain rates. This makes it possible to achieve large permanent deformation in a single operation rather than in successive operations.

6.11 Deformation Defects

Besides the defects that can plague sheet metal forming (such as earing, wrinkling, and thinning), there are a number of other potential bulk-deformation defects. These are summarized in the following list and in Figure 6.19. Some of the defects are a direct result of the prior casting history.

The definitions apply to both ferrous and nonferrous metals, and in some instances, the same defect can result from either hot or cold deformation.

1. *Segregation.* Inclusions and alloying elements segregate during solidification and are rolled into bands.

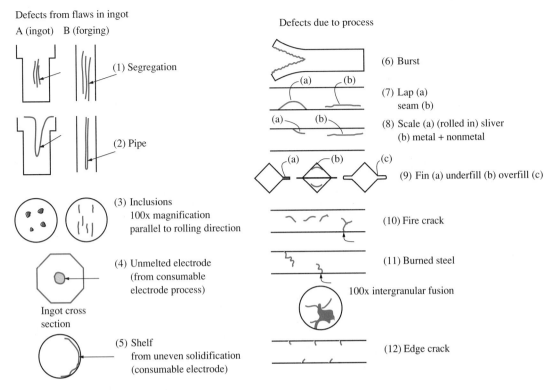

FIGURE 6.19 Typical defects found in forming processes

2. *Pipe (shrinkage cavity).* If the material is oxidized, a shrinkage cavity does not weld. If the material is forged, a shrinkage cavity produces a fissure in the rolled product.

3. *Inclusions.* In the ingot, inclusions are uniformly distributed; after forming, they elongate in the forming direction.

4. *Unmelted electrodes.* During the process of melting metal via electrodes, pieces of the electrode fall into the ingot.

5. *Shelf.* When solidification occurs in an uneven wave, there are lap-like defects.

6. *Burst (also called alligatoring).* When surface stresses at rolled edges are higher than in the interior, bursting occurs.

7. *Lap.* Folding over of rolled surface at one region.
   Seam. Continuous folding over of rolled surface.

8. *Scale.* Oxidized scale can be rolled into the surface, or internal deoxidation products can be strung out in the direction of rolling (in the form of slivers).

9. *Fin.* An extrusion of metal that occurs when the fit of the die is poor.
   Underfill. Insufficient metal between dies.
   Overfill. Excess metal between dies.

10. *Fire cracks.* These are caused by cracking of rolls and imprinting of striations on the piece being worked.
11. *Burned steel.* A defect that occurs when steel is heated to temperatures above the melting point of one of the phases (normally the sulfide phase); also called hot shortness. Similar defects can occur in nonferrous metals.
12. *Cracks.* These are caused by improper forging temperature or improper hammer force.

Hot Working

Defects encountered during hot working (above the recrystallization temperature) can be caused by the following factors:

1. Working temperature that is too high
2. Working temperature that is too low
3. Pre-existing defects in the ingot
4. Residual stresses
5. Defective working procedure

Burning, or Excessive Working Temperature If the metal is heated at or near the solidus, the material at grain boundaries melts and forms voids, which are not sealed by subsequent working. It is not necessary to reach the solidus of the pure alloy if other elements, such as sulfur in the form of metal sulfides, form a liquid with a melting point below the solidus of the pure alloy. Segregation of some of the desired elements may lead to melting below the estimated equilibrium temperature. Overheating is related to burning, but occurs at slightly lower temperatures, and is not really a well-defined condition. It may be evidenced by loss of ductility and cracking at high temperatures without definite evidence of fusion, as in the case of burning. Heating at temperatures that are too high for the particular composition may also lead to excessive grain growth.

Low Working Temperatures It is generally advisable to work at as high a temperature as possible because of the greater ductility, lower energy requirements, and ease of flow of the metal, provided that excessive temperatures are avoided. Many hot processes are designed with high ductility in mind. When the temperature standards are not met, *cracking* takes place. There are several special cases. In forging and rolling, corner and edge cracks develop. Another rolling defect, called a *center burst*, is quite striking because it takes place inside the bar being rolled. It arises from the fact that the material next to the rolls is being rapidly deformed relative to the undeformed metal at the center.

Pre-existing Defects in the Ingot An ingot is, after all, a casting and is subject to casting defects such as gas holes, shrinkage porosity, and inclusions. Fortunately, in most cases the gas holes are welded shut by rolling.

One grade of steel, called *rimming steel*, is made deliberately with gas holes that are not open to the atmosphere and that prevent shrinkage cavities from developing. Shrinkage porosity, however, results in a *pipe* in the ingot that is compressed but not welded shut during rolling because the internal surfaces are oxidized. If this portion of the ingot is not cut off, piping defects occur. *Inclusions* are either elongated, as in the case of manganese sulfide, or fractured, as in the alumina type. When the metal is tested in the rolling direction, the effect of inclusions is minimal, but across the rolling direction, ductility is lowered. Actually the inclusions may be treated as cracks from the point of view of fracture toughness. *Segregation* in the ingot may lead to *banding*; for example, in hypoeutectoid steel, there may be alternating layers of ferrite and pearlite.* *Seams* are regions of inhomogeneity that may open up during forging. They can be caused by segregation.

Residual Stresses   Residual stresses develop when a part that has severe temperature differences (thermal gradients) between different sections is cooled from the hot-working range. When the overall temperature of the part reaches the elastic range, hotter sections contract more than cooler sections, resulting in residual stresses. These stresses may crack the part during the cooling cycle or, if unrelieved, may lead to premature failure. When this condition is encountered, the part should be reheated within a reasonable time to a temperature at which creep occurs, to relieve the elastic strain. Generally this temperature is near the recrystallization temperature. The part should be slowly cooled to prevent redevelopment of the thermal gradient.

Defective Working Procedure   Improper manipulation of the piece during forging or poor die design leads to the development of a lap, also called a *cold shut.* This defect is caused when one surface region is forced over an adjacent region. The overlap traps a portion of the usually oxidized and scaled surface beneath the surface, producing a plane of weakness.

Cold Working

During cold working, which is generally conducted at about room temperature, the ductility of the metal is much lower and recrystallization does not take place to remove work hardening (except in low-melting-point metals such as lead). As a result, the part is more subject to cracking when defects such as inclusions are present, especially if they have been elongated by previous hot working. If harmful residual stresses are present, the condition may be aggravated by cold working and the part may actually break. The same terms are used to describe typical defects, except that the phrase *cold short* denotes a material that cracks particularly easily during operations at room temperature, most often because of the presence of brittle phases.

---

*See chapter 8 for an explanation of steel microstructures.

6.12 Powder (Particulate) Processing

In this chapter we will limit our discussion to the processing of metal powders, or powder metallurgy (P/M). However, many of the concepts we will discuss are also applicable to ceramic components.

There are three basic steps in producing a P/M part: mixing of elemental or alloy powders, compaction of the powders into a shape, and finally *sintering*, or heating of the compact to produce a metallurgical bond between the powder particles.

As an example, consider the production of a small 70Cu–30Zn brass spur gear with the hardened steel tooling shown in Figure 6.20. The lower punch is inserted into the bottom of the die, and the core rod is placed through the center of the lower punch. A measured quantity of prealloyed powder is placed into the die cavity, and the upper punch is positioned in the die. An organic lubricant such as 0.25–0.50% each of lithium and zinc stearate is also present in the powder to aid in compaction and to provide higher strength after compaction.

The actual compaction in the rigid tooling may occur by several techniques, as shown schematically in Figure 6.21. The more complex tool motions are required as the thickness-to-diameter ratio increases, because friction between particles and at the die walls can cause variation in density within the "green compact." Normal compaction pressures are 60,000–100,000 psi (414–690 MPa).

The compacted gear shape is then sintered at 815°C (1500°F) for 30 minutes in a controlled-atmosphere furnace. The atmosphere is low in oxygen to prevent the formation of scale between the particles, which would inhibit a good metallurgical bond. Dry hydrogen, dissociated ammonia, nitrogen, and carbon monoxide–carbon dioxide mixtures all have minimum oxygen.

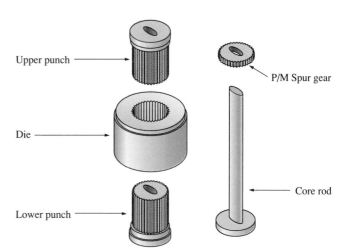

FIGURE 6.20 Tooling for forming a small spur gear by P/M

[*P/M Design Guidebook*, Metal Powder Industries Federation, Princeton, N.J. 1983, p. 4. Used by permission.]

Upper punch

Die

Lower punch

P/M Spur gear

Core rod

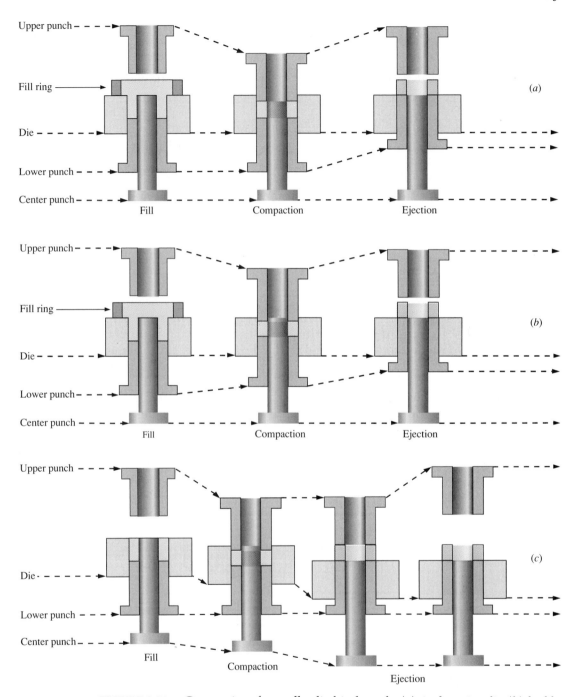

FIGURE 6.21   Compaction of a small cylindrical part by (*a*) single-acting die, (*b*) double-acting dies, and (*c*) floating dies. (*Note*: Dotted lines show the progressive motion of dies and punches.)

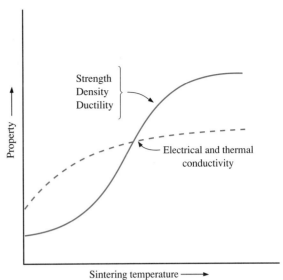

FIGURE 6.22 Schematic representation of the variation in mechanical and physical properties with solid-state sintering temperature. The sintering time and compaction pressure are assumed constant. (The origin is not zero for either the measured property or the sintering temperature.)

Two concepts are very important to the sintering process: phase diagrams and diffusion. The sintering temperature for our brass sample has been selected as 100°C (180°F) below the solidus of the 70Cu–30Zn alloy. At higher temperatures we risk liquid formation and *higher shrinkage.* At lower temperatures the densification process takes too long and the optimal properties may not be achieved, as shown schematically in Figure 6.22.

The closure of porosity can be represented by a model of three spherical particles and their response to sintering (Figure 6.23). The diffusion of atoms — and correspondingly the diffusion of point defects (atom holes) — is important to the model. Shrinkage takes place and is normally 0.5% to 3.0%, depending on the characteristics of the powder, on compaction pressures, and on sintering time and temperature. The shrinkage is predictable, however, once the processing variables are controlled, and dimensional tolerances of 0.004 in. (0.10 mm) in our small brass gear are typical.

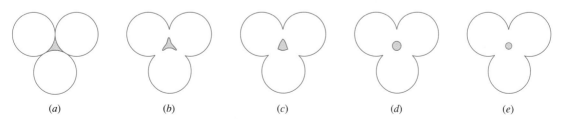

(a)   *(b)*   *(c)*   *(d)*   *(e)*

FIGURE 6.23 A three-sphere sintering model, showing (a) original point contact, (b) formation of a neck, (c) initial pore rounding, (d) a spherical pore, and (e) spherical pore shrinkage

6.13 Further Principles Involved in Metal Powders Processing

In our brass gear example, we have introduced several concepts that require further explanation, such as powder production, packing in solids, and types of sintering.

Packing in Solids

In P/M, the sintered properties depend on the final void space, which can be determined from density measurements. However, we must define more specifically what kind of density and porosity we are talking about before the property variation can be explained. Consider the placement of odd-shaped particles in a box as shown in Figure 6.24. The following definitions will apply:

1. *Porosity*
   Open pores: Spaces into which water can penetrate
   Closed pores: Spaces into which water cannot penetrate
   True porosity: Volume of (open + closed pores)/total volume of box
   Apparent porosity: Volume of open pores/total volume of box
2. *Volume*
   Bulk volume: True volume of particles + (volume of open + closed pores)
   Apparent volume: True volume + volume of closed pores only
3. *Density*
   Bulk density: Mass of particles/bulk volume
   True density: Mass of particles/true volume
   Apparent density: Mass of particles/apparent volume or Mass/(true volume + volume of closed pores)

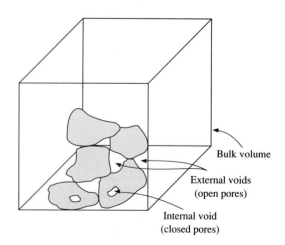

Bulk volume

External voids
(open pores)

Internal void
(closed pores)

FIGURE 6.24   Relationship among bulk volume, external voids, and internal voids

4. *Packing factor*
   True packing factor: True volume/bulk volume or bulk density/true density
   Apparent packing factor: Apparent volume/bulk volume or bulk density/apparent density
5. *Archimedes's Principle*
   Weight in fluid = dry weight − buoyant force, where buoyant force = weight of fluid displaced

The normal engineering problem is to find methods of optimizing or controlling the packing factor or, more important, the porosity, which is 1 minus the packing factor.

EXAMPLE 6.3 *[ES]*

In the packaging industry, container shape is important. What is the apparent porosity of the package if basketballs are shipped inflated so that they just fit into the box?

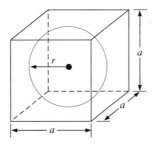

Answer For a radius r and a cubic box of side length a,

$$2r = a$$

$$\text{Packing factor} = \frac{\frac{4}{3}\pi r^3}{a^3} = \frac{\frac{4}{3}\pi r^3}{8r^3}$$

$$= \pi/6 = 0.524$$

$$\text{Porosity} = 47.6\%$$

The answer would be the same for any spherical shape placed in a cubic container into which it just fits.

The next question is "How can the amount of porosity be changed?" There are essentially three different ways to change the porosity:

1. *Shape changes.* Bricks can be packed with less porosity than spheres. A

box can hold a greater weight of pencils of hexagonal cross section than of dowels. Even the honeybee has optimized the shape of its honeycomb.

2. *Compression and impregnation.* The exclusion of air pockets by compression, as happens when one stuffs a sleeping bag into a tote bag or compresses garbage with a garbage compactor, increases the amount that can be packed. The use of wood fibers and resins to make plywood is an example of impregnation. Here the resins fill (impregnate) the void space both within and between the wood fibers.

3. *Mixed sizes.* Small spheres can be used to fill the spaces between large spheres. This principle is used in the manufacture of concrete. Similarly mixed spherical particles are used to control the porosity in packed beds for chemical reaction control and filtering processes.

EXAMPLE 6.4 *[ES]*

Pure iron powder is to be compacted and sintered into a complex shape, the bulk volume of which is not easily measured. The following weights have been experimentally determined for the sintered product:

Weight in air: 95.00 g
Weight suspended in water: 82.65 g
Weight saturated with water: 96.80 g

Calculate the following values:

A. Open pore volume
B. True volume
C. Apparent volume and closed-pore volume
D. Percentage of theoretical density, defined as (bulk density/true density) × 100

Answer

A. Water fills only the open pores, so we need saturated weight minus the dry weight, or

$$96.80 - 95.00 = 1.80 \text{ g}$$

Because water has a density of 1.0 g/cm^3,

$$\text{Open-pore volume} = 1.80 \text{ g}/(1.0 \text{ g/cm}^3) = 1.80 \text{ cm}^3$$

B. True volume = true weight / true density

$$= 95.00 \text{ g}/(7.87 \text{ g/cm}^3) = 12.07 \text{ cm}^3$$

(*Note:* True density is obtained from the table on the inside front cover.)

C. Apparent volume = true volume + closed-pore volume. We can find the apparent volume by using Archimedes' principle:

Weight in fluid = dry weight − buoyant force
= dry weight − weight of fluid displaced
= dry weight − (density × volume) of fluid displaced,

where volume of fluid displaced = apparent volume of the product. Therefore,

$$82.65 \text{ g} = 95.00 \text{ g} - (1.0 \text{ g/cm}^3 \times \text{apparent volume})$$

or

$$\text{Apparent volume} = 12.35 \text{ cm}^3$$

Then

$$\text{Closed-pore volume} = \text{apparent volume} - \text{true volume}$$
$$= 12.35 \text{ cm}^3 - 12.07 \text{ cm}^3$$
$$= 0.28 \text{ cm}^3$$

D. In order to calculate the percentage of theoretical density, we must first determine the bulk volume and bulk density.

$$\text{Bulk volume} = \text{apparent volume} + \text{open-pore volume}$$
$$= 12.35 \text{ cm}^3 + 1.80 \text{ cm}^3$$
$$= 14.15 \text{ cm}^3$$

$$\text{Bulk density} = \text{dry weight} / \text{bulk volume}$$
$$= 95.00 \text{ g}/(14.15 \text{ g/cm}^3)$$
$$= 6.71 \text{ g/cm}^3$$

$$\text{Percentage theoretical density} = (\text{bulk density} / \text{true density}) \times 100$$
$$= (6.71 \text{ g/cm}^3)/(7.87 \text{ g/cm}^3) \times 100$$
$$= 85.3\%$$

The mechanical properties for this density could be determined from the data given in Figure 6.26.

Powder Characterization

Metal powders are produced by a number of methods, such as the mechanical grinding of brittle metals and the atomization of liquid alloys by blasting liquid-metal streams. Whatever method is used, the range of particle size and shape must be controlled to optimize compaction and sintering. Several important considerations are the following:

1. Mixed particle sizes give a higher apparent density (20–50% of the theoretical density in the loose, noncompacted state).
2. Fine grain size within individual particles and fine powders give more uniform shrinkage, higher mechanical properties, and faster sintering.
3. Annealing a powder makes possible higher compaction pressures without particle fracture.
4. In high production, the flow rate of powder into the dies must be rapid. Excess lubricant such as the stearates can inhibit the flow.

Figure 6.25 illustrates that above a certain compaction pressure, density cannot be increased significantly. This is due to work hardening of the powder particles and to increased friction between particles and at the die walls. However, with careful control of powder characteristics and compaction pressure, it is possible to attain a green compact density 85% of the theoretical density.

Types of Sintering

In our brass gear example, sintering was accomplished by diffusion in the solid; hence the name *solid-state sintering*. Another common thermal treatment is *liquid-state sintering*. The carbide cutting tools are an excellent example of the liquid process.

Here a tungsten carbide powder is blended with several percent of pure cobalt plus a lubricant and is compacted into a desired shape. Sintering is done at such temperatures that the cobalt melts and the entire shape is bonded. We then have tungsten carbide particles that are completely surrounded by cobalt with the elimination of voids. (See Fig. 11.1)

In any sintering process, several things occur.

1. *Chemical changes.* The powder surfaces desorb any previously adsorbed gases. Organic lubricants burn off.

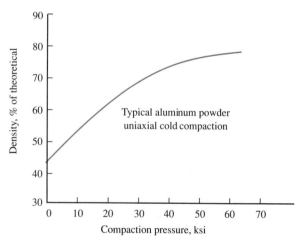

FIGURE 6.25 The compacted density of a powder does not increase linearly with pressure.

2. *Relief of internal stresses.* The particles lose their work-hardened structure from compaction, and elastic strains are relieved
3. *Dimensional changes.* Shrinkage occurs.
4. *Alloying.* In non-prealloyed systems, the individual elemental species diffuse to form an overall bulk composition.
5. *Phase changes.* Examples are liquid formation in liquid-state sintering and any phase changes predicted by the appropriate phase diagram.

6.14 Other Powder Processes and Powder Defects

Most of the variations in the processing of P/M parts employ methods that overcome some of the potential deficiencies inherent in the traditional die compaction shown in Figures 6.20 and 6.21. The objective is usually to develop higher properties through higher density in the sintered shape. An example of these effects is shown for pure iron powder in Figure 6.26.

Repressing after sintering not only provides closer dimensional tolerances but can also close porosity, thereby achieving an outcome closer to the theoretical density. A similar process is the *forged-powder method*, wherein the forging operation closes porosity in the sintered product.

Hot-pressing the powder in heated dies removes the work-hardening limitation on compaction pressure and gives higher sintered density.

Because uniaxial dies may give nonuniform compacted density (and hence nonuniform response to sintering), especially in tall components, *isostatic compaction* offers more uniform compaction. In this case, the powder is encapsulated in a flexible, impermeable mold and is submerged in a fluid that is then pressurized. The hydraulic pressure from all sides gives more uniform density, but achieving dimensional control in even a shaped mold is more difficult than with conventional dies. A variation is *hot isostatic pressure* (HIP), wherein uniform pressure and heated fluid give high compacted density.

Other methods for shaping metal powder are rolling, extrusion, and slip casting. These techniques will be discussed more fully in Chapter 12 because they are also important in the processing of ceramic powder.

Finally, we might consider porosity a defect because the properties vary with density, as shown in Figure 6.26. On the other hand, open porosity can be an advantage for specific applications. Consider the *impregnation* of the voids in a P/M bearing with up to 30 vol. % oil. When the bearing heats by friction, the oil expands and flows to lubricate the surface. Upon cooling, the oil returns to the open porosity by capillary action.

In another variation, open porosity can also be filled with another metal with a lower melting point; this is called *infiltration*. A composite metal structure results that is not dependent on phase equilibrium, and unique properties can be developed.

As with any primary metal process, P/M parts can receive such traditional secondary treatments as machining, heat treatment, welding/brazing, and plating.

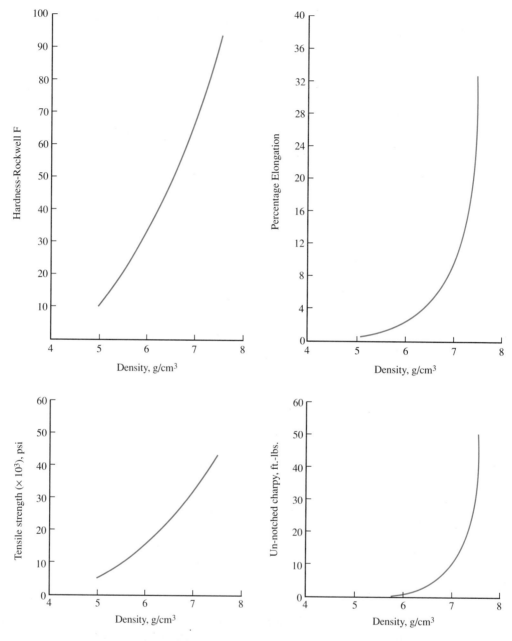

FIGURE 6.26 Mechanical properties of sintered iron related to the sintered density. The data represent six different powders sintered at 1100°C (2012°F) for 1 hr. The theoretical density of iron is 7.87 g/cm³.

[Data reprinted with permission from *Transactions of the Metallurgical Society*, Vol. 171, pp. 485–505 (1947), a publication of The Metallurgical Society, Warrendale, Pennsylvania 15086 USA.]

6.15 Joining Metals

Many methods are available for joining metallic materials, but in this section we will limit our discussion to welding. Mechanical joints, such as those created by bolting and riveting, are commonly treated in mechanical design. And, the use of adhesives is certainly an important and rapidly expanding field. However, adhesives are organic in nature, so we will defer their discussion until we cover polymeric materials (Chapter 14).

Welding generally implies bonding by a fusion process and, in most cases, can be thought of as a "fast-casting" technique. Therefore all of the principles we noted when we discussed casting, such as phase diagrams, nucleation/growth, and residual stresses, apply also to welding. The major difference is that nonequilibrium structures are more likely to occur in welding because the process is so rapid.

Before proceeding with a simple example of welding a cold-worked metal, we must differentiate between welding and the similar processes soldering and brazing.

Soldering utilizes a filler metal with a relatively low melting point (below 1000°F or 538°C) compared to the solidus temperatures of the metals to be joined. Common fillers are lead–tin alloys. Safe stresses in design are less than 1000 psi (6.9 MPa), and continuity is more important than the ability to withstand high operating stresses. Because of the low melting point of the filler, the alloys are also subject to creep.

Brazing uses filler metals that melt at a temperature that is above 1000°F (538°C) but is less than the melting range of the base metals to be joined. The joint strength can be higher than that of the base metal as long as the filler thickness is small. The greater the filler thickness, the greater the tendency for the joint strength to approach that of the filler. Copper–zinc alloys and silver are common fillers for joining ferrous alloys.

Fluxes are used in both soldering and brazing. They perform two essential functions. They clean the surfaces of oxides and coatings, and they prevent further reaction with the atmosphere. This provides the necessary wettability between the filler and the base metal.

One key to deciding whether to use soldering or brazing rather than welding and to choosing the composition of the filler metal is recognizing that many base metals are joined in the heat-treated condition. The objective is to select a joining process that will least modify the microstructure of the base metal.

Of course, achieving higher joint strength requires trade-offs. However, we can see the effect of changes in microstructure by considering an example of the electric arc (or so-called stick welding) of a cold-worked, single-phase metal (Figure 6.27). If we now cut across the weld, view the microstructure, and conduct a hardness traverse, we find variation in the hardness shown schematically in Figure 6.28. The following areas or zones can be identified:

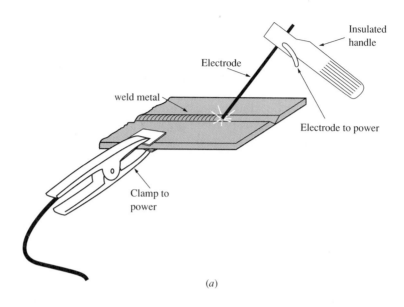

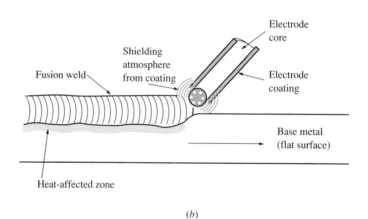

FIGURE 6.27 (*a*) Essential elements of shielded metal arc welding. (*b*) Enlargement of the electrode area in "stick welding."

1. *Unaffected zone.* The parent metal remains cold-worked and retains its cold-worked hardness.
2. *Transition, refined, and coarsened zones.* In combination, these are called the *heat-affected zone* (HAZ). The hardness gradually falls from the transition zone (partial recrystallization) to the area where grain growth occurs (the coarsened zone).
3. *Fusion and deposited-metal zones.* The fusion zone shows the structure characteristic of metal that has exceeded the solidus but remains below

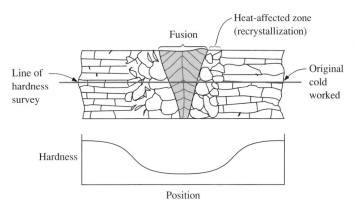

FIGURE 6.28 Schematic representation of fusion welding of a cold-worked metal

the liquidus. The melted-metal filler then constitutes the deposited metal zone. Here it is assumed that the parent and filler metals have the same chemical analysis.

A similar variation in hardness occurs in precipitation- or martensite-hardened microstructures that are welded, although the plots may be more complex.

6.16 Welding Methods

We can divide the welding processes into three general categories based on the nature of the energy they employ: mechanical energy alone, mechanical and thermal energy, or thermal energy alone. Here we will broaden the definition of welding to recognize that filler metals and the external application of heat need not always be present to obtain a good metallurgical bond. A brief description of welding processes and a few examples will make it easier to understand our new definition.

1. *Mechanical energy.* Although in this method we may begin with cold materials, mechanical energy is converted to heat energy to produce the joint. An example is the *friction welding* of shafts with no addition of filler metal, as shown in Figure 6.29.
2. *Mechanical and thermal energy.* Again, in this case, filler metals are not used. The closing of internal voids in cast ingots by hot rolling is an example we have encountered before. In the common *spot weld*, shown in Figure 6.30, pressure and heat from the flow of an electric current produce the joint.
3. *Thermal energy.* Filler metals are often used in this process, such as in the shielded metal arc welding (stick welding) shown in Figure 6.27. *Gas welding* is a similar process wherein oxygen and acetylene gases are used for

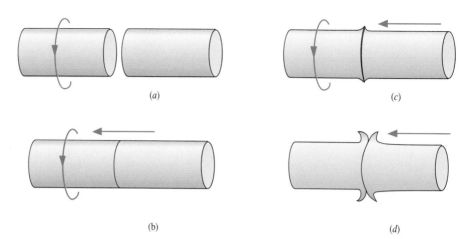

FIGURE 6.29   Successive steps in friction welding. When the parts come in contact, heat is generated and the amount of pressure determines the flash, as shown in (*d*).

[R. A. Lindberg and N. R. Braton, *Welding and Other Processes*, Allyn and Bacon, Boston, Mass., 1976, p. 47]

the heat source. However, filler metals need not always be used in these thermal processes. A local heat source can quickly melt the adjacent base metal, and be recast as the heat source is moved to a new position. *Electron-beam* and *laser welding* fall into this category.

6.17 Welding Defects

The objective in welding is to produce a continuous and homogeneous component with minimum disruption of the parent microstructure. This generally means that the heat-affected zone (HAZ) must be as narrow as possible. Therefore, a gas-welded joint with its wider HAZ would locally soften a cold-worked metal moreso than arc welding.

Heat is very localized in laser and electron-beam welding because of their small HAZ. However, these processes are expensive due to the special equipment needed. In diffusion welding (Figure 6.31) a very narrow concentration gradient at the interface of the dissimilar metals may be the only equivalent of the heat-affected zone.

In view of many new discoveries, new welding processes have been developed that utilize unique combinations of heat and mechanical energy sources. An example is *explosion welding*, wherein solid-state fusion is accomplished by a controlled detonation that provides a high-velocity contact between parts to be joined. *Ultrasonic welding* is also a solid-state joining process wherein external pressure is applied in conjunction with local high-frequency vibration energy.

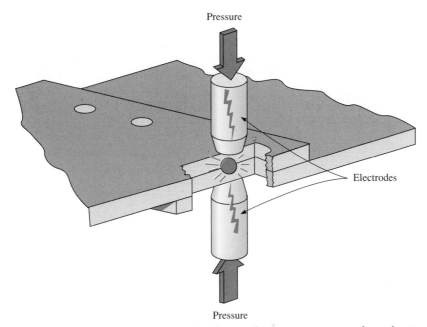

Pressure

Electrodes

Pressure

FIGURE 6.30 Pressure and heating by electrical resistance are used to obtain a fusion spot weld.

[J. W. Geachino, W. Weeks, and G. S. Johnson, *Welding Technology*, 2nd ed., p. 110. American Technical Publishers, Inc. Reprinted by permission.]

There are, of course, defects that can occur in any welding process that can influence the service performance of components. Some of these defects—such as gas porosity, shrinkage, hot tearing, and inclusions—are similar to those discussed for casting and need not be considered further. However, other defects are peculiar to the welding process.

Figure 6.32 illustrates defects that can be introduced during welding.

1. *Cracks.* Several types of cracks are possible: crater, transverse, longitu-

FIGURE 6.31 A schematic representation of multilayer diffusion welding or merely bonding by a diffusion process. As with all diffusion, the rate increases exponentially with temperature.

[R. A. Lindberg and N. R. Braton, *Welding and Other Processes*, Allyn and Bacon, Boston, Mass., 1976, p. 50]

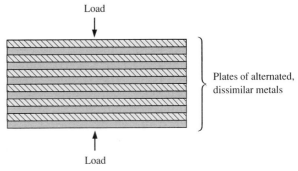

Load

Plates of alternated, dissimilar metals

Load

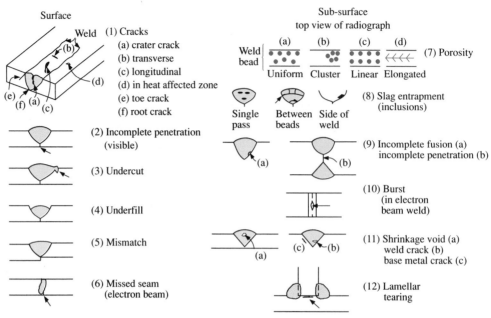

FIGURE 6.32   Typical defects found in welding

dinal, in heat-affected zone, toe crack, and root crack (due to stresses on heating and cooling).

2. *Incomplete penetration.* This defect is due to too little heat.
3. *Undercut.* This is due to poor control of weld rod.
4. *Underfill.* This is due to poor deposition of weld rod.
5. *Mismatch.* This is a shift of mating parts in a fixture.
6. *Missed seam.* In an electron-beam weld, this defect is due to poor alignment of the parts to be welded.
7. *Porosity.* Porosity results from evolution of gas from the weld metal; also due to shrinkage.
8. *Slag entrapment.* This is due to improper control and poor removal of slag between passes.
9. *Incomplete fusion.* This defect is due to poor manipulation of weld rod or a dirty joint.
10. *Burst void.* This is due to shrinkage and/or gas in an electron-beam weld.
11. *Shrinkage void.* These internal defects are caused by shrinkage during solidification and by weld stresses.
12. *Lamellar tearing.* These cracks are due to weld stresses.

Lack of fusion and penetration types of defects occur when the welding conditions do not produce a hot enough weld bead to dissolve (penetrate) the base metal. This leads to a weak joint. It is quite evident compared with a normal weld.

Undercutting takes place when the weld melts the base metal at the side wall but does not fill in the groove. This gives a longitudinal notch that can lead to failure.

Hydrogen cracking occurs if hydrogen from the materials of the weld-rod coating is dissolved in the bead; it may precipitate later at inclusion sites at very high pressure and generate cracks.

Chapter 18 discusses certain types of corrosion peculiar to welds.

EXAMPLE 6.5 *[E]*

What is the sign of the stress (tension or compression) in a weld, and why might the weld be peened with a hammer?

Answer Because the weld filler metal is molten, it undergoes both solidification shrinkage and contraction during cooling. The parent metal is much cooler and larger in mass: this inhibits the contraction of the weld material. Therefore the weld is in tension.

Peening of the weld introduces compressive stresses that tend to relieve the built-up tension. It also removes the surface oxides and residual flux present on the weld surface. Although hand peening may be used, shot blasting can produce more uniform cleaning and stress removal. A residual stress in compression that is developed by peening may even be an advantage in a weld, because fatigue initiation is often enhanced by residual tensile stresses.

A number of defects can result from the incorrect use of shielding gases and fluxes associated with a given process. Normally the gases and fluxes clean the metal surfaces and prevent oxidation during welding. However, gases also serve other important purposes.

In *Heliarc welding*, electrons from the electric arc interact with helium ions to produce a metastable excited state within the gas. When the helium reverts to its more stable unexcited state, energy is released as heat. If argon is used as the gas, the excited state has a lower energy difference with respect to the stable state. Less heat is therefore evolved and lower temperatures are achieved. The mixing of gases, then, can provide a wide range of heat input to the components to be welded.

Finally, residual stress in a weld can cause premature failure or, in thinner sheet metal, can result in buckling or wrinkling. Preheating a joint gives a lower thermal gradient between the parent and filler metal, whereas postheating can be an effective anneal for stress relief.

6.18   The Machining Process

In a very general sense, *machining* means the removal of the material to produce a desired shape. As such, it can compete with other processes like casting or forming, yet more often is a secondary operation following these processes.

There are many variations in machining equipment, but they all include the simple components shown schematically in Figure 6.33.

1. A *workpiece* is a component whose shape is to be modified.
2. A *workholder* is necessary to prevent the workpiece from moving when cutting forces are applied.
3. The *cutting tool* modifies the workpiece geometry through the removal of material.
4. The *toolholder* prevents undesired motion of the tool.
5. Finally, there must be relative *motion* between the workpiece and cutting tool, and there must be a *force* to provide the cutting action.

It is important to realize that these five components are present in all machining processes, from handsawing a piece of wood to the most complex, multitask, computer-controlled machining of a space-age alloy.

Several principles introduced earlier should be reviewed. For example, the simple cutting of a material suggests study of wear and abrasion (Section 2.19). The wear occurs on both the workpiece and the cutting tool. As a result, microstructures at the contact surface may be modified.

The important considerations then become cold working (or other microstructural change), tool wear, and process economics while achieving control over dimensions and surface finish. Several of these points are discussed further in the next section.

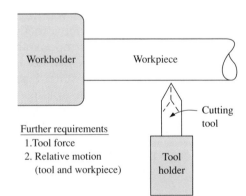

FIGURE 6.33   The components essential to all conventional machining processes

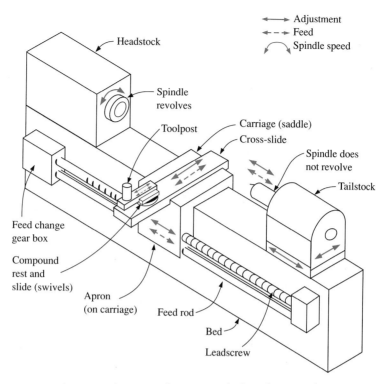

FIGURE 6.34 Schematic drawing of an engine lathe, showing the major movements

[Ludema/Caddell/Atkins, *Manufacturing Engineering: Economics and Processes.* © 1987, p. 216. Reprinted by permission of Prentice-Hall, Inc., Englewood Cliffs, N.J.]

6.19 Variables in Machining

In all of the machining processes, the relative speed, feed, and depth of cut control the life of the tool and the tendency for a microstructural change. The schematic drawing of a common engine lathe in Figure 6.34 shows how speed and feed can be controlled. The depth of cut, of course, depends on the relative positions of the tool and the workpiece (Figure 6.35).

The amount of work to be done in machining can be determined by finding the area under the stress–strain curve (see Figure 6.14), realizing that local frac-

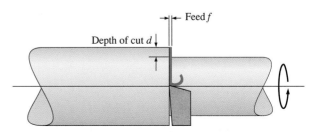

FIGURE 6.35 The depth of cut in simple turning

[Ludema/Caddell/Atkins, *Manufacturing Engineering: Economics and Processes.* © 1987, p. 217. Reprinted by permission of Prentice-Hall, Inc., Englewood Cliffs, N.J.]

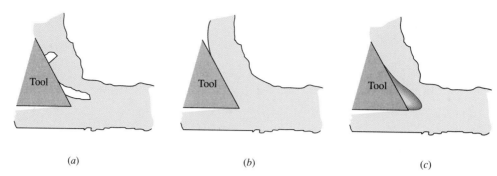

FIGURE 6.36   Three types of chip formation: (*a*) segmented, (*b*) continuous, (*c*) continuous with built-up edge

[R. A. Lindberg, *Processes and Materials of Manufacture*, 2nd ed., Allyn and Bacon, Boston, Mass., 1977, p. 127]

ture of the workpiece at the tool interface must occur. The strain-hardening exponent, the tensile strength, and reduction in area are important to determination of the work to be done.

As the area under the stress–strain curve increases, the amount of heat generated increases, and this heat must be carried away by *both* the workpiece and the tool. There is then an opportunity for a surface change in microstructure due to the local heating.

All materials do not machine in precisely the same manner and some insight can be gained by observing the three types of chips that can be formed (Figure 6.36).

1. *Discontinuous chips.* The chips break up because of the inherent limited ductility of the workpiece. An example is gray cast iron.
2. *Continuous chips.* A continuous ribbon of material is produced as a result of the high ductility of the workpiece. There is a flow of metal along the tool face and little or no tendency for the chip to adhere to the tool.
3. *Continuous chips with built-up edge.* In this case there is more contact area between workpiece and tool, and local "welding" occurs on the tool surface. This condition generally leads to poorer surface finish because the tool geometry is changing.

The wear of the tool material depends on the difference in inherent hardness between workpiece and tool *and* on the type of chip that is formed. If welding occurs between workpiece and tool, the strength of the weld determines the relative wear, as we noted in Section 2.19.

EXAMPLE 6.6   *[ES/EJ]*

A *built-up edge* on a tool is considered detrimental. Explain why each of the following might minimize the built-up edge.

1. Using a cutting lubricant
2. Decreasing the feed rate and depth of cut
3. Increasing the cutting speed

Answer

1. The lubricant inhibits contact welding through lower friction and can provide another vehicle for heat removal.
2. When the thickness of the chip is decreased, there is less energy to be dissipated as local heating.
3. At higher speeds the system runs hotter, which would seem to contradict the foregoing arguments. The strength of the weld is important, however, and higher temperatures can anneal the weld, making it *weaker* such that failure is through the weld. A built-up edge occurs when the weld is strong, i.e., not annealed.

Tool materials are normally hard, and their ability to retain this hardness at high temperatures is important. Their strength must be high because large forces are applied to them over a very small area. Several materials that meet these criteria are discussed in Chapter 9 and Chapter 10.

6.20 Defects in Machining, and Other Machining Processes

A few of the defects that can occur in machining are

1. *Limited dimensional control.* Although a major objective of machining is to control dimensions, there are limitations to tolerances. For example, when a material is heated, it expands locally, and its dimensions in the heated condition are not precisely those found when it is cooled. As pointed out in Chapter 1, the coefficient of expansion is related to the bond energy, which is likely to be different for the workpiece and the tool.
2. *Residual stress.* The forces of machining coupled with heating and cooling can impose a residual stress in the material. Also, nonsymmetrical machining can cause dimensional variation and distortion. A casting that may have a residual stress when machined on one surface can distort as a result of redistribution of the stress.
3. *Surface finish.* Speed, feed, and depth of cut can modify the finish. Tool marks may simply affect appearance, may influence wear resistance, or may become points where fracture begins (such as when fatigue is a design consideration).
4. *Metallurgical changes.* It is possible for machining to heat the workpiece so much that it modifies the surface microstructure. High-speed grinding of a hardened steel can destroy the surface hardness by tempering the hard

martensite structure. Similarly, a precipitation-hardened material can be overaged on its surface.

It should be pointed out that all machining processes do not have an equivalent tendency to produce the preceding defects. We have used the example of turning (engine lathe) in our discussion, and there is general familiarity with such other processes as sawing, drilling, threading, and grinding. These are conventional machining processes and several more should be identified. A summary of a few of these processes appears in Figure 6.37.

In *milling* the cutting tool rotates, and the workpiece, clamped to a table, moves in a linear fashion. The rotating tool has multiple cutting edges. In *shaping* a single cutting edge is moved back and forth across a clamped, non-rotating workpiece surface. Material is removed only on the forward stroke. If we now provide a tool with multiple cutting edges and the same back-and-forth motion, we have a *broaching* operation.

A number of nonconventional machining processes have also evolved to overcome some of the difficulties attributed to the conventional methods. Examples include *laser machining* and *electron-beam machining*, which are analogous to their welding counterparts. Very fine cuts are attainable by these means, but the thickness of the workpiece is usually limited to 0.250 in. (6.35 mm). Heat damage to the workpiece is minimal.

Another example of welding technology applied to machining is *torch* or *arc machining*. In this case, however, the heat-affected surface depth is greater than in conventional machining.

Minimal microstructural damage together with very low surface residual stresses can be obtained by still other processes.

In *electrical-discharge machining* (EDM), a current is passed between the workpiece and the tool through electrical discharge in a dielectric fluid. Small particles of metal are removed from the workpiece during the discharge, and the shape of the tool controls the shape of the hole. We can "drill" square holes in very hard materials by this method.

A similar process is *electrochemical machining* (ECM), wherein the workpiece is made an anode in an electrolytic cell. An electrolyte such as a sodium chloride solution is pumped between the cathode (tool) and the anode to remove material by accelerated corrosion. In *ultrasonic machining*, abrasive particles suspended in a fluid are vibrated at high frequency between the tool and the workpiece.

These techniques are expensive, but they can be used on hard materials where irregularly shaped cavities are necessary.

6.21 Thermal Treatment

We have spent considerable time in earlier chapters discussing heat treatment as a method of controlling the metal microstructure. We will now review a few of these concepts from a more general standpoint.

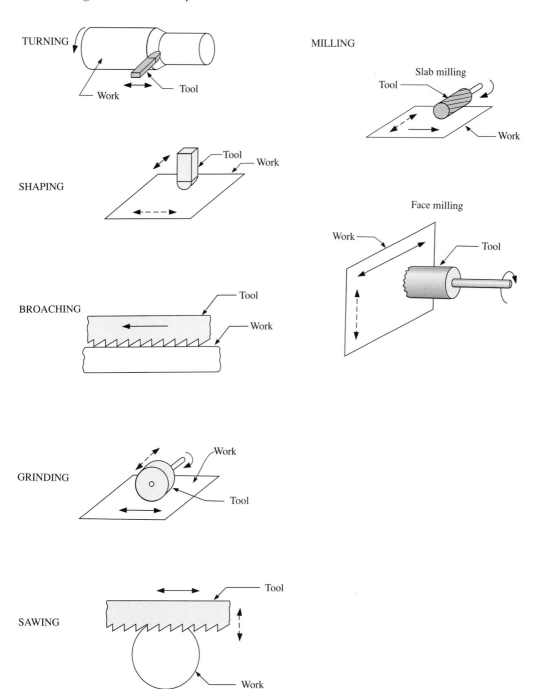

FIGURE 6.37   Schematic diagrams of several machining processes

[Reprinted with permission of Macmillan Publishing Company from *Materials and Processes in Manufacturing*, 6th ed., by E. P. DeGarmo, J. T. Black and R. A. Kohser, Copyright © 1984 by Macmillan Publishing Company.]

There are essentially three reasons to apply heat *only* to a metal — that is, reasons why an external force is not also applied, as it would be in hot forming.

1. *To enhance a diffusion process.* The diffusion rate increases exponentially with an increase in temperature. We also find that chemical reactions occur more readily at higher temperatures because of the greater mobility of the reacting species.
2. *To homogenize a structure.* This implies that a nonequilibrium or metastable structure is present and that heating allows equilibrium to be achieved. An example is the heating of a cast structure to obtain stress relief and more uniform microstructural chemistry or phase distribution.
3. *To obtain a nonequilibrium structure.* Here it is assumed that the nonequilibrium condition is more desirable. Examples include age hardening and martensite heat-treatment cycles.

It should be emphasized that these thermal treatments are not simply temperature-controlled but are also dependent on time. Most defects that occur in thermal treatment are due to poor control of the temperature, of the time at temperature, and of heating or cooling rates. A few examples will illustrate the importance of these controls.

It is obvious that we must achieve an *optimal temperature* in order to obtain the desired microstructure. For example, the "window" for obtaining all α in the solution treatment of a 4% Cu–aluminum alloy is 75°C, as shown in Figure 6.38. Above 575°C we would have a liquid phase and a destroyed component. At 500°C or below, we would have some θ phase that would appear as a continuous brittle network at the α grain boundaries. (Although high temperatures might be suggested where the thermal window is large, there is a greater risk of reaction with the furnace atmosphere and, consequently, the production of the surface scales.

Time at temperature is another important control. In our example, 10 min at, say, 550°C may not be long enough for all of the θ phase to go in solution. Upon quenching and aging, the hardness may be too low. As a part becomes thicker, longer times at temperature are required because it takes longer to achieve the desired temperature at the center of the component.

Heating and cooling rates are related to optimal temperature and time at temperature. A high cooling rate may be necessary to retain supersaturated α in the age-hardening sequence. On the other hand, a high thermal gradient (surface to center), as in a water quench, leads to residual stress. In some systems (such as forming martensite in steels), the phase-transformation strain coupled with a high cooling rate can lead to cracking. Specifying the appropriate thermal treatment cycle then becomes a matter of selecting a heat source and the corresponding cooling rate to control the reaction kinetics. Special names are often given to these cycles. They will be described further and applied to both nonferrous and ferrous metals in Chapters 7–9.

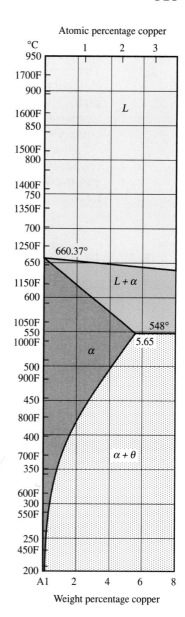

FIGURE 6.38 Aluminum-rich end of the aluminum–copper phase diagram

[Reprinted with permission from *Metals Handbook*, 8th ed., Vol. 8, *Metallography, Structures and Phase Diagrams*, American Society for Metals, Metals Park, OH, 1973, p. 259.]

EXAMPLE 6.7 *[ES/EJ]*

Why might controlling the furnace atmosphere be as important as controlling the temperature, time at temperature, and heating and cooling rates during heat treatment?

Answer   The entire thermal cycle is intended to accomplish a specific objective, but reaction with the furnace atmosphere can undo the best efforts. The formation of surface scale (such as an oxide) or the selective removal of elements at the metal surface (as in the *decarburization* of steel) can lead to undesired service performance. Either lower surface strength or the introduction of stress-raising surface phases can be disastrous in a highly stressed component.

However, the use of more desirable atmospheres with low oxidizing potential adds to the processing cost. It is therefore important to determine the end use for the product before specifying the atmosphere requirements and necessary degree of control.

6.22   Finishing

A steel automobile fender may begin as a cast slab that is rolled to sheet stock, coated with zinc for corrosion protection, formed to the desired shape, and finally painted before or after assembly. During all of these processes, measurements of "quality" are conducted and compared to standards that can be broadly categorized as "specifications."

The difficulty of precise specifications in commercial products will be treated in the next section. Here we will describe some of the finishing processes that modify both the appearance and the function of a component.

Surface finishing operations fall into four major categories: mechanical, thermal, and chemical surface modification and surface coatings.

Mechanical surface modification.   After primary processing such as casting or forming, burrs, fins, and scratches may have to be removed by abrasive blasting, tumbling, wire brushing, or buffing. This is generally done for appearance, although service performance and safety may also be reasons for the treatment.

Thermal surface modification.   This may be a consequence of a normal heat-treatment cycle. An example is quenching a steel to form hard martensite at the surface while the core or center cools too slowly to form the metastable martensite.

Chemical surface modification.   Here the chemistry of the surface is changed, normally by diffusion. Examples include *carburization* (carbon diffusion) and *nitriding* (nitrogen diffusion) into the surface of steel to provide better wear resistance. A similar process is *ion implantation*, wherein a metal surface is bombarded by an accelerated beam of other metal ions. The ions wedge into the surface atom layers, giving higher resistance to wear and corrosion.

Surface coatings.   There are numerous coatings that can be applied to metal surfaces. Paint is an example of a coating applied for corrosion resistance and appearance. Metallic coatings perform the same functions and also boost resistance to wear. Methods of applying metal range from electrode-position to flame spraying.

It is important to recognize that the finishing operations can be expensive and are utilized mainly when service performance dictates their necessity. As in any process, defects in production can occur and can lead to undesirable consequences. For example, simple pinhole porosity in a painted surface can lead to shortened product life, because it reduces resistance to corrosion.

6.23 Quality Control

The history of engineered components bears out the dictum that "nothing lasts forever." On the other hand, life cycles can at least be made predictable through the designation of minimum specifications. The guarantee of these minimum specifications that can be used in design is one definition of "quality." In a broader sense, it is also assumed that the defined level of quality can be "controlled."

Throughout the processing of a component, checks must be made to determine whether the final specifications can be met. These checks may be dimensional, microstructural, chemical, electrical, tests of strength, or whatever is necessary to ensure adherence to "anticipated performance standards."

We have two questions to answer after the level of desired quality is defined:

1. How do we check for quality?
2. And how often?

The first question has as many answers as there are properties that have to be controlled and for which reliable tests are available. And the problem becomes even more complex when we realize that some tests are destructive whereas others do not harm the part.

Thus the different test methods are divided into nondestructive and destructive tests. Both are covered more thoroughly in Chapter 19.

1. *Nondestructive tests (NDT).* A dimensional analysis is likely to fit into this category, as does a simple visual inspection. For example, a weldment that is only partially intact may not be considered "good quality." More sophisticated nondestructive methods include x-ray inspection for internal porosity and ultrasonic inspection (sound transmission depends on material thickness, porosity and microstructure). The level of technology associated with NDT is changing rapidly because of the development of sensitive instruments that interface with computers for the rapid acquisition and comparison of data (see Section 6.24).

2. *Destructive tests.* In a tensile test, the sample is destroyed. Similarly, cutting a sample for microexamination and microstructural control is destructive, though the sample may be taken from an area that is ultimately to be removed by machining. A hardness test, though it leaves an indenta-

tion and is therefore destructive, can also be performed in a noncritical area. Whereas NDT may be applied to all components, (though there is generally a high cost involved), destructive tests are carried out at selected intervals to preserve the maximum number of undamaged parts.

This brings us to the question of how often we must check the quality. In practice, we can test a number of random samples and establish a statistical distribution from which "control limits" for a given process can be determined.

This practice, known as *statistical process control* (SPC), makes it possible to check any given portion of a process to find out whether it is "in" or "out of" control in terms of prior statistical experience with the process.

We will not pursue SPC concepts further here. However, you should appreciate how difficult it is to deal with absolute numbers in specifications and design. To illustrate, let us briefly consider hardness data for ½-in.-diameter bolts (Figure 6.39).

Hardness is used because there is a reasonable correlation between Rockwell C hardness and the tensile strength of a bolt. We need the following data in order to interpret Figure 6.39.

1. A single heat of 1038 steel was used to manufacture the bolts. (This material is commonly used for threaded fasteners.)
2. All bolts were heat-treated in one plant to eight prescribed hardness levels. (Only five levels have been included in the figure.)
3. Hardnesses on the finished bolts were taken at eight different laboratories.
4. The hardness location was the same at all laboratories, as indicated by the sketch.

If we look at the data for a nominal or aim hardness of 30 HRC, we find that the hardness range is approximately 25 to 32.5 HRC. This would translate into a bolt tensile strength of 125,000 to 150,000 psi, as determined from actual tests on bolts. If the data are typical, the engineer using bolts has to consider the statistical variation.

We will not discuss the possible reasons for the data spread here, but you might consider the following potential sources of error:

1. Manufacturing contributions such as metal chemistry, rod stock production, bolt production, and consistency of the heat treatment.
2. Testing contributions such as sample preparation, hardness location, machine standardization, and test technique.

In Chapter 23 we will again discuss the relationship between "typical" properties for engineering materials and the control of product quality.

6.24 Computer Use in Metal Processing

Computers can be used effectively in any of the unit processes employed in the manufacture of engineering components. The objective in all cases is to produce a more consistent product of higher quality at a lower cost. However,

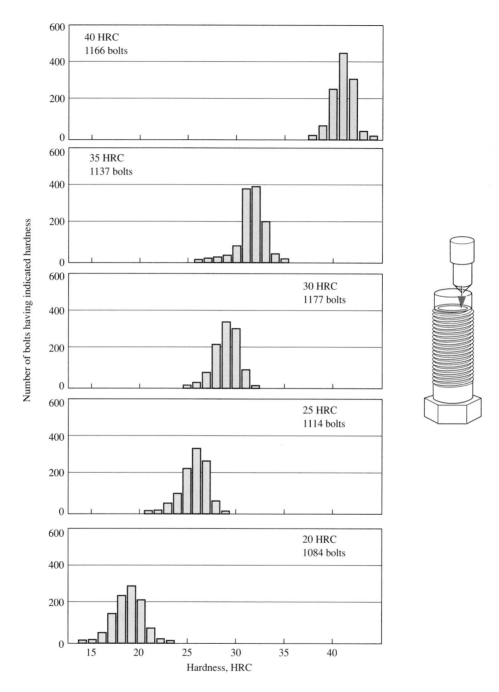

FIGURE 6.39 Hardness distribution for several lots of ½-in.-diameter 1038 steel bolts

[Adapted with permission from *Metals Handbook*, 9th ed., Vol. 1, *Properties and Selection: Irons and Steels*, American Society for Metals, Metals Park, OH, 1978, p. 281.]

there is no guarantee that this objective can be met, especially if the problems are incorrectly identified.

The scope of computer use can best be appreciated by considering simple examples in which computers are incorporated into the casting, machining, and welding processes.

Metal Casting

It is necessary to make several important decisions in order to produce a sound casting. The size of risers, their position on the casting, and the use of chills make up one set of variables that must be optimized to produce a casting free of shrinkage defects.

Theoretical and empirical heat-transfer models for metal castings are available as software packages from several sources. These programs—together with graphics support—make it possible to study solidification patterns and the tendency toward shrinkage porosity. By selecting several alternative riser designs, one can determine an optimum without having to pour a series of test castings.

The technique just outlined is one form of *computer-aided design* (CAD), and using it can save considerable time. Similar approaches can be used for design of the gating system for a casting.

One note of caution: Any so-called computer solution is only as valid as the program and the supporting data base. Therefore, a prototype casting is necessary to verify the final CAD solution and to provide assurance that all of the variables have been properly identified.

Machining

A typical example of computer applications in machining is the use of a *computer numerical control* (CNC) vertical milling machine operation. A schematic drawing of the machine is shown in Figure 6.40. If we secure the head and spindle from vertical motion, we can determine x, y, and z coordinates for the revolving tip of the cutter with respect to a workpiece clamped to the table.

We then program a series of x and y coordinate motions for the table, together with z coordinates for the knee of the machine. Upon selection of speed and feed for the cutting tool, contoured shapes can be machined. The cutter shape must also be included in the program, because cutter diameter limits the minimum internal radii.

We might well want to use our digitized CAD solution for the final casting design to produce the necessary pattern shape on the milling machine. The machining process would then be called *computer-aided manufacturing* (CAM), and the total process CAD/CAM.

Although CNC machine programming takes time, it may make a complex machining operation faster than having a machinist perform the same task on

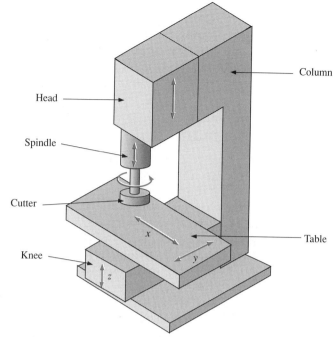

FIGURE 6.40
Schematic drawing of a vertical milling machine that could be adapted to computer control

[Adapted and reprinted with permission of Macmillan Publishing Company from *Materials and Processes in Manufacturing*, 6th ed., by E. P. DeGarmo, J. T. Black and R. A. Kohser, Copyright © 1984 by Macmillan Publishing Company.]

a non-CNC machine. As the number of duplicate components increases, the advantages the computer bestows become more obvious, especially when we consider that the machine can be reprogrammed for other tasks.

Welding

In electric-resistance spot welding (see Figure 6.30), it is difficult to determine the integrity of a weld except by visual examination and destructive testing of random samples. Another method of inspection employs a computer to both inspect and modify the weld parameters while making a series of weld buttons.

When a spot weld solidifies, strain energy develops and has characteristics that change with discontinuities such as porosity and with the amount of residual strain (see Example 6.5). This energy can be detected at the surface as an "acoustic event" that is referred to as *acoustic emission*. The profile of the emissions can be detected by a transducer attached to the surface (see Chapter 20).

We now determine the emission characteristics generated by a suitable spot weld and store the profile in a computer memory as an "ideal weld." Upon making other spot welds, we compare their emission characteristics to those of the ideal as a test of weld integrity.

Although this technique can be used simply as an inspection method for determining acceptance or rejection, it is also possible to devise an active feedback loop that will modify the weld parameters and produce a button closer to the ideal on the next cycle. The computer makes possible a very rapid response to the comparison between the profiles of acoustic emissions and facilitates the implementation of any necessary change.

The level of technology will continue to evolve in manufacturing processes, and the computer is likely to play an increasingly important role.

Summary

We now have an overview of the nature of individual operations and their relation to different metal structures. Let us sum up these points before proceeding to examine the processing of the different alloy families in the next three chapters.

Casting is the first process that practically every metal component undergoes. In only about 20% of the time is an effort made to cast the part near the final shape; this is done in foundries (casting shops). In the remaining 80% of all casting, the metal is cast into ingots, slabs, or rounds that are then processed further.

In any alloy family, it is important to distinguish between the alloy compositions that can only be cast and those that can be worked (wrought alloys). In the aluminum alloys, the high-silicon alloys can only be cast because of the large amounts of the high-silicon phase. In the ferrous alloys, the cast irons are not forged either, this time because of large amounts of nonductile graphite or large grains of hard and brittle iron carbide.

In specifications, therefore, it is essential to know whether we are dealing with a *cast* or a *wrought* alloy. *Powder metallurgy* products are yet a third option.

In *deformation*, we distinguish between the processes that produce gross bulk deformation (such as reducing a large ingot to a shaft or plate) and the processes that don't (such as forming a sheet into an automotive fender without much change in material thickness). In general, the hot-working processes are used to accomplish bulk deformation, and cold working is used for near-finished parts.

Particulate processing, or *powder metallurgy*, is a technique for producing parts to near net shape and also for forming mixtures of structures that cannot be obtained by melting and casting, such as cobalt-bonded tungsten carbide. Limitations of die design, press size, and die costs make this process appropriate primarily for large numbers of small parts.

Joining is valuable for attaching complex shapes and for the automated production of simple shapes. Components made by different methods, such as castings and forgings, can be joined with caution. To prevent disastrous failures, the structures in the fusion zone and the adjacent heat-affected zone must be understood, and residual stresses must be accounted for.

Machining seems superficially the simplest and easiest way to obtain a desired shape. However, modern engineering design attempts to minimize machining by casting or forming parts to near their desired (net) shapes. This saves the cost of the material otherwise devoted to producing chips that have low value, and it reduces the labor and energy costs of the process. Also, the soundest material is often near the surface of a part, and machining removes this. Another precaution is to avoid the change in structure and even cracking that surface heating can cause during the machining process.

Thermal treatment is one of the most powerful methods for improving the properties of a component. Two vital considerations must always be borne in mind when using thermal treatment: (1) what structural changes are taking place and whether the test-bar properties are representative of these and (2) how the pattern of residual stress is being changed. The build-up of harmful residual stress in a complex component can overshadow the beneficial effects displayed in test bars!

Finishing, or surface treatments such as carburizing, plating, and shot peening, can greatly enhance the life of a component and may often be used to produce the desired structure where it is needed rather than throughout the part. As an example, a gear needs a shallow "case" only on its surface to enhance wear resistance. The hazards associated with these final treatments include hydrogen embrittlement in plating and the redistribution of surface stress when hard surfaces are applied.

Finally we should note that control of quality during all processing stages is essential to successful product performance. The control can be enhanced by appropriate use of the computer.

With this background in processing, we may now proceed to investigate the metal structures and how they are used or specified.

Definitions

Anisotropy   The directionality of properties in a material that results from orientation of phases and crystallographic orientation.

Apparent density   The mass divided by the apparent volume of a material. The apparent volume accounts for the open pores but does not recognize any closed pores. The true volume is equal to the apparent volume when there are no closed pores.

Brazing   Fusion with a filler metal that has a liquidus temperature above 1000°F but lower than the solidus temperature of the parent metals to be joined.

Built-up edge   Local welding between a machine tool and a workpiece that results in modification of the tool geometry.

Bulk deformation   A deformation process wherein there is extensive change in cross-sectional thickness, such as in rolling or forging.

Bulk density   The mass of a loosely packed container of material per unit volume of the container.

Burning   Heating of steel or other alloy above the solidus temperature during heat treatment, producing intergranular voids.

Carburization   The diffusion of carbon into a metal surface. The diffusion of carbon out of a metal surface is called *decarburization*.

Castings   Parts made by pouring liquid metal into a mold of the proper dimensions. The mold may be made of sand bonded with clay or resin (sand castings) or metal (die or permanent-mold castings).

Chill   An insert such as graphite in a sand mold that extracts heat more rapidly than the sand.

Closed pores   The internal cavities in a material that are not penetrated when the material is immersed in a liquid.

Cope flask   The top form for holding the sand in a casting.

Dendrites   Fern-like solid spikes growing into the liquid phase during the solidification process.

Drag flask   The bottom form or box for holding the sand in a casting.

Earing   A scalloped edge at the rim of formed cup shapes that results from material anisotropy. Trimming is usually necessary.

Fluidity   The inverse of viscosity. Superheat is the temperature difference between pouring and the liquidus, that is, it is an index of fluidity.

Fluxes   Cleansing and atmosphere-control agents used during the fusion processes.

Gas porosity   Rounded voids created in a casting, ingot, or weldment by the evolution of dissolved gas during solidification.

Gating   Channels necessary for the entry of molten metal into the casting cavity.

Heat-affected zone (HAZ)   That portion of the parent metal interface that undergoes metallographic modification during a fusion process.

Hot tear   A fracture in casting due to restraint (often by the mold) when the temperature is between the liquidus and the solidus.

Hydrogen cracking   In welding, cracking produced by the dissolving of excessive amounts of hydrogen during welding, as a result of high hydrogen content of the weld rod.

Ion implantation   The placement of ions in a metal surface via bombardment with an accelerated beam of other metal ions.

Isostatic compaction   A cold- or hot-compaction method by which a solid is consolidated under hydrostatic pressure.

Machining   Removal of material to produce a desired shape.

Nitriding   The formation of metal nitrides at a surface by nitrogen diffusion.

Nondestructive tests (NDT)   The examination of a component by methods that do not change the surface or center of the part. Examples include x-ray radiography and ultrasonic inspection.

Nonmetallic inclusions   Brittle oxides or silicates and less brittle sulfides that result from the entrapment of foreign matter in liquid metal during pouring or from reactions within the liquid metal.

Open pores The spaces between particles of a material held in a container that are filled when a liquid is added.

Packing factor A true or apparent volume divided by bulk volume; or a bulk density divided by true or apparent density. True and apparent packing factors are equal when there are no closed pores.

Parting line The surface between the cope and the drag where the mold is separated to remove the pattern.

Residual stresses Stresses that result from the presence of elastic strains remaining in a component as a result of prior working or treatment.

Riser A liquid-metal reservoir to supply molten metal to a casting and thus prevent shrinkage.

Sheet forming Forming in which there is little change in cross-sectional thickness but extensive change in shape.

Shrinkage In the liquid-to-solid transformation in casting, a volumetric contraction that can lead to gross porosity or microporosity.

Sintering The fusion of particles by diffusion bonding (solid state) or by the formation of a small amount of a liquid phase (liquid state).

Soldering A fusion process that uses a filler metal with a liquidus temperature less than 1000°F (538°C).

Thinning Local change in thickness during a sheet-forming operation.

True density The mass per unit volume of a material in the absence of internal voids.

Undercutting In welding, a defect that involves the melting away of the base metal at the side of the weld without filling in of the welding groove.

Welding A fusion process where melting at the interface is common. A filler material may or may not be used.

Problems

6.1 *[ES/EJ]* Indicate how a continuous-cast product may contain some shrinkage porosity and yet the final commercial product can be pore-free. (Sections 6.1 through 6.6)

6.2 *[ES/EJ]* A large, complex casting is to be redesigned. One of the engineers suggests that casting closer to final service dimensions will not only save money but also result in better service performance. Explain whether you agree with this comment. (Sections 6.1 through 6.6)

6.3 *[ES]* Explain why the following statement is correct: "The castability of an alloy is dependent on its solidification temperature range." (Sections 6.1 through 6.6)

6.4 *[ES/EJ]* The maximum sound length L of an unchilled bar cast in a 70Cu–30Ni alloy is governed by the equation $L = 2T + 4$, where T is the bar thickness in inches. [A bar is defined as having a width W less than $3T$.] (Sections 6.1 through 6.6)
a. What happens to L when chills are used?
b. Why might the use of a chill potentially add to casting cost?
c. Why do we differentiate between bars and plates?

d. How does the material of which the mold is made influence the value of *L*?

6.5 *[EJ]* Actual data such as those shown schematically in Figure 6.5 make it possible to calculate a riser size that will provide the precise amount of liquid metal necessary to accommodate the solidification shrinkage of a casting. Explain why risers larger than those that calculation of the ideal suggests are employed. (Sections 6.1 through 6.6)

6.6 *[ES/EJ]* Consider Example 6.1. (Sections 6.1 through 6.6)
   a. What changes in wheel design would you recommend to increase the radial compression in the spokes?
   b. What changes in wheel design would you recommend to increase the radial tension in the spokes?

6.7 *[ES/EJ]* With a mold-cooled wheel as in Example 6.1 and residual radial compression in the spokes, suggest heat-treatment processes for the following results. [Hint: There is an expansion when FCC → martensite (Section 5.6).] (Sections 6.1 through 6.6)
   a. Increase the residual radial compression in the spokes.
   b. Develop residual radial tension in the spokes.
   c. Develop a wheel free of residual stress.

6.8 *[ES/EJ]* Give reasons for designing a riser attachment that is smaller than that shown in Figure 6.6. Point out how you might determine the minimum size of such a riser attachment. (Sections 6.1 through 6.6)

6.9 *[ES/EJ]* A crack is noted in a casting, and there is some confusion about whether it is a hot tear or a crack that resulted from the build-up of a residual stress during cooling to room temperature. How do cracks caused by these two mechanisms differ in appearance? [Consider both macroscopic and microscopic appearance.] (Sections 6.1 through 6.6)

6.10 *[EJ]* Gas porosity in castings can occur even when the molten metal delivered from the furnace is low in dissolved gas. What are the other sources of gas porosity in castings? (Sections 6.1 through 6.6)

6.11 *[EJ]* Superclean steels have very small inclusions and low inclusion counts. The processing costs are high for such steels. What are the advantages? [*Hint:* Consider both further processing and mechanical properties.] (Sections 6.7 through 6.14)

6.12 *[EJ]* One of your associates comments that an aluminum-, magnesium-, or copper-based alloy sheet can be formed with the same dies as long as the ductility of the alloy is not exceeded. Explain why this might not be true, especially if close dimensional tolerances are to be maintained. [*Hint:* Consider the characteristics of a stress–strain curve when unloading occurs in the plastic region.] (Sections 6.7 through 6.14)

6.13 *[EJ]* Wrinkling in sheet metal products is generally not advantageous. Explain why wrinkling in aluminum baking pans purchased in the grocery store is not considered detrimental. (Sections 6.7 through 6.14)

6.14 *[EJ]* Obtain a twisted galvanized decking nail and suggest the sequence of operations in its manufacture. (Sections 6.7 through 6.14)

6.15 *[EJ]* Sketch a cross section of a ball for towing a trailer and indicate the most desirable texture, as in Figure 6.15. (Sections 6.7 through 6.14)

6.16 *[ES]*
   a. Determine the constants in the equation $\sigma = K\epsilon^n$ for the 0.05 carbon annealed steel (alloy 1) and the 2014 annealed aluminum alloy (alloy 9) in Figure 2.11.
   b. The following identities apply to cold-formed products:

$$\epsilon = \ln\left(\frac{100}{100 - \%\,\mathrm{CW}}\right)$$

$$\text{Work} = VK \left[\frac{\epsilon^{(n+1)}}{(n + 1)} \right] \qquad \text{where } V = \text{volume of metal deformed}$$

Determine the work necessary to cold-work a slab 1 in. × 1 in. × 10 in. 25% if it is made of the two alloys described in part a. (Sections 6.7 through 6.14)

6.17 *[ES]* Anisotropy is measured in a sheet metal 0.040 in. thick by the standard true plastic elongation of 15% ($\epsilon_l = 0.1500$). Specimens were 0.5000 in. wide × 2.0000 in. gage length. The widths after test were as follows: 0° to rolling direction, 0.4662 in.; 45°, 0.4753 in.; 90°, 0.4592 in. [*Hint:* Assume conservation of volume to determine the sample thicknesses.] (Sections 6.6 through 6.14)

a. Determine the ΔR value and indicate whether earing might occur.

b. Determine the average R value, defined as $\overline{R} = (R_{0°} + 2R_{45°} + R_{90°})/4$, and the possibility of thinning. ($\overline{R}$ values differing from 1.0 determine thinning resistance.)

6.18 *[ES]* A metal tube is made by compaction of a powder. In the "green" state it is 5 cm OD by 3 cm ID by 5 cm long and is allowed to absorb water and become saturated. It absorbs 35 g of water. If the final porosity after sintering is 2%, what will be the final dimensions if the shrinkage is uniform in all directions? (Sections 6.7 and 6.14)

6.19 *[E]* The following statements relate to compacting a metal powder at room temperature. Explain why they are true. (Sections 6.7 and 6.14)

a. The particles may be coated with a small amount of paraffin wax.

b. Pressures beyond a certain level do not significantly increase the bulk density of the compact.

c. Careful control of the particle sizes is necessary to maximize the bulk density of both the dry powder and the compact.

6.20 *[ES]* A self-lubricating bearing is made by sintering a cylinder of brass powder followed by impregnation with an oil. Determine the percentage open and closed porosity (open-pore volume and closed-pore volume, respectively, divided by the bulk volume) in the bearing from the following data. (Sections 6.7 through 6.14)

Dimensions—2.00 cm in diameter × 6.00 cm high

Weight after sintering—123.85 g

Weight after impregnation with oil of specific gravity 0.90 is 126.80 g

True density of the brass—8.47 g/cm³

6.21 *[ES/EJ]* Why might the actual method of producing metal powder be important to the design of the processing sequence for a P/M product? (Sections 6.7 through 6.14)

6.22 *[ES/EJ]* Explain how the properties might be different for the same P/M component produced by the three compaction-die designs shown in Figure 6.21. (Sections 6.7 through 6.14)

6.23 *[ES/EJ]* A small piece of seam-welded aluminum pipe is tested for residual stress by application of a strain gauge, as shown in the accompanying figure. The original strain-gauge reading is 2100 μin./in. The pipe is then cut as indicated to remove the

Strain gauge

Seam weld

Slot cut to relieve stress

stress, and the strain gauge reading is 1345 μin./in. Calculate the residual stress, tell whether it is tension or compression, and state why the stress might have been anticipated. (Sections 6.15 through 6.21)

6.24 *[EJ]* Explain why, when two pieces of cold-formed alloy are butt-welded (abutting one another in the same plane), the strength is not likely to be as high as in the unwelded condition. (Sections 6.15 through 6.21)

6.25 *[ES/EJ]* Indicate whether the following statements are correct or incorrect, and justify your answer. (Sections 6.15 through 6.21)

a. Brazing has less influence on mechanical properties than welding.

b. Preheating a material prior to welding may be advantageous.

c. In designing products, it is appropriate to assume that the welding efficiency is less than 100%.

6.26 *[ES/EJ]* In machining, why might it be more difficult to hold dimensional tolerances in a polymer than in a piece of steel? [*Hint:* Consider properties that change with bond strength.] (Sections 6.15 through 6.21)

6.27 *[EJ]* Criticize the statement that both the milling and the shaping (use of a shaper) of a flat surface result in the same quality. (Sections 6.15 through 6.21)

6.28 *[EJ]* A machinist will generally let you know when your product is too hard for his liking. Explain why he may also become concerned when the product is too soft. (Sections 6.15 through 6.21)

6.29 *[ES/EJ]* Why would you probably not grind a wood chisel on a high-speed electric grinder? (Sections 6.15 through 6.21)

6.30 *[ES/EJ]* A number of thermal-treatment processes affect only the surface of a component. Why may the surface be more important than the entire cross section? (Sections 6.15 through 6.21)

6.31 *[ES]* Technicians were giving a material a homogenizing heat treatment, and because long times are beneficial from a diffusion standpoint, they left the parts in the furnace for several days. What defects might occur from the extended heat treatment? (Sections 6.15 through 6.21)

6.32 *[EJ]* In nondestructive ultrasonic testing, how do we differentiate between variation in microstructure and defects in a material? (Sections 6.22 through 6.24)

6.33 *[EJ]* Some components are sold as being of "x-ray quality." That is, radiographs are used to determine the existence of flaws. Does this necessarily mean that we need not worry about brittle fracture in high-strength materials that have passed an x-ray examination? Explain. (Sections 6.22 through 6.24)

6.34 *[ES/EJ]* At the top of the next page is a distribution of yield strengths for sand and permanent mold test bars of AZ91C (9 Al; 1 Zn) magnesium that have been *cast separately* from the production castings. Also given are data for specimens cut from production sand castings. Both T4 (natural aging after solution treatment) and T6 (artificial aging after solution treatment) heat-treatment data are given.

a. Why are there variations in both range of yield strength and average values of yield strength for differences in heat treatment, type of casting, and separately poured bars versus samples cut from production castings?

b. Explain how data such as these distributions of yield strength are used in the development of specifications.

6.35 *[ES]* Give an approximate arithmetic average for the five aim hardness lots shown in Figure 6.39. (Sections 6.22 through 6.24)

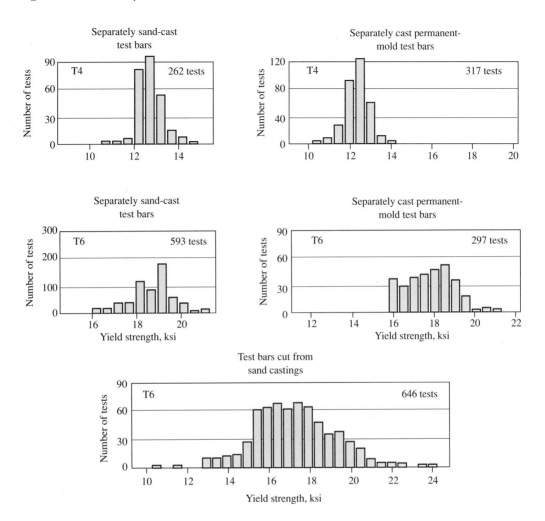

6.36 *[ES/E]* The text suggests several reasons for the data spread shown in Figure 6.39. Indicate how each of the contributions attributed to testing could influence the hardness. (Sections 6.22 through 6.24)

6.37 *[E]* Suggest several ways in which a computer could be used in a rolling process and in metal powder processing. (Sections 6.22 through 6.24)

CHAPTER 7

Nonferrous Alloys: Aluminum, Magnesium, Copper, Nickel, Titanium, Zinc, and Others

The hang glider provides a good illustration of the use of high-strength lightweight alloys. For centuries people have tried to make wings with which they could glide like a bird. The success of the present design is related to the use in the ribs of high-strength aluminum alloy 6061-T6, discussed in this chapter, and the high-strength polymer fibers described in Chapter 14.

In this chapter we will use the concepts of structure and structural control that we discussed in the earlier chapters. We will review the structures of the important nonferrous alloys in order to understand the interrelation of structure, mechanical properties, and processing.

7.1 Overview

We come now to one of the most important concepts of the entire course, the *specification*. Just as we say in a broad sense that the engineer is the bridge between science and society, we will show that the specification is the bridge between the scientific ideas about structure that we introduced in the first five chapters and the application of those ideas in obtaining real material for components.

Let us illustrate. In Chapter 5 we developed the concept of using 4% copper in aluminum and age hardening to obtain high tensile strength. Now in every engineering drawing, there is a box at the lower right-hand corner in which to specify material for the purchasing department. Suppose you wrote in "Material: 4% copper–aluminum alloy with θ' precipitate." You would be the laughing stock of the Engineering Department! On the other hand, the basic knowledge you have obtained puts you in a superior position to look at available materials and heat treatments like a shopping list of structures *if only you remember to think in terms of structures*.

For example, you will find in handbooks of standards and in our short tables of specifications the following data:

	Cu	Si	Mn	Mg	Al	Yield Strength, psi
Aluminum alloy 2014	4.5	0.8	0.8	0.5	balance	14,000 annealed 40,000 heat-treated T6

Looking up the T6 heat treatment, you will find that it corresponds to the solution treatment and artificial aging discussed in Chapter 5. You may ask, "How about the other elements—silicon, manganese, and magnesium?" Although the copper is the principal alloy used to form the $CuAl_2$ precipitate, other elements may be added to degasify the melt, control grain size, or provide additional age hardening or solid-solution hardening. Sometimes the designation (max.) is attached to an element to indicate the highest tolerable level of an undesirable impurity.

Now let us go on to another concept: thinking in terms of structure to simplify our understanding of the thousands of available alloys. This chapter is concerned with the nonferrous alloys. For simplicity in classification, metallic materials are divided into *nonferrous alloys* (non-iron-base alloys, such as aluminum and copper) and *ferrous alloys* (alloys with iron as the base element, such as wrought iron, steel, and cast iron). We find that over 90% of the nonferrous alloys are based on the metals aluminum, magnesium, copper, nickel, titanium, and zinc. These elements are more expensive than iron, but each finds application because it has unique properties, such as corrosion resistance or the ease of processing that zinc die castings offer.

We can get an excellent overview of the properties of these alloys by classifying them as follows:

1. *Single-phase alloys.* Review first the properties of the pure element and then consider how these are modified by cold-working, recrystallization, and solid-solution strengthening.
2. *Polyphase alloys.* Review first the polyphase equilibrium structures and then consider how these are modified by controlling the second phase via treatments such as age hardening, other dispersion hardening, and martensite reactions.

It is also important to note whether a given alloy is available in the wrought or the cast condition, because this will affect component design. In practice, some alloys may be available only in a cast form because of the occurence of brittle phases.

Let us begin by applying this classification to the aluminum alloys.

7.2 Aluminum Alloys

The first group of alloys whose actual engineering applications we will consider are the aluminum alloys. As you read through this discussion, note how even complex specifications are based on the fundamental factors you have learned. For each alloy, it is important to understand why certain alloying elements are used and why the processing results in the properties given.

Because this family has an aluminum base, let us review briefly what we know about pure aluminum. The atomic structure is $1s^2$, $2s^2$, $2p^6$, $3s^2$, $3p^1$, or, more simply, a tightly bonded core surrounded by three valence electrons that are readily given up. This structure results in a light, reactive metal with good electrical and thermal conductivity. The unit cell is FCC, which leads to a number of plane and direction combinations for easy slip, {111} and ⟨110⟩, so we expect excellent ductility and formability. All these predictions are borne out by our everyday experiences with aluminum foil and wire.

Unalloyed or relatively pure aluminum has many uses. We can raise its yield and tensile strengths by cold working, as discussed in Chapter 2. Table 7.1 gives two sets of properties for commercially pure aluminum (1060), for conditions marked 0 and H18. We shall discuss these symbols in detail shortly, but for the present you may consider them as denoting annealed (recrystallized) and cold-worked conditions, respectively. The microstructures are similar to those of the cold-worked and annealed α brasses (Cu-Zn solid solutions) discussed earlier. As in the case of brass, we can obtain different combinations of strength, elongation, and hardness between the extremes listed in Table 7.1 by controlling the sequence and amount of annealing and cold working.

Now let us advance to the specifications involving the second method for improving the strength of a single-phase alloy, namely solid-solution strengthening. This is the mechanism operating in alloys 3003 and 5052, listed in

TABLE 7.1 Typical Properties of Aluminum Alloys

Alloy Number	Chemical Analysis, percent*	Condition	Tensile Strength,† psi × 10³	Yield Strength,† psi × 10³	Percent Elongation	HB	Typical Use
		Single-Phase Wrought Alloys					
1060	99.6 minimum Al	0	10	4	42	19	Sheet, plate, tubing
		Hard H18	19	18	6	35	
3003	1.2 Mn	0	16	6	30	28	Truck panels, ductwork
		Hard H18	29	27	4	55	
5052	2.5 Mg, 0.2 Cr	0	28	13	25	47	Bus bodies, marine applications
		Hard H38	42	37	7	77	
5050	1.2 Mg	0	21	8	24	36	Sheet, trim, gas lines
		Hard H38	32	29	6	63	
		Two-Phase Wrought Alloys					
2014	4.5 Cu 0.8 Si 0.8 Mn 0.5 Mg	Annealed	27	14	18	45	Airplane structures
		Heat-treated T6	70	60	13	135	
6061	1 Mg 0.6 Si 0.2 Cr 0.3 Cu	Annealed	18	8	30	30	Transportation equipment, pipe
		Heat-treated T6	45	40	12	95	
7178	7 Zn, 0.3 Mn, 3 Mg, 0.3 Cr, 2 Cu	Annealed	33	15	16	60	Structural parts in aircraft
		Heat-treated T6	88	78	10	160	

(Continued)

TABLE 7.1 (Continued)

Alloy Number	Chemical Analysis, percent*	Condition	Tensile Strength,† psi × 10³	Yield Strength,† psi × 10³	Percent Elongation	HB	Typical Use
		Two-Phase Cast Alloys					
296.0	4.5 Cu	Solution heat-treated (T4)	37	19	9	75	Aircraft fittings, pump bodies
		Aged (T6)	40	26	5	90	
356.0	7 Si, 0.4 Mg	Aged T5	25	20	2	60	Auto transmission casings, wheels
		Aged T6	38	27	5	80	
712.0	5.5 Zn 0.6 Mg 0.5 Cr 0.15 Ti	Aged T5	35	25	3	75	Machine parts
208.0	3 Si, 4 Cu	As cast (F)	21	14	2	55	General
380.0	8 Si, 3.5 Cu	As cast (F)	47	23	4	80	Die casting
390.0	17 Si, 1 Fe, 4.5 Cu, 0.5 Mg	As cast (F)	41	35	<3	120	Die casting

*Balance aluminum; analysis given as weight percent in this table and those that follow.
†Multiply psi by 6.9×10^{-3} to obtain MPa or by 7.03×10^{-4} to obtain kg/mm².

Table 7.1. From the aluminum–magnesium phase diagram (Figure 7.1), we find the 2.5% magnesium of alloy 5052 is trapped in solid solution in the α (FCC) phase at room temperature. Similarly, the manganese in 3003 is essentially in solid solution. The tensile and yield strengths of both these alloys in the annealed (0) condition are much higher than those of 1060 aluminum; the percent elongation is somewhat lower. However, the microstructures still show the simple grains of a single phase. Basically, the solute atoms have raised the stress required for slip. Note that in the cold-worked conditions (marked H18 and H38), the alloys are still proportionately higher in strength than the cold-worked pure aluminum. This means that the effects of solid-solution strengthening and work hardening can be used together; that is, the effects are additive. This method of alloy design is used in all the other systems as well. In other words, when we wish to retain the ductility and formability of the base metal, we add elements that dissolve in solid solution to raise the strength in the annealed condition and then cold-work to the desired properties. In general, however, the solid solution will exhibit lower ductility than the pure metal.

Now let us examine the two-phase wrought alloys shown in Table 7.1. The most important element in 2014 is the 4.5% copper. Turning to the aluminum–copper phase diagram (Figure 7.2), we see that we have a single-phase alloy at 550°C (1022°F) but a two-phase material at room temperature. As discussed in Chapter 5, this alloy is a candidate for age hardening if the precipitate is coherent, a condition that is found to exist. To develop a fine coherent precipitate, we give the alloy a two-step treatment. We first heat the annealed material to 500–550°C (932–1022°F) to dissolve the θ phase and then quench

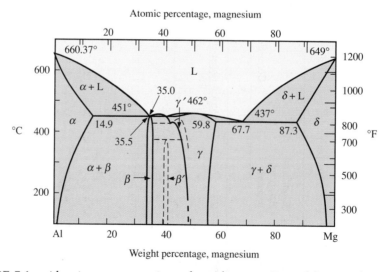

FIGURE 7.1   Aluminum–magnesium phase diagram. Dotted lines indicate uncertain solubility limits.

[Reprinted with permission from *Metals Handbook*, 8th ed., Vol. 8, *Metallography, Structures and Phase Diagrams*, American Society for Metals, Metals Park, OH, 1973, p. 261.]

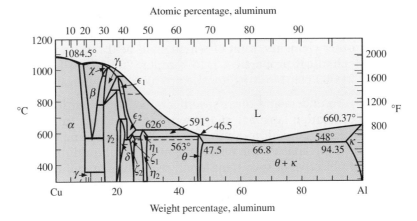

FIGURE 7.2   Aluminum–copper phase diagram. Dotted lines indicate uncertain solubility limits.

[Reprinted with permission from *Metals Handbook*, 8th ed., Vol. 8, *Metallography, Structures and Phase Diagrams*, American Society for Metals, Metals Park, OH, 1973, p. 259.]

it. This gives supersaturated κ. We then produce a fine precipitate by aging at 170°C (338°F) for 10 hr; this precipitate is so fine that it is not visible by conventional microscopy. A tensile strength of 70,000 psi (483 MPa) develops (the condition marked T6), compared with 27,000 psi (186 MPa) in the annealed condition.

The third section of Table 7.1 shows the casting alloys. Casting alloys are usually two-phase. (A notable exception is the casting of relatively pure aluminum cooling fins around iron laminations in electric rotors, where it is desired to capitalize on the high thermal conductivity of aluminum.) In one group of heat-treatable casting alloys, such as 356.0, age hardening is produced as in the wrought alloys. In some cases the solution-treating step may be omitted if the cooling rate of the casting is fast enough to produce a supersaturated solid solution. In this case we need only age the casting. This condition is denoted by T5. (See the code given later in this section.) In other cases the second phase acts simply as a hard dispersion to improve hardness and wear resistance. This is the case with the silicon-rich phase in the 390.0 alloy for die-cast engine blocks. The size and shape of the dispersion are controlled by the cooling rate and by nucleation with phosphorus (forming AlP). Fast cooling rates and abundant nuclei give a dispersion of fine equiaxed crystals of silicon rather than long blade-like crystals, thereby improving strength and facilitating machining. The alloy can be further hardened with a solution heat treatment and aging because of the presence of 4% copper, as we discussed in connection with alloy 2014.

Now let us consider some details of the actual engineering specifications, which may seem complex after our simple discussions of working and heat treatment. For example, the specification for an ordinary aluminum frying pan would read:

Alloy 1100: 99.0% minimum Al, 1.0% maximum (Fe + Si), 0.20% maximum Cu, 0.05% maximum Mn, 0.10% maximum Zn, 0.05% maximum each of other elements, total of which shall be 0.15% maximum.

Mechanical properties: H16 temper, 20,000 psi (138 MPa) yield strength, 21,000 psi (145 MPa) tensile strength, 6% elongation.

To understand these specifications, you must learn to see through all the specification language and determine what structure we are talking about.

First, pure aluminum is an FCC structure and therefore has good ductility. If we consult the phase diagrams for aluminum (Figure 7.3), we see that the objective of specifying small maximum amounts of impurities is to avoid the formation of any quantities of hard second phases that would reduce ductility and increase corrosion (see Chapter 18). One could specify a purer material, but this would raise the cost.

Next we note the specification of "H16 temper." This denotes a certain amount of cold working, such as cold rolling, which raises the yield strength and prevents denting of the frying pan. (The temperature encountered during normal heating of the pan would be too low for recrystallization.) In essence, then, the specification is written to give a cold-worked single-phase structure of commercially pure aluminum.

Let us now review the general specifications. We shall see that the alloys can be divided into two general categories: single-phase and polyphase alloys. Only working and annealing are used to control the properties of single-phase alloys. For the polyphase alloys, combinations of working and precipitation hardening are employed.

The principal alloying elements used with aluminum are copper, manganese, silicon, magnesium, zinc, nickel, and tin. For the wrought alloys, a code has been developed to make it easy to recognize the type of alloy.

Code	Type of Alloy (Major Element)	Example
1XXX	Essentially pure Al	1060 (99.6% minimum Al)
2XXX	Cu (two-phase)	2014 (4.5% Cu)
3XXX	Mn (one-phase)	3003 (1.3% Mn)
4XXX	Si (two-phase)	4032 (12.5% Si)
5XXX	Mg (one-phase)	5050 (1.2% Mg)
6XXX	Mg and Si (two-phase)	6063 (0.4% Si, 0.72% Mg)
7XXX	Zn (two-phase)	7075 (5.6% Zn)

Unfortunately, no simple difference in numbering systems separates the single-phase and two-phase alloys, so it is necessary to recognize that the 2XXX, 4XXX, 6XXX, and 7XXX series are two-phase alloys. The type of processing is covered by the following *aluminum alloy designations:*

H1X, cold-worked. The higher the number X, the greater the cold working. For example, 1060-H14 denotes a 1060 alloy cold-worked about half of the

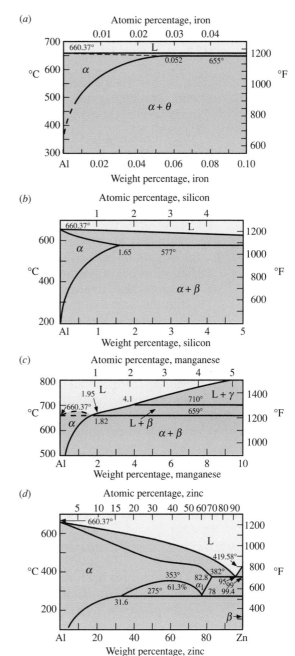

FIGURE 7.3 Portions of aluminum-rich phase diagrams. (*a*) Aluminum–iron. (*b*) Aluminum–silicon. (*c*) Aluminum–manganese. (*d*) Aluminum–zinc. Dotted lines indicate uncertain solubility limits.

[Reprinted with permission from *Metals Handbook*, 8th ed., Vol. 8, *Metallography, Structures and Phase Diagrams*, American Society for Metals, Metals Park, OH, 1973, pp. 260, 262, 263, 265.]

total possible amount, whereas 1060-H19 represents the maximum amount of cold working.

H2X, cold-worked and annealed. The X still represents the extent of cold work.

H3X, cold worked and stabilized. Again, the X represents the amount of cold work. Stabilizing means heating to 50 to 100°F (30 to 55°C) above the maximum service temperature so that the material does not soften in service.

Specifications with the letter T involve heat treatment to produce age hardening. These are used only for alloys that develop coherent precipitates: the 2XXX, 6XXX, and 7XXX series. The code is as follows:

T3, solution treatment followed by strain hardening and then natural aging (that is, holding at room temperature).

T4, solution treatment plus natural aging.

T5, aging only. In special cases where the part is cooled quickly enough from the forging or casting temperature, the solution heat treatment is omitted.

T6, solution heat treatment plus artificial aging.

T7, solution heat treatment plus stabilization.

T8, solution heat treatment plus strain hardening, followed by artificial aging.

T9, solution heat treatment plus artificial aging, followed by strain hardening.

T10, cooled from an elevated-temperature shaping process, cold-worked, and artificially aged.

0, the annealed condition in wrought alloys.

T2, cooling from the elevated-temperature shaping process, followed by cold working and natural aging.

F, the as-fabricated (as-rolled, etc.) condition in wrought alloys and the as-cast condition in castings.

The code for casting alloys is quite different:

Numeral	1XX.X	2XX.X	3XX.X	4XX.X	5XX.X	6XX.X	7XX.X	8XX.X
Family	Al 99. min	Al-Cu	Al-Si Cu, Mg	Al-Si	Al-Mg	Unused series	Al-Zn	Al-Sn

The codes for heat treatment are used also where applicable. The T2, T3, T8, T9, and T10 designations, which call for strain hardening, apply only to wrought products, however.

EXAMPLE 7.1 *[EJ]*

The T5 designation means aging only. How can alloy 712.0-T5 attain properties close to those of 356.0-T6 if no solution treatment is used after casting? (See Table 7.1)

Answer   A thin section in a casting cools rapidly enough to provide a solution treatment, that is, supersaturated α is obtained. Next, at room temperature, natural aging takes place and the presence of a coherent precipitate results in an age-hardened alloy. In a casting with grossly varying section sizes, large sections may not develop the physical properties of lighter sections unless the material is solution-treated. Furthermore, the T5 heat treatment does not always mean that natural aging will take place. The casting may have to be reheated to artificially age, depending on the alloy system.

EXAMPLE 7.2 *[ES/EJ]*

How much $CuAl_2$ (θ phase, Figure 7.2) is present as a function of temperature in a 4.5% copper–aluminum alloy cooled under equilibrium conditions? It is found experimentally that there is more $CuAl_2$ present in an annealed alloy than in a naturally aged alloy but that the yield strength of the latter is greater. Why?

Answer

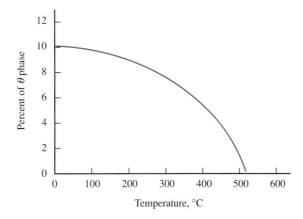

The approximate phase fraction chart is shown here. It is assumed from the phase diagram that no copper is soluble in aluminum at room temperature and that 53% by weight copper is soluble in the θ phase at all temperatures. In terms of the formula $CuAl_2$, this corresponds to 33.3 at.% copper. The higher strength in the naturally aged alloy is a result of the more uniform distribution of the second phase; the annealed alloy has continuous $CuAl_2$ at the grain boundaries, which is not effective in preventing slip. Furthermore, the aging process produces coherency strains that more effectively block slip and dislocation movement when compared to the annealed material.

7.3 Magnesium Alloys

Magnesium alloys are competitive with aluminum alloys because the density of magnesium is two-thirds that of aluminum. Therefore, many aircraft parts are made of magnesium alloys. Also, an interesting competition has evolved between magnesium and aluminum alloys for the Volkswagen engine block, because both materials can be used successfully.

We discussed the basic atomic structure of magnesium $(1s^2, 2s^2, 2p^6, 3s^2)$ in Chapter 1. Beneath the two valence electrons is a stable ring of eight, making magnesium a very active element. Although fine magnesium powder burns in air, a melt of liquid magnesium can be poured in air into castings or ingots if simple precautions are taken. Magnesium parts do not ignite during heat treatment. However, magnesium corrodes more rapidly than aluminum in many environments, such as sea water, so outboard motor parts, for example, are not made of magnesium.

Another major difference between magnesium and aluminum is that the unit cell of magnesium is HCP. Therefore it has only three slip systems at room temperature, compared with the twelve of aluminum. However, additional systems become operative at higher temperatures, so magnesium alloys are usually hot-worked rather than cold-worked.

Table 7.2 lists some magnesium alloys. The *magnesium alloy designations* for heat treatment (T4, T6, etc.) and the code systems for cold working (H24, etc.) are the same as those for aluminum alloys. However, the methods for indicating the composition of the alloys are different. The following letters signify the most important alloying elements according to the code indicated.

A = aluminum   K = zirconium   M = manganese
E = rare earths (such as cerium)   Q = silver   S = silicon
H = thorium   Z = zinc   T = tin

The numbers following the letters give the amounts of the elements. The first number gives the percentage of the element indicated by the first letter, and the second number the percentage of the second. For example, AZ92 means an alloy containing 9% Al, 2% Zn.

Magnesium alloys are used principally in aircraft and spacecraft, machinery, tools, and materials-handling equipment. The strength increases with aluminum level through solid-solution strengthening and the precipitation of $Mg_{17}Al_{12}$ (the phase γ in Figure 7.1). In both the wrought and cast alloys, rare earths are added to minimize flow (plastic deformation) at elevated temperatures (150 to 200°C, 302 to 392°F). The rare earths produce a rigid grain-boundary network in the microstructure, such as Mg_9Ce, which resists deformation.

Careful control of the network is necessary, because the second phases also limit room-temperature ductility. If they are present in excessive amounts, they may even decrease the tensile strength of the alloy.

Magnesium alloys should play an increasing role as engineering materials, because the amount of magnesium in sea water is tremendous, whereas

TABLE 7.2 Typical Properties of Magnesium Alloys

Alloy Number	Chemical Analysis, percent	Condition	Tensile Strength,* psi × 10³	Yield Strength,* psi × 10³	Percent Elongation	HB	Typical Use
Wrought Alloys							
AZ31B	3 Al, 1 Zn 0.2 Mn	F H24	38 42	28 32	9 15	55 73	Sheet and plate
AZ80A	8 Al, 0.2 Zn, 0.2 Mn	T5	50	34	6	72⎫	Forgings and extrusions
ZK60A	6 Zn, 0.5 Zr	T5	44	30	16	82⎬	
HK31A	3 Th, 1 Zr	H24	37	29	8	—	Elevated temperatures
Cast Alloys							
AZ63A	6 Al 3 Zn 0.13 Mn	As cast Solution heat-treated T6	29 40 40	14 13 19	6 12 5	55 50 73	General sand castings
AZ91C	9 Al 1 Zn 0.13 Mn	As cast Solution heat-treated T6	24 40 40	14 12 19	2 14 5	52 53 66	High-strength castings
QE22A	2 Ag, 2 R.E.†	T6	40	30	4	77	Highest-strength uses
EZ33A	3 R.E., 3 Zn	T5	23	16	3	50⎫	Elevated temperatures
EK31A	3 R.E., 1 Zr	T6	31	16	6	55⎬	

*Multiply psi by 6.9×10^{-3} to obtain MN/m^2 [MPa] or by 7.03×10^{-4} to obtain kg/mm^2.
†R.E. = rare earth elements.

supplies of other important metallic elements, which are usually mined as complex oxides and sulfides, are dwindling.

EXAMPLE 7.3 *[EJ]*

We have now discussed two families of nonferrous alloys, aluminum-based and magnesium-based. What are the essential differences between alloy compositions that might be used for a forging or a casting?

Answer In general, a component manufactured by forging is competitive with one manufactured by casting. However, some alloy compositions exhibit low ductility and therefore may only be cast to a final shape. (Whereas only alloys with good ductility can be used to produce a forged component, both low- and high-ductility alloys may be cast.) In practice, forging stock is produced from cast ingot, bar, or slab.

However, even though the room-temperature properties might indicate that an alloy has low ductility, it is possible to heat a multiphase alloy to a higher temperature at which a single phase suitable for forging exists. Therefore we must differentiate between hot- and cold-forming processes.

It is not uncommon to find the same nominal composition available in both cast and wrought shapes. If an alloy is available only as cast products, this suggests an inability to achieve sufficient ductility for forming at all temperatures. For a specific alloy, the appropriate phase diagram and microstructure provide the necessary data for explanation.

EXAMPLE 7.4 *[ES/EJ]*

We can define the *specific strength* of an alloy as the yield strength ($lb/in.^2$) divided by the density ($lb/in.^3$). Show that aluminum- and magnesium-based alloys are more competitive with one another than the simple comparison of yield strength indicates.

Answer An average specific gravity for aluminum alloys is 2.70, and for magnesium alloys, 1.75. Knowing that specific gravity is the ratio of the material density to the density of water (0.0361 $lb/in.^3$), we calculate the approximate densities of aluminum and magnesium alloys to be 0.0975 $lb/in.^3$ and 0.0632 $lb/in.^3$, respectively. The accompanying table summarizes the properties of the strongest wrought and cast alloys in the two systems.

Alloy	Yield Strength, $lb/in.^2$	Specific Strength, in.
7178-T6 (Al-wrought)	78,000	8.0×10^5
390.0 (Al-cast)	35,000	3.6×10^5
AZ80A (Mg-wrought)	34,000	5.4×10^5
QE22A (Mg-cast)	30,000	4.7×10^5

Although the yield strengths may vary by more than a factor of 2, the specific strengths are much closer to each other. Because many applications are dependent on component weight, the specific strength becomes an important specification. Further consideration of these factors is given in Chapter 23.

7.4 Copper Alloys in General

The copper alloys have a unique combination of characteristics: high thermal and electrical conductivity, high corrosion resistance, generally high ductility and formability, and interesting color for architectural uses. Although the hardness and strength of these alloys do not equal those of the hardest steels, some copper alloys reach tensile strengths of 150,000 psi (1.035×10^3 MPa).

The atomic structure of copper is $1s^2$, $2s^2$, $2p^6$, $3s^2$, $3p^6$, $3d^{10}$, $4s^1$. Note that the outer electron $4s^1$ does not have a shell of eight beneath it, as do the *s* electrons of aluminum and magnesium. The energy of this electron is very close to that of the $3d$ electrons. Therefore the $3d$ and $4s$ electrons are equivalently attracted to the positively charged nucleus and so are tightly bonded. For this reason copper, instead of being an active metal similar to aluminum, is considered a noble (that is, corrosion-resistant) metal in the same vertical group in the periodic table as silver and gold. As we will discuss in Chapter 20, the unique red color of copper is due to selective absorption of the spectrum of white light by interaction with the $3d$ electrons.

At first glance it is easy to be confused by the variety of copper alloys. Over thousands of years the names bronze and brass have been used differently, and other names such as gun metal, admiralty metal, gilding bronze, manganese bronze, and ounce metal have added to the confusion.

We shall take a simple approach to classification that is based on the microstructures of the alloys involved. Like the light metals, all the copper alloys can be divided into two classes: single-phase and polyphase alloys. We would expect the single-phase alloys to exhibit good ductility because the unit cell is FCC. Table 7.3 confirms this. The mechanisms for strengthening the single-phase wrought alloys are the usual solid-solution hardening and combinations of cold work and annealing. In the two-phase alloys, age hardening and other hardening by precipitates or second-phase dispersions are used.

In Table 7.3 we have chosen a few popular alloys to illustrate these points. We begin with ETP (electrolytic tough pitch) copper. This is copper that has been refined electrolytically. The term *tough pitch* refers to the oxygen level of about 0.04% (present as copper oxide), which gives the ingot its unique appearance. This grade of copper is widely used, although it embrittles if it is heated in an atmosphere containing hydrogen, because the hydrogen diffuses through the copper and encounters copper oxide at grain boundaries. A reaction takes place and water vapor is generated. These water molecules are large compared to the hydrogen molecules. They do not diffuse

TABLE 7.3 Typical Properties of Copper Alloys

Alloy Number	Chemical Analysis, percent	Condition	Tensile Strength,* $psi \times 10^3$	Yield Strength,* $psi \times 10^3$	Percent Elongation	Hardness	Typical Use
		Single-Phase Wrought Alloys					
C11000	ETP, 99.9 Cu	Annealed	32	10	45	40 HRF	Architectural, electrical
		Cold-worked	50	40	6	85 HRF	
C26800	65 Cu, 35 Zn Yellow brass	Annealed	46	14	65	88 HRF	Plumbing, Grill work
		Cold-worked	74	60	8	80 HRB	
C61400	91 Cu, 7 Al, 2 Fe Aluminum bronze	Cold-worked	82	40	35	90 HRB	Condenser tubing
C71500	70 Cu, 30 Ni Cupronickel	Annealed	44	20	40	37 HRB	Desalinization tubing
		Cold-worked	75	68	12	85 HRB	
		Polyphase Wrought Alloys					
C17200	98 Cu, 2 Be Beryllium copper	Annealed	70	30	42	57 HRB	Springs, tools
		Precipitation-hardened	175	140	7	38 HRC	
		Cast Alloys					
C81100	Cu	As cast	25	9	40	HB 44	Electrical conductors
C83600	85 Cu, 5 Sn, 5 Zn, 5 Pb	As cast	37	17	30	HB 60	Valves, bearings
C93700	80 Cu, 10 Sn, 10 Pb	As cast	35	18	20	HB 60	Bearings, pumps
C96400	70 Cu, 30 Ni	As cast	68	37	28	HB 140	Marine valves
C82400	98 Cu, 2 Be	Hardened	150	140	1	38 HRC	Dies, tools
C90500	88 Cu, 10 Sn, 2 Ni	As cast	44	22	6	HB 85	Gears
C95300	89 Cu, 10 Al, 1 Fe	As cast	75	27	25	HB 140	Gears, bearings
		Heat-treated	85	42	15	HB 174	

*Multiply psi by 6.9×10^{-3} to obtain MPa or by 7.03×10^{-4} to obtain kg/mm².

out of the metal, but segregate to form voids at the grain boundaries, leading to embrittlement. For applications in which embrittlement would be a problem, two other grades of copper are available: phosphorus deoxidized copper (C12200), in which the dissolved oxygen is eliminated by a phosphorus addition to the melt, and OFHC (oxygen-free high-conductivity) copper (C10200), which is melted under special reducing conditions to eliminate oxygen. The mechanical properties of all grades are comparable. The change in yield strength from 10,000 to 40,000 psi (69 to 276 MPa) brought about by cold working is especially important.

7.5 Solid-Solution Copper Alloys

An important solid-solution effect is obtained in alloys with a silver content specified as 10 to 25 troy oz/ton (12 troy oz = 1 lb). Although this is only on the order of 0.05% by weight silver, it raises the softening temperature of the cold-worked copper by more than 100°C (180°F), permitting the fabricator to soft-solder (with a lead-tin alloy) cold-worked copper without lowering the strength by recovery and recrystallization. The presence of silver does not change the electrical conductivity.

The most widely used solid-solution alloys are those that contain zinc. These alloys are called *brass*. The most common alloys range from 65% Cu, 35% Zn to 70% Cu, 30% Zn. Because copper costs about $1.40/lb and zinc about $0.82/lb (1989 prices), the higher-zinc brasses are cheaper. The phase boundary of the α field is quite important because the higher-zinc alloys (above 35%) contain the β phase (Figure 7.4). Although this phase is stronger than α, it is more susceptible to a particular type of corrosion called *dezincification*, which is discussed in Chapter 18. The cold working and recrystallization of brass were discussed in Chapter 3.

Considerable strengthening is also accomplished by alloying copper with aluminum and nickel, as shown in the C61400 and C71500 alloys. The C71500 alloy is especially important in applications involving sea water, as in desalinization equipment.

7.6 Polyphase Wrought Copper Alloys

The highest-strength copper alloy is produced by age hardening a 2% beryllium alloy (C17200). Heating to 800°C (1472°F) gives a single-phase α solid solution (Figure 7.5). After being quenched in order to obtain supersaturated α, the alloy is aged for 3 hr at 315°C (599°F) to precipitate the γ_2 phase CuBe, which gives a coherent precipitate. This alloy is used widely for springs, nonsparking tools, and parts that require good strength plus high thermal and electrical conductivity. Other precipitation-hardening alloys contain silicon and zirconium.

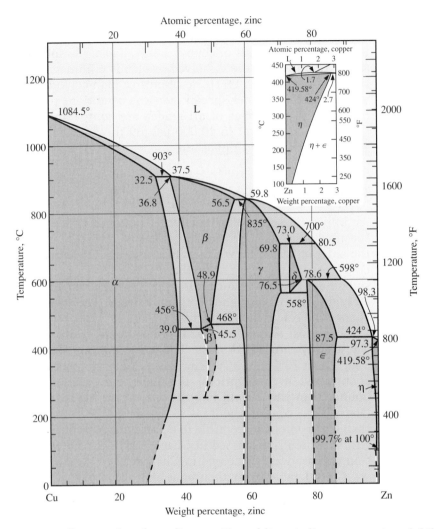

FIGURE 7.4 Copper–zinc phase diagram. Dotted lines indicate uncertain solubility limits.

[Reprinted with permission from *Metals Handbook*, 8th ed., Vol. 8, *Metallography, Structures and Phase Diagrams*, American Society for Metals, Metals Park, OH, 1973, p. 301.]

7.7 Cast Copper Alloys

The cast alloys offer a wider range of structures than the wrought alloys because high ductility is not required for working. Of great importance are the high-lead alloys for bearings and the high-tin alloys for gears.

Copper itself is used in castings requiring good electrical and thermal conductivity. As an example, although copper melts at 1084°C (1983°F), water-

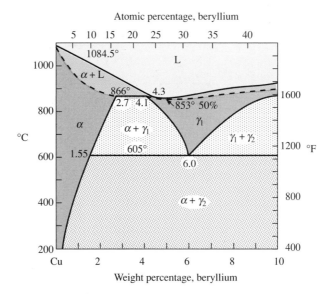

FIGURE 7.5   Copper–beryllium phase diagram. Dotted lines indicate uncertain solubility limits.

[Reprinted with permission from *Metals Handbook*, 8th ed., Vol. 8, *Metallography, Structures and Phase Diagrams*, American Society for Metals, Metals Park, OH, 1973, p. 271.]

cooled copper tuyères and lances can be used in the processing of steel, where temperatures reach 1750°C (3182°F) in the atmosphere above the liquid metal. As with the wrought alloys, copper forms strong solid solutions with zinc, nickel, and aluminum.

The lead alloys are of interest because liquid copper can dissolve lead in unlimited amounts (Figure 7.6). Upon cooling, the lead precipitates as metallic lead because it is insoluble in solid copper. Alloys such as 85% Cu, 5% Sn, 5% Zn, 5% Pb are valuable for bearings because of the lubricating effects of the lead droplets (Figure 7.7a).

The alloys with tin contain a hard intermetallic compound δ ($Cu_{31}Sn_8$). The presence of δ in a ductile α matrix gives a *bronze* that is excellent for gears because it provides a good mating surface against hardened steel gears. (See also Sections 2.19–2.21.) The photomicrograph in Figure 7.7b shows that the δ phase cracks only after considerable deformation of the surrounding α.

The two-phase alloys include aluminum bronze, in which the aluminum exceeds the solid solubility in α and a hard γ_2 phase is formed on cooling (Figure 7.2). Under stress, failure occurs through γ_2 regions (Figure 7.7c). The alloy is also used in the heat-treated condition, which is obtained by heating to the β region and quenching. The β transforms on quenching to a structure called *martensite* (Figure 7.7d), which is not shown on the equilibrium diagram. Chapter 5 discussed the nature of this transformation. Other polyphase alloys include cast beryllium–copper, copper–silicon, and manganese–bronze.

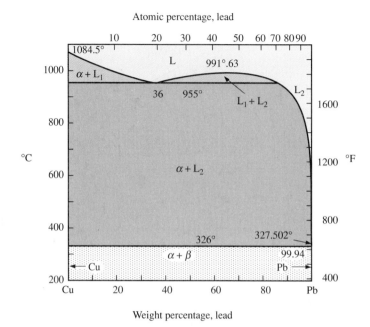

Atomic percentage, lead

Weight percentage, lead

FIGURE 7.6 Copper–lead phase diagram. Copper and lead are essentially insoluble in each other at room temperature. The three-phase equilibrium at 955°C (1751°F) is a monotectic (introduced in Section 4.9).

[Reprinted with permission from *Metals Handbook*, 8th ed., Vol. 8, *Metallography, Structures and Phase Diagrams*, American Society for Metals, Metals Park, OH, 1973, p. 296.]

7.8 Nickel Alloys in General

Nickel is an element somewhat similar to iron in strength, but its alloys have exceptional corrosion resistance and elevated temperature strength, as well as important magnetic properties. This section will discuss only the corrosion-resistant alloys; we will reserve discussion of the complex heat-resistant nickel–chromium–aluminum superalloys for Chapter 9.

The atomic structure of nickel is related to those of copper and iron: $1s^2$, $2s^2$, $2p^6$, $3s^2$, $3p^6$, $3d^8$, $4s^2$. As in copper, there is little difference in energy between the $3d$ and $4s$ electrons, so the metal is relatively noble and corrosion-resistant. The unit cell is FCC, and the lattice parameter a_0 is close to that of copper: 3.52 Å vs. 3.62 Å (0.352 nm vs. 0.362 nm).

The principal wrought alloys are *Monel* and *Inconel* (Table 7.4). The nickel–copper alloy Monel is an extension of the copper–nickel alloys to the high-nickel side of the phase diagram (Figure 7.8).

The rules of solid solubility suggest that copper and nickel should form an extensive series of solid solutions (Chapter 1), and this is the case. The fact that the intermediate nickel–copper alloy (Monel) has a higher yield strength than nickel, although nickel is stronger than copper, is another illustration of solid-solution strengthening. This shows that the element added in solid

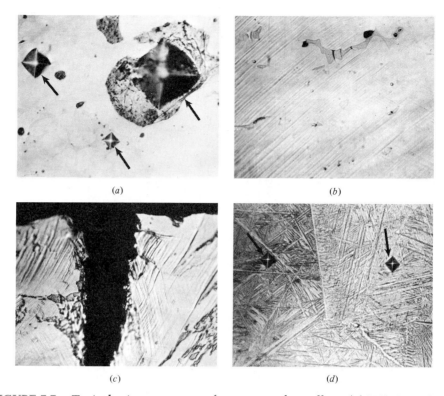

FIGURE 7.7 Typical microstructures of cast copper-base alloys. (*a*) 85% Cu, 5% Sn, 5% Pb, 5% Zn alloy. Lead phase (dark gray), HV 16; matrix (white), HV 98; δ phase (light gray), HV 293, 500×, chromate etch. (*b*) 88% Cu, 8% Sn, 4% Zn. Islands of δ phase are in the FCC copper matrix. The specimen was stressed after polishing. The ductile matrix shows slip, while the brittle δ phase cracked; 500×, chromate etch. (*c*) 89% Cu, 10% Al, 1% Fe (as cast). Gray γ_2 phase is in the FCC copper matrix. During stress the ductile matrix showed slip, but failure occurred through the γ_2; 500×, chromate etch. (*d*) An alloy of (*c*) heated to 900°C (1650°F) for 1 hr and water-quenched to form a martensitic structure. The specimen was not stressed; 500×, chromate etch. (Arrows point to microhardness indentations.)

solution can harden and strengthen the solvent metal even though the solute element is soft itself. Both Monel and Inconel may be age-hardened, as shown by the alloys *Duranickel 301* and *Monel K500*.

7.9   Cast Nickel Alloys

The cast alloys are parallel to the wrought alloys, with two exceptions — Monel 505 and Inconel 705 (Table 7.4). These materials contain 4 to 6% silicon, which produces the hard intermetallic compound Ni_3Si. This dispersion hardens the cast alloy, and the material is further hardened by aging. Because the elongation is only 3%, these alloys are not available in wrought form.

TABLE 7.4 Typical Properties of Nickel Alloys

Alloy Number	Chemical Analysis, percent*	Condition	Tensile Strength,[†] psi × 10³	Yield Strength,[†] psi × 10³	Percent Elongation	Hardness	Typical Use
Single-Phase Wrought Alloys							
Nickel 200	99.5 Ni	Annealed	65	22	47	HB 75	Corrosion-resistant parts
		Cold-worked	120	92	8	HB 230	
Monel 400	66 Ni, 32 Cu	Annealed	72	35	42	HB 110	Corrosion-resistant parts
		Cold-worked	120	110	8	HB 241	
Inconel 600	78 Ni, 15 Cr, 7 Fe	Annealed	100	50	35	HB 170	Corrosion-resistant parts
		Cold-worked	150	125	15	HB 290	
Polyphase Wrought Alloys							
Duranickel 301	94 Ni, 4.5 Al, 0.5 Ti	Annealed	105	42	40	90 HRB	Corrosion-resistant parts
		Cold-worked, age-hardened	200	180	8	40 HRC	High-strength parts
Monel K500	65 Ni, 2.8 Al, 0.5 Ti, 30 Cu	Annealed	97	52	35	85 HRB	Corrosion-resistant parts
		Cold-worked, age-hardened	185	155	7	34 HRC	High-strength parts
Single-Phase Cast Alloys							
Nickel 210	95 Ni, 0.8 C	As cast	52	25	22	HB 100	Condensers
Monel 411	64 Ni, 32 Cu, 1.5 Si	As cast	77	38	35	HB 135	Paper mill equipment
Inconel 610	68 Ni, 15 Cr, 2 Nb, 10 Fe	As cast	82	38	20	HB 190	Dairy equipment
Polyphase Cast Alloys							
Monel 505	63 Ni, 29 Cu, 4 Si	Aged	127	97	3	HB 340	Valve seats
Inconel 705	68 Ni, 9 Fe, 6 Si, 15 Cr	Aged	110	95	3	HB 340	Exhaust manifolds

*These represent the major elements. Other elements may be present.
[†]Multiply psi by 6.9×10^{-3} to obtain MPa or by 7.03×10^{-4} to obtain kg/mm².
Note: Nickel-base superalloys are discussed with other superalloys in Chapter 9.

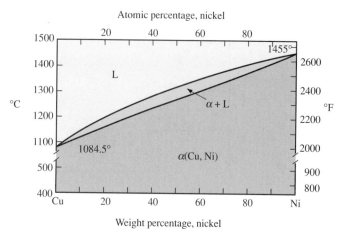

FIGURE 7.8 Copper–nickel phase diagram

[Reprinted with permission from *Metals Handbook*, 8th ed., Vol. 8, *Metallography, Structures and Phase Diagrams*, American Society for Metals, Metals Park, OH, 1973, p. 294.]

7.10 Titanium Alloys

Titanium and *titanium alloys* are used in two quite distinct applications: corrosion-resistant parts and highly stressed parts from low temperatures up to about 650°C (1200°F). In the latter case the strength-to-weight ratio is competitive with that of aluminum.

The alloys and their structures are well adapted to these applications. Single-phase, relatively pure, unalloyed titanium is used for the first of these applications, because it has sufficient strength and better corrosion resistance than the alloyed types.

In the second case, where optimal strength is important, alloying and processing for solid-solution strengthening and second-phase dispersion control are used. To understand these compositions, it is best to start with the element itself.

The atomic structure of titanium is $1s^2$, $2s^2$, $2p^6$, $3s^2$, $3p^6$, $3d^2$, $4s^2$. The $3d$ and $4s$ electrons are bound rather lightly, and a stable shell of eight remains ($3s^2$, $3p^6$). Titanium is therefore quite active. For example, there is no known crucible material that does not react with liquid titanium, so it is melted in a water-cooled copper crucible. This produces a shell of solid titanium between the liquid titanium and the copper. However, upon exposure to air and a number of chemicals, titanium develops a very adherent TiO_2 protective film (similar to the Al_2O_3 film in aluminum), which gives it good corrosion resistance.

The structure of titanium is HCP and is called alpha (α) up to 880°C (1616°F), where it transforms to BCC, beta (β). Alloys produce the two types

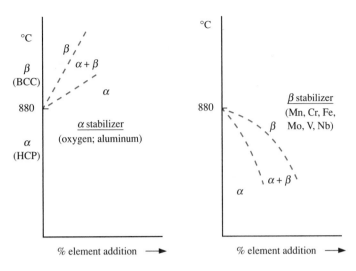

(No effect: Zirconium, tin)

FIGURE 7.9   Effects of different alloy elements on α-β equilibrium in titanium alloys

of phase diagrams shown in Figure 7.9. In the first case, the *α-forming elements* raise the temperature of the $\alpha \rightarrow \beta$ transformation; in the second, the *β-forming elements* lower the $\alpha \rightarrow \beta$ temperature. Note that in both cases, instead of a sharp $\alpha \rightarrow \beta$ transformation we pass through an $\alpha + \beta$ region, following the phase rule. These two-phase fields are important because in some cases, hot working in such a field is easier than in a single-phase field, and because the final properties are better. A third type of diagram (not shown) exhibits a reduced α field and the presence of a eutectic. A martensite type reaction (see Chapter 5) is also encountered. Typical properties of different structures are shown in Table 7.5.

TABLE 7.5   Typical Properties of Titanium Alloys (Wrought)

Material	Chemical Analysis, percent	Structure	Tensile Strength,* psi $\times$ 10^3	Yield Strength,* psi $\times$ 10^3	Percent Elongation	HB
Ti	99 Ti[†]	α	38 to 100	22 to 85	17 to 30	115 to 220
Ti, Al, Sn	5 Al, 2.5 Sn[‡]	α	125	117	18	360
Ti, Al, V	6 Al, 4 V[§]	α-β	170	155	8	380

*Multiply psi by 6.9 $\times$ 10^{-3} to obtain MPa or by 7.03 $\times$ 10^{-4} to obtain kg/mm^2.
[†]Variations in properties depending on whether cold-worked or annealed
[‡]Annealed sheet
[§]Solution heat-treated and aged

TABLE 7.6   Typical Properties of Zinc Alloys (Die Cast)

Alloy Number	Chemical Analysis, percent	Nominal Tensile Strength,* psi $\times 10^3$	Percent Elongation	HB
AG40A	4 Al, 0.05 Mg	41	10	82
AC41A	4 Al, 1 Cu, 0.05 Mg	48	7	91

*Multiply psi by 6.9×10^{-3} to obtain MPa or by 7.03×10^{-4} to obtain kg/mm^2.

7.11 Zinc Alloys

The principal engineering components made of zinc are die castings. Zinc alloys are ideally suited for this process because of their low melting point and because they do not corrode steel crucibles or dies. As a result, simple automotive parts, toys, and building hardware are often made of zinc alloys (Table 7.6).

EXAMPLE 7.5 *[EJ]*

The tables of mechanical properties have included a column for yield strength. Why are these data missing for the zinc alloys in Table 7.6?

Answer   An inadvertent omission is always a possibility, but a review of reference texts would show that these data are also missing from those tables. This suggests experimental difficulty in determining the modulus of elasticity, which is necessary for using the 0.2% offset method to determine yield strength.

Because of the low melting point of zinc, the material can adjust to stress at room temperature in the few minutes required to conduct a tensile test. This phenomenon is actually creep, as discussed in Chapter 3. The result is that the stress–strain curve for zinc is nonlinear, and we are unable to define a modulus of elasticity. Therefore a "pseudomodulus" is defined, on the basis of the time of loading and the amount of allowable permanent elongation. The design is thus based on creep rather than yield strength. Similar difficulties might be expected for other low-melting alloys such as lead and tin.

7.12 The Less Common Metals and Precious Metals

We are entering a period of greater growth in the use of the rarer metals. The era began with the discovery that in piston-driven aircraft each part is literally worth its weight in gold. That is, removing one pound of dead weight makes it possible to carry one more pound of cargo each trip. Over the life of the plane, this extra revenue certainly exceeds the value of a pound of gold. In space travel, the value of a pound of weight is orders of magnitude greater. As a result, parts made of beryllium, rhenium alloys, molybdenum alloys, platinum, gold, silver, and iridium are used freely in commerce. A common example is the use of expendable fine wire platinum/platinum–10% rhodium thermocouples to check each heat of steel. Years ago a thermocouple of this type was a carefully guarded laboratory tool. A detailed discussion of the properties of the less common metals is beyond the scope of this text, but we should mention that the following metals are readily available:

Refractory metals for high-temperature service: molybdenum, tantalum, tungsten, rhenium
Metals for use in nuclear reactors: zirconium, hafnium
Precious metals: gold, silver, platinum, palladium, rhodium, ruthenium, osmium, iridium
Tin and its alloys
Rare earths

Data on the analyses and physical properties of these materials are available in the American Society for Metals handbooks and *The Materials Selector* (see References).

EXAMPLE 7.6 *[EJ]*

The discussion in this chapter suggests that most commercial alloys have compositions close to the extremities of any phase diagram. In other words, compositions such as 50%X, 50%Y are seldom used. Speculate on why.

Answer   Alloys are classified as either single-phase or polyphase. Single-phase alloys contain relatively small amounts of other elements to prevent second phases from forming, except in systems such as the copper–nickel system where there is complete solid solubility (Figure 7.8). Here we stay near the copper end because of the higher cost of nickel, unless we absolutely require a particular property of nickel, such as high-temperature resistance.

In polyphase alloys the center portion of the phase diagram usually shows hard and brittle compounds, that, although wear- and abrasion-resistant, are difficult to maintain in a desirable distribution to maximize strength.

Summary

In this chapter we considered how the properties of a wide range of nonferrous alloys can be understood in terms of their structures.

In the aluminum alloys the wrought materials can be divided into two groups: the single-phase materials, which are strengthened by solid-solution and work hardening, and the polyphase alloys, which are age-hardened and may also be solid-solution or work-hardened. The cast alloys are strengthened by age hardening or by controlling the composition to produce large amounts of a hard dispersed phase such as silicon.

The magnesium alloys are similar in many ways to the aluminum alloys, but the single-phase alloys are not so ductile because of limited slip in the hexagonal structure.

The copper alloys have great ductility in the pure metal and in the single-phase alloys. The polyphase alloys may be age-hardened or strengthened with a martensite reaction.

The nickel and other alloys show similar relationships.

Definitions

Aluminum alloy designations   HX indicates the amount of cold work and annealing for single-phase alloys; TX indicates the combination of age hardening and cold work for two-phase alloys. (See text.).

Copper Alloys
    Brass   An alloy of copper and zinc.
    Bronze   An alloy of copper and a specified metal, such as tin, aluminum, or silicon.
    Cupronickel   An alloy of copper and 10 to 30% nickel.

Ferrous alloys   Alloys that have iron as the base element, such as wrought iron, steel, and cast iron.

Magnesium alloy designations   The same H and T designations used for aluminum alloys, but with a different code for composition.

Nickel alloys
    Monel   An alloy of nickel and 30% copper.
    Inconel   An alloy of nickel and 15% chromium.
    Duranickel   An alloy of nickel and 4.5% aluminum.

Nonferrous alloys   Alloys that do not have iron as the base element, such as the alloys of aluminum, copper, magnesium, nickel, zinc, and titanium.

Titanium alloys   Expensive alloys that have important applications where thermal and corrosion environments are severe or where strength-to-weight ratios must be high.

Problems

7.1   *[EJ]* Below are several applications for metal components. The alloys under consideration are 1060 aluminum, 5050 aluminum, 380.0 aluminum, AZ31B magnesium, and ZK60A magnesium. Determine which alloy would probably be specified for each application. There may be more than one appropriate alloy, but only one is to be se-

lected. When making your choice, consider such factors as strength, ductility, weight, cost, corrosion, conductivity, and method of fabrication. (Sections 7.1 through 7.3)
a. Home outdoor television antenna
b. Chalk tray for a blackboard
c. Metal frame for a computer
d. Accessory bracket on an air-cooled automotive engine
e. Corrosion-resistant nail

7.2 *[ES/E]* A 7075 T6 aluminum (5.6% Zn, 2.5% Mg, 1.6% Cu, 0.23% Cr) is exposed 10,000 hours at the temperatures indicated below and then tested at 75°F. The T6 specification is to solution-treat at 870–900°F, quench, and age at 250°F. (Sections 7.1 through 7.3)
a. Plot the data.
b. Explain how the data might be useful in specifications for particular applications.

Temperature, °F	Tensile Strength, psi	Yield Strength, psi	Percent Elongation
−112	90,000	79,000	11
−18	86,000	75,000	11
75	83,000	73,000	11
212	70,000	65,000	14
300	31,000	27,000	30
400	16,000	13,000	55
500	11,000	9,000	65
600	8,000	6,500	70
700	6,000	4,600	70

7.3 *[ES/E]* Calculate the parameter called the "strength-to-weight ratio" or "specific strength" for several of the higher-strength aluminum alloys in Table 7.1, and compare these to high-strength steel with a yield strength of 250,000 psi (1.725×10^3 MPa). This value is equal to yield strength divided by density (in pounds per cubic inch). What is the significance of this value? Take 2.7 as the specific gravity of aluminum and 7.8 as that of steel. Density = specific gravity $\times$ 0.0361 lb/in.^{3}. (Sections 7.1 through 7.3)

7.4 *[ES/E]* An aerospace engineer selects a 2014 T6 aluminum over a steel for a structural airframe part. The yield strength of the steel is 160,000 psi; that of the aluminum alloy is 60,000 psi. (Sections 7.1 through 7.3)
a. Give reasons why selection of the aluminum alloy is justified.
b. Recommend further processing that will enable the aluminum alloy to exhibit a yield strength in excess of 60,000 psi.

7.5 *[ES/E]* The density of magnesium is 0.064 lb/in.^{3}. What yield strength would be required to enable a magnesium alloy to compete with the best aluminum alloy on a strength-to-weight basis? [See Problem 7.3.] (Sections 7.1 through 7.3)

7.6 *[ES]* A modern lightweight engine block contains 16% silicon in aluminum. What percentage of the area of the cylinder wall contains the β silicon phase? The specific gravity of silicon is 2.33 and that of aluminum is 2.70. [*Hint:* Area % = Volume %.] (Sections 7.1 through 7.3)

7.7 *[EJ]* Assume that a large aluminum structure is to be made from an age-hardening alloy. (Sections 7.1 through 7.3)
  a. At what stage in the aging process would cold working be done? Why?
  b. Why would an aluminum alloy that exhibits natural aging be selected?
  c. Why might welding be difficult?

7.8 *[EJ]* A student tested two tensile specimens of 1060 aluminum in the annealed and H18 conditions, respectively, and one specimen of 5052 aluminum in the H18 condition, and then mixed up the data. Place the following values in their logical places in the accompanying table. (Sections 7.1 through 7.3)

Tensile strength:	19,000	10,000	41,000 psi
Yield strength:	4,000	18,000	39,000 psi
Percent elongation:	42	6	4

Analysis	Condition	Tensile Strength, psi	Yield Strength, psi	Percent Elongation
1060 99.6% min Al	0 (annealed)	_____	_____	_____
1060 99.6% min Al	H18	_____	_____	_____
5052 2.5% Mg, 0.2% Cr, Bal. Al	H18	_____	_____	_____

7.9 *[ES]* The specification for the T6 condition reads "solution heat-treated plus artificially aged." (Sections 7.1 through 7.3)
  a. Using the phase diagram shown, draw a time–temperature chart specifying the heat treatment you would use to attain this condition in a 6% Cu, 94% Al alloy, assuming that your starting material was slow-cooled from 900°F (482°C).
  b. Choose the description that best describes the microstructure after the T6 treatment of the 6% Cu alloy: (1) supersaturated κ with separate grains of θ, (2) supersaturated κ, (3) κ with fine precipitate of θ, (4) κ with fine precipitate of θ plus coarse grains of θ.

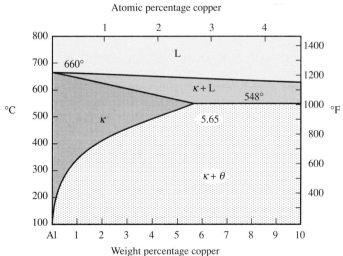

7.10 *[ES/EJ]* Referring to Figure 7.1, determine whether each of the following alloys is a likely candidate for age hardening. (Sections 7.1 through 7.3)
  a. 90% Al, 10% Mg
  b. 97% Mg, 3% Al
  c. 80% Al, 20% Mg

7.11 *[ES/EJ]* Shown here is a series of age-hardening curves obtained at different temperatures for an aluminum alloy. Answer the following questions, using the letters A, B, and C to indicate the three curves. There may be more than one answer for each question. (Sections 7.1 through 7.3)

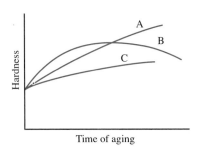

  a. Which curve shows aging at the lowest temperature? _____
  b. Which shows overaging? _____
  c. Which shows solution treatment? _____
  d. If an age-hardening curve showed two hardness peaks, what might this curve mean?

7.12 *[EJ]* In the age hardening of aluminum alloys, what does the T7 condition mean? [*Hint:* Stabilization generally means overaging. Think of how an age-hardening alloy might react to elevated-temperature service.] (Sections 7.1 through 7.3)

7.13 *[ES/EJ]* Typical uses for copper alloys are given in Table 7.3. What unique characteristics of these alloys suggest their continued use, even though the cost may be high? (Sections 7.4 through 7.12)

7.14 *[ES/EJ]* Use phase diagrams to explain the following phenomena. (Sections 7.4 through 7.12)
  a. Copper–nickel alloys are used above 330°C, but copper–lead alloys are not.
  b. Commercial copper–lead alloys generally contain less than 30 wt % lead.

7.15 *[EJ]* Aged copper–beryllium alloys are often used in thin electrical switch gear. Why is it that these alloys are used rather than other copper alloys such as the brasses? [*Hint:* Alloying elements also have an effect on the electrical conductivity.] (Sections 7.4 through 7.12)

7.16 *[EJ]* Why is the yield strength of wrought Inconel greater than that of wrought Monel 400 even though the alloy content is lower? (Sections 7.4 through 7.12)

7.17 *[EJ]* Indicate why a Ti-6Al-4V alloy might be competitive with stainless steel for a leg bone replacement. [*Hint:* Corrosion is not the only consideration.] (Sections 7.4 through 7.12)

7.18 *[EJ]* In cast-nickel–based alloys, Monel 505 and Inconel 705 are said to form Ni_3Si, which dispersion-hardens. The text indicates that further hardening can take place through aging. What is the difference between dispersion hardening and age hardening, if any? Recalling that the alloys are cast, speculate on the significance of the text statement? (Sections 7.4 through 7.12)

7.19 *[ES/EJ]* Refer to Figure 7.9 and the particular elements that stabilize the β phase in titanium. (Sections 7.10 and 1.21 and Chapter 4)

a. What characteristic of these alloys suggests that they should stabilize β?

b. Indicate what might limit the amounts that we could add and still have β or α + β mixtures.

c. Why does the text indicate that an α–β region should exist according to the phase rule?

7.20 *[ES/EJ]* The text indicates that 10–25 troy ounces per ton of silver are often added to copper [12 troy ounces = 1 lb]. (Sections 7.4 through 7.12)

a. Convert the silver content to a weight percent range.

b. Why might silver in these small amounts be added to a copper tea kettle?

7.21 *[ES/EJ]* Twenty-four-karat gold is 1000 fine gold, or 100% gold (995 fine means 99.5% gold). (Sections 7.4 through 7.12)

a. Calculate the "net worth" of a 50-g piece of jewelry that is 18 karat (18/24 of 100% gold). Look up the current price of gold bullion in the commodity quotes in your newspaper. Commercially traded gold bullion is at least 995 fine, and 12 troy ounces = 1 lb.

b. Why might the actual net worth be higher or lower than your calculation?

c. Why is gold jewelry seldom 24-karat gold?

7.22 *[ES/EJ]* We are going to produce a new solid-solution alloy by adding 5 at. % balonium to copper. Indicate how the physical properties might vary by placing an X in the appropriate column. (Section 7.4 and earlier chapters)

	Increase	*Decrease*	*No Change*
Tensile strength	_____	_____	_____
Ductility	_____	_____	_____
Electrical conductivity	_____	_____	_____
Hardness	_____	_____	_____

Data you may need:

	Pure Copper	*Pure Balonium*
Tensile strength, psi	35,000	30,000
Ductility (percent elongation)	50	50
Electrical conductivity relative to copper, %	100	90
Hardness	90 HB	85 HB

7.23 *[ES/EJ]* A particular zinc alloy with 4% Al, 0.50% Cu, 0.09% Fe, and 0.03% Mg is tested in tension at room temperature for different loading times required to give

0.10% elongation. The data for the elastic modulus, the stress level, and the time required to produce the elongation follow. (Section 7.11 and earlier chapters)

a. What would be the effects of higher temperatures and lower allowable elongations on the modulus and stress values for a given test time?

b. How are these data useful in design and specifications?

Test Time	Modulus, 10^6 psi	Stress, 10^3 psi
1 day	7.2	7.5
100 days	3.0	3.0
1 year	2.1	2.2
5 years	1.2	1.2
10 years	1.0	1.0

7.24 *[EJ]* In the following table, list by code letter *all* the hardening mechanisms that have raised the strength and hardness of each material above that of the pure element (copper or aluminum) paying attention to the microstructure given. [*Code:* A = solid-solution strengthening; B = age hardening; C = cold working; D = dispersion hardening; E = martensite formation.] (Review of Chapter 7)

Material	Chemical Analysis	Microstructure	Hardening Mechanisms (Use Code)
(1) Cartridge brass	Cu 70%, Zn 30%	Elongated grains of α showing slip	
(2) Gun metal	Cu 88%, Sn 10%, Zn 2%	Equiaxed α + δ phase of CuSn compound (hard)	
(3) Aluminum bronze	Cu 90%, Al 10%	Grain boundary precipitate of α plus α + γ_1 eutectoid. γ_1 is CuAl compound (hard)	
(4) Aluminum alloy	4% Cu, balance Al, condition 0 (as annealed)	Particles of $CuAl_2$ + equiaxed α	
(5) Same alloy as in (4)	Condition T9: solution heat-treated, aged, cold-worked	Elongated α showing slip plus very fine (coherent) precipitate of $CuAl_2$ (at 60,000×)	

7.25 *[ES/EJ]* List by code letter the one hardening mechanism that is most important for each of the following alloys: [*Code:* A = work hardening; B = dispersion hardening;

C = age hardening; D = solid-solution hardening; E = martensite reaction.] (Review of Chapter 7)
a. 1060 aluminum (99.6% min Al)
b. 2014 aluminum (4.5% Cu, 0.8% Si, 0.8% Mn, 0.5% Mg)
c. C11000 ETP 99.9% Cu
d. C71500 70% Cu, 30% Zn brass
e. C17200 98% Cu, 2% Be

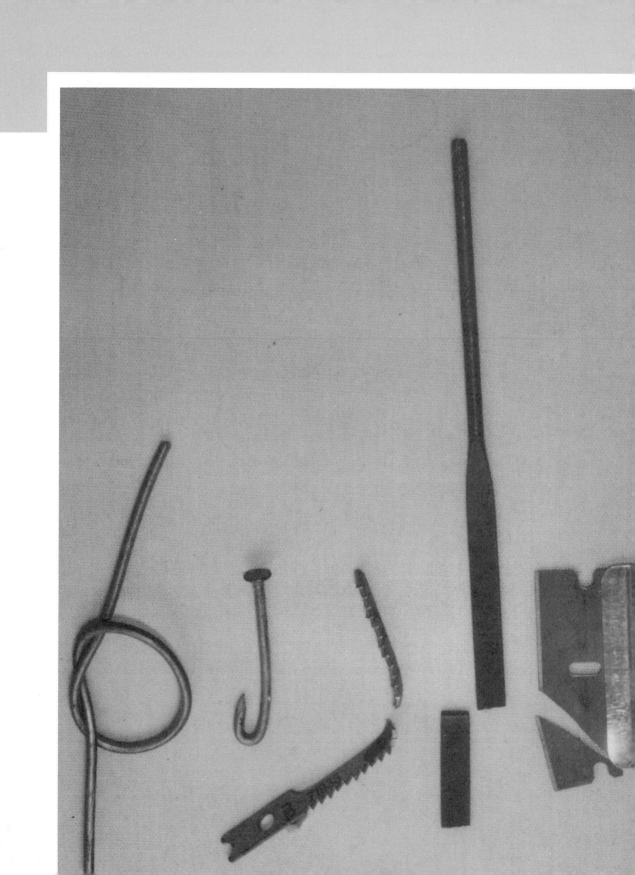

CHAPTER 8

Plain-Carbon and Low-Alloy Steels

Steel is not a single material, but a family of alloys of iron and carbon. Some components such as files and razor blades, are hard with low ductility; others, such as steel wire and nails, are softer and quite ductile. An intermediate group of parts, such as screwdrivers, bolts, and shafts, have intermediate strength and toughness.

The variation in properties is due to the amount and distribution of the carbon present in hard particles of iron carbide. In the hardest components, the iron carbide is in a *high concentration* of *fine particles*, which block dislocation movement. In the soft components there is less carbide, and it is present in a coarse dispersion. The amount of carbide is controlled by chemistry; the distribution is varied by heat treatment.

8.1 Overview

We tend to take ferrous (iron-base) alloys for granted because they are so common in everyday applications. In fact, approximately 85% of all of the metal tonnage used for engineering applications in the United States is based on iron. This of course suggests that ferrous metallurgy retains the importance that it first assumed in the "iron age" approximately 3000 years ago.

Today, we find many iron alloys when we look at the components in an automobile, even though the nonferrous alloys, polymers, and composites are used in a number of items that were made of ferrous alloys just a few years ago. It is significant, however, that iron alloys are still used for those applications that require heavy load-bearing capability or must transmit power.

We shall discuss the plain-carbon and the low-alloy steels separately in this chapter; they account for the bulk of the ferrous alloys. In Chapter 9, we will discuss the high-alloy steels such as the stainless steels, the superalloys used for many high-temperature applications, and the cast irons, wherein the presence of graphite provides many unique properties.

The *plain-carbon steels* can be defined as pure iron to which carbon has been added. There are other elements present, but their amounts are insufficient to affect the equilibrium materially. When we add alloying elements to the iron–carbon alloy under a total of approximately 5 wt %, we call the products *low-alloy steels*.

The major reason for adding alloying elements is to obtain a microstructure that will result in more desirable mechanical properties even though an alloy addition adds to the cost.

We shall begin with the iron–carbon equilibrium phases and follow this with a discussion of the nonequilibrium structures that are so important to steel metallurgy. Finally, we will be able to appreciate why we need alloy additions in order to produce engineered components that can tolerate high stresses and still perform reliably over many years in service.

8.2 The Iron–Iron-Carbide Diagram and its Phases

The iron–iron-carbide diagram is essential to understanding the basic differences among iron alloys and the control of properties. Figure 8.1 shows the diagram that is commonly used. We note first that the carbon scale only goes up to 6.7% carbon, where we encounter iron carbide, Fe_3C. This should cause us no concern, because we have already used a part of another diagram, the portion of the copper–aluminum diagram extending to $CuAl_2$, (θ), in our discussion of age hardening.

The first step is to examine the individual phases. Beginning with pure iron, we see three solid phases, alpha (α), gamma (γ), and delta (δ), shown at the left side. If we begin with α at 68°F (20°C) and heat slowly, we can expect

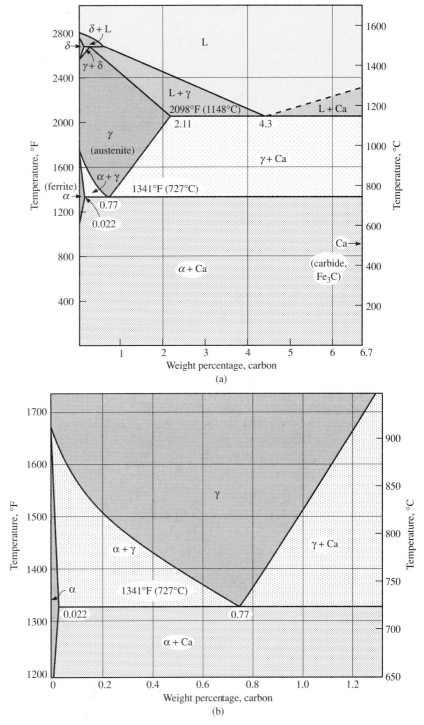

FIGURE 8.1 (*a*) Iron–iron-carbide phase diagram. (*b*) Enlarged eutectoid portion of the iron–iron-carbide phase diagram.

α to transform to γ at 1674°F (912°C), γ to transform to δ at 2541°F (1394°C), and δ finally to melt at 2802°F (1538°C).

We shall look in vain for β iron; it was lost in antiquity. Naturally, because of its importance, the iron–iron-carbide diagram was the first to receive attention. Early investigators noted that near 1418°F (770°C) iron lost its ferromagnetism. They considered this a phase change and gave the symbol β to iron in the range 1418 to 1674°F (770 to 912°C). It was found, however, that this magnetic effect is not a phase change—it is due to a shift in alignment of the atoms. This point will be considered further in Chapter 21.

This left metallurgists with α, γ, and δ. The picture was further simplified when x-ray diffraction evidence showed α and δ to have the same crystal structure. Both these phases have BCC structures, and the lattice parameter a for δ is the same as that for α if allowance is made for expansion with temperature (Figure 8.2).

The γ phase is FCC and therefore more densely packed. There is a contraction of about 1% in volume in the $\alpha \rightarrow \gamma$ transformation and an expansion of 0.5% in volume in the $\gamma \rightarrow \delta$ change.

According to the phase diagram (Figure 8.1a), the amount of carbon that can be dissolved in γ is 2.11% maximum (2098°F, 1148°C). This value is many times greater than the maximum solubility in α, 0.022% carbon (1341°F, 727°C). We can see the basis for this difference if we compare the sizes of the interstitial holes in BCC and FCC unit cells where the carbon atoms dissolve. The α phase is also called *ferrite* and the γ phase containing carbon is called *austenite*.

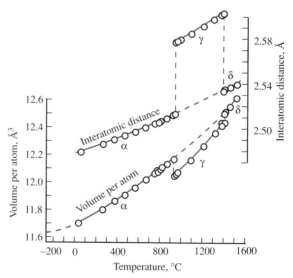

FIGURE 8.2   Lattice parameter of iron as a function of temperature

[Z. S. Basinski, W. Hume-Rothery, and A. L. Sutton, "The Lattice Expansion of Iron," *Royal Society of London: Proceedings*, Vol. A229, 1955, p. 459. Used by permission.]

EXAMPLE 8.1 *[ES]*

Compare the sizes of the largest interstitial holes in the BCC and FCC structures found in iron. How does the size of the carbon atom compare? [The atomic radius of iron is 1.27 Å (0.127 nm) in FCC and 1.24 Å (0.124 nm) in BCC.] Calculate the atomic packing factor in both iron structures.

Answer   A horizontal plane through the center of an FCC structure contains the largest interstitial hole (the same hole can be found in the face of the FCC).

$$a = \frac{4R}{\sqrt{2}} = 2R + 2r \quad \text{or} \quad r = 0.414R = 0.414 \times 1.27$$

$$= 0.52 \text{ Å } (0.052 \text{ nm})$$

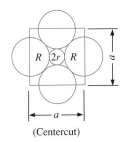

(Centercut)

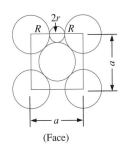

(Face)

In the BCC structure the largest interstitial hole occurs at the coordinates $\frac{1}{2}, 0, \frac{3}{4}$.

$$(r + R)^2 = (\tfrac{1}{2}a)^2 + (\tfrac{1}{4}a)^2$$

$$(r + R)^2 = \frac{5}{16}a^2 = \frac{5}{16}\left(\frac{4R}{\sqrt{3}}\right)^2$$

$$(r + R)^2 = \frac{5R^2}{3}$$

$$r + R = R\sqrt{\tfrac{5}{3}}$$

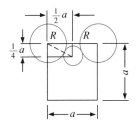

or

$$r = 0.291R = 0.291 \times 1.24 = 0.36 \text{ Å } (0.036 \text{ nm})$$

Because the atomic radius of carbon is 0.71 Å (0.071 nm), it has a much better chance of forming an interstitial solution in FCC iron. (Note that the radius of the iron atom is slightly different in the BCC and FCC configurations.) The atomic packing factor (APF) is defined as (volume of atoms per unit cell)/volume of unit cell.

$$\text{BCC:APF} = \frac{2 \text{ atoms} \times \frac{4}{3}\pi(1.24 \text{ Å})^3}{\left(\dfrac{4 \times 1.24 \text{ Å}}{\sqrt{3}}\right)^3} = 0.68$$

$$\text{FCC:APF} = \frac{4 \text{ atoms} \times \frac{4}{3}\pi(1.27 \text{ Å})^3}{\left(\dfrac{4 \times 1.27 \text{ Å}}{\sqrt{2}}\right)^3} = 0.74$$

Even though the packing of atoms is more efficient in the FCC, the interstitial holes in the FCC are larger than those in the BCC.

In addition to the solid solutions of iron and carbon, the other solid phase in the diagram is iron carbide, Fe_3C, also called *cementite*. This curious name arose from an ancient process in which carbon was diffused into iron by making up a package of layers of iron and carbon compounds, then heating it in a furnace. The term is derived from this so-called cementation process. It is evident from the diagram that cementite will be encountered at 20°C in any alloy from 0.006 to 6.7% carbon. In sharp contrast to the BCC and FCC iron structures, which are both ductile (over 40% elongation) and soft (HB 100 to 150), cementite is brittle (0% plastic elongation) and hard [HB over 700 (HV 1200)]. We can summarize the phases and their several names as follows:

Temperature Range, Pure Fe	Phase	Crystal Structure	a, Å	HB	Percent Elongation
Room temperature to 1674°F (912°C)	α, ferrite	BCC	2.86 (0.286 nm)	~150	~40
1674 to 2541°F (912 to 1394°C)	γ, austenite	FCC	3.60 (0.360 nm)	~150	~40
2541 to 2802°F (1394 to 1539°C)	δ, delta ferrite	BCC	2.89 (0.289 nm)	~150	~40
	Iron carbide (cementite, Fe_3C)	Orthorhombic	—	~700	~0

EQUILIBRIUM STRUCTURES OF PLAIN-CARBON AND LOW-ALLOY STEELS

In general, steel is an iron–carbon alloy that contains less than 2.1% carbon, and this is the material we shall concentrate our attention on. We shall discuss the steel structures formed under equilibrium conditions before turning to the nonequilibrium hardening reactions.

8.3 Hypoeutectoid Steels

We shall consider first *hypoeutectoid steels* (up to 0.77% carbon) (Figure 8.3) and then *hypereutectoid steels* (0.77 to 2.11% carbon).* The low-carbon steels are by far the most important of the group, mainly because of their high ductility both hot and cold, which enables them to be readily fabricated into shapes of excellent toughness and strength. All structural and automobile body steels fall into this group. Most steel castings, such as railroad-car parts, are also in this composition range.

Let us follow the cooling of a typical steel of this type (0.3% carbon) from the liquid state, using the phase diagram as a map and drawing a fraction chart (Figure 8.4) as discussed in Chapter 4.

The most important changes during cooling occur in the range 1600 to 1300°F (871 to 704°C). Ferrite (α) begins to precipitate at the austenite (γ) grain boundaries at 1475°F (802°C) (Figure 8.3). This precipitation continues until the temperature reaches 1342°F (728°C), and the γ changes from 0.3 to 0.77% carbon as a result (tie line changes in the α + γ region upon cooling). At 1341°F (727°C) the remaining γ transforms to a mixture of α + carbide through a constant-temperature eutectoid reaction, γ → α + carbide. The important point is that an interleaved, or *lamellar*, mixture of ductile α and hard carbide is formed. This composite is called *pearlite.* It has higher strength but lower ductility than ferrite. Note that the name *pearlite* is given to the unique *lamellar mixture* of the two phases α + carbide.

Let us now return to our discussion of the fraction chart (Figure 8.4). At 1342°F (728°C) we have approximately 63% ferrite and 37% austenite. We shall refer to this ferrite as *primary ferrite* so as not to confuse it with the ferrite that results from the formation of pearlite. At 1341°F (727°C) all the remaining austenite (of 0.77% carbon) transforms to pearlite, and we therefore obtain *eutectoid ferrite.* If we had started out with 100% austenite at 1341°F (727°C) (which would have required an initial carbon content of 0.77%), we would have obtained (6.69 − 0.77)/(6.69 − 0.022), or 89%, ferrite and 11% carbide. However, because we had only 37% austenite, we obtain 0.89 × 37 = 33% ferrite upon slow cooling. The *total* ferrite would therefore be 96% (63 + 33) at 1340°F (726°C). The fact that Figure 8.4 shows no change in the amount of ferrite or carbide when the mixture is cooled from 1340 to 68°F (726 to 20°C) indicates that a slight decrease in carbon solubility in ferrite (0.022 to <0.01) has a negligible effect on the calculation.

In order to achieve equilibrium, slow cooling must be used. More rapid cooling, such as air cooling, results in a finer pearlite and is called *normalizing.* Slow cooling, such as furnace cooling, through the eutectoid is called *full annealing.* Full annealing results in slightly lower strength but higher ductility than does normalizing. We can get still another variation in the pearlite by

---

*Hypoeutectoid means **below eutectoid composition**—that is, less than 0.77% carbon. Hypereutectoid means **above eutectoid composition**—that is, greater than 0.77% carbon.

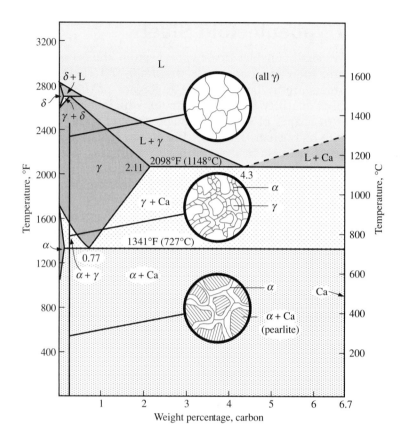

FIGURE 8.3 Changes that occur in the microstructure of a hypoeutectoid steel when it is cooled. The photomicrograph shows white ferrite and the lamellar structure of pearlite; 500×, 2% nital etch.

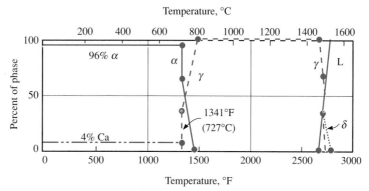

FIGURE 8.4 Phase fraction chart for 0.3% carbon steel

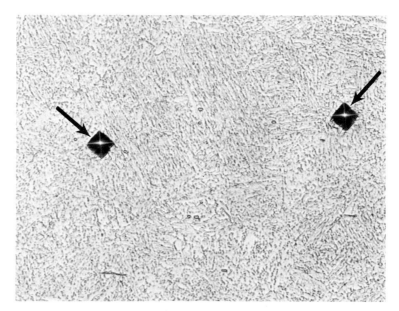

FIGURE 8.5 Spheroidized structure in 0.5% carbon steel. There are very fine spheroids of iron carbide in the α matrix; HV 170, 500×, 2% nital etch.

reheating or holding for extended periods just below the eutectoid (approximately 1300°F, 704°C). The carbides in the pearlite change from the normal plates to spheres (Figure 8.5). This microstructure is called *spheroidite* or *spheroidized pearlite*. It has lower hardness than lamellar pearlite but higher ductility and toughness.

8.4 Hypereutectoid Steels

High-carbon steels naturally contain more of the hard carbide phase. They are useful where higher strength, hardness, and wear resistance are needed, as in knife blades, other cutting tools, and bearings. Let us follow the cooling of a 1% carbon steel on the phase diagram and plot the results as a fraction chart (Figure 8.6).

The important temperature range for this material is 1550 to 1300°F (843 to 704°C). It is vital to observe that a brittle phase, iron carbide, precipitates at the austenitic grain boundaries from 1520 to 1341°F (827 to 727°C) (Figure 8.7), although this *primary carbide* amounts to only approximately 3.5%. This is in contrast to the ductile primary ferrite that precipitated in the 0.3% carbon steel. The eutectoid reaction is exactly the same in both cases; austenite with 0.77% carbon in solid solution forms the α + carbide in pearlite.

The properties of this high-carbon material in the slow-cooled condition are very poor. Elongation, for instance, is less than 5% because of the brittle carbide network. However, if we heat the material to 1550°F (843°C) to dissolve the carbide in the austenite, cool rapidly to 1300°F (704°C) to allow transformation to a fine mixture of α + carbide, and then hold at 1300°F (704°C), the coarse grain-boundary carbide does not have time to form on cooling, and we obtain a better carbide shape and distribution with higher ductility. This material is also called *spheroidite*. In general, the toughness of these hypereutectoid steels is still lower than that of the hypoeutectoid type because of the greater amount of carbide.

8.5 Specifications

We can now discuss a few specifications and uses of these materials (Table 8.1). Wrought iron and ingot iron both show maximum ductility. *Wrought iron* is essentially pure iron with slag fibers rolled into the structure. The use of these materials in pipe and architecture is well known. Present-day wrought iron components are usually made of a low-carbon steel. Next we have a group of steels called the plain-carbon steels, to distinguish them from alloy steels. All of these contain about 0.5% manganese, however. The code symbol for these steels is 10XX, where XX designates the percentage of carbon;

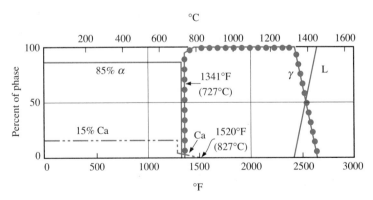

FIGURE 8.6 Phase fraction chart for 1% carbon steel

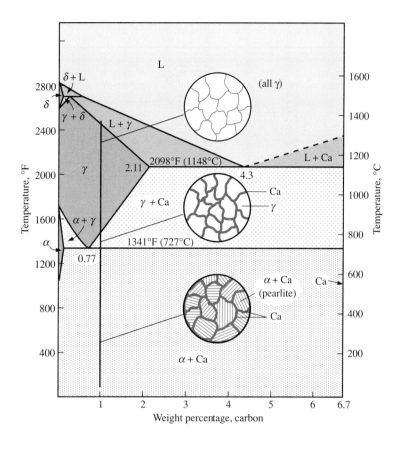

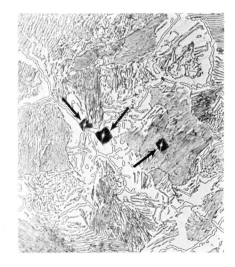

FIGURE 8.7 Changes that occur in the microstructure of hypereutectoid steel when it is cooled. Photograph is an actual photomicrograph; 500×, 2% nital etch. Pearlite (gray), HV 235; some ferrite, HV 160, precipitated next to the carbide (white).

TABLE 8.1 Typical Properties of Plain-Carbon Steels

Steel Number	Carbon Content, percent	Condition	Tensile Strength,* psi × 10³	Yield Strength,* psi × 10³	Percent Elongation	HB	Typical Use
Ingot iron†	0.02	Annealed	42	19	48	69	Pipe, architecture
		Hot-rolled	44	23	47	83	
		Cold-drawn	73	69	12	142	
1010	0.10	Hot-rolled	47	26	28	95	Car fenders
		Cold-drawn	53	44	20	105	
1020	0.20	Hot-rolled	55	30	25	111	Structural forms
		Cold-drawn	61	51	15	121	
1040	0.40	Hot-rolled	76	42	18	149	Crankshaft
		Cold-drawn	85	71	12	170	
1060	0.60	Hot-rolled	98	54	12	201	Chisel
		Cold-drawn‡	90	70	10	183	
1080	0.80	Hot-rolled	112	62	10	229	Wear-resistant parts
		Cold-drawn‡	98	75	10	192	
1095	0.95	Hot-rolled	120	66	10	248	Cutting blades
		Cold-drawn‡	99	76	10	197	

* Multiply psi by 6.9 × 10⁻³ to obtain MN/m² (MPa) or by 7.03 × 10⁻⁴ to obtain kg/mm².
† Wrought iron has mechanical properties similar to those of ingot iron.
‡ Spheroidized, then cold-drawn

for example, 1020 steel contains approximately 0.20% carbon. As we increase the amount of carbon, the percentage of pearlite rises and the strength increases from 44,000 psi (304 MPa) at 0.02% carbon to 112,000 psi (774 MPa) at approximately 0.8% carbon. The cold-drawn material shows still higher strength and hardness because of work hardening. However, above 0.5% carbon it is necessary to spheroidize the steel prior to cold drawing to attain sufficient ductility. The yield strength of the cold-worked spheroidized material is higher than that of the corresponding hot-rolled pearlitic structure.

EXAMPLE 8.2 *[EJ]*

Why would automobile fenders be made of a 1010 steel, whereas railroad rails might be made of a 1080 steel?

Answer Automobile fenders are relatively thin, and their strength is developed by cold working. Therefore, a low carbon content is required to obtain the necessary ductility for cold forming. Railroad rails, which are rolled sections, must be hot-formed. The high carbon content gives the required wear resistance; however, it limits the cold-working capability and the steel is formed while it is all austenite.

A higher-carbon steel fender might seem desirable because of the added strength; however, it would have to be hot-formed. The hot-worked steel would show excessive oxide scales (Chapter 18) and might not be so strong in the annealed condition as a lower-carbon cold-worked steel. (Compare a cold-drawn 1010 steel and a hot-worked 1020 steel in Table 8.1.)

NONEQUILIBRIUM REACTIONS

8.6 Steel Hardening

If we were limited to the equilibrium structures and the plain-carbon steels of the iron–iron-carbide diagram, we could not make a great many critical tools and components. Imagine the problems we would have without hardened drills, files, lathe tools, gears, chisels, plows, ball bearings, rolls, razor blades, knives, saws, dies, and hundreds of other parts in which we need hardness and strength! As an example of the hardening operation, let us consider the changes in the properties of 1080 steel that we can accomplish by nonequilibrium cooling. See the table below. It is not the *act* of quenching but the *effect* of quenching on the structure that produces the change. For example, if we performed the same treatment on a piece of pure iron or on an 18% Cr, 8% Ni stainless steel (see Chapter 9), there would be no change in hardness.

Before we discuss how this change in hardness takes place, we should recall that all reactions take time, because the nucleation of new structures or phases and the diffusion of atoms to allow growth do not occur instantaneously. We shall see in the next section that the rate of transformation of austenite to ferrite and carbide (as in pearlite) depends on nucleation and growth.

	1080 Steel			
Condition	Tensile Strength, psi × 10³	Yield Strength, psi × 10³	Percent Elongation	HB
Slow-cooled from 1550°F (843°C)	112 (773 MPa)	62 (428 MPa)	10	192
Water-quenched from 1550°F (843°C)	>200 (>1380 MPa)	200 (1380 MPa)	1	680

8.7 Austenite Transformation

The key to understanding the variation in hardness of different steels is a knowledge of the austenite transformation. Let us consider first the simplest reactions—those that can occur with eutectoid (0.8% nominal carbon) austenite.* Later we shall take up the effects of different compositions. So far we have discussed only the transformation of austenite under equilibrium conditions to a relatively coarse, plate-like mixture of carbide and ferrite called pearlite.

What if we avoid the reaction at 1341°F (727°C) by cooling the austenite rapidly from 1500 to 1200 or 800°F (815 to 649 or 427°C) or lower and allowing it to react at these temperatures?

Let us perform an experiment to determine the effects. First we machine a thin specimen of 0.8% carbon steel with two holes that serve as reference points for measurement. Then we test the specimen, using an instrument called a *dilatometer* (Figure 8.8). We hang the specimen from a hook *A* on a tube of fused silica. The tube itself is rigidly held from platform *B*. Next we place the hook of an inner tube into the lower specimen hole *C*. The specimen is now in light tension, supporting the inner tube, which slides smoothly in the outer tube. Finally we fasten a dial gage onto the top of the inner tube with the point bearing on the support of the outer tube.

This instrument is called a dilatometer because it measures change in length, or dilation, of the specimen. Whether the specimen expands or contracts, the dial shows the movement faithfully. The use of fused silica, a very-low-thermal-expansion material, reduces any errors that might be caused by expansion of the tubes.

Now let us change the structure of the specimen to austenite by heating to 1600°F (871°C). We do this simply by immersing the lower part of the dilatometer and the specimen in a pot of liquid lead at 1600°F (871°C). The fused silica is inert and has excellent thermal shock resistance.

8.8 Pearlite Formation

To observe transformation at 1300°F (704°C), we quickly replace the 1600°F (871°C) pot with another pot of lead at 1300°F (704°C). We record the changes in length of the specimen as time passes and obtain the graph shown in Figure 8.9*a*.

We find the microstructural changes in the specimen by interpreting the graph. Stage 1 is a contraction, as the austenite cools. The magnitude of this contraction corresponds exactly to the value predicted via the coefficient of expansion (or contraction) of austenite. The specimen is at constant temperature during period 2. No change in microstructure is taking place, so no change in length occurs. At time 3 the length of the specimen is increasing, although it is in a bath at constant temperature. This change must be due to the change in structure, austenite → pearlite. The change in length is not surprising given the data showing the volume change when FCC iron transforms

---

*Although the equilibrium eutectoid is 0.77% carbon, a nominal carbon content of 0.8%, as in a 1080 steel, will be used to represent a eutectoid steel, in which we might expect 100% pearlite under equilibrium cooling.

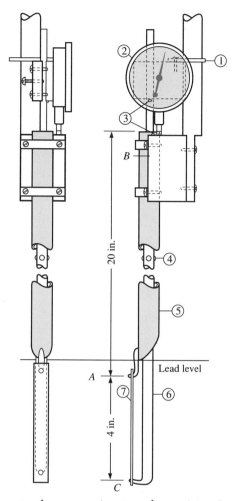

FIGURE 8.8 Dilatometer for measuring transformations. 1 = dial guide, 2 = dial gage, 3 = fused silica lugs, 4 = inner-tube guides (fused silica, two sets), 5 = outer fused silica tube, 6 = inner fused silica tube, 7 = specimen ($4\frac{1}{2}$ by $\frac{1}{2}$ by $\frac{1}{32}$ in.; 114.3 by 12.7 by 0.8 mm).

to BCC. The graph tells us, therefore, the time at which transformation begins and the time at which it ends, at this temperature, for this steel: 30 and 1000 sec. After the transformation there is no change in dimension as long as the temperature remains constant.

If we repeat the experiment using a temperature of 1200°F (649°C) for the transformation, we obtain different transformation times: 3 and 25 sec (Figure 8.9b).

We now graph the time from the beginning to the end of the transformation as a function of temperature, using a logarithmic scale to accommodate the wide variation in time (Figure 8.9c). This graph can be called either an

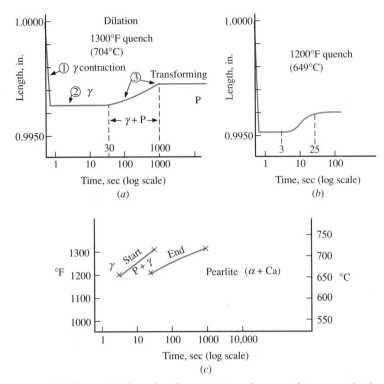

FIGURE 8.9 (*a*) Change in length when 0.8% carbon steel is quenched to 1300°F (704°C). (*b*) Change in length when 0.8% carbon steel is quenched to 1200°F (649°C). (*c*) TTT curve [derived from data from (*a*) and (*b*)].

isothermal transformation curve or a TTT (time–temperature–transformation) curve.

If we measure the hardness of the steel at room temperature after transformation, we find that the hardness increases as the transformation temperature decreases. The spacing between the carbide plates in the pearlite is finer, and the hardness and strength are related to the distance between the carbides (Figure 8.10).

8.9 Bainite Formation

Below 1000°F (538°C) the isothermally transformed specimen changes in structure from the alternating α and carbide plates of pearlite to a feathery or acicular (needle-like) structure called *bainite* (Figure 8.11). The hardness continues to increase because the carbide is becoming increasingly fine, and as a result the distance over which slip can take place in the ferrite is becoming shorter. The starting and ending times of the transformation are longer because the diffusion rate is slower. From these data we can obtain more of the isothermal transformation curve (Figure 8.12).

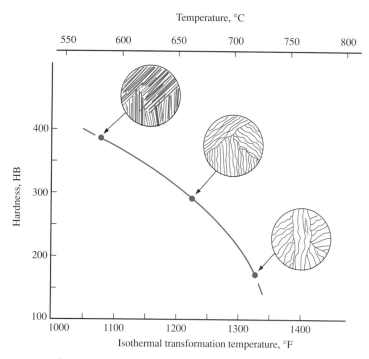

FIGURE 8.10 Relationship between the hardness of 100% pearlite (0.80% carbon) and the isothermal transformation temperature. Lower temperatures result in finer pearlite.

FIGURE 8.11 Bainite produced by the isothermal transformation of an 0.8% C austenite at 600°F (316°C); 500 × , 2% nital etch

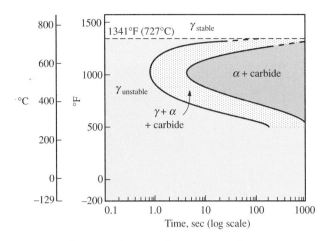

FIGURE 8.12   A partially completed isothermal transformation curve for 1080 steel

8.10   Transformation to Martensite

Below 420°F (216°C) we encounter a type of transformation quite different from the isothermal transformations we have just discussed, and the resulting structure is different (Figure 8.13). As we decrease the temperature below 420°F (216°C), a fraction of the new structure, called *martensite*, forms *instantaneously* with each decrease in temperature. If we halt the cooling and hold the sample at 300°F (149°C), for example, no further martensite is produced until we resume cooling. The sample is completely transformed when we reach lower temperatures. We call the temperature at which this transfor-

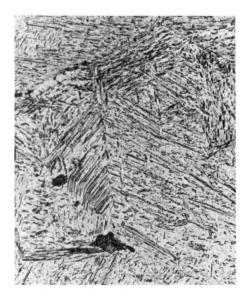

FIGURE 8.13   Martensite produced by rapid quenching of an 0.4% C steel from 2300°F (1260°C). The high austenitizing temperature gives a coarse martensite. The two dark spots are pearlite; 500 × , 2% nital etch.

mation begins the *martensite start temperature*, M_s, and the temperature at which it ends the *martensite finish temperature*, M_f.

The dilatometer curves show this effect clearly (Figure 8.14). If we quench to just above the M_s, we obtain a normal transformation curve to bainite. If we quench to below the M_s (300°F, 149°C), the decrease in length *on quenching* is less than it would be if we had quenched to 420°F (216°C). Therefore, some expansion resulting from transformation took place *during* cooling. We may now complete our isothermal transformation curve for 1080 steel to include the M_s and M_f, as shown in Figure 8.15.

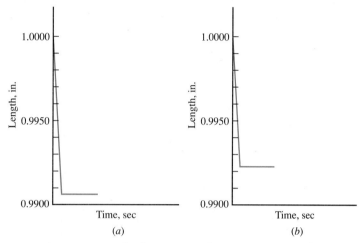

FIGURE 8.14   Dilation curves for lower quenching temperatures (0.8% carbon, 1080 steel). (*a*) Quenched to 420°F (216°C), M_s. (*b*) Quenched to 300°F (149°C), γ + martensite.

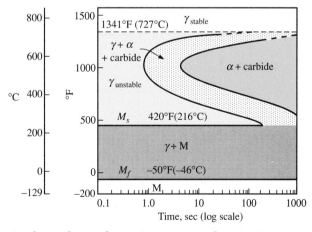

FIGURE 8.15   Isothermal transformation curve and austenite–martensite transformation (0.8% carbon, 1080 steel)

Martensite is important because it is the hardest structure formed from austenite.* X-ray diffraction evidence indicates that the structure is a distorted BCC (tetragonal). The distortion is caused by trapped atoms of carbon, as shown in Figure 8.16.

We can explain the difference between the nucleation and growth of pearlite and bainite and the rapid transformation to martensite as follows. In the case of pearlite and bainite, nucleation of the two new phases ferrite and carbide has to occur, and diffusion and long-range movement of atoms must take place. In the case of martensite, the structure can be formed by short shearing movements. In Figure 8.17 we see the similarity between martensite and the parent austenite. The end product, martensite, forms by atomic movements that result in an expansion of the *a* direction in the tetragonal cell, which is sketched within the FCC lattice of the austenite. Note that the carbon atoms are trapped along the sides, so the *c* axis is finally longer than *a*. At the maximum carbon level, the *c/a* ratio is 1.08, as shown in Figure 8.16.

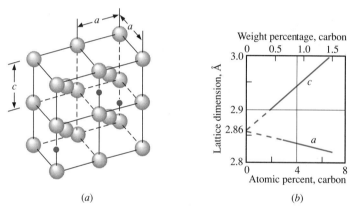

(a) (b)

FIGURE 8.16 (a) Location of carbon atoms in martensite. (b) Carbon atoms expand the BCC lattice, giving rise to strain and higher hardness.

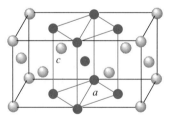

FIGURE 8.17 Relation of a unit cell of martensite to the parent austenite. Two unit cells of austenite (FCC) and the resultant BCT lattice of martensite are shown.

---

*When cooling carbon-saturated austenite from the eutectic to the eutectoid, iron-carbide, which is harder than martensite, can precipitate. We will discuss this further in Chapter 9.

Let us summarize the austenite transformation of an 0.8% carbon steel shown in Figure 8.15:

1. Above 1341°F (727°C): Austenite is the stable phase, as we see from the equilibrium diagram.
2. From 1341 to 1050°F (727 to 566°C): (a) Austenite isothermally transforms to α + carbide as pearlite. (b) As the temperature decreases, the time required for transformation *decreases* because nucleation is easier. (c) A more rapid nucleation rate at lower temperatures leads to finer pearlite (HB 170 to 400).
3. From 1050 to 420°F (566 to 216°C): (a) Austenite isothermally transforms to α + carbide as bainite. (b) As the temperature decreases, the time required for transformation *increases*, because although nucleation rates are high, diffusion is slower. (c) High nucleation results in finer bainite as the temperature decreases (HB 400 to 580).
4. From 420 to −50°F (216 to −46°C), M_s to M_f: Austenite transforms to martensite upon cooling. The hardness of martensite is HB 680.

Note that once the austenite has transformed to one of these structures, such as pearlite, it cannot be transformed to another unless the sample is reheated to the equilibrium austenite temperature [above 1341°F (727°C)]. However, a sample can be transformed to a mixture of transformation products if it spends a short interval in the pearlite and bainite transformation ranges and then is cooled through the martensite range.

EXAMPLE 8.3 *[ES/EJ]*

Draw schematic time–temperature diagrams of the processes involved in obtaining the following microstructures in 1080 steel: (a) coarse pearlite, (b) 50% pearlite and 50% bainite, (c) 80% martensite and 20% pearlite. Indicate the microstructures on the diagram at each stage of the treatment.

Answer

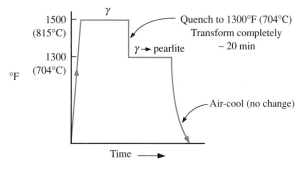

(a)

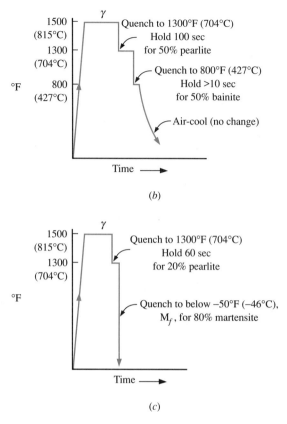

(b)

(c)

Note: These partial transformations at the indicated temperatures can be accomplished only in thin sections that respond rapidly to the quench.

During this discussion you probably recognized that Figure 8.15 is similar to the figure developed in Example 5.4. The concept of nucleation and growth is very important to the control of microstructures and mechanical properties of materials. The next several sections will apply these concepts to other varieties of steel.

8.11 Uses of the Time–Temperature–Transformation (TTT) Curve

Now that we have discussed austenite transformation, let us consider some of the uses of the *TTT curve.*

Suppose that we wish to machine a knife blade from a piece of 1080 steel and then harden it. Suppose further that the hardness of the piece of steel we have on hand is HB 370. This steel is too hard to be machined conveniently.

We therefore heat the steel to the austenite range, 1500°F (815°C)*, cool it to 1250°F (677°C) in the furnace, allow it to transform in the furnace for 1 hr, and then air-cool it. The result is coarse pearlite with a hardness of HB 250. We could have obtained a still softer pearlite by holding it for a longer time at 1300°F (704°C), but what we have is adequate. Also, holding for a long period of time at 1300°F (704°C) would cause the plates of carbide to spheroidize, giving a still softer structure, similar to Figure 8.5 except at a higher carbon level.

Now we machine the blade and place it in the furnace at 1500°F (815°C) again. (It is customary to hang the part on a wire or to use some support so that thin sections do not sag at the high temperature when the steel is soft.) After austenite is obtained (in less than $\frac{1}{2}$ hr), the blade is quenched in rapidly circulating oil at 68°F (20°C). In order for martensite to be produced, the quenching must be rapid enough to avoid transformation to either pearlite or bainite. Lower quench temperatures increase the amount of martensite.

8.12 Tempering

If we test the knife blade, we find that it is at the hardness we wanted, HB 680 or 64 HRC. However, it is also quite brittle. It will break with slight bending, and the edge will chip with impact. We recall that martensite is a highly stressed supersaturated solid solution of carbon in a distorted ferrite. Therefore, we heat the blade just enough to cause the body-centered tetragonal structure to collapse to the BCC and to allow the carbon atoms to migrate and form some very fine iron carbide crystals. Admittedly the hardness may decrease slightly, but the ductility should improve.

We find the relationships shown in Figure 8.18. The martensite transforms to ferrite plus fine carbide. The higher the tempering temperature, the coarser the carbide, the lower the hardness, and the greater the ductility.[†] Fur-

FIGURE 8.18 Relationship between the hardness of tempered martensite and tempering temperature and time (1080 steel). Because tempering is a diffusion phenomenon, one might anticipate the dependence on time and temperature.

[Lawrence Van Vlack, *Elements of Materials Science,* © 1964, Addison-Wesley Publishing Co., Inc., Reading, Massachusetts. Reprinted with permission of the publisher.]

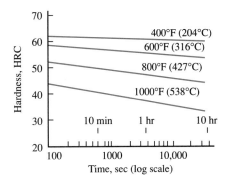

*Although 1341°F (727°C) is adequate to obtain austenite, higher austenitizing temperatures are required to be assured of homogeneous austenite in a reasonable period of time.

[†] In some steels certain tempering temperatures are avoided because secondary reactions in the structure cause temper embrittlement.

thermore, the sample may be quenched from the tempering temperature without changing the effect on hardness.

In many cases we want a balance between hardness and ductility. For example, after the head of a typical claw hammer is quenched, the striking face is tempered to 60 HRC, but the claws are tempered to 48 HRC because greater toughness is needed in this region. This is accomplished by selective heating with a flame or induction coil.

EXAMPLE 8.4 *[ES]*

What are the different microstructures that are mixtures of ferrite and carbide, and how are they obtained in a eutectoid steel?

Answer

1. *Pearlite* is produced by normal equilibrium cooling of austenite or isothermal holding of austenite above the nose of the TTT diagram. (The *nose of the TTT curve*, which is also called its *knee*, represents the shortest transformation time for austenite.)
2. *Bainite* is produced by isothermal holding of austenite below the nose of a TTT diagram. (Bainite may also be obtained by continuous cooling if two noses are present, with the bainite nose out ahead of the pearlite nose. See Section 8.16 and Figure 8.29)
3. *Spheroidite* is produced by spheroidization of pearlite by extended holding at approximately 1300°F (704°C) or lower.*
4. *Tempered martensite* is produced by reheating the distorted body-centered tetragonal martensite to precipitate the carbon as very fine carbide and allow the microstructure to return to the BCC ferrite plus carbide.

8.13 Marquenching (Martempering)

One of the hazards of quenching is the possibility of distorting and cracking the part. Let us consider these effects in a simple knife blade (Figure 8.19). To simplify the discussion, let us assume that the thin portion *A* of the blade cools first and transforms to martensite, while the thicker section *B* remains austenite (above M_s) at, say, 420°F (216°C). Portion *B* is still soft and hot, and it responds to the changes in portion *A*. Later, when portion *B* cools enough to transform, its expansion causes the knife to bow severely and cracks the thinner layer *A*, which is brittle. [This effect is found even in a round bar (Figure 8.20), where surface cracking is a greater problem than bowing.]

---

* Bainite and martensite can also be spheroidized. Whereas it may take 6 to 12 hr to spheroidize pearlite, bainite and martensite can be spheroidized in 1 hr or less.

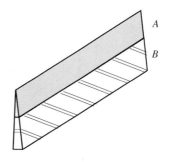

FIGURE 8.19 Schematic representation of cooling and the build-up of stress in a knife blade

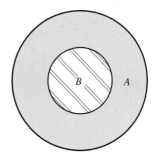

FIGURE 8.20 Schematic representation of cooling and the build-up of stress in a round bar

The solution to both the bowing problem and the cracking problem is to have the austenite–martensite transformation take place in *A* and *B* at the same time. This is done by *marquenching* (martempering) as follows:

1. Austenitize (heat to 1500°F, 815°C).
2. Quench in hot oil, liquid metal, or molten salt at just *above* the M_s. Equalize the temperature throughout the part.
3. Air-cool through the M_s-to-M_f range.

The quenching in liquid is necessary to avoid the formation of pearlite or bainite. The part is removed from the hot bath *before* bainite begins to form and is allowed to transform throughout to martensite. Tempering is still used *after* cooling to below the M_f (Figure 8.21). In the case of a 1080 steel, the use of dry ice may be necessary, the M_f is well below room temperature. Austempering as discussed in the next section may be a more reasonable alternative.

8.14 Austempering

Austempering is another strengthening process for avoiding distortion and cracking. In this process, the part is simply transformed to bainite at the desired level of the bainite temperature range. The part is austenitized, quenched in a salt bath above the M_s, allowed to transform, and then air-cooled. Naturally the as-quenched hardness is lower than if martensite were formed. However, martensite is usually tempered to a lower hardness level, so the results are comparable. It is often not practical to use austempering with some alloy

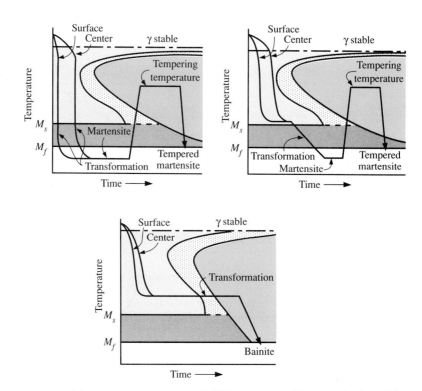

FIGURE 8.21 Schematic comparison of different quenching methods, with cooling curves superimposed on TTT diagrams. (*a*) Conventional process. (*b*) Martempering (marquenching). (*c*) Austempering.

[Reprinted with permission from *Metals Handbook*, 8th ed., Vol. 2, *Heat Treating, Cleaning and Finishing*, American Society for Metals, Metals Park, OH, 1964]

steels because of the long time required to form bainite, as we will discuss later. On the other hand, the advantages of austempering are that the extra tempering step is avoided and in general the distortion is less. We can compare the TTT curves of conventional quenching and tempering, martempering, and austempering as shown in Figure 8.21.

8.15 Effects of Carbon on Transformation of Austenite and Transformation Products

Up to this point we have considered only the simple eutectoid steel 1080. The heat treatment of hypoeutectoid and hypereutectoid steels is also important, and the use of alloying elements permits the successful hardening of many complex designs that could not be produced otherwise.

The first step in attaining maximum hardness in a *hypoeutectoid steel* is to austenitize completely. The austenitizing temperature for 1045 steel is approximately 100°F (56°C) above that for 1080 steel (Figure 8.22). The steel may then be hardened by quenching to martensite at rapid cooling rates.

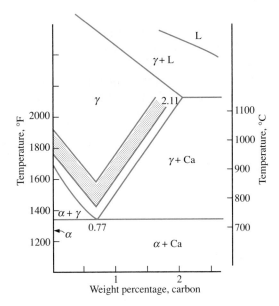

FIGURE 8.22 Austenitizing temperatures for steels containing different amounts of carbon (shaded area)

If the 1045 steel is cooled slowly from 1500°F (815°C), ferrite forms in the range from 1430 to 1341°F (777 to 727°C), and the austenite changes from 0.45 to 0.77% carbon. Then at 1341°F (727°C) pearlite forms. Now let us see what happens with isothermal transformation. If a sample is quenched from 1500 to 1400°F (815 to 760°C), we encounter start and end times for the formation of α at equilibrium with austenite in this temperature range (Figure 8.23). But

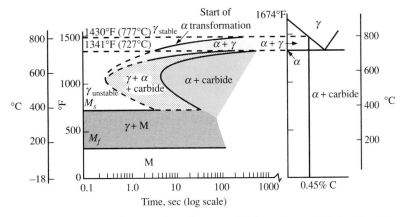

FIGURE 8.23 TTT curve for 0.45% carbon steel. There is an additional region above the nose of the curve that is not found with 1080 steel. A portion of the iron–iron-carbide diagram is included to show why primary α occurs.

[Lawrence Van Vlack, *Elements of Materials Science*, © 1964, Addison-Wesley Publishing Co., Inc., Reading, Massachusetts. Reprinted with permission of the publisher.]

until the reaction temperature reaches 1341°F (727°C), we have isothermal transformation only to ferrite. If we quench from 1500°F (815°C) to below 1341°F (727°C) to as low as 1000°F (538°C), we observe first a period of ferrite formation, then a period of pearlite formation.

Also, the time required for transformation in the pearlite range is shorter than that for 1080 steel. The practical consequence is that martensite or bainite can be obtained only in thin sections, which must be quenched rapidly.

If the steel is transformed below 1000°F (538°C), separate formation of ferrite does not take place. Therefore the TTT curve shows only the start and end of bainite, just as for 0.8% carbon.

The hardness of the transformation products at any temperature is lower than that of 0.8% carbon steel, because the amount of the hard carbide phase is smaller.

If the steel is quenched rapidly enough to avoid prior transformation to pearlite or bainite, martensite forms from the M_s down to the M_f. Note that as the carbon content decreases, the M_s increases (Figure 8.24) and the hardness of the martensite decreases (Figure 8.25).

Now let us consider the *hypereutectoid steels*, which usually contain from 0.8 to 1.2% carbon. If we cool a 1.2% carbon steel slowly from the austenite field, hypereutectoid carbide forms first. Then at 1341°F (727°C) the 0.77% carbon austenite transforms to pearlite. Therefore, the TTT curve for a hypereutectoid steel shows curves for the start and end of the carbide formation and then shows curves for pearlite formation (Figure 8.26). It is interesting to compare the dilation curve for a hypereutectoid steel with the curve for a hypoeutectoid steel; in the former, the precipitation of carbide alone causes

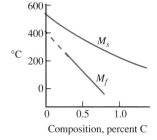

FIGURE 8.24   The effect of carbon content on the M_s and M_f

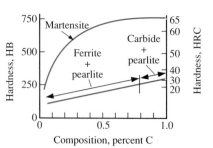

FIGURE 8.25   The hardness of martensite and annealed structures as a function of carbon content

[Lawrence Van Vlack, *Elements of Materials Science*, © 1964, Addison-Wesley Publishing Co., Inc., Reading, Massachusetts. Reprinted with permission of the publisher.]

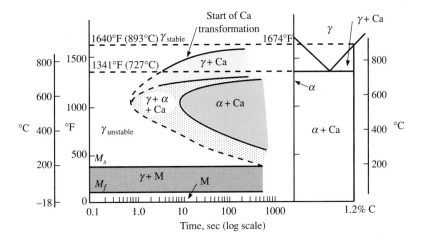

FIGURE 8.26 TTT curve for hypereutectoid steel (1.2% carbon)

contraction of the material (Figure 8.27), followed by expansion when the pearlite transforms.

The TTT curve for the hypereutectoid steels differs from the one for 0.8% carbon steel in that more time is required for transformation, the M_s is lower, and there is an additional curve for carbide precipitation.

We can summarize the effects of carbon in the TTT curve as follows:

1. Above 1000°F (538°C), the formation of ferrite in hypoeutectoid steel and of carbide in hypereutectoid steel precedes the transformation to pearlite.
2. Upon isothermal transformation in the bainite range, steel containing less than 0.8% carbon is softer than steel containing more carbon.

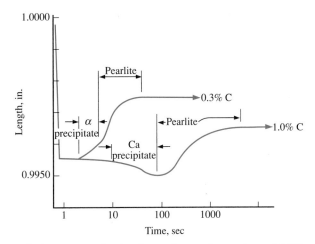

FIGURE 8.27 Dilation curves for isothermal transformation of 0.3% carbon and 1% carbon steels. The transformation temperature is 1300°F (704°C).

3. In the martensite range (an athermal transformation), the M_s and M_f decrease with increased carbon, as shown in Figure 8.24.
4. The hardness of the martensite is a function of the carbon content (Figure 8.25).

EXAMPLE 8.5 *[ES/EJ]*

When TTT diagrams are used in conjunction with the equilibrium diagram, errors commonly arise because of a misunderstanding of how the diagrams should be used. Several sequential thermal treatments are listed below, along with predictions of the resulting microstructures, some of which are partially incorrect. What would be the correct final microstructure after each step?

1. 1.2% carbon steel   *Microstructure*
   a. Heat to 1500°F (815°C), hold 1 hr   Austenite
   b. Quench to 0°F (−18°C), hold 1 hr   Martensite
   c. Allow to return to room temperature   Martensite + austenite
2. 1.2% carbon steel
   a. Heat to 1700°F (927°C), hold 1 hr   Austenite
   b. Quench to 1050°F (565°C), hold 5 sec   Austenite + ferrite
                                                          + carbide
   c. Quench to 500° F (260°C), hold 300 sec   Austenite + bainite
   d. Quench to room temperature   Martensite
   e. Heat to 800°F (427°C), hold 1 hr   Tempered martensite

Answers

1. See Figures 8.1(*b*) and 8.26.
   a. *Austenite + carbide.* In 1 hr we would achieve equilibrium. Therefore,

   $$\% \ Ca \approx \frac{1.2 - 1.0}{6.7 - 1.0} \times 100 \approx 4\% \quad \text{and} \quad \% \ \gamma = 96\%$$

   b. *Martensite + carbide.* Only the austenite transforms to martensite; the carbide is unaffected. Note that the temperature must be below the M_f for there to be complete transformation to martensite.
   c. *Martensite + carbide.* Raising the temperature only a few degrees does not temper the martensite. We cannot obtain austenite again unless we reheat to above 1341°F (727°C). *Once the austenite has transformed, the TTT diagram is not used.*
2. See Figure 8.26.
   a. *Austenite* is correct, because we are well into the γ region.
   b. *Austenite + fine pearlite.* Although the *phases* are α + Ca, the *microstructure* would be pearlite. It is fine pearlite, because the transformation is near the nose of the curve. Had the transformation temperature been 1100°F (593°C), we would have passed through the γ + Ca region,

and after 5 sec we would have had primary carbide + austenite + fine pearlite.

c. *Austenite + fine pearlite + acicular bainite.* Some of the remaining austenite transforms to bainite, but we are still in a $\gamma + \alpha + Ca$ region after 300 sec. The pearlite remains unchanged.

d. *Martensite + fine pearlite + acicular bainite*, assuming that the room temperature is below the M_f. If not, some austenite may remain.

e. *Tempered martensite + fine pearlite + spheroidized bainite.* Again, nothing will happen to the pearlite because 800°F is too low a temperature to spheroidize it in 1 hr. The martensite will temper, and the bainite will begin to spheroidize. Cooling to room temperature will not change the structure.

8.16 Quenching and the Advantages of Alloy Steels

In previous sections we found that we could produce a hard, strong, wear-resistant structure by heating to form austenite, then transforming the austenite to martensite. We also established that the austenite must be cooled fast enough to avoid transformation to pearlite or bainite.

Figure 8.28 shows the effect of the type of quenchant on cooling of 1-in.-diameter bars. Water provides a more rapid quench than liquid salt or oil, because the conversion of water to steam at the surface of the bar absorbs a great deal of heat (as a result of the high latent heat of vaporization). On the other hand, quenches in oil and salt lead to less distortion in the material because of lower thermal gradients.

In many cases in which the section is thick, it is not possible to obtain

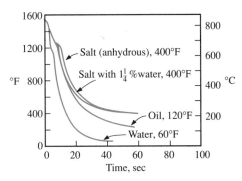

FIGURE 8.28 The effect of various quenchants on cooling rates at the center of 1040 steel bars, 1 in. in diameter by 4 in. long (25.4 mm × 101.6 mm)

[Reprinted with permission from *Metals Handbook*, 8th ed., Vol. 2, *Heat Treating, Cleaning and Finishing*, American Society for Metals, Metals Park, OH, 1964.]

martensite even with a water quench. And with other parts, a water quench cannot be used because of cracking or distortion. The remedy is to increase the time for transformation of austenite to pearlite, so that even with a slower cooling rate we still avoid austenite transformation to α + Ca. We do this by adding such alloying elements as nickel, molybdenum, chromium, and manganese to the steel.

The TTT curves of Figure 8.29 demonstrate the basic effects of alloys. Note that the time required for pearlite formation is much greater for these steels containing alloying elements than for plain-carbon steels.

These alloy steels lend flexibility to heat treatment in two ways. First, a thin or intricate part can be cooled more slowly from the austenitic field, lessening the susceptibility to cracking. Second, a thick section that would cool slowly when quenched can still be produced with a martensitic structure.

Because alloy steels are more expensive than plain-carbon steels, some of the effects of alloys acting in combination are worth noting. In general, a

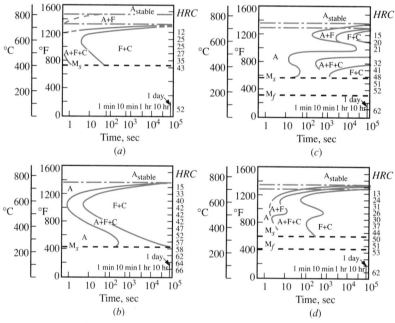

FIGURE 8.29   The effect of alloys on TTT curves. The hardness values to the right of each diagram are those obtained after complete isothermal transformation at the temperatures indicated. (*a*) TTT diagram for 1034 steel, 0.34% C. (*b*) TTT diagram for 1090 steel, 0.9% C. (*c*) TTT diagram for 4340 steel, 0.4% C, 1.7% Ni, 0.8% Cr, 0.25% Mo. (*d*) TTT diagram for 5140 steel, 0.4% C, 0.8% Cr. (The text uses the symbols γ, α, and Ca instead of A, F, and C.)

[Reprinted with permission from *Metals Handbook*, 8th ed., Vol. 2, *Heat Treating, Cleaning and Finishing*, American Society for Metals, Metals Park, OH, 1964, figs. 4 and 5, p. 38.]

triple-alloy steel, such as NiCrMo 8640, has a better TTT curve—that is, a longer pearlite formation time—than a single-alloy steel of the same cost. The alloys also affect the M_s. An empirical formula provides an estimate of this point. (Note that all the alloys lower the M_s.)

$$M_s = 930°F - [540 \times (\% \ C) + 60 \times (\% \ Mn) + 40 \times (\% \ Cr)$$
$$+ 30 \times (\% \ Ni) + 20 \times (\% \ Mo)] \tag{8-1}$$

It should be emphasized that the alloys do not change the hardness of the pearlite, bainite, or martensite. They merely increase the transformation time for pearlite and bainite and lower the temperature range for martensite.

In alloy steels the eutectoid temperature and the composition are somewhat different from 1341°F (727°C) and 0.77% carbon, and the differences are reflected in the data shown in Figure 8.30. The effects at low alloy levels are approximately additive.

Finally, Figure 8.29 shows that two noses may appear in some of the TTT diagrams for alloy steels (4340, for example). The lower nose is called the *bainite nose*, and the upper one is called the *pearlite nose*. In 4340 steel, for example, it is possible to cool at such a rate as to miss the pearlite transformation completely and therefore obtain bainite by continuous cooling instead of by the isothermal treatment we discussed previously.

EXAMPLE 8.6 *[ES/EJ]*

Calculate the eutectoid temperature, percent eutectoid carbon, and M_s for a steel of the following alloy composition: 2% Ni, 1% Cr, 0.5% C.

Answer   From Figure 8.30 we find:

Alloying Element	Change in Eutectoid Temperature	Change in Eutectoid Carbon, percent
2% Ni	−30°F	−0.05
1% Cr	+50°F	−0.10
	+20°F	−0.15

Approximate eutectoid temperature = 1341°F + 20°F = 1361°F (738°C)
Approximate percent eutectoid carbon = 0.77 − 0.15% = 0.62%
M_s = 930 − 540 × 0.5 − 40 × 1 − 30 × 2 = 560°F (293°C)

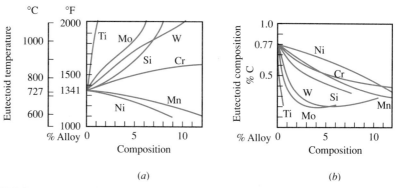

FIGURE 8.30   The effects of additions of alloys on (a) the temperature of the eutectoid reaction, and (b) the carbon content of the eutectoid

[Lawrence Van Vlack, *Elements of Materials Science*, © 1964, Addison-Wesley Publishing Co., Inc., Reading, Massachusetts. Reprinted with permission of the publisher.]

CHOOSING THE PROPER STEEL
FOR A GIVEN PART

In the previous sections we established the importance of the time and temperature of austenite transformation in determining the hardness of the structures produced—pearlite, bainite, and martensite. The basic effects of the carbon and alloy content were shown by TTT curves. With this background, we can now come to grips with the important engineering problem of selecting the steel for a given part and heat treatment.

8.17 Hardenability Evaluation

Many years ago steel producers developed extensive handbooks showing the hardness profiles that could be developed in different steels with different quenching conditions in different section thicknesses. To take into account the variables of composition, section thickness, and quenching medium, thousands of curves were needed to cover all the possibilities. In addition, there arose the issue of how to check a steel for its *ability to harden* within commercial variation in chemical analysis between ingots. True, the TTT diagram could be used, but determining it is time-consuming.

To solve the problem of evaluating the *hardenability* of the millions of tons of steel used by the automotive industry, Walter Jominy and his associates developed the *hardenability bar (Jominy bar)*. The idea behind this test is to produce in one bar a wide variety of known cooling rates. Then, by measuring the hardness along the bar, we can find the hardnesses obtainable with different cooling rates from the austenitizing temperature.

Figure 8.31 shows the test method. A bar of the steel to be tested is machined to produce a cylinder 4 in. (101.6 mm) long and 1 in. (25.4 mm) in diameter with an upper lip. The bar is then austenitized in standard fashion in

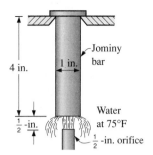

FIGURE 8.31 Jominy end-quench hardenability test

[Guy, *Elements of Physical Metallurgy*, © 1974, Addison-Wesley Publishing Company, Inc., Reading, Massachusetts. Page 484. Reprinted with permission of the publisher.]

a furnace and placed in the fixture. A stream of water is quickly turned on, positioned so as to provide a smooth film striking only the end of the bar. The quenched end cools rapidly. The regions away from the end cool at rates proportional to their distance from the quenched portion. After the bar is cool, it is removed from the fixture, one side is ground flat, and HRC hardness readings are taken every $\frac{1}{16}$ in. (1.6 mm). Figure 8.32 shows a typical graph of the results of hardness testing along a bar. The greatest hardness is at the quenched end, where martensite is formed. The lower hardness farther away from the quenched end is due to softer transformation products.

The most significant information revealed by this testing method is not the hardness at a given point along the specimen but the hardness at a given cooling rate. (The cooling rates have been measured at the different points and are quite constant for different steels.) A little region in the bar does not know it is in a Jominy end-quench test; it only feels the effect of a given cooling rate. For this reason, the hardness of a position on the Jominy bar is equivalent to the hardness that would be obtained at a point in an oil- or water-quenched part with the *same cooling rate*. Figure 8.33 gives the relationships

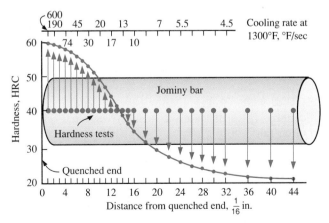

FIGURE 8.32 Typical distribution of hardness in Jominy bars

[Guy, *Elements of Physical Metallurgy*, © 1974, Addison-Wesley Publishing Company, Inc., Reading, Massachusetts. Page 484. Reprinted with permission of the publisher.]

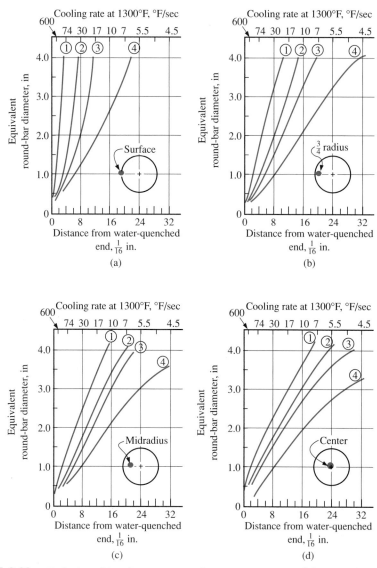

FIGURE 8.33 Relationships between cooling rates in round bars and in Jominy locations. 1 = still water; 2 = mildly agitated oil; 3 = still oil; 4 = mildly agitated molten salt.

between given Jominy positions and different locations in actual bars quenched in several media.

Armed with these data and a few *Jominy curves (hardenability curves)* for representative steels (Figure 8.34), we can now choose the proper steel for a given part.

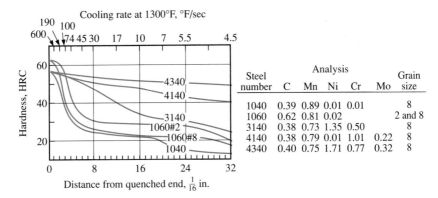

FIGURE 8.34 Hardenability curves for several steels with different compositions and austenite grain sizes. (A higher number indicates a finer grain size.)

[Lawrence Van Vlack, *Elements of Materials Science*, © 1964, Addison-Wesley Publishing Co., Inc., Reading, Massachusetts. Reprinted with permission of the publisher.]

EXAMPLE 8.7 *[ES/EJ]*

Suppose that we have a simple shaft 2 in. (50.8 mm) in diameter and we wish to obtain at least a hardness of 50 HRC to $\frac{1}{2}$ in. (12.7 mm) beneath the surface. We wish to use a still-oil quench for hardening, having experienced cracking in tests with a water quench. Which steel should we use?

Answer Figure 8.33c reveals that the cooling rate at this position (mid-radius) is 20°F/sec, and this corresponds to a Jominy position of $\frac{11}{16}$ in. from the quenched end of the bar. We then draw a vertical line at this position on the Jominy curve of Figure 8.34. We see that, of this group, only 4340 steel would provide hardness of above 50 HRC (4140 would be almost 50 HRC).

EXAMPLE 8.8 *[ES/EJ]*

We have a complex shape made of 1040 steel. After austenitizing and quenching in agitated oil, the hardness $\frac{1}{8}$ in. below the surface is 30 HRC. This steel is too soft for our application; 50 HRC at $\frac{1}{8}$ in. below the surface is necessary. What steel from Figure 8.34 would you recommend if the austenitizing treatment and quenching are to remain the same?

Answer In the Jominy curves (Figure 8.34), 30 HRC for 1040 steel appears at $\frac{1}{4}$ in. from the water-quenched end. This means that the material $\frac{1}{4}$ in. from the water-quenched end of a Jominy test cools at the *same* rate (74°F/sec) as the material $\frac{1}{8}$ in. *below the surface* of our complex shape quenched in oil. Therefore, we look at the hardness obtained for other steels at the same cooling rate ($\frac{1}{4}$ in. position or 74°F/sec) and see that 4340, 4140, and 3140 steels all give over 50 HRC under these quenching conditions.

8.18 Analysis and Properties of Typical Low-Alloy Steels

To simplify the specification of low-alloy steels, the Society of Automotive Engineers (SAE) and the American Iron and Steel Institute (AISI) have issued a joint specification of SAE-AISI steels (Table 8.2). The key is as follows. The last two digits in the code number for each type of steel give the carbon content, as in plain-carbon steels; that is, 4340 steel contains 0.40% carbon. In rare cases five digits are used, where the last three represent the carbon; in 52100 ball-bearing steel, for example, the carbon content is 1.00%. The chemical analysis of any given steel deviates somewhat from that given in Table 8.2 because of manufacturing variables. The first two digits indicate alloy content; the table includes the plain-carbon steels for completeness. Figure 8.35 shows the combinations of strength, ductility, and hardness that

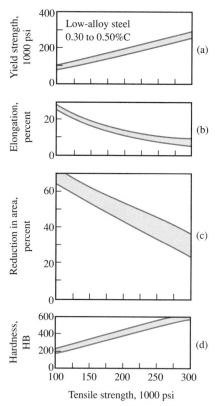

FIGURE 8.35 Properties of tempered martensite (low-alloy steel, 0.3 to 0.5% carbon). (*a*) Yield strength. (*b*) Elongation. (*c*) Reduction in area. (*d*) Brinell hardness. [To obtain MPa, multiply psi by 6.9×10^{-3}. To obtain kg/mm^2, multiply by 7.03×10^{-4}.]

[Reprinted with permission from *Metals Handbook*, 8th ed., Vol. 1, *Properties and Selection of Metals*, American Society for Metals, Metals Park, OH, 1961, Fig. 3, p. 109.]

TABLE 8.2 SAE-AISI Chemical Specifications

			Nominal Chemical Analysis, percent								
Type	Name	Example	C	Mn	P Maximum	S Maximum	Si	Ni	Cr	Mo	Other
10xx	Plain carbon	1020	0.2	0.4	0.04	0.05	0.3				
11xx	Free machining	1111	0.1	0.7	0.09 average	0.12 average	0.3				
13xx	Mn	1330	0.3	1.7	0.04	0.04	0.3				
3xxx	NiCr	3140	0.4	0.8	0.04	0.04	0.3	1.0	0.6		
40xx	Mo	4042	0.42	0.8	0.04	0.04	0.2			0.25	
41xx	CrMo	4140	0.4	0.8	0.04	0.04	0.3		1.0	0.20	
43xx	NiCrMo	4340	0.4	0.7	0.04	0.04	0.2	1.8	0.8	0.25	
46xx	NiMo	4620	0.2	0.6	0.04	0.04	0.3	1.8		0.25	
47xx	NiCrMo	4720	0.2	0.6	0.04	0.04	0.3	1.0	0.4	0.20	
48xx	NiMo	4820	0.2	0.6	0.04	0.04	0.3	3.5		0.25	
50xx	Cr	5015	0.15	0.4	0.04	0.04	0.3		0.4		
52xx	Cr	52100	1.0	0.4	0.02	0.02	0.3		1.4		
61xx	CrV	6120	0.2	0.8	0.04	0.04	0.3		0.8		0.10 V
81xx	NiCrMo	8115	0.15	0.8	0.04	0.04	0.3	0.3	0.4	0.10	
86xx	NiCrMo	8650	0.5	0.8	0.04	0.04	0.3	0.5	0.5	0.20	
87xx	NiCrMo	8720	0.2	0.8	0.04	0.04	0.3	0.5	0.5	0.25	
88xx	NiCrMo	8822	0.22	0.8	0.04	0.04	0.3	0.5	0.5	0.35	
92xx	Si	9260	0.6	0.8	0.04	0.04	2.0				
93xx	NiCrMo	9310	0.1	0.6	0.02	0.02	0.3	3.0	1.2	0.10	
94xx	NiCrMo	94B30*	0.3	0.8	0.04	0.04	0.3	0.4	0.4	0.10	0.0015 B
98xx	NiCrMo	9840	0.4	0.8	0.04	0.04	0.3	1.0	0.8	0.25	

*"B" refers to the presence of boron.

can be obtained with these steels. From a simple hardness test after quenching and tempering, it is possible to determine the other properties. In general, the amount of carbon determines the hardness of the martensite in the quenched structure. The alloy content is what determines the structure at a given cooling rate — that is, the hardenability.

Cast Steel

In the preceding discussions we have generally referred to specifications for wrought steel — that is, for material that is cast and then rolled or forged to shape. However, many important parts are made as steel castings. By using steel castings we can easily obtain many shapes that would be difficult to fabricate. To illustrate this point, let us consider several applications.

A cast low-carbon steel (0.05 to 0.15% carbon) is used for electrical machinery such as generators and motors to obtain optimal magnetic properties. Steel in the range of 0.2 to 0.5% carbon is used for structural parts for the railroad industry, such as side frames, to attain a good combination of strength and ductility. For railroad car wheels and mining equipment components, which require good wear resistance, eutectoid (0.8% carbon) steel is used. Other parts that are to be hardened, such as large gears, are made from low-alloy steels similar to the SAE-AISI grades already discussed.

8.19 Tempering of Alloy and Noneutectoid Steels

We discussed the tempering of 0.8% carbon steel in Section 8.12. The tempering of other steels follows the same principles.

In general terms, tempering is a reheating operation that leads to precipitation and spheroidization of carbide. There are notable side effects, such as the change in martensite from a stressed tetragonal structure to BCC and, often, the transformation of retained austenite (untransformed during cooling in high-carbon alloys), but the basic effect on the carbide phase is the most important.

Although we normally refer to the tempering of martensite, the term is also used for the softening of bainitic and even pearlitic structures. If samples of martensite, bainite, and pearlite of the same steel are heated at 1200°F (649°C), for example, all will contain spheroidite (spheroidal carbide plus ferrite) after sufficient time has passed.

Some quantitative relations among alloy content, tempering temperature, and hardness are shown in Figure 8.36a. The alloy steels generally retain their hardness more than the plain-carbon steels. The steels shown in Fig. 8.36a are all 0.45% carbon; the effect of other percentages of carbon is illustrated in Fig. 8.36b.

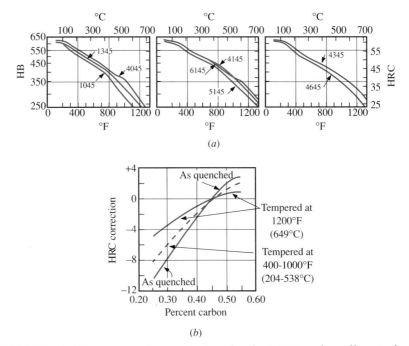

FIGURE 8.36 (*a*) Tempering characteristics of eight 0.45% carbon alloy steels. Duration of tempering = 1 hr. (*b*) Effect of carbon content on the hardness of quenched and tempered steel, with HRC units to be added or subtracted from the value for 0.45% carbon for different temperatures.

[Reprinted with permission from *Metals Handbook*, 8th ed., Vol. 1, *Properties and Selection of Metals*, American Society for Metals, Metals Park, OH, 1961, Fig. 2, p. 109.]

EXAMPLE 8.9 *[ES/EJ]*

Find the tempered hardness of a 4330 steel after 1 hr at 600°F (315°C).

Answer We read the hardness for 4345 at 50 HRC from Figure 8.36*a*. Then, from Figure 8.36*b*, we find that a subtraction of 6 HRC is needed, which gives 44 HRC as the final reading. (*Note:* These corrections should be applied only to steels of similar alloy content.)

8.20 High-Strength Low-Alloy Steels (HSLA Steels)

One step that auto manufacturers have taken to improve gasoline mileage is to reduce the car weight. Because a good percentage of the weight is in the steel body, intensive effort has been directed toward producing higher-strength, very low-alloy steels that can be used in thinner sections. The conventional "low-alloy" steels that we have just discussed are not so weldable and formable as the usual low-carbon sheet steels with 0.15 to 0.25% carbon. Therefore, in the 1960s, low-carbon sheet steels were developed that were "microalloyed" and rolled under controlled conditions to produce high strength and good weldability. Because of the careful control of small amounts of alloy and the processing, these steels are largely proprietary compositions; that is, the precise chemical analysis is not published. They are sold on the basis of their mechanical properties rather than their chemical analysis.

Although there are a number of HSLA steels for specific purposes, the properties of two grades of a specific type given below indicate the properties that might be anticipated for HSLA steels in general.

Alloy Designation	% C	% Mn	% Si	% Nb	% V	% N
A633-Grade A	0.18	1.00/1.35	0.15/0.30	0.05	—	—
A633-Grade E	0.22	1.15/1.50	0.15/0.50	0.01/0.05	0.05/0.15	0.01/0.03

Alloy Designation	Minimum Tensile Strength* (psi × 10³)	Minimum Yield Strength* (psi × 10³)	Minimum Elongation[†] (%)
A633-Grade A	63/83	42	23
A633-Grade E	75/100	55/60	23

*Multiply psi by 6.9×10^{-3} to obtain MN/m² (MPa).
[†]Elongation in 2 in. The value is 18% in 8 in. (a more common gage length for sheet samples).

It is interesting that these steels have their own price structure that is based not on alloy content but rather on the minimum mechanical properties illustrated in the table. Because mechanical properties can be modified in many ways, the HSLA steels have been divided into four broad categories:

1. *As-rolled pearlitic structural steels.* Properties are enhanced by the addition of one or more elements other than carbon. Quenching and tempering are generally not done, although normalizing or annealing may be carried out.
2. *Microalloyed high-strength low alloy steels.* Here elements such as niobium and vanadium are added to a steel to be hot-rolled without changing the carbon and manganese. Heat treatment is again not required, but rather the properties are obtained by controlled hot rolling.
3. *High-strength structural carbon steels.* These are basically carbon–manganese–silicon steels that can be used in the normalized or quenched-and-tempered condition.
4. *Heat-treated structural low-alloy steels.* The higher alloy than in the carbon grades results in better hardenability — and hence higher properties — in thicker sections.

The selection of a specific steel within a particular family of HSLA steels depends on the mechanical properties that are required. Although we usually think of strength or ductility requirements, fracture toughness, fatigue and impact properties are other characteristics that may have to be optimized.

Summary

The ferrous, or iron-based, alloys make up a large group of metals that are used for numerous engineering applications. Among these alloys, the plain-carbon and low-alloy steels are the most widely used; their applications range from nails to highly stressed gears and shafts.

We began with a discussion of possible structures when iron and carbon are in equilibrium. We then progressed to the time required to achieve equilibrium and found that new structures such as martensite can be obtained by rapid quenching. Alloying elements modify the equilibrium structures but, more important, make it easier to obtain martensite at lower quenching rates.

Three important equilibrium phases are involved in these steels: ferrite (α, BCC with up to 0.02% carbon in solid solution), austenite (γ, FCC with up to 2.11% carbon in solid solution), and iron carbide, Fe_3C (6.7% carbon). The majority of plain-carbon steels, such as construction and automobile-body grades, consist simply of ferrite with increasing amounts of carbide as the carbon content is increased. When the carbon content reaches 0.77%, the ferrite and carbide form a lamellar structure called pearlite that, though strong, is not easily formed at room temperature. Hot rolling in the austenite region is then necessary.

When the carbon content in an unalloyed steel is above 0.3% it is possible to harden thin sections by heating to produce austenite, followed by rapid quenching in water to form martensite. However, low-alloy steels are used for most heat-treated parts, because the hard martensitic structure can be obtained with a less drastic quench and in thicker sections.

A key point to remember is that the *hardness* of martensite depends on the carbon content, whereas its ease of formation (*hardenability*) depends on the alloy content (including carbon).

Accurate measures of the kinetics of austenite transformation and the relationship between cooling rate and martensite formation are provided by referring to data derived from TTT curves and from a Jominy hardenability bar. These same tests are employed in the control of other microstructures, such as pearlite and bainite.

Definitions

Austempering   A hardening process involving austenitizing, quenching in a liquid bath above the M_s, holding in the bath to produce bainite, and then cooling.

Austenite γ iron, FCC, with a maximum of 2.11% carbon in solid solution.

Austenitizing   Heating to the austenite temperature range (which varies with the particular steel) to produce austenite.

Bainite   A mixture of α and very fine carbides that shows a needle-like structure and is produced by transformation of austenite below 1000°F (538°C) and above 500°F (260°C).

Carbide   In the context of iron-based alloys, iron carbide, Fe_3C; also called *cementite*.

Dilatometer   An instrument for measuring the change in the length of a sample during heating and cooling.

Eutectoid composition   In the iron–carbon system, 0.77% carbon. The eutectoid reaction is γ (austenite) $\rightarrow \alpha$ (ferrite) + iron carbide at 1341°F (727°C). The α + iron carbide mixture is called *pearlite*.

Ferrite α iron, BCC, with a maximum of 0.022% carbon in solid solution.

Full annealing   Heating a ferrous alloy to the austenite region, followed by slow cooling such as furnace cooling to produce coarser pearlite.

Hardenability   A general term indicating the ability of a given steel to harden. Hardenability depends on the ease with which pearlite or bainite transformation can be avoided so that martensite can be produced. High hardenability is not the same as high hardness. The maximum hardness attainable is a function of the carbon content.

Hardenability curve, Jominy curve   A graph showing the hardness attained for a given steel via the Jominy end-quench test.

Hypereutectoid steels   Steels with carbon content of from 0.77 to 2.11% (plain carbon).

Hypoeutectoid steels   Steels with carbon content below 0.77% (plain carbon).

Jominy bar, hardenability bar   A bar 1 in. (25.4 mm) in diameter and 4 in. (101.6 mm) long that is austenitized, water-quenched on one end, and then tested for hardness along its length. The bar is used as a quantitative measurement of hardenability.

Lamellar   Made up of alternating layers of phases. For example, pearlite is lamellar because it has alternating plates of ferrite and carbide.

M_f   Temperature at which austenite-to-martensite transformation finishes.

M_s   Temperature at which austenite-to-martensite transformation starts.

Marquenching   (martempering) A hardening process involving austenitizing, quenching in a liquid bath at above the M_s, and then air-cooling through the martensite transformation range.

Martensite   A nonequilibrium phase consisting of body-centered tetragonal iron with carbon in supersaturated solid solution.

Normalizing   Heating a ferrous alloy to the austenite region followed by air-cooling to produce fine pearlite.

Nose of TTT curve   The point on the TTT curve that indicates the temperature range in which isothermal transformation is fastest.

Pearlite   A mixture of α and carbide phases in parallel plates, produced by the transformation of austenite between 1341°F (727°C) and 1000°F (538°C).

Primary phase   A phase that appears first upon cooling. For example, at 0.4% carbon, ferrite is a primary phase occurring before the pearlite reaction.

Proeutectoid phase   Literally, "before the eutectoid." At 1.0% carbon, carbide is the proeutectoid phase. (Compare with *primary phase.*)

Spheroidite α + spheroidized carbide, produced by heating pearlite, bainite, or martensite at elevated temperatures.

Steel An iron–carbon alloy with 0.02 to 2.11% carbon. The most common range is 0.05 to 1.1% carbon. Steel can be hot- or cold-worked.

Tempered martensite Martensite that has been heated to produce BCC iron and a fine dispersion of iron carbide.

TTT curve A time–temperature–transformation curve that indicates the time required for austenite of a given composition to transform to α + carbide at different temperatures (isothermally).

Wrought iron Iron that has a very low carbon content (0.02%) and slag fibers and that is easily worked into intricate forgings while it is hot.

Problems

8.1 *[ES]* What would be the maximum solubility, in percent by weight, if all the positions like $\frac{1}{2}, \frac{1}{2}, \frac{1}{2}$ in γ iron were occupied by carbon? Why is this level of solubility not attained? (If 2.1% by weight carbon is soluble in the FCC iron, what is the number of unit cells of iron per carbon atom?) [*Hint:* See Figure 1.30.] (Section 8.1 through 8.5)

8.2 *[ES]* Draw fraction charts showing the amount of each phase present as a function of temperature for the following alloys. (Sections 8.1 through 8.5)
a. SAE-AISI 1010 steel (0.1% carbon)—neglect peritectic
b. SAE-AISI 1080 steel (0.8% carbon)

8.3 *[ES]* Fe_3C is listed in various text and reference books as having a composition of 6.67% C, 6.69% C, and 6.70% C. Show by calculation which value is most likely to be correct. (Sections 8.1 through 8.5)

8.4 *[EJ]* From the steels listed in Table 8.1, what compositions would you select for a hammer and a nail? [*Hint:* A nail is generally cold-formed, and a hammer head is hot-formed.] (Sections 8.1 through 8.5)

8.5 *[EJ]* An automobile fender is dented. Why would you be likely to "hammer out" the dent at room temperature rather than using a torch to help remove the dent? (Sections 8.1 through 8.5)

8.6 *[ES]* Use the accompanying phase fraction chart for a plain-carbon steel to answer the following questions. (Sections 8.1 through 8.5)
a. Is the steel hypo- or hypereutectoid?
b. What is the carbon content?
c. What is the amount of pearlite?

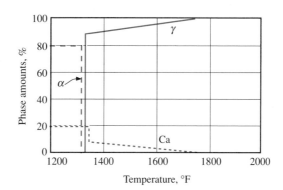

8.7 *[ES/EJ]* Determine the percentage of pearlite at equilibrium in a 1010, 1020, 1040, 1060, and 1080 steel. What is the engineering significance of these values? (Sections 8.1 through 8.5)

8.8 *[ES]* Figure 8.2 gives the lattice parameter for pure iron as a function of temperature. (Sections 8.1 through 8.5)

a. Using the interatomic distance values (twice the atomic radii), calculate the volume change when α iron transforms to γ iron.

b. Estimate the linear coefficient of thermal expansion (in./in. $\cdot$ °C) for α and γ iron.

8.9 *[ES/EJ]* Strains and accompanying stresses can be developed by thermal expansion or contraction while a component is being restrained from movement. These strains and stresses are usually elastic, where the strain is determined from the coefficient of thermal expansion and the stress is dependent on the modulus of elasticity. A railroad track of 1080 steel is welded at its ends and laid in the winter (32°F or 0°C). What magnitude of stress can be expected in the middle of the summer sunlight (132°F or 55.6°C)? The linear coefficient of thermal expansion is 6.0×10^{-6} in./in. $\cdot$ °F (10.8×10^{-6} m/m $\cdot$ °C). (Sections 8.1 through 8.5)

8.10 *[EJ]* The following table gives the hardness of isothermally transformed austenite for a 1080 steel at the indicated temperatures. Plot the data and explain the shape of the curve on the basis of the transformation product characteristics. (Sections 8.6 through 8.14)

Temperature, °F	Hardness, HRC
1200	32
1100	38
1000	40
900	40
800	41
700	43
600	50
500	55

8.11 *[ES/EJ]* Using the TTT curve for 1080 steel, give the microstructure present in a thin [0.010-in. (0.254-mm)] strip after each step of the following treatments. The original material is hot-rolled. [See Figure 8.15.] (Sections 8.6 through 8.14)

a. 1600°F (871°C), 1 hr; quench in lead at 1300°F (704°C), hold 20 min; quench in lead at 900°F (482°C), hold 1 sec; water-quench.

b. 1300°F (704°C), 6 hr; water-quench.

c. 1600°F (871°C), 1 hr; quench in salt at 800°F (427°C), hold $\frac{1}{2}$ hr; heat rapidly to 1100°F (593°C), hold 1 hr; water-quench.

d. 1600°F (871°C), 1 hr; quench in salt at 500°F (260°C), hold 1 min; air-cool. What is the name of this treatment?

e. 1600°F (871°C), 1 hr; quench in salt at 600°F (316°C), hold 1 hr; air-cool. What is the name of this treatment?

8.12 *[ES/EJ]* From the following group, choose heat-treating cycles to produce the following microstructures in a 0.010-in. (0.254-mm)-diameter wire of 1080 steel. (Sections 8.6 through 8.14)

a. 100% pearlite
b. 100% bainite
c. 50% pearlite, 50% bainite
d. 50% martensite, 50 bainite

Code:

A = 1100°F (593°C), 1000 sec; air-cool.

B = 1600°F (871°C), 500 sec; quench in lead bath at 700°F (371°C); hold 1000 sec; air-cool.

C = 1100°F (593°C), 10 sec; quench in lead at 700°F (371°C); hold 1000 sec; air-cool.

D = 1600°F (871°C), 500 sec; quench in lead bath at 1250°F (677°C); hold 100 sec; quench in lead bath at 800°F (427°C); hold 1000 sec; water-quench.

E = 1600°F (871°C), 500 sec; quench in water to 0°F (−18°C); heat to 800°F (427°C); hold 5 sec; water-quench.

F = 1600°F (871°C), 400 sec; quench in lead at 600°F (316°C); hold 300 sec; water-quench.

G = 1600°F (871°C), 500 sec; quench in lead at 1100°F (593°C); hold 1000 sec; air-cool.

8.13 *[ES]* Using the data given in Figure 8.18, develop separate equations at 1000°F (538°C) and 600°F (316°C) that relate tempered hardness to tempering time. (Sections 8.6 through 8.14)

8.14 *[ES/EJ]* In Figure 8.15, the knee or nose of the curve (1000°F or 538°C) has a beginning transformation at approximately 0.6 sec. Explain whether a finer prior austenite grain size would increase, decrease, or not affect the time required to initiate the transformation. (Sections 8.6 through 8.14)

8.15 *[ES/EJ]* It may be necessary to approximate the 50% transformation for austenite in Figure 8.15. Sketch the beginning and the end of the TTT curve between stable γ and the M_s temperatures, and then indicate with a dotted curve where you might expect the 50% transformation value to be. [*Hint:* Note that the time scale is logarithmic.] (Sections 8.6 through 8.14)

8.16 *[ES/EJ]* Indicate whether the following statements about a 1080 steel are correct or incorrect and justify your answer. (Sections 8.6 through 8.14)
a. The hardness of pearlite is a fixed value.
b. Martensite is obtained by the isothermal transformation of austenite.
c. The isothermal transformation curve is an equilibrium diagram.

8.17 *[ES/EJ]* Draw schematic time–temperature diagrams for obtaining the following structures in an SAE 1045 steel, giving the structure at each step. (*Percentages are approximate.*) [See Example 8.3] (Sections 8.15 through 8.16)
a. 45% ferrite, 55% pearlite
b. 20% ferrite, 80% martensite
c. 100% martensite
d. 100% spheroidite
e. 90% bainite, 10% ferrite
f. Calculate the approximate M_s temperature for the martensite in part b.

8.18 *[ES/EJ]* The accompanying figure gives the TTT curve for an alloyed 0.30% carbon steel. The hardness data are for fully transformed structures. (Sections 8.15 through 8.16)
a. A foundry finds that castings made of this steel are hard and unmachinable (400 HB) in the as-cast condition. Name two microstructures that could be responsible.
b. The same foundry hears that a competitor is annealing its castings with a short cycle called an *isothermal anneal*. This involves heating of the castings, followed by isothermal transformation to a structure of 250 HB max. Draw a time–

temperature chart giving this result, *labeling temperatures* and isothermal transformation *time* accurately.

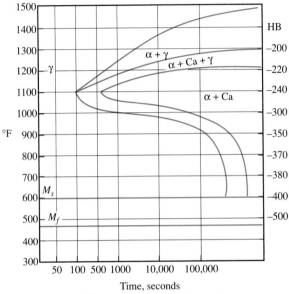

TTT curve for 0.30% C alloy steel

8.19 *[ES/EJ]* An isothermal transformation curve for a 4130 steel is shown below. Describe the microstructure after each of the following heat treatments, using only the following words: austenite, ferrite, liquid, pearlite, bainite, martensite, tempered martensite, spheroidite. (Sections 8.15 through 8.16)

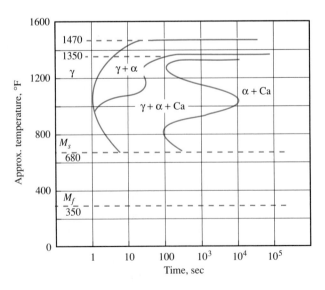

   a. A thin section of hot-rolled 4130 steel is heated to 1400°F, held 1 hr, and water-quenched to room temperature.

   b. A thin section of hot-rolled 4130 is heated to 1550°F, held 1 hr, quenched in a molten salt bath at 800°F, held 1000 sec, and slowly cooled in air.

   c. A thin section of hot-rolled 4130 is heated to 1550°F, held 1 hr, quenched in a salt bath at 1200°F, held 1000 sec, and water-quenched.

   d. Which material (a, b, or c) will be softest?

8.20 *[ES/EJ]* Indicate whether the following steels might be hypoeutectoid or hypereutectoid. [*Hint:* The alloy content changes both the percent carbon and the temperature associated with the eutectoid.] (Sections 8.15 through 8.16)

   a. 0.6% C, 1% Mo, 1% Cr

   b. 0.6% C, 1% Ni, 1% Cr

8.21 *[ES]* Calculate the amount of proeutectoid ferrite and the amount of pearlite for the alloy steel in Example 8.6 at 1360°F (737°C). (Sections 8.15 through 8.16)

8.22 *[EJ]* You are given three plain-carbon steels, 0.45% C, 0.80% C, and 1.2% C. For each steel, specify a heat treatment; that is, estimate the time and temperature that would provide the following room-temperature microstructures. (Sections 8.15 through 8.16)

   a. The hardest microstructure

   b. The microstructure that would be most ductile

8.23 *[ES/EJ]* Indicate whether the following statements are correct or incorrect and justify your answer. (Sections 8.15 through 8.16)

   a. It is possible to obtain 100% bainite by continuous cooling.

   b. Martensite always has very high hardness.

   c. It is possible to have 0.6% C in a steel and have proeutectoid carbide.

   d. The austenitizing treatment prior to quenching to martensite is important to the final hardness.

8.24 *[EJ]* Although a single specimen of steel gives a single line as a hardenability curve, results of the tests of commercial steel fall within a hardenability band such as that shown for the 8600 series steels in Figure 8.37 on page 414 and the table on page 415. (Sections 8.17 through 8.20)

   a. Why is the hardness band at the $\frac{2}{16}$-in. position higher in 8650H than in 8630H? What is the microstructure in each case?

   b. How will the microstructure of 8620H at the $\frac{18}{16}$ position compare with that of 8640H? Are there several structures that might be encountered at the $\frac{18}{16}$ position in 8640H?

   c. What is the principal element responsible for the bandwidth at the $\frac{2}{16}$ position in all cases?

8.25 *[EJ]* A midradius hardness of minimum HRC 50, maximum HRC 62 is required in a 2-in. (50.8-mm) bar. Which steels would meet the specification and how should the steel be quenched (Figure 8.37)? (Sections 8.17 through 8.20)

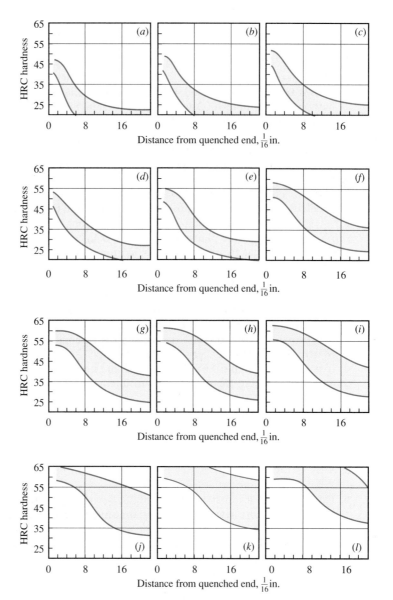

FIGURE 8.37   Hardenability of 8600 series steels: (a) 8620H, (b) 8622H, (c) 8625H, (d) 8627H, (e) 8630H, (f) 8637H, (g) 8640H, (h) 8642H, (i) 8645H, (j) 8650H, (k) 8655H, (l) 8660H

[Reprinted with permission from *Metals Handbook*, 8th ed., Vol. 1, *Properties and Selection of Metals*, American Society for Metals, Metals Park, OH, 1961.]

8.26   *[ES/EJ]* The distance from the water-quenched end in a hardenability bar is equivalent to a specific cooling rate. In Figure 8.32 why is cooling at 1300°F (704°C) chosen? Would the cooling be faster or slower at 1000°F (538°C)? Justify your answers. (Sections 8.17 through 8.20)

TABLE for Figure 8.37

| SAE-AISI Steel | Chemical Analysis, percent | | | | | | Normalizing Temperature, °F | Austenitizing Temperature, °F |
	C	Mn	Si	Ni	Cr	Mo		
8620H	0.17 to 0.23	0.60 to 0.95	0.20 to 0.35	0.35 to 0.75	0.35 to 0.65	0.15 to 0.25	1700	1700
8622H	0.19 to 0.25	0.60 to 0.95	0.20 to 0.35	0.35 to 0.75	0.35 to 0.65	0.15 to 0.25	1700	1700
8625H	0.22 to 0.28	0.60 to 0.95	0.20 to 0.35	0.35 to 0.75	0.35 to 0.65	0.15 to 0.25	1650	1600
8627H	0.24 to 0.30	0.60 to 0.95	0.20 to 0.35	0.35 to 0.75	0.35 to 0.65	0.15 to 0.25	1650	1600
8630H	0.27 to 0.33	0.60 to 0.95	0.20 to 0.35	0.35 to 0.75	0.35 to 0.65	0.15 to 0.25	1650	1600
8637H	0.34 to 0.41	0.70 to 1.05	0.20 to 0.35	0.35 to 0.75	0.35 to 0.65	0.15 to 0.25	1600	1550
8640H	0.37 to 0.44	0.70 to 1.05	0.20 to 0.35	0.35 to 0.75	0.35 to 0.65	0.15 to 0.25	1600	1550
8642H	0.39 to 0.46	0.70 to 1.05	0.20 to 0.35	0.35 to 0.75	0.35 to 0.65	0.15 to 0.25	1600	1550
8645H	0.42 to 0.49	0.70 to 1.05	0.20 to 0.35	0.35 to 0.75	0.35 to 0.65	0.15 to 0.25	1600	1550
8650H	0.47 to 0.54	0.70 to 1.05	0.20 to 0.35	0.35 to 0.75	0.35 to 0.65	0.15 to 0.25	1600	1550
8655H	0.50 to 0.60	0.70 to 1.05	0.20 to 0.35	0.35 to 0.75	0.35 to 0.65	0.15 to 0.25	1600	1550
8660H	0.55 to 0.65	0.70 to 1.05	0.20 to 0.35	0.35 to 0.75	0.35 to 0.65	0.15 to 0.25	1600	1550

8.27 *[ES/EJ]* A $3\frac{1}{2}$-in. (88.9-mm)-diameter automatic transmission gear is sketched below. Specifications call for oil quenching to give as-quenched hardness values of (1) at least HRC 50 near the surface (point A) and (2) no greater than HRC 45 at point B. A trial gear was made from AISI 1060 steel, austenitized at 1550°F, and quenched in oil. Hardnesses were recorded as HRC 42 at point A and HRC 34 at point B. (Sections 8.17 through 8.20)

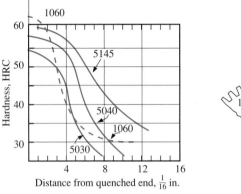
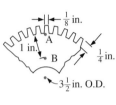

Distance from quenched end, $\frac{1}{16}$ in.

   a. AISI steels 5145, 5040, 5030, and 1060 are under consideration for this part. Which steel or steels would be acceptable?

   b. What would be the hardness at point A if the selected steel were used?

   c. What would be the hardness at point B if the selected steel were used?

8.28 *[ES/EJ]* In a certain chisel the desired hardness is *minimum* HRC 50 at $\frac{1}{8}$ in. (3.2 mm) from the point (in the center of the chisel), *maximum* HRC 30 at 2 in. (50.8 mm) from the point. An investigator tries to meet these specifications with steel A with an oil quench, but the hardness is HRC 45 at $\frac{1}{8}$ in. (3.2 mm) from the chisel point and HRC 35 at 2 in. (50.8 mm) from the point. To meet the specification, would you use steel B, C, or D? Determine the hardness you would obtain at the points specified. The Jominy end-quench curves for these steels are shown. Which steel has the highest carbon content? (Sections 8.17 through 8.20)

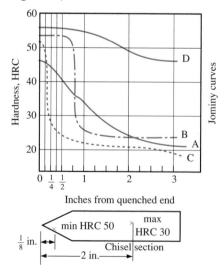

8.29 *[ES/EJ]* An unidentified bar of steel is known to be from one of four lots. The hardenability curve for each of the four lots is given. The unknown bar has the hardness traverse shown. (Sections 8.17 through 8.20)
   a. Which of the four steels is the unknown bar? (Explain your reasoning.)
   b. Estimate the carbon content of these four steels.
   c. Sketch on the first figure a possible hardenability curve for steel A when it has a coarser γ grain size (no change in composition).
   d. Draw the hardness traverse of the unknown bar if it were reheated and quenched in still water.

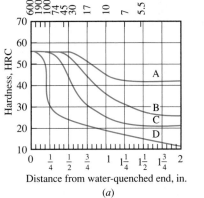

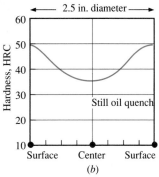

(a) (b)

8.30 *[ES/EJ]* Figure 8.36 gives data for the tempering of 0.45% C steels. (Sections 8.17 through 8.20)
   a. Which steel would be most resistant to a hardness loss in service at 1000°F (538°C)?
   b. Copy the curve for the 4045 steel, and sketch the tempering curves for a 4035 and a 4055 steel.

8.31 *[ES/EJ]* A quenched and tempered 4145 steel has a hardness of HRC 35. What should we anticipate for the tensile strength, yield strength, and ductility for this steel? Give a range of values. (Sections 8.17 through 8.20)

8.32 *[ES/EJ]* Elongation values for the HSLA steels are often given for an 8-in. gage length rather than a 2-in. gage length. Why might this be the case, and why are elongation values generally less as the gage length increases? (Sections 8.17 through 8.20)

8.33 *[ES/EJ]* Why is it difficult to obtain some grades of HSLA steels in the cold-rolled condition? (Sections 8.17 through 8.20)

8.34 *[ES/EJ]* One of the reasons for specifying an HSLA steel is to obtain better weldability. List some applications wherein good weldability is necessary, and indicate how a pre and post heat may be required for some HSLA grades. (Sections 8.17 through 8.20)

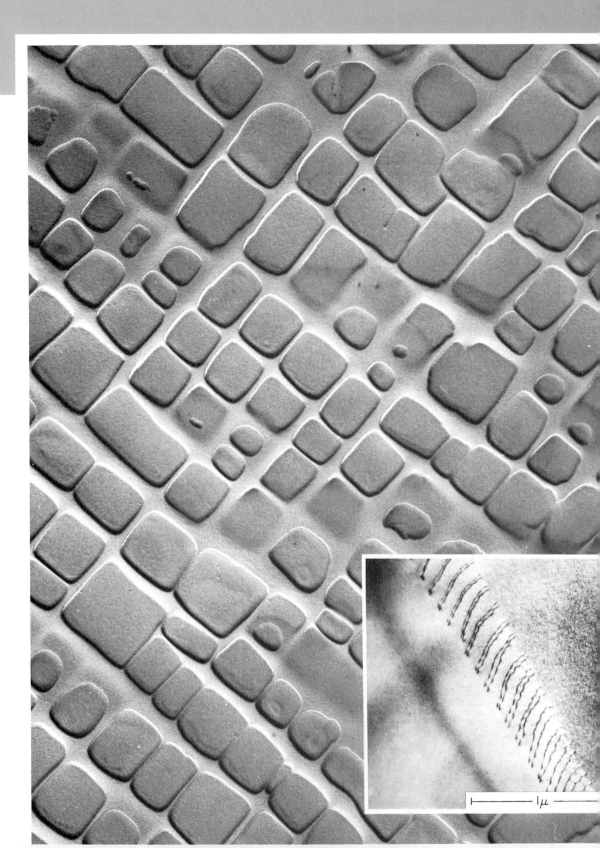

CHAPTER 9

High-Alloy Steels, Superalloys, Cast Irons

In this chapter we shall take up the second principal group of metallic materials, those based on iron and a few related high-nickel alloys called the "superalloys." Only a few structures are involved: the two crystal forms of iron, BCC and FCC, discussed previously, and a hard compound, iron carbide (Fe_3C). By altering the amount, shape, and distribution of the carbide, we can obtain a wide range of properties.

The electron micrograph shows another special type of precipitation in a nickel-based superalloy. The matrix is FCC, but the precipitate is a cubic intermetallic compound, $Ni_3(Al, Ti)$. This combination gives the exceptional blend of high-temperature strength and oxidation resistance required for gas turbine blades.

The large photomicrograph shows Ni_3Al (γ') precipitates in a nickel alloy (γ) matrix (approximately 10,000×). The small insert shows dislocations cutting through the γ' precipitate (approximately 50,000×).

9.1   Overview

If we were limited to using plain-carbon and low-alloy steels only, a number of common products such as cookware, drill bits, and jet engines would not exist. Let us examine some of the reasons why highly alloyed steels, the very high-alloy metals called superalloys, and the cast irons are so important.

The *high-alloy steels* may be defined as having an alloy content of 10% or more, and one major family within this group consists of the *stainless steels*. This suggests that one of the reasons for adding the large quantity of alloying elements is to provide better corrosion resistance. However, these are not just random alloy additions but rather are carefully selected elements, such as chromium and nickel, chosen to give a desirable microstructure and mechanical properties while they enhance the steel's ability to withstand a hostile environment.

We should also observe that the presence of more alloy generally lowers the diffusion rate, which can be advantageous in other applications. For example, a martensitic microstructure might be suggested for a cutting tool, yet when heating occurs in service, tempering of the martensite can occur and the cutting edge is lost. The high-speed *tool steels*, then, resist tempering because the high alloy content slows diffusion rates and coarsening of the carbide.

Similarly, we know from our discussion in Chapter 3 that creep is diffusion-dependent. Thermodynamic principles tell us that engines operate more efficiently at high temperatures, so we produce highly alloyed materials to resist the creep strains at these higher temperatures. Although we may simply add elements such as chromium, nickel, and molybdenum to iron to achieve thermal resistance, there are upper temperature limits at which creep again becomes a problem. Fortunately, beginning with another base metal such as nickel, cobalt, or titanium and alloying it with other elements extends the resistance to creep at the upper temperatures.

These materials are classified as *superalloys* and have evolved from a principal development requirement in high alloy steels: the need to extend the corrosion, oxidation, and thermal resistance of metallic materials.

We may question how we can include the seemingly mundane cast irons in the same chapter as the more exotic high-alloy steels and superalloys. The reason is that the cast irons can have any of the background microstructures (matrices) found in the plain-carbon, low-alloy, and high-alloy steels—plus an added high-carbon phase of carbide or graphite. The metallurgy can therefore be complex, and we had to develop the steel metallurgy first.

Recall from the iron–iron-carbide phase diagram in Figure 8.1 that when the carbon content exceeds 2.11%, we can no longer obtain 100% austenite on freezing. The occurrence of the hard and brittle iron carbide requires that the material be cast rather than formed; hence its classification as *cast iron*. More specifically, we call it *white cast iron* when iron carbide is embedded in a steel matrix, because the fracture surface appears white.

With the addition of certain alloying elements such as silicon, the iron carbide becomes unstable and pure carbon in a flake shape (graphite) is formed, giving us *gray cast iron* (so called because of the appearance of the fracture surface). Another addition agent such as magnesium changes the flake graphite to spheroids, yielding what we call *ductile cast iron*.

We can then alloy the cast iron matrices (steel background structure) to produce microstructures such as ferrite, pearlite, bainite, martensite, and austenite, just as we modified the low- and high-alloy steels. The numerous applications for such materials range from water mains (ductile cast iron) to engine blocks (gray cast iron) and corrosion-resistant pump bodies (white cast iron).

With this general background on the commercial importance of these ferrous materials, we will now consider each classification in greater detail.

HIGH-ALLOY STEELS

9.2 General Characteristics

In Chapter 8 we saw that relatively small amounts of alloy are used to change the TTT curve and hardenability of steels. There is another group of steels in which well over 5% alloy is added for various special purposes. These high-alloy steels may be divided into the groups shown in the accompanying table.

A number of these steels are called austenitic, because one of the effects of the alloy is to produce a structure that is austenite at room temperature. This can be explained only by a radical alteration of the iron–iron-carbide diagram we have studied. The phase relationships in the high-alloy steels (see the table) must therefore be our first consideration.

Type	*Common Chemistry, percent*
Stainless steel	
Austenitic	18 Cr, 8 Ni, balance Fe
Ferritic	16 Cr, 0.1 C
Martensitic	17 Cr, 1 C
Precipitation-hardened stainless steel	17 Cr, 7 Ni, 1 Al
Maraging steel*	18 Ni, 7 Co
Tool steel; high-speed steel	18 W, 4 Cr, 1 V
Manganese steel, austenitic	12 Mn

*See Section 9.5.

9.3 Phase Diagrams of the High-Alloy Steels

The most important effect of the presence of alloys is the change in the austenite field. We can classify all alloys on the basis of their effects on the size and shape of the γ field. Examples are shown in Figure 9.1. Note that

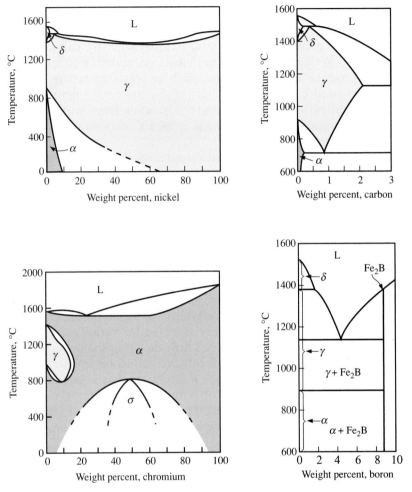

FIGURE 9.1   Effect of alloys on the γ field of iron. Some alloying elements **narrow** the field; others widen it.

nickel widens the field until, at 32% nickel, γ is encountered at room temperature under *equilibrium* conditions. Conversely, chromium narrows the field. At 13% chromium we have only α as the solid phase at all temperatures. Although the interstitial elements such as carbon and boron do not produce completely open austenitic or ferritic fields, they expand or contract the γ field. Figure 9.2 summarizes the effects of various elements. It is interesting that all of the Group 8 elements and manganese expand the γ field, whereas the other transition elements contract it.

We shall now discuss the structures encountered in the stainless steel groups.

Period	1a	1b	2a	2b	3a	3b	4a	4b	5a	5b	6a	6b	7a	7b	8a	8b
1														1 H		2 He
2	3 Li ▲		4 Be ●			5 B ○		6 C □		7 N □	8 O ▲ ?		9 F			10 Ne
3	11 Na ▲		12 Mg ▲			13 Al ●		14 Si ●		15 P ●	16 S ▲		17 Cl			18 A
4	19 K ▲		20 Ca ▲	21 Sc		22 Ti ●		23 V ●		24 Cr ●	25 Mn ■		26 Fe	27 Co ■	28 Ni ■	
4		29 Cu □		30 Zn □ ?		31 Ga ●		32 Ge ●		33 As ●		34 Se		35 Br		36 Kr
5	37 Rb ▲		38 Sr ▲	39 Yt		40 Zr ○		41 Cb ●		42 Mo ●		43 Tc	44 Ru ■	45 Rh ■	46 Pd ■	
5		47 Ag ▲		48 Cd ▲		49 In		50 Sn ●		51 Sb ●		52 Te		53 I		54 Xe
6	55 Cs ▲		56 Ba ▲	58 Ce ○	72 Hf			73 Ta ●		74 W ●		75 Re	76 Os ■	77 Ir ■	78 Pt ■	
6		79 Au □		80 Hg ▲		81 Tl ▲		82 Pb ▲		83 Bi ▲		84 Po		85 At		86 Rn
7	87 Fa		88 Ra ▲	89 Ac		90 Th		91 Pa		92 U						

■ Open γ-field   □ Expanded γ-field   ▲ Insoluble
● Closed γ-field   ○ Contracted γ-field

FIGURE 9.2   Periodic table of the elements, showing the effects of various elements on the γ field of iron

[C. Barrett, *Structure of Metals*, 2d ed., McGraw-Hill, New York, 1952, Fig. 16, p. 251, © 1952. Reprinted by permission.]

9.4 The Stainless Steels

The stainless steels fall into three principal groups—ferritic, martensitic, and austenitic. Each is named for its predominating structure.

Ferritic Steels

We see from the iron–chromium diagram (Figure 9.1) that the formation of austenite is completely suppressed in a pure iron–chromium alloy above 13% chromium. However, when carbon is present, the chromium forms chromium carbide. In this case 1% carbon combines with 17% chromium. This carbide precipitates and deprives the iron matrix of a corresponding amount of chromium. The commercial steel 430 listed in Table 9.1 is made with 16% chromium and low carbon so that the ferritic structure will be retained. The advantage of this material over unalloyed ferrite is its corrosion resistance, as we will discuss in Chapter 18. Steel 430 is used in the cold-worked condition for automotive trim and kitchen utensils.

Martensitic Steels

When the carbon content of the high-chromium steels is high, as in 440C (Table 9.1), we can heat to the γ field and quench to produce martensite. With

TABLE 9.1 Typical Properties of Stainless Steels

Number	Chemical Analysis, percent	Condition	Tensile Strength,* psi × 10³	Yield Strength,* psi × 10³	Percent Elongation	HB	Typical Use
		Austenitic Steels					
301	17 Cr, 7 Ni	Annealed	110	40	60	160	Lightweight, high-strength transportation equipment
		Cold-worked	185	140	9	388	
304	19 Cr, 10 Ni	Annealed	85	35	60	149	General chemical equipment
		Cold-worked	110	75	12	240	
347	18 Cr, 11 Ni†	Annealed	90	35	45	160	Welded construction
		Ferritic Steels					
430	16 Cr, <0.1 C	Annealed	80	55	25	140	Automobile trim, kitchen equipment
		Cold-worked	90	80	20	200	
		Martensitic Steels					
410	12 Cr, 0.15 C	Annealed	70	40	30	155	General-purpose springs, rules
		Quenched and tempered	140	100	20	300	
440C	17 Cr, 1 C	Annealed	110	65	14	230	Instruments, cutlery, valves
		Quenched and tempered	285	275	2	580	
		Precipitation-hardened Steels					
17-7PH	17 Cr, 7 Ni, 1 Al	Hardened	235	220	6	400	Airframe parts
		Maraged Steels					
Maraging steel	18 Ni, 7 Co‡	Maraged	275	268	11	500	Aircraft components

* Multiply psi by 6.9×10^{-3} to obtain MPa or by 7.03×10^{-4} to obtain kg/mm^2.
† Contains niobium (columbium) to prevent weld embrittlement. Percent Nb = 10 × percent C. Percent C is about 0.08 in usual grades.
‡ 0.025% C, 0.1% Mn, 0.1% Si, 0.22% Ti, 0.003% B.

424

lower chromium, as in type 410, we can produce martensite with only 0.15% carbon. Because of their high alloy content, these steels have high hardenability. In some cases it is necessary only to cool them in air to form martensite, and thus pearlite and bainite formation are easily avoided. These steels are excellent for cutlery and dies.

We can summarize the carbon–chromium relationships for the ferritic and martensitic types as follows:

1. When [% Cr − (17 × % by weight C)] > 13, only ferrite is present.
2. When [% Cr − (17 × % by weight C)] < 13, austenite can be formed and quenched to martensite.

Figure 9.3 depicts this relationship.

Austenitic Steels

In the austenitic steels, we find substantial amounts of nickel as well as chromium, as in the familiar 18% Cr, 8% Ni types. The nickel enlarges the austenite field to such an extent that it is stable at room temperature. To illustrate this effect, we can use a horizontal section of the iron–nickel–chromium ternary diagram (Figure 9.4) of the type discussed in Chapter 4. This section represents the phases present at 650°C (1202°F). (We use this temperature to avoid the complications of transformation of the austenite in the low-alloy steels that occur at lower temperature.) We note that the composition 18% Cr, 8% Ni is close to the boundary between the $\alpha + \gamma$ and the γ fields. If we want to be sure to have an all-γ structure at room temperature, we use a type 304 composition (Table 9.1), with 10% nickel. However, there is an advantage to using type 301 steel, in which some ferrite is encountered: A comparison of its yield strength with that of type 304 steel shows that this steel is more responsive to cold working. Type 347 steel is used where there will be welding, to avoid the combination of carbon with chromium that leads to poor corrosion resistance. In type 347 steel the carbon combines with the niobium, instead of with chromium, preventing the corrosion due to chromium depletion of the matrix, as we will discuss in Chapter 18.

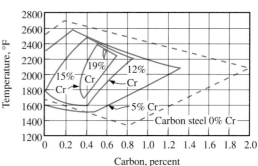

FIGURE 9.3 Closing of the gamma region by the addition of chromium and carbon

[Reprinted with permission from *Alloying Elements in Steel*, 2nd ed., by Bain, American Society for Metals, Metals Park, OH, 1961.]

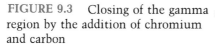

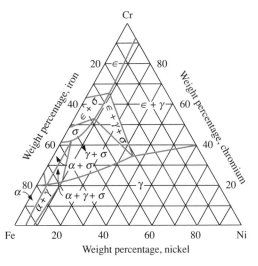

FIGURE 9.4 Iron–nickel–chromium ternary phase diagram at 650°C (1202°F)

[Reprinted with permission from *Metals Handbook*, 8th ed., Vol. 1, *Properties and Selection of Metals*, American Society for Metals, Metals Park, OH, 1961.]

9.5 Precipitation Hardening and Maraging Steels

Where maximum strength and some corrosion resistance are required, 17-7PH and maraging 18% nickel steels are used. In the 17-7PH type steel, a solution treatment at 1925°F (1052°C) is followed by aging at 900°F (482°C) to produce a precipitate. In the *maraging steels*, the material is austenitized at 1500°F (815°C) to dissolve precipitated phases; relatively soft, low-carbon martensite is formed on cooling. The material can be worked or machined in this condition. The material is strengthened by heating the alloy to 900°F (482°C) for 3 hr, which causes aging. Thus the hardening in such steels is due to both martensite and aging precipitates. There is little distortion.

9.6 Tool Steels

The tool steels encompass a wide range of compositions. It is possible to produce a hard cutting edge in a high-carbon plain-carbon steel. The advantages of the high-alloy steels such as 18% W, 4% Cr, 1% V are twofold. With higher hardenability we can use slower cooling rates during quenching, which protects tools from cracking and distortion. Second, the highly alloyed martensite and excess-alloy carbides retain their hardness at the high temperatures generated by fast cutting speeds; that is, they resist tempering.

The various grades of tool steels are represented in Table 9.2 in order of increasing cost. The plain-carbon W1 grade is heat-treated, in the same manner as the iron–carbon alloys already discussed. The final hardness depends on the carbon content and the tempering temperature. The same hardness can be attained in this steel as in the highly alloyed types. The next grade, O1, can be oil-hardened in simple geometric shapes such as small bars

TABLE 9.2 Typical Characteristics of Tool Steels

Number	Chemical Analysis, percent*					Austenitizing Temperature, °F (°C)	Quenching Medium	Tempering Temperature, °F (°C)	HRC	Typical Use
	C	Cr	Mo	W	V					
W1	0.6/1.4					1550 (843)/1400 (760)	Water	350 (177)/650 (343)	64/50	Tools and dies used below 350°F (177°C)
O1	0.9	0.5		0.5		1500 (815)	Oil	350 (177)/500 (260)	62/57	Tools and dies requiring less distortion than W1
D1	1.0	12.0	1.0			1800 (982)	Air	400 (204)/1000 (538)	61/54	Wear-resistant, low-distortion tools
H11	0.35	5.0	1.5		0.4	1850 (1010)	Air	1000 (538)/1200 (649)	54/38	Hot-working dies
M1	0.8	4.0	8.0	1.5	1.0	2200 (1204)	Air	1000 (538)/1100 (593)	65/60	High-speed tools
T1	0.7	4.0		18.0	1.0	2350 (1288)	Air	1000 (538)/1100 (593)	65/60	High-speed tools

* The balance is Fe.

because of the alloy and shows less distortion. The DI grade, with 12% chromium, resembles the martensitic stainless steels, with a high level of carbon, which we have just discussed. The H grade is different from all the others because of the lower level of carbon. It is used for hot-working dies and is therefore exposed to thermal shock. A higher-carbon steel would crack under these conditions. To prevent softening of the martensite, a considerable amount of alloy is used: 5% Cr, 1.5% Mo, 0.4% V. The M1 and T1 steels can be considered together as high-speed steels. The T1 type, with 18% tungsten, was developed first; the molybdenum modification was developed in response to the shortage of tungsten in World War II.

The most significant feature of the heat treatment of these steels is the need for a very high austenitizing temperature. The phase diagram for the T1 type (Figure 9.5) illustrates this point. During the early research on these steels, the hardness obtained after quenching was disappointingly low. This was because, with an austenitizing temperature of 900°C (1652°F), the carbon content of the γ is only 0.2%. Therefore a low-carbon martensite with a hardness of about 50 HRC is obtained. However, when the austenitizing temperature is raised to 1250°C (2282°F), the carbon content of austenite is 0.6%, which gives martensite of 64 HRC on quenching.

The tempering of high-speed steels has received extensive study because of an effect called *secondary hardening*. The hardness rises slightly after tempering at 1000°F (538°C). The as-quenched structure contains some retained austenite, and upon its heating to 1000°F (538°C) some alloy carbides precipitate. This effect raises the M_s of the retained austenite (because carbides take the alloy out of solution), and the structure is transformed to martensite upon cooling from the tempering temperature. Therefore, although the original

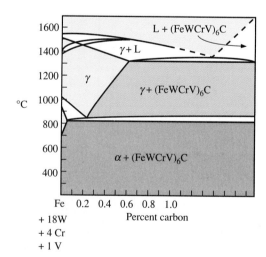

FIGURE 9.5 Phase diagram for T1 type tool steels

[R. M. Brick and A. Phillips, *Structure and Properties of Alloys*, Copyright 1949. Used by permission of McGraw-Hill Publishing Co.]

martensite was softened at 1000°F (538°C), the new martensite raises the overall hardness. Retempering often improves the toughness of the structure.

9.7 Hadfield's Manganese Steel

The presence of manganese in a steel, like that of nickel, produces an enlargement of the γ field. We take advantage of this fact in a very tough, abrasion-resistant austenitic steel containing 12% manganese and 1% carbon. The as-cast structure contains carbide, austenite, and some pearlite. However, after heating to 2000°F (1093°C) and water quenching, the structure is completely austenitic. A water quench may be used even with complex castings because there are no stresses from transformation of austenite to martensite. Typical properties are as follows: tensile strength, 120,000 psi (828 MPa); yield strength, 50,000 psi (345 MPa); percent elongation, 45; HB, 160. This steel is widely used in mining and earth-moving equipment and in railroad-track components.

EXAMPLE 9.1 *[EJ]*

The discussion of Hadfield's manganese steel (12% Mn, 1% C) indicates that a water quench from 2000°F (1093°C) gives an austenitic microstructure. A slower cooling rate, such as would occur with an air quench, can result in martensite, carbides, and austenite. Explain why martensite might be obtained with slow cooling rates rather than a rapid quench. [*Hint:* The alloy content of the austenite determines the M_s temperature.]

Answer   Although the calculation of M_s temperature from the empirical relationship performed in Section 8.16 may not be completely appropriate for this high-alloy content, it does show the low M_s for the alloy:

$$M_s = 930 - 540\,(1\%\ C) - 60\,(12\%\ Mn) = -330°F$$

Therefore, when all of the carbon and manganese is in solution, as when quenching occurs from 2000°F (1093°C), martensite is not obtained because of the low M_s temperature.

However, slow cooling allows precipitation of complex iron–manganese carbides from the austenite. This lowers the alloy content of the austenite and raises the M_s temperature above room temperature, thus providing some martensite. Because the M_f temperature is still below room temperature, complete transformation to martensite may not occur. Hence the microstructure of austenite, martensite, and carbides can exist. Similar effects are observed in stainless steels and tool steels that require double-quench cycles, as the text explains.

SUPERALLOYS

9.8 Superalloys in General

The name *superalloys* has been given to materials that exhibit far greater strength at high temperatures — 1500 to 2000°F (815 to 1093°C) — than conventional alloys. Certain of these alloys are even called "exotic" because of the high cost of some of the elements used. These alloys were developed largely for use in gas turbine rotors, because the efficiency of a gas turbine increases with the temperature at which the rotor can be operated. However, strength at elevated temperatures is necessary not merely in the rotor itself but also in the vanes, the combustion chamber, and even the compressor section.

We look for two principal qualities when developing materials for parts operating at high temperatures: oxidation resistance and strength. The need for oxidation resistance is related primarily to the problem of gas corrosion, which will be taken up in Chapter 18. At this point we shall consider only the need for strength.

Simply running a hot tensile test (testing the conventional 0.505-in.-diameter specimen in a furnace) does not give the information we need to evaluate strength at high temperatures. Suppose that the conventional test gives a yield strength of 50,000 psi (345 MPa). Let us load the specimen to 40,000 psi (276 MPa) and observe it over a period of time at the elevated temperature. First, the sample continues to elongate although it is at constant load. As we discussed in Chapter 3, this phenomenon is called *creep*. Second, the specimen ruptures after a certain time, say 20 hr. These are two important factors in design. Therefore, if we design a rotor for a gas turbine merely on the basis of yield strength during a tensile test, it will fail when it is in service, because the blades will heat up and either stretch excessively and contact the housing or break after a period of time (see also Sections 3.18 and 3.19). Figure 9.6 shows typical deformation or creep-rupture data for an austenitic stainless steel.

9.9 High-Temperature Properties of Typical Superalloys

At first glance the names and analyses of superalloys seem random and complex (Table 9.3). There are, however, certain common features correlated with analysis, structure, and performance.

First, all the important alloys for high temperature use have an FCC structure based on a combination of nickel, iron, and chromium. Therefore, we can view these structures as simple extensions of the austenitic field of the ternary diagram (Figure 9.4) to higher-nickel compositions. The chromium, though not added to obtain austenite, is essential to provide oxidation resis-

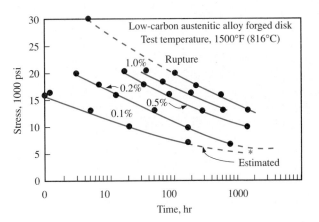

FIGURE 9.6 Stress-deformation data for a low-carbon austenitic stainless steel turbine disk. Percentages refer to elongation due to creep.

[Reprinted with permission from *Metals Handbook*, 8th ed., Vol. 1, *Properties and Selection of Metals*, American Society for Metals, Metals Park, OH, 1961.]

tance. Thus we have accounted for the iron, nickel, and chromium content. Cobalt is used to replace part of the iron or nickel. It dissolves in solid solution, strengthening the material.

Aluminum and titanium have a special use in providing a very fine precipitate called γ' (this is read "gamma prime"). The name is derived from the fact that the structure is close to the γ solid solution. γ' has the formula Ni_3Al, and some of the aluminum can be replaced with titanium with only small changes in a_0. The γ' structure is an ordered substitutional solid solution; aluminum or titanium atoms occupy the corners of the unit cell and nickel atoms are in the face centers. The precipitate is in the form of cubes. The magnification of the electron microscope is necessary to resolve it (see the photo on page 418 at the beginning of this chapter). This precipitate is essential to high strength in these alloys.

Molybdenum adds to the strength of the matrix in superalloys and also forms complex carbides that reduce the creep rate. Small amounts of zirconium and boron are added for their beneficial effects on carbide size refinement. Usually the carbon content is about 0.1 to 0.2%, and the carbides are carefully controlled.

Change comes rapidly in the field of superalloys, because the rewards are high for developing a new material or technique that results in just moderately better performance. One recent development was the addition of 1.5% hafnium to some superalloys. Another was the directional solidification of turbine blades using an induction field to develop the desired temperature gradient in the mold and also the "single crystal" blades described in Chapter 6.

Practically all grades of superalloys are cast, but only a limited number are wrought. The reason is that with increased creep strength, hot working becomes difficult and die life short. Also, some of the alloys have low ductility that is adequate for service but not for rolling. The investment casting process is the most widely used method. It is capable of producing intricate shapes.

TABLE 9.3 Superalloys: Iron and Nickel-Based Alloys

Code and Name	Chemical Analysis, percent								Stress to Rupture in 100 hr,[*] psi × 10³, at Indicated Temperature, °F							Availability	
	Fe	Ni	Cr	Co	Al	Ti	Mo	Other	1200	1350	1400	1500	1600	1700	1800	Cast	Wrought
HF (18/8 type)	Balance	10	20					0.3 C	30		14		6			X	X
Incoloy 901	Balance	43	13		0.2	3.0	6	0.01 B		50							X
Inconel 718	18	52	19		0.6	0.8	3	5.0 Nb			45					X	X
Waspalloy	2	Balance	20	13	1.3	3.0	4	†				40	25			X	X
Inco 713	2	Balance	13		6.0	0.8	4	2.0 Nb					44			X	
Udimet 500	4	Balance	18	18	3.0	3.0	4	†				45	32	19		X	X
Udimet 700	1	Balance	15	18	4.5	3.5	5						42			X	X
In 100	1	Balance	10	15	5.5	5.0	3						56			X	
René 41	5	Balance	19	10	1.5	3.0	10					45	28	17		X	X

[*]Multiply psi by 6.9 × 10⁻³ to obtain MPa or by 7.03 × 10⁻⁴ to obtain kg/mm².
†Plus small amounts of boron and/or zirconium.

EXAMPLE 9.2 *[ES/EJ]*

Refer to Table 9.3 and indicate why the *stress to rupture* in 100 hr, rather than the creep rate, is given for the different alloys. Is the creep rate not important?

Answer   Certainly the creep rate is important, as the complete set of data in Figure 9.6 suggests. However, a review of 100-hr rupture stresses at different temperatures for the alloys listed in Table 9.3 makes possible rapid comparison and selection of a suitable alloy.

For example, if the design conditions require a minimum 100-hr rupture stress of 35,000 psi at 1600°F, a number of the alloys in Table 9.3 are unsuitable. Once we select an alloy, such as Udimet 700, we must then obtain creep data to determine whether the high-temperature plastic deformation design criteria can be met. It would be difficult to present all of the creep-rupture data in a single table, because we are interested in temperature, stress to rupture, and creep elongation.

Data such as those shown in Figure 9.6 exist for many alloys at different temperatures and would fill several volumes — especially because we may also superimpose the effect of atmosphere.

CAST IRONS

9.10   Importance of High-Carbon Alloys

Gray cast iron is one of the oldest alloys; ductile iron, which is being used more and more, is one of the newest. Together they account for 95% of the weight of a typical automotive engine, though aluminum alloys are competing successfully in the trend to decrease vehicle weight. Typical gray iron parts of a car engine are the engine block, head, camshaft, piston rings, lifters, and manifolds. The crankshaft and rocker arms are ductile iron. Malleable iron and ductile iron are also used in many other parts of the automobile, such as the differential housings.

The annual production of castings in this group of high-carbon iron alloys (about 11 million tons) is second only to that of steel in the field of metallic materials. Nevertheless, the structures and properties of these alloys are not well understood because of a few fundamental differences from steel. To use these alloys properly and to understand their heat treatment, it is important to focus first on the portion of the structure that is different from steel:

White iron:	massive carbide	
Gray iron:	flake graphite	
Ductile iron:	spheroidal graphite	+Steel matrix (ferrite,
Malleable iron:	clump-like graphite	pearlite, martensite, etc.)
	(temper carbon)	

Figure 9.7 shows examples of these microstructures.

We are already familiar with the control of austenite transformation to ferrite and other products, so let us first explore the factors affecting the massive carbide and graphite structures.

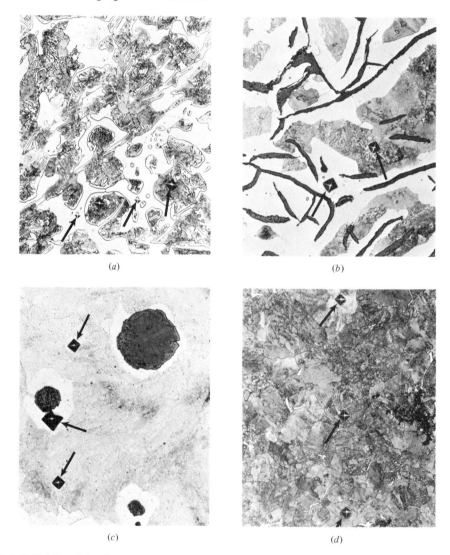

(a)   (b)   (c)   (d)

FIGURE 9.7   (*a*) White cast iron: 3.5% C, 0.5% Si. Massive carbide (white islands), HV 1300. Pearlite (lamellar), HV 300 as cast; 500 ×, 2% nital etch. (*b*) Gray cast iron: 3.5% C, 2% Si. Flake graphite. Pearlite (lamellar), HV 285. Ferrite (white), HV 195 as cast; 500 ×, 2% nital etch. (*c*) Ductile cast iron: 3.5% C, 2% Si, 0.05% Mg. Spheroidal graphite. Pearlite (lamellar), HV 300. Ferrite (white), HV 190 as cast; 500 ×, 2% nital etch. (*d*) Malleable cast iron: 2.5% C, 1% Si. Temper carbon. Pearlite (lamellar), HV 345. Heat-treated, 1750°F (954°C), 12 hr, air-cooled; 500 ×, 2% nital etch.

9.11 Relationships Among White Iron, Gray Iron, Ductile Iron, and Malleable Iron

White Iron

We see from the fraction chart of white iron (Figure 9.8) that carbide precipitates during three important periods (3% carbon iron):

1. 2098°F (1148°C): the eutectic reaction, $L \rightarrow \gamma$ + carbide
2. From 2098 to 1341°F (1148 to 727°C): from the eutectic to the eutectoid, $\gamma \rightarrow \gamma^*$ + carbide
3. At 1341°F (727°C): the eutectoid reaction, $\gamma \rightarrow \alpha$ + carbide (as pearlite)

In the first reaction large, massive carbides are formed from the liquid. In the second reaction the carbide crystallizes onto the existing massive carbide, and in the third reaction pearlite is formed, resulting in the microstructure shown in Figure 9.7a. The final product, therefore, contains a high percentage of massive carbide and is hard and brittle.

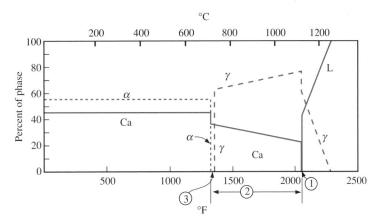

FIGURE 9.8   Fraction chart for white cast iron

---

*The analysis of the austenite changes from 2.11% carbon to 0.77% carbon with decreasing temperature.

Gray Iron

To understand gray cast iron, you need to understand that iron carbide is not basically a stable phase. With abnormally slow cooling (or in the presence of certain alloys such as silicon), graphite (pure carbon) and iron crystallize (Figure 9.7b). Furthermore, if we heat iron carbide for extended periods, it decomposes: iron carbide → iron + graphite.

In other words, the true equilibrium diagram is the iron–graphite system, which is superimposed on the iron–iron-carbide diagram in Figure 9.9. We do not need to learn a new diagram. For all practical purposes, all we need to do is substitute graphite for carbide in the two-phase fields, as shown in Figure 9.9, and move the right-hand vertical line to 100% carbon. There are, however, slight differences in eutectic and eutectoid carbon contents and temperatures.

Using this diagram, let us follow the formation of graphite in a ferritic gray cast iron with 3% carbon (ferrite matrix with flake graphite). To produce graphite in place of carbide, we either cool very slowly or add silicon, which also promotes graphite formation. We obtain the fraction chart shown in Figure 9.10. The following table contrasts the formation of graphite in gray cast iron with that of carbide in white cast iron of eutectic composition (4.3% carbon).

Temperature, °F (°C)	Gray Iron*	White Iron	Comments
2098 (1148)	L → γ + graphite	L → γ + carbide	
2098 to 1341 (1148 to 727)	γ → γ + graphite	γ → γ + carbide	Graphite (carbide) precipitates from γ on existing graphite (carbide). Carbon solubility is decreasing.
1341 (727)	γ → α + graphite	γ → α + carbide (as pearlite)	The analysis of the austenite is 0.77% carbon.

*We shall continue to use the iron–iron-carbide equilibrium temperatures in this introductory section for simplicity. Later we shall modify these remarks when we discuss the actual commercial alloys containing, for example, silicon.

Basically, in place of the carbide precipitation in white cast iron, we obtain graphite in the gray cast iron. (We shall modify this statement somewhat when we discuss other gray cast iron microstructures. It is possible, for example, by fast-cooling gray iron from 1500°F (815°C), to obtain pearlite or the other austenite transformation products we found in the austenite transformation of steel.)

The final structure in ferritic gray cast iron is therefore flake graphite plus ferrite. The flakes formed at 2098°F (1148°C) grew during cooling to 1341°F (727°C), and at the eutectoid transformation γ → α + Gr.

"Gray" and "white" not only refer to the microstructure but also describe the appearance of the fracture of the cast irons. Therefore, a fracture test of cast iron reveals at least whether the high-carbon phase is carbide or graphite.

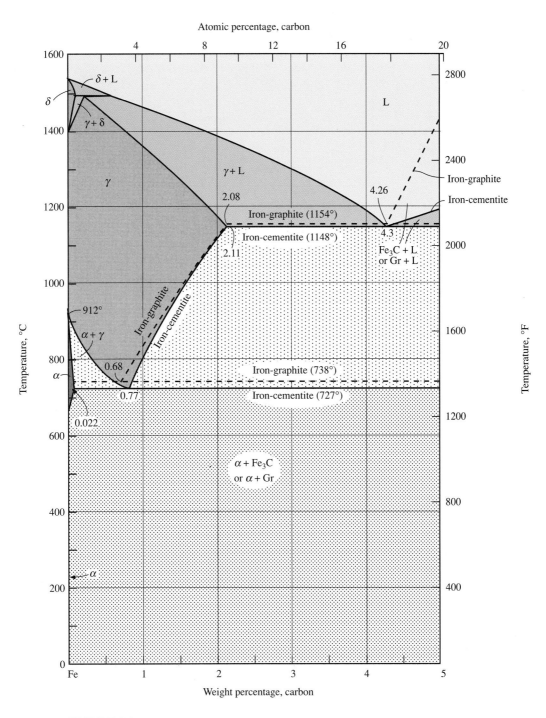

FIGURE 9.9 Iron–iron-carbide and iron–graphite phase diagrams

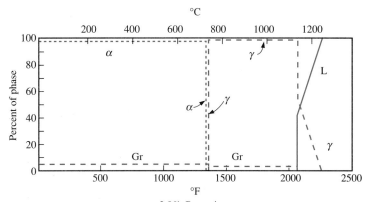

FIGURE 9.10   Fraction chart for gray cast iron

Ductile Iron

Although gray iron is excellent for parts such as engine blocks because the graphite flakes give good machinability, a spheroidal graphite shape is better for strength. In 1949 Millis and Gagnebin discovered that dissolving a small (0.05%) amount of magnesium in the liquid iron causes the formation of spheroidal graphite instead of flake graphite from the melt. Then the graphite that precipitates from 2098 to 1341°F (1148 to 727°C) and at the eutectoid continues to crystallize in the spheroidal form. The final structure in ferritic *ductile iron* is therefore *spheroidal* graphite plus ferrite. The phase amounts and analyses are the same as for ferritic gray cast iron, but the different shape of the graphite gives it twice the tensile strength and 20 times the ductility.

Malleable Iron

The difference between *malleable iron* and gray or ductile iron is that the *as-cast* structure of malleable iron is the same as that of white cast iron and contains no graphite. In other words, the cooling rate and composition produce white cast iron in the original casting. The cold casting is then heated in a malleablizing furnace at 1750°F (954°C) to change the iron carbide to the equilibrium structure of iron plus graphite. At this point you might wonder why manufacturers don't produce the graphite during the original cooling and avoid the expense of heat treatment. The answer lies in the fact that malleable iron was being produced for many years before ductile iron was discovered. Malleable iron had been found to be stronger and more ductile than gray cast iron because during the heat treatment, graphite was formed in clump-like nodules (something like spheroidal graphite); this graphite is called *temper carbon*. After Millis and Gagnebin discovered how to form ductile iron in a one-step process, many producers of malleable iron changed to the new material, but malleable iron is still used for many thin-section castings.

It is interesting to follow the structural changes in the heat-treatment cycle for malleable iron. There are two stages:

STAGE 1.   Graphitization. Heat at 1750°F (954°C) for 12 hr. The structure changes from austenite plus carbide to austenite plus temper carbon.

STAGE 2.   Cool from 1750 to 1400°F (954 to 760°C). Slow-cool at 10°F/hr (5.5°C/hr) to 1200°F (649°C).

Whereas austenite would normally transform to α + carbide, the slow-cooling rate in stage 2 gives α + graphite in the eutectoid temperature range. The graphite crystallizes on the existing temper carbon formed at the higher temperatures. The final structure is then ferrite plus temper carbon.

EXAMPLE 9.3   *[EJ]*

Why is it difficult to produce malleable cast iron in large sections?

Answer   Malleable cast iron must first be cast white (with carbide as the high-carbon constituent). In large sections the cooling rate is slow enough for graphite to form at the high temperatures during solidification. Because this graphite is in flake form, the desirable properties of the temper carbon structure cannot be obtained. Although it is possible to stabilize the carbide by using a lower percentage of silicon or by adding carbide-stabilizing elements such as chromium, the time that would be required for stage 1 makes this alternative uneconomical.

Now let us consider the variations in properties in these different families as we change the structure of the steel matrix (Table 9.4). We have austenite at 1600°F (871°C), so we can produce any of the transformation products found in steel: ferrite, pearlite, bainite, martensite, and even retained austenite. We shall now discuss these variations for each type of cast iron.

9.12   White Cast Iron

The largest tonnage of *white cast iron* is made with a pearlitic matrix—that is, a structure of pearlite and massive carbide. If alloys are added to suppress the pearlite transformation, martensitic white irons are obtained. The following table shows some examples.

TABLE 9.4   Minimum Properties of White Iron, Gray Iron, Ductile Iron, Malleable Iron, and Special Alloys

Name and Number	Chemical Analysis, percent	Condition	Tensile Strength,[a] psi × 10³	Yield Strength,[a] psi × 10³	Percent Elongation	HB	Typical Use
White cast iron, unalloyed	3.5 C, 0.5 Si	As cast	40	40	0	500	Wear-resistant parts
Gray Iron							
Ferritic class 25	3.5 C, 2.5 Si	As cast	25	20	0.4	150	Pipe sanitary ware
Pearlitic class 40	3.2 C, 2 Si	As cast	40	35	0.4	220	Machine tools, blocks
Quenched martensitic	3.2 C, 2 Si[b]	Quenched	80	80	0	500	Wearing surfaces
Quenched bainitic	3.2 C, 2 Si[c]	Quenched	70	70	0	300	Camshafts
Ductile Iron							
Ferritic (60-40-18)	3.5 C, 2.5 Si	Annealed	60	40	18	170	Heavy-duty pipe
Pearlitic (80-55-06)	3.5 C, 2.2 Si	As cast	80	55	6	190	Crankshafts
Quenched (120-90-02)	3.5 C, 2.2 Si	Quenched and tempered	120	90	2	270	High-strength machine parts
Malleable Iron							
Ferritic (35018)	2.2 C, 1 Si	Annealed	53	35	18	130	Hardware, fittings
Pearlitic (45010)	2.2 C, 1 Si	Annealed	65	45	10	180	Couplings
Quenched (80002)	2.2 C, 1 Si	Quenched and tempered	100	80	2	250	High-strength yokes
Special Alloy Irons							
Austenitic gray	20 Ni, 2 Cr[d]	As cast	30	30	2	150	Exhaust manifolds
Austenitic ductile	20 Ni[d]	As cast	60	30	20	160	Pump casings
High-silicon gray	15 Si, 1 C	As cast	15	15	0	470	Furnace grates
Martensitic	4 Ni, 2.5 Cr[e]	As cast	40	40	0	600	} Wear-resistant parts, liners
white	20 Cr, 2 Mo, 1 Ni[f]	Heat-treated	80	80	0	600	}

[a]Multiply psi by 6.9×10^{-3} to obtain MPa or by 7.03×10^{-4} to obtain kg/mm².
[b]+1% Ni, 1% Cr, 0.4% Mo
[c]+1% Ni, 1% Mo
[d]+3% C, 2% Si
[e]+3.2% C, 0.8% Si
[f]+2.7% C, 0.8% Si

	Analysis, %					
Material	*C*	*Si*	*Cr*	*Ni*	*Mo*	*HB*
Typical pearlitic white cast iron	3.2	0.5	1.0			450
Martensitic white iron	3.2	0.5	2.5	4.0		600
Martensitic high-chromium iron	2.5	0.5	20.0	1.0	2.0	600

The high alloy content pushes the nose of the TTT diagram so far to the right that martensite can even be obtained in as-cast sections. These alloys are used chiefly in liners and balls, for abrasion resistance in grinding mills, in cement manufacture, in mining equipment, and in rolls for finishing steel.

9.13 Gray Cast Iron

As in the case of white cast iron, it is possible to vary the matrix of *gray cast iron*. The range of gray iron structures in actual use is wider than that of white iron structures. It includes ferrite, pearlite, bainite, martensite, and even austenite. In all cases the structure depends on the alloy content and cooling rate of the austenite.

Pearlite is easily obtained by cooling fast enough through the eutectoid range to avoid the transformation of austenite to α + graphite. An equally important factor is the percentage of silicon and other elements that affect the stability of carbide. For example, a pearlitic gray iron engine block is made with 3.2% carbon and 2% silicon and is air-cooled through the eutectoid after pouring. If the block were removed from the mold at 1400°F (760°C) and cooled slowly in a furnace at 10°F/hr (5.5°C/hr), the structure would be ferritic. The presence of small amounts of tin or copper results in a pearlite matrix even with cooling in the sand molds.

Bainite can be obtained either by an isothermal heat treatment or by casting an alloy combination that produces a TTT curve with a pronounced bainite nose. For example, the following analysis is used to produce a bainite structure in a 1-in.-diameter crankshaft: 3.2% C, 2% Si, 0.7% Mn, 1% Ni, 1% Mo. The HB is 300. The hardness is lower than that of a bainitic steel because of the presence of graphite.

Martensite is usually obtained by heat treating—austenitizing plus oil quenching. A typical analysis for an automotive camshaft is 3.2% C, 0.7% Mn, 2% Si, 1% Ni, 1% Cr, 0.4% Mo. The HB is 550. Figure 9.11 shows a section of an induction-hardened shaft. In this case, because of the selective heating, only the surface layers were austenitized before quenching.

To obtain austenite, enough alloy is added to prevent pearlite transformation *and* to lower the M_s below room temperature. This can be done with a 20% Ni, 2% Cr alloy or with a 14% Ni, 6% Cu, 2% Cr alloy. This type of gray cast iron is used in high-performance exhaust manifolds, in which we want a stable heat-resistant structure, and in pump parts, for which we want the corrosion resistance of a high-nickel chromium iron.

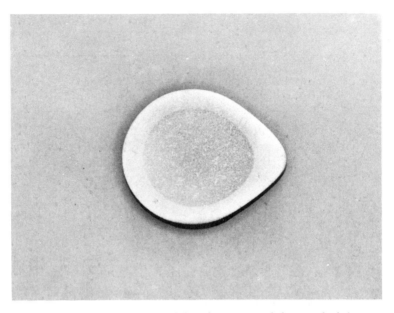

FIGURE 9.11   Cross section of a cam lobe of an automobile camshaft (gray iron). The camshaft was heated on the surface by induction, then quenched. The hardened layer shows as a light etching structure. Actual size, 5% nital etch.

EXAMPLE 9.4   *[ES/EJ]*

A solid slug of gray cast iron, 1 in. (25.4 mm) in diameter by 2 in. (50.8 mm) long is to have a hardness of 400 HB on one end, free of graphite to a depth of $\frac{1}{4}$ in. (6.35 mm). The remainder of the casting is to be 250 HB maximum and must be machinable. Indicate how this might be accomplished in a class 40 gray cast iron.

Answer   The key here is to pour the slug in a sand mold where one end has a chill (such as metal or graphite) that can remove the heat rapidly. The chilled end then has carbide as its high-carbon phase, while flake graphite occurs in the more slowly cooled portions away from the chill where sand surrounds the casting. In each case, pearlite is the matrix because cooling through the eutectoid is fairly rapid. Close control of the chemical analysis and pouring temperature are required in order to guarantee he $\frac{1}{4}$-in. (6.35-mm) depth for the white cast iron end.

   This technique has been used for a wear-resistant surface that requires machining on the remainder of the component. An example is a valve lifter that contacts the camshaft shown in Figure 9.11.

9.14 Ductile Iron

From Table 9.4 we see that the tensile strengths of the common non-heat-treated grades of gray iron vary from 25,000 to 40,000 psi (173 to 276 MPa) and that the elongation is very low. By contrast, the ductile irons have strengths up to 120,000 psi (828 MPa) and high elongation.

The most common grade of ductile iron is the as-cast pearlitic type 80-55-06. The analysis is controlled to avoid massive carbide, but in addition a pearlitic matrix is formed on cooling. For this alloy, the silicon is maintained at under 2.5% and the manganese is above 0.4%. Small amounts of copper and tin are also used to promote pearlite.

The ferritic grade is obtained by slow-cooling from the casting temperature or by reheating to 1600°F (871°C) and holding, then slow-cooling near 1300°F (704°C).

The quenched and tempered grade of ductile iron is heat-treated in a manner similar to that used for 0.8% carbon steel. It is austenitized at 1600°F (871°C), quenched, and tempered to the desired hardness.

The austenitic grade has higher strength and toughness than the corresponding gray iron, as we could have predicted from the differences in graphite shape.

A more recent matrix development in ductile cast is *austempered ductile iron* (ADI), which derives its name from an isothermal treatment in the bainite region. When a ductile iron is austenitized, the microstructure is graphite spheres plus an alloyed austenite (primarily silicon). Upon quenching and isothermal holding to form bainite, the microstructure first becomes ferrite plus austenite plus graphite. That ferrite forms first might be expected because bainite is nucleated by ferrite.

What is not anticipated is that 1–4 hr at the isothermal temperature may be required before the carbide forms and combines with more ferrite to produce bainite. This window provides us with an option. We can quench the austenite–ferrite–graphite mixture and produce martensite that can then be tempered to produce a tough microstructure. Or we can allow complete transformation to bainite, which results in higher strength but lower toughness and ductility.

Considerable interest exists in the use of both the tempered martensite and bainite austempered ductile irons for highly stressed wear applications such as are found in gears.

9.15 Malleable Iron

The principal grades of malleable iron are ferritic, pearlitic, and quenched. The ferritic grade is obtained via the two-stage graphitization previously described. As with ductile iron, the ferritic grade of malleable iron has the highest ductility (Table 9.4).

The pearlitic grade is made by avoiding the stage 2 graphitization. After stage 1 the casting is air-cooled, and the austenite transforms to pearlite. If the pearlite is too hard, it is subsequently tempered (spheroidized).

The quenched grade of malleable iron is also called pearlitic, but this is a misnomer. A martensitic structure is produced by quenching from austenite and then tempering to spheroidite.

EXAMPLE 9.5 *[EJ]*

Suggest possible iron-based alloys that might be used for the following applications. (Data from Chapter 8 are also required.)

A. road grader blade
B. file cabinet
C. wood chisel
D. metal mixing bowl
E. fire hydrant

Answers

A. *Road grader blade.* Because of the wear that a grader blade is subjected to, abrasion resistance is required, so a work-hardening material is desirable for this application. A quenched and tempered steel would not have sufficient impact resistance. A Hadfield steel (12% Mn, 1% C) would meet the requirements. An austenitic stainless steel, though highly work-hardenable, would be too expensive.

B. *File cabinet.* Given the need for excellent cold formability and a good surface quality for subsequent painting, a low-carbon steel such as a 1010 steel would be appropriate for this application.

C. *Wood chisel.* A tool steel is probably not necessary for this application; the chisel does not heat in service. A quenched and tempered low-alloy steel would be adequate. (Care should be exercised in regrinding a sharp edge, because a high-speed noncooled grinding operation will further temper the steel.)

D. *Metal mixing bowl.* Because the bowl will contain food that might be contaminated by corrosion by-products, a stainless steel is suggested. The most inexpensive stainless steel is ferritic (such as type 430); however, an austenitic grade (type 304) provides better corrosion resistance.

E. *Fire hydrant.* Given the complex shape of the fire hydrant, a casting would be appropriate. The most inexpensive iron-based alloy would be a gray cast iron such as class 25. Another advantage of using gray cast iron may be that the hydrant would shear more easily if an automobile should collide with it, and the accident might therefore do less physical harm to the driver.

Summary

We have covered three special classes of metals containing iron: high-alloy steels, superalloys, and cast irons. Although some of the superalloys may not be iron-based, their development has been a natural extension of the structures of high-alloy steels. The incorporation of elements such as chromium, nickel, cobalt, aluminum, molybdenum, and titanium into the superalloys slows diffusion and increases the resistance to creep.

The presence of large amounts of alloy in the high-alloy steels, substantially alters the iron–iron-carbide diagram. Elements in one group, such as nickel, give the open γ (austenite) field, leading to alloys that are austenitic at room temperature, such as austenitic stainless steels. By contrast, the elements that give a closed γ field, such as chromium, tend to produce an α structure at all temperatures, as in the ferritic stainless steels. An intermediate condition occurs in the martensitic stainless steels. Ternary diagrams indicate the quantitative relationships. Tool steels and other specialty grades depend on similar principles.

The cast irons and related materials all contain over 2.11% carbon. Some of this carbon may be present as graphite; to analyze these relationships, we use the iron–graphite diagram. Each of these materials can be considered a steel matrix (pearlite, bainite, etc.) plus carbide or graphite. No graphite is present in white cast iron, and the product is hard and brittle because of the presence of massive particles of iron carbide. Gray cast iron is soft and machinable because the iron carbide is replaced by graphite flakes, but the ductility is low because of the sharp notch shape of the individual graphite flakes. Ductile iron exhibits better ductility than gray cast iron because the shape of the graphite is spheroidal. Malleable iron also is ductile because of the temper carbon produced during heat treatment of white cast iron.

Definitions

Creep Plastic elongation as a function of time (and temperature) in stressed material.

Ductile iron An iron–carbon alloy with 3.5 to 4% carbon and graphitizing elements such as silicon. It differs from gray cast iron in that the graphite is in the form of spheres and the ductility is high. The spheroidal graphite is produced by the addition of 0.05% magnesium just before the liquid metal is poured into castings.

Gray cast iron An iron-carbon alloy with 2.1 to 4% carbon and graphitizing elements such as silicon. Most of the carbon is in the form of graphite flakes, so the material is easily machined but low in ductility.

High-alloy steel A steel that has an alloy content of 10% or more.

Malleable iron An iron–carbon alloy with 2.1 to 3% carbon. Malleable iron is produced when castings of white cast iron are heat-treated at 1750°F

(954°C). The massive iron carbide in the white cast iron is converted to nodules of graphite called *temper carbon* in an iron matrix. The properties are similar to those of ductile iron.

Maraging steel   A highly alloyed steel that is hardened both by martensite and by age hardening.

Stainless steel   A steel that is highly alloyed with chromium and often nickel to improve corrosion resistance.

Stress to rupture   The amount of stress needed to cause rupture in a specified time at a given temperature.

Superalloys   Alloys used for high-temperature service. FCC alloys with nickel, cobalt, iron, and chromium are usually strengthened with an $Ni_3(Al,Ti)$ precipitate.

White cast iron   An iron-carbon alloy with 2.1 to 4% carbon. The most common range is 3 to 3.5% carbon. Because of the presence of large amounts of brittle iron carbide, the material is not hot-worked or machined.

Problems

9.1   *[ES]* Here are three compositions of stainless steel and their potential applications. Indicate for each whether ferritic, austenitic, or martensitic stainless steel should be expected. (Sections 9.1 through 9.9)

   a. 16% chromium; 0.5% carbon (cutter for a compost grinder)

   b. 16% chromium; 10% nickel; 0.03% carbon (vessel to hold corrosive fluids)

   c. 16% chromium; 0.1% carbon (decorative trim)

9.2   *[EJ]* List by code letter all the mechanisms you could use to harden each of the following steels. [*Code:* A = work hardening; B = quenching from austenitizing temperature and tempering; C = solution heat treating and age hardening.] (Sections 9.1 through 9.9)

   a. Hadfield manganese steel; 12% Mn, 1% C, equiaxed grains

   b. Austenitic stainless steel; 18% Cr, 8% Ni, 0.02% C

   c. Ferritic stainless steel; 14% Cr, 0.05% C

   d. Martensitic stainless steel; 14% Cr, 0.7% C, 200 HB

   e. 18% W, 4% Cr, 1% V tool steel

   f. Which steel(s) would be more sensitive to cracking by the heat treatment 1900°F, 2 hr, water-quench?

9.3   *[ES/EJ]* A nominal 17 wt % Cr stainless steel may be purchased with three different microstructures. Identify the three microstructures and indicate differences among them in chemical composition. (Sections 9.1 through 9.9)

9.4   *[ES/EJ]* Why might a tool steel be air-quenched, tempered at an intermediate temperature, and then requenched and retempered? (Sections 9.1 through 9.9)

9.5   *[EJ]* The cost of stainless steels and tool steels is high, and not all of the high cost is attributable to increased alloy content. What processing variables contribute to the cost? (Sections 9.1 through 9.9)

9.6   *[EJ]* Which steel(s) listed in Table 9.1 would you specify for each of the following applications? Explain the deficiencies of the others. (Sections 9.1 through 9.9)

   a. A kitchen knife of intricate blade shape for preparing grapefruit

   b. A blade for carving that would hold its edge best
   c. Automobile trim
   d. Corrosion-resistant aircraft parts with the best strength-to-weight ratio
   e. A cast alloy for exhaust valves contains 15% Ni, 15% Cr, 1% C. From the point of view of structure, why is this alloy more resistant to deformation than one of the austenitic steels listed?

9.7 *[ES]* Manganese and nitrogen have found use in austenitic stainless steels. Discuss their role and importance. (Sections 9.1 through 9.9)

9.8 *[ES]* From the creep data given in Fig. 9.6, calculate the following. (Sections 9.1 through 9.9)
   a. Minimum creep rate (percent strain per hour) at a stress of 15,000 psi (103 MPa). (It may be necessary to replot the data. Use only the straight-line portion of the curve.)
   b. Stress to rupture in 1000 hr
   c. Total creep in 1000 hr at stresses less than 15,000 psi

9.9 *[ES]* As mentioned in the text, nickel combines with aluminum and titanium to form the compound Ni_3X, where X is aluminum or titanium or both. From the analysis given in Table 9.3 for Udimet 700, calculate what percent by weight of this compound is present, assuming all aluminum and titanium are in the compound. Now convert to percent by volume. At what content of aluminum weight percent or titanium weight percent would the structure be 100% γ'? (Sections 9.1 through 9.9)

9.10 *[EJ]* The superalloys listed in Table 9.3 often contain many elements and may exhibit age hardening. Suggest why these alloys might not overage in service. (Sections 9.1 through 9.9)

9.11 *[ES/EJ]* Find the metal price quotation in one of the newspapers that list commodity prices, and compare the costs of 100 lb of HF and Udimet 500 superalloys on the basis of element content. Indicate why the costs of superalloys can be easily justified. (Sections 9.1 through 9.9)

9.12 *[ES/EJ]* We austenitize an M1 tool steel by heating for 1 hr at 2200°F, and upon air quenching, we obtain a hardness of 55 HRC rather than the value given in Table 9.2. What might have happened, and what experiments would you suggest performing before calling up the supplier to complain that the wrong material was sent? (Sections 9.1 through 9.9)

9.13 *[ES/EJ]* Why might a Hadfield's steel be more similar to a tool steel than to a stainless steel? (Sections 9.1 through 9.9)

9.14 *[ES]* Seventy pounds (31.8 kg) of an iron–carbon alloy containing 3.0% carbon is equilibrated at 2111°F (1155°C). (Sections 9.10 through 9.15)
   a. What phase(s) is (are) present? What is the percent of carbon in each phase? What is the amount of each phase? Show all calculations.
   b. The alloy is held at 2000°F (1093°C) until all the Fe_3C has graphitized. How many pounds (kg) of graphite will the 70-lb (31.8-kg) casting then contain? Show your calculations.

9.15 *[ES/EJ]* The effect of appreciable silicon (over 0.5%) on the iron–iron-carbide diagram is important. Figure 9.12 shows the phase diagram for 2% silicon. (Sections 9.10 through 9.15)
   a. According to the phase rule, why can $\alpha + \gamma + Ca$ (three phases) exist over a range of temperature?
   b. The eutectic composition is no longer 4.3% carbon, but follows this formula: eutectic = 4.3 − (0.3)(% Si). Verify this for the diagram shown in Fig. 9.12.
   c. What is the carbon content of pearlite in a 2% silicon cast iron?

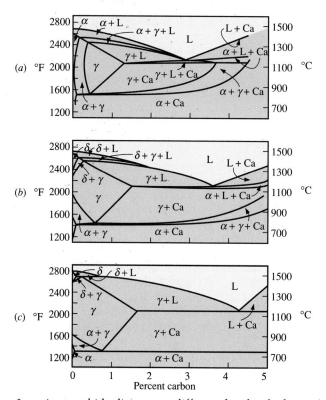

FIGURE 9.12 Iron–iron-carbide diagram at different levels of silicon. (a) 4% silicon. (b) 2% silicon. (c) 0% silicon.

[Reprinted from Greiner, *Alloys of Iron and Silicon.* Copyright 1933. Used by permission of McGraw-Hill Publishing Co.]

    d. How does the austenitizing temperature prior to quenching compare with that of an 0.8% carbon steel?

9.16 *[ES/EJ]* Assume that we have a cast iron of chemistry (3.0% C, 2.0% Si) that is all liquid and can be treated in several ways. Indicate the final matrix and high-carbon phase expected for each treatment. [*Matrix:* ferrite or pearlite; *high-carbon phase:* carbide, flake graphite, spheroidal graphite, or temper carbon.] (Sections 9.10 through 9.15)

Treatment	Matrix	High-Carbon Phase
1. Cast in a large section, slowly cooled in the mold to 1300°F (704°C), then air-cooled	_____	_____
2. Cast in a thin section and air-cooled from 1500°F (816°C)	_____	_____
3. Cast in a large section after treatment with magnesium and air-cooled from 1500°F (816°C)	_____	_____

9.17 *[EJ]* Explain why it takes so long for carbide to appear in austempered ductile cast iron. [*Hint:* The austenite is alloyed with silicon, and carbide stability is important.] (Sections 9.10 through 9.15)

9.18 *[EJ]* Not only is gray cast iron one of the most inexpensive of the iron base alloys, but it also offers other advantages. What are some of these other desirable characteristics? (Sections 9.10 through 9.15)

9.19 *[EJ]* Evaluate the following statement: Nodular cast iron can be cast to a rough shape, and then the final component dimensions can be achieved by a forming operation such as forging or rolling. (Sections 9.10 through 9.15)

9.20 *[ES]* Determine the phases present, the phase analyses, and the amounts of phases at 70°F (21°C) for the following materials. (Sections 9.10 through 9.15 and Chapter 8)
   a. SAE 1080 steel, annealed (1600°F or 871°C) 5 hr, slow-cooled
   b. Martensitic gray iron (3% C), quenched from 1500°F (816°C) (fully martensitic)
   c. White cast iron (3% C) (pearlite plus massive carbide)
   d. Ductile cast iron (3.5% C, 2.0% Si), [ferrite plus nodular (spheroidal) graphite]

9.21 *[EJ]* A machinist says that gray cast iron is easier to machine in thinner than in thicker sections. Explain. [*Hint:* Recall that nucleation rates change with the cooling rate.] (Sections 9.10 through 9.15)

9.22 *[EJ]* Below are several applications for metal components and, for each, a choice of iron-based alloys. Determine the alloy that would be most likely to be specified for each application. There may be more than one appropriate alloy among the choices, but you are to select only one. When making your choice, consider such factors as strength, ductility, weight, cost, corrosion, conductivity, and method of fabrication. [*Note:* There may be other alloys that could be used but are not included in the choices.] (Chapters 8 and 9)
   a. Painted wastebasket. (Alloys: 1010 steel, 1080 steel, quenched and tempered 4345 steel, 430 stainless steel)
   b. Lawn mower blade. (Alloys: class 40 gray cast iron, pearlitic ductile cast iron, 1010 steel, 1080 steel)
   c. Outside door handle on a commercial building. (Alloys: 1040 steel, 120-90-02 ductile cast iron, 301 stainless steel, A633 HSLA steel)
   d. Cold chisel. (Alloys: quenched and tempered 1040 steel, white cast iron, 440C stainless steel, D1 tool steel)
   e. Paper clip. (Alloys: 430 stainless steel, 1010 steel, A633 HSLA steel, W1 tool steel)

9.23 *[ES/EJ]* A 1020 steel bolt is inserted in a 304 stainless steel tube, as shown in the accompanying figure, and the nut is hand-tightened to give negligible stress in the bolt at 68°F (20°C). The following table gives diameters and mechanical properties.

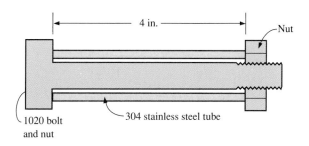

	Coefficient of Expansion, 10^{-6} in./in. °F	Elastic Modulus, 10^6 psi
Bolt (1020: 0.500 in. thrd. root dia.)	7.0	30
Tube (304: 0.800 in. OD × 0.600 in. ID	11.0	28

a. Calculate the stress in the tube and bolt if the assembly is heated up to 200°F (93°C).

b. In another experiment, the assembly is unfastened and then tightened lightly upon reaching equilibrium at 1368°F (742°C). It is then heated at 10°F (5.5°C) per hour until it reaches 1500°F (815°C). Compare the stress level qualitatively with the answer to part *a* and explain any differences. [See also Problem 8.9] (Chapters 8 and 9; Section 3.3)

9.24 *[ES/EJ]* The figure gives several schematic heating and cooling curves that would be useful in obtaining the given microstructures. Indicate *by letter* the correct heat-treatment cycle for each microstructure. Do not worry about actual times and temperatures; the curves are relative and more than one microstructure may be obtainable from the same type of cycle. [*Note:* A vertical line means quench.] (Chapters 3, 5, 8 and 9)

(1) bainite in plain-carbon steel
(2) martensite
(3) tempered martensite
(4) age hardening
(5) normalizing a plain-carbon steel
(6) ferritic malleable cast iron
(7) recrystallization of cold-worked steel

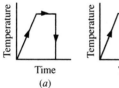

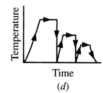

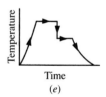

Time   Time   Time   Time   Time
(a)   (b)   (c)   (d)   (e)

PART III

Ceramic Materials

Ceramics are not just pottery any more! For a long time engineers have been aware that ceramics include the highest-strength, hardest, most wear- and creep-resistant materials. These properties originate in the strong covalent and ionic bonds characteristic of ceramic materials.

At the same time, however, this type of bonding does not allow the extensive plastic deformation (ductility) of the metallic bond. Ceramics are therefore generally considered brittle and subject to catastrophic failure. For example, although ceramic blades would enable aircraft gas turbines to operate at higher temperatures and therefore higher efficiencies, they cannot pass the "bird test" that simulates the accidental ingestion and impact of a bird or other foreign object.

Great progress has been made in the use of ceramics in many engineering applications. For example, they are currently used in the tiles on the space shuttle, hard machine tools, superconductors, optical fibers, and selected structural components such as turbine stators.

It is also important for the engineer to understand such traditional ceramics as those widely used in construction—brick, pipe, concrete, and glass—and to be aware of the potential of the new engineering ceramics that are under development and are already finding wide use in composites in glass, carbon, and boron fibers.

We believe the most understandable way to present these two aspects of ceramics is to begin with traditional ceramics, with which you are more familiar, in Chapter 10 and then take up the advanced ceramics in Chapter 11. These two chapters focus on the relationship among structures, properties, and applications. In Chapter 12 we take up processing in the same order because many of the processes and principles for traditional materials are used for the advanced ceramics along with some important innovations. We shall see that processing concepts are even more important for ceramics than for metals, because variations can affect strength and toughness by over 100%, as well as the dimensions of the product.

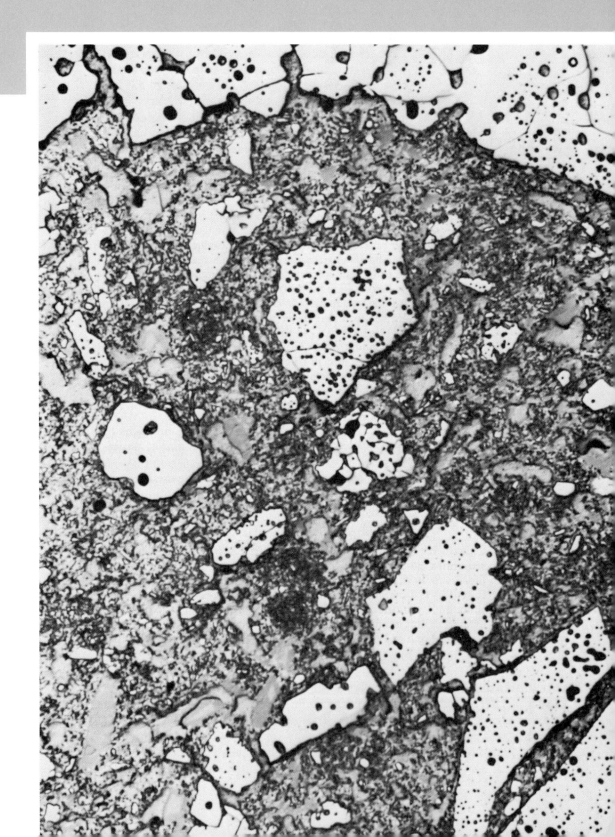

CHAPTER 10

Conventional Ceramic and Glass Structures

In this chapter we will concentrate on traditional ceramics such as brick. Advanced structural ceramics will be discussed in the next chapter.

This illustration shows a typical 90% alumina refractory brick, which is used in critical applications such as ladles for liquid steel. The structure consists of coarse synthetic alumina (white) bonded by a glassy alumina silicate matrix. Magnification 100 ×.

GLASS AND GLASS PROCESSING

10.1 Overview

At a recent annual meeting of the American Ceramic Society, the book of abstracts alone contained 377 pages! We use the divisions into which these abstracts were broken down as the subjects for our three chapters on ceramics.

Traditional Ceramics	New Ceramics	Processes
Glass (traditional)	Engineering ceramics	Traditional ceramics
Structural clay products	Electronics	New ceramics
Whitewares	Nuclear	
Refractories	Basic science	
Cements	Glasses (new)	

Although these topics seem very diverse, there are certain unifying concepts. We will explain these in this overview and then discuss them in detail in the chapters ahead. We have placed the discussion of traditional materials first because some of the concepts are easier to illustrate with familiar materials. For example, many of the basic principles that guide the fabrication of inexpensive fiberglass insulation are also important in producing a sophisticated optical fiber.

The path we will take in this chapter and the concepts to be discussed are as follows:

We will begin with a discussion of glass itself, because a glassy phase is used as a bond for crystalline phases in most ceramics. Also, it is better to include the production of glass here because the process is very different from that used to produce other ceramics and is intimately related to the product. To illustrate this, compare glass processing with ceramic processing.

	Starting Materials	Processing
Glass	Silica Calcium oxide Sodium oxide, etc.	*Melt*, then cast, blow, or roll
Ceramic 1	Clay base	Add H_2O, press solids, then fire in furnace to sinter for glassy bond
Ceramic 2	Cement base	Add H_2O, then allow to set (hydraulic-bond) at room temperature
Ceramic 3	New ceramics	In addition to the processes listed for ceramics 1 and 2, special powder preparation, forming of compacts, gaseous reaction, vapor deposition, and other methods

After discussing glass we take up traditional ceramics.

To understand and use these materials in components, we must develop a new way of thinking about them compared to the way we think about metals. Instead of involving plastic flow on slip planes, as in metals, brittle fracture occurs on well-defined *cleavage* planes (see also Chapter 2) or by irregular fracture, as in glass. And with the exception of glass, we do not melt and cast but instead press, cement, or sinter the grains of a ceramic into a shape. Because this product is usually hard and unmachinable, a good deal of control is needed to attain the desired properties in the final shape. Finally, we must recognize that the structure we begin with is often found in the final part, along with new structures formed during firing.

For these reasons it is important to begin by investigating the materials we have as building blocks, the *starting structures.* These may occur in nature as minerals, or they may be produced as synthetic materials. In general, in the traditional materials such as brick and whiteware covered in this chapter, natural, lower-price starting materials are involved, whereas in the high-strength products discussed in the next chapter, high-purity synthetics are needed. We have already discussed the unit cells of ceramic structures in simple fashion in Chapter 1. We reserve a more extensive discussion for Chapter 20, in which electrical properties are related to more complex structures.

10.2 Glass Products in General

Glass has certain properties quite different from those of metals and alloys. These properties are very useful in processing. Recall that, in the case of a pure metal, when the liquid cools to the freezing point a crystalline solid precipitates. With a glassy material, however, as the liquid cools it becomes more and more viscous, turns to a soft plastic solid, and finally becomes hard and brittle. There is no sharp melting or freezing point. When the specific volume (1/density) of a glass is plotted as a function of temperature, there is a point where the curve changes slope (Figure 10.1). This is the temperature at which the material becomes more like a solid than a liquid. It is called the *glass transition temperature.*

To express these relations in more detail, let us review a typical curve of glass viscosity versus temperature (Figure 10.2). Four important levels are defined as the glass cools:

1. *Working point*, viscosity = 10^4 poises (10^3 N · s/m^2). In this temperature range the glass is readily drawn or pressed.
2. *Softening point*, viscosity = 10^8 poises (10^7 N · s/m^2). At this temperature the glass still deforms under its own weight.
3. *Anneal point*, viscosity = 10^{13} poises (10^{12} N · s/m^2). Above this point the glass still creeps, and stress can be relieved through the conversion of elastic strain to plastic strain.
4. *Strain point*, viscosity = $10^{14.5}$ poises ($10^{13.5}$ N · s/m^2). Below this temperature the behavior of the glass is essentially elastic. No permanent plastic deformation can occur without fracture.

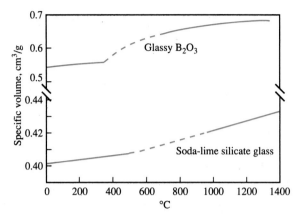

FIGURE 10.1 Relationship of the temperature of two glasses to their specific volume. The point at which the curve changes slope is called the glass transition temperature.

[After R. H. Doremus, *Glass Science*, John Wiley, New York, 1973, Fig. 1, p. 115]

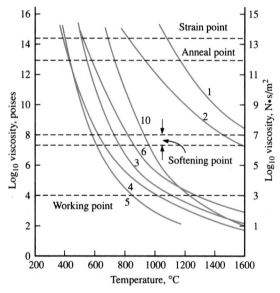

FIGURE 10.2 Change in viscosity as a function of temperature for several varieties of glass. The curves are numbered to correspond to the chemical compositions of glasses given in Table 10.1.

[O. H. Wyatt and D. Dew-Hughes, *Metals, Ceramics, and Polymers*, Cambridge University Press, 1974, p. 259. Reprinted by permission.]

Although Figure 10.2 indicates exact points for working a glass (the working point), we can use higher or lower viscosities for specific processes. For example, rapid processing, such as blowing a glass shape, may require lower viscosity (or higher fluidity, which is the inverse of viscosity). Conversely, slower processing, such as pressing, can be done at higher viscosity and therefore lower temperatures.

The exact viscosity requirement for a specific process is determined by experience. It is related to temperature by an equation not unlike the diffusion relationship:

$$\text{Fluidity} = \frac{1}{\text{viscosity}} = Ae^{-B/T} \tag{10-1}$$

where A and B are constants and T is absolute temperature in kelvin.

Figure 10.2 and Table 10.1 show compositions of different types of glass and their viscosity–temperature relationships. The figure demonstrates the important influence of the additions. If we compare 96% silica glass (Vycor; item 2 in the table) with fused silica (item 1), we see that the softening point of Vycor is about 200°C (360°F) lower. Moving to window glass, we see that the softening point is about 750°C (1382°F), compared with 1400°C (2552°F) for the Vycor. As a result, window glass can easily be rolled into large plates, whereas Vycor is available only in smaller shapes and is far more costly.

EXAMPLE 10.1 *[ES]*

A particular glass is known to have a viscosity of $10^{1.7}$ N · s/m^2 at 1000°C (1273 K) and 10^3 N · s/m^2 at 835°C (1108 K). It is necessary to limit the viscosity of a particular forming operation to 2×10^2 N · s/m^2 in order to ensure that failure does not occur during forming. What is the minimum temperature that can be used in the forming operation?

Answer We find the temperature by first deriving a general relationship between viscosity and temperature using Equation 10-1:

$$\frac{1}{10^{1.7}} = Ae^{-B/1273} \quad \text{and} \quad \frac{1}{10^3} = Ae^{-B/1108}$$

Dividing the first equation by the second, we get

$$10^{1.3} = e^{-B(1/1273 - 1/1108)} \quad \text{or} \quad \ln 10^{1.3} = -B\left(\frac{1}{1273} - \frac{1}{1108}\right)$$

from which we find that $B = 2.56 \times 10^4$. Substituting back into the first equation gives

$$\frac{1}{10^{1.7}} = Ae^{-2.56\times10^4/1273} \quad \text{and} \quad A = 1.08 \times 10^7$$

Therefore,

$$\frac{1}{\text{viscosity}} = 1.08 \times 10^7 \, e^{-2.56\times10^4/T}$$

Substituting to find our unknown temperature yields

$$\frac{1}{2 \times 10^2} = 1.08 \times 10^7 \, e^{-2.56\times10^4/T} \quad \text{and} \quad T = 1191 \text{ K, or } 918°C$$

TABLE 10.1 Composition of Some Glasses

Glass	SiO_2	Na_2O	K_2O	CaO	MgO	BaO	PbO	B_2O_3	Al_2O_3	Remarks
1 [Fused] silica	99.5+									Difficult to melt and fabricate but usable to 1000°C (1832°F). Very low expansion and high thermal shock resistance.
2 (Vycor) 96% silica	96.3	<0.2	<0.2					2.9	0.4	Fabricate from relatively soft borosilicate glass (high in B_2O_3); heat to separate SiO_2 and B_2O_3 phases; acid-leach B_2O_3 phase; heat to consolidate pores.
3 Soda-lime: plate glass	71–73	12–14		10–12	1–4				0.5 to 1.5	Easily fabricated. Used in slightly varying grades, for windows, containers, and electric bulbs.
4 Lead silicate: electrical	63	7.6	6	0.3	0.2		21	0.2	0.6	Readily melted and fabricated with good electrical properties. High lead absorbs x rays; high refractive index used in achromatic lens. Decorative crystal glass.
5 high-lead	35		7.2				58			
6 Borosilicate: low expansion (Pyrex)	80.5	3.8	0.4					12.9	2.2	Low expansion, good thermal shock resistance, chemical stability. Used in chemical industry.
7 low electrical loss	70		0.5				1.2	28	1.1	Low dielectric loss.
8 Aluminoborosilicate: standard (apparatus)	74.7	6.4	0.5	0.9		2.2		9.6	5.6	Increased alumina, lower boric oxide improves chemical durability.
9 low alkali (E-glass)	54.5	0.5		22				8.5	14.5	Widely used for fibers in glass resin composites
10 Aluminosilicate	57	1.0		5.5	12			4	20.5	High-temperature strength, low expansion
11 Glass-ceramic (low expansion) plus 4.5 TiO_2 2.6 Li_2O	65				10				18	Crystalline ceramic made by devitrifying glass. Easy fabrication (as glass), good properties. Various glasses and catalysts.

From O. H. Wyatt and D. Dew-Hughes, *Metals, Ceramics, and Polymers*, Cambridge University Press, 1974, p. 261; and W. D. Kingery, *Introduction to Ceramics*, 2nd ed. John Wiley, New York, 1976 p. 372. Reprinted by permission.

The reason for glassy behavior is related to the structure of the material. If we fuse pure silica (SiO_2), upon cooling it forms a glass called *vitreous silica*. This glass is very useful for chemical glassware because it resists thermal shock. The basic unit of the structure is the silica tetrahedron (Figure 10.3), composed of a silicon nucleus (valence = 4+) surrounded by four equidistant oxygen atoms. Because each oxygen atom has a valence of 2−, the charge is shared with adjacent SiO_4^{4-} tetrahedra (Figure 10.4), producing a network in space of chains of silica tetrahedra. At high temperatures these chains slide easily past each other because of the thermal vibrations. However, as the melt cools, the structure becomes rigid. Note that silica, SiO_2, is found in nature in the nonglassy crystalline state also shown in Figure 10.4. The most common form of silica is the mineral quartz, found in sandstone and silica sand.

Silica is the most important constituent of glass, but other oxides are added to lower the melting point in order to simplify processing or to change the physical characteristics, such as the index of refraction for optical or

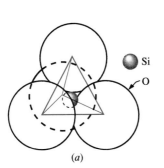

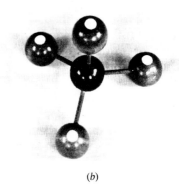

(a)	(b)

FIGURE 10.3   Silica tetrahedron. (a) Sketch. (b) Photograph of a model of a silica tetrahedron. Light-colored dots indicate unsatisfied oxygens to which other ions or molecules may attach.

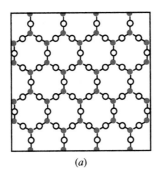

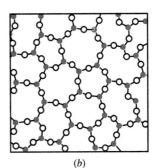

(a)	(b)

FIGURE 10.4   Two-dimensional sketch of a silica network in (a) crystalline and (b) low-order glass form

[After Zachariasen and Warren]

decorative glass. There are three major constituent groups of oxides that may be found in glass:

1. Other *glass formers*, such as boron oxide, B_2O_3. The valence of the metal ion is usually 3 or greater, and the ion is small.
2. *Modifiers*, or oxides of low-valence elements such as sodium and potassium. They tend to break up the continuity of the chains, but they can be added only in limited amounts (Figure 10.5). The addition of such oxides leads to lower melting temperatures and simplifies processing.
3. The intermediate oxides, or *intermediates*. These oxides do not form glasses by themselves but join the silica chain to maintain a glass (Figure 10.6). An example is lead oxide. It is added in large amounts (up to 60%), producing an ornamental glass of great brilliance. Lead oxide can both become a part of the chain and modify the structure at internal positions. This explains why we can add large quantities and still maintain a glassy structure, as shown in Fig. 10.6.

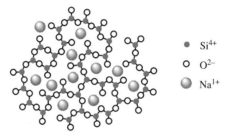

FIGURE 10.5 The effect of a modifier (Na^+) in breaking up the continuity of silica glass

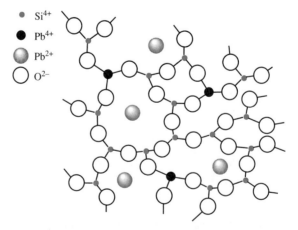

FIGURE 10.6 Lead oxide, considered an intermediate oxide, occurs both in the silica network (as Pb^{4+}) and as a modifier in internal positions (as Pb^{2+}).

Before we continue with our discussion of glass ceramics, let us digress a moment to point out that glass—or what might more appropriately be called amorphous material—can also exist in metal and polymer systems. From earlier discussion we know that, in order to inhibit the crystalline form and retain the amorphous state at room temperature, it is only necessary to cool rapidly enough from the liquid region.

In the ceramic materials the complex ion combinations at unit cell sites slow the crystallization rate. This effect is chemistry-dependent, so it should not surprise you to learn that modifiers and intermediates are found in ceramic materials. Furthermore, time–temperature–transformation (TTT) curves exist for predicting the cooling rates necessary to retain the glassy state and the temperatures required to promote isothermal conversion to the crystalline form.

Chapter 1 explores how amorphous metals and polymers can be produced.

10.3 Glass Processing and Glass Products

We will now continue our discussion of glass by considering its processing as a natural extension of its physical characteristics. Because of the unique properties of glass, it can be cast, rolled, drawn, and pressed like a metal. In addition, it can be blown.

Casting is accomplished by pouring the liquid into a mold. A famous case is the pouring of a 200-in.(5.08-m)-diameter telescope disk by Corning Glass Works. Under normal circumstances, if a disk of this size were to reach the elastic range with temperature gradients present, the material would crack later on as these temperature gradients equalized. Also, as material was removed in grinding and polishing, the surface would distort. Therefore, the disk was cooled over a very long period of time to avoid temperature differences in the mass.

Rolling is widely used to produce window glass and plate glass. The raw materials are melted at one end of a large furnace called a *tank furnace,* and the liquid flows to the other end over a period of time long enough to allow bubbles to float out. The temperature at the end where the rolls are located is controlled so that the glass is of the right viscosity to be rolled into a sheet. The sheet then passes through a long annealing furnace called a *lehr,* where residual stresses are removed. In its original form the rolled material can be used for ordinary window glass, but extensive grinding and polishing are required to produce plate glass.

A novel method for forming plate glass has become quite important (Figure 10.7). In this method the glass flows from the melting furnace onto a float bath of liquid tin, which is covered by a refractory roof. A controlled, heated atmosphere is maintained to prevent oxidation. In the float chamber both surfaces of the glass become mirror-smooth. The sheet then passes into the annealing furnace, which has smooth rollers that will not harm the

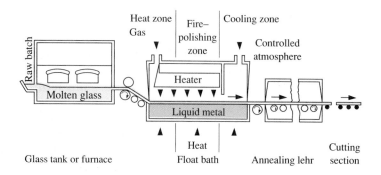

FIGURE 10.7   Method of making "float" glass (Pilkington process)

surface finish. This process has replaced the costly grinding and polishing operations of the old plate-glass roller method.

Centrifugal casting is used to make the funnels at the back sides of television tubes. A gob of glass is dropped into a metal mold. The mold is then rotated so that the glass rises by centrifugal force. The upper edge is trimmed, and then the glass faceplate is sealed to the funnel through the use of a special low-melting-point solder glass.

Drawing of glass tubing is similar to rolling. Glass of the proper viscosity flows directly from the melting furnace around a ceramic tube or mandrel pulled by refractory-covered rollers. Air blowing through the mandrel keeps the tube from collapsing after it passes the mandrel. Annealing is necessary, as it is for plate glass.

Pressing is accomplished by metering a gob of glass into a metal mold, compressing it, and removing it for annealing.

The *press-and-blow* method shown in Figure 10.8 is widely used to make containers. A gob is fed into a mold and pressed. Then the bottom half of the mold is removed and a mold of the final shape is substituted. The blow operation gives the desired contour. The partly formed glass is called a *parison*.

The fiber for fiberglass is now a very important product. One method of making it involves remelting glass marbles and directing the flow of molten glass through a heated platinum plate with orifices to produce filaments. Traction is provided by rotating the winding tube at surface speeds of up to 12,000 ft/min (61 m/sec). Sizing material is applied to separate and lubricate the fibers as they are wound. The bonding of glass fibers with plastic is discussed in Chapter 16.

The specifications for glass products vary widely, depending on the end use. For window and plate glass the chief requirements are flatness, transparency, and freedom from bubbles and harmful stresses that may cause not only breakage but also distortion. For containers, accuracy of volume is usually important. In chemical ware, compositions that will not corrode are necessary. When thermal shock is a consideration, the coefficient of expansion is important and thus pure silica or high-silica glass is specified. In optical

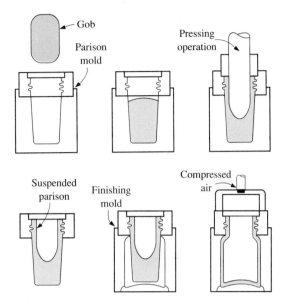

FIGURE 10.8   Press-and-blow method for forming a wide-mouthed container

[W. D. Kingery, *Introduction to Ceramics*, John Wiley, New York, 1960, Fig. 3.25, p. 67. By permission of John Wiley & Sons, Inc.]

glasses the index of refraction is most important, (Chapter 20) whereas in the electrical industry the dielectric constant is of great importance (Chapter 20).

The current interest in recycling has been readily applied to glass products. Because of the fabrication methods, it is advantageous to begin with a "prealloyed" glass—that is, one with a controlled chemistry and viscosity. Rather than starting with raw silica and adding other materials to reduce the softening point, manufacturers need only remelt recycled glass and reform it to the necessary shape. This can result in a considerable saving.

10.4   Glass Ceramics

This topic fits naturally between our discussions of glass and of ceramics. Though developed for advanced engineering uses in spacecraft, glass ceramics are now commonly used in the household under such trade names as *Pyroceram*. They are noted for resistance to thermal shock and for good mechanical strength. Let us see what structural characteristics lead to these properties and how that structure is produced.

The frontispiece of Chapter 5 shows a mixture of long, blade-like crystals (high in titanium oxide) and a background of fine grains (lithium aluminum silicate, LAS). The resistance to thermal shock is due to the low coefficient of expansion of the LAS phase, which is actually negative parallel to the c-axis of the crystal to slightly positive in other directions. The good mechanical properties are due to the fine, uniform grain size and to the lack of porosity.

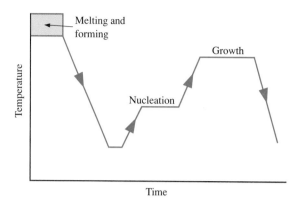

FIGURE 10.9   Schematic temperature–time cycle for the controlled crystallization of a glass–ceramic body

[*Introduction to Ceramics*, 2nd ed., by W. D. Kingery, H. K. Bowen and D. R. Uhlmann, p. 369, Copyright © 1976 John Wiley & Sons, Inc. Reprinted by permission of John Wiley & Sons, Inc.]

This type of material has a unique advantage: It can be processed as a glass, and then, after attaining the desired shape, the crystalline structure can be produced by heat treatment (Figure 10.9).

After melting and forming, the part is cooled and (if desired) trimmed and chemically treated or glazed. Next it is heated to nucleation temperature (typically 800°C, 1472°F) and then to above 1000°C (1832°F) for more rapid growth. Postcrystallization treatments to strengthen the surface via cladding or "stuffing" by ion exchange (see the Problems) can also be added.

Finally, note that recent electron microscopic investigation has shown that separation of a second glassy phase can be produced by the heat treatment, offering new combinations of structures.

10.5   Traditional Ceramics in General

Before discussing traditional ceramic materials and their applications, we need to take a look at the common structures on which they depend: silica, silicates, and other minerals. These are listed in Table 10.2. (The engineering ceramics are listed in Table 11.1) In some cases, such as alumina, a structure is used in both traditional and new structural ceramics. In the latter case, the structure is refined further, as is discussed in the chapter on processing. For general interest, the occurrence of any structure as a natural mineral is noted.

We should also review several concepts introduced in Chapter 1 before proceeding to the discussion of specific traditional ceramics. Although we could refer to the complex unit cells of ceramics, it is easier to consider the coordination number and the use of interstitial sites within a close-packed structure such as an FCC.

Furthermore, we must realize that solid solutions are just as significant in ceramics as they are in metals, although the "rules of solid solution" are slightly different. Also, the occurrence of lattice vacancies can have a pro-

TABLE 10.2 Traditional Ceramics

Oxides	Structure	Natural (N) Synthetic (S)	Traditional	Structural	Electrical	Optical	Magnetic	Composite
Silica, SiO_2 cristobalite	Cubic	N	X					
tridymite	Complex	N	X					
quartz	Complex	N	X					
vitreous silica	Glass	S	X	X		X		X
Silicates								
Kaolinite and other clays	Hexagonal	N	X					
mullite	Complex	N, S	X	X				X
Spinel	Cubic	N, S	X	X	X		X	
Alumina, Al_2O_3	Hexagonal	N, S	X	X		X		
Magnesia, MgO	Cubic	N, S	X					
Dolomite, $MgO \cdot CaO$	Complex	N	X					
Olivine, $MgO \cdot SiO_2$	Complex	N	X					
Cordierite, $MgO \cdot Al_2O_3 \cdot SiO_2$	Complex	N, S	X					
Zircon, $ZrO_2 \cdot SiO_2$	Cubic	N	X					
Zirconia, ZrO_2 (CaO stabilized)	Cubic	N, S		X				
Graphite	Hexagonal	N, S	X	X	X			

465

nounced influence on the mechanical properties of the ceramic, just as they did with metals.

Therefore, let us consider these topics in greater detail.

10.6 Coordination Number; Interstitial Sites

The ionic radii of the positive and negative ions result in very specific types of ion packing. This packing can be treated quantitatively via the *coordination number* relationship. Furthermore, there exist in ceramic unit cells *interstitial sites* that result in more complex structures. Ionic radii, as given in Table 1.6, become very important in the discussion of coordination number and interstitial sites.

Note that in general the negative (nonmetallic) ions have the larger radii. This is because the positive (metallic) ions give up outer electrons, and the ionic radius therefore becomes smaller than the atomic radius. In contrast, negative ions add electrons to the outer shell. This reduces the force of the nucleus on the individual electrons, so the ionic radius is greater than the atomic radius. See, for example, Fe^{2+} versus Fe^0 and Cl^- versus Cl^0.

Coordination Number (CN)

To achieve a stable grouping of larger ions around a smaller ion—that is, to achieve a given CN, the smaller ion must touch all the larger ones. The greater the ratio of the radius r of the smaller ion to the radius R of the larger (the closer the ratio is to 1), the greater the number of ions that can be grouped around the smaller ion. In the unit cells found in nature the CN can equal 2, 3, 4, 6, 8, or 12. Figure 10.10 illustrates the minimum radius ratio, r/R, for each of these groupings.

EXAMPLE 10.2 *[ES]*

Calculate the minimum r/R for CN = 3.

Answer   Note that in Figure 10.10*b* the smaller ion has just reached a radius at which it is touching all three larger ions. We can calculate r/R by geometry as follows. From the figure on the next page we see that

$$AD = r + R \quad \text{and} \quad \tfrac{1}{2}AC = R$$

Because *ABC* is an equilateral triangle (each angle = 60°) and the smaller ion is at the center, α must equal 30°.

Then

$$\cos 30° = \frac{\tfrac{1}{2}AC}{AD}$$

$$0.866 = \frac{R}{r + R}$$

$$r/R = 0.155$$

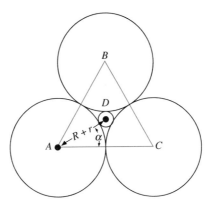

We can calculate the minimum radius ratios for other atomic configurations in a similar way, beginning by finding $r + R$ as a function of R from the geometry of the figure.

When we consider the effect of the radius ratio on the shape of the unit cell, we find that in boron nitride, for example, we have a simple triangle of larger N^{3-} ions with a Ba^{3+} ion in the center (CN = 3). If we formed a tetrahedron with four touching N^{3-} ions (CN = 4), the Ba^{3+} ion would rattle around. Therefore the configuration would be unstable ($R_{N^{3-}} = 1.17$ Å, $r_{B^{3+}} = 0.2$ Å).

When we consider the radius ratio of silicon to oxygen, we find that the important SiO_4^{4-} complex ion is made up of a silicon atom at the center of four oxygen atoms. Another electronically balanced compound is ZnS. (In general practice the words "ion" and "atom" are used interchangeably, just as they are in discussions of metal structures. However, we must use the *ionic* radius in calculations for ceramic structures.)

Now let us look specifically at the ionic radius ratios for sodium chloride and cesium chloride, which have CN's of 6 and 8, respectively. For sodium chloride we have 0.98 Å/1.81 Å = 0.54, which is in the range 0.732 to 0.414 (CN = 6). For cesium chloride we have 1.65 Å/1.81 Å = 0.91, which is in the range 1 to 0.732 (CN = 8). The sodium chloride coordination is described as octahedral. Although the Cl^- ions form an octahedral figure around the Na^{+}, there are only six ions, not eight, and the CN = 6.

Let us sum up the effect of CN on the unit cell structure. For a given CN the radius ratio is usually large enough so that the smaller atom touches all the larger atoms indicated by the CN. For example, for CN = 8, r/R should be *greater* than 0.732. However, we may have a structure in which the CN is lower than that predicted by the radius ratio. In such a case the smaller atom

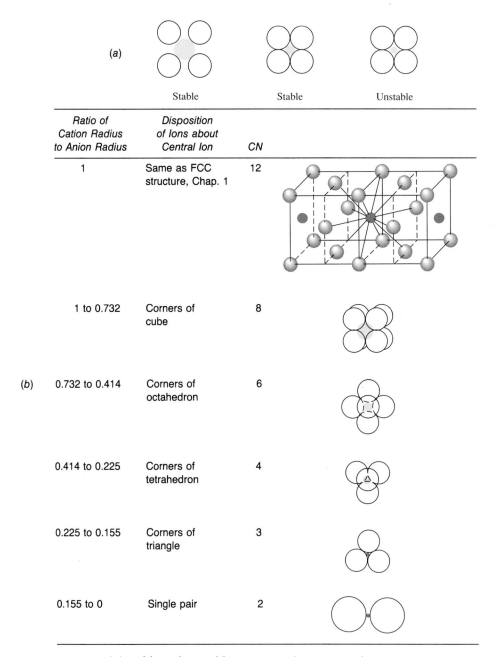

Ratio of Cation Radius to Anion Radius	Disposition of Ions about Central Ion	CN	
1	Same as FCC structure, Chap. 1	12	
1 to 0.732	Corners of cube	8	
0.732 to 0.414	Corners of octahedron	6	
0.414 to 0.225	Corners of tetrahedron	4	
0.225 to 0.155	Corners of triangle	3	
0.155 to 0	Single pair	2	

FIGURE 10.10 (*a*) Stable and unstable ionic coordination configurations. Instability arises when the ion (a cation in this instance) "rattles around" between the anions. (*b*) Radius ratios for various atom arrangements in ionic bonding. The ratios are cation/anion (r^+/R^-).

[Barrett, Nix, Tetelman, *Principles of Engineering Materials*, ©1973, p. 340. Reprinted by permission of Prentice-Hall, Inc., Englewood Cliffs, N.J.]

is indeed touching all the larger atoms, but additional larger atoms could be packed around the smaller atom.

EXAMPLE 10.3 *[ES/EJ]*

Compute the coordination number for KCl and compare your computed value with the actual coordination number of 6.

Answer From Table 1.6 we have $r = 1.33$ Å for K^+ and $R = 1.81$ Å for Cl^-. Thus

$$\frac{r^+}{R^-} = \frac{1.33}{1.81} = 0.735$$

Because the lower limit for CN = 8 is 0.732, our calculation does not accurately predict the true CN of 6. Apparently the calculation and the model break down close to the limiting values.

Ionic radii vary to some degree depending on the neighboring ions and on how many of these ions are in contact with the ion in question. The values given in Table 1.6 for the radii are for a coordination number of 6 in the ionic state. The radii can vary in the last significant figure.

A slight change of ± 0.01 Å in ionic radius would change the radius ratio, so the calculation (and thus the prediction of a coordination number) near the limiting values is not foolproof.

Interstitial Sites

While we have before us the models of tetrahedral and octahedral groups of ions around a central ion for CN = 4 and CN = 6, respectively, let us discuss the terms *octahedral site* and *tetrahedral site*. These are voids in the unit cell in which we can place different atoms. We recall that inside the FCC structure of iron we could place a carbon atom at $\frac{1}{2}, \frac{1}{2}, \frac{1}{2}$ (Example 8.1). This is called an octahedral site because the six nearest atoms, the iron atoms at the centers of the faces, form an octahedron around the void (Figure 10.11). The size of the site (or void) is expressed as the diameter of the sphere that would just touch the iron atoms (0.414 times the radius of the iron atom). There are also tetrahedral sites in the same structure (Figure 10.12). An atom at $\frac{1}{4}, \frac{1}{4}, \frac{1}{4}$ is equidistant from four iron atoms, making up a tetrahedron, and $r = 0.225R$. [r = small atom, R = large (iron) atom.]

In a larger region of the space lattice, including adjacent unit cells, we find that there are octahedral sites at the center of each cube edge. On the average there are four octahedral sites per FCC unit cell, or one site for each FCC atom, and eight tetrahedral sites per cell, or two sites per atom. Thus

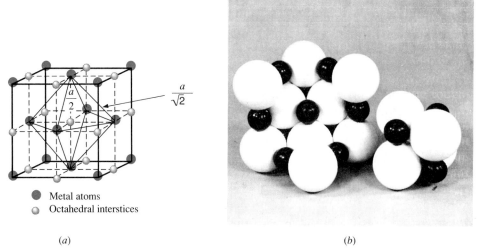

Metal atoms
Octahedral interstices

(a)   (b)

FIGURE 10.11   Octahedral sites (CN = 6). (a) Sites in an FCC structure. (b) Photograph of a hard-sphere model of an FCC structure with octahedral sites filled. Corner atoms have been removed to show the site in the center of the structure.

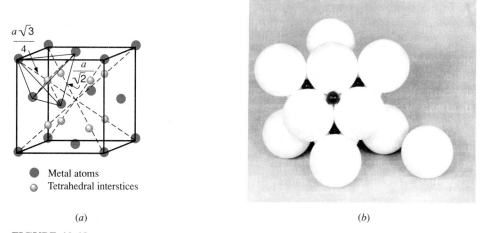

Metal atoms
Tetrahedral interstices

(a)   (b)

FIGURE 10.12   Tetrahedral sites (CN = 4). (a) Sites in an FCC structure. (b) Photograph of a hard-sphere model of an FCC structure, showing only four of the eight tetrahedral sites. The corner atom has been removed. Note the small size of these sites compared to the size of the octahedral sites (Figure 10.11).

close-packed structures such as the FCC have *twice* as many tetrahedral as octahedral sites.

HCP structures also have octahedral and tetrahedral sites in the same ratios of sites to atoms as FCC structures. This concept of sites is very useful in understanding the ferrimagnetic ceramics. We shall see in Chapter 22 that magnetic field strength is directly related to the way in which atoms are placed in these sites.

10.7 Solid Solutions

Up to this point we have discussed only pure compounds. We also encounter solid solutions in ceramics, as in metal systems, although additional principles affect their formation. For example, a series of minerals important in ceramics is that which occurs between pure magnesia, MgO, and iron oxide, FeO. Both of these compounds have the sodium chloride structure (Figure 1.32), and they form a continuous series of solid solutions similar to the copper–nickel system in metals. However, other characteristics, such as similar valence, are equally important to the existence of a continuous series of solid solutions.

EXAMPLE 10.4 *[ES/EJ]*

Olivine sands are usually given the formula $(Mg, Fe)_2SiO_4$, which assumes a complete solid solution of magnesium and iron. Justify the high solid solubility and indicate whether lithium ions might also substitute.

Answer Mg^{2+} (0.78 Å) and Fe^{2+} (0.87 Å) have the same charge and are very close in size, fulfilling the criteria for a high degree of substitutional solid solubility. Although Li^+ (0.78 Å) is also the same size, it does not have the same charge. It is still possible to achieve some solid solution, however, in the following way:

1. $Li^+ + Fe^{3+}$ (0.67 Å) substitute for two Mg^{2+}.
2. Two Li^+ substitute for one $Mg^{2+} + \square$. (The symbol $\square$ signifies a cation vacancy assumed to exist in the olivine structure; see Section 10.8.)

Types of Transformations

As in the metals, in the ceramics we encounter transformations of the nucleation and growth type and of the diffusionless type. When the transformation of a structure involves the breaking of bonds, it is called a *reconstructive transformation*. When only small shifts in position are involved, the transformation is called a *displacive transformation* and is similar to the martensite type of transformation in metals. Examples of displacive transformation are $\alpha \rightarrow \beta$ quartz and cubic $\rightarrow$ tetragonal $BaTiO_3$. When a large volume change accompanies displacive transformation, as in quartz, severe cracking can take place, because there is no ductility. (See also Chapter 5.)

$$O^{2-}\ Fe^{2+}\ O^{2-}\ Fe^{2+}\ O^{2-}\ Fe^{2+}\ O^{2-}\ Fe^{2+}$$
$$Fe^{2+}\ O^{2-}\ Fe^{2+}\ O^{2-}\ Fe^{2+}\ O^{2-}\ Fe^{2+}\ O^{2-}$$
$$O^{2-}\ Fe^{3+}\ O^{2-}\ Fe^{2+}\ O^{2-}\ \blacksquare\ \ O^{2-}\ Fe^{2+}$$
$$Fe^{2+}\ O^{2-}\ \blacksquare\ \ O^{2-}\ Fe^{3+}\ O^{2-}\ Fe^{3+}\ O^{2-}$$
$$O^{2-}\ Fe^{3+}\ O^{2-}\ Fe^{2+}\ O^{2-}\ Fe^{2+}\ O^{2-}\ Fe^{2+}$$
$$Fe^{2+}\ O^{2-}\ Fe^{2+}\ O^{2-}\ Fe^{2+}\ O^{2-}\ Fe^{2+}\ O^{2-}$$

FIGURE 10.13 The defect structure of $Fe_{(1-x)}O$. The symbol $\square$ denotes an ion vacancy. The value of x depends on the fraction of iron ion vacancies.

Also as in metals, the kinetics of reactions in ceramic structures are very important. TTT diagrams exist for ceramics and have been used to determine the crystallization times of glasses. As we would expect, adding "alloying elements" shifts the nose of the diagram, as it does in metals. In general, most reactions are more sluggish in ceramics because several ions must move in combination to maintain electroneutrality, a problem not encountered in metals. These unusual features of ceramics will be discussed further in subsequent sections.

10.8 Defect Structures, Lattice Vacancies

The point defects or vacancies in the lattices that we found in metals also occur in ceramic crystals. Vacancies are especially important when a foreign ion of different valence is dissolved in the standard structure, and some compensation is needed to obtain a balanced charge over the crystal as a whole. A typical case is iron oxide crystals in which both Fe^{2+} and Fe^{3+} are present. The normal FeO structure is like that of NaCl, but if we make up the crystal with some Fe^{3+} ions present, one Fe^{2+} must be absent to balance two Fe^{3+} ions (Figure 10.13).

The symbol $\square$ signifies an ion vacancy where an Fe^{2+} was omitted to balance the charge surplus caused by two Fe^{3+}. Such a vacancy is called a *cation vacancy*. If, on the other hand, we substitute a cation of lower than normal valence, say K^+, then *anion* vacancies are needed to balance the charge. (In this case some O^{2-} would be left out.) Defect structures are used in voltage rectifiers, as we will discuss in Chapter 20.

EXAMPLE 10.5 *[ES]*

In wüstite (Figure 10.13) what fraction of the cation (iron) sites will be vacant if there are 10 Fe^{3+} ions to every 100 Fe^{2+} ions? *Note:* Wüstite does not have a fixed stoichiometry, but it can have cation vacancies (as in the example) or excess O^{2-} ions at interstitial positions.

Answer In a problem of this sort, we must first decide whether the anion or the cation lattice will be perfect. In this case it is the anion (oxygen) lattice that will be perfect, because there are two valency states for the cations.

$$100 \ Fe^{2+} + 10 \ Fe^{3+} = 110 \ \text{cation sites filled}$$

For every two Fe^{3+} there must be one Fe^{2+} vacancy for the charge to balance. Therefore there are five Fe^{2+} vacancies because we have 10 Fe^{3+} ions.

The anion sites are all filled, however, with oxygen, and there are equal numbers of anion and cation sites in FeO.

$$100 \ Fe^{2+} + 10 \ Fe^{3+} + 5 \ \square = 115 \ \text{cation sites total}$$

or

$$\text{Anion sites} = 115 \ \text{(all filled)}$$
$$\text{Percent vacant cation sites} = \tfrac{5}{115} \times 100 = 4.35$$

An alternative method of solution is as follows:

$$100 Fe^{2+} = +200$$
$$10 Fe^{3+} = \underline{+ \ 30}$$
$$+230 \quad \text{(charge to be balanced by } O^{2-})$$

Therefore,

$$\frac{-230}{-2} = 115 \ O^{2-} \ \text{ions required}$$

Stoichiometry in FeO is 1:1, so

$$115 \ \text{anions} = (100 + 10) \ \text{cations} + 5 \ \text{cation vacancies}$$

We cannot select a few common properties as a basis for comparison of ceramics in the same way we did for metals. The tensile test, a common means of comparing metals, is rarely performed on ceramics. In many cases the insulating or optical properties are more important than the mechanical properties. Therefore, we shall discuss some of the properties that are important in ceramics, as we take up the individual materials.

10.9 Silica and Silicate Structures

Silica (SiO_2) in the form of quartz and the mineral silicates, such as clay, are the most abundant starting materials and also the commonest phases in finished ceramic products. Let us therefore examine silica with special care, just as we studied in detail the element iron when we considered ferrous alloys.

An important simplifying principle is that whether we deal with silica itself or silicates such as clay (which is a complex aluminum silicate with other elements), the silicon atom is always at the center of a tetrahedron of four oxygen atoms. Hence we always start with tetrahedral building blocks that have an unsatisfied electron bond at each oxygen (see Figure 10.3).

Let us first concentrate on silica itself, in which all the oxygen bonds are attached to adjacent silicon atoms, giving a *framework* structure in space. Then we will see how the basic tetrahedron is found in the different forms of silica (*silica polymorphs*). This is not merely an academic exercise; these transformations are intimately related to the shattering of furnace roofs and the weakening of "fused quartz" components.

At first glance, Figure 10.14 may be a little intimidating because it shows that silica can exhibit *seven* different structures.* If we review the simpler case in the iron alloys, we recall that austenite can change structure in two ways. Under equilibrium conditions, pearlite is formed by nucleation and growth. In ceramics this is called a *reconstructive* transformation. With fast cooling of steel, however, martensite is formed with only small shear-like movements. This is called a *displacive* transformation, as we noted in Chapter 5 and Section 10.7.

Let us see how these mechanisms operate for cristobalite, the crystalline phase formed when liquid silica is cooled. The cristobalite structure is relatively open (Figure 10.15) and has an FCC array of oxygen atoms. When the

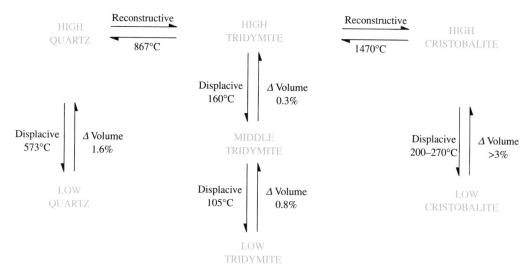

FIGURE 10.14 Transformations and volume changes for SiO_2 polymorphs. Liquid forms at 1710°C (3110°F). All volume changes are negative on cooling.

[Reprinted from D. W. Richerson, *Modern Ceramic Engineering*, 1982, p. 20, courtesy of Marcel Dekker, Inc.]

---

*And there are still other structures; coesite is found in meteors, for example.

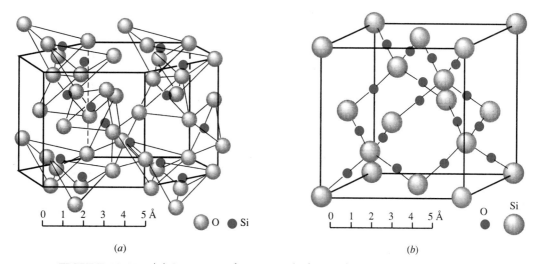

FIGURE 10.15 (a) Structure of α quartz (a form of SiO_2). (b) Structure of β cristobalite (a form of SiO_2).

[Part (a) after W. L. Bragg and Gibbs. Part (b) after Wyckoff.]

cristobalite is cooled slowly to provide equilibrium conditions, a more complex structure, tridymite, is formed by nucleation and growth (a reconstructive transformation). On the other hand, when cristobalite is cooled more rapidly, it transforms to "low" cristobalite via a displacive transformation at 200–270°C (392–518°F) with a large change in volume (over 3%). If we follow the tridymite phase in similar fashion, we see that we have the option of forming "high" quartz by slow cooling or going through two displacive transformations upon faster cooling. Finally, if "high" quartz is formed, a rapid displacive transformation to low quartz (Figure 10.15a) occurs. It should be emphasized that the reconstructive transformations are slow whereas the displacive transformations are not suppressed by fast cooling, just as we found for the austenite–martensite transformation that started at the M_s temperature regardless of cooling rate.

EXAMPLE 10.6 *[ES/EJ]*

Silica bricks are excellent inexpensive refractories for lining furnace roofs where the flames heat the roof to 1500°C (2732°F). If we start with quartz, the brick expands 1.8% by volume upon passing through 573°C (1063°F). This produces a tolerable level of compressive stress. Furthermore, when temperatures around 1500°C are attained, the stress is relieved by plastic flow. Upon cooling through 573°C, however, the high tensile stresses due to transformation shatter the roof.

It was found that the addition of 2% CaO to silica brick increased the rate of transformation to tridymite during firing of the brick. With proper maintenance this eliminated the cracking problem. Explain.

Answer   The change of structure to tridymite eliminated the large change in volume that occurred upon cooling at 573°C (1063°F). However, because there are appreciable volume changes in tridymite at 160°C (320°F) and 105°C (221°F), it is necessary to maintain the roof above this temperature during shut-down periods. This is easily done by "banking the furnaces," using a small amount of fuel.

It is intriguing that we have this complex set of transformations to consider in the simple compound SiO_2. We may well ask whether any general rules of behavior govern these transformations. Indeed, there are two:

1. In transforming to a low-temperature structure from a high-temperature variety, a decrease in volume occurs.
2. The higher-temperature structure is more ordered, more open, and has lower specific gravity. Compare the simpler cristobalite unit cell with quartz (Figure 10.15). The low-temperature structure has a collapsed structure attained by rotating rows of $(SiO_4)^{4-}$ tetrahedra.

The properties of the silicates are also intimately related to their structures. Returning to the basic unit, the silica tetrahedron, we can bond foreign atoms at from one to four of the oxygen atoms. If we place foreign atoms at all four positions, the tetrahedron becomes an "island." If we bond only to

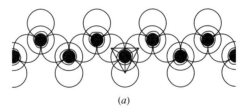

(a)

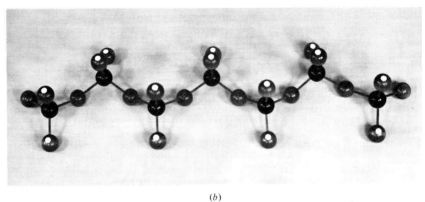

(b)

FIGURE 10.16   A single-chain silicate structure. (a) Sketch. (b) Photograph of silicate model. Oxygen atoms with white dots have bonds that are still unsatisfied.

[Part (a) from Lawrence Van Vlack, *Elements of Materials Science,* © 1964, Addison-Wesley, Publishing Co., Inc., Reading, Massachusetts. Reprinted with permission of the publisher.]

two or three oxygen atoms, the remaining O–Si bonds lead to chain structures (Figure 10.16). If more Si–O bonds are retained, a sheet structure develops (Figure 10.17). The consequences are clear in the properties of the minerals (Figure 10.18). When the chain structure is present we have fibrous cleavage; with the sheet structure we have plate-like cleavage. In the island structure we usually have cleavage in three dimensions.

The clays deserve special attention because of their wide use in traditional ceramics and their abundance. The structure of one of the common

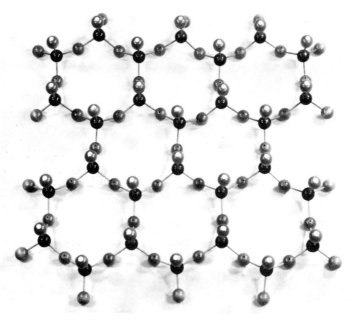

FIGURE 10.17 Photograph of a model of a sheet-structure silicate. Only the fourth (upper) oxygen atoms have unsatisfied bonds (light-colored dots).

FIGURE 10.18 Three types of cleavage in silica minerals. Left: asbestos (one direction). Center: mica (two directions). Right: feldspar (three directions).

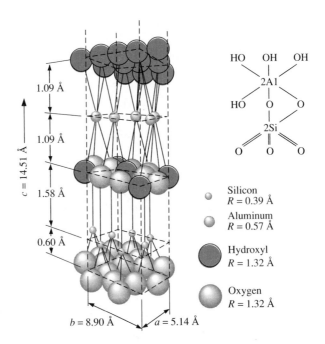

FIGURE 10.19   Sheet structure of kaolinite clay. (Note the planes of potential easy cleavage.)

[After W. E. Hauth]

clays, kaolinite, is shown in Figure 10.19. Note that there are no primary bonds extending vertically from the ends of the unit cell. Therefore, the cleavage develops readily parallel to the hexagonal face. When water is added, the H_2O molecules attach by van der Waals forces and the platelets of clay slide easily over each other, giving excellent moldability.

Alumina (mineral name, corundum) is more expensive than silica, but it has a variety of uses and a wide range of cost. At one end of the spectrum it is used in the "pure" form as jewelry or in lasers: Sapphire and ruby are alumina crystals with very small amounts of impurities. At the other extreme we have alumina brick of different grades, many of which are used as furnace refractories. The structure of alumina is shown in Figure 10.20.

Mullite

Mullite is one of a group of important ceramics composed of various percentages of silica and alumina (see the phase diagram Figure 10.21). When a fire clay (kaolin) is heated, water is given off and other complex changes take place, so above 1595°C (2903°F) the mineral mullite and a liquid are present. This limits the use of alumina–silica refractories with less than 72% Al_2O_3 to temperatures below about 1550°C (2822°F). However, above 72% Al_2O_3 we have mullite, $3Al_2O_3 \cdot 2SiO_2$, or mullite plus corundum, Al_2O_3, with a solidus at 1840°C (3344°F). Therefore, the high-alumina refractories are widely used

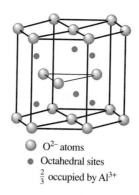

○ O²⁻ atoms
● Octahedral sites
$\frac{2}{3}$ occupied by Al³⁺

FIGURE 10.20   Structure of corundum (alumina)

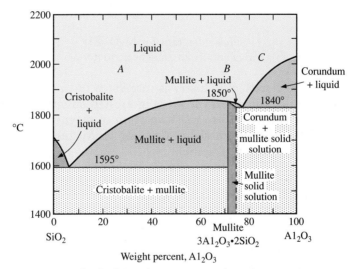

FIGURE 10.21   SiO₂–Al₂O₃ phase diagram. *A* = fire clay composition; *B* = mullite refractory; *C* = high-alumina refractory.

for steel making (1600°C, 3020°F). It should be emphasized that a relatively slight change in Al₂O₃—from 70 to 80%, for example—changes the solidus by 255°C (459°F) in a very important temperature range. For this reason mullite is an important refractory.

Spinels

Spinels, which have the form A²⁺B₂³⁺O₄ (where A and B are metal ions), are used as refractories and in the electrical industry. In refractories the phase MgAl₂O₄ is found in the MgO–Al₂O₃ phase diagram (Figure 10.22) at a 1:1 molar ratio of MgO and Al₂O₃, just as we find mullite in the Al₂O₃–SiO₂ system at a 3:2 molar ratio. The spinel has a greater resistance to thermal shock than MgO. It also has good resistance to most slags.

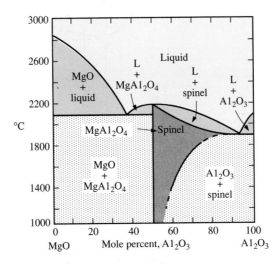

FIGURE 10.22 Phase diagram for $MgO–Al_2O_3$. The spinel-type structure is important in refractories and in ferrimagnetic materials.

The electrical industry is interested in the spinel structure because of its magnetic properties; thus it is interested only in those spinels with transition elements that lead to the formation of magnetic materials such as $CoFe_2O_4$, which will be discussed in Chapter 21. Magnetite, $FeFe_2O_4$, or lodestone, is a naturally occurring spinel.

CERAMIC PRODUCTS

10.10 Brick and Tile

We can now discuss the special features of ceramic products, beginning with brick and tile. For construction brick, a low-cost, easily fused clay is used as the base. This clay contains high-silica, high-alkali, high-FeO, gritty materials found in natural deposits. The bricks are formed by either dry or wet pressing. They are fired at relatively low temperatures.

The specifications are simple, involving only compressive strength ranging from 2000 to 8500 psi (13.8 to 58.6 MPa) depending on grade and dimensional tolerances.

Building tile (unglazed), clay pipe, and drain tile are processed similarly.

10.11 Refractory and Insulating Materials

Furnace and ladle linings use either brick or monolithic (rammed in place) linings. In dealing with liquid metals and slag, it is essential to distinguish among acidic, neutral, and basic refractories. In general, refractories composed of MgO and CaO are called "basic" because the water solutions are basic. Refractories based on SiO_2 form very weak acid solutions. The distinctions are important because acid slags attack basic refractories and basic slags attack acid refractories.

The most important characteristics of refractory bricks are resistance to slag, resistance to temperature effects, and insulating ability. Figure 10.23 shows the relation of thermal conductivity to temperature for various materials.

Acidic bricks are less expensive than basic bricks, but in many furnaces slags high in CaO and MgO are used to refine the metal (remove phosphorus and sulfur in ferrous alloys). These slags react with SiO_2 to form low-melting-point materials that erode the bricks. Therefore, when basic slags are used, the refractories must be basic. In intermediate cases alumina and chromite brick are used. The furnace or ladle linings can be formed from brick or from stiff refractory mud that has been rammed in place and fired with a gas flame or coke fire.

Insulating brick contains a great deal of pore space. It is therefore not so resistant to slag and it is generally used behind the working lining. Insulating brick for lower temperatures has been made of asbestos and plaster mixtures although the toxicity of asbestos limits such use.

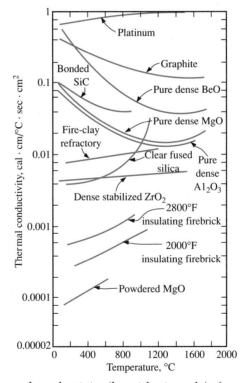

FIGURE 10.23 Thermal conductivity (logarithmic scale) of ceramic materials over a wide range of temperatures

[W. D. Kingery, *Introduction to Ceramics*, John Wiley, New York, 1960, Fig. 14.37, p. 507. By permission of John Wiley & Sons, Inc.]

10.12 Earthenware, Stoneware, China, Ovenware, and Porcelain

Earthenware, stoneware, china, ovenware, and porcelain are used for a variety of products, from simple earthenware shapes to fine china and electrical ceramics. We shall review first the raw materials used to make each product, then the reactions that occur during processing.

Earthenware is made of clay (kaolin, for example). In some cases silica, SiO_2, and feldspar, such as $KAlSi_3O_8$, are present. The important feature of earthenware is that it is fired at a lower temperature than the other products in this group and thus has a relatively porous, earthy fracture. For example, the cup shown in Figure 10.24a is made today by Indians in Chile from clay mined from the nearby hillside and fired in a wood-fired kiln. Although the cup is quite serviceable, it is not high in strength because there has been very little fusion or sintering. Much of the clay "soil pipe" widely used for drains has similar characteristics.

In finer grades of earthenware, called *semivitreous earthenware*, clay–silica–feldspar mixtures are used. Because of the three constituents, these earthenwares are called *triaxial*. The use of a higher firing temperature results in the formation of some glass and gives the earthenware lower porosity and higher strength. All grades may be either unglazed or coated with a separate glassy-surface-forming material that gives a *glazed* surface.

Stoneware differs from earthenware in that a higher firing temperature gives a porosity of less than 5%, compared with 5 to 20% for earthenware (Figure 10.24b). The composition of stoneware is usually controlled more carefully than that of earthenware, and the unglazed product has the matte finish of fine stone. In some variations, such as the well-known Wedgwood jasper stoneware, barium compounds are added. Stoneware is an excellent material for ovenware or chemical tanks and coils. It is essentially unattacked by most acids but is corroded by alkalis.

China is obtained by firing either the triaxial mixture mentioned above or other mixtures to a high temperature in order to obtain a translucent object. (The translucence of a piece of china is an indication of its quality. This is why experts hold a plate toward the light to see how clearly the shadow of a hand comes through.) China is translucent because a large share of the mixture of quartz, clay, and feldspar crystals has been converted to a clear glass. The expression "soft paste" porcelain is sometimes used. The firing temperature of such porcelain is lower than that of "hard" porcelain because a small amount of CaO is present as a flux. English bone china (Figure 10.24c) is different in composition from regular china. It contains about 45% bone ash (from cattle bones) and 25% clay; the rest of the mixture is feldspar and quartz. The calcium phosphate from the bones gives a lower-melting-point material, and the phosphate group substitutes for part of the silica as a glass former. In another well-known china, "Beleek," glass is added to the original mixture as a flux, giving the final product high translucency.

(a)

(b)

(c)

FIGURE 10.24 (a) Earthenware cup made in Chile. Potters obtain the clay from nearby hills and shape it, then bake it in a charcoal-fired kiln. Beneath the glaze, the structure is relatively porous. (b) Stoneware mug made in Annapolis, Md. The clay–silica mix is fired in an electrically heated kiln at 2100°F (1149°C). The product, which is partly fused, is stronger than earthenware. (c) English bone china (Minton) made in England. The mixture of flint, clay, and bone ash is fired at a temperature high enough to produce a glassy phase, giving it a translucent appearance.

The specifications for *ovenware* and *flameware* are interesting. It has been found that ceramics with a coefficient of expansion of about $4 \times 10^{-6} (°C)^{-1}$ $[2.2 \times 10^{-6} (°F)^{-1}]$ withstand heating in an oven and cooling in air, but utensils that are in direct contact with a flame, such as a frying pan, need a coefficient below $2 \times 10^{-6} (°C)^{-1} [1.1 \times 10^{-6} (°F)^{-1}]$. As a result, some triaxial compositions with coefficients of about $4 \times 10^{-6} (°C)^{-1} [2.2 \times 10^{-6} (°F)^{-1}]$ are used in the oven. To obtain the lower coefficients, it is necessary to add Li_2O from such sources as the minerals spodumene or cordierite.

Porcelain, which is closely related to china, is fired at higher temperatures than the other members of the group. In general, the avoidance of fluxes
and the higher temperatures make the product very hard and dense.

Because the clay–silica–feldspar combination is of such great importance
in all these products, the ternary phase diagram containing these constituents
should be reviewed briefly. Figure 10.25 is essentially a contour map showing
the decrease in liquidus temperature as we go from hard porcelain, which is
high in silica and alumina, to dental porcelain, which is high in alkali (K_2O).
The low melting point of the *dental porcelain* makes it possible to fuse most of
an artificial tooth and convert it to a glassy translucent material similar to a
real tooth.

10.13 Large-Scale Ceramics for Electrical Use

Electrical ceramic products may be divided into large insulators, such as
those used for power lines, and small electronic components, such as the capacitors and magnets to be discussed in Chapters 20 and 21.

Large insulators (Figure 10.26) are produced from a triaxial porcelain such
as 60% kaolin, 20% feldspar, and 20% silica, as shown on the ternary diagram in Figure 10.25, through either specialized slip casting or plastic forming. Special glazes are used for three reasons. First, the glaze strengthens the
surface because its coefficient of expansion is lower than that of the porcelain, and thus the glaze is put in compression upon cooling from the glaze
sintering temperature. Second, a semiconducting glaze is used in some cases
to equalize the charge. Finally, a glaze lowers the surface porosity, which is

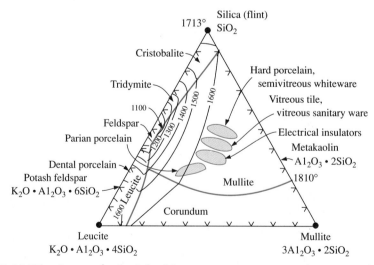

FIGURE 10.25   Areas of triaxial whiteware compositions shown on the silica–
leucite–mullite phase equilibrium diagram

[W. D. Kingery, *Introduction to Ceramics*, John Wiley, New York, 1960, Fig. 13.10, p. 419. By permission of John Wiley & Sons, Inc.]

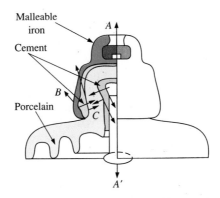

FIGURE 10.26 Cross section of a disk insulator. The mechanical strength of porcelain, like that of most ceramic materials, is greater in compression than in tension. Therefore, disk insulators, which are exposed to tension during service, must be designed to ensure that the main stresses on the porcelain component are of a compressive nature.

[Courtesy of Doulton Insulators, Ltd.]

an advantage when the insulator will be used where water might collect and cause an electrical short.

EXAMPLE 10.7 *[EJ]*

The tensile strength of porcelain is far lower than its compression strength—and also less reliable. Porcelain is needed, however, to insulate high-voltage wires from the towers. The design shown in Figure 10.26 takes advantage of the ductility of malleable iron and the compressive strength of porcelain. If such a design were not used, the porcelain would have to be in tension. Analyze the direction of stresses (tension or compression) at points A, B, and C. The cement bonds firmly to the insulator and may be considered part of it. Why is it used? Why does the ceramic have a large outside diameter and deep sheds, or "petticoats" (ripples), on the bottom?

Answer   The part is designed so that only the metal is in tension and the ceramic is in compression. The use of the cement permits the metal part to be inserted into the insulator first. The wire cable attachment is through A-A' (tension), which also results in tensile stresses in the malleable iron at B because of its apron shape. The knob shape of the insulator as cemented to the malleable iron gives compression in the ceramic insulator, point C, because of the pinching action of the malleable iron apron. The large outside diameter gives an electrical discharge a long path to travel on the surface and therefore prevents discharge. The deep sheds are to avoid build-up of a continuous sheet of water or ice that could cause a short circuit.

10.14 Abrasives

In producing abrasive wheels or papers, the idea is to grip the hard particles firmly enough so that they do not leave the wheel until they have become rounded. This is accomplished by two means: using a soft matrix that is worn away and providing porosity to weaken the support. The porosity is also important in conveying coolant to prevent "burning" of the part while it is being ground by the abrasive wheel. In the case of steel, for example, heavy grinding heats the surface to the austenitic range. The final structure of the steel depends on the cooling rate and the composition; many a carefully heat-treated part has been ruined by lack of coolant during finish grinding, because a brittle fresh martensite layer formed.

About 85% of the abrasive wheels manufactured are synthetic alumina; 15% are silicon carbide. The carbide is harder but more fragile, so for certain applications there is a great deal of competition between the two. The matrix to which the given material is bonded may be either a ceramic glass, polymer resin, or rubber. The softer bonds wear faster, but the wheels can be operated at higher speeds and produced in thinner sections, such as cutoff wheels.

10.15 Cements and Cement Products

In many ceramic shapes the crystalline phases are held together by the glass phase, so we can consider glass a special high-temperature cement. In the field of common cements and plasters, however, a mixture may be formed into a shape at low temperature. Because of interaction with water, a hydraulic bond develops. In general, when water is added to these cements, the existing minerals either decompose or combine with the water, and a new phase grows throughout the mass. Examples are the growth of gypsum crystals in plaster of Paris and the precipitation of a silicate structure from Portland cement. It is usually quite important to control the amount of water to prevent the presence of an excess that would not be part of the structure and would therefore weaken it.

10.16 Portland Cement, High-Alumina Cement, and Other Cements

There are several grades of common cements, as shown in the ternary diagram in Figure 10.27. A typical *Portland cement* contains 19 to 25% SiO_2, 5 to 9% Al_2O_3, 60 to 64% CaO, and 2 to 4% FeO, whereas a high-alumina cement is largely CaO and Al_2O_3. Both types are prepared by grinding a proper mixture of clays and limestone, firing the mixture in a kiln, and then regrinding it.

Several minerals are present in Portland cement: tricalcium silicate C_3S ($3CaO \cdot SiO_2$); dicalcium silicate, C_2S; and tricalcium aluminate, C_3A

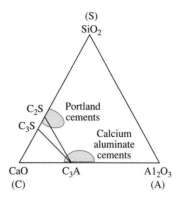

FIGURE 10.27   Cement compositions

(see Figure 10.27). Tetracalcium aluminum ferrite, $4 \, CaO \cdot AL_2O_3 \cdot FeO$ is also present.

The reaction by which cement is formed involves solution, recrystallization, and precipitation of a silicate structure.* To retard setting, 2% gypsum, $CaSO_4 \cdot 2H_2O$, is added. The gypsum reacts with the tricalcium aluminate and reduces shrinkage. The heat of hydration (heat of reaction in the adsorption of water) in setting can be large and can damage massive structures. Low-heat cements are made by reducing the amount of tricalcium aluminate.

High-alumina cement has approximately 35 to 42% CaO, 38 to 48% Al_2O_3, 3 to 11% SiO_2, and 2 to 15% FeO. Setting is produced by the formation of hydrated alumina crystals from the tricalcium aluminate. The cement is quick-setting and attains in 24 hr the same strength that Portland cement attains in 30 days.

The effects of Portland cement and of a lime reaction are combined in the mortar used in bricklaying. When lime, CaO, is mixed with water, it forms hydrated lime, $Ca(OH)_2$. This hydrated lime reacts with oxides of carbon in air to form $CaCO_3$, and it also reacts with sand to form a calcium silicate. The formation of these new crystals causes bonding. A typical mortar is one part Portland cement, two parts lime hydrate, and eight parts sand by volume. One of the most important uses of Portland cement is as a binder in the manufacture of concrete. (Concrete will be considered as a composite material in Chapter 17.)

It should be pointed out that Portland cement does not have a fixed composition. Five different grades of Portland cement are used in the manufacture of concrete. The composition is modified to control the setting time and the heat of hydration. The lower the water-to-cement ratio, the higher the compressive strength, which is considerably greater than the tensile strength. Portland cement sets through the formation of a gel structure.

---

*The silicate structure is a sheet structure of silicate groups with calcium and oxygen ions in the interstices. Water molecules separate the sheets.

A *gel* is a two-phase colloidal mixture of a liquid and a solid. The liquid can be in capillaries or in voids in the solid network. The gel network can be either crystalline or noncrystalline and either rigid or nonrigid. A companion material is a *sol*, which is a suspension of fine solid particles in a liquid. A sol is usually the first stage in the formation of a gel; hence all the members of this class of materials are often referred to as sol-gels.

Portland cement is a rigid gel network that begins as a noncrystalline material and then crystallizes as complex hydrates of the oxides of silicon, calcium, aluminum, and iron. Water is required for formation of the rigid gel; however, excess water added for workability results in voids that lower the cement strength.

The rigid network gels do not generally weaken as the temperature is increased. However, when the temperature of a material such as Portland cement is increased, the excess water is driven off and the chemically combined water is lost. If the excess water is trapped in discontinuous voids, heating may cause the cement to explode. The loss of the chemically combined water results in a complete loss of strength via a process commonly called *calcining*.

A weakly bonded network gel is called an elastic gel. A familiar example is common gelatin. In elastic gels an increase in temperature is accompanied by a decrease in strength. This decrease in strength is analogous to the weakening of van der Waal bonds with increasing temperature. The rigid and elastic network gels have many similarities to the thermoset and thermoplastic polymers, which will be discussed in Chapters 13 and 14.

Silicate cements are made from sodium silicate and fine quartz. The quartz is bonded by the formation of a silica gel that results from the breakdown of the sodium silicate. The process is used in the manufacture of molds and cores in the foundry. The reaction is accelerated by passing carbon dioxide gas through the sand to speed up the formation of sodium carbonate and silica.

Gypsum cement is used widely in wall construction. The mineral gypsum is heated to change the double hydrate to the hemihydrate: $CaSO_4 \cdot 2H_2O \rightarrow CaSO_4 \cdot \frac{1}{2}H_2O + \frac{3}{2}H_2O$ (gas). The hydrogen and oxygen atoms are an integral part of the crystal structure. When the hemihydrate is mixed with water, the gypsum phase reforms and hardens the shape.

The same type of reaction is involved in cements made from magnesium oxychloride and phosphate minerals.

10.17 Molds for Metal Castings

Although not formally recognized as a branch of ceramics, the annual preparation of ceramic molds for 20 million tons of metal castings deserves attention because of both the volume of ceramics used and the intricacy of the problems of metal-mold reactions. The problems are really at the interface between ceramics and metallurgy. The goal of the ceramist is to produce an expendable mold that will provide adequate surface finish and dimensional accuracy. From the metallurgical point of view, the action of the metal on the mold should be as neutral as possible. For example, the result of the reaction of metal with a mold containing excess water is a solution of hydrogen in the metal and gas holes in the casting. The combination of aluminum dissolved in liquid steel with SiO_2 in the mold wall can affect the casting surface, as shown in Figure 10.28. A wide variety of ceramic materials have been developed, including clay-bonded green (undried) sand, dry sand, oil-bonded sand, and sodium-silicate–bonded sand, as well as mixtures using alumina, zirconia, and olivine $(Mg, Fe)_2SiO_4$. The different "mixes" provide variation in casting surface finish and dimensional tolerances.

10.18   Residual Stresses and Contraction

An understanding of how residual stresses develop and how they can be controlled is more important in relation to ceramics than in relation to metals, because in ceramics there is no ductility to accommodate plastic deformation. But residual stresses, if properly controlled, can improve performance, as we pointed out in previous sections.

In glasses we can control stress by both mechanical and chemical methods. The compressive strength of glass is many times greater than its tensile strength. Therefore, in a sheet of glass that is subjected to bending, we want

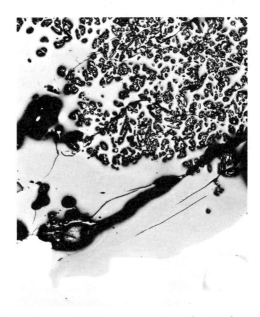

FIGURE 10.28   Photomicrograph of a cross section of an inclusion attached to a steel casting. Aluminum dissolved in the liquid steel reacted with SiO_2 from the silica sand mold, producing Al_2O_3 (corundum) crystals as well as a glassy iron silicate material; 100 ×, unetched.

[After G. A. Colligan]

to have residual compression in the surface layers. We accomplish this in *tempered glass* by quenching the surfaces (normally by an air jet) while the glass is in the plastic state. The surfaces are at a lower temperature as a result of the quench. Immediately after the quench there is no residual stress, because the core of the sheet of glass is plastic. However, when it cools, the core contracts more than the surface because it falls through a greater temperature range. When the glass reaches room temperature, there is tension in the core and compression in the surface (Figure 10.29).

The chemical method of controlling stress is to expose the surface of a glass containing sodium ions to a solution of (larger) potassium ions. Chemical exchange takes place, and the "wedging in" of the larger ions causes surface compression.

Present government safety regulations require tempered glass for many applications. Windows other than windshields in automobiles are tempered glass. When such a window fails, it breaks up into many small, almost dust-like pieces rather than large splinters that could become projectiles in an accident. Windshields, on the other hand, are laminated with a polymer that retains glass splinters, because the use of highly tempered glass might result in a total loss of transparency during an accident.

In reinforced concrete, prestressed beams are used to take advantage of the high compressive strength of the concrete. Before the concrete is poured, steel rods are positioned in the mold on the side of the beam under tension. The steel rods are stressed in tension by jacks, then the concrete is poured. After the concrete is set, the jacks are removed. The steel contracts, producing residual compression in the concrete. When the beam is loaded, the first portion of the load merely reduces the stress in the concrete on the tension side of the beam to zero. Therefore, the total load can be greater before the capacity of the beam is exceeded. (See also Chapter 17.)

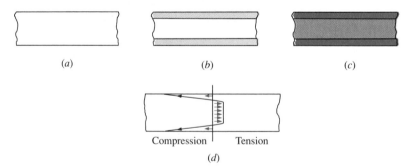

FIGURE 10.29   Changes in the dimension of glass as it is tempered. (a) Hot glass has no stresses. (b) The surface cools quickly; as the surface contracts, the center adjusts, and there are only minor stresses. (c) When the center cools and contracts, the surface is compressed and the center is in tension. (d) These compressive stresses due to the surface compression of tempered glass must be overcome before the surface can be broken in tension.

[Lawrence Van Vlack, *Elements of Materials Science*, © 1964, Addison-Wesley Publishing Co., Inc., Reading, Massachusetts. Reprinted with permission of the publisher.]

Stresses due to thermal expansion and *spalling* are related. When the surface layers of a refractory are heated, they expand and tend to spall off. Two techniques have been developed to combat this problem. In one method, glazes that have a lower coefficient of expansion than the base material are applied and fired. Because the base material tends to contract more than the glaze, the glaze is in compression, and heating merely lowers the stress level of the glaze. The other method is to use a glass with a very low or negative coefficient of expansion, such as Pyroceram.

These examples show that although ceramics inherently have low tensile strengths, they can still be used in tensile applications if residual surface compressive stresses can be imposed on them.

Summary

In Chapter 10 we discussed the structure, properties, and applications of traditional ceramics. Chapter 11 will address the new structural ceramics; and Chapter 12, ceramic processing.

We began this chapter with the study of glass, because glass itself has great importance and because a glass bond appears in the majority of ceramics. We found that the glass structure is made up of silica tetrahedra with fluxes (such as calcium and sodium oxide) and intermediate oxides (such as alumina) to lower the melting point and modify properties. The outstanding feature of the glass structure is that gradual softening occurs upon heating, rather than there being a sharp melting point. This allows for substantial shaping above T_g, the glass transition temperature.

The traditional ceramics are principally related to the minerals silica (quartz) and clay. The transformations and accompanying changes in volume that silica undergoes are of great importance in the performance of siliceous materials upon heating and cooling.

Ceramic products range from china to abrasives and cements, but in general their hardness and resistance to corrosion are of primary importance, whereas their brittleness is a drawback. Often the low cost, like that in cement, is most significant.

Definitions

Cleavage The fracture path that results when a material is broken. For example, sodium chloride cleaves into cubes, whereas glass has no regular cleavage.

Coordination number, CN The number of equidistant nearest neighbors to a given atom. The sodium ion, for instance, has six equidistant chloride ions in sodium chloride; therefore, CN = 6. The CN is related to the ratio of radii:

$$\frac{\text{Ionic radius of the smaller atom}}{\text{Ionic radius of the larger atom}}$$

Displacive transformation   Small angular displacement between ions to produce new dimensions within the same crystal structure.

Gel   A liquid and solid mixture in which the solid is present as a rigid or elastic network and the liquid is trapped in capillaries or voids.

Glass   A noncrystalline solid that softens upon heating rather than showing a sharp melting point. The structure is composed of small units such as SiO_4^{4-} tetrahedra, linked to each other or to other ions in a mostly random three-dimensional network.

Glass characteristics

   Working point   The temperature at which glass is easily formed, 10^4 poises.

   Softening point   The temperature at which glass sags appreciably under its own weight, 10^8 poises.

   Anneal point   The temperature at which locked-up stresses can be relieved, 10^{13} poises.

   Strain point   The temperature at which glass becomes rigid, $10^{14.5}$ poises.

   Glass transition temperature   The temperature at which the rate of contraction of cooling glass changes to a lower value.

Intermediate   An oxide, such as lead oxide, that can be added in large amounts to a glass former, such as silica, while still providing a glass structure.

Modifier   An oxide, such as sodium oxide, that is added to a glass in limited amounts to break up the continuity of the chain structure.

Mullite   A refractory compound formed between alumina and silica $(3\ Al_2O_3 \cdot 2SiO_2)$.

Octahedral site   The void within a group of *six* atoms that, when connected, form an octahedron. For example, the hole in the center of an FCC unit cell is an octahedral site.

Portland and high-alumina cements   Fine powders produced by sintering mixtures of clays and limestone and then grinding the mixture. A strong sheet structure or network is formed when water is added.

Pyroceram   A strong ceramic produced through formation of a crystalline phase via heat treatment of a glass structure.

Reconstructive transformation   Formation of a new crystal structure as a result of the complete rearrangement of ions.

Refractory   A material that resists exposure to high temperatures, as in a furnace lining or a gas turbine blade.

Silica polymorphs   Crystalline phases of silica (quartz, tridymite, and crystobalite).

Silicate   Any of the inorganic silicon–oxygen compounds based on the $(SiO_4)^{4-}$ tetrahedron.

Silicone   A family of silicon–oxygen compounds that have organic molecules as side groups.

Sol   A suspension of fine particles in a liquid. The sol state often precedes the gel state.

Spalling   The cracking and breaking away of surface layers resulting from exposure to heat or cold, sometimes combined with a chemical action.

Spinel A compound of two oxides characterized by $A^{2+}B_2^{3+}O_4$. An example is $MgAl_2O_4$ ($MgO \cdot Al_2O_3$).

Tempered glass A glass structure that is differentially cooled on its surface such that residual surface compressive stresses are present.

Tetrahedral site The void at the center of *four* atoms that, when connected, form a tetrahedron.

Viscosity coefficient The force required to slide one layer of a liquid past another under specified conditions. The viscosity coefficient affects the flow rate of liquid glass in a mold or tube, because the layer next to the tube is stationary and the inner layers must flow past it. The unit is the poise (1 poise = 0.1 Pa-sec = 0.1 N $\cdot$ s/m^2).

Vitreous silica Pure silica that has been cooled to retain the glassy rather than the crystalline state. It has a low coefficient of thermal expansion and hence exhibits good thermal spalling resistance. This is also called *fused quartz* to indicate its crystalline origin.

Problems

10.1 *[ES]* Determine the annealing temperature for the glass in Example 10.1. (Sections 10.1 through 10.4)

10.2 *[ES]* We said in Section 10.2 that the viscosity–temperature relationship for glass was similar to the diffusion coefficient–temperature relationship.

$$D = D_o e^{-Q/RT} \quad \text{and} \quad \frac{1}{\text{viscosity}} = Ae^{-B/T}$$

For glass 6 in Figure 10.2 (low-expansion borosilicate), develop the general viscosity–temperature equation by using the work point and the annealing point. Check the equation by using the upper softening point (10^7 N $\cdot$ s/m^2) to determine the softening temperature. (Sections 10.1 through 10.4)

10.3 *[EJ]* From the discussion of a nonfixed working point for glass as illustrated in Figure 10.2, suggest why the annealing point may also not be a fixed point. (Sections 10.1 through 10.4)

10.4 *[ES/EJ]* Refer to the following data to answer the questions below. Code numbers of the compositions may be used twice. (Sections 10.1 through 10.4)

Type	SiO_2	$Na_2O + CaO$	Al_2O_3	B_2O_3	PbO
A	99.9	—	—	—	—
B	96.	—	—	4	—
C	73.6	25.4	1.0	—	—
D	35.0	7.0	—	—	58

a. Which glass will have the lowest cost in a finished part (not necessarily the lowest raw material cost)?

b. Which glass will have the highest melting point?

c. Which glass will have the highest index of refraction?

d. Which glass will have the best thermal shock resistance?

e. Which compositions have only glass formers? List them.

f. Which compositions have only modifiers? List them.

g. Which compositions have only intermediates? List them.

10.5 *[ES]* Replot the data for glasses 2 and 3 in Figure 10.2 in terms of $\log_{10}$ of viscosity versus the reciprocal of absolute temperature. Why is the original curve not shaped like the replotted curve? (Sections 10.1 through 10.4)

10.6 *[ES]* The Na_2SiO_3–SiO_2 phase diagram is shown in Figure 10.30. (Sections 10.1 through 10.4)

a. Does the diagram show the existence of glass or crystalline structures? Explain.

b. What would be the liquidus and the solidus of a composition containing 15 wt % Na_2O?

c. Explain whether the composition in part (b) would be easier or more difficult to process than glass 3 in Table 10.1.

10.7 *[ES]* The text indicates that modifiers can be added to glass only in small amounts. What might happen if large amounts were added? (Sections 10.1 through 10.4)

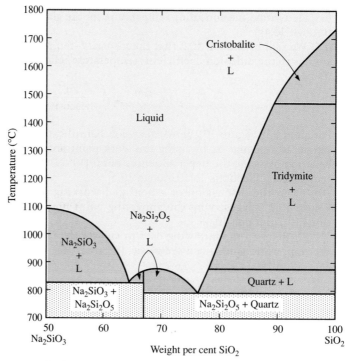

FIGURE 10.30   The binary system Na_2SiO_3–SiO_2. After F. C. Kracek, *J. Am. Chem. Soc.*, **61**, 2869 (1939).

[W. D. Kingery, *Introduction to Ceramics*, John Wiley, New York, 1960, p. 259]

10.8   *[ES]* Explain the logic behind each step in the time–temperature cycle for the glass ceramic crystallization shown in Figure 10.9. (Sections 10.1 through 10.4)

10.9   *[EJ]* a. How can one tell the difference between a cast and a blown glass shape? b. In the production of float glass (see Figure 10.7) why is molten tin used rather than another molten metal, such as aluminum? (Sections 10.1 through 10.4)

10.10   *[ES]* Many minerals and glasses with complex formulas, such as $LiAlSiO_4$, are called "stuffed derivatives" of SiO_2. The ions Li^+ and Al^{3+} substitute for one Si^{4+} ion. The Al^{3+} ion takes the place of the Si^{4+} ion in the tetrahedron, and the Li^+ ion is "stuffed" into a hole. In the allotropic form of silica called tridymite, important stuffed derivatives are $KNa_3Al_4Si_4O_{16}$ (nepheline) and $KAlSiO_4$. Which ions replace silicon in the tridymite lattice, and which are stuffed?

One grade of Pyroceram has the formula of β spodumene, or $Li_2O \cdot Al_2O_3 \cdot 4SiO_2$, plus quartz. Which are the stuffed ions? (Sections 10.1 through 10.4)

10.11   *[ES]* The NaCl crystal structure is shown in Figure 10.31; it is the same structure found in titanium nitride, TiN. Calculate the coordination number of TiN and the unit cell side length if its density is 5.41 g/cm³. (Sections 10.5 through 10.8)

10.12   *[ES]* Strontium oxide, SrO, has the same structure as the CsCl shown in Figure 10.32. Show all calculations for the following questions. (Sections 10.5 through 10.8)
a. What is the coordination number?
b. What is the unit cell size?
c. Determine the density.
d. List two other elements that might substitute for the strontium.

10.13   *[EJ]* Outline a general experimental procedure by which one could determine ionic radii and hence coordination numbers. (Sections 10.5 through 10.8)

10.14   *[ES]* Zirconia, ZrO_2, is often stabilized with calcium to produce an important refractory. The basic cell is ZrO_2 with 1 Ca^{2+} ion present for every 10 Zr^{4+} ions. Will the vacant sites be anion or cation? What percentage of the total number of all sites will be vacant? (Sections 10.5 through 10.8)

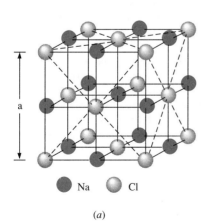

Na   Cl

(a)

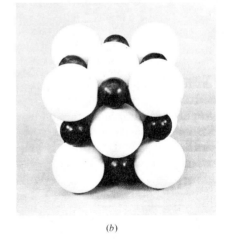

(b)

FIGURE 10.31   Sodium chloride structure. (*a*) The ion configuration. (*b*) Photograph of a hard-sphere model. Note that ions do not touch along what might be considered a facial diagonal. (Larger spheres are Cl^-.)

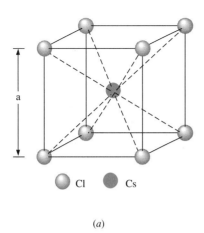

(a)   (b)

FIGURE 10.32   Cesium chloride structure. (a) The ion configuration. (b) Photograph of the hard-sphere model. (Larger ions are Cl^-.) The structure looks like two interpenetrating simple cubics. The ions touch along the body diagonal.

10.15 *[ES]* At high temperatures, the spinel $MgAl_2O_4$ shows solid solubility with Al_2O_3. For the spinel containing 80 mol % Al_2O_3 (Figure 10.22), determine the type and percentage of the vacancies. [*Hint:* Al^{3+} substitutes for Mg^{2+}, and the spinel stoichiometry indicates 50 mol % with no vacancies.] (Sections 10.5 through 10.8)

10.16 *[ES/EJ]* The text refers to ion defects that can occur in the iron oxide structure, FeO, giving values other than 50 mol % of oxygen and iron. (Sections 10.5 through 10.8)
a. Show with an appropriate sketch in the vicinity of 50 mol % oxygen how the compound chemistry occurs over a composition range rather than being fixed.
b. Explain how these defects contribute to diffusion rates, strength, and electrical conductivity.

10.17 *[ES]* From the following description, determine the coordination number, the stoichiometry between the ions, and the name of a similar ionic arrangement. "Oxygen ions are in a close-packed arrangement with magnesium ions at the octahedral positions with all octahedral sites filled." (Sections 10.5 through 10.8)

10.18 *[ES]* From the following data, calculate the ionic radius of barium in cubic $BaTiO_3$, density = 5.97 g/cm³. Assume Ba^{2+} at cube corners; Ti^{4+} at $\frac{1}{2}, \frac{1}{2}, \frac{1}{2}$; and oxygen at face centers. (*Hint:* The [110] direction cuts two radii of barium and oxygen ions, which is equal to $a_0\sqrt{2}$. See also Figure 1.34) (Sections 10.5 through 10.8)

	Atomic Weight	Ionic Radius, Å
Ba	137.3	X
O	16	1.32
Ti	47.9	0.64

$N_A = 6.02 \times 10^{23}$ atom/at. wt

10.19 *[ES/EJ]* Indicate whether the following statements are correct or incorrect and justify each answer. (Section 10.9)

a. $FeCr_2O_4$ is likely to have a spinel structure.

b. All silicates are noncrystalline.

c. All ceramic materials exhibit both covalent and ionic bonding.

10.20 *[EJ]* The common silicate structures are summarized below. (Section 10.9)

Structure	Identification	O/Si Ratio
Island	$(SiO_4)^{4-}$	4.0
Single chain	$(SiO_3)^{2-}$	3.0
Double chain	$(Si_4O_{11})^{6-}$	2.75
Sheet	$(Si_2O_5)^{2-}$	2.5
Framework	SiO_2	2.0

a. Show with a sketch what type of silicate structure is present in $PbSiO_3$.

b. Explain the melting point of this compound from the following data:

Compound	Thermal Properties
$PbSiO_3$	Melts at 766°C
SiO_2	Melts at 1710°C
PbO	Melts at 888°C
PbO_2	Decomposes at 290°C
Pb_2O	Decomposes at 500°C

[*Hint:* What might the phase diagram look like?]

10.21 *[ES]* Indicate the type of silicate structure (island, sheet, etc; see Problem 10.20) for each ceramic material listed below. Also show the stoichiometry of the simple oxides that make up the complex. For example, the spinel $MgAl_2O_4$ can be treated as $MgO \cdot Al_2O_3$. (Section 10.9)

a. Fayalite — Fe_2SiO_4 (common iron oxides are FeO, Fe_2O_3, and Fe_3O_4)

b. Enstatite — $MgSiO_3$

c. Talc — $Mg_3(OH)_2(Si_2O_5)_2$

10.22 *[EJ]* Pure alumina (corundum) has a melting point that is higher than that of mullite and a thermal coefficient of linear expansion that is three times higher than that of mullite. (Section 10.9)

a. Under what circumstances might you consider using mullite rather than alumina for a tube furnace?

b. What difficulties are encountered if excess SiO_2 is present in the mullite?

c. Why is a pure silica coating, SiO_2, on pure alumina, Al_2O_3, inappropriate for high-temperature service?

10.23 *[ES/EJ]* The introduction to Section 10.9 uses the term *fused quartz*, which we will now define as *vitreous silica*. Explain how the two terms can mean the same thing. (Section 10.9)

10.24 *[ES/EJ]* Indicate a thermal test that can be used to approximate the amount of tridymite present in a silica brick composed of tridymite, quartz, cristobalite, and vitreous silica. (Section 10.9)

10.25 *[ES/EJ]* Indicate whether the following statements are true of pure silica structures, and justify each answer. (Section 10.9)
a. Beach sand is more likely to be cristobalite than quartz.
b. Displacive transformations are faster than reconstructive transformations.

10.26 *[ES]* From the following data, estimate the amount of airborne asbestos in pounds per day at a busy intersection. (Section 10.9)

The rate of cars passing through is 50,000 cars/day.
Each car loses approximately 0.0001 in. of brake lining at each wheel in passing through the intersection (brake shoe area per wheel = 10 in^2).
The density of the brake shoe is 2.5 g/cm^3, and it is composed of 20% asbestos. Assume 75% of the asbestos from each wheel becomes airborne.

10.27 *[EJ]* Concrete does not burn. Explain how well its integrity is maintained after exposure to a hot fire. (Sections 10.10 through 10.17)

10.28 *[EJ]* Explain the meaning of the claim that sol–gel technology can produce ceramic shapes more inexpensively than traditional processing methods. (Sections 10.10 through 10.17)

10.29 *[ES]* A brick is made from sand (SiO_2) and 10% sodium metasilicate ($Na_2SiO_3 \cdot 9H_2O$). Sodium metasilicate is known to lose $6H_2O$ at 100°C. The brick weighs 3 lb at room temperature. How much will it weigh after heating at a temperature slightly above 100°C? (Sections 10.10 through 10.17)

10.30 *[ES/EJ]* In the setting of plaster of Paris, the following reaction takes place:

$$CaSO_4 \cdot \tfrac{1}{2}H_2O + \tfrac{3}{2}H_2O \longrightarrow CaSO_4 \cdot 2H_2O$$

Indicate why plaster of Paris molds work well for making aluminum castings but do not work for making steel castings. (Sections 10.10 through 10.17)

10.31 *[EJ]* A student places a steel sample in a vitreous silica boat and places the boat in a heat-treat furnace at 1800°F (982°C). Leaving it at that temperature over the weekend, he returns to find the sample fused to the boat. Explain what happened. (Sections 10.10 through 10.17)

10.32 *[ES/EJ]* Compare MgO, SiO_2, and Al_2O_3. Indicate which has the highest melting point, the highest modulus of elasticity, and the highest thermal conductivity. Give reasons for your choices. (Sections 10.10 and 10.17)

10.33 *[ES]* An investigator observes a sample of refractory brick under the microscope and determines that 70% of the area is mullite and 30% is corundum. Assuming equilibrium conditions, determine the lowest temperature at which the brick will liquefy completely. Assume that the density of corundum is 4.0 g/cm^3 and that of mullite 3.6 g/cm^3. (Sections 10.10 and 10.17)

10.34 *[ES/EJ]* A very thin glaze is placed on a ceramic underbody and is fired at 600°C, at which point there is zero stress. Because of differences in the thermal coefficient of linear expansion (glaze, 3.7×10^{-6}/°C; underbody, 4.5×10^{-6}/°C), a stress is present at room temperature, or 25°C. (Sections 10.10 and 10.17)
a. Calculate the maximum stress present in the glaze if the modulus of elasticity of both the underbody and the glaze is constant at 20×10^6 psi (13.8×10^4 MPa).
b. Why would this maximum value not be realized?

10.35 *[EJ]* A student has a choice of two glazes to use with a given underbody. The coefficients of expansion are glaze 1, $4 \times 10^{-6}/°C$; glaze 2, $6 \times 10^{-6}/°C$. The underbody has a coefficient of $5 \times 10^{-6}/°C$. What type of stress (tensile or compressive) is present in each glaze? Which glaze is more likely to crack, and why? (Sections 10.10 and 10.17)

10.36 *[EJ]* A ceramic tile used on the walls of a shower stall is generally glazed on only one side. (Sections 10.10 and 10.17)

    a. Should the glaze have a higher or a lower thermal coefficient of expansion than the underbody? Give reasons.

    b. The tile contains both open and closed porosity. What are the advantages of each type of porosity?

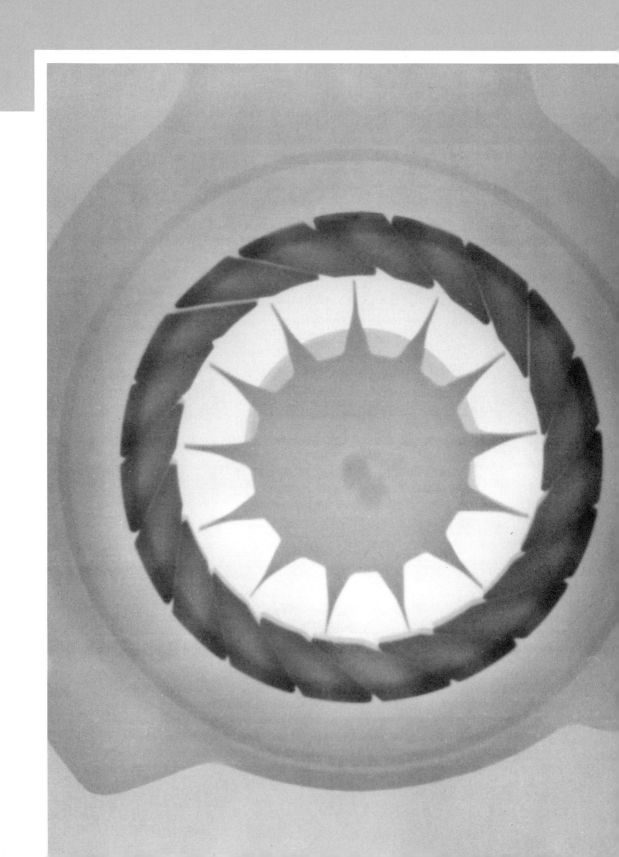

CHAPTER 11

Advanced Structural Ceramics

For many years design engineers have sought materials for turbine blades that could operate at higher temperatures than metal blades because of higher engine efficiency. Only by developing ceramics of greater fracture toughness has this goal recently been realized.

The photograph shows a ceramic rotor for a gas turbine engine developed by Garrett Turbine Engine Division and Ford Motor Company for the U.S. Department of Energy and NASA.

11.1 Overview

We have known for a long time that the ceramic materials have the hardest and strongest structures, that they offer exceptional resistance to wear, and that some are capable of withstanding higher temperatures than the best metal alloys. The reason why ceramics have not been used more widely for demanding applications such as aircraft gas turbine and auto engine components is their brittleness, which can be related to plane strain fracture toughness, K_{IC}. The values of K_{IC} for traditional ceramics are around 1–2 MPa$\sqrt{m}$ (0.9–1.8 ksi$\sqrt{in.}$) or about 1/50 of those of metals. The structural reason for this difference is that metals undergo substantial plastic deformation (slip) at the tip of a crack, thereby absorbing energy and retarding fracture. By contrast, a conventional ceramic fractures catastrophically without significant plastic deformation.

Recently, considerable effort has been devoted to finding mechanisms for increasing the fracture toughness of ceramics. The stakes are very high; components for spacecraft, as well as for conventional aircraft and automobiles, are involved.

Appreciable success has been attained with $K_{IC} \sim 15$ MPa$\sqrt{m}$ (13.6 ksi$\sqrt{in.}$), and though the field is still experimental, it seems poised for explosive growth. There are two paths for reaching the required toughness:

1. The composites (either metal–ceramics as in tungsten-carbide–cobalt tools or polymer–ceramics as in fiberglass); see Chapter 16. These are well-established materials, but their use at elevated temperatures is limited by the properties of the matrix.
2. The toughened ceramic materials that we will discuss here. Investigating the ceramics in their own right serves a dual purpose. If sufficient toughness can be developed, we will have a new class of materials. If not, we will still have better materials for use in composites and will understand their behavior better.

In this chapter, rather than concentrating only on the materials popular today, we will take a broad perspective of this rapidly changing field by covering the wide range of possible materials and the general mechanisms for strengthening and toughening. However, we will use materials currently in widespread use as examples in the following outline of the chapter contents.

1. A survey (Table 11.1) of the properties and structures of ceramics that are candidates for a wide range of engineering applications. These include oxides such as alumina and zirconia, hard carbides, nitrides, borides and silicides, pyrolitic graphite, and glass ceramics.
2. The hardness and modulus of elasticity of these materials.
3. Toughening mechanisms for ceramics (which are quite different from those we studied for metals). These include transformation toughening, microcrack development, and crack deflection.
4. The properties of ceramics at elevated temperatures.

TABLE 11.1 Characteristics of Engineering Ceramics

	Mohs* Hardness	Vickers Hardness	Structure	Use:				
				Structural	Electrical	Optical	Magnetic	Composites
Oxides								
Silica, vitreous silica	7	650	glass	X		X		X
Silicates								
Li, Al glass ceramics	6+	600	glass + crystalline	X				
Alumina	9	2000	hexagonal	X	X	X		X
Zirconia	8+	1150	cubic (stabilized) monoclinic (unstabilized)	X				
Carbides								
Metallic				X				
Ti, V, Cr, Y, Zr, Mo Hf, Ta, W	9+	$\frac{2000}{3000}$	cubic					X
Nonmetallic								
Si, B	9+	$\frac{2000}{3000}$	cubic	X				X
Nitrides								
Metallic				X				X
Ti, V, Cr, Y, Zr, Mo	9+	2000+	cubic or hexagonal, other					
Nonmetallic				X	X			X
Al, B, Si also Si, O, N								
Borides								
Ta, V, Cr, Zr, Mo, W also Ti, C	9+	2000+	cubic + other	X				X
Silicides								
Mo, transition element	~8	$\frac{1000}{1500}$	tetrahedral		X			
Carbon								
Pyrolytic, Type I Type II	1	<20	hexagonal	X				
Diamond	10	$\frac{5500}{7000}$	cubic (hexagonal)	X				
Spinel	7+	800	cubic				X	
Garnet	7+	800	cubic				X	

*See Figure 11.2 for comparison between hardness scales. The Mohs hardness uses a scratch method of ten minerals ranging from hard diamond (10) to soft talc (1). A specimen of a higher number will scratch a specimen of a lower number.

11.2 Key Structures

The important engineering ceramics are listed and characterized in Table 11.1. Some of these were discussed when we considered traditional ceramics, but with improved methods of processing they also qualify as structural ceramics and for use in composites.

1. *Silica and Silicates.* The most important silicates are the glasses, which we discussed earlier. Low-expansion lithium aluminum silicate (LAS) and vitreous silica are used alone or in composites. Synthetic garnets such as YAG (yttrium aluminum garnet) have important electrical uses. The garnet structure will be discussed in this context in Chapter 20.

2. *Alumina.* Alumina has a hexagonal structure and very strong ionic–covalent bonding. Because of the lack of ductility, it is essential to avoid serious flaws in stressed applications. Their interference with light transmission also makes flaws undesirable in optical devices. A great deal of effort has been directed toward production of high-strength alumina, and the most promising material developed so far is zirconia-toughened alumina. The mechanism of toughening in this material by stress-induced transformation is discussed in the next section.

3. *Zirconia.* Zirconia in the pure form is cubic at high temperatures and transforms upon cooling to a monoclinic form between 800 and 1100°C (1472–2012°F). However, the addition of the order of 16% of other oxides called "stabilizers" such as CaO, MgO, or Y_2O_3 results in a stable cubic phase which does not transform. The fully stabilized material has important electrical properties, and a partially stabilized form which transforms under stress has exceptional toughness.

4. *Carbides.* Hard carbides are well-established commercial products (such as tungsten carbide) and may be used in composites or alone. The microstructure of a typical cutting tool is shown in Figure 11.1. In general, the elements of Groups 4, 5, and 6 of the periodic table form cubic or hexagonal carbides. The Hägg rule states that when the ratio of the radius of the carbon atom to the radius of the metal atom is below 0.59, a simple structure is to be expected. This rule is based on the principle that to form a stable structure, the interstitial atom must touch the surrounding atoms. Table 11.2 summarizes the radius ratios of the common carbides. The more important applications are covered when composites are discussed in Chapter 16.

5. *Nitrides.* The structures of the hard nitrides are close to those of the carbides and follow the same Hägg relationship (Table 11.2).

6. *Borides.* The borides are also related to the carbides, but they contain boron–boron bonds as well as metal–boron bonds.

7. *Silicides.* The silicides exhibit high hardness, resistance to oxidation, and high melting points.

8. *Carbon.* Until recently graphite was considered a soft, low-strength material suitable only for crucibles, electrodes, and antifriction materials. The

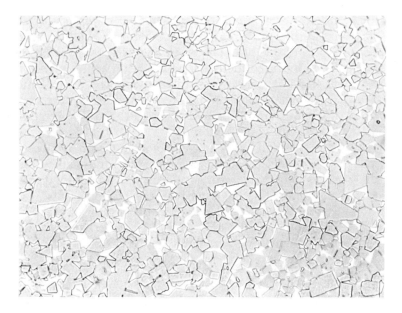

FIGURE 11.1 Cobalt-bonded tungsten carbide material for making tools. The gray crystals are tungsten carbide; the matrix is cobalt. To make this structure, tungsten carbide crystals were placed in a ball mill with cobalt powder. After milling, 1% paraffin was added. Then the mixture was pressed in a die and sintered for 0.5 hr at 2650°F (1452°C). The cobalt melted and the entire mass was sintered; 1500×, alkaline ferrocyanide etch.

TABLE 11.2 Atomic Radius of Transition Metals and Radius Ratios of Carbides and Nitrides *

Element (Me)[†]	Ti	V	Cr	Mn	Fe	Co	Ni
Atomic radius Å	1.467	1.338	1.267	1.261	1.260	1.252	1.244
C/Me ratio	0.526	0.576	0.609	0.611	0.612	0.616	0.620
N/Me ratio	0.504	0.553	0.584	0.587	0.587	0.591	0.595
Element (Me)	Y	Zr	Nb	Mo	Ru	Rh	Pd
Atomic radius Å	1.797	1.597	1.456	1.386	1.336	1.342	1.373
C/Me ratio	0.429	0.483	0.530	0.556	0.577	0.574	0.561
N/Me ratio	0.418	0.463	0.508	0.534	0.554	0.551	0.539
Element (Me)	La	Hf	Ta	W	Os	Ir	Pt
Atomic radius Å	1.871	1.585	1.457	1.394	1.350	1.355	1.385
C/Me ratio	0.412	0.486	0.529	0.553	0.571	0.569	0.557
N/Me ratio	0.396	0.467	0.508	0.531	0.548	0.546	0.534

*Radius of carbon, 0.77 Å; Radius of nitrogen, 0.74 Å. Metal radii may differ slightly from those given in Table 1.6.
[†]*Me* stands for the metal atom.
Reprinted by permission of the publisher from W. Baumgart, A. C. Dunham, and G. C. Amstutz, eds., *Process Mineralogy of Ceramic Materials*, p. 177. Copyright 1984 by Elsevier Science Publishing Co., Inc.

recent development of high-strength graphite fibers is due entirely to realizing that the strength in the plane of the hexagonal plates is very high. Fibers of this orientation are widely used. *Pyrolitic graphite* fibers are prepared by three methods, which begin with polyacrylonitrite (PAN), pitch, and rayon, respectively. In each case the organic molecules are heated to remove noncarbon atoms in order to provide carbon chains from the original hydrocarbon chain structure. Ribbon-like crystallites about 6 nm thick and 45 nm wide are formed.

Diamond has a long-established record as a cutting material. In Chapter 4 we discussed the phase diagram and diamond formation at high pressures and temperatures. Recently, as a result of many years of research, diamond films have been deposited from a plasma containing hydrogen and methane. This development is of great interest. It is applied to lengthen the life of cutting tools and to improve thin film insulation in electronic components.

11.3 Hardness and Modulus of Elasticity of Ceramics

We will begin our discussion of the properties of advanced ceramics with hardness, because many uses are related to this characteristic and it gives us an overall view of the field. Because of the brittle nature of many ceramics, the usual standard is Vickers hardness (HV), obtained with a diamond indenter.

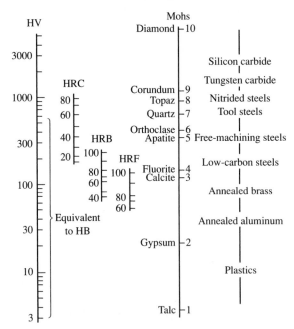

FIGURE 11.2   Interrelationships of different hardness scales. (Readings on carbides were obtained with special indenters.)

Figure 11.2 shows typical values for different materials. The hardness of the different carbides is shown as a function of temperature in Figure 11.3. Whereas TiC is hardest at room temperature, tungsten carbide is hardest at elevated temperatures, a significant consideration in the manufacture of cutting tools that will operate at high speeds. Diamond is the hardest material; next hardest are boron carbide and silicon carbide.

In general, the covalent bond or a combination of covalent and ionic bonds gives the greatest hardness, followed by the metallic bond and finally by the weaker intermolecular bonds of the polymers.

The modulus is related to the hardness as shown in Table 11.3. We would expect diamond, with perfect covalent bonding, to show the highest value. The hard carbides and nitrides follow, exhibiting values far above those for iron and steel. The glasses are below iron because of the lower order of the molecular structure. Softer structures (such as calcite) with ionic bonds only also have lower moduli.

11.4 Fracture Toughness

This property is the one that most radically limits the use of ceramics in highly stressed applications. The basic problem is that in order to reach a

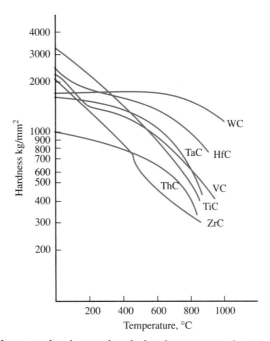

FIGURE 11.3   The microhardness of carbides decreases with increasing temperature. At room temperature, TiC is the hardest binary carbide; at 1000°C, WC is the hardest binary carbide.

[Reprinted by permission of the publisher from W. Baumgart, A. C. Dunham, and G. C. Amstutz, eds., *Process Mineralogy of Ceramic Materials*, p. 179. Copyright 1984 by Elsevier Science Publishing Company, Inc.]

TABLE 11.3   Modulus of Elasticity of
            Selected Ceramics

Material	Modulus	
	GN/m^2	$10^6 psi$
Diamond	1000	145
Tungsten carbide	450/650	65/94
Borides: Ti, Zr, Hf	500	72
Silicon carbide	450	65
Alumina	390	57
Titanium carbide	379	55
Silicon nitride	340	49
Magnesia	250	36
Zirconia	170	25
(Iron and steel)	207	30
Silica glass	94	14
Soda glass	69	10
Concrete, cement	47	6.8
Calcite (marble)	31	4.5
Graphite	27	3.9

level of toughness similar to that of metals (over 20 MPa$\sqrt{m}$, 18.2 ksi$\sqrt{in}$.),
we need to find a substitute for the plastic deformation that absorbs energy at
the root of a crack in the metals. Three methods of accomplishing this have
been developed: transformation toughening, microcrack formation, and crack
deflection. They are illustrated in Figure 11.4.

Transformation Toughening

This method has resulted in the highest values of K_{IC}. We first develop a two-
phase structure in which one of the phases transforms under stress to a new,
less dense phase. This produces compressive stress in the region of the crack
tip and arrests further propagation of the crack.

Let us consider an example in a prominent material, zirconia-toughened
alumina (ZTA). By processing and heat treatment, a two-phase structure consisting of a tetragonal zirconia phase in an alumina matrix is produced. The
tetragonal zirconia transforms under stress by a martensite-type transformation, with an increase in volume of about 5%. This produces compressive
strain around the crack tip, arresting further cracking. The progress of a crack
and of transformation were followed by M. Ruhle (Figure 11.5). It is important
to note that transformation occurred ahead of the crack and that regions to
the sides of the crack, as well as in front of it, were affected. The effect of the
side transformation is to impede crack opening and to further absorb energy.

Transformation toughening is also used in *partially stabilized zirconia*
(PSZ), in which the microstructure consists of cubic and tetragonal phases.
The tetragonal phase transforms under stress to monoclinic.

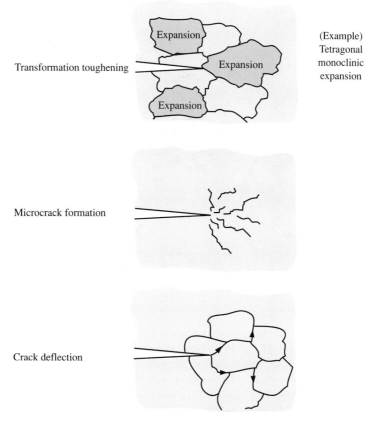

FIGURE 11.4　Toughening mechanisms in ceramic materials

Microcrack Formation

We instinctively regard crack formation as leading to failure because the remaining area of sound material undergoes higher stress. However, this analysis applies only to *macrocracks*; *micro*cracking can absorb energy and raise K_{IC}. As a matter of fact, it is believed that in the transformation toughening just discussed, a good deal of the increase in K_{IC} is caused by microcracking of the tetragonal phase (Figure 11.6). This mechanism of energy adsorption is very important in composites, as we will discuss later.[*]

Crack Deflection

If the progress of a crack can be changed from a rapid catastrophic type through uniform material to a slower, irregular type via the positioning of a

---

[*]We will also have to differentiate between this *plane stress* fracture toughness and the *plane strain* fracture toughness that is discussed in detail in Chapter 19.

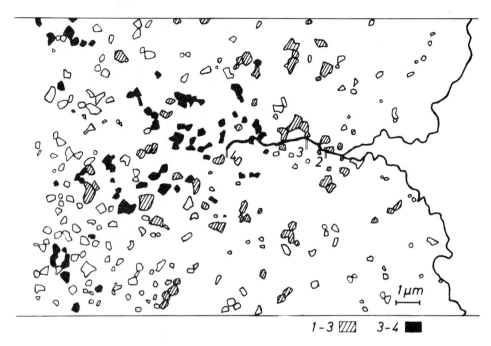

1-3 ▨   3-4 ◼

FIGURE 11.5 Schematic representation of transformation zone from *in situ* TEM experiment on Al_2O_3–15% ZrO_2 material. The specimen was stressed so that the crack propagated from the opening 1 (not marked) to (successively) 2, 3, and 4. All ZrO_2 grains are marked on the drawing. The cross-hatched areas are grains transformed during progress of the crack from position 1 to position 3. The dark areas are grains transformed during crack movement from position 3 to position 4.

[Courtesy of Professor M. Ruhle, University of California, Santa Barbara]

second phase across the crack path, which deflects the crack, energy can be absorbed and higher K_{IC} obtained. This mechanism is used on a fine scale in the development of two-phase microstructures and on a coarser scale in composites. This approach, in which a crack follows grain boundary facets instead of proceeding across grains, seems quite contrary to the metallurgical concept of avoiding brittle phases at the grain boundary and grain boundary fracture. The difference is that a crack proceeding through ductile metal absorbs greater energy, whereas little or no plastic deformation occurs during intragranular rupture in a ceramic. Therefore, the intergranular crack path in a ceramic absorbs greater energy.

11.5 High-Temperature Applications of Ceramics

There are very strong incentives for developing ceramics that can be used safely at high temperatures. The strength and creep resistance of ceramics are much higher than those of metals, which permits the design of engines and

Engineering Material Uses

FUNDAMENTALS

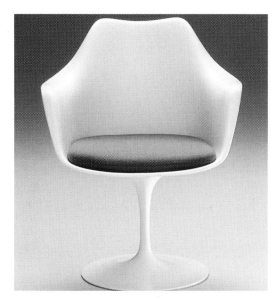

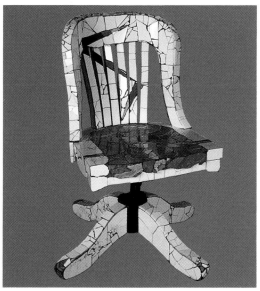

All homogeneous materials may be classified into three groups: metals, ceramics, and polymers. Frequently, any one of the three types of materials may be used for the same application. The chairs above are plastic (upper left), ceramic tile (lower left) and welded steel (upper right). The fourth chair (lower right) is made of fiberglass, polyurethane foam, and wood, which are classified as composite materials. A composite is combined of two or more materials, either naturally or synthetically. (Plastic and steel chairs courtesy Knoll International; ceramic chair "Swivel" © 1987 Nina Yankowitz; composite chair "Pacifica Lounge" by Brian Kane for Metro furniture)

METALS

Shown above, compact strip production is a process for casting and rolling steel strips. The good surface quality and uniform temperature of continuously cast thin steel slabs allow direct charging of the slab into the hot rolling mill. This strip is 50 mm. thick. (Courtesy SMS-Schloemann Siemag AG)

On the left, a hardenability test to determine the depth of a hard martensite layer obtained by quenching a steel. A standard-size bar is heated to produce the high-temperature structure of iron and then quenched on one end. The three photographs show the gradual cooling of the bar. After cooling, the hardness is measured along the bar, and the depth of hardening is determined.

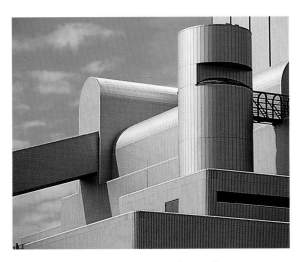

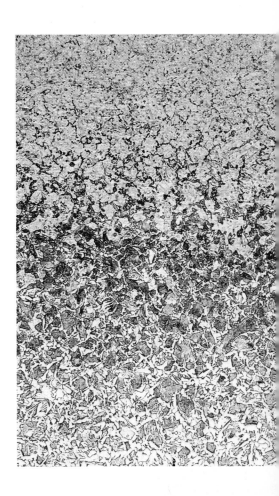

Above, a power station made from aluminum in Bonn, West Germany. (Aluminium, cover photo, Issue 64, Dec. 12, 1988. Aluminium-Verlag, Düsseldorf. Vereinigte Aluminium-Werke AG, Bonn)

The microstructures below and to the right are developed by chemically tinting polished specimens. Upper right (200×): carburized steel (Courtesy Leco® Corporation). Below, left (100×): a two-phase structure in air-cooled α-β brass. Below, right (100×): grains in an annealed α brass (70% Cu, 30% Zn) (George Vander Voort, Carpenter Technology Corp.)

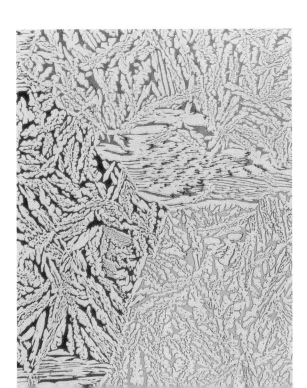

CERAMICS

Left, technicians inspect the world's largest window for transmitting infrared energy. Made of zinc selenide, the window represents advances in optical processing and coating techniques. (Shot at Perkin-Elmer Corp. by Gabe Palmer, © Palmer/Kane (1987)

Above, production of a glass optical wave guide. First a glass glob (called a boule) is made separately so that the outer layers (due to differing composition) prevent light from escaping from the core. The boule is then heated, and its weight pulls the glass like a strand of taffy down through the drawer tower. The final diameter is about 0.1 mm, yet the outer layer is preserved. (Courtesy Corning Inc.)

Left, catalytic converter. Catalytic activity creates a glow that outlines the cellular honey-comb structure of a Corning Celcor-brand ceramic monolith used as a substrate in automotive catalytic converters. (Courtesy Corning Inc.)

POLYMERS

Right, testing Monsanto Santo-prene® thermoplastic rubber. Elastomers such as this are engineered to make rubber parts for many industries, including automobile, architecture, business machine, hardware, and fluid delivery.

Below, defeating the "oil and water" problem. Shown are colorized electron micrographs of nylon/rubber blends in a Dupont study. Left, the blend without a grafting agent; center, the blend with a small amount of grafting agent; and right, with a larger percentage of grafting agent. (Photos reproduced with permission of Plastics Technology Magazine, New York)

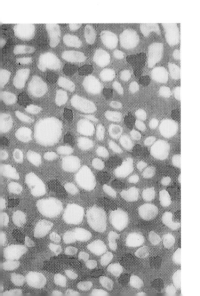
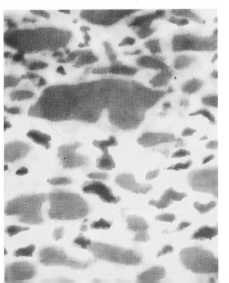
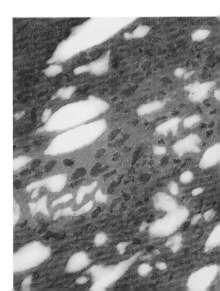

COMPOSITES

The objective of hull design for a windsurfer is to attain the lightest hull with adequate strength, rigidity, balance, and wear resistance. The core is epoxy-polystyrene foam wrapped in a low-density glass mat. The outer skin (yellow) is ABS/Polycarbonate with a UV-resistant outer layer. The footbed area (gray) is made of lightweight glass mat underlaid by high-density three-layer glass cloth with unidirectional and multiaxial layers. Linear carbon fiber strips (black) are used as top deck reinforcement. Woven carbon fiber cloth (curved section) is used for reinforcement at the bottom of the hull and at the rails. The rectangular section shown as a cutaway is a high-density, hard composite foam stringer (yellow) wrapped with glass mat (blue) and resin soaked (pink) for positive adhesion. A lightweight woven glass cloth is used overall as a sock to keep laminates in place and to add integrity to the hull. (Courtesy Roy van Oostendorp, Fanatic, Robin Hill Corporate Park, Patterson, N.Y.)

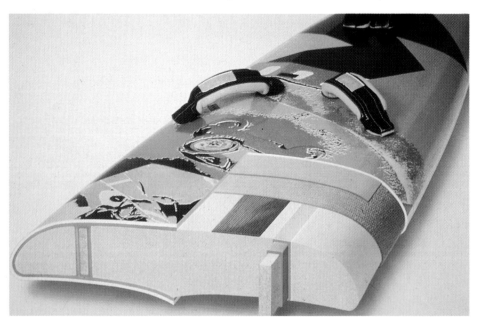

ELECTRICAL/OPTICAL

Above, fiber optics are widely used for information, for communications and for medical diagnosis. Optical fibers are comprised of two layers of glass. The light beam is guided through the fiber with minimal loss because the density of the inner layer is more than that of the outer layer. (Courtesy Siecor Corporation, Hickory, NC)

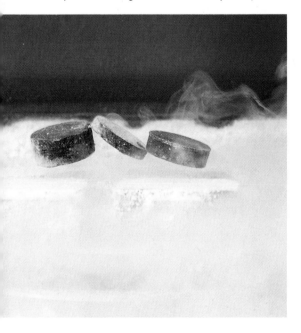

Above, laser driller, one of the many ways in which we now use lasers. (Gil C. Kenny/The Image Bank)

Left, levitation of 12 mm disks of the high temperature superconductor $YBa_2CU_3O_{7-8}$ above a permanent magnet, which rests in a bowl with liquid nitrogen. (Science cover photo, vol. 243, Jan. 20, 1989, "Levitation in Physics" by E. H. Brandt, Max-Planck Institüt für Festkörperforschung, Stüttgart, FDR)

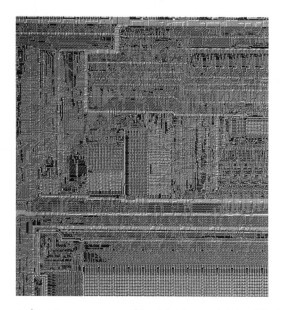

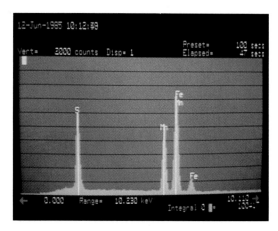

Left, microprocessor chip. The base of this chip is a thin slice of a single crystal of silicon. Selective masking and diffusion permit the development of complex circuits. (Alfred Pasieka/Taurus Photos)

Right, the x-ray spectrum produced by an inclusion in steel as determined by a modern scanning electron microscope. The x-ray peaks show the structure to be a manganese sulfide inclusion in an iron matrix.

Left, gallium arsenide semiconductor laser operates at an unusually short wavelength of 650 nm in the red visible light range. The indium gallium phosphide laser is on a gallium arsenide substrate and attached to a 2 mm copper tube for cooling. (Courtesy of Philips Research Laboratories, Einhoven, The Netherlands)

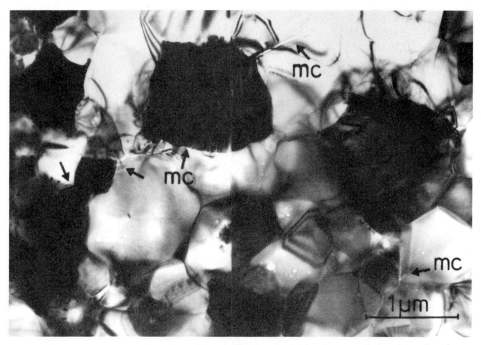

FIGURE 11.6 A TEM micrograph revealing a typical microcrack (mc) in the stress zone of 15% ZrO_2 (dark grains) Al_2O_3 (light grains)

[Courtesy of Professor M. Rühle, University of California, Santa Barbara]

turbines that can operate at higher temperatures and hence more efficiently. In many cases, such as with refractory oxides, the oxidation resistance of ceramics is also superior. In other instances, their greater hardness at elevated temperatures gives seals and bearings better resistance to wear.

All of these properties depend fundamentally on the strength of the covalent bond compared to those of the metallic bond. However, the same bonding and structures that lead to these excellent properties make ceramics deficient in slip compared to the metals and lead to their fracturing in brittle fashion.

Despite this handicap, extensive efforts are being made to develop ceramics of sufficient ductility and to realize a greater potential of these new materials. We will review the principal structural concepts.

In considering a ceramic material, we must evaluate three points:

1. Resistance to thermal shock
2. The effect of temperature on stress-strain relationships (creep).
3. The effect of different atmospheres at high temperatures. (Chapter 18)

We will discuss the first two points in detail.

Resistance to Thermal Shock

When a ceramic is heated uniformly to an elevated temperature and then the surface is quenched, the surface layers undergo *thermal shock* and they are subjected to tensile stress. It can be shown in physics that the surface stress S in a slab is

$$S = \frac{E\alpha}{(1 - \nu)}\Delta T \qquad (11\text{-}1)$$

where

E = Young's modulus

α = coefficient of thermal expansion

ν = Poisson ratio

ΔT = temperature difference between surface layer and interior

This a logical relation, because the tensile strain developed in the surface is $\alpha\Delta T$ and we convert this to stress by multiplying by the modulus. The $(1 - \nu)$ term corrects for the condition that the surface layer is not free but is constrained by the subsurface layers.

EXAMPLE 11.1 *[ES]*

Compare the surface stresses and predict whether ordinary (sodium) bottle glass and borosilicate glass (Pyrex) would fracture when quenched from boiling water into ice water (100 to 0°C).

	E	*Fracture stress*	ν	α
Bottle glass	10^7 psi	10,000 psi	0.20	$10 \times 10^{-6}/°C$
Pyrex	10^7 psi	10,000 psi	0.20	$3 \times 10^{-6}/°C$

Answer

$$S = \frac{E\alpha}{(1 - \nu)}\Delta T \qquad (11\text{-}1)$$

Bottle glass:

$$S = \frac{10^7 \text{ psi}(10 \times 10^{-6}/°C)}{(1 - 0.2)} \ (100°C) = 12{,}500 \text{ psi}$$

Pyrex:

$$S = \frac{10^7 \text{ psi}(3 \times 10^{-6}/°C)}{(1 - 0.2)} \ (100°C) = 3750 \text{ psi}$$

The bottle glass would break; the Pyrex wouldn't. Often the quantity $S(1 - \nu)]/\alpha E$, or ΔT in Equation 11.1, is used as an index of resistance to thermal shock. This ΔT value is called R and is shown in Table 11.4.

TABLE 11.4   Calculated Values of the Thermal Shock Parameter R for Various Ceramic Materials, Using Typical Property Data

Material	Strength,[a] S (psi)	Poisson's ratio, ν	Thermal expansion, α $\left(in./in.\cdot{}^{\circ}C^{-1}\right)$	Elastic modulus, E (psi)	$R = \dfrac{S(1-\nu)}{\alpha E}$ (°C)
Al_2O_3	50,000	0.22	7.4×10^{-6}	55×10^6	96
SiC	60,000	0.17	3.8×10^{-6}	58×10^6	230
RSSN[b]	45,000	0.24	2.4×10^{-6}	25×10^6	570
HPSN[b]	100,000	0.27	2.5×10^{-6}	45×10^6	650
LAS[b]	20,000	0.27	-0.3×10^{-6}	10×10^6	4860

[a]Flexure strength rather than tensile strength.
[b]RSSN, reaction-sintered silicon nitride; HPSN, hot-pressed silicon nitride, LAS, lithium aluminum silicate (β-spodumene).
[Reprinted from D. W. Richerson, *Modern Ceramic Engineering*, 1982, p. 141. Courtesy of Marcel Dekker, Inc.]

Creep

The first rule in designing with ceramics for use at high temperatures is to appreciate the difference in the stress–strain curves at those temperatures compared to the curve at room temperature and to understand the structural reasons for this difference.

At room temperature and for an appreciable range above room temperature, the curve shows no plastic elongation, and brittle fracture will occur (Figure 11.7). The criteria for fracture are the fracture toughness K_{IC} and the

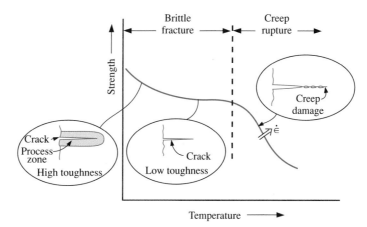

FIGURE 11.7   A schematic illustrating trends in strength with temperature: the trends at lower temperature, in the brittle range, reflect the temperature dependence of the toughness; the trends at high temperature involve creep and creep rupture.

[A. G. Evans and B. J. Dalgleish, "Some Aspects of High-Temperature Performance of Ceramics and Ceramic Composites,"*Cer. Eng. and Sci.* Proceedings, Vol. 7 No. 9–10, 1986, p. 1083]. Reprinted by permission of the American Ceramic Society.

maximum allowable crack length. We will call this *crack-propagation-dominated fracture.*

At elevated temperatures we encounter plastic strain for a number of reasons that we will discuss. We will call this *creep-dominated* fracture.

At an intermediate temperature range, such as 1300°C (2372°F) for alumina, we encounter the possibility of fracture by either of these mechanisms. (This is an important commercial range, so the contrast is significant.) To illustrate the great difference, let us review the results of an experiment by Evans *et al.* Polycrystalline specimens of alumina were subjected to three-point bend testing in air at 1250 and 1300°C (2282 and 2372°F) at several strain rates. The results fell into two distinct groups, as shown in Figure 11.8. The pronounced difference was in total strain at fracture, ϵ_f. In one group this was about 10^{-2}; in the other it was 10^{-1}. The first group showed crack-propagation-dominated failures. Cracks were nucleated at coarse-grained regions, at impurity concentrations such as nickel atoms, and at amorphous silica regions. In each case the inhomogeneity led to stress concentration, development of cracks above a critical size, and *more* rapid fracture.

In the case of the specimens exhibiting greater strain at fracture, several structural features were important. The cracks followed grain boundaries, and there was evidence that cavity formation (cavitation) took place ahead of the principal crack tip. The formation of these cavities reduces the stress concentration at the crack tip and impedes its progress. Finally the cavities joined, and the crack extended over a facet of a grain. At higher stress intensities, the cavitation is less and the crack grows directly along the grain boundaries.

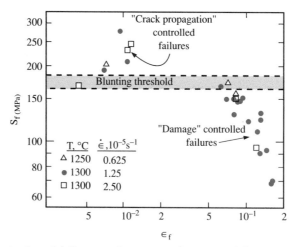

FIGURE 11.8 A plot of failure strain. ε_f, as a function of fracture stress, S_f, indicating that data exhibit two distinct regimes of behavior: one at small failure strains and the other at large failure strains.

[B. J. Dalgleish, E. Slamovich, and A. G. Evans, "High-Temperature Failure of Ceramics," *Journal of Materials for Energy Systems*, Amer. Soc. For Metals, Vol 8, No. 2, September 1986, p. 223. Reprinted by permission.]

This process still results in slower crack propagation, however, because the crack tip is blunted by surface-tension effects at the grain boundary.

Other evidence shows that crack blunting and even crack healing can occur at elevated temperatures when the effects of diffusion are greater than the growth rate of the crack. This leads to the concept of a threshold value of stress intensity, K_{th}, below which cracks do not propagate in a homogeneous material.* Furthermore, when crack-blunted regions develop, shear bands form between cracks before final failure. This absorbs additional energy (Figure 11.9).

These concepts are summarized in Figure 11.10. In the left-hand region below K_{th}, there is no crack growth (unless inhomogeneities are present). At low values of stress intensity, growth proceeds slowly by cavitation. At higher values, crack growth by direct creep takes place. Note that catastrophic rates of growth are not obtained even if K_{IC} is exceeded, because the

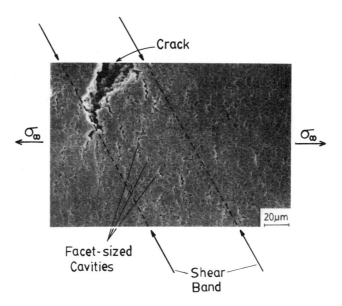

FIGURE 11.9   A scanning electron micrograph showing a high density of facet-sized cavities within a shear band in polycrystalline alumina

[B. J. Dalgleish, E. Slamovich, and A. G. Evans, "High-Temperature Failure of Ceramics," *Journal of Materials for Energy Systems,* Amer. Soc. for Metals International, Vol. 8, No. 2, 1986, p. 221. Photo by Geoffrey Campbell.]

*Derived from the equation developed in Chapter 2: $\sigma_f = K/Y\sqrt{\pi a}$, where $Y = 1$ and $K = K_{th}$, or the greatest stress intensity at which a crack will heal itself. See also Chapter 19.

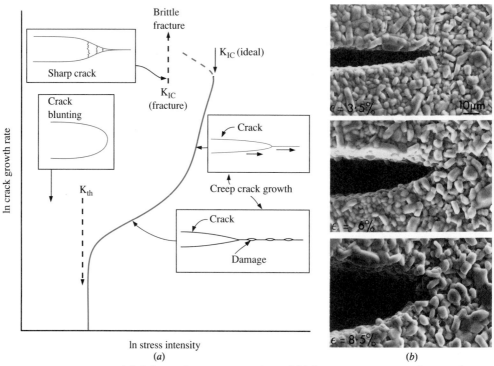

FIGURE 11.10 (a) Schematic representation of high-temperature crack growth rate as related to the stress intensity in polycru. (b) A crack blunting sequence in SiC (greater blunting, higher strain at lowest stress intensity).

[A. G. Evans and B. J. Dalgleish, "Some Aspects of the High-Temperature Performance of Ceramics and Ceramic Composites,"*Cer. Eng. and Sci. Proceedings*, Vol. 7, No. 9–10, 1986, p. 1086. Photos by Geoffrey Campbell. Reprinted by permission of the American Ceramic Society.]

crack tip is blunted by diffusion of material to the tip. The competing mechanism, crack propagation and brittle fracture, is shown for sharp crack tips and high rates of crack growth.

EXAMPLE 11.2 *[ES/EJ]*

Figure 11.8 shows that the specimens with lower total strain exhibited higher *stresses* at failure in the bend test than did the more ductile samples (which exhibited higher total strain). Explain.

Answer The specimens were subjected to a constant strain rate. In the more ductile specimens, the strain required was provided by cavitation and grain boundary cracking, so a considerably higher strain was present at fracture. In the crack-dominated samples no plastic strain occurred, so the stress was a function of elastic strain. Although the values of stress were higher for the brittle specimens, they could not be used in safe design. In summary, the data indicate the greater need for achieving uniformity and structural control in ceramics for high-temperature applications.*

Summary

In this chapter we focused our attention on advanced engineering ceramics. These materials are not yet used widely by themselves, but a number of important composites incorporate ceramic fibers, so it is vital to know the characteristics when the ceramics are tested alone. Also, thanks to the development of toughening mechanisms (transformation toughening, controlled microcrack formation, and crack deflection), we may soon be able to use these materials alone in applications where high hardness and resistance to elevated temperatures are called for.

The other fields wherein high-technology ceramics are important— electrical, optical, and magnetic materials—will be covered in later chapters.

Definitions

Crack deflection   A toughening mechanism in which the progression of a crack is diverted by a grain boundary or a second phase.

Creep-dominated fracture   Failure due to plastic strain and cavitation plus grain boundary sliding ahead of the crack.

Microcrack formation   A toughening mechanism in which the progression of a crack is slowed by dissipation into several paths.

Partially stabilized zirconia (PSZ)   Zirconia alloyed with CaO or the like to produce a tetragonal phase that transforms under stress.

Pyrolitic graphite   Graphite grown with a controlled crystal orientation so that the direction of greatest strength and modulus is parallel to the direction of applied stress—for example, the longitudinal direction in a fiber.

Thermal shock   A sudden temperature change, such as results from quenching a hot specimen in water or heating a specimen rapidly. The difference in temperature between the outer surface and the interior causes unequal expansion or contraction, in the case of cooling, and rupture can occur in brittle materials.

Transformation toughening   Production of a phase that transforms to a more voluminous phase under the stress field of a crack. The transformation produces a compressive stress field on the sides and ahead of the crack, limiting the crack's progress.

---

* A pertinent discussion of the use of Weibull graphs for ceramics versus their use for metals is given in D. W. Richardson, *Modern Ceramic Engineering*, Dekker, New York, 1982.

Zirconia-toughened alumina (ZTA)   Alumina containing zirconia in a two-phase structure with an alumina matrix and zirconia-rich grains, which provide transformation toughening.

Problems

11.1 *[ES/EJ]* From the nature of the bonding, explain why ceramic materials are brittle whereas metals are ductile. (Section 11.1 through 11.3)

11.2 *[ES/EJ]* The ceramic compound TiC is even harder than martensite. It is therefore very valuable as a cutting tool in industry. TiC is cubic. The sketch shows the atom arrangements on the (100), (010), and (001) planes (facial planes). The center-to-center distance between the closest titanium and carbon atoms is 2.16 Å (0.216 nm). (Sections 11.1 through 11.3)

   a. Is the arrangement of the titanium atoms in this structure simple cubic, BCC, or FCC?

   b. What is the coordination number for titanium in this unit cell?

   c. What is the unit cell volume in cubic angstroms?

   d. What is the closest center-to-center approach of two titanium atoms in angstroms? Each titanium atom is related to how many other titanium atoms at this same distance?

   e. Is the packing similar to CsCl or NaCl? (See Figures 10.31 and 10.32.)

   f. Why is the material so hard?

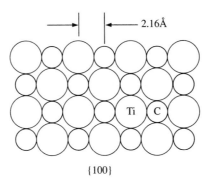

{100}

11.3 *[ES/EJ]* Determine the metal radii for Ti, Cr, Fe, Mo, and W from the data given in Table 11.2. Check your results with the radii given in Table 1.6 and suggest why differences might occur. (Sections 11.1 through 11.3)

11.4 *[ES]* The explanation of modulus differences shown in Table 11.3 is based on bond strengths. What would be the order of materials in the table if we considered their melting points or coefficients of thermal expansion? (Sections 11.1 through 11.3)

11.5 *[ES/EJ]* Why are we interested in the values of the modulus of elasticity given in Table 11.3 when the materials are so brittle? (Sections 11.1 through 11.3)

11.6 *[ES/EJ]* The text says that the modulus of elasticity is related to hardness. However, in steel we can change the hardness drastically (ferrite + pearlite versus martensite) and the modulus remains approximately the same at 30×10^6 psi (207 GPa). Explain. (Sections 11.1 through 11.3)

11.7 *[ES/EJ]* Table 11.1 shows that carbides, nitrides, and borides have approximately the same Mohs hardness of 9^+. Explain whether all of these materials can be used for cutting tools because they are similar in hardness. (Sections 11.1 through 11.3)

11.8 *[EJ]* Give several reasons why the carbides shown in Figure 11.3 do not all decrease in hardness with temperature at the same rate. (Sections 11.1 through 11.3)

11.9 *[ES/EJ]* Explain the statement that crack formation in a ceramic material can actually toughen it. (Sections 11.4 through 11.5)

11.10 *[ES/EJ]* In earlier chapters on metals, we developed the multiphase relationships: nature, size, shape, amount, distribution, and orientation. How do these relationships apply to crack deflection in ceramic materials? (Sections 11.4 through 11.5)

11.11 *[EJ]* There are other methods of treating thermal shock in ceramic materials in addition to that given in the text. For example, in refractory materials there is concern about developing surface stresses than can result in spalling off of surface layers. A spalling resistance index for refractories is given by the formula

$$\text{Spalling resistance} = \frac{kS}{\alpha E C_p \rho}$$

where k = thermal conductivity, S = tensile strength of the material, α = coefficient of thermal expansion, E = Young's modulus (modulus of elasticity), C_p = specific heat, and ρ = density. Explain how each term contributes to a reduction or increase in the tendency to spall. (Sections 11.4 through 11.5)

11.12 *[ES/EJ]* LAS is known to have excellent thermal shock resistance. (Sections 11.4 through 11.5)
a. What major physical property of LAS contributes most to its thermal shock resistance.
b. Explain why the value of R in Table 11.4 is not negative, even though the coefficient of thermal expansion is negative.

11.13 *[ES/EJ]* What variables that make up the thermal shock parameter R have the greatest influence on maximizing the value of R? (Sections 11.4 through 11.5)

11.14 *[ES/EJ]* In Example 11.2, the comment is made that the brittle samples in Figure 11.8 could not be used in safe design. What other information about the design would we need before we could assess the validity of this statement? (Sections 11.4 through 11.5)

11.15 *[ES/EJ]* Crack healing is one method of lowering the crack growth rate in the high-temperature creep of ceramics. Identify and explain some of the variables that affect the success of crack healing. (Sections 11.4 through 11.5)

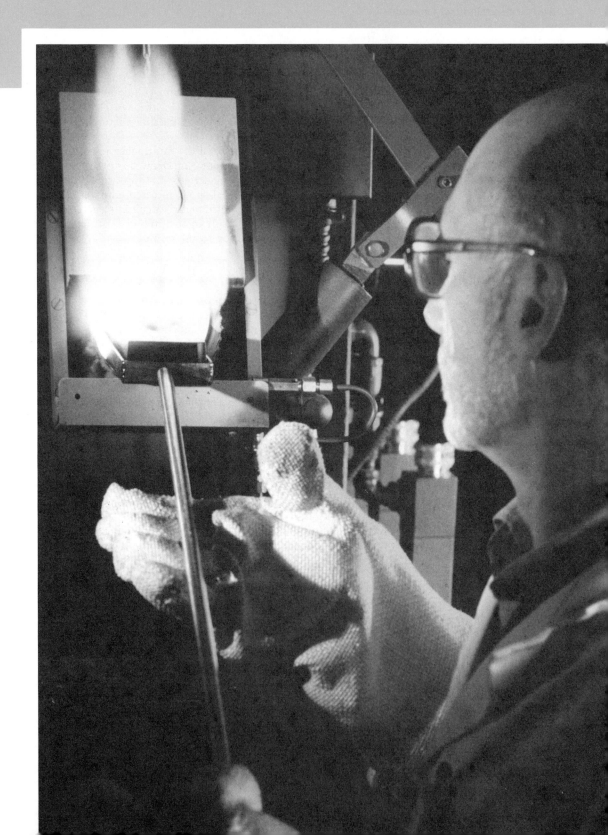

CHAPTER 12

Processing of Ceramics

The processing of conventional high-volume ceramics and the processing of advanced structural ceramics have many features in common.

In the illustration, a high-quality ceramic component has been made by advanced techniques in powder manufacture, binderless compaction and dimensional control, but it is still subjected to firing in a kiln by long-established methods.

12.1 Overview

The processing operations for a ceramic need more attention in design than those for a metallic or polymer component for two reasons.

1. The ceramic part is usually hard and brittle, and machining is costly. Therefore, the initial shape should be close to the final shape.
2. The sizes of the flaws and their population are related to the raw materials because usually, in contrast to metals, ceramic material is not melted to eliminate pores and dissolve foreign particles.

We will investigate these concepts by examining processing first in traditional ceramics such as those used for making bricks and pipe, where prices are in dollars per *ton*, and then in the engineering ceramics such as those used for making cutting tools, integrated circuit bases, turbine parts, and ceramic magnets, where prices are in dollars per *pound*.

Obviously, the processing costs for the latter group can be much higher. Typical building bricks are made from sand and clay fresh from the dunes and quarries, whereas high-strength parts are made of alumina and zirconia that have been chemically extracted, purified, and precipitated to a controlled grain size.

The well-educated engineer should understand the processing of both types of materials. The consumption of one traditional material—concrete, which is based on Portland cement—is greater than that of all metals and woods combined. And we need only glance around at the cracks that form in relatively new concrete building structures to realize that a need for further research still exists. The new ceramics, too, confront us with many challenges, and the key to the growth of this field lies as much in the refinement of processing to reduce flaws as in the development of new materials.

In this chapter we will discuss the processing techniques for traditional ceramics first, because many of them are also used in processing the new ceramics. In general, our discussion in both cases will follow this sequence:

1. Preparing the powder or other raw material
2. Shaping
3. Developing the bond (usually by firing but also by the formation of a hydraulic bond)
4. Finishing

Before beginning our discussion, however, let us review some of the concepts developed in Chapter 10 by considering the following example.

EXAMPLE 12.1 *[EJ]*

The following components are often made of ceramics: (*a*) "oven-proof" casserole, (*b*) coffee cup, (*c*) cap for a tooth, (*d*) fibrous insulating material that will be exposed to 2000°F (1093°C). Indicate the characteristics you would consider desirable for each application, and suggest a material that would meet these requirements. (Several answers may be appropriate for each application; indeed, normally there are competitive materials or a given component.)

Answers

A. *"Oven-proof" casserole.* Good thermal shock resistance, high thermal conductivity, and resistance to thermal spalling are desirable characteristics. From Table 10.1, we can see that fused silica or 96% silica would be excellent; however, processing costs for these materials are high. Therefore, we might consider borosilicate glass. Pyroceram is competitive and has the advantage of decorative finishes.

B. *Coffee cup.* At first it would seem that the materials discussed in the answer to part (a) would be the correct selections. However, the temperatures to which a cup is exposed are not as high as those found in an oven, and high thermal conductivity may be undesirable if the coffee is to be kept hot. An earthenware cup that has some porosity and provides lower thermal conductivity may be a better choice, although it has low strength.

C. *Cap for a tooth.* Strength, ability to be cemented to a natural tooth, and coloration are most important. Figure 10.25 gives some compositions of dental porcelain.

D. *Fibrous insulating material.* This application requires a fibrous material that packs in a container such that dead air space is utilized to provide maximum insulation. Asbestos, though fibrous, is toxic. It is also a hydrate, so it may lose weight by dehydration and become more compacted. We must determine the availability of ceramic fibers with decomposition or softening temperatures above 2000°F (1093°C). Pure silica, alumina, or any of their intermediate compositions would be an excellent choice (see the phase diagram in Figure 10.21).

TRADITIONAL CERAMICS

12.2 Preparation of Powder or Other Material

We will not dwell long here on the preparation of the starting powder, because it is more important in the engineering ceramics and will be discussed later. However, there is a relatively wide spectrum of preparation techniques.

For building brick, pipe, and inexpensive china, material from native deposits is used after being screened to remove large particles. To produce silica brick to be used as a refractory, it is necessary to mine and mill a rock called quartzite because, as a result of certain catalyst minerals, the transformation to cristobalite and tridymite (discussed earlier) takes place more rapidly than with ordinary silica sand. At the far end of the cost spectrum, high-alumina brick is obtained via chemical separation of the alumina from the ore to eliminate iron oxide and silica prior to fusion and grinding.

12.3 Shaping and Binding

In the following sections, we will discuss the manufacture and production of materials that are not glass and are competitive with glass — brick, refractories, artware, cookware, dinnerware, tile, chemical ware, plumbing fixtures, and ceramic bonded abrasives. There is a group of standard processes used for the nonglasses, including pressing and sintering. The fabrication methods used for glasses were discussed in Chapter 10.

Table 12.1 lists the different ceramic processing methods for a few sample products. Note that the same processes are used for a variety of products. Let us now discuss the processes listed, not in great mechanical detail, but to obtain a feel for how they can affect the final structure and shape.

12.4 Molding Followed by Firing

In the first group of ceramic processing methods, a shape is formed by one of several different methods and then fired to give it strength.

TABLE 12.1   Processes Used for Different Ceramic Products

Product	Molding plus Firing Processes						
	Slip Casting	Wet Plastic Forming	Dry Powder Pressing	Hot Pressing	Viscous Fluid*	Chemical Bonding	Single Crystal
Cemented products		X				X	
Brick		X	X		X	X	
Refractory and insulation	X	X	X	X	X	X	
Whiteware	X	X					
Vitreous enamelware		X					
Abrasive wheels			X				
Molds for metal castings	X	X				X	
Special (magnets, laser crystals, etc.)			X	X			X

*A second liquid phase is formed that bonds the primary grains.

Slip casting is an interesting and rather unusual method of forming ceramic products. A suspension of clay in water is poured into a mold (Figure 12.1). The mold is usually made of plaster of Paris *with porosity controlled* so that some of the water of the suspension enters the mold wall. As the water content of the suspension decreases, a soft solid forms. The remaining fluid is poured out, and then the hollow form is removed from the mold. The bond at this point is clay–water. Next the part is fired. Although in the past this method was used only for production on a small scale, it is now being employed with fine dispersions of refractory *metal* powders on curved surfaces. A similar technique is used to form a refractory shell on the outside of wax patterns. In this case the pattern is melted away and the mold is fired, leaving a precise cavity in the refractory for subsequent casting of metal.

Wet plastic forming is done by several methods. In some cases a wet or damp refractory is rammed into a mold and then ejected to create the required shape. In other cases extrusion is used for simple shapes such as brick or pipe (Figure 12.2). The plastic mass is forced through a die to produce a long shape, which is then sliced to the desired lengths. When circular shapes such as plates are to be formed, a mass of wet clay is placed on a rotating potter's wheel and shaped by a tool. This old process has now been highly mechanized.

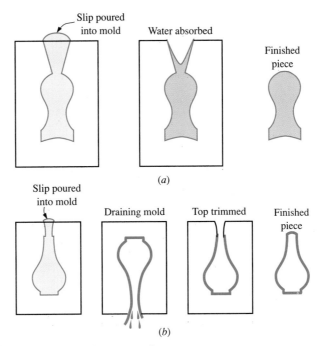

FIGURE 12.1   Slip casting of ceramic shapes

[W. D. Kingery, *Introduction to Ceramics*, John Wiley, New York, 1960, Fig. 3.17, p. 52. By permission of John Wiley & Sons, Inc.]

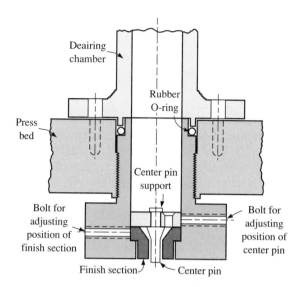

FIGURE 12.2   Piston-extrusion die for forming tubing

[W. D. Kingery, *Introduction to Ceramics*, John Wiley, New York, 1960, Fig. 3.14, p. 48. By permission of John Wiley & Sons, Inc.]

Dry powder pressing is accomplished by loading a die with a charge of powder and pressing the powder. The powder usually contains some lubricant such as stearic acid or wax. A variety of shapes can be made through the ingenious design of multipart dies (Figure 12.3).

After any of the preceding processes, the "green" part is placed in a kiln and fired. During firing, the water and volatile binders are driven off, low-

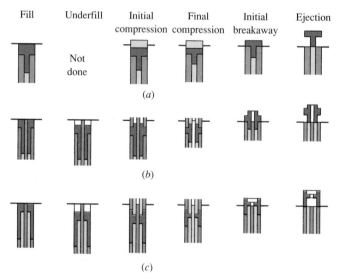

FIGURE 12.3   Die motions required for three different shapes

[W. D. Kingery, *Introduction to Ceramics*, John Wiley, New York, 1960, Fig. 3.9, p. 42. By permission of John Wiley & Sons, Inc.]

melting-point fluxes melt and bond the refractory, and sintering of the refractory grains takes place at 700 to 2000°C (1292 to 3632°F). We mentioned *sintering* in our discussion of powder metallurgy in Chapter 6. Remember that sintering is a diffusion process. It is especially important in ceramics, because the high melting point of the material often makes it impossible actually to melt the material to bond the grains together. Sintering provides the solution, because the diffusion of material results in true bonding between grains.

Hot pressing involves simultaneous pressing and sintering. The advantages it offers over cold powder pressing are greater density and finer grain size. The problem is to obtain adequate die life at elevated temperatures. Protective atmospheres are often used.

Isostatic compaction is a special way of pressing powders surrounded by a compressible fluid to overcome the uneven compaction sometimes observed with dies. The powder is encapsulated in a compressible can (made of rubber, for example) and immersed in the pressurized fluid. The shapes of the can and the removable cores determine the pressed shape. The pressing can be either hot or cold.

Figure 12.4 shows the tools used in cold powder pressing and sintering and some products of these processes. Here it was necessary to produce a crucible with an internal thermocouple protection tube that could withstand superheated liquid magnesium under an inert gas pressure to prevent its boiling. An MgO–MgF$_2$ mixture was selected for the crucible. The shrinkage was very severe and had to be accounted for in the original design. The use of a graphite external die allowed for induction heating to accomplish the sintering.

(a)   (b)

FIGURE 12.4   Objects used in making ceramic shapes (MgO) by pressing and sintering. (a) Components of a crucible die and fabricating tools. (b) A powdered compact in the graphite mold-susceptor (center). At left, a sintered crucible after it has been removed from the mold. At right, a crucible after it has been resintered in an oxidizing atmosphere. Note the shrinkage of the crucible as it went from its "green" state in the mold-susceptor to its sintered shape.

[Courtesy of Guichelaar, Flinn, and Trojan, University of Michigan]

12.5 Chemical Bonding

The advantages of using *chemical bonding* to process ceramics are that it is a cold process and that it can produce precise dimensions. The most common and important bond is the one involved in the setting of Portland cement, but other processes involve plaster of Paris, ethyl silicate, or phosphate cements. These structures were described in Chapter 10.

ENGINEERING CERAMICS

12.6 Processing Restrictions of Structural Ceramics*

The principal barrier to the wider use of structural ceramics for applications requiring resistance to high temperatures and corrosive environments is the presence of flaws combined with low fracture toughness. We recall that the stress required for fracture is a function of the fracture toughness divided by the square root of the largest flaw. When we select a material, we fix the maximum value of K_{IC} such that the allowable stress is in the hands of the processors via their control of flaw size, a. That is,

$$\sigma_f = K_{IC}/\sqrt{\pi a}. \qquad (12\text{-}1)$$

However, the processes are limited by flaw size control.

In contrast to metals, direct casting of ceramics from the melt usually produces a friable object because of uncontrolled grain growth and shrinkage, except in the case of glass. Important glass ceramic structures have been developed, as we noted earlier, but they are limited to specific glass-forming compositions. Vapor deposition methods are important principally for wires and thin surfaces.

The most efficient way to produce a variety of ceramic structures is with powder compacts made dense by heat treatment, so we shall discuss this method in some detail. To process ceramics successfully, we must study and overcome the sources of flaws in each step. Often the sources of flaws can be found by examination of the fracture, as we shall discuss later.

12.7 The Manufacture of Powder

Either mechanical or chemical methods can be used to obtain powder. The ball mill, in which the ceramic is placed inside a rotating barrel containing abrasive balls, is one of the commonest. Average particle sizes can be reduced below 5 μm. The disadvantages include possible contamination from the container walls and the ball material giving rise to inclusions that can act as flaws. Small surface fractures also may occur in the ball-milled product.

---

*A major portion of this discussion is adapted from F. F. Lange's Sosman Mineral Lecture, American Ceramic Society, 1987.

One of the preferred chemical methods is freeze drying, a process that consists of four steps:

1. Water-soluble salts are weighed out to give the desired final concentration as an oxide, for example.
2. After the salts dissolve, the solution is formed into very small droplets (0.1–0.5 mm in diameter) and rapidly frozen to avoid segregation.
3. The water is removed from the droplets as vapor in a vacuum chamber. No liquid phase forms.
4. The product is heated to decompose the salt, which forms fine oxides. The most common beginning salts that are used when forming oxides are sulfates.

This method is not universally applicable. For example, it is difficult to find proper salts for some ceramics.

Another popular method is precipitation from solution. An important case is the preparation of pure alumina by the Bayer process. The alumina is dissolved from the bauxite (high aluminum-bearing) ore to eliminate iron and silica, which remain insoluble and are filtered out. Aluminum hydroxide is then formed by precipitation and is transformed to alumina by heating (calcining) the $Al(OH)_3$.

The advantages of both of these methods is that they avoid the large aggregates of material that later in processing lead to separation from the adjacent small grains and they minimize the segregation of constituents.

Powder Preparation Prior to Compaction

Let us assume we have produced a fine, uniform powder by one of the methods described. Inevitably, weak aggregates of particles develop by van der Waals attractions. At every stage in the process, large aggregates must be dispersed to prevent the eventual formation of flaws. One of the most effective dispersion methods is the use of small amounts of organic chemicals that are called *surfactants* because they develop a charge on the surfaces of particles. If the charges are opposite, the particles are attractive and the phenomenon is called *flocculation* (the root of this word means "tuft of wool"; the fibers in a tuft of wool adhere quite tightly together). The reverse process, developing like charges, leads to repulsion and is called *de*flocculation.

To break up the weak aggregates, a deflocculant is added. This is called colloidal preparation, and once it is started, the powders should not be dried thereafter. The next step is to remove large, hard aggregates and large inclusions. This is done by *sedimentation*. The powder is shaken in a columnar vessel, and after a short interval to allow the larger particles to settle out, the balance is decanted or centrifuged. If several powders of different density (such as alumina and zirconia) are mixed, it is important to flocculate the mixture to avoid segregation due to gravity. To accomplish this, another surfactant or a change in pH is used.

12.8 Other Consolidation Methods

In structural ceramics, the need to avoid variations in density and the occurrence of flaws is much more acute than in traditional ceramics. The processes used include slip casting, doctor-blade forming, injection molding, and filtration.

1. *Slip casting,* which we discussed earlier, offers the advantage that after the desired slurry is formed, there is no need for an intermediate drying stage to prepare a powder. Therefore, this process is particularly desirable for a thin-walled cylindrical object.
2. *Doctor-blade forming (tape forming)* (Figure 12.5) is a method of obtaining thin sheet such as is needed for electrical bases and plate-fin heat exchangers. A slurry is placed on a moving carrier surface such as cellophane and spread to the desired thickness by an adjustable *doctor blade.* The product is then dried carefully and yields a thin, flexible tape that can be cut or stamped to the desired shape before firing. The desired flexibility is attained by adding plasticizers such as polyethylene glycol, which is eliminated during firing.
3. In *injection molding,* a heated liquid polymer is used as a carrier for the ceramic powder. The mixture is injected into a die (a step we will describe when we examine polymer processing in Chapter 15) and allowed to solidify. The polymer is driven off during sintering. More complex shapes can be produced by this method than by tape forming.
4. In *filtration* (Figure 12.6), the die contains a filter and a slurry is added. The liquid is eliminated by either applying a vacuum on the outside of the filter or applying pressure above the slurry in the chamber.

After the die is filled, a number of methods can be used to harden the compact to permit removal without cracking. In slip casting, the compact is allowed to dry to a leather-like consistency. In cases where a deflocculated slurry has been used, the addition of a flocculant stiffens the compact.

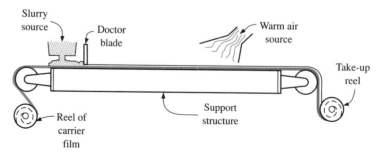

FIGURE 12.5   Schematic representation of the doctor-blade forming of ceramic tape

[Reprinted from D. W. Richerson, *Modern Ceramic Engineering,* 1982, p. 210. Courtesy of Marcel Dekker, Inc.]

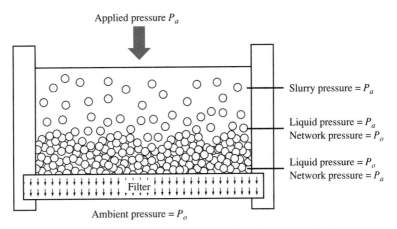

Applied pressure P_a

Slurry pressure = P_a

Liquid pressure = P_a
Network pressure = P_o

Liquid pressure = P_o
Network pressure = P_a

Filter

Ambient pressure = P_o

FIGURE 12.6   Producing a green ceramic shape by filtration of a slurry

[F. F. Lange, "Powder Processing Science and Technology for Increased Reliability", Sosman Memorial Lecture, Amer. Ceramic Soc., 1987, Figure 7. Reprinted by permission of the American Ceramic Society.]

12.9 Sintering

The *sintering* process is to ceramic processing what the solidification process is to metallurgy. The green shape has little strength, and the sintering process serves two purposes: (1) bonding between adjacent grains and (2) densification and, in the optimal case, elimination of voids. To understand the process, it is necessary to examine how material can be transported at the sintering temperature (Figure 12.7).

In *vapor-phase sintering* (Figure 12.7a), particles enter the atmosphere as a result of the vapor pressure of the solid; then they recondense to form a neck. A similar neck is formed by surface movement. These necks lead to bonding but do not decrease porosity.

On the other hand, volume diffusion or *bulk diffusion*, (mass transfer of the material away from the neck throughout the volume of the sphere), leads to a movement of a centerline (Figure 12.7b). This is accompanied by overall contraction of the compact and by densification. The complete elimination of pores may not be attained, even with a very long sintering time. The extent to which pores are eliminated is a function of the ratio of the energy per area of grain boundary to the energy per area of particle surface. This varies with different materials. Also, grain growth occurs in the later stages of densification, which may be harmful.

Another way of densifying is to add some material of different composition that liquefies. This can provide capillary forces attracting the particles with pressures as high as 7 MPa (1000 psi). The amount of liquid must be carefully controlled or the compacts can fall apart in the furnace from gravitational forces. A related method, *reactive liquid sintering*, is somewhat safer.

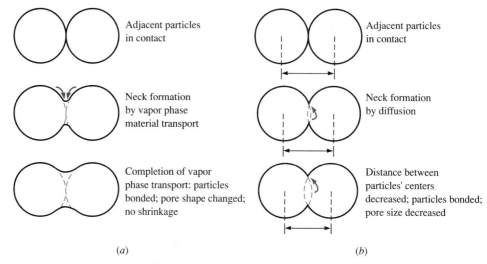

FIGURE 12.7   Material transport during sintering. (*a*) Vapor-phase transport. (*b*) Solid-state transport.

[Reprinted from D. W. Richerson, *Modern Ceramic Engineering*, 1982, pp. 219 and 221. Courtesy of Marcel Dekker, Inc.]

Material is added to provide a short-lived liquid for densification, which is then absorbed (dissolved by going into solid solution or by forming a solid compound with the matrix).

Still another method of densification is *hot pressing.* This is accomplished either in a press with heated dies operating with uniaxial force or by hot *isostatic pressing* (HIP); see Chapter 6. The major objective is to deform the sintering network to form new particle contacts that are centers for densification. The neck region is a major region for mass transfer from the interior, and increasing this area leads to more shrinkage.

12.10   Effect of Processing Heterogeneities on Strength

In Figure 12.8 we have schematically summarized the effects of processing variables on the strength of a sintered ceramic body. These potential defects are as follows:

Soft agglomerates — Low strength due to inadequate pressure during powder consolidation.
Hard agglomerates — Locally higher strength because nonuniform attractive forces of the powders give some agglomeration prior to compaction.
Organic inclusions — Arise from organic binders.
Inorganic inclusions — Occur as impurities in the original powder or can be produced by reaction with the atmosphere.

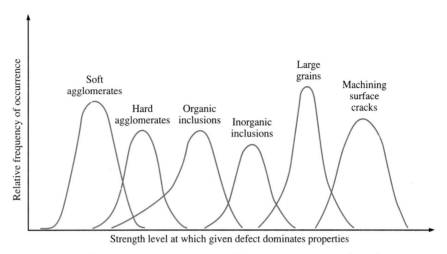

FIGURE 12.8 Schematic representation of heterogeneities on the relative strength and occurrence in sintered ceramics

[F. F. Lange, "Powder Processing Science and Technology for Increased Reliability," Sosman Memorial Lecture, Amer. Ceramic Soc., 1987, Figure 1. Reprinted by permission of the American Ceramic Society.]

Large grains — Grain growth after initial sintering where heating is continued to reduce the size of voids.

Machining surface cracks — Flaws in the sintered surface that are introduced by secondary processing such as machining.

Each distribution shown in Figure 12.8 can be treated as a population that has a maximum frequency of occurrence and a range of effects on the strength of the sintered ceramic body. For example, soft agglomerates give a lower mean strength (average within the frequency distribution) than hard agglomerates. Therefore, even if we should solve the problem of soft agglomerates, the properties of hard agglomerates become the next flaw problem that limits the strength, but at a higher mean-strength level.

Correspondingly, solving the difficulties of agglomerates and inclusions means that the strengths may still be limited by large grain size and surface cracks, but again at a higher strength level.

In viewing our schematic representation, do not assume that the order of flaw population and frequency of occurrence have the same position or relative values for all ceramics. The actual material and its processing history exert considerable influence.

An example of the effects of these flaws on the three-point bending strength of an yttrium-containing alumina–zirconia ceramic is shown in Figure 12.9. The advantage of having only surface flaws is apparent. Although the data are for several different chemical analyses, the major effect is still the presence of the indicated kind of flaw.

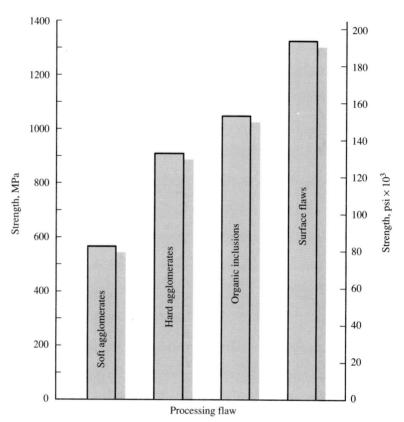

FIGURE 12.9   The strength of an yttrium-containing alumina–zirconia ceramic related to the presence of specific processing flaws

[Adapted from data of F. F. Lange, "Powder Processing Science and Technology for Increased Reliability," Sosman Memorial Lecture, Amer. Ceramic Soc., 1987. Reprinted by permission of the American Ceramic Society.]

12.11   Ceramic Processing Compared to Metal Processing

In Chapter 6 we defined the unit processing operations for metals and indicated that they could be applied to all engineering materials. Now let us review these operations and see how they can be applied to ceramics.

1. *Casting.*   In the classical sense of liquid transforming to a solid, glass compositions are cast when we heat to provide sufficiently low viscosity for flow. The melting temperatures of the advanced ceramics are normally too high for casting. However, slip casting is one method of particle agglomeration prior to sintering. Therefore, the term *casting* in ceramics has been expanded to mean flow into a mold without the necessity of melting, and further processing may still be required. Within our new definition, chemical bonding after pouring cemented products such as Portland-cement-bonded aggregate into a mold can also be classified as casting (see Chapter 17).

2. *Deformation.* Here we have shape control by forming, but it can be done only to ceramics that exhibit sufficient plasticity. Wet-bonded ceramic grain such as we produce when making bricks is one example. However, firing or sintering must again follow. Hot glass compositions are blown and rolled, which follows our prior concept of how metals are formed, but we must remember that cold ductility does not exist in ceramics.

3. *Particulate processing.* The processing of ceramic powders by compaction and sintering is, of course, one of the primary methods of component fabrication. We have discussed these concepts at some length, but take special note of the fact that a key to success in engineered components is the reduction of porosity.

4. *Joining.* Welding as in metals does not seem appropriate in ceramics. However, glass components are welded together to produce larger structures, as you have probably recognized if you have observed a glassblower. We can use polymer adhesives to join ceramic components; however, doing so usually impairs their load-bearing capability. Chemical bonding with cement products is also a common joining method, as in the classic "bricks and mortar" construction.

5. *Machining.* Ceramics can be machined in the green or leathery state that precedes final sintering. As pointed out earlier, however, the machining of fired ceramic shapes is difficult because of their inherent high hardness and brittleness. More important, grinding a ceramic may produce surface flaws. These stress raisers can then lead to premature failure of an engineering ceramic.

6. *Thermal treatment.* In ceramics as in metals, we use heat to enhance diffusion, to homogenize the structure, or to obtain a nonequilibrium structure. Temperatures may be higher or holding times longer in ceramics because more sluggish reactions are found in these ionic- and covalent-bonded materials.

7. *Surface treatment.* Here we may polish surfaces, which in reality is similar to machining or what may be called mechanical surface treatment. We may thermally or chemically treat the surface to provide compressive surface stresses, as in tempered glass. The application of a glaze can provide lower porosity and more desirable residual surface stresses.

In summary, the unit operations developed for metals in Chapter 6 can also be used for ceramic materials, although we need to modify our terminology to recognize the inherent differences in bonding that distinguish these two classes of materials.

EXAMPLE 12.2 *[ES/E]*

Sintering is time- and temperature-dependent. Why may it not be desirable to shorten the time by going to a high temperature?

Answer   A number of defects can arise from sintering at a high temperature, or what is sometimes called "overfiring."

1. Higher temperatures result in more shrinkage, which, if nonuniform, may be observed as warpage.
2. Excessive grain growth with lower strength is more likely at high temperatures.
3. New, unwanted phases can occur at high temperatures.
4. When a piece is brought up to temperature too quickly, binders may be driven off rapidly enough to cause surface cracks.

Therefore, optimal sintering times and temperatures are necessary to provide the most serviceable component. End use, processing, and the characteristics of the powder are also important considerations.

Summary

The processing of ceramics is especially important for two reasons. In all types of ceramics, we are usually dealing with hard, brittle materials, so it is necessary to form them into the desired final shape in the initial operation. In other words, the sequence of processing operations used for metals and polymers (such as hot working, cold forming, and machining) is difficult or not available. Second, because of inherent brittleness, the attainment of maximum toughness, K_{IC}, is essential in engineering ceramics. This is accomplished via the preparation of fine homogeneous powders, careful powder preparation, pressing, and sintering. Despite all the special care that is taken in their processing, ceramic products exhibit much more variation in properties than metallic products.

Definitions

Chemical bonding   A process in which a ceramic shape is bonded by the formation of a new structure between the grains (usually a hydraulic bond).

Doctor-blade forming   Spreading of a uniformly thick ceramic powder slurry over a moving surface.

Dry powder pressing   A process in which ceramic powder plus lubricant and binder are pressed into a die.

Flocculation   Attraction of particles as a result of charges of opposite sign on their surfaces. *Deflocculation* occurs when the surface charges are of the same sign.

Hot pressing   A method similar to dry pressing, except that the heated die causes sintering of the part.

Isostatic compaction A cold or hot compaction method by which a solid is consolidated under hydrostatic pressure.

Sintering The fusion of particles by diffusion bonding (solid state) or by the formation of a small amount of a liquid phase (liquid state).

Slip casting A shape-forming method in which a suspension of a solid, such as clay in water, is poured into a porous mold. The water diffuses from the layers next to the mold surface, leaving a solid shape. The liquid is then poured out of the interior of the mold.

Surfactants Organic chemicals used in small amounts to develop a charge on the surfaces of particles.

Wet plastic forming A shape-forming method in which a wet plastic ceramic mix is shaped by extrusion or other methods.

Problems

12.1 *[EJ]* Indicate why a slip-cast ceramic shape might be prefired at 95°C before being placed in a sintering furnace. (Sections 12.1 through 12.5)

12.2 *[EJ]* In the discussion of dry powder pressing in the text, there is mention of using a lubricant such as stearic acid or wax. Explain how these materials might be added to the ceramic powders and what the disadvantage of adding too much might be. (Sections 12.1 through 12.5)

12.3 *[ES/EJ]* If you were to purchase bricks, you might be asked whether you wanted common or vitreous bricks. How are the two varieties likely to differ in processing and ultimate use? (Sections 12.1 through 12.5)

12.4 *[ES/EJ]* Give reasons why protective atmospheres might be required in hot pressing. (Sections 12.1 through 12.5)

12.5 *[ES/EJ]* A building brick measures 9 in. × 4 in. × 2 in. (228.6 mm × 101.6 mm × 50.8 mm), weighs 5.25 lb (2.38 kg), and has a specific gravity of 2.49. When saturated with water, the brick weighs 5.60 lb (2.54 kg). (Sections 12.6 through 12.11 and 6.13)
a. Determine the true volume, apparent open-pore volume, and apparent closed-pore volume of the brick.
b. Why are the pore volumes apparent rather than true values?

12.6 *[ES]* An insulating brick (true specific gravity = 2.58) weighs 3.90 lb (1.77 kg) dry, 4.77 lb (2.17 kg) when the open pores are saturated with kerosene, and 2.59 lb (1.18 kg) when suspended in kerosene (specific gravity = 0.82). (Sections 12.6 through 12.11 and 6.13)
a. What are the true volume, apparent pore volume, and bulk volume?
b. What is the closed porosity?

12.7 *[EJ]* A buyer comparing some small pure Al_2O_3 crucibles notes that there is a significant difference in physical properties among manufacturers. Suggest why this might occur. Also, why might one manufacturer have size tolerances of ±2%, whereas another's might be twice that high? (Sections 12.6 through 12.11)

12.8 *[EJ]* After a ceramic shape has been sintered, its density is found to be too low. It is suggested that hot isostatic compaction be used to close the porosity. Why would this not be likely to be a good process for densification of the sintered product? (Sections 12.6 through 12.11)

12.9   *[ES]* You want to produce a ceramic tube by compacting and sintering MgO with the addition of 10% by weight CaF_2. (Sections 12.6 through 12.14 and portions of Chapter 10)

a. What is the structure of MgO? Support your answer.

b. The objective in sintering is to have the CaF_2 break down according to the reaction

$$CaF_2 + \tfrac{1}{2}O_2 \longrightarrow F_2(gas) + CaO$$

(1) With the initial addition of 10 wt % CaF_2, what is the weight percentage of CaO in the MgO–CaO mixture?

(2) From the accompanying CaO–MgO phase diagram, find the liquidus and solidus temperatures. Are these changed significantly from those of pure MgO?

(3) Will the Ca^{2+} ions replace the Mg^{2+} ions in the MgO structure? Support your answer.

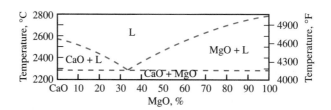

12.10   *[EJ]* The text indicates that preparing ceramic powder by precipitation is preferable to using mechanical ball milling. Explain whether this is correct for all ceramic components. (Sections 12.6 through 12.11)

12.11   *[ES/EJ]* Refer to Figures 12.8 and 12.9. (Sections 12.6 through 12.11)

a. Why do soft agglomerates have a maximum strength of 550 MPa, whereas hard agglomerates show a maximum strength of 900 MPa?

b. Why do surface flaws appear to have less effect on strength than internal flaws have?

12.12   *[ES/EJ]* Explain the problems that we might encounter if we attempted to weld two pieces of ceramic together. Would the ceramic pieces be more or less difficult to weld if they were of different chemical composition? (Sections 12.6 through 12.11)

PART IV

Polymeric Materials

The field of plastics or, more properly, "high polymers" has been growing at an amazing rate. There are three main reasons:

1. New polymers with enhanced properties are continuously being developed because the field of synthesis is unbounded.
2. The use of polymers in new *composites* is of major importance.
3. New high-strength *fibers* (such as Kevlar), made of *polymers* not ceramics, for use in *composites* have been synthesized.

To use polymers properly in components, we need a general view of their advantages and disadvantages. Let's begin with the advantages. Polymers are light in weight (specific gravity 1–2), easily cast or formed and machined, and generally corrosion-resistant. Many are transparent, electrical insulators, and semiconductors or conductors (by the addition of fillers). They exhibit a wide range of properties from the rigid, higher-strength types to highly plastic or elastic types such as rubbers.

Now let's look at the disadvantages. Although the low softening temperatures of the thermoplastics suit the producer just fine—only low melting temperatures are needed, and die life is excellent—the engineer has to be sure the parts aren't going to be heated to a softening temperature in service. Worse yet, we have to be sure that creep at room temperature won't be a factor. Although the bonding is strong in the main chain of the polymer, there are many van der Waals bonds holding the long-chain molecules together. There is usually no real modulus, as we know it for metals and ceramics, because the molecules straighten out and slip as a function of time at room temperature. Even apart from the effect of time, the modulus is several orders of magnitude lower than for metals and ceramics, reflecting the lower bond strength.

In Chapter 13 we consider the fundamental properties of polymers and survey the high-usage (commodity) polymers. Then in Chapter 14 we investigate how high-performance polymers are developed, much like metals, by alloying of the simple polymers. In Chapter 15 we discuss the different processing methods and show how they are related to the structure and properties of the product. Composite materials are covered in Chapters 16 and 17.

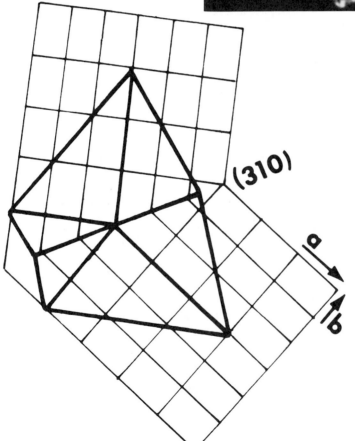

(310)

a

b

CHAPTER 13

Traditional Polymers: Fundamentals and High-Volume Varieties

In earlier discussions we have spoken of more or less random arrangements of molecules in the thermoplastic polymers. However, in linear polymers of fairly simple structure such as polyethylene the long molecules may form crystals. In this case, the long molecule forms a space lattice of regular dimensions. In this way a number of crystalline grains are formed.

The accompanying photograph shows a polyethylene crystal with a twin similar to a deformation twin in metals. The crystal was extracted from a solution of 0.05 wt % polyethylene in xylene. The crystal structure of polyethylene is orthorhombic, with the a and b dimensions in the accompanying sketch 0.253 nm and 0.493 nm, respectively.

The properties of the crystalline polymer are different from those of amorphous material of the same chemical analysis. In general, the crystalline material will be translucent rather than transparent because of the effect of grain boundaries on transmission of light. Other physical properties will vary, as discussed in the text.

13.1 Overview

We mentioned in the introduction to this part that we would concentrate on the fundamentals in this chapter. What are these basic concepts?

The first concept to master is that we must revise our notions of crystalline structure as it is found in the metals and ceramics and of the regular motions of slip and cleavage. The basic unit of polymers is the *molecule* or, more accurately, the macromolecule, which is composed of thousands of atoms. To complicate the picture further, crystallization can occur within the molecule!

The spine of these molecules is the carbon chain of atoms, and the three different forms that this chain can take lead to the three basic divisions of polymers (Figure 13.1).

1. The *linear structure*, a chain that is not a straight line but more like a piece of spaghetti. These molecules slide by each other upon heating and form the *thermoplastic* polymers. Note in Figure 13.1 that the carbon bonds are not at 180° but show a tetrahedral arrangement, with 109° between carbon atoms. For simplicity, however, we will draw them linearly rather than as a three-dimensional tetrahedron.
2. The *space network structures*, which are rigid and make up the *thermosetting polymers.*,
3. The coiled structures, which exhibit *elastic* extension of as much as 1000% and are aptly termed the *elastomers*.

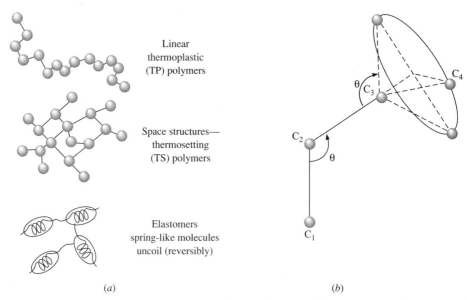

Linear thermoplastic (TP) polymers

Space structures—thermosetting (TS) polymers

Elastomers spring-like molecules uncoil (reversibly)

(*a*)   (*b*)

FIGURE 13.1   (*a*) Structures of thermoplastic polymers, thermosetting polymers, and elastomers. (*b*) Detail of the formation of the carbon backbone. Atoms C_1, C_2, and C_3 define a plane; atom C_4 may lie at any one of several preferred positions on the circle shown.

We will consider each division of the polymers separately, beginning with their formation from simple chemicals and the bonding they exhibit, and progressing through their variations in structure and the correlation of structure with properties. Having developed correlations between structure and property with one group at a time, we will then end with overall comparisons. Because over 70% of the polymer tonnage is thermoplastic and because this is the simplest group, we will begin with the thermoplastics.

13.2 Formation of Polymer Structure

One of the great contributions of chemistry to engineering materials is the process called *polymerization* in which we change a substance such as ethylene into the useful continuous, solid polymer known as *polyethylene*, (Figure 13.2). An individual C_2H_4 molecule is called a *monomer*, but when present as a basic building block, the C_2H_4 molecule is called a *mer*. *Polymer* therefore means "many mers." The basic mechanism lies in unlocking one of the double bonds between carbon atoms in each monomer and linking it to a bond in a carbon atom in an adjacent molecule (Figure 13.3).

Addition Polymerization

Recall from our discussion of the double bond between carbon atoms that the first bond was sigma type, a stronger bond than the second, pi type bond. If we add a material called an *initiator*, such as hydrogen peroxide, it will supply free radicals — that is, groups of atoms with a free electron. Thus

$$H_2O_2 \rightarrow 2OH\cdot$$

To initiate polymerization, an $OH\cdot$ combines with one pi electron in a mer, leaving the other free to bond with a pi electron in an adjacent ethylene mer. This initiates a fast, zipper-like reaction of bonding to other mers (Figure 13.3), which ends when a *terminator* is added.[*]

Addition polymerization is the simplest case, and we observe that we need two or more potential points of attachment on the mer in order to form a polymer. We therefore say that ethylene is *bifunctional*.

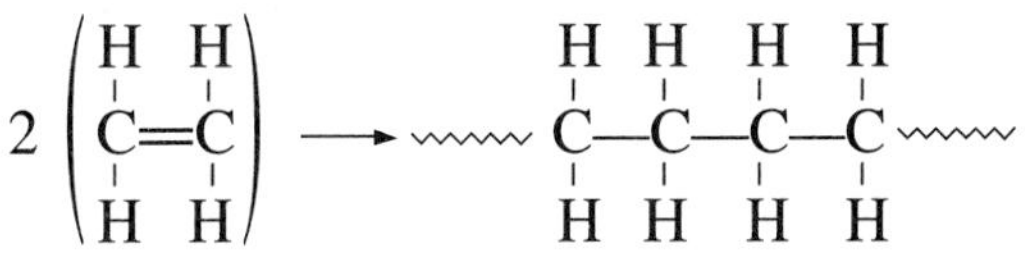

$$2 \begin{pmatrix} H & H \\ | & | \\ C{=}C \\ | & | \\ H & H \end{pmatrix} \longrightarrow \text{\Large\char`\~\char`\~} \begin{matrix} H & H & H & H \\ | & | & | & | \\ C{-}C{-}C{-}C \\ | & | & | & | \\ H & H & H & H \end{matrix} \text{\Large\char`\~\char`\~}$$

FIGURE 13.2   An example of the thermoplastic linear polymer, polyethylene

---

[*]Other effects that may terminate the reaction are impurities and reversion to a double bond between carbon atoms in a molecule, which is called *disproportionation*.

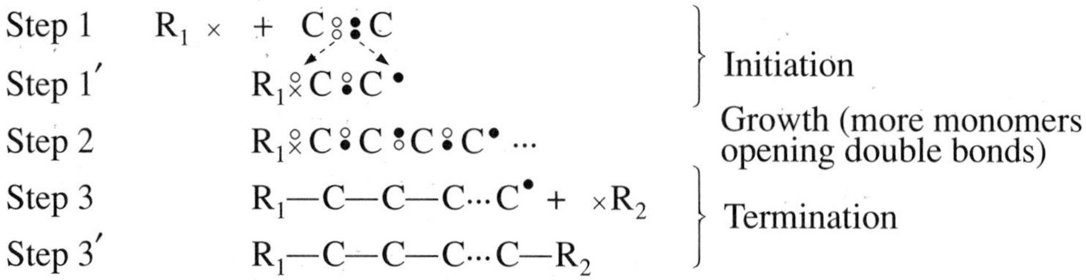

Step 1 R_1 ×   +   C $\overset{\circ}{\underset{\circ}{:}}$ C   ⎫
Step 1′ $R_1 \overset{\circ}{\underset{\times}{}} C \overset{\circ}{\underset{\circ}{}} C \,^\bullet$ ⎬ Initiation

Step 2 $R_1 \overset{\circ}{\underset{\times}{}} C \overset{\circ}{\underset{\circ}{}} C \overset{\circ}{\underset{\circ}{}} C \,^\bullet$ ...   Growth (more monomers opening double bonds)

Step 3 R_1—C—C—C···$C^\bullet$ + ×R_2 ⎫
Step 3′ R_1—C—C—C···C—R_2 ⎬ Termination

FIGURE 13.3   Atom bonding in addition polymerization. For simplicity, the hydrogen or other atoms bonded to the sides of the carbon chain are not shown. R_1 and R_2 are the initiator and terminator groups, respectively.

To obtain different properties, we can use other bifunctional molecules such as propylene to make *polypropylene.*

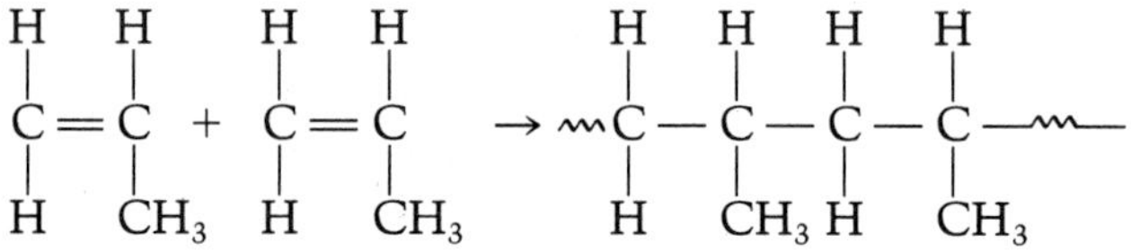

$$
\begin{array}{ccccc}
\text{H} & \text{H} & \text{H} & \text{H} & \\
| & | & | & | & \\
\text{C} = \text{C} & + & \text{C} = \text{C} & \rightarrow & \text{\tiny ww}\text{C} - \text{C} - \text{C} - \text{C} -\text{\tiny ww} \\
| & | & | & | & \\
\text{H} & \text{CH}_3 & \text{H} & \text{CH}_3 & \\
\end{array}
$$

(The symbol $\sim\!\!\sim$ means that the mer is attached to adjacent mers.) Note that this structure is different from that of polyethylene because of the extra or bulky side groups. These provide extra bonding (van der Waals forces) between molecules. For example, polyethylene will soften in a dishwasher but polypropylene will not.

Let us go on to another common thermoplastic, *polystyrene.* The term *styrene* simply means that we have a carbon ring of six carbon atoms (shown by the hexagonal symbol) attached instead of the CH_3 groups of polypropylene.

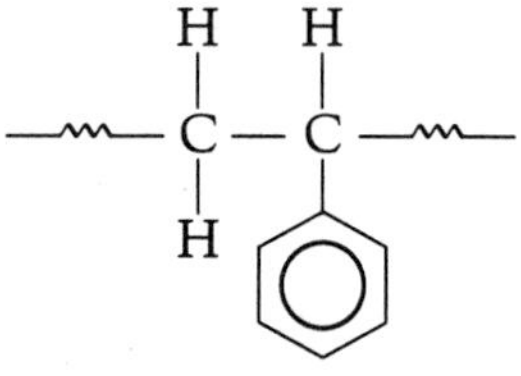

In addition to using side groups to improve properties, we may build foreign atoms into the carbon backbone. For example, Kevlar has the structure

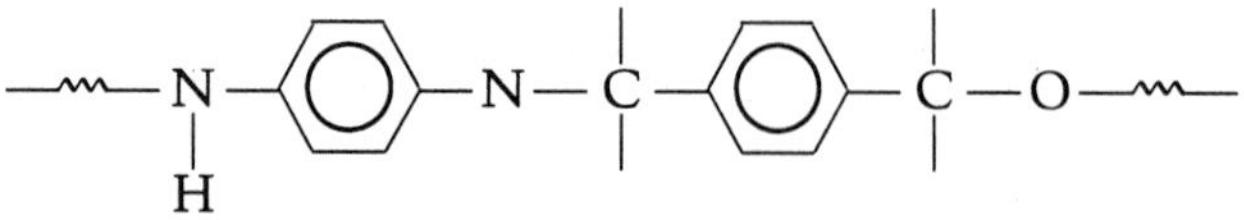

which seems quite complex compared to the simpler structures we have studied. However, this is only because the repeating unit, the mer, has oxygen and nitrogen as well as carbon rings in the backbone.

Condensation Polymerization

Condensation polymerization is another method of forming polymers. In this case, instead of breaking carbon double bonds, we bring together molecules that interract and form a *byproduct* (such as water or ammonia), which *condenses*. In the important case of nylon 6/6:

$$\begin{array}{c} H \\ \diagdown \\ N-(CH_2)_6-N \\ \diagup \qquad\qquad \diagdown \\ H \qquad\qquad\qquad H \end{array} \quad + \quad HO-\overset{O}{\overset{\|}{C}}-(CH_2)_4-\overset{O}{\overset{\|}{C}}-OH$$

(hexamethylene diamine + adipic acid)

$$\begin{array}{c} H \\ \diagdown \\ N-(CH_2)_6-N-\overset{O}{\overset{\|}{C}}-(CH_2)_4-\overset{O}{\overset{\|}{C}}-OH \ + \ H_2O \\ \diagup \qquad\qquad\quad | \\ H \qquad\qquad\quad H \end{array}$$

Note that both molecules are *bifunctional* at the N—H bonds and the C—OH bonds and that the reaction can continue. (It might appear that the hexamethylene diamine is tetrafunctional, but there is no space for additional molecules to attach around the nitrogen atom.)

13.3 Nomenclature

Before we investigate the bonding forces between different atom groupings, it may be helpful to review the names of a few common ones. (We will also provide some advice on how to deal with the mouth-filling ones, such as the hexamethylene diamine and adipic acid that we just encountered.)

It is important to know the following four prefixes:

Meth-1 Carbon	*Eth*-2 Carbons	*Prop*-3 Carbons	*But*-4 Carbons
$-\overset{\|}{\underset{\|}{C}}-$	$-\overset{\|}{\underset{\|}{C}}-\overset{\|}{\underset{\|}{C}}-$	$-\overset{\|}{\underset{\|}{C}}-\overset{\|}{\underset{\|}{C}}-\overset{\|}{\underset{\|}{C}}-$	$-\overset{\|}{\underset{\|}{C}}-\overset{\|}{\underset{\|}{C}}-\overset{\|}{\underset{\|}{C}}-\overset{\|}{\underset{\|}{C}}-$
*Meth*ane	*Eth*ane	*Prop*ane	*But*ane

When only single bonds between carbons are present, the suffix is *ane* as in *propane*. When a double bond is present, the suffix is *ene* as in *ethylene*. These are the important monomers for polymers when hydrogen atoms are present at each bond end.

We form a simple alcohol by substituting an OH group for an H; we form a dialcohol by substituting two OH groups.

The term *vinyl* means that we have substituted another atom for a hydrogen atom in *ethylene*, and that atom will be named. Thus *vinyl chloride* monomer is

$$
\begin{array}{cc}
\text{H} & \text{Cl} \\
| & | \\
\text{C} & = \text{C} \\
| & | \\
\text{H} & \text{H}
\end{array}
$$

The term *styrene* represents a special kind of vinyl in which a benzene ring instead of a single atom is substituted for a hydrogen atom.

The term *amine* denotes an organic group substitution for hydrogen atoms in NH_3, and *amide* denotes a

$$
\begin{array}{c}
| \quad\quad\quad \text{H} \\
-\text{C}-\text{N} \big\langle \\
| \quad\quad\quad \text{H}
\end{array}
$$

group in a polymer (polyamide).

These simple guidelines enable us to understand the structure of over two thirds of the polymers used. In the following list, the abbreviations of the polymers are given, along with the percentage of the market for polymers that each currently fills.

	Use % of market
Polyethylene, PE	32
Polypropylene, PP	11
Polystyrene, PS	9
Polyvinylchloride, PVC	15
	67

These structures and others are illustrated in Table 13.1.

In writing structural formulas, we use the following valences or sharing of electron bonds per atom: C = 4, N = 3, O = 2, and S = 2, and there is one shared bond for H, Cl, F, Br, and I.

Now what can we do to understand the others? Often the chemical name is not given and a trade name, such as nylon, is used instead. It is perfectly proper to ask the producer for the structural formula, which immediately reveals whether we are dealing with a thermoplastic or a thermoset (discussed later). Then, by examining the backbone structure and the side groups, we can make some predictions about bonding and substructure, which we will now discuss.

TABLE 13.1 Summary of Important Polymers

Group I. Thermoplastics		
Polymer	Percentage of Market	Monomer(s) Used
Polyethylene	32	$\begin{array}{c} \text{H} \quad \text{H} \\ | \quad | \\ \text{C}=\text{C} \\ | \quad | \\ \text{H} \quad \text{H} \end{array}$
Polyvinylchloride	15	$\begin{array}{c} \text{H} \quad \text{H} \\ | \quad | \\ \text{C}=\text{C} \\ | \quad | \\ \text{H} \quad \text{Cl} \end{array}$
Polystyrene	9	$\begin{array}{c} \text{H} \quad \text{H} \\ | \quad | \\ \text{C}=\text{C} \\ | \\ \text{H} \end{array}$ (⬡ is benzene, C_6H_6)
Polypropylene	11	$\begin{array}{c} \text{H} \quad \text{H} \\ | \quad | \\ \text{C}=\text{C} \\ | \quad | \\ \text{H} \quad \text{CH}_3 \end{array}$
ABS	3	$\begin{array}{c} \text{H} \quad \text{H} \\ | \quad | \\ \text{C}=\text{C} \\ | \quad | \\ \text{H} \quad \text{C}\equiv\text{N} \end{array}$ Acrylonitrile (graft) $\begin{array}{c} \text{H} \quad \text{H} \quad \text{H} \quad \text{H} \\ | \quad | \quad | \quad | \\ \text{C}=\text{C}-\text{C}=\text{C} \\ | \quad\quad\quad | \\ \text{H} \quad\quad\quad \text{H} \end{array}$ Butadiene (chain) $\begin{array}{c} \text{H} \quad \text{H} \\ | \quad | \\ \text{C}=\text{C} \\ | \\ \text{H} \end{array}$ Styrene (graft)
Acrylics (examples: polymethyl methacrylate, Lucite)	1	$\begin{array}{c} \text{H} \quad \text{CH}_3 \\ | \quad\quad | \\ \text{C}=\text{C} \\ | \quad\quad | \\ \text{H} \quad \text{C}=\text{O} \\ \quad\quad | \\ \quad\quad \text{O} \\ \quad\quad | \\ \quad\quad \text{CH}_3 \end{array}$

TABLE 13.1 Summary of Important Polymers (*Continued*)

Group I. Thermoplastics (Continued)		

Polymer	Percentage of Market	Monomer(s) Used
Cellulosics	<1	(mer of cellulose)
Acetals	<1	(mer)
Nylons	1	
Polycarbonates	<1	(mer)
Fluoroplastics (example: polytetrafluoroethylene)	<1	
Polyester, thermoplastic type [example: polyethylene-terephthalate (dacron)]	2	

Group II. Thermosetting Polymers		

Polymer	Percentage of Market	Monomer(s) Used
Phenolics (example: phenol formaldehyde)	6	

TABLE 13.1 Summary of Important Polymers (*Continued*)

Group II. Thermosetting Polymers (Continued)

Polymer	Percentage of Market	Monomer(s) Used	
Amino resins (example: urea formaldehyde)	4	(urea-formaldehyde monomer structure: H_2N—$C(=O)$—NH—CH_2—NH—$C(=O)$—NH_2 with CH_2O)	
Polyesters, thermoset type	3	H_2C—OH, HC—O(H), H_2C—OH and HO—$C(=O)$—$(CH_2)_x$—$C(=O)$—OH	
Epoxies	1	epoxide structure: H_2C(—O—)CH—$C(=O)$—R—$C(=O)$—CH(—O—)CH_2 and H_2N—R′	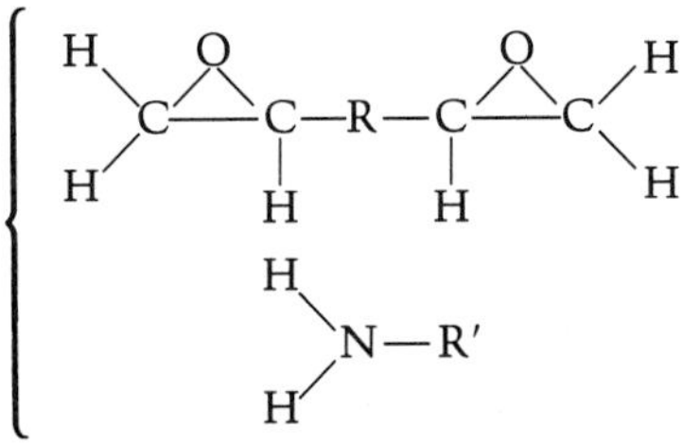 (R and R′ are complex polyfunctional molecules)
Polyurethane, also thermoplastic	4	OCN—R—NCO (diisocyanate) + HO—R′—OH	(R and R′ are complex polyfunctional molecules)
Silicones	1	Cl—$Si(CH_3)(Cl)$—Cl Trichlorosilane; H—O—$Si(CH_3)(OH)$—OH Trihydroxy silane	

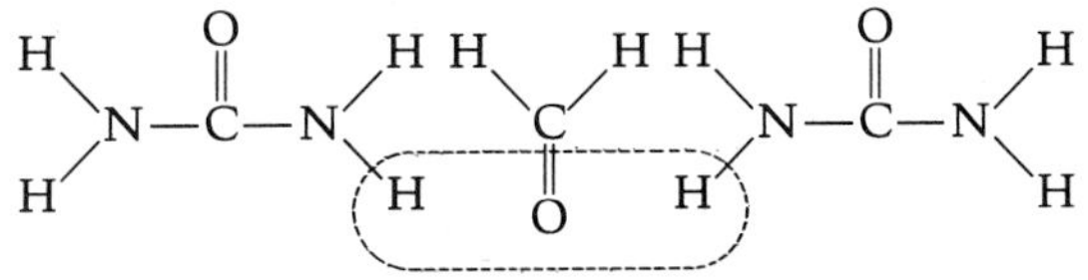

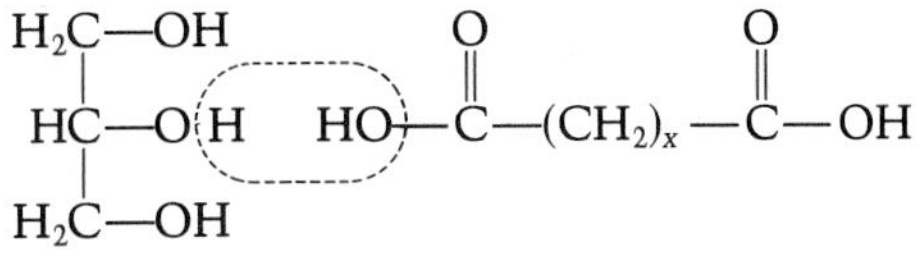

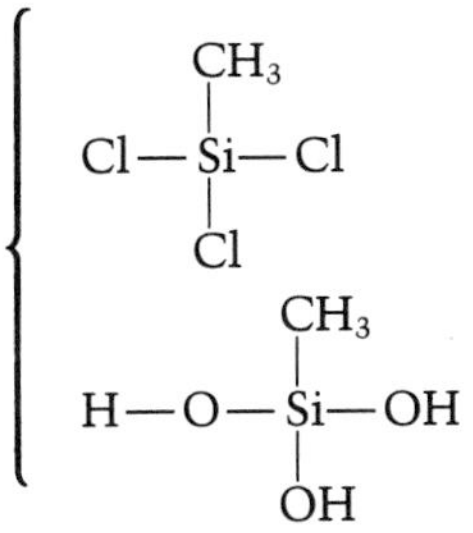

13.4 Bond Strengths

Let us consider what is involved in bonding the basic structures into polymers. First, however, take a moment to review the table of *bond strengths*, Table 13.2 (page 551).

EXAMPLE 13.1 *[ES]*

If we were to make polyethylene from ethylene, how much energy would be absorbed or evolved? Would the reaction tend to be spontaneous?

Answer   The reaction can be written in the following way:

$$
\begin{array}{cccccc}
& H \ H & \ \ \ H \ H & \ \ \ \ \ H \ \ H \ \ H \ H \\
& | \ \ | & \ \ \ \ | \ \ | & \ \ \ \ \ | \ \ | \ \ | \ \ | \\
& C{=}C \ + & C{=}C & \longrightarrow \ {-}C{-}C{-}C{-}C{-} \\
& | \ \ | & \ \ \ \ | \ \ | & \ \ \ \ \ | \ \ | \ \ | \ \ | \\
& H \ H & \ \ \ H \ H & \ \ \ \ \ H \ \ H \ \ H \ H
\end{array}
$$

mer

or

$$2 \ C{=}C \text{ bonds} \longrightarrow 4 \ C{-}C \text{ bonds} \quad (3 \text{ full bonds } + 2 \times \tfrac{1}{2} \text{ bonds}$$
because of sharing with adjacent mers)

[Note that if more than two monomers ($C{=}C$ bonds) are used, twice as many $C{-}C$ bonds will always be formed. Note also that, in accordance with our definitions, C_2H_4 with the unsaturated or double bonds is called the monomer, whereas C_2H_4 within the chain is called the mer.] Therefore,

$$2(145 \text{ kcal/mole}) \longrightarrow 4(80 \text{ kcal/mole})$$

or

$$290 \text{ kcal/mole} \longrightarrow 320 \text{ kcal/mole}$$

Because we obtain 30 kcal/mole more for the final products than for the reactants, we get more energy out of the bonds formed than we put in to break the bonds. Therefore, the reaction tends to be spontaneous.

We shall use the same reasoning later in evaluating more complex reactions to decide what process is the most probable.

In addition to the covalent bonds within a molecule, there are van der Waals bonds between molecules and in overlapping parts of the same molecule. These are much lower in strength than the covalent bonds, but they are very important because in most cases, the stress required for fracture is related to the force needed to *separate* molecules rather than to that required to break bonds *within* the molecule. The bonds within the long-chain molecule do not break during fracture; they are covalent. The hydrogen bond deserves special attention because it is very strong in a number of cases, especially in cellulose (cotton) and polyamides (nylon and protein). Examples of strong hydrogen bonds follow.

Bond	$-H...X-$ Bond Length, Å	Energy, kcal/g-mole[*]
O—H...O—	2.7	3 to 6
N—H...O—	2.9	4
N—H...N—	3.1	3 to 5

[*]To obtain J/g-mole, multiply kcal/g-mole by 4.186×10^3.

TABLE 13.2 Bonding Energies in Organic Molecules

Bond	Bond Energy* (kcal/g-mole)	Wavelength for Corresponding Energy
1. C≡N nitrile	209	
2. C≡C	200	
3. C=O	174	
4. C=C	145	
5. C≡S	129	
6. C—C aromatic	124	
7. C—H acetylene	121	
8. C—F	119	
9. O—H	110	2537 Å (short-wavelength UV)
10. C—H ethylene	106	
11. C—H methane	98	
12. Si—O	89	
13. C—O	87	
14. S—H	87	
15. N—H	84	3650 Å (long-wavelength UV)
16. C—C aliphatic	80	
17. C—O ether	79	
18. C—Cl	78	
19. S=S	76	
20. Si—H	75	
21. Si—C	70	4000 Å (blue light)
22. C—N nitromethane	68	
23. C—S	66	
24. O—O peroxide	64	5500 Å (yellow light)
25. N—N hydrazine	37	

*The values can vary with the type of neighboring bonds. Note in particular the C—H bond in 10 and 11. To obtain J/g-mole, multiply kcal/g-mole by 4.186×10^3.
From *Plastics*, 6th ed., by J. H. DuBois and F. W. John. Copyright © 1981 by Van Nostrand Reinhold Company Inc. Reprinted by permission of the publisher.

The hydrogen bond does not have the strength of a covalent bond, however, because no electrons are shared. The bond arises from the fact that at the H side of an OH radical, for example, there is a positive polarity because the electron from the hydrogen is attracted strongly to the oxygen. Similarly, an oxygen that has attracted an electron from another source is of negative polarity.

These strong hydrogen bonds can result in exceptions to our traditional definition of thermoplastics, insofar as it includes the easy flow of chains when the temperature is raised. In materials such as cellulose and polyte-trafluoroethylene, these strong secondary bonds are not all broken at the same time when the temperature increases. Furthermore, high chain stiffness (as in cellulose) prevents easy motion between chains. The result is better thermal resistance than might be anticipated for a thermoplastic. (See the column showing *heat-distortion temperature* in Table 13.3.)

13.5 Bonding Positions on a Monomer, Functionality

A monomer must have bonds that can be opened up for attachment to other monomers of the same or different formula. The number of bonds the monomer has is called the *functionality*.* The ethylene monomer is bifunctional, because the $C=C$ bond can react with two neighboring monomers. For polymerization the monomers must be at least bifunctional. Not only is a double carbon bond bifunctional, but we can split the $N-H$ bond in an amine, the $O-H$ bond in an alcohol, and the $C-OH$ bond in an acid to form another bond. There are two or more of these bonds in amines and dialcohols. For example, in ethylene glycol (a dialcohol of ethylene), $H-O-CH_2-CH_2-O-H$, we find bifunctionality for polymerization because both OH groups can be split.[†] Therefore, we can form polymers without a carbon double bond in the monomer. When the monomers are trifunctional or greater, we can form network or thermosetting polymers.

The *degree of polymerization* (DP) is the molecular weight of the polymer divided by the molecular weight of the mer. It gives the number of mers in the molecule, or its average length. It is important because, as we have already seen, the presence of larger molecules results in higher bond strengths and therefore higher melting or softening points. There is always a variation in the DP in a given batch of polymers, because not all chains start growing at the same time. The time of growth of individual chains is also variable. This variation can be measured and expressed statistically.

The molecules in a polymer are of different lengths and therefore of different molecular weights, so it is important to assess this variation. Two calculations are employed for this purpose: the number average and the weight average (see also the Problems).

EXAMPLE 13.2 *[ES/EJ]*

Assume we wish to produce polyethylene with a "number average" molecular weight of 20,000 g/mol. What is the average degree of polymerization and how much H_2O_2 is required to polymerize the ethylene if the hydrogen peroxide acts as both an initiator and a terminator?

---

*This discussion is so greatly simplified that it is not possible to predict many cases of functionality from these simple premises. See a text on organic chemistry for a fuller explanation.

[†]This happens provided there are present other bifunctional molecules *that have groups that can react with dialcohols in this manner.*

Answer

$$DP = \frac{(MW)_{poly}}{(MW)_{mer}}$$

$$= \frac{20{,}000 \text{ g/mole}}{(2 \times 12 + 4 \times 1) \text{ g/mole}} = 714$$

The average length of the chains is 714 mers; some will be longer and some shorter.

$$H_2O_2 + x(C_2H_4) \longrightarrow H-O\left(\begin{array}{cccccc} H & H & H & H & H & H \\ | & | & | & | & | & | \\ C & C & C & C & C & C \\ | & | & | & | & | & | \\ H & H & H & H & H & H \end{array}\right)_x O-H$$

If we ignore the contribution of OH to the molecular weight of the polymer and use a 100-g basis, we get

$$\frac{100 \text{ g}}{20{,}000 \text{ g/mole}} = \text{moles polymer} = \text{moles } H_2O_2 \text{ required}$$

or

$$\frac{100 \text{ g}}{20{,}000 \text{ g/mole}} = \frac{y \text{ g } H_2O_2}{(2 \times 1 + 2 \times 16)}$$

$$y = 0.17 \text{ g } H_2O_2 \text{ required}$$

(It is unlikely that all of the H_2O_2 will break down to form OH groups; thus more than the calculated ideal amount of H_2O_2 may be necessary.)

13.6 Variations in Thermoplastic Polymer Structures

As we have already suggested, the bonding between molecules determines the properties, so it is important to understand the variations in structure that affect bonding. These include branching, side group orientation, crystallinity, copolymers, and plasticizers.

Branching

Branching can take place in a linear, thermoplastic polymer as well as in the network type. It is important in polyethylene. One way in which branching can occur is if an addition agent removes a hydrogen atom from the side of a chain, whereupon growth can occur at this point. The difference between a branching structure and a network structure is shown in Figure 13.4.

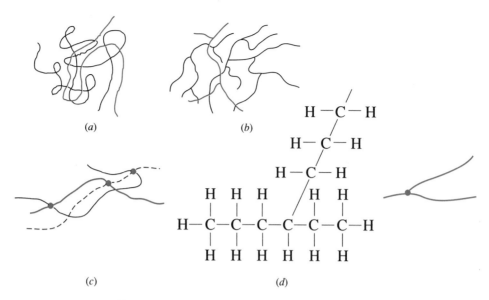

FIGURE 13.4 Different types of carbon backbone structures: (*a*) linear, (*b*) branched, (*c*) loose network, (*d*) branching in polyethylene

It is important to recognize that branching results in side arms that create more mechanical entanglements than occur without the side arms. A network, on the other hand, has covalent bonding *between chains* and is hence more rigid. We will discuss networks more extensively when we consider the thermosetting materials.

Location of Atom Groups

Another factor that provides for variation in polymer structures is the location of different atom groups along the chain. It is important to control the symmetry or distribution of a given group or element on the sides of the chain. There are three possibilities, exemplified by the three types of polypropylene (Figure 13.5). In polypropylene the *atactic* (random side group) material is a wax-like product of little use, softening at 74°C (165°F). However, both the *isotactic* (same side of chain) and the *syndiotactic* (opposite side of chain) types are useful, tough, partly crystalline plastics that melt at about 175°C (347°F). Bulky side groups may also cause *stereo interference* that limits the symmetry configuration. In other words, when the side groups, such as CH_3 in Figure 13.5, become large, they may not fit on the same side of the chain. The strain on carbon atoms within the chain may not allow some stereo combinations, such as isotactic and syndiotactic.

Other Important Configurations

Other configurations that are important in polymer structures are *trans* and *cis* structures, exemplified by gutta-percha and natural rubber, two different configurations of the same chemical formula (Figure 13.6).

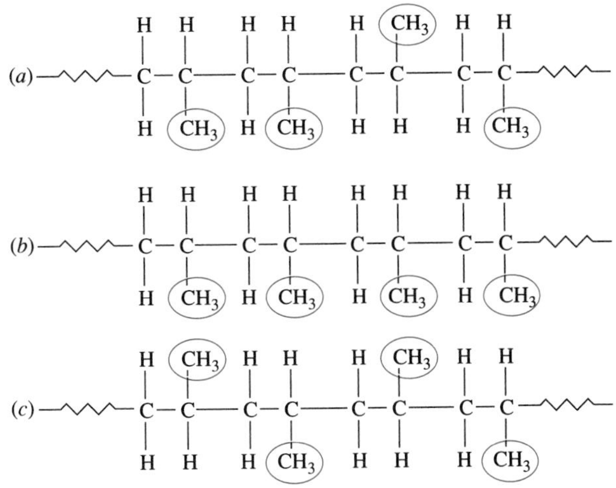

FIGURE 13.5   (a) Atactic (random sides), (b) isotactic (same side), and (c) syndiotactic (opposite side) structures for polypropylene

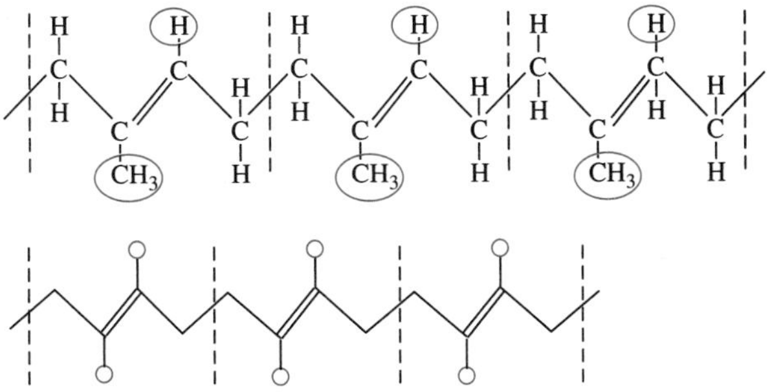

(a) *trans*–polyisoprene
[Possible rotations about indicated C——C bonds (- - - - -)
retain the linear structure of the molecule.]

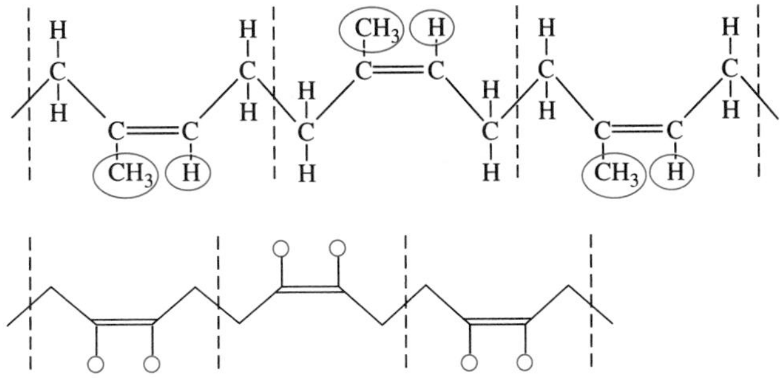

(b) *cis*–polyisoprene
[Possible rotations about indicated C—C bonds (- - - - -)
allow the molecule to coil up upon itself.]

FIGURE 13.6   *Trans* and *cis* configurations: (a) *trans:* gutta-percha, (b) *cis:* natural rubber. [*Note:* The structure cannot pivot around a C=C bond.]

In gutta-percha, the *trans* structure, the CH_3 and H are on opposite sides of the double bond, leading to a relatively straight carbon chain. In rubber, the *cis* configuration, the chain curves (there cannot be later rotation of the *double* bond), leading to a helical molecule that makes it rubber or spring-like under tension compared with gutta-percha, which is brittle.

Crystallinity

To this point we have considered only the random arrangement of linear molecules. A second type of structure, the crystalline, is important in the linear polymers: it leads to stiffer, stronger materials and usually makes an amorphous material translucent or opaque because light scatters at grain boundaries.

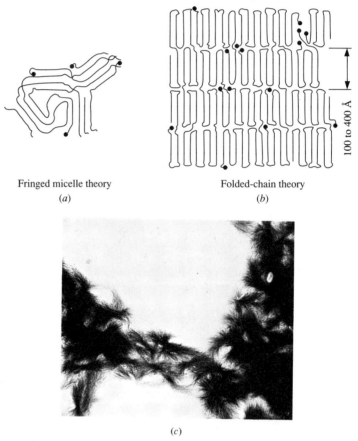

Fringed micelle theory
(a)

Folded-chain theory
(b)

100 to 400 Å

(c)

FIGURE 13.7 (a) Fringed micelle model of crystallinity. (b) Crystallinity created by the folding back of chains. The growth mechanism involves the folding over of the planar zigzag chain on itself at intervals of about every 100 chain atoms. A single crystal may contain many individual molecules. (c) Spherulites (spherelike areas of crystallinity) in polycarbonate polymer; transmission electron micrograph, 66,000×.

[Parts (a) and (b) from L. E. Nielsen, *Mechanical Properties of Polymers*, Litton Educational Publishing, Inc., New York, 1962. Reprinted by permission of Van Nostrand Reinhold Company. Part (c) courtesy Jim de Rudder. University of Michigan.]

The earliest model of the crystalline structure is called the *fringed micelle* (Figure 13.7a). In this model the molecules were considered to be side by side in some crystalline regions (the micelles) and to be randomly arranged in others. This model is rather crude, but it does help to explain the strengthening that occurs when a fiber such as nylon is drawn: The yield strength increases because the molecules in the amorphous regions become stretched out and aligned in the direction of the applied stress, forming crystals. Recent research using electron microscopy has given more detailed information on the crystalline regions. Researchers have grown single crystals of a number of polymers and have found that the chain structures of the molecules fold back and forth in regular fashion to build up the crystal (Figure 13.7b). In polycrystalline materials, the crystal size can be increased either by annealing at elevated temperatures or by increasing the temperature of crystallization, as in the crystals of metals. For example, if a polyethylene crystal that has a typical thickness of 100×10^{-7} mm at 100°C (212°F) is heated at 130°C (266°F) for several hours, its thickness will increase to 400×10^{-7} mm. Figure 13.7c shows the formation of spherulitic crystals. Relatively large crystals can be grown from liquid solutions, and x-ray diffraction techniques can identify crystallographic planes, as shown in the chapter frontispiece.

The tendency to crystallize is very important and is closely related to the structure and polarity of the molecule. Regular molecules without bulky side groups or branches show strong tendencies to crystallize. It should be emphasized that partial rather than complete crystallization is obtained at best.

EXAMPLE 13.3 *[ES/EJ]*

Explain why the following polymers have little or no crystallinity whereas related polymers have much more: (a) branched polyethylene, (b) atactic polypropylene, (c) random copolymers of linear polyethylene and isotactic polypropylene, and (d) random *cis* and *trans* forms.

Polymer	Percent Crystallinity
Linear polyethylene	90
Branched polyethylene	40
Isotactic polypropylene	90
Atactic polypropylene	0
Random copolymers of linear polyethylene and isotactic polypropylene	0
Trans-1,4-polybutadiene	80
Cis-1,4-polybutadiene	80
Random *cis* and *trans* forms of above	0

Answers

A. The branches interfere with regular arrangement and thus reduce the crystallinity.

B. The random arrangement of the CH_3 side groups in the atactic form leads to noncrystalline material.

C. Introducing a nonregular assortment of ethylene and polypropylene mers leads to a random spacing of the CH_3 side groups of the polypropylene and thus to a noncrystalline material.

D. The situation in the random case is similar to that of atactic polypropylene in part (b).

Copolymers, Blending, Plasticizers

It is possible to synthesize useful polymers by joining different mers. For example, if we alternate mers of ethylene and vinylchloride, we form the ethylene-vinylchloride copolymer:

$$-(CH_2)_2-C_2H_3Cl-(CH_2)_2-C_2H_3Cl-$$

This is an alternating copolymer. Other types of copolymers are random, block (in which a number of mers of one kind are inserted into a chain as a block), and graft (in which the copolymer is added as a branch) (Figure 13.8). Crystallization in copolymers is only partial at best. *Blending* or alloying is the combining of two or more distinct polymer molecules to form a new product with different characteristics. In a case such as the blending of polyethylene and nylon the new product has unusual permeability.

Blending with a low-molecular-weight (approximately 300) material is called using a *plasticizer* because this is a common way to soften a polymer or make it flexible. Just as adding water to clay plasticizes it, adding a small polar molecule can affect the van der Waals forces between the polymer chains. If we add too much plasticizer, the bond will be chiefly between the small molecules and we will have a liquid. In a common paint, excess plasticizer is added to make the paint liquid and allow brushing. After the paint is applied, the plasticizer evaporates and the paint dries. This evaporation is usually accompanied by some polymerization and crosslinking with oxygen.

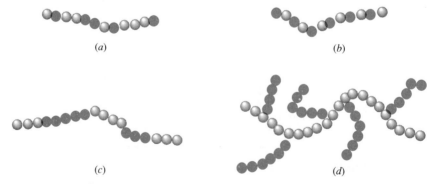

(a)   (b)

(c)   (d)

FIGURE 13.8   Different types of copolymers: (a) random, (b) alternating, (c) block, (d) graft

The residual plasticizer makes the film of paint tough and flexible. With time, however, the combination of further oxidation and loss of plasticizer can result in brittleness and flaking of the paint.

It should be pointed out that plasticizers may be either internal or external; that is, they can be part of the chain or separate small polar molecules that allow chains to move relative to one another. The external plasticizers are more readily lost with time than the internal plasticizers, which resist evaporation and thus enable the polymers to retain their flexible characteristics over longer periods.

A *filler* is somewhat the opposite of a plasticizer. It is usually added to improve strength and dimensional stability while decreasing the cost of the polymer. A more complete discussion of fillers appears in Chapter 14.

EXAMPLE 13.4 *[ES/EJ]*

Draw a possible structure of ABS (acrylonitrile-butadiene-styrene), given that it is described as a "graft of styrene and acrylonitrile on a butadiene backbone" (Table 13.1). Indicate whether the material is thermosetting, thermoplastic, or crystalline and whether it could form a network.

Answer

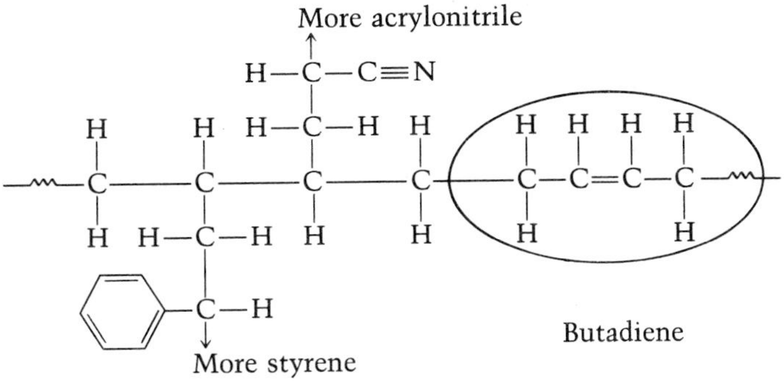

The material is thermosetting or thermoplastic, depending on whether or not it crosslinks to form a network. Because there can still be unsaturated carbon positions in the butadiene (as outlined in the second mer), crosslinking is possible. The material is too irregular to be crystalline (copolymers show little crystallinity). In practice, separate regions of butadiene are present. Crosslinking is discussed more fully in Section 13.11 as it applies to elastomeric materials and in terms of how it can lead to network structures.

13.7 Properties of Thermoplastics at Ambient Temperatures

Table 13.3 lists the structures and properties of some important thermoplastics, along with their abbreviated structures. We have already discussed

TABLE 13.3   Properties of Thermoplastic Polymers

Name and Structure[a]*	Use,[b] per-cent	Price[c] dollars/lb	Tensile Strength[d] psi	Percent Elonga-tion	Rockwell Hardness, HRR
Polyethylene					
High density	13	0.32	4,000	15 to 100	40
Low density	19	0.40	2,000	90 to 800	10
Polypropylene	11	0.40	5,000	10 to 700	90
Polystyrene	9	0.50	7,000	1 to 2	75
Polyvinylchloride (rigid)	15	0.33	6,000	2 to 30	110
Polytetrafluorethylene (Teflon)	<1	5.00	2,500	100 to 350	70
ABS, acrylonitrile-butadiene-styrene copolymer[h]	3	0.80	4,000 to 7000	20 to 80	95
Polyamides (6/6 nylon)	1	2.00	11,800	60	118

*See page 562 for table footnotes.

Impact[e] izod, ft-lb	Modulus,[d] psi $\times 10^3$	Specific Gravity	Coefficient of Expansion, $(°F)^{-1} \times 10^{-6}$ $[(°C)^{-1} \times 10^{-6}]$	Heat-Distortion Temperature,[f] °F (°C)	Burning Rate[g] in./min	Typical Applications
1 to 12	120	0.95	70 (126)	120 (49)	1	Clear sheet,
16	25	0.92	100 (180)		1	bottles
1 to 11	200	0.91	50 (90)	150 (66)	1	Sheet, pipe, coverings
0.3	450	1.05	38 (68.5)	180 (82)	1	Containers, foams
1	400	1.40	30 (54)	150 (66)	<1	Floors, fabrics
4	60	2.13	55 (99)	270 (132)	0	Chemical ware, seals, bearings, gaskets
1 to 10	300	1.06	50 (90)	210 (99)	1	Luggage, telephones
1	410	1.10	55 (90)	220 (104)	Low	Fabric, rope, gears, machine parts

TABLE 13.3 Properties of Thermoplastic Polymers (*Continued*)

Name and Structure[a]	Use,[b] per-cent	Price[c] dollars/lb	Tensile Strength[d] psi	Percent Elonga-tion	Rockwell Hardness, HRR
Acrylics (Lucite)	1	0.91	8,000	5	130
Acetals	0.3	0.91	10,000	50	120
Cellulosics	<1	0.55 to 0.70	2,000 to 8,000	5 to 40	50 to 115
Polycarbonates	0.7	2.50	9,000	110	118
Polyesters	2	0.45	8,000	300	117

[a]A continuing bond is indicated by —ᴧᴧ—. R may be any complex molecule. In general, hydrogen atoms are not shown.
[b]Total tonnage 1987, 24 million tons.
[c]1987 prices.
[d]Multiply by 6.9×10^{-3} to obtain MN/m^2 (MPa) or by 7.03×10^{-4} to obtain kg/mm^2.
[e]Multiply by 0.138 to obtain kg-m.
[f]Loaded at 264 psi, 1.82 MN/m^2 (MPa), gives severe deflection.
[g]Multiply by 25.4 to obtain mm/min.
[h]The Butadiene is present as fine spheres as a second phase.

polyethylene, polypropylene, polystyrene, and polyvinyl chloride. We see that as the side groups become more complex, we gain in strength and lose ductility.

In Teflon all the hydrogen atoms of ethylene are replaced with fluorine. The C—F bond is so strong that there is little bonding to external substances, which accounts for the low coefficient of friction. In the lower part of the table involving the more complex structures such as polyamides (nylon) and polycar-

Impact[e] izod, ft-lb	Modulus,[d] psi × 10³	Specific Gravity	Coefficient of Expansion, (°F)⁻¹ × 10⁻⁶ [(°C)⁻¹ × 10⁻⁶]	Heat-Distortion Temperature,[f] °F (°C)	Burning Rate[g] in./min	Typical Applica- tions
0.5	420	1.19	40 (72)	200 (93)	1	Windows
2	520	1.41	44 (79)	255 (124)	1	Hardware, gears
2 to 8	500 to 4,000	1.25	75 (135)	115 to 190 (46 to 88)	1.4	Fibers, films, coatings, explosives
14	350	1.2	25 (45)	275 (135)	<01	Machine parts, pro- pellers
1	340	1.3	33 (60)	130 (54)	Low	Magnetic tape, fibers, films

bonate, the added atoms give stronger secondary bonding forces. For example, the tensile strength of nylon is 11,800 psi compared to 2000–4000 psi for polyethylene. The modulus of elasticity is also four times higher.

13.8 Effect of Time and Temperature on Properties of Thermoplastics

When tensile tests are conducted on specimens of cellulose acetate at differ- ent temperatures (Figure 13.9*a*), high strength and brittle behavior are found at low temperatures, and lower strength and ductile performance at higher temperatures. A change in modulus with temperature for a group of polymers is shown in Figure 13.9*b*. We may represent this behavior, which is typical of

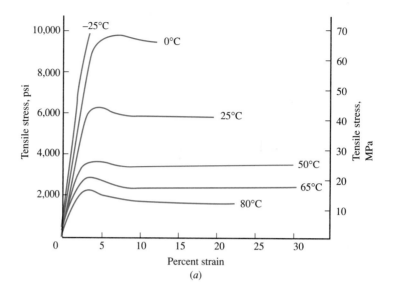

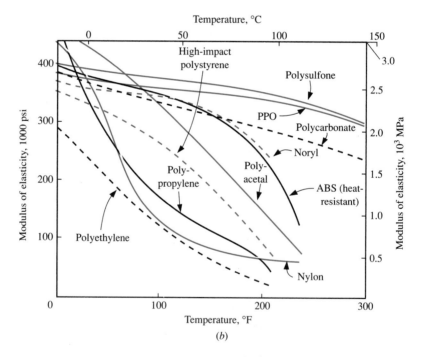

FIGURE 13.9 (*a*) Effect of temperature on the stress–strain curve of cellulose acetate. (*b*) Effects of temperature on properties of various resins.

[Part (*a*) after T. S. Carswell and H. K. Nason. Part (*b*) from *Modern Plastics Encyclopedia*, McGraw-Hill, New York, 1968.]

a thermoplastic polymer, with a schematic graph (Figure 13.10). At low temperatures characteristic for the particular material, the behavior is glassy and Hookian (that is, stress is proportional to strain). As higher temperatures are reached, the bonds between molecules are overcome by thermal agitation and the flow becomes viscous.

Now let us consider the effects of speed of testing or stressing, again using the elastic modulus as a measure of bond strength. When strain takes place, molecules must move, and such movement takes time. Therefore, Figure 13.10 applies to one strain rate. It is possible to modify the results by changing the strain rate. (Also, the effects noted at higher temperatures in Figure 13.10 would be obtained at lower temperatures with lower strain rates.) Thus at a given temperature a short-time test results in a higher modulus and brittle or glassy behavior for a given polymer. On the other hand, the same specimen would be rubbery or would even flow if the material were given sufficient time to react to the applied stress. Silly Putty is an excellent example of a material exhibiting this varied behavior. This familiar material bounces like a ball (short-time stress) but flows like a liquid if left for a long time on a table top. In an extreme case, with very rapid application of stress (such as a hammer blow), it will shatter.

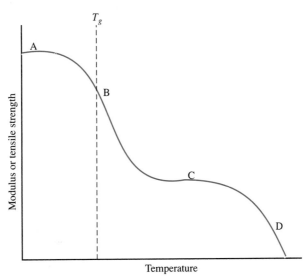

Period	
A	Hooke's law is obeyed: material is glassy and has low ductility.
B	Side chains loosen, some main chain movement occurs, material shows mixed elastic and viscous behavior.
C	Main chain loosen; material is rubbery.
D	Thermal excitation overcomes bonds, material flows.

FIGURE 13.10   Variations of modulus and tensile strength as a function of temperature for a thermoplastic (schematic). T_g is the glass transition temperature.

The term that describes the interdependence of time, strain, and temperature in polymers is *viscoelasticity*. The relationship is literally a combination of elastic and viscous flow characteristics. This is not the first time we have encountered the phenomenon; recall our discussion of the viscosity of glass in Chapter 10. We also introduced T_g, or the glass transition temperature, in Chapter 10.

Variables Affecting T_g

The *glass transition temperature* is such an important engineering property that it is worthwhile to consider some of the variables that affect it.

1. The *free volume, V_f,* of the polymer is the volume of the polymer not occupied by the molecules. The free volume is then the volume of the polymer mass, V, minus the volume of solidly packed molecules. We find that the higher the percentage of free volume, the lower T_g becomes. In other words, we have to cool the polymer to have sufficient attractive forces between molecules for the material to exhibit solid properties. In many polymers V_f/V is approximately equal to 0.025 at T_g.
2. Higher *attractive forces* between molecules increase T_g.
3. *Chain mobility.* The more difficult the rotation about bonds and the stiffer the chains, the more difficult it will be for molecules to separate and assume the properties of a liquid. Therefore, the lower the chain mobility, the higher the T_g.
4. The greater the *chain length*, the more difficult the separation for melting. An interesting empirical relationship has been developed.

$$T_g = T_g^\infty - \frac{C}{x} \tag{13-1}$$

where T_g is the glass temperature for the chain length x, T_g^∞ is the extrapolated glass temperature for chains of very great or infinite length, and C is a constant for the particular polymer. When x is large (greater than 500 mers), $T_g \approx T_g^\infty$.

There is a useful structural correlation between T_g and T_m, the melting point. For polymers with a symmetrical repeating unit, such as polyethylene, $T_g/T_m = 1/2$. For polymers with unsymmetrical repeating units, such as polypropylene,

$$-\underset{\underset{\text{H}}{|}}{\overset{\overset{\text{H}}{|}}{\text{C}}}-\underset{\underset{\text{H}}{|}}{\overset{\overset{\text{CH}_3}{|}}{\text{C}}}-$$

the ratio $T_g/T_m = 2/3$. The irregular structure restricts molecular movement at a higher temperature; that is, T_g is closer to T_m.

Effect of Strain Rate

In an ideal solid material, application of a load within the elastic region gives the result shown in Figure 13.11a. When stress is applied at $t = 0$, a corresponding elastic strain occurs (point 1). The value of the strain depends on the stress and the modulus of elasticity (strain = stress/E). The elastic strain remains constant with time (point 2), and upon unloading, all the elastic strain is recovered (point 3). Steel, for example, exhibits such behavior at room temperature even if it has been under load for several years.

Now let us consider a polymer that may undergo some viscoelastic flow at room temperature (Figure 13.11b). Upon loading at $t = 0$, the strain increases to point 1. The material reacts in an elastic fashion because *immediate* unloading would return the sample to zero strain. However, over time the strain increases to point 2 even though the load is held constant. Upon removal of the load, we find that there has been some permanent or plastic deformation (point 3). We would never have observed this had we not held the sample under load for an extended period. We may now make the following observations:

1. The modulus of elasticity, stress divided by strain, depends on the strain value, which may change with time.
2. The plastic strain component varies with temperature, because the increase in plastic strain is a response of the material to the load. Atoms or molecules must move, and their motion is temperature-dependent as well as load-dependent.

Therefore the strain-rate–temperature interdependency of polymers is explained by the viscoelastic response of these materials.

Figure 13.11b has been somewhat idealized for our discussion. In actual practice we would find varying degrees of curvature between points 1 and 2. Also, upon unloading, we may find exponential decay to point 3. It is possible

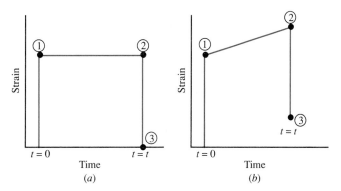

FIGURE 13.11 When a material is loaded elastically, it may return to its original zero strain condition, as in (a), or it may exhibit viscoelastic flow with a permanent residual strain, as in (b).

that all the strain may be recovered, but only after a long period of time. These observations have been treated mathematically by using analogs of springs and dashpots in various arrangements. The springs are used to represent the elastic strain component, whereas dashpots represent viscous flow.

We may think of the viscoelastic response as reflecting the ability of the material to adjust to its stress. That is, if we had maintained a constant strain on our material, the stress would decrease with time. A number of polymers show stress decay to be exponential according to the relationship

$$\sigma = \sigma_0 e^{-t/\lambda} \tag{13-2}$$

where σ_0 = stress at time 0, σ = stress at time t, and λ = relaxation time. The relaxation time is therefore the time necessary to reduce the stress to $1/e$ of its original value. The relationship can be derived from a single spring and dashpot in series.

In the example in Figure 13.11a the relaxation time approaches infinity and we have a perfectly elastic material. However, in polymers the relaxation time may be finite. The reduction of stress (or, correspondingly, the increase in strain) then becomes important to the selection of materials.

EXAMPLE 13.5 *[ES]*

The relaxation time for a particular polymer is known to be 1 year at 20°C. How long would it take to reduce the original stress on a component made of that polymer by 25%? How might we determine the stress relaxation at 100°C if the relaxation time is 8 months at 50°C?

Answer

$$\sigma = \sigma_0 e^{-t/\lambda} \qquad 0.75\sigma_0 = \sigma_0 e^{-t/1} \qquad \ln 0.75 = -t$$

or

$$t = 0.29 \text{ yr (approximately } 3\tfrac{1}{2} \text{ months)}$$

The dependence of relaxation on temperature can be expected to follow the same relationship as that for diffusion and viscosity, or

$$\frac{1}{\lambda} = Ae^{-B/T} \qquad \frac{1}{12} = Ae^{-B/293} \qquad \frac{1}{8} = Ae^{-B/323}$$

Solving simultaneously, we have $A = 6.58$ and $B = 1280$. Therefore,

$$\frac{1}{\lambda} = 6.58e^{-1280/T} \text{ for } \lambda \text{ in months} \quad \text{and} \quad \frac{1}{\lambda_{373\,\text{K}}} = 6.58e^{-1280/373} = 0.213$$

or

$$\lambda_{373\,\text{K}} = 4.7 \text{ months and } 0.75\sigma_0 = \sigma_0 e^{-t/4.7}$$

$$t = 1.35 \text{ months}$$

Relaxation requires a much shorter time period at the higher temperatures.

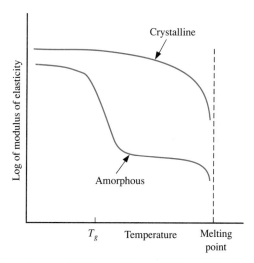

FIGURE 13.12   Effect of crystallinity on the modulus of elasticity

Figure 13.12 shows schematically the effect of crystallinity on a thermoplastic material. With increased crystallinity the polymer becomes stronger but more brittle. Toughness can be increased by having partial crystallinity at a sacrifice in strength.

13.9 Formation of the Thermosetting (TS) Structure

The distinction between linear — thermoplastic (TP) — polymers and TS polymers is that the latter have a space network structure. A bowling ball made of a thermosetting polymer is essentially one giant molecule made up of mers linked in space. For a space network structure to be obtained, the monomers must be at least trifunctional, or, if several monomers are present, one must be trifunctional. Either addition or condensation polymerization may be used, but in the largest tonnage (phenol-formaldehyde), a condensation reaction occurs (Figure 13.13). In the reaction shown, only two monomers of phenol and one of formaldehyde are used. To develop a space network structure, two other bonds can be made to the carbon ring at alternating carbon positions. The phenol is therefore the trifunctional monomer, and the formaldehyde is bifunctional, tying together the phenol groups.

This does not mean that a complete three-dimensional *network* must be formed. For example, we may stop the reaction in phenol-formaldehyde so that the product appears linear, as shown in Figure 13.13. This stage is called the B stage and is used as a starting point in production molding where it can be softened by later heating. (The network stage is called the C stage, and the beginning monomers A.)

The amount of this network structure will dictate the polymer's properties. A tightly networked structure is said to be highly crosslinked, and a

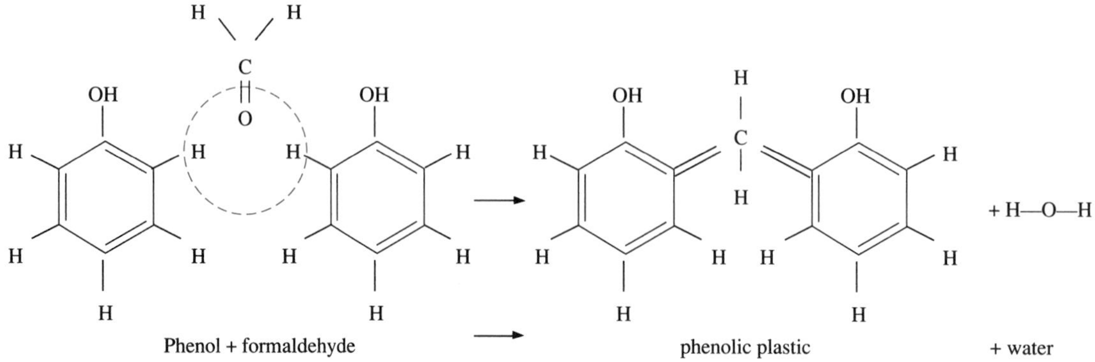

FIGURE 13.13 Formation of phenol-formaldehyde polymer where further polymerization gives a network. [Note: The condensation reaction for phenol-formaldehyde is more complex than what is shown; it involves several intermediate steps.]

loose network (as shown in Figure 13.4c) is lightly crosslinked. We will discuss chemical *crosslinking* methods in Section 13.11 on elastomers.

Ring Scission

Ring scission is a related phenomenon in that two molecules are joined by a third. However, in this case a ring structure such as an epoxide group,

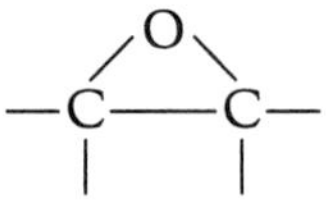

is broken by combination with the linking reagent.

EXAMPLE 13.6 *[ES/EJ]*

As many people know, epoxy cement comes in two tubes. In one tube is the epoxy itself, and in the other is a material that will crosslink the epoxy molecules. Show how polymerization could take place through the breaking of one of the rings (Table 13.1). Also explain how the reaction may lead to a tough, crosslinked network.

Answer

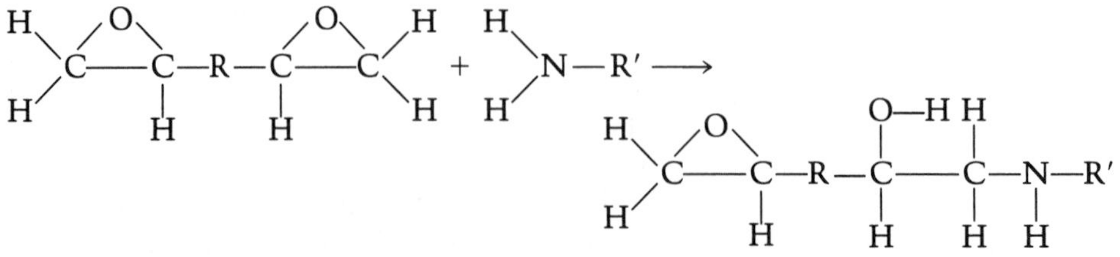

If we looked only at the

ring (there are two per molecule), we could not explain crosslinking or the formation of a network because we need three bond sites. However, crosslinking may take place between molecules at the R and R' groups, which are usually multifunctional. Another possibility is to use a monomer that has more than two rings per molecule. Several epoxy grades actually do this.

EXAMPLE 13.7 *[ES/E]*

The reaction between urea and formaldehyde is somewhat like the phenol-formaldehyde reaction (Figure 13.13) but more complex. It takes place in two steps: first the formation of a methylol compound and then the condensation step.

Urea + formaldehyde ⟶ methylol compound

How does the condensation step take place, and is the product thermoplastic or thermosetting?

Answer

(The reaction is repeated at other N—H bonds.)

A complete network is formed, which makes the polymer thermosetting. The network would also inhibit crystallinity.

13.10 Typical Thermosetting Polymers

In Table 13.4 a number of TS polymers are listed, with their structures and properties. It is important to recognize ways in which they differ from the thermoplastics.

TABLE 13.4 Properties of Thermosetting Polymers

Name and Structure[a]	*Use, per-cent*	*Price,*[b] *dollars/lb*	*Tensile Strength,*[c] *psi*	*Percent Elonga-tion*	*Rockwell Hardness, HRR*
Phenolics (phenol-formaldehyde) 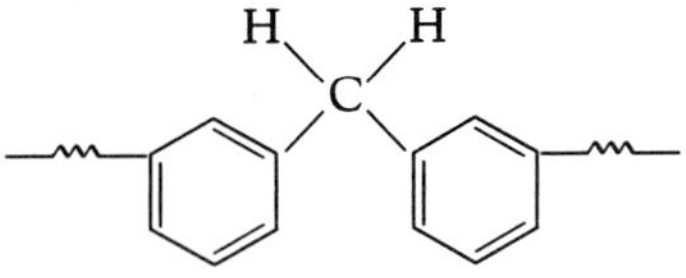	6	0.60	7,500	0	125
Urea-melamine 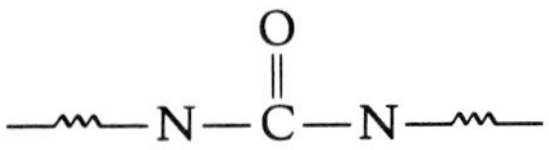	4	0.64	7,000	0	115
Polyesters 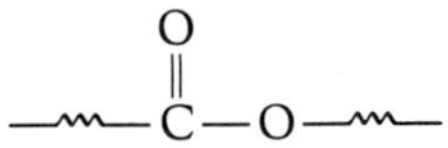	3	0.50	4,000	0	100
Epoxies 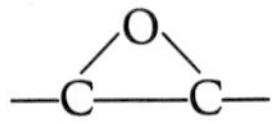	1	1.5	10,000	0	90
Urethanes 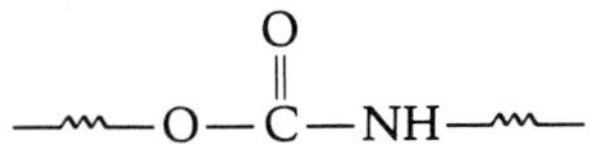	4	1.10	5,000	—	—
Silicones 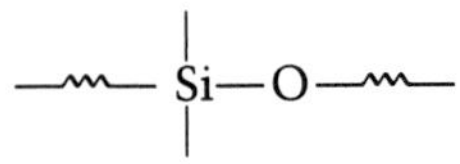	<1	1.50	[1,000] 3,500	[400] 0	89
	Properties are for glass-filled silicone.				

[a]A continuing bond is indicated by —∿—. R may be any complex molecule. In general, hydrogen atoms are not shown.
[b]1987 prices
[c]Multiply by 6.9×10^{-3} to obtain MN/m^2 (MPa) or by 7.03×10^{-4} to obtain kg/mm^2.
[c]Multiply by 6.9×10^{-3} to obtain MN/m^2 (MPa) or by 7.03×10^{-4} to obtain kg/mm^2.
[d]Multiply by 0.138 to obtain kg-m.
[e]Loaded at 264 psi 1.82 MN/m^2 (MPa).
[f]Multiply by 25.4 to obtain mm/min.

Impact,[d] izod, ft-lb	Modulus,[c] psi × 10³	Specific Gravity	Coefficient of Expansion, (°F)⁻¹ × 10⁻⁶ [(°C)⁻¹ × 10⁻⁶]	Heat Distortion,[e] °F (°C)	Burning Rate,[f] in./min	Typical Applications
0.3	1,000	1.4	45 (81)	300 (149)	<1	Electrical equipment
0.3	1,500	1.5	20 (36)	265 (129)	0	Dishes, laminates
0.4	1,000	1.1	42 (75.5)	350 (177)	1.4	Fiberglass composite, coatings
0.8	1,000	1.1	40 (72)	350 (177)	1	Adhesives, fiberglass composite, coatings
—	—	1.2	32 (57.5)	190 (88)	<1	Sheet, tubing, foam, elastomers, fibers
0.3	1,200	[1.25] 1.75	[139 (250)] 20 (36)	360 to 900 (177 to 482)	<1	Gaskets, adhesives, elastomers

Elastomer properties are shown in brackets.

1. They have heat-distortion temperatures 100–200°C (180–360°F) higher than TP polymers.
2. The effect of heat is breakdown (degradation) rather than melting.
3. They have higher tensile strength and much lower elongation (Figures 13.14 and 13.15).
4. Their modulus of elasticity is generally several times higher.

Thermosets are less widely used than thermoplastics, chiefly because TP materials are easier to form. This difference will be covered when we discuss production methods. In general, the TS polymers are used where lower creep rates and resistance to higher temperatures are needed.

However, the possibility of degradation (in which the polymer undergoes chemical decomposition, often with the evolution of poisonous gases) must be avoided. It should be noted that thermoplastic polymers may suffer degradation at high temperatures or even at excessive liquefying temperatures via "unzipping" of the structure to revert to the monomer.

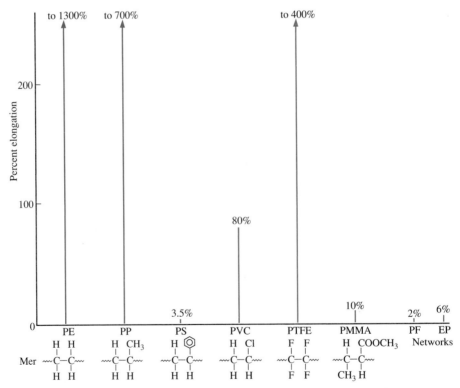

FIGURE 13.14   Range of percent elongation in tension of selected polymers at room temperature

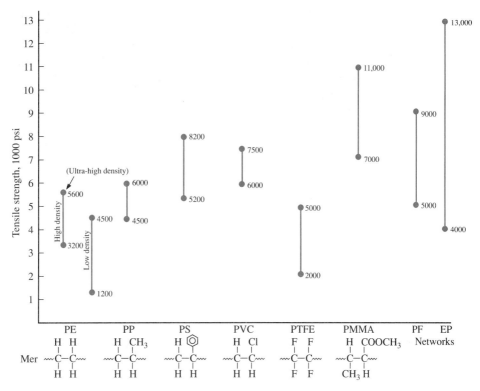

FIGURE 13.15   Range of tensile strength in selected polymers at room temperature

13.11 Elastomers

This fascinating group of polymers provides the greatest elastic strain of any group of materials (witness the common rubber band). We arbitrarily define an *elastomer* as a polymer that will return to its original length after being stretched repeatedly to twice its original length. The structural basis for this behavior is the elastic uncoiling and coiling of the molecules that we discussed when we examined *cis* and *trans* configurations. It is very important to distinguish between such *elastic* strain and the high elongations of over 400% observed for the thermoplastics (Figure 13.14); these *plastic* elongations are permanent extensions.

Although most early research was devoted to elastomers for rubber tires, more recent developments have been applied in making conveyors, foams, seals, and chemical-resistant coatings. To illustrate the importance of seals, it is only necessary to mention the O-ring failure that apparently caused the 1986 space shuttle disaster. One of the demonstrations in the Senate hearings showed that after immersion in a glass of ice water (freezing temperatures prevailed at launch time), the O-ring lost a great deal of its elasticity.

It is important, therefore, to examine one of the prominent methods used to modify the properties of elastomers: crosslinking by vulcanization. Up to the time of Charles Goodyear (1800–1860), natural rubber was little more than good pencil eraser material. Goodyear found that rubber heated with sulfur became harder and stronger and still retained its elasticity. The structural change is shown in Figure 13.16, and the effect on modulus in Figure 13.17.

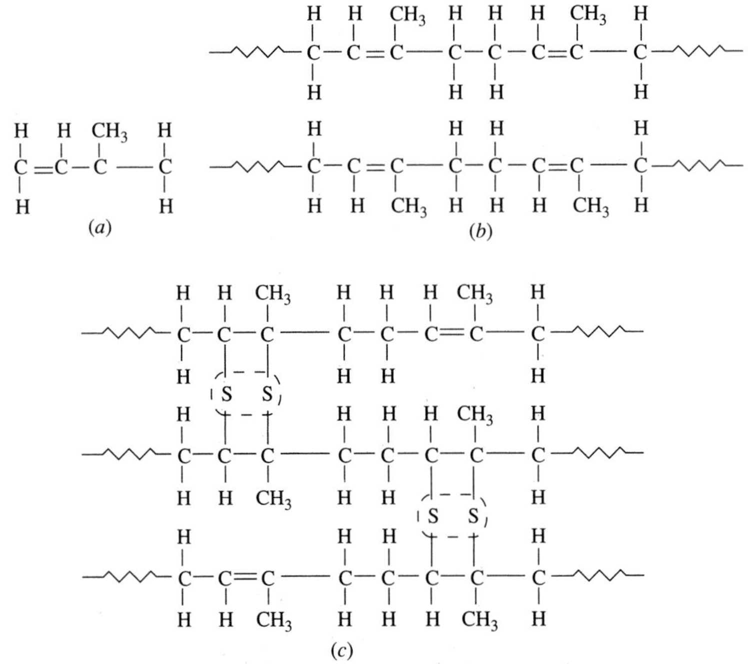

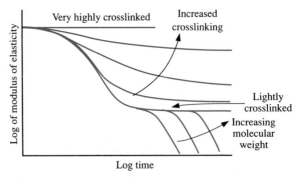

FIGURE 13.16   Chemical crosslinking of rubber by sulfur (vulcanization): (*a*) isoprene monomer, (*b*) isoprene polymer, (*c*) three polymer coil-like chains crosslinked with sulfur (outlined with dashed lines)

FIGURE 13.17   Effect of crosslinking on the modulus of elasticity

The range of properties varies from the soft rubber used in sponges and balls through the intermediate condition in tires to hard-rubber combs and battery cases. Other polymers have replaced the rubber battery case, because the extensive elasticity of an elastomer is not needed in this application, and other polymers may be used to produce lighter components.

Recently the labor-intensive vulcanization treatment has been eliminated in some thermoplastic elastomers. The process depends on the development in the polymer chains of a combination of stiff and flexible regions called domains. These materials include styrene and polyurethane copolymers; annual production in the United States is one-quarter of a million tons.

Typical commercial rubbers are listed in Table 13.5. Natural rubber, polyisoprene, and butadiene-styrene copolymer are widely used but have poor resistance to oil and gasoline and a limited temperature range. The more complex types such as nitrile have extended temperature ranges and better resistance to oil and gasoline. Elastomer foams will be discussed later, in Chapter 14.

Finally, it should be noted that as long as they are not heavily crosslinked, elastomers are similar in behavior to thermoplastics. They become brittle at low temperatures (below T_g) and can age in service as a result of further crosslinking with oxygen.

TABLE 13.5   Properties of Different Rubbers (Elastomers)

Common Name	Chemical Name	Tensile, Strength,* psi	Percent Elongation	Resistance to Oil, Gas	Useful Temperature Range, °F (°C)
Natural rubber	cis-Polyisoprene	3,000	800	Poor	−60 to 180 (−51 to 82)
GR-S or Buna S	Butadiene-styrene copolymer	250	3,000	Poor	−60 to 180 (−51 to 82)
Isoprene	Polyisoprene	3,000	400	Poor	−60 to 180 (−51 to 82)
Nitrile or Buna N	Butadiene-acrylonitrile copolymer	700	400	Excellent	−60 to 300 (−51 to 149)
Neoprene (GR-M)	Polychloroprene	3,500	800	Good	−40 to 200 (−40 to 93)
Silicone	Polysiloxane	700	300	Poor	−178 to 600 (−117 to 315)
Urethane	Diisocyanate polyester	5,000	600	Excellent	−65 to 240 (−54 to 115)

*Multiply psi by 6.9×10^{-3} to obtain Mn/m^2 (MPa) or by 7.03×10^{-4} to obtain kg/mm^2.

EXAMPLE 13.8 *[ES]*

Why is it typical for a thermoplastic to have a *higher* modulus of elasticity than an elastomer (rubber)?

Answer   The modulus of elasticity is defined as the load divided by the strain within the "elastic region." In rubbers a small load provides a high elastic strain and therefore the modulus is low. Confusion usually arises from an emphasis on the word *elasticity* rather than on the word *modulus* in the definition. Highly crosslinked elastomers show an increase in the modulus of elasticity with a decrease in elasticity (the maximum amount of elastic strain).

13.12   Other Properties of Polymers: Specific Gravity, Transparency, Coefficient of Expansion

The low specific gravity of the plastics compared with other classes of materials is important. It leads to favorable strength-to-weight and stiffness ratios. Some sample ratios are shown in the following table.

Material	Tensile Strength,* psi	Strength/Weight,[†] psi/(lb/in.^{3}) × 10^3
Cold-drawn SAE 1010 steel	53,000	155
ABS	6,500	170
6/6 nylon	11,800	297
Polyethylene (high density)	4,000	117

* Multiply by 6.9 × 10^{-3} to obtain MN/m^2 (MPa) or by 7.03 × 10^{-4} to obtain kg/mm^2.
[†] Yield strength of metals or tensile strength of plastics (in psi) divided by density (in lb/in.^{3})

Specific gravity is a function of the weight per volume of the individual molecules and the way they are bonded. Hydrocarbons are made of light atoms, so their density is generally low. The density of simple hydrocarbon polymers — polyethylene, polypropylene, polystyrene — is 0.9 to 1.1 g/cm^3. Substituting relatively heavy atoms such as chlorine or fluorine for hydrogen gives polyvinylchloride, with a density of 1.2 to 1.55 g/cm^3, or Teflon, with a density of 2.1 to 2.2 g/cm^3. The acetals are even denser because of the C—O—C chain packing.

The crystalline form of a plastic is always denser than the amorphous form because of more efficient packing. The difference in density is important in determining transparency, because the index of refraction is proportional to density. If, on the one hand, the densities of the amorphous and crystalline forms of a material are close, there is little dispersion as the light passes through a mixture of the two forms, and the material is transparent. On the other hand, if there is a substantial difference between the densities, the material is opaque.

As an example, let us consider the series polyethylene, polypropylene, and polypentene. In polyethylene the difference in density is great (specific gravity is 0.85 for the amorphous form and 1.01 for the crystalline). Therefore, where substantial crystallization takes place, the material is opaque. With polypropylene the difference is smaller (0.85 amorphous versus 0.94 crystalline), and parts are at least translucent. With polypentene $(-C_5H_{10}-)$ the densities are similar, and the material is transparent.

It is important to remember, however, that a crystalline material may be transparent if it is cooled rapidly so that the crystallites formed are very small—shorter than the wavelength of the light to be transmitted. Also, although amorphous polymers may be transparent, if fillers such as asbestos and carbon black are used the product is opaque.

The most important point concerning the coefficient of expansion is that in plastics it is from 2 to 17 times as great as it is in a typical metal such as iron. Thus due allowance should be made when a plastic is substituted for a metal in a part with close tolerance. Furthermore, when a composite metal-plastic part is molded of a brittle plastic, separation and cracking may take place. Such cracking is often noted on older automobile steering wheels that have been through severe temperature changes.

Summary

The backbone of the structure of high polymers consists of chains of carbon atoms. In thermoplastic materials the chains have long linear structures, and as a result the materials soften and flow upon heating. In thermosetting materials the chains are connected in a network structure and do not separate and yield a liquid.

The backbone structure is made by joining small structural elements called monomers. When there are only two bonding positions in the monomer, the linear thermoplastics are formed; if at least one variety of monomer in the mix is more than bifunctional, a network forms. When the monomers have been combined, as in polymerization, the basic building block is called a mer.

The polymer can crystallize if the mers are linear and possess no side groups on the chain or if they have only small, regularly oriented side groups (isotactic or syndiotactic).

Amorphous polymers exhibit behavior similar to that of glasses, passing from a viscous to a rigid structure on cooling through the glass transition temperature T_g. They have a rubbery range above T_g. When a linear polymer is made of several different monomers, it is called a *copolymer;* generally co-polymers do not crystallize.

Polymers exhibit a wide range of mechanical properties depending on their structure. Their low density gives them a good strength-to-weight ratio that makes them competitive with metals. The amorphous structures such as Plexiglas and polystyrene are transparent, whereas the crystalline polymers are translucent to opaque.

The elastomers have structures that uncoil readily and thus exhibit substantial elastic elongation. This property is modified by crosslinking with sulfur or oxygen.

Definitions

Addition polymerization   Bonding between similar and/or different monomers by linking at functional positions without the formation of a condensation product; also called *chain-reaction polymerization.* When monomers are different, the process is called *copolymerization.*

Atactic structure   A structure in which the side groups (such as CH_3) of a molecule are arranged randomly or nonpreferentially on the sides of the chain.

Bifunctional   Having two bonding positions. ("Tri" signifies three positions and "tetra" signifies four.)

Blending   An intimate mechanical mixture of several polymers.

Bond strength   The energy required to break a particular bond.

Branching   Structure wherein a linear polymer has forked branches.

Chain stiffening   The inclusion of foreign atoms in or along the carbon chain, resulting in increased bonding.

Cis **structure**   A curved carbon backbone produced, for example, by positioning the H and CH_3 groups on the same side of the chain.

Condensation polymerization   Bonding between mers that is effected by a reaction in which the bonding takes place with the emission of a by-product such as H_2O or NH_3 gas. For example, phenol + formaldehyde gives phenol-formaldehyde resin + water.

Crosslinking   The formation of connections between polymer molecules by a crosslinking agent. For example, butadiene rubber can be crosslinked with sulfur (vulcanization) or oxygen (oxidation).

Crystallinity   The alignment of a molecule or molecules in a regular array for which a unit cell can be described. Crystallinity is generally not complete.

Degree of polymerization   The molecular weight of a polymer divided by the molecular weight of the mer.

Elastomer   A high polymer with a coiled structure and rubber-like characteristics.

Filler   Foreign material that is used to strengthen or modify properties of polymers.

Functionality   The number of positions on a monomer at which bonding to another monomer can take place.

Glass transition temperature, T_g The temperature at which a high polymer becomes rigid. T_g is similar to the transition temperature for glasses.

Glassy range The temperature range in which the polymer acts glass-like or brittle and the strength and modulus are high.

Heat-distortion temperature The temperature at which a polymer softens severely.

Initiator A material that, when added to a monomer, acts to initiate polymerization.

Isotactic structure A structure in which the side groups of a molecule are all arranged on one side of the chain.

Linear structure The polymer structure in which the mers are joined in a line rather than in a network.

Mer A unit consisting of relatively few atoms joined to other units to form a polymer.

Monomer A mer standing alone—that is, not part of a polymer.

Plasticizer A chemical of lower molecular weight that is added to a polymer to soften or liquefy it.

Polymer A molecule made up of repeating structural groups, or mers. For example, polyethylene is made up of $-CH_2-CH_2-$ groups.

Ring scission The breakup of a ring structure to provide bonds for polymerization.

Rubbery range The temperature range in which very great elastic elongation is obtained. The polymer chains uncoil but then return to their original positions.

Solvent A liquid that dissolves a polymer—for example, in a paint.

Space network structure The structure formed when one or more of the monomers is tri-or polyfunctional.

Syndiotactic structure A structure in which the side groups of a molecule are regularly arranged on alternating sides of the chain.

Terminator A material that reacts with the end of a growing polymer chain to terminate growth.

Thermoplastic polymer A high polymer that flows and melts when heated. Scrap may be recovered by remelting and remolding.

Thermosetting polymer A high polymer that sets into a rigid network. Such a polymer does not melt when heated but chars and decomposes in a process called *degradation*. These polymers are not reusable.

***Trans* structure** A relatively straight carbon backbone produced, for example, by positioning the H and CH_3 on alternating sides of the chain.

Viscous range The temperature range in which a polymer shows permanent flow or deformation if given enough time. The relationship among time, strain, and temperature in polymers is called *viscoelasticity*.

Problems

13.1 *[ES]* Explain why propylene produces a good polymer but propane does not. (Sections 13.1 through 13.4)

13.2 *[ES]* Determine the coordination number for carbon atoms in diamond, and indicate why diamond is classified as a thermoset rather than a thermoplastic. (Sections 13.1 through 13.4)

13.3 *[ES]* Show with a sketch how the following monomer can polymerize by addition while still providing an unsaturated polymer. (Sections 13.1 through 13.4)

$$\overset{\displaystyle |}{C}=\overset{\displaystyle |}{C}-\overset{\displaystyle |}{C}=\overset{\displaystyle |}{C}$$

13.4 *[ES]* Sketch hypothetical monomers that could be classified as (a) trialcohol, (b) triacid, (c) diamine. (Sections 13.1 through 13.4)

13.5 *[ES/EJ]* Polytetrafluoroethylene (PTFE) can be used on a component that may be exposed to high temperatures. Explain why, as a thermoplastic, PTFE does not flow readily in high-temperature service. Why does PTFE not burn readily? (Sections 13.1 through 13.4)

13.6 *[ES/EJ]* The most common acrylic is polymethylmethacrylate, sold as Lucite or Plexiglas. Table 13.1 gives the methylmethacrylate structure. Sketch the structure of the polymer. Is it linear or network? Is it crystalline? (See Section 13.6) Is it thermoplastic? (Sections 13.1 through 13.4)

13.7 *[ES]* The most important nylons are 6/6, 6, and 6/10. The numbers refer to the numbers of carbon atoms in the mers that are joined by a condensation reaction. Let us take 6/6 nylon, which is formed from adipic acid and hexamethylene diamine. Actually the condensation takes place as the H of the NH_2 group combines with an OH to form water. Calculate the energy of the bonds broken and of the new bonds formed. Compare the alternative possibility of breaking off an NH_2 group and the H of the OH group to form NH_3 gas as a condensation product. Is the nylon thermoplastic or thermosetting? (Sections 13.1 through 13.4)

13.8 *[EJ]* Example 13.1 gives the calculation of the energy exchange for the polymerization of polyethylene. Explain whether this means that ethylene monomers placed in a container will polymerize. (Sections 13.1 through 13.4)

13.9 *[ES]* From the calculation of bond energies, what is the possibility of the chemical reaction $CH_4 + H_2O \rightarrow CH_3OH + H_2$? (H—H bond energy is 104 kcal/g-mole.) (Sections 13.1 through 13.4)

13.10 *[ES/EJ]* In Table 13.2, why is it that the values of the bond energy vary with the type of neighboring bonds? What is the significance of this variation for a calculation of an energy exchange to determine the polymerization potential? (Sections 13.1 through 13.4)

13.11 *[ES]* A researcher has suggested that the following condensation polymerization occurs between ethylene glycol and acetone. (Sections 13.1 through 13.4)

Ethylene glycol   Acetone

a. Show by calculating the energy involved whether the proposed reaction may be expected to take place.

b. Calculate how many pounds of water are formed per pound of mer (outlined in the diagram).

13.12   *[ES]* Indicate how hydrogen bonds can contribute to the strength of 6/6 nylon. (Sections 13.1 through 13.4)

13.13   *[ES/EJ]* Criticize the statement that addition polymerization produces a thermoplastic material, whereas condensation polymerization results in a thermosetting polymer. (Sections 13.5 through 13.6)

13.14   *[ES/EJ]* Polytetrafluoroethylene (PTFE, or Teflon) is an important plastic. What is the structure of the mer? Is the polymer thermoplastic or thermosetting? Suggest how it is attached to frying pans. (Sections 13.5 through 13.6)

13.15   *[ES]* The basis of the *silicone* structure is silane, SiH_4, which is analogous to methane, CH_4. Silicon hydrides are named according to the number of silicon atoms in the chain; thus Si_3H_8 is trisilane. When chlorine replaces hydrogen, we have chlorosilane. It is possible to produce a backbone structure analogous to the acetal structure consisting of alternating silicon and oxygen atoms. These are the silicones. They are prepared by first reacting chlorosilanes with water to form the trihydroxy silane given in Table 13.1. Show how these hydroxyl compounds can undergo condensation polymerization yielding a silicone. (Sections 13.5 through 13.6)

13.16   *[EJ]* The frequency distribution for two polymerization treatments of the same monomer is given below. (Sections 13.5 through 13.6)

   a. Is the degree of polymerization (DP) the same for the two polymers? Explain.

   b. Explain why we might be interested in the distribution rather than just the numerical value of DP.

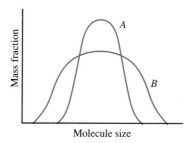

13.17   *[ES/EJ]* As pointed out in Section 13.5, there are several ways to measure the degree of polymerization (DP). There is no difficulty in determining the mer weight. Determining the polymer molecular weight is complicated, however, because there is a distribution of sizes.

   Finding the number average, $\overline{M}_n$, and finding the weighted average, $\overline{M}_w$, are two methods for determining the polymer molecular weight. The number average is mathematically defined as follows:

$$\overline{M}_n = \sum_{i=1}^{n} \left(\frac{n_i M_i}{\sum_{i=1}^{n} n_i} \right)$$

where n_i is the number of molecules with molecular weight M_i within an interval i.

   Although seemingly complex, the relationship merely says that each molecule-size interval contributes to the average in proportion to the fraction it represents of the total molecule sizes present. In other words, a numerical average is the sum of all the individual elements divided by the total number of elements.

Using a weighted average is another approach wherein each element is counted in proportion to the fraction of the total that it represents. With respect to the polymer molecular weight, the weighted average $\overline{M}_w$ is defined as follows:

$$\overline{M}_w = \sum_{i=1}^{n} \left(\frac{n_i M_i}{\sum\limits_{i=1}^{n} n_i M_i} \right) M_i$$

$\overline{M}_w$ is always greater than $\overline{M}_n$. The ratio of $\overline{M}_w / \overline{M}_n$ is called the *polydispersity index* (PDI) and can have values up to 30 for the commercial polymers. Large PDI values indicate a large proportion of smaller molecules. (Sections 13.5 through 13.6)

a. Give reasons why we would like to have a high value of $\overline{M}_n$.
b. What effect does a large PDI value have on the thermal resistance of a polymer?
c. Determine the values of $\overline{M}_n$, $\overline{M}_w$, and PDI from the following simplified data (more intervals and more molecules are usually present).

i	M_i	n_i
1	5,000	2
2	15,000	4
3	25,000	6
4	50,000	2

13.18 *[ES]* Calculate the energy exchange for the branching of polyethylene if hydrogen gas is a by-product. The H—H bond energy may be taken as being approximately 104 kcal/mol. (Sections 13.5 through 13.6)

13.19 *[ES]* (Sections 13.5 through 13.6)
a. Sketch the polymerization process in the formation of polyacrylonitrile.
b. The average molecular weight of the polymer is 24,350 g/mol. wt. What is the average degree of polymerization?
c. What would be the advantage if the degree of polymerization were doubled?

13.20 *[ES/EJ]* Explain why a high degree of crystallinity may be difficult to achieve in branched polymers, in atactic polymers, and in copolymers. (Sections 13.5 through 13.6)

13.21 *[EJ]* Anyone who has ever owned a car will recognize the following situations. Suggest what might be occurring in the structure of the material in each case. (Sections 13.5 through 13.6)
a. The insides of the windows develop a haze that resembles smoke, but it is not smoke.
b. Vinyl seats crack after a period of time. Those areas exposed to sunlight are especially susceptible.

13.22 *[ES]* For each case that follows, explain in one sentence how the addition of a filler might change the characteristic of the polymer as indicated. (Sections 13.5 through 13.6)
a. Reduces crystallinity
b. Inhibits crosslinking
c. Increases rigidity

13.23 *[ES/EJ]* Why might someone conclude that a thermoplastic material is weaker than a thermoset one? (Sections 13.7 through 13.10)

13.24 *[EJ]* Phenol–formaldehyde (Figure 13.13) is used for mounting metal samples to be polished and etched for examination under the microscope. Explain why a firmer mount is obtained when higher compaction pressures are used and when pressure is maintained until the mount has cooled to 90°C. The initial temperature may be 140°C. (Sections 13.7 through 13.10)

13.25 *[ES/EJ]* Indicate whether the following statements are correct or incorrect and justify each answer. (Sections 13.7 through 13.10)
   a. The degree of polymerization is a more appropriate measurement for a thermoplastic polymer than for a thermoset one.
   b. Polyesters are thermoplastic.
   c. Bulky side groups in a linear polymer are likely to result in higher strength.

13.26 *[ES]* Figure 13.13 shows the polymerization of phenol–formaldehyde. Show by calculation whether the following polymerization is also possible. (Sections 13.7 through 13.10)

13.27 *[ES]* By calculating energy values, show that polymerization of epoxy, as shown in Example 13.6 will indeed take place. (Sections 13.7 through 13.10)

13.28 *[ES/EJ]* Explain the following effects with respect to the glass transition temperature, T_g. (Sections 13.7 through 13.10)
   a. An increase in pressure on a polymer increases its glass transition temperature.
   b. The ratio T_g/T_m increases as the percentage of crystallinity decreases in a copolymer.

13.29 *[ES/EJ]* Using the data from Example 13.5, determine the time necessary to reduce the stress by 50% and by 75% of the original value at 50°C. Suppose that the polymer is used as bonding to hold together a wooden crate. What is the significance of your results? (Sections 13.7 through 13.10)

13.30 *[ES]* A stringed musical instrument uses polymer strings that can relax with time. What is the minimum relaxation time for the polymer if 5% stress relaxation in 1 hr can be tolerated before it becomes necessary to retune the instrument? (Sections 13.7 through 13.10)

13.31 *[EJ]* What would be the significance of a material exhibiting viscoelastic characteristics, such as that shown in Figure 13.11*b*, if we were to use it as a fishing line? (Sections 13.7 through 13.10)

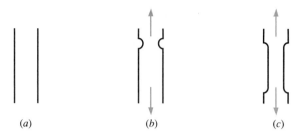

FIGURE 13.18   Behavior of polyethylene: (a) unstressed, (b) necking down begins, (c) necking down continues, with more stress.

13.32   *[ES/EJ]* Suppose that we established a "relaxation modulus" as follows:

$$E_t = E_0 e^{-t/\lambda} \tag{13-3}$$

where E_0 is the modulus at $t = 0$, E_t is the modulus at $t = t$, and λ = relaxation time. How might the relaxation modulus vary with temperature? What effect would it have on the data given in Figure 13.9b? (Sections 13.7 through 13.10)

13.33   *[EJ]* Explain the behavior of polyethylene in the tensile specimen shown in Figure 13.18: Why doesn't failure take place at the reduced section in (b) instead of the necking down continuing? (Sections 13.7 through 13.10)

13.34   *[ES/EJ]* In Tables 13.3 and 13.4 only tensile strengths are listed for the polymers. Why are the yield strengths not given? (Sections 13.7 through 13.10)

13.35   *[ES]* The relationship $T_g = T_g^\infty - C/x$ is given in the text (Equation 13-1). Plot T_g versus $1/x$ schematically where x might range from 50 to 1000. (Sections 13.7 through 13.10)

13.36   *[ES/EJ]* Explain why a thermoplastic polymer with high crystallinity is desirable for conditions of exposure to high temperatures. (Sections 13.7 through 13.10)

13.37   *[EJ]* When nylon reinforcing cord was first used in rubber tires, the tires developed "flat spots" as the vehicle remained stationary overnight. What accounts for this phenomenon, and why is the return to the round shape more rapid in the summer than in the winter? (Sections 13.7 through 13.10)

13.38   *[EJ]* Rubber products sometimes undergo degradation over extended periods of time. Explain why the following phenomena occur in rubber. (Sections 13.11 and 13.12)
   a. A rubber band fails after having been wrapped very tightly around an object for several months.
   b. An automobile heater hose bursts.
   c. A rubber washer in a water faucet no longer seals.

13.39   *[EJ]* For the outdoor storage of a recreational vehicle, the tires are often covered with an opaque material such as plywood, a tarp, or even aluminum foil. Why might this be a good idea? (Sections 13.11 and 13.12)

13.40   *[ES/EJ]* Indicate whether the following statements are correct or incorrect and justify each answer. (Sections 13.11 and 13.12)
   a. A 50% crystalline polymer is likely to be more transparent than an amorphous polymer.
   b. The specific gravity of a linear polymer increases in conjunction with an increase in its crystallinity.
   c. A polymer firmly attached to a piece of steel at 25°C shows residual tension when cooled to 0°C.

13.41 *[EJ]* The strength-to-weight ratio is the strength divided by the density (units of lb/in.^{3}) of the material involved. Determine this ratio for the following materials, using yield strength for the metals and tensile strength for the polymers. In addition, indicate the significance of these values in component design. (Sections 13.11 and 13.12, Chapters 7 through 9)
  a. High-density polyethylene
  b. 6/6 nylon
  c. Epoxy
  d. Annealed type 304 stainless steel
  e. Cold-drawn type 1020 steel
  f. Cold-worked 65-35 brass
  g. 7178-T6 aluminum

13.42 *[ES]* If all the isoprene in Figure 13.16 is crosslinked with sulfur, what percent by weight sulfur will have been added (weight of sulfur/weight of sulfur + isoprene)? (Sections 13.11 and 13.12)

13.43 *[ES/EJ]* (Sections 13.11 and 13.12)
  a. Show by a sketch how butadiene rubber, $CH_2(CH)_2CH_2$, can polymerize and crosslink with oxygen.
  b. Show by *calculation* whether the crosslinking with oxygen would be a spontaneous reaction (O═O bond energy is 118 kcal/mole).
  c. What would be the increase in weight if oxygen crosslinking were to occur at all of the available positions? (Use units of pounds increase in weight per pound of butadiene.)
  d. Explain how oxygen crosslinking is minimized.

13.44 *[ES]* Glyptal is a commercial sealer that is familiar to any experimentalist who has tried to produce a leak-proof joint. The condensation reaction is shown schematically below.

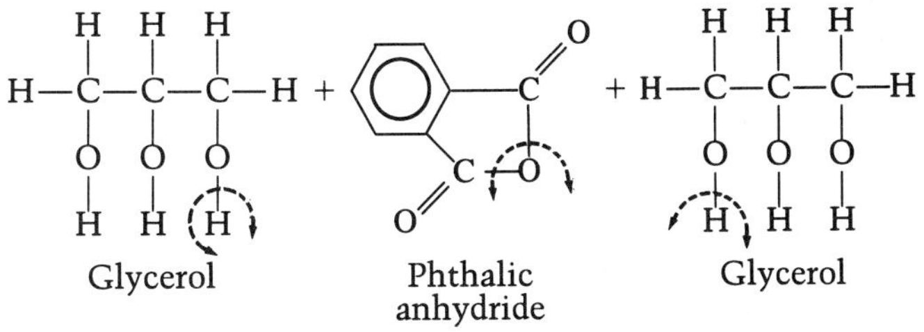

Glycerol   Phthalic   Glycerol
                anhydride

By-product = H—O—H

Show by making an appropriate sketch how the material can be either a thermoplastic or a thermoset, and also calculate the energy exchange for the thermoplastic form. (Chapter 13)

13.45 *[EJ]* Using only the following chemicals, illustrate with *structural formulas* [*] the following reactions. Be sure that all bonds satisfy valences. Show the unsatisfied bonds of a continuing polymer as follows: —ᴠᴠ—C═C═C—ᴠᴠ— (Chapter 13)

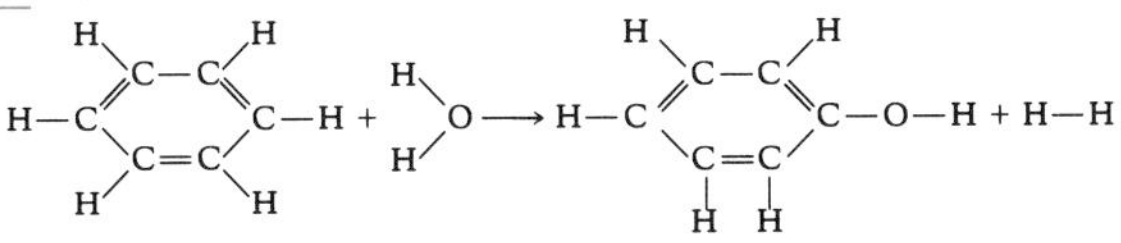

Chemicals:

$$CH_2O, \qquad C_6H_5OH, \qquad C_2H_4, \qquad C_2H_3Cl, \qquad CH_3NH_2, \qquad H_2O_2,$$

a. Formation of copolymer by addition polymerization
b. Formation of thermosetting resin by condensation polymerization
c. Polymerization by ring scission
d. Formation of a *cis* (isotactic) structure
e. Formation of a syndiotactic copolymer

13.46 *[ES/EJ]* The chemicals listed below are to be used alone or in combination to form the following polymers (a through e). Draw the structures of the components and indicate how they combine to form each of the following: (Chapter 13)

a. A transparent thermoplastic copolymer
b. A translucent thermoplastic linear polymer
c. A highly elastic, crosslinked elastomer
d. An opaque network polymer (thermoset)
e. A thermosetting *polyester* made by condensation polymerization

1. Formaldehyde, CH_2O
2. Sulfur
3. Carbon black
4. Ethylene, C_2H_4
5. Vinylchloride, C_2H_3Cl
6. Ethyl alcohol, C_2H_5OH
7. Styrene, $C_2H_3(C_6H_5)$
8. Phenol, C_6H_5OH
9. Tetrafluoroethylene, C_2F_4

10. A trialcohol,

11. A diacid,

12.

```
      H         H
      |         |
      C         C      H
    ⌇  |       ⌇|      |
   /   H    C         C  ⌇
            |         |
            CH₃       H
```

13.

```
      H              H
      |              |
      C              C
    ⌇ |            ⌇ |   ⌇
      H    C = C    H
      |   /     \   |
      CH₃        \  H
                    H
```

CHAPTER 14

Special Polymer Products

After seeing the high values of covalent bond strength along the axis of polymer molecules, the question may be asked: Why can these materials not be formed into fibers competitive with glass, graphite, and boron? By incorporating benzene rings and amide (NH_2) groups into the carbon backbone, the aramid fibers (Kevlar) have been developed. These fibers exhibit mechanical properties equivalent to or higher than other fibers with sufficient ductility to permit weaving into fabric.

In a typical case the use of Kevlar fiber results in a canoe weighing 50 pounds compared to 70 pounds for an equivalent fiberglass product. This substantial reduction is obviously of great advantage in portaging.

14.1 Overview

In the previous chapter we concentrated on the widely used materials called *commodity polymers*, such as polyethylene, polystyrene, polyvinylchloride, and the phenolic types. In this chapter we will investigate advanced *engineering polymers* that offer greater strength, creep resistance, and abrasion resistance. In the second part of the chapter we will take up special applications such as adhesives, fibers, foams, and coatings and their relationship to structure.

Returning to our first topic, advanced engineering polymers, let us start with the question "How do we proceed to obtain improved properties in a polymeric material?" There are four ways:

1. We can synthesize a *new polymer* structure. For example, Kevlar fiber is a highly crystalline polymer in which the backbone of carbon atoms is strengthened by the introduction of nitrogen and oxygen primary bonds. The rigidity of the chain prevents the folding, as that which occurs in the typical crystal structure, and leads to rod-like molecules with high strength in the longitudinal direction.*

Synthesis of a new structure is the most expensive way to achieve improved properties. Developing the material and an adequate market can take as long as 15 years.

2. We can develop a *copolymer* by introducing foreign monomers during polymerization. Depending on the type of copolymer desired (block, alternating, or other type), this can require considerable effort. But it is less costly and time-consuming than synthesizing a new structure.

3. We can *blend* two or more polymers. This process, which is also called alloying, should be clearly distinguished from developing a copolymer. In this case we start with two sets of large (polymer) molecules and bond them by van der Waals forces via mixing.

4. We can produce a composite such as fiberglass, as discussed in Chapter 16.

We will investigate approaches 1, 2, and 3 in order.

---

*The symbols ⬡ and ⬡ are equivalent and refer to an aromatic ring with cyclic bonding.

14.2 Advanced Engineering Polymers

One way to classify engineering polymers is by considering price levels:

Commodity polymers — low thermal resistance $0.30/0.75 per lb
Transition polymers — better thermal properties $0.75/1.25
Engineering polymers — good up to 350°F (177°C) $1.25/2.50
Specialized polymers — for applications above 350°F $2.00/20.00

We have considered the criterion of thermal stability, but other considerations (such as higher strength and toughness, corrosion resistance, ease of processing, or electrical properties) may justify a higher-cost material.

Table 14.1 describes the characteristics of several engineering polymers for comparison with nylon 6/6,* the "work horse" engineering polymer we discussed in Chapter 13, whose properties are listed first in the table.

Polyphenylene Sulfide

In the case of *polyphenylene sulfide* (PPS), we have a phenyl group (benzene ring with sulfur) as the mer. This *hetero*chain carbon bonding provides excellent strength and a high modulus, compared to *homo*chain carbon bonding. But how can we justify a price of $4.50/lb? This is a crystalline material and, as we shall see later when we discuss processing, provides a very fluid, easily injected liquid above the melting point. The lower coefficient of expansion and higher softening temperature are useful in many components. In addition, it is flame-resistant. PPS is used in a wide variety of applications, from large, thick-walled mechanical parts such as pump housings to delicate electronic parts such as for encapsulating integrated circuits on silicon chips.

Polysulfone

Polysulfone (PSU) is another polymer in which the carbon backbone is stiff. Here this strength is achieved by incorporating sulfur and oxygen and adding methyl side groups:

Compared to PPS, polysulfone is amorphous rather than crystalline because of the bulky side groups. However, the van der Waals forces between molecules are strong and the glass transition temperature is high; PSU has a

---

*6/6 indicates that there are six carbons in the diamine and six carbons in the diacid.

TABLE 14.1 Engineering Polymers

Name and Structure	Use, Tons/yr ×10³	Price,[a] $/lb	Tensile Strength,[b] ×1000 psi	% elongation	Modulus,[b] ×10⁶ psi	Coefficient of Expansion, 10⁻⁶/°C	Heat Distortion,[c] °F (°C)	Advantages
Polyamide (PA, nylon type); semicrystalline	200	1.75	11.8	60	0.4	90	220 (104)	Strength, thermoplastic
Polyphenylene sulfide (PPS); crystalline	8	4.50	9.5	1.6	0.5	49	275 (135)	Melt viscosity controlled, solvent-resistant, flame-resistant
Polysulfone (PSU); amorphous	n.a.	5.00	10.2	75.	0.36	51	345 (174)	Hydrolytic stability, resistance to high temperature, inert
Ultrahigh-molecular-weight polyethylene (UHMWPE); crystalline	n.a.	1.00	7.0	350.	0.1	200	n.a.	High resistance to impact and abrasion, low coefficient of friction
Ionomers; amorphous	70	1.75	30	400.	0.04	1.5	110 (43)	High strength and elongation

[a]1987 prices
[b]Multiply by 6.9×10^{-3} to obtain MN/m² (MPa) or by 7.03×10^{-4} to obtain kg/mm².
[c]Loaded at 264 psi, 1.82 MN/m² (MPa), gives severe deflection.

heat-distortion temperature of 345°F (174°C). This material has high hydrolytic stability (resistance to water) and therefore can be used in medical and food service applications requiring repeated sterilization. It is also used in electronic circuit boards.

Ultrahigh-Molecular-Weight Polyethylene

UHMWPE has the highest abrasion resistance and impact strength of all polymers. A slab 1 inch thick can stop a 38-caliber pistol slug fired from 6 inches! The name UHMWPE arises from the fact that this polymer's average molecular weight of 4×10^6 is ten times that of high-density polyethylene. Because it also exhibits self-lubricating, non-stick properties, UHMWPE is widely used for handling bulk materials (such as grain, cement, and gravel) in trucks, silos, and conveyors.

Ionomers

The key factor in the ionomer structure is the use of sodium or zinc ions to control crosslinking between typical polymer molecules—for example, in copolymers with ethylene. This results in high levels of resilience and impact resistance. As shown in Table 14.1, these materials have very high strength and elongation. Examples of severe use include golf ball covers and bowling pins. In the automotive industry, foamed bumper guards are a typical application. Although the T_g and softening temperatures are much lower than those of the other materials we have discussed, the lower T_g results in higher values of strength and ductility at subzero temperatures.

14.3 Copolymerization

In Chapter 13 we defined different types of copolymers; here we will go into greater detail. It is easy to get the impression that all we need do to form a copolymer is mix two monomers.

To produce a *random copolymer*, it is necessary to begin with a mixture of the *monomers* and induce concurrent growth. However, in many cases the composition of the growing molecule favors the more reactive mer, as shown in Figure 14.1.

We can think of a copolymer as being composed of two species, A and B. Since it is improbable that the reactivity of both monomers is exactly the same, the growing polymer favors the more reactive monomer. This can be treated quantitatively as a reactivity ratio r. For example:

r_A is a measure of the A monomer's tendency to react with itself compared to its tendency to react with B.

r_B is a measure of the B monomer's tendency to react with itself compared to its tendency to react with A.

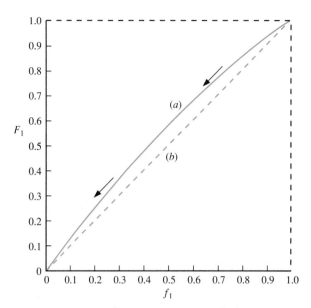

FIGURE 14.1   Instantaneous copolymer composition (F_1) versus monomer composition (f_1). (*a*) butadiene (A), styrene (B), 60C; $r_A = 1.39$, $r_B = 0.78$. The direction of composition drift in a batch reactor is indicated by arrows. The ideal path is shown by (*b*).

[From *Fundamental Principles of Polymeric Materials* by Stephen L. Rosen. Copyright © 1971 by Barnes & Noble, Inc. Reprinted by permission of Harper & Row, Publishers, Inc.]

We will now consider only two of several possibilities:

1. $r_A = r_B = 0$

   Each monomer has difficulty polymerizing with itself, yet it can exist quite well with the other monomer. We therefore obtain a perfectly alternating copolymer, ABABABABAB.... Polymerization ends when one of the species is depleted.

2. $r_A r_B = 1.0$

   This is a case of ideal copolymerization, wherein each monomer shows no preference for reacting with itself or with the opposite species, as in random copolymerization. However, the polymerization is driven in the direction from the more to the less reactive. Figure 14.1 is an example of this case; here butadiene is the A species and styrene the B monomer. The reactivity ratios of 1.39 for butadiene and 0.78 for styrene yield a result modestly above 1.0 when multiplied together, which is why the copolymerization path is above the dotted line that indicates the ideal.

   More complex plots than that shown in Figure 14.1 exist; they follow the general concepts developed in chemical kinetics.

   Either cationic or anionic polymerization can be used to form a block copolymer. In both cases the size of the molecules (DP) is confined to a narrow range, because each molecule is nucleated by an ionic catalyst. Therefore, after the molecules are grown to a given size via monomer A, another monomer, B, may be added to produce the desired length of B. Then the A monomer is added again to produce a block copolymer. Ions are used to control these events. It is even possible to grow three different mers in blocks.

14.4 Phase-Behavior Diagram for Copolymers

Though it is not an equilibrium diagram, a phase diagram can be developed for copolymers (Figure 14.2). In one case where the mers are similar in structure, a single-phase copolymer is formed. This is much like the formation of a continuous series of solid solutions that we find in the copper–nickel diagram. Just as in metals, however, two phases with different structures may separate. This phase diagram is more complicated than that for metals because a glassy phase may also exist. In other words, if the overall composition is between about 30% and 50% A, the copolymer passes gradually from a melt to a rubbery condition and then to a low-ductility glass below T_g (region 1 to region 5). At the A or B side of this region, crystalline polymers of the A or B type will develop, and the balance of the structure will be rubbery above T_g and brittle below (regions 2 and 4 for high B and regions 3 and 6 for high A). In both cases the melting point, T_m, is depressed from that of the pure polymer, just as the liquidus decreases in the diagram for a metal or ceramic.

Products that fall in the various regions shown in Figure 14.2 have various uses. A copolymer in region 5 (such as a homopolymer like polystyrene) is transparent and strong but is low in ductility below its T_g. A polymer in region 2 or 3 is rubbery but strengthened by the crystalline particles. These

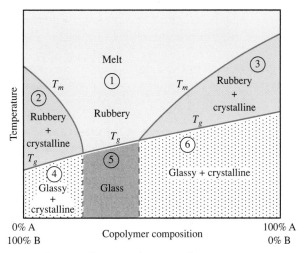

FIGURE 14.2   Phase diagram for a random copolymer system

[From *Fundamental Principles of Polymeric Materials*, by Stephen L. Rosen. Copyright © 1971 by Barnes & Noble, Inc. Reprinted by permission of Harper & Row, Publishers, Inc.]

properties resemble those of a pure polyethylene squeeze bottle, which is between T_m and T_g at room temperature. In regions 4 and 6 the mixture of crystals and glass gives a hard but low-ductility material such as nylon 6/6.

The advantages of the copolymers include not only the potential increase in mechanical properties but also the flexibility in processing provided by control of the melting point and the glass temperature.

Examples of Changes Made in Polymer Structure

To illustrate how the structure of a given polymer can be changed to attain different properties, we may consider the unsaturated polyesters. The essential structure of an unsaturated polyester is the presence of the ester group,

$$-O-\overset{\overset{\displaystyle O}{\|}}{C}-$$

and the existence of one or more double bonds that are functional (shown by the arrow).

$$-\!\!\wedge\!\!\!\wedge\!\!-\overset{\overset{\displaystyle O}{\|}}{C}-O-\underset{\underset{\displaystyle H}{|}}{C}=\underset{\underset{\displaystyle H}{|}}{C}-\!\!\wedge\!\!\!\wedge\!\!-$$

In a conventional polyester, a monomer such as styrene crosslinks the ester chains.

Resins of this type are used widely for applications such as boats, shower stalls, and swimming pools. For greater toughness, however, the polyester chains are synthesized to have double bonds at the ends, rather than a random distribution of styrene links. This gives greater resiliency after crosslinking.

In another variation, a low-shrinkage polyester is used to avoid severe shrinkage pockets that can cause depressions in the surface of a part. In this case, a thermoplastic polymer that is only partially soluble is added. Upon

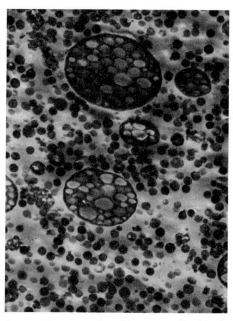

FIGURE 14.3   Typical ABS plastic with rubber particles (dark areas) dispersed in the SAN matrix (light background areas)

[Reprinted with permission from Engineered Materials Handbook, Vol. 2, *Engineered Plastics*, ASM International, 1988, p. 110.]

cooling, any shrinkage is dispersed as microvoids nucleated at the interface between phases.

Acrylonitrile–butadiene–styrene (ABS) is another prominent copolymer (Figure 14.3). The acrylonitrile and styrene form a true copolymer and a single phase. However, there is a second phase composed of dispersed rubber particles (butadiene) that have a layer of styrene-acrylonitrile grafted on the surface. The rubber particles bond to the SAN (*styrene-acrylonitrile*) phase. Each component contributes to the overall properties. Acrylonitrile provides heat resistance and surface hardness. Styrene contributes strength and makes processing easier. And the butadiene (rubber) improves toughness and impact resistance.

14.5 Blending

This method of improving materials is less expensive and gives more predictable properties than the development of a new polymer. The producer is generally making one or both polymers already, so it is only necessary to provide mixing equipment. In the usual case, the melted polymers are intimately blended in a shear intensive extruder. The mixture is extruded in strands and cut into pellets.

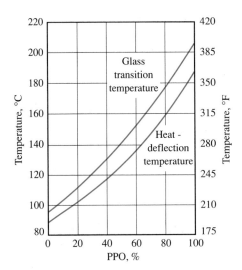

FIGURE 14.4 Heat-deflection temperature and glass transition temperature as a function of composition for PS–PPO blends

[Reprinted with permission from *Engineered Materials Handbook*, Vol. 2, ASM International, Metals Park, OH, 1988.]

It is important to analyze the miscibility of the polymers being blended. In the rather exceptional case of polysulfone–polypropylene oxide (PS–PPO), the amorphous polymers are completely miscible and only one T_g, which depends on the percentage of PPO, is obtained (Figure 14.4). In this way the desired processing characteristics are obtained *and* the chemical resistance of PS is improved.

When crystalline and amorphous materials that are miscible are blended, a crystalline phase can be produced at the proper cooling rate, which stiffens the overall structure and raises the T_g. Miscible blends of two crystalline structures have also been developed.

When the polymers are immiscible, inferior properties can be encountered if lack of mixing between phase boundaries produces high surface tension and phase separation. However, materials called *compatibilizers* can be used to reduce phase separation and also to produce finer dispersions when more than one phase is desired.

EXAMPLE 14.1 *[ES/EJ]*

Figure 14.4 shows that the heat-deflection or -distortion temperature is less than the glass transition temperature for all PPO contents in a PS–PPO blend. This seems to imply that the material can distort at temperatures less than those at which it reacts in a glassy or brittle fashion. Explain.

Answer   The glass transition temperature (T_g) is a change in the slope of the specific volume-versus-temperature plot. It reflects an inability of entangled polymer chains to respond rapidly to a change in temperature. If we assume slow cooling, a more-or-less equilibrium T_g can be determined. However, more rapid cooling rates would give a different glass transition temperature. Furthermore, a polymer can be strained rapidly above its published T_g and still react in a brittle fashion.

The heat-distortion temperature is measured as deflection of 0.010 in. (0.25 mm) at the center of a 5 in. × 0.5 in. × 0.5 in. (127 mm × 12.7 mm × 12.7 mm) beam with an outer fiber stress of 264 psi (1.82 MPa). Normally a heating rate of 2°C (3.6°F)/min is used, as in ASTM Standard D648.

Note that the heat-distortion temperature and the glass transition temperature are experimentally obtained by totally different methods. Therefore, we should think of them as separate indices appropriate only for making comparisons between polymers.

SPECIAL PRODUCTION TECHNIQUES AND PRODUCTS: FIBERS, FOAMS, ADHESIVES, COATINGS

14.6 Fibers

There are many important engineering uses for such well-known *fibers* as nylon, polyester, and glass fibers and the newer high-strength graphite and boron fibers. Although natural fibers such as cotton are used as fillers in polymers, we will emphasize new developments in synthetic fibers in our discussion.

The plastic fibers, such as nylon and polyester, are produced by melting chips of the polymer and then forcing the liquid through orifices in an extremely fine die called a *spinneret*. The extruded material is cooled either by liquid (in the wet process) or by air. If the fiber is coiled continuously, it is called monofilament; if it is broken into lengths for later use, it is called staple. The process for making nylon fibers incorporates an extra step. The filaments leaving the spinneret are gathered into yarn, and then the yarn is drawn four to five times its length. This orients the long axis of the polymer molecules along the length of the yarn, increasing strength at some sacrifice in elongation. The strengths of different polymer fibers are shown in Figure 14.5. The wide range for nylon is due to differences in processing.

The standard measure of strength in the textile industry is grams per denier. The denier is defined as follows:

$$\text{Denier} = \frac{4,464,528}{\text{yd/lb of fiber}} \tag{14-1}$$

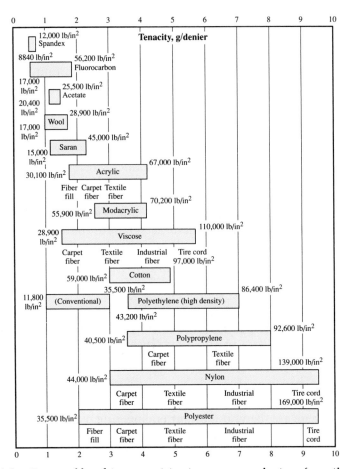

FIGURE 14.5   Range of breaking tenacities in grams per denier of textile yarns

[C. A. Harper, ed., *Handbook of Plastics and Elastomers.* Copyright © 1975. Reprinted with permission of McGraw-Hill Publishing Company.]

Another definition for *denier* is the weight in grams of a fiber that is 9000 m long. Whatever units are used, a denier is proportional to the density of the fiber and to its cross section. The strength of a fiber (Figure 14.5) is generally given not in lb/in² or MPa but rather in terms of *tenacity*, which has units of g/denier. Two fibers that have the same denier and the same breaking strength in grams have the same tenacity. If the two fibers have the same density, they also have the same tensile strength. (See Example 14.2.)

EXAMPLE 14.2   *[ES]*

Develop a mathematical formula relating tensile strength (lb/in²) to the tenacity of a fiber. The density will also have to be considered as a variable.

Answer

$$\text{Tenacity} = \frac{\text{g (force)}}{\text{denier}} = \frac{\text{g (force)}}{\text{g (mass)}/9000 \text{ m}}$$

Therefore,

$$\text{Tenacity} \times \text{density} = \left(\frac{\text{g (force)}}{\text{g (mass)}/9000 \text{ m}}\right)\left(\frac{\text{g}}{\text{cm}^3}\right)\left(\frac{100 \text{ cm}}{\text{m}}\right) = \frac{\text{g}}{\text{cm}^2}$$

This can be converted to lb/in^2:

$$(9000 \text{ m})\left(\frac{\text{g}}{\text{cm}^3}\right)\left(\frac{100 \text{ cm}}{\text{m}}\right)(2.54 \text{ cm/in})^2\left(\frac{1}{454 \text{ g/lb}}\right) = 12{,}790 \text{ lb/in}^2$$

Finally,

$$\text{Tensile strength} = \text{tenacity}\left(\frac{\text{g}}{\text{denier}}\right) \times \text{density}\left(\frac{\text{g}}{\text{cm}^3}\right) \times 12{,}790 \qquad (14\text{-}2)$$

$$\text{(in units of lb/in}^2)$$

Polymer fibers can be used alone, as in rope or fabric, or as fillers, as we will discuss later. Common synthetic fillers include nylon, orlon, rayon, and teflon; many natural organic fillers, such as jute, sisal, and wood flour, are also used. (See Table 14.6.)

14.7 Foams

The average engineer is aware mostly of low-stress applications of *foam*, such as in packing material, drinking cups, inexpensive coolers, and home insulation. However, the performance of elastomer foam in higher-stress applications, such as in place of steel springs in upholstery and mattresses, demands respect. We shall see that strong foamed engineering structures can also be produced. Table 14.2 shows various properties of typical polymer foams.

The finished foam product consists of a gaseous phase dispersed in the polymer. Dispersion is obtained by one of three methods: (1) injecting gases directly under pressure, (2) adding volatile liquids to the polymer, or (3) adding chemical agents that decompose into gas and other by-products. For example, in the production of polyurethane foam, the isocyanate reacts with water to produce CO_2 gas. In addition, silicone copolymers are incorporated to control cell size and stabilize the foam. Another method of producing foams, which ensures constant cell size, employs the dispersion of hollow spheres of glass or plastic in the polymer matrix.

Over twenty polymers, thermosets, thermoplastics, and elastomers are available in many foamed modifications. The closed-cell type of foam is favored for structural, insulating, or flotation applications, whereas the open-cell type is used for filtration and absorption.

TABLE 14.2   Properties of Typical Polymer Foams

Polymer Name	Type of Foam	Density, lb/ft^3	Tensile Strength, psi	Max. Service, °F	% H$_2$O absorption*
ABS	Injection-molded pellets	31/56	1800/4100	180	0.4/0.6
Cellulose acetate	Boards (closed cell)	6/8	170	350	13/17
Epoxy	Syntactic sheet	23	2100	500	1.8
Phenolics	Foam in place	7/10	80/130	300	10/15
Polyethylene	High-density, molded	35	1200	230	0.2
Polypropylene	High-density, molded	35	1600	—	0/3
Polystyrene	Molded beads	5	150/170	165/185	0/<3
Polystyrene	Extruded sheet	10	600/1000	175	nil
Polyurethane	Molded foam in place	1/40	10/1350	250	—
Polyurethane	Rigid, closed-cell (low-density)	1.5/3.0	15/95	250	0.1/5
Polyurethane	Rigid, closed-cell (high-density)	41/70	3000/8000	300	—
Polyvinylchloride	Flexible, molded, open-cell	10 and up	10/200	125/225	—

* % by volume, ASTM Specification D2842, 100% relative humidity for 96 hr
Data adapted from C. A. Harper, ed., *Handbook of Plastics and Elastomers*, McGraw-Hill, New York, 1975. Reprinted with permission.

It is possible to control the density of foams over a wide range, as illustrated in Table 14.2. As we would expect, the tensile strength of a foam increases with density (Figure 14.6). Many foams are used as structural beam-like

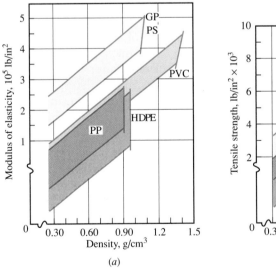
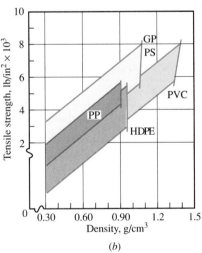

(a)   (b)

FIGURE 14.6   Properties of selected foamed thermoplastics as related to density. (*a*) Modulus of elasticity. (*b*) Tensile strength. GP-PS: general-purpose polystyrene; PVC: polyvinylchloride; PP: polypropylene; and HDPE: high-density polyethylene.

[Weir, C. L., "These Data Show Why the Action Is Swinging to Structural Foams," *Plastics Technology*, April 1972, p. 37. Used by permission.]

members, because a beam made of foam can be lighter in weight but have the same deflection as a solid beam of the same plastic with a smaller cross section. (See Example 14.3.)

EXAMPLE 14.3 *[ES]*

The formula for the deflection of the beam shown is

$$\delta = \frac{WL^3}{48EI} \tag{14-3}$$

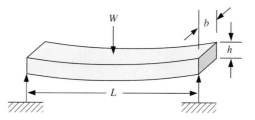

where W = concentrated load at center

L = length of span

E = modulus of elasticity

I = moment of inertia, which in this case is $bh^3/12$

where b = width of beam and h = depth of beam. Thus, for the rectangular beam that is supported at its ends and has a single concentrated load at its center,

$$\delta = \frac{WL^3}{4Ebh^3} \tag{14-4}$$

Consider two beams of the same material and of equal width b and length L. One beam is solid and the other is foamed with 90% air by volume. Determine the relative thickness and relative weight of the two beams if the deflections and the loads are equal.

Answer

$$\delta_f = \frac{WL^3}{4E_f bh_f^3} = \delta_s = \frac{WL^3}{4E_s bh_s^3}$$

where subscripts f and s refer to foamed and solid, respectively.

Then

$$\frac{h_f^3}{h_s^3} = \frac{E_s}{E_f}$$

From Figure 14.6 we see that the modulus for a given polymer varies linearly with density:

$$E = K \times \text{density}$$

where K is a constant.

Therefore,

$$\frac{h_f}{h_s} = \sqrt[3]{\frac{d_s}{d_f}}$$

where d is the density. Because the same material is used for both beams and one has 90% air,

$$\frac{d_s}{d_f} = \frac{1}{0.1} = 10$$

or

$$\frac{h_f}{h_s} = \sqrt[3]{10} = 2.15$$

Thus the foamed beam is 2.15 times thicker than the solid beam.

Furthermore,

$$\frac{Wt_f}{Wt_s} = \frac{(Lbh_f)d_f}{(Lbh_s)d_s} = \left(\frac{h_f}{h_s}\right)\left(\frac{d_f}{d_s}\right) = 2.15\left(\frac{1}{10}\right) = 0.215$$

The weight of the foamed beam is 0.215 that of the solid beam, representing a 78.5% saving in weight and material.

(It is assumed that the application will tolerate the added bulk of the foamed beam.)

In practice, techniques have been developed for coating beams or sheets with high-strength dense plastic on the upper and lower surfaces where the stress is highest.

Foamed polymers also have many other useful properties. For example, foam may be blown into existing, irregularly shaped cavities for insulating and sound-proofing, as well as for strengthening. The valuable electrical properties of foams are discussed in Chapter 20.

When elastomers are foamed, their properties are distinctly different from those of the structural foams. The materials are highly flexible, compressible, and soft. However, depending on the nature of the elastomer, the foams exhibit different degrees of resistance to heat and chemicals.

The processing of elastomer foams is different from that of the structural foams, because the vulcanization (crosslinking) reaction takes place at the same time as the gas bubbles are being generated in the foam. If the gas is generated too rapidly, the foam will collapse. This is somewhat like the technique of using sodium bicarbonate to preserve a risen cake or soufflé. If an open-cell structure is desired, as in fabric coating, inorganic salts that liberate CO_2 are used to make the CO_2 diffuse rapidly. If a closed-cell structure is desired, as for a gasket, nitrogen is used as the cell former because it diffuses more slowly.

The specifications for elastomer foams in the ASTM standards are quite different from those for structural materials. They include psi for 25% compression deflection, effect of overaging on compression deflection, and low-temperature testing. These criteria reflect the intended end uses.

14.8 Adhesives in General

For thousands of years people have used natural *adhesives* (glues) made from animal sources. However, with advances in polymers in recent years, synthetic adhesives with superior characteristics have appeared. There is even a startling TV ad showing that one drop of adhesive can support an autombile.

One of the fascinating aspects of this rapidly growing field is that thermosets, thermoplastics, and elastomers have all found applications as adhesives, and so-called alloys have been developed in which more than one type of polymer is used.

We will consider first the theories of adhesion, then the general advantages and disadvantages of adhesive bonding, and finally the properties of individual adhesives.

14.9 The Adhesive Bond

Despite the wide use of adhesives, a good deal of controversy surrounds the nature of the bond. Five theories have some support; each seems to be particularly useful in explaining certain phenomena associated with adhesive bonding. It is worthwhile to review these because they indicate procedures commonly followed for optimal bonding.

The *mechanical interlock theory* points out that surfaces on a micro scale are very rough. Therefore, when a liquid adhesive is placed between two surfaces, it penetrates the crevices and pores and then solidifies. Thus a cement interlocks with the surface layers on both sides and provides a mechanical bond. The fact that fresh, roughened surfaces provide the best bond supports this theory.

The *absorption theory* states that to be successful, an adhesive must wet the surface to be bonded (called the adherend). This theory has led to the development of materials with lower surface tension than that of the adherend. Supporting this theory is the fact that epoxy wets steel and provides a good bond, whereas it does not wet the olefins PE, PP, and PTFE and does not bond them.

The *electrostatic theory* postulates that as a result of the interaction of the adhesive and the adherend, an electrostatically charged double layer of ions develops at the interface. The fact that electrical discharges are observed when an adhesive is peeled from a substrate is cited as evidence of these attractive forces.

The *diffusion theory* is particularly applicable to cases in which the adhesive contains a solvent for the adherend. A type of bonding similar to diffusion bonding in metals develops, and molecules pass across the interface. This diffusion can obliterate the mechanical plane of the interface and its weakness.

It is generally agreed that the highest bonding strength is indicated when, upon stressing, the fracture occurs in the body of the adherend or within the adhesive, not at the interface. The *weak boundary layer theory* holds that for an adhesive to perform satisfactorily, the weak boundary layer should be eliminated. For example, in the case of metals with a scaly oxide layer, failure takes place at the boundary. The problem does not exist for aluminum, which has a coherent oxide layer. Similarly, in the case of polyethylene, a weak, low-molecular-weight additive is present throughout the structure, and this leads to a weak interface. In both cases the potentially weak layers can be removed by surface treatments.

14.10   Requirements for Satisfactory Bonding

As a result of theoretical considerations and extensive practical testing, the following recommendations have been developed for attaining satisfactory joints.

1. *Cleanliness of bond surfaces.* Not only is a clean bond surface in the conventional sense required, but fresh cleaning to avoid adsorbed gases is often useful. Activated inert gases are sometimes used. Detailed procedures for different material–adhesive systems are given in *Handbook of Plastics and Elastomers* (C. A. Harper, McGraw-Hill, New York, 1975).
2. *Adhesive choice.* The adhesive should wet the adherend and solidify under proper conditions of time, temperature, and pressure. Often the desired production conditions narrow the choice of adhesive. A variety of liquids, pastes, and solids is available.
3. *Joint design.* Adhesive joints are generally more resistant to shearing, compressive, and tensile stresses than they are to stress systems due to peeling. For example, it is easier to remove adhesive tape from a surface by peeling than by any other method of applying stress.
4. *Service conditions.* In general, the coefficient of expansion of the polymer in adhesives is greater than that of metals (Figure 14.7). If severe temperature changes are to be encountered, this effect and the required accomodations of the adhesive must be considered. Weathering and solvents that may be encountered in service are also important considerations. (See Chapter 18.)

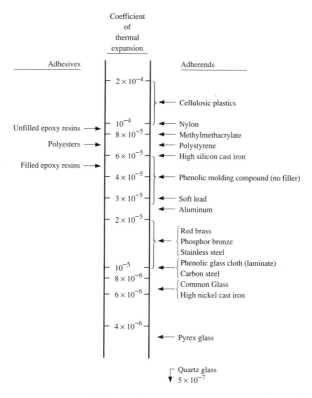

FIGURE 14.7 Coefficients of thermal expansion, in./in./°C, for selected adhesives/adherends

[Adapted from Perry, H. A., "Room Temperature Setting Adhesives for Metals and Plastics," in J. E. Rutzler and R. L. Savage (eds.), *Adhesion and Adhesives: Fundamentals and Practices*, Society of Chemical Industry, London, 1954. Used by permission.]

14.11 Available Adhesives

The principal adhesives and their uses are given in Table 14.3. In general, the most useful materials for structural stressed applications are the thermosets and their alloys. Because the molecules of thermosets are densely cross-linked, their resistance to heat and solvents is good and they show less elastic deformation under load than the thermoplastics.

Typical shear strengths (ASTM Specification D1002) for both stainless steel and aluminum adherends are given in Table 14.4. Because cryogenic temperatures may be encountered in aircraft and spacecraft, low-temperature strength is also important. Examples of strength at low temperature are shown in Figure 14.8. It is apparent that at lower temperatures, some polymer adhesives increase in strength whereas others decrease. For this reason, adhesives with little temperature sensitivity must be selected when the service temperature range is large.

TABLE 14.3 Characteristics and Uses of Adhesives

Adhesive Classification	Examples	Common Physical Form	Cure Required	Bond Characteristics	Major Use	Materials Most Commonly Bonded
Thermoplastic	Cellulose acetate, polyvinyl acetate, polyvinyl acetals, polyamide, acrylic	Liquid, some dry film	No	To 150–200°F; Poor creep strength, fair peel strength	Unstressed joints	Nonmetals such as wood, paper
Thermosetting	Cyanoacrylate, urea formaldehyde, melamine formaldehyde epoxy, polyimide, acrylic acid diester	Liquid and others	Most	To 200–500°F; Good creep strength, fair peel strength	Stressed joints	Most materials employed for structural uses
Elastomeric	Natural rubber, butyl, nitrile, polyurethane, polysulfide, silicone, neoprene	Liquid, some film	Most	To 150–400°F; Low strength, high flexibility	Unstressed or flexural joints	Rubber, leather, modifications with synthetic resins are used for most materials
Alloys	Epoxy-phenolic epoxy-nylon, neoprene-phenolic, Vinyl-phenolic	Liquid, paste, film	Varies	Wide range depending on the type of adhesive	Some combinations give highest joint strength but at high cost	Metals, ceramics, glass, thermoplastics

610

TABLE 14.4   Shear Strength of High-Temperature Structural Adhesives

Property	Modified Epoxy[a]	Epoxy-phenolic[b]	Nitrile-phenolic[b]	Poly-imide[a]	Poly-benzimid-azole[a]	Vinyl Phenolic[b]
Initial room-temperature shear strength, lb/in^2	3800	3240	3200–4100	3300	2920	3800–4500
% of original shear strength after 10-min soak:						
At 180°F	83	94	35–70		99	35–106
At 250°F	74	81	31–42		99	12–79
At 350°F		66		64	98	6–11
At 500°F	28	54		61	83	
At 650°F		43		45	74	
At 800°F		15			68	
After 1000-hr soak:						
At 180°F		96	62–78			60–122
At 250°F		81	52–66			33–82
At 350°F		43	27–56		110	13–30
At 500°F				63	6–7	
At 600°F				61		

[a] Stainless steel used as adherend material
[b] Aluminum used as adherend material

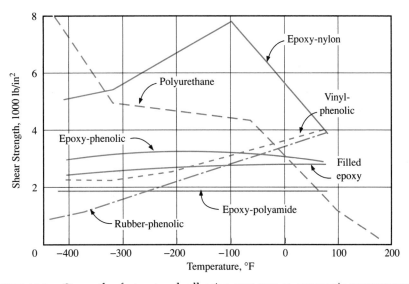

FIGURE 14.8   Strength of structural-adhesive systems at cryogenic temperatures

[Adapted from Kausen, R. C., "Adhesives for High and Low Temperatures," *Materials Engineering*, August–September, 1964. Used by permission.]

The gluing of wood is an important application for polymer adhesives. Although many polymers have been used successfully, in home crafts the "white glue" is often used. This adhesive is a polyvinyl acetate latex containing a high percentage of solids. Polymerization is by emulsion whereby the monomers coalesce and polymerize upon loss of the plasticizer (usually water that is drawn into the woodgrain).

EXAMPLE 14.4 *[EJ]*

Which type of adhesive listed in Table 14.3 is most likely to include the one called "model airplane glue"? What are some of its disadvantages?

Answer "Model airplane glue" is most likely to be a thermoplastic adhesive, though all the varieties listed would work on models, especially the plastic models. The cement is a polymer dispersed by a solvent. (Recall from Chapter 13 that a plasticizer carried to an extreme is a solvent.) The solvent not only wets the adherend but also reacts with coatings such as paint or varnish on wood surfaces.

Because polymerization takes place by evaporation of the excess plasticizer (solvent), there is considerable shrinkage of the adhesive, which can cause stresses that weaken the joint. Continued loss of plasticizer over time will make the polymer adhesive more brittle and more subject to impact failure.

14.12 Advantages and Disadvantages of Adhesive Bonds Compared to Metallic Bonds

Advantages

In thin-metal honeycomb structures such as those used in aircraft, stress is transmitted across a joint bonded with an adhesive better than it is across a welded or riveted construction.

Heat-treated structures, such as age-hardened alloys, are weakened by processes such as welding and brazing. With the use of adhesives, which do not require heat, we do not risk changing the microstructure.

Structures bonded with adhesives generally exhibit less stress concentration and lower residual stresses than those bonded by welding or riveting.

An adhesive bond undergoes less corrosion, and the adhesive can act as an insulator between different metallic alloys (see Chapter 18 on galvanic corrosion).

Any two solids, regardless of size or shape, can be bonded together with proper adhesives.

Disadvantages

Usually the temperature range within which adhesive bonds remain intact is lower, and creep takes place.

Adhesive bonds are susceptible to peeling.

Inspection for joint integrity is different and often difficult when adhesives are used.

14.13 Coatings

Coatings have been used since the earliest times. Even the Biblical account of the construction of Noah's Ark speaks of coating with pitch, inside and out. In modern times, the development of synthetic polymers has led to great advances; the thermoplastics, thermosets, and elastomers all find uses as coatings. The plastic is either dissolved in a solvent that evaporates, sprayed as a powder on a hot surface, or, in the case of the thermosets, baked to produce final polymerization.

In addition to the conventional ways of applying coatings, several newer methods are important in production. Electrostatic spraying consists of charging the particles of the coating to a high negative potential as they pass an electrode. The surface to be coated is grounded and the particles are attracted to it. Recently, dry powder has been applied with the same method. The part is preheated to 450–650°F (232–343°C) to fuse the particles. This eliminates the solvent hazard. Another coating method, called electrodeposition, involves precipitation from a water solution onto a charged object. Variations on these techniques are used to provide high-integrity coatings in factory-applied automotive exterior paints and to provide corrosion-resistant finishes in nonvisible areas.

The fluidized bed application process also operates without a solvent. The coating is suspended as a powder in a controlled upward stream of gas. The part to be coated is preheated and immersed in the stream. Particles strike the part and fuse with it to provide the desired coating.

Properties of Coatings

The properties of a number of TP, TS, and elastomer coatings are summarized in Table 14.5. It is quite satisfying to find that even when we investigate thin films of polymers, we find that the processing and properties are predictable from the basic principles covered earlier. For example, the TS materials generally require baking for curing, whereas the TP materials are cured by evaporation of the solvent or by flame spraying. The TS coatings can be used at higher temperatures than the TP coatings, and their moisture resistance is also superior.

TABLE 14.5   Properties and Application of Typical Polymeric Coatings

Name	Application Method	Curing Method	Moisture Resistance	Max. °F (continuous)	Adhesion to Metals	Flexibility	Typical Uses
Acrylic	Spray, brush, dip	Air-dry	Good	180	Good	Good	Auto lacquer, circuit boards
Alkyd[†]	Spray, brush, dip	Air-dry	Poor	200	Excellent	Fair	Auto enamel
Epoxy-phenolic	Spray, dip	React, bake	Excellent	400	Excellent	Good[*]	Electric insulation
Fluorocarbon	Spray, dip	Fusion from solvent	Excellent	500	Excellent with primer	Excellent	High-temperature cookware
Phenolics	Spray, dip	Heat	Excellent	350	Excellent	Fair[*]	High-bake coatings
Nylon	Flame, spray	None	Fair	250	Excellent	Fair[*]	Low-friction bearings
Polystyrene	Spray, dip	Evaporation of solvent	Good	160	Excellent	Fair	Coil coating
Polyurethane	Spray, brush, dip	Air-dry	Good	250	Poor	Good[*]	Furniture
Silicone	Spray, brush, dip	Air-dry to bake	Excellent	500	Good with primer	Excellent	Heat-resistant coating
Vinyl chloride	Spray, dip, roller	Air-dry	Good	150	Excellent	Excellent[*]	Can coatings, furniture
ELASTOMERS							
Butyl rubber	Usually spray, dip, brush for all elastomers	Air-dry	Excellent	—	Good to excellent for all elastomers	Elastomers used where high flexibility is required	Superior replacement for natural rubber
Chloroprene rubber	Air-dry	Excellent	190				Protection for sportswear
Fluorelastomer	Air-dry	Excellent	450				Maintenance coating in hostile environments
Neoprene	Air or heat	Excellent	—	Primer required			

[*] Poor at low temperature
[†] Alkyds are essentially polyesters (acid and alcohol) made with naturally occurring chemicals.
Data adapted from C. A. Harper, ed., *Handbook of Plastics and Elastomers*, McGraw-Hill, New York, 1975. Reprinted with permission.

EXAMPLE 14.5 *[ES]*

A piece of iron 100 mm × 10 mm × 1 mm is coated on one side with an epoxy that is 0.10 mm thick. Polymerization has taken place at 100°C to accelerate curing. What is the stress in each material if the coated steel is kept flat between clamps while it is cooled to 0°C? The properties are as follows:

	Modulus, 10^6 psi	Coefficient of Expansion, in./in./°C
Iron	30	1.4×10^{-5}
Epoxy	1.0	8.0×10^{-5}

Answer   From mechanics we know that the sum of the forces equals zero.

$$\sum F = 0 = F_{ep} + F_{Fe} \qquad (14\text{-}5)$$

or

$$S_{ep} A_{ep} + S_{Fe} A_{Fe} = 0$$

Because $A_{Fe} = 10 \text{ mm} \times 1 \text{ mm} = 10 \text{ mm}^2$ and $A_{ep} = 10 \text{ mm} \times 0.1 \text{ mm} = 1 \text{ mm}^2$ we know that

$$S_{ep}(1) + S_{Fe}(10) = 0 \quad \text{or} \quad S_{ep} = -10 S_{Fe} \qquad (1)$$

The materials are bonded together and are kept straight, so the total strain in the epoxy must equal the total strain in the iron. However, the total strain is made up of a thermal strain plus a strain to straighten the material, or

$$e_{\text{thermal}-ep} + e_{\text{restore}-ep} = e_{\text{thermal}-Fe} + e_{\text{restore}-Fe}$$

Now,

$$e_{\text{thermal}-ep} = (8.0 \times 10^{-5})(0 - 100°C) = -8 \times 10^{-3} \text{ in./in. or mm/mm}$$

$$e_{\text{restore}-ep} = \frac{S_{ep}}{E_{ep}} = \frac{S_{ep}}{1 \times 10^6} = 1.0 \times 10^{-6} S_{ep}$$

$$e_{\text{thermal}-Fe} = (1.4 \times 10^{-5})(0 - 100°C) = -1.4 \times 10^{-3} \text{ in./in. or mm/mm}$$

$$e_{\text{restore}-Fe} = \frac{S_{Fe}}{E_{Fe}} = \frac{S_{Fe}}{30 \times 10^6} = 3.33 \times 10^{-8} S_{Fe}$$

Therefore,

$$-8 \times 10^{-3} + 1.0 \times 10^{-6} S_{ep} = -1.4 \times 10^{-3} + 3.33 \times 10^{-8} S_{Fe} \qquad (2)$$

We now have two equations, (1) and (2), and two unknowns, S_{ep} and S_{Fe}. Solving them simultaneously gives us the stress in each material:

$$S_{Fe} = -658 \text{ psi} \quad \text{(negative sign means compression)}$$

$$S_{ep} = +6580 \text{ psi} \quad \text{(tension)}$$

Because epoxy has a tensile strength of 10,000 psi (see Table 13.4), failure might be induced if the temperature difference were greater. When the part is heated to induce polymerization and curing, creep or viscoelastic flow may take place. This example thus points out the problem with metal–polymer combinations.

14.14 Fillers and Laminates

One of the most important uses of plastics is as a cement or matrix in which particles or fibers of a second phase, such as glass, are suspended. Some of the most advanced engineering materials are composed of high-strength fibers of glass, graphite, or boron in a plastic matrix. Although the fibers have high strength, they are too brittle and notch-sensitive to be used alone. But if the fibers are surrounded with plastic, the composite has a very high strength-to-weight ratio. In another recent development, whiskers (small, high-strength crystals that are practically flaw-free) have been bonded in the same way.

Currently, the term *filler* is used to refer to any foreign solid—from wood flour to graphite—that is incorporated in a plastic. *Laminates* are made by bonding together two or more sheets of reinforcing fibers, usually with heat or pressure. Although both filled plastics and laminates are composites, in the aircraft industry the word *composite* signifies high-performance constructions obtained from high-strength, high-modulus fibers and specialty resin systems. We will discuss fillers and laminates in this chapter and the high-strength composites in Chapter 16.

Fillers

A summary of fillers and their uses is given in Table 14.6. Most of the properties are familiar, but two may need explanation. Lubricity is not expressed quantitatively but refers to the low-friction characteristics at a bearing surface. Graphite, kaolin, mica, molybdenum disulfide, nylon, PTFE, and talc all increase lubricity. Processability refers to the ease of molding or forming; many of the same materials are used to increase the processability of both thermoplastics and thermosets.

The effects of fillers on properties are presented in Table 14.7. Glass fiber is commonly used as a filler for the thermoplastics, whereas a wider variety of materials are used for the thermosets. To reduce the expense, a number of additives, such as "micoal," wood flour, and fabric, are also used. The data show that for the thermoplastics, the addition of glass raises strength and modulus at the expense of a severe loss in ductility. Of the several fillers shown for phenol formaldehyde, only the glass improves tensile strength.

However, other advantages are gained from the fillers, such as increased processability and heat resistance.

Laminates

A variety of materials can be used for the fibrous sheet or web of a *laminate:* cellulose, glass fibers, or synthetic plastics. The binders are usually based on thermosetting plastics: phenolic, melamine, epoxy, or silicone. The resin is dissolved in a suitable solvent, and the mats are impregnated with the solution. Then the solvent is removed in a treating tower. The sheets are pressed at 250–400°F (121–204°C) at 200/3000 psi (1.4/2.07 MPa) until they are cured. Tubes and formed shapes can also be produced. Table 14.8 compares the properties of different mats used to laminate melamine–formaldehyde.

The automotive industry in particular has become much interested in *reinforced plastics* (RP) because of their excellent strength-to-weight ratios and corrosion resistance. Effort is currently being applied to the compression molding of *sheet molding compound* (SMC) for use in exterior automotive body panels. The SMC can be produced continuously as a mat and cut to the size needed for the mold. The final cure by compression molding is by application of heat and pressure. Surface finish and coating adhesion become important final considerations.

EXAMPLE 14.6 *[E]*

What are some of the advantages and disadvantages of using a preimpregnated mat, called *prepreg,* rather than a separate mat and polymer in the manufacture of reinforced plastic components? (An example is a fender extension on an automobile.)

Answer   The earliest RPs used polyester resins and glass fibers and were produced by hand layup on an open die shape. This procedure was costly in terms of both labor and time. More recently, thermoplastic sheets have been used to sandwich a prewoven mat formed in a heated mold. Two difficulties encountered were incomplete adhesion between the polymer and the mat and displacement of the mat because of the difference in density between the mat and the thermoplastic sheets. A similar process has been used for thermosets, but only with the polymer in stage B.

A prepreg overcomes some of these difficulties, because adhesion between mat and fiber has already taken place and the major requirement is polymer flow to eliminate the prepreg porosity. A disadvantage of using prepregs is the cost of the intermediate processing required to produce the prepreg.

TABLE 14.6 Some Fillers and Reinforcements and Their Contributions to Plastics*

Filler or reinforcement	Chemical resistance	Heat resistance	Electrical insulation	Impact strength	Tensile strength	Dimensional stability	Stiffness	Hardness	Lubricity	Electrical conductivity	Thermal conductivity	Moisture resistance	Processability	Recommended for use in[a]
Alumina tabular	•	•				•								S/P
Alumina trihydrate, fine particle			•				•					•	•	P
Aluminum powder		•	•			•				•	•			S
Asbestos	•	•		•		•	•	•						S/P
Bronze							•	•		•	•			S
Calcium carbonate[b]		•				•	•	•					•	S/P
Calcium metasilicate	•	•				•	•	•				•		S
Calcium silicate		•				•	•	•						S
Carbon black[c]		•				•	•			•	•		•	S/P
Carbon fiber					•		•			•	•			S
Cellulose				•		•	•	•						S/P
Alpha cellulose			•		•	•								S
Coal, powdered	•											•		S
Cotton (macerated/chopped fibers)			•	•	•	•	•	•						S

618

Filler	Symbol[a]
Fibrous glass	S/P
Fir bark	S
Graphite	S/P
Jute	S
Kaolin	S/P
Kaolin (calcined)	S/P
Mica	S/P
Molybdenum disulphide	P
Nylon (macerated/chopped fibers)	S/P
Orlon	S/P
Rayon	S
Silica, amorphous	S/P
Sisal fibers	S/P
PTFE-fluorocarbon	S/P
Talc	S/P
Wood flour	S

The chart does not show differences in degrees of improvement; calcined kaolin, for example, generally gives much higher electrical resistance than kaolin. Similarly, differences in characteristics of products under one heading, such as talc (which varies greatly from one type to another and from one grade to another), are not distinguished.

[a]Symbols: P — in thermosets only; S — in thermoplastics only; S/P — in both thermoplastics and thermosets.

[b]In thermosets, calcium carbonate's prime function is to improve molded appearance.

[c]Prime functions are imparting coloring and resistance to ultraviolet light, also is used in crosslinked thermoplastics.

From *Plastics*, 6th ed., by J. H. DuBois and F. W. John. Copyright © 1981 by Van Nostrand Reinhold. Reprinted by permission of the publisher.

TABLE 14.7   Effects of Fillers on Mechanical Properties

Material	Filler	Amount, %	Tensile Strength, psi × 10³*	Percent Elasticity	E × 10³ psi*
ABS	0	0	4.5/8.5	20/80	300
	Glass	20/40	8.5/19	2.5/3	590/1030
Acetal	0	0	8.8/10	25/75	520
	Glass	20	8.5/11	2/7	1000
Nylon 6/6	0	0	11.2/13.1	60	410
	Glass	33	22/28	4/5	—
Phenol formaldehyde	0		7/8	1.0	750/1000
	Wood flour	—	5/8	0.4/0.8	800/1700
	Asbestos	—	4.5/7.5	0.2/0.5	1000/3000
	Mica	—	5.5/7	0.1/0.5	2500/5000
	Glass	—	4/18	0.2	1900/3300

*To obtain MPa, multiply psi by 6.9×10^{-3}. To obtain kg/mm², multiply psi by 7.03×10^{-4}.

TABLE 14.8   Effect of Mat Material on the Properties of a Melamine–Formaldehyde Laminate

	No Mat	Cellulose Paper	Cotton Fabric	Asbestos Paper	Glass Fabric	Glass Mat
Tensile strength, psi × 10³*	7/8	10/25	7/17	6/12	25/63	16/25
E × 10³ psi*	700/1000	—	1000/1900	1600/2200	2000/2500	—

*To obtain MPa, multiply psi by 6.9×10^{-3}. To obtain kg/mm², multiply psi by 7.03×10^{-4}.

Summary

The engineering polymers, which offer higher levels of strength, particularly at elevated temperature, have been developed by making basic changes in the carbon backbone structure of the simple polymers such as polyethylene. Even in the case of PE, the change to an ultrahigh-molecular-weight material has yielded a structure of exceptional abrasion resistance. Similarly, research directed toward the controlled synthesis of copolymer structures and toward the blending of polymer molecules has resulted in new, valuable combinations of properties.

The varied and unique combinations of structures found in polymers lend themselves to a wide range of applications in fibers, foams, adhesives, paints, and other coatings.

Polymer fibers, which have applications ranging from rope to fabrics, are generally manufactured from thermoplastics. The strength of polymer fibers is usually expressed in terms of tenacity rather than in terms of the measures of tensile strength commonly given for the bulk polymers.

Foam products are not limited to drinking cups and insulation; they include a variety of engineered components produced from over twenty thermoplastics, thermosets, and elastomers. Structural forms can give a resistance to deflection equivalent to that of solid plastics at a considerable saving in weight.

Polymer adhesives are also available in the three polymer types. They offer many inherent advantages over other joining processes such as welding, because almost any engineering material can be bonded by adhesive. However, the same environmental considerations that apply to the bulk polymers also apply to the adhesives.

Coatings such as paint are really polymers. The monomers and the chains are polar, so electrostatic application can enhance the coating adhesion and minimize porosity.

Fillers are added to polymers to modify their properties; the addition of fillers can also make the product less expensive. Glass fibers, for example, can be used as fillers or can be woven into a mat and processed with a number of polymers to produce the laminates.

Definitions

Adhesive A molecular structure that bonds two other materials by either physical or chemical means.

Alloy adhesive An adhesive made up of more than one type of polymer.

Blend An intimate mixture of polymers obtained by melting and mechanically blending two or more polymers.

Coating A thin film of a polymer that is applied to another material.

Commodity polymers Widely used polymers such as polyethylene, polystyrene, and polyvinylchloride that have low thermal resistance.

Copolymer A chemical bonding of two or more monomers by controlled polymerization to produce molecules. These are called random, alternate, block, or graft types depending on the arrangement of the monomers.

Denier A measure of fiber size defined as the weight in grams of a 9000-meter length.

Engineering polymers High-cost polymers that have thermal resistance up to 350°F (177°C).

Fiber A polymer in a finely drawn form.

Filler A foreign solid incorporated into a polymer.

Foam A polymer containing porosity to add to the insulating value or sponginess or to decrease the weight.

Heterochain bonding The incorporation of more than one type of monomer in the carbon chain as in polyphenylene sulfide (PPS). The use of one species only is called *homochain bonding*.

Ionic polymerization The use of an anion or cation to aid in the formation of block copolymers.

Laminate A polymer-impregnated mat set and formed by heat and pressure.

Prepreg A mat preimpregnated with a polymer to simplify further processing.

Reinforced plastics (RP) The reinforcement is usually provided by fibers or mats.

Sheet molding compound (SMC)   A prepreg used for panels such as exterior automotive components. It is formed by compression molding.
Spinneret   Orifice die used to produce polymer fibers.
Tenacity   The strength of fibers, expressed in grams per denier.

Problems

14.1   *[ES/EJ]* Table 14.1 indicates that several polymers are flame-resistant and possibly heat-resistant. What characteristics determine whether a material is flame and/or heat-resistant? (Sections 14.1 through 14.4)

14.2   *[ES/EJ]* Refer to the definitions given in Problem 13.17 with respect to the measurement of molecular weight in a polymer. The following table provides data for several polyethylenes. (Sections 14.1 through 14.4)

	$\overline{M}_n$
LDPE	20,000
HDPE	200,000
UHMWPE	4×10^6

a. Determine the degree of polymerization for each polyethylene.
b. Indicate why toughness is not solely dependent on $\overline{M}_n$ but also increases when the molecular weight distribution narrows.

14.3   *[ES]* In Figure 14.1, why are we interested in the direction of composition drift during the polymerization of copolymers? (Sections 14.1 through 14.4)

14.4   *[ES]* Assume that we have two monomers with the reactivity ratios $r_A = 0.01$ and $r_B = 50$. What sort of copolymer (random, block, alternating, etc.) would we obtain? Explain. (Sections 14.1 through 14.4)

14.5   *[ES/EJ]* Referring to the phase diagram given in Figure 14.2, we might be able to determine the phase amounts by applying the inverse lever law. (Sections 14.1 through 14.4)
a. Why are the properties not determined even if we assume equilibrium?
b. Is it possible to change this equilibrium? Explain.

14.6   *[ES/EJ]* ABS is discussed in the text as normally having dispersed butadiene particles in the copolymer. What controls whether these particles are present? (Sections 14.1 through 14.4)

14.7   *[EJ]* Differentiate between copolymerization and blending and indicate why one might be preferred over the other. (Sections 14.5 through 14.7)

14.8   *[ES/EJ]* Compare the tensile strengths for the bulk polymers in Table 13.3 with the fiber strengths in Figure 14.5. (Sections 14.5 through 14.7)
a. Why is the range so broad for the fibers?
b. Why are the strengths higher for the fibers?

14.9   *[ES/EJ]* The strength-to-weight ratio for a polymer is the tensile strength divided by the density. Explain how this ratio is used in polymer fibers. (Sections 14.5 through 14.7)

14.10   *[ES/EJ]* Why are the most successful polymer fibers partially crystalline? Dyes are usually polar in nature. Where do they attach in partially crystalline polymers, and how might the chain geometry be modified to better accept dyes? (Sections 14.5 through 14.7)

14.11   *[EJ]* Cotton fiber is 95% cellulose and absorbs moisture readily by polar attraction of water, which acts as a plasticizer. Explain why cotton clothes wrinkle readily on a hot, humid day and why polyester-cotton blends are used for wash-and-wear fabrics. Check a piece of your clothing for the normal blend composition. (Sections 14.5 through 14.7)

14.12 *[EJ]* The 10-lb test fishing line obtained from several manufacturers shows monofilaments that are not of the same diameter. They also exhibit different resistance to kinking, especially as the temperature is varied. Explain the variation in diameter and in kink resistance. (Sections 14.5 through 14.7)

14.13 *[ES]* Refer to Example 14.3. Two foam beams of different material are to have the same width, length, and deflection and to support the same concentrated load. What will be the ratio of thickness range for two such beams of general-purpose polystyrene and high-density polyethylene if their densities must also be the same at 0.8 g/cm^3? (Sections 14.5 through 14.7)

14.14 *[ES]* A foam beam is to be 2 in. wide × 6 in. high × 8 ft long and is to be made of polypropylene with a density of 0.6 g/cm^3. (Sections 14.5 through 14.7)
  a. What will be the maximum beam deflection at its center with a 500-lb concentrated load?
  b. What maximum tensile load will the beam support if it is longitudinally loaded?

14.15 *[EJ]* Most adhesives cannot be used in all types of environments. Explain. What type of glue could be used where it would be exposed to weathering? (Sections 14.8 through 14.14)

14.16 *[EJ]* Although thermoset adhesives are usually stronger than thermoplastic ones, why are the latter identified particularly to be used for wood in Table 14.3? (Sections 14.8 through 14.14)

14.17 *[ES/EJ]* Why would it be better to specify a filled rather than an unfilled polyester adhesive for bonding aluminum that will be subject to thermal cycling? (Sections 14.8 through 14.14)

14.18 *[ES]* Why is shear strength rather than tensile strength used in tables of mechanical properties of adhesives? (Sections 14.8 through 14.14)

14.19 *[ES/EJ]* Two 1-in.-wide strips of aluminum are overlapped $\frac{1}{4}$ in. and bonded with epoxy-phenolic adhesive. (Sections 14.8 through 14.14)
  a. From data in Table 14.4 plot the *tensile load* (shear in the joint) that this arrangement can support as a function of cure temperature and time.
  b. How might such a plot be useful in automobile design? Consider different environments within the vehicle.

14.20 *[ES/EJ]* Indicate whether the following statements are correct or incorrect and justify each answer. (Sections 14.8 through 14.14)
  a. Polyethylene is difficult to glue.
  b. Masking tape does not adhere well at low temperatures.
  c. Plywood for indoor use and plywood for outdoor use incorporate the same adhesive between the layers.

14.21 *[ES/EJ]* Repeat Example 14.5 for a rigid polyvinylchloride coating on aluminum. Supplementary data may be found in several different sections in the text. (Sections 14.8 through 14.14)

14.22 *[EJ]* The paint on an older automobile often appears "checked," especially if the car was not kept in a garage. What is the mechanism of this effect? Suggest why newer automotive finishes may be more resistant to the deterioration. (Sections 14.8 through 14.14)

14.23 *[EJ]* Both epoxy paint and latex (water-base) paint show temperature sensitivity. (Sections 14.8 through 14.14)
  a. What controls the minimum application temperature for each type of paint?
  b. Which type can more readily be applied in direct summer sunlight? Why?

14.24 *[ES/EJ]* Carbon black is widely used as a filler in thermoplastic and thermoset polymers. Determine its effect on their properties of hardness, heat resistance, conductivity, color, dimensional stability, and resistance to ultraviolet (UV) radiation. (Sections 14.8 through 14.14; see also Chapter 13)

14.25 *[EJ]* Why is stage B used in a thermosetting prepreg? What must be controlled in processing the prepreg to provide maximum strength and a good surface finish? (Sections 14.8 through 14.14)

14.26 *[ES/EJ]* A printed circuit board is to be prototyped from a melamine formaldehyde laminate (Table 14.8). (Sections 14.8 through 14.14)
   a. Why is this material suggested?
   b. What filler would you suggest? Why?

14.27 *[EJ]* Table 13.1 lists a number of monomer structures and how they polymerize. For each of the following applications, select one of the polymers as the most likely candidate. (Chapters 13 and 14)
   a. Frying pan handle
   b. Children's toy truck
   c. Container resistant to organic solvents
   d. Window for a storm door
   e. Outdoor trash container

14.28 *[EJ]* (Chapters 13 and 14)
   a. From the following polymers (as sketched), select the one that best fits each of these applications: (1) rigid electrical outlet; (2) aircraft window; (3) nonflammable chemical tubing; (4) squeeze bottle.

(a)   (b)

(c)   (d)

   b. Which of these polymers are thermoplastic?
   c. Which of them formed a by-product during polymerization?

14.29 *[EJ]* Several monomers are listed here. (Chapters 13 and 14)
   1. C_2H_4
   2. C_2F_4
   3. $\left.\begin{array}{l} C_6H_5OH \\ CH_2O \end{array}\right\}$ (Added together to form a polymer)

4.
$$
\text{HO} - \overset{\overset{\displaystyle O}{\|}}{\text{C}} - \overset{\overset{\displaystyle H}{|}}{\underset{\underset{\displaystyle H}{|}}{\text{C}}} - \overset{\overset{\displaystyle H}{|}}{\underset{\underset{\displaystyle H}{|}}{\text{C}}} - \overset{\overset{\displaystyle H}{|}}{\underset{\underset{\displaystyle H}{|}}{\text{C}}} - \overset{\overset{\displaystyle H}{|}}{\underset{\underset{\displaystyle H}{|}}{\text{C}}} - \overset{\overset{\displaystyle O}{\|}}{\text{C}} - \text{OH}
$$

(Added together to form a polymer)

$$
\overset{H}{\underset{H}{\diagdown}}\text{N} - \overset{\overset{\displaystyle H}{|}}{\underset{\underset{\displaystyle H}{|}}{\text{C}}} - \overset{\overset{\displaystyle H}{|}}{\underset{\underset{\displaystyle H}{|}}{\text{C}}} - \overset{\overset{\displaystyle H}{|}}{\underset{\underset{\displaystyle H}{|}}{\text{C}}} - \overset{\overset{\displaystyle H}{|}}{\underset{\underset{\displaystyle H}{|}}{\text{C}}} - \overset{\overset{\displaystyle H}{|}}{\underset{\underset{\displaystyle H}{|}}{\text{C}}} - \overset{\overset{\displaystyle H}{|}}{\underset{\underset{\displaystyle H}{|}}{\text{C}}} - \text{N}\overset{\diagup H}{\underset{\diagdown H}{}}
$$

a. Sketch the polymer structure of each.

b. From the following suggested applications, indicate a typical use for each polymer: rope, electrical outlets, chemical valves, squeeze bottles, baking dishes, microscope lenses, ball bearings.

14.30 *[ES/EJ]* For each description in the right-hand column, choose the monomer(s) that best fit(s) it or whose corresponding polymer best fits it. (Chapters 13 and 14)

a. Ethylene

b. Vinylchloride

c. Propylene

d. Styrene

e. Butadiene

f. Urea formaldehyde

1. One that would form the strongest polymer
2. Example of one that might form various stereo arrangements
3. A combination that might form a copolymer
4. One that might crosslink with another atom
5. One that might crystallize
6. One that would be most difficult to glue
7. One that might be easy to recover and reuse as scrap
8. One that would have the lowest glass transition temperature
9. One that could not be plasticized
10. One that might be classified as an elastomer

14.31 *[EJ]* Five polymers and five ceramics are listed below. (Chapters 10, 11, 13, 14)

Polymers	Ceramics
1. Polyethylene	6. Soda-lime glass
2. Polyvinylchloride	7. Vitreous silica
3. Polymethyl-methacrylate	8. Alumina (Al_2O_3)
4. Phenol formaldehyde	9. Pyroceram
5. Chloroprene rubber	10. Portland cement

Choose *one* of each (ceramic *and* polymer) to meet the indicated material requirement. (There may be more than one correct answer, but only one is requested. The first requirement is given as an example.)

Requirement	Polymer	Ceramic
Processing uses more than one component	4	10 (or 9)
a. Is most transparent		
b. Is most crystalline		
c. Is most temperature-resistant		
d. Is easiest to recycle		
e. Is the hardest		

CHAPTER 15

Processing of Polymers

The body and other parts of this experimental convertible are made of polymers. The purpose of the development was primarily to determine the cost savings in a relatively low production car (40,000/yr) which could be effected because of the lower cost of dies and the ability to form more complex shapes compared to metal.

The polymers used fall into three divisions: (1) Structural fiber-reinforced plastics (FRP) are vinyl ester resin with 55–60% fiberglass in a mixture of random and oriented placement. The front panel which serves as the radiator support is an example. (2) Nonstructural FRP is polyester resin with 50% filler such as talc used as the relatively inexpensive matrix. 25/27% fiberglass is added in a random orientation. (3) Complex parts are made of conventional resins such as reaction injection molded polyurea.

The low investment vehicle (LIV) passed the federal 30 mph crash test. Cost of tooling was 60% lower than for steel stamping dies. Adhesive bonding fixtures were 25–40% less costly than traditional welding fixtures.

15.1 Overview

In preparing previous editions of this text, we justifiably assumed that the producer of finished parts would obtain the polymer from the chemical manufacturer either in the completely polymerized condition or close to it (in the case of thermosets). Recently, however, the automotive industry has taken the lead in *producing the polymer from its reacting materials in the same process in which the part is formed.* This trend is spreading to other high-production operations. In order to understand the characteristics of parts made in this way, the engineer should know more about the polymerization process than the pencil and paper syntheses. For these reasons, we will discuss the following two topics here:

1. Reactions occurring in the polymerization chamber or kettle for addition and condensation reactions
2. Production processes for forming components, which we will compare, when appropriate, to the processing of metal parts

15.2 Polymerization Reactions — Addition Polymerization

Let us investigate two important concepts related to addition polymerization in a reaction chamber: (1) the rate of formation and the size of the molecules produced, and (2) the temperature generated by the reaction, which can limit the size of the product.

Rate of Formation

To understand this feature, we need to examine the rates of several reactions going on in the chamber in the initiator and the molecules of the monomer.

First, in order for the initiator to break the double bond of the monomer, it must decompose to form free radicals. In general terms,

$$I \xrightarrow{k_d} 2R \cdot \tag{15-1}$$

Example: $H_2O_2 \longrightarrow 2OH \cdot$

where $\cdot$ signifies the free electron needed to break the $C{=}C$ bond. We use the symbol k_d to signify the rate of decomposition in units of sec^{-1}.

Next we have the rate at which the *free radical* bonds to the *monomer*, which is assumed to be constant regardless of the size of the molecule. This rate is represented as k_a.

$$R \cdot + M \xrightarrow{k_a} M \cdot \tag{15-2}$$

After this the monomer, which is now a free radical, bonds to another monomer.

$$M\cdot + M\cdot \xrightarrow{k_p} M_2^{\cdot} \qquad (15\text{-}3)$$

This rate, k_p, is used for all molecules regardless of their size.

Finally, a growing chain can be terminated by running into another chain with a free electron (instead of opening up a $C=C$ bond in a new monomer). That is, we have

$$M\cdot + \cdot M \xrightarrow{k_{tc}} M—M \qquad (15\text{-}4)$$

where M—M is a standard shared electron bond, instead of

$$M_x^{\cdot} + M_y \longrightarrow M_x—M_y^{\cdot}$$

The symbol k_{tc} is used to express rate of *termination by combination.*

Another method is to reform a $\overset{\displaystyle |}{C} = \overset{\displaystyle |}{\underset{\displaystyle |}{C}}$ bond by *disproportionation.*

Molecule I   Molecule II

The hydrogen with one electron moves to molecule II. The free electron and the residual hydrogen electron reconstitute a double bond in molecule I. This reaction is

$$M_x^{\cdot} + M_y^{\cdot} \xrightarrow{k_{td}} M_x + M_y \qquad (15\text{-}5)$$

Note that there are no longer free electrons in the products. The symbol k_{td} is used to express the rate of *termination by disproportionation.*

Now we can simplify all these reaction k's as follows:

1. Eliminate k_a because once the initiator is decomposed, it reacts very quickly; therefore this rate is not limiting.
2. Lump the k's for different terminations together as k_t.

We need one more constant, however, because not all the free radicals created by decomposition react with monomers. Some recombine, and others are involved in side reactions. Accordingly, we define f as the fraction of initiator molecules that react with monomers.

Using these relations and rate theory, the expression can be developed:

$$\ln\frac{[M]}{[M_0]} = -k_p\left(\frac{fk_d[I_0]}{k_t}\right)^{1/2}t \tag{15-6}$$

where

$[I_0]$ = concentration of initiator

$[M]$ = total concentration of chain radicals (polymer), regardless of length (moles per liter)

$[M_0]$ = original concentration of monomer (moles per liter)

k_d, k_p, k_t = constants for reactions

f = fraction of free radicals from the decomposition of initiator that react to form polymer (Some are lost in side reactions.)

t = time of reaction

(In this approximation, a constant level for $[I_0]$ is assumed.) A more complex solution is given in the literature.[*]

EXAMPLE 15.1 *[ES]*

Calculate the conversion of styrene to polystyrene as a function of time, using the following data for isothermal conditions (60°C).

$$\frac{k_p^2}{k_t} = 1.18 \times 10^{-3}\ \frac{\text{liter}}{\text{mol}\cdot\text{sec}}$$

Styrene density = 0.907 g/cm³

$$k_d = 0.96 \times 10^{-5}\ \text{sec}^{-1}$$

$$[I_0] = 0.05\ \text{mol/liter (azobisisobutylnitrile)}$$

$$f = 1.0$$

Answer   Because

$$\frac{k_p^2}{k_t} = 1.18 \times 10^{-3}\frac{\text{liter}}{\text{mol}\cdot\text{sec}}$$

we know that

$$k_p = 0.034\sqrt{k_t}\left(\frac{\text{liter}}{\text{mol}\cdot\text{sec}}\right)^{1/2}$$

[*]S. L. Rosen, *Fundamental Principles of Polymeric Materials*, Barnes & Noble, New York, 1971, p. 102.

Then

$$\ln\frac{[M]}{[M_0]} = -k_p\left(\frac{(f)k_d[I_0]}{k_t}\right)^{1/2} t \tag{15-6}$$

$$= -0.034(k_t)^{1/2}\left(\frac{\text{liter}}{\text{mol}\cdot\text{sec}}\right)^{1/2}$$

$$\times \left(\frac{(1)(0.96\times 10^{-5}\ \text{sec}^{-1})(0.05\ \text{mol/liter})}{k_t}\right)^{1/2} t$$

or $\quad \ln\dfrac{[M]}{[M_0]} = -2.35\times 10^{-5}t$

where t is in units of seconds. We could calculate an actual concentration [M] by assuming a time value and knowing the molecular weight of styrene (104).

$$[M_0] = (0.907\ \text{g/cm}^3)(1000\ \text{cm}^3/\text{liter})(1\ \text{g}\cdot\text{mol}/104\ \text{g})$$

$$= 8.72\ \text{g mol/liter}$$

Or we could calculate the *conversion fraction*, which is defined as $(1 - [M]/[M_0])$ for a given time value.

When $t = 10$ hr $= 36,000$ sec,

$$\ln\frac{[M]}{[M_0]} = (-2.35\times 10^{-5})(36,000) = -0.846$$

$$\frac{[M]}{[M_0]} = 0.43 \quad \text{and} \quad \left(1 - \frac{[M]}{[M_0]}\right) = 0.57 \quad \text{or} \quad 57\% \text{ conversion}$$

This is plotted for several time values in Figure 15.1.

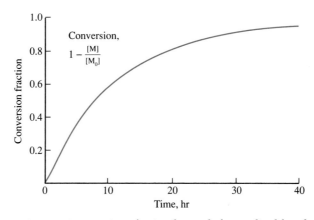

FIGURE 15.1   Conversion vs time for isothermal, free-radical batch polymerization.

[S. L. Rosen, *Fundamental Principles of Polymeric Materials*, Barnes & Noble, New York, 1971, p. 105. Reprinted by permission of Harper & Row, Publishers, Inc.]

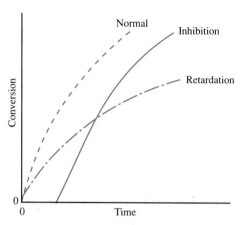

FIGURE 15.2   Inhibition and retardation effects on batch polymerization conversion
[From *Fundamental Principles of Polymeric Materials* by Stephen L. Rosen. Copyright © 1971 by Barnes & Noble, Inc. Reprinted by permission of Harper & Row, Publishers, Inc.]

We chose a relatively slow reaction for our example. At higher temperatures and with other monomers, much faster reactions can be produced.

In commercial practice an inhibitor may be added to avoid polymerization during shipment, and a retarder may be employed to reduce the reaction rate. The effects are shown schematically in Figure 15.2.

Temperature Effects Due to the Reaction

In principle it seems simple to start with a monomer, stir in an initiator, and form the desired polymer. However, problems in heat dissipation plague the production of either larger components or masses of polymer. For example, the important vinyl polymers have large exothermic heats of formation (-10 to -21 kcal/g $\cdot$ mol). The organic materials have only about half the heat capacity of aqueous solutions, and because of high viscosity it is hard to achieve heat transfer to the vessel by mixing. The temperature increase raises the reaction rate and, to quote Schildknecht's famous comment on laboratory bulk polymerization, "It may lead to loss of the apparatus, the polymer, or even the experimenter."

EXAMPLE 15.2 *[ES]*

Estimate the adiabatic (no temperature loss to surroundings) temperature increase in the bulk polymerization of styrene. The heat liberated, ΔH (enthalpy change), for polymerization = 16.4 kcal/g $\cdot$ mol; molecular weight = 104; heat capacity = 0.5 cal/g $\cdot$ °C.

Answer By definition $dH = C_p dT$, or $\Delta H = C_p \Delta T$, when the heat capacity is assumed to be a constant.

$$\Delta T_{max} = \Delta H/C_p = (16{,}400 \text{ cal/g} \cdot \text{mol})/[(0.5 \text{ cal/g} \cdot \text{c})(104 \text{ g/g} \cdot \text{mol})]$$

$$\Delta T_{max} = 315°C$$

If we began at 20°C, the new temperature would be 335°C—and the boiling point of styrene is 146°C!

In order to avoid the disastrous result described in Example 15.2, we take one of the following precautions.

1. Produce small, thin sections for adequate heat transfer.
2. Suspend or dissolve the monomer and some partially polymerized product in an inert solvent to absorb heat. This produces an interesting side effect called auto-acceleration. The reaction rate is much greater in concentrated solutions, because at high viscosities the *termination rate* decreases. (It is harder for the polymer molecules to move and combine.) The *combination rate* is still high, however, because the small monomer molecules can diffuse easily.

15.3 Polymerization Reactions—Condensation

We will not discuss these reactions in detail, because the effects we noted for addition reactions are either absent or not so potentially disastrous. For example, the exothermic effect is usually less when it is accompanied by the formation of a condensation product. The polymerization step is different because, in contrast to the zipper-like formation of free radicals of increasing length in addition reactions, here the positioning of two polymer molecules of increasing length is needed.

An interesting relation can be developed from probability theory (and confirmed by experiment). It relates the ratio of the number of molecules of a given size x to the time for reaction (Figure 15.3). The fraction of molecules of a given size is expressed by the ratio n_x/N, where n_x is the number of molecules of size x, and N is the total number of molecules in the mix.

The symbol p denotes the fraction of original material reacted (conversion), and $\overline{x}_n$ denotes the average size of the molecules (number of mers in the chain). After only a short time ($p = 0.90$, $\overline{x}_n = 10$), there are practically no molecules above a degree of polymerization of 60; at higher conversions, there is a wide range of sizes.

To achieve the average size of 100 that is needed to obtain useful mechanical properties, 99% reaction must take place. Therefore it is necessary to start with pure monomers and, if two are involved, to use exact proportions.

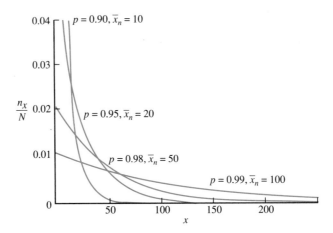

FIGURE 15.3   Number (mole) fraction distributions for linear poly-condensations

[From *Fundamental Principles of Polymeric Materials* by Stephen L. Rosen. Copyright © 1971 by Barnes & Noble, Inc. Reprinted by permission of Harper & Row, Publishers, Inc.]

15.4   Production Processes

Now that we have discussed polymerization techniques, we can review the principal processes used in the production of polymer components. They are summarized in Table 15.1. The thermoplastics and thermosets require different approaches for the fabrication of components, and the same equipment may not be appropriate for both classes of materials.

Let us first discuss the methods by which three dimensions may be varied (rather than only two, as in the case of rolled or extruded parts.) These methods are casting, compression molding, transfer molding, and injection molding, followed by slush molding, rotational molding, and blow molding.

Casting generally employs a thermosetting resin without external heating. The components are mixed and poured into the mold, an exothermic reaction occurs, and the polymer sets. (Some thermoplastics can be cast as films if they are deposited in solution or as a hot melt against a cold polished surface.) Casting is used extensively for potting—that is, encapsulating—electrical circuits. It is not a high-production process but has practically unlimited geometry.

In *compression molding* (Figure 15.4), powder or prepared pellets are placed in the die cavity. The die is closed and heated, and the plastic (usually a thermoset) softens to take the shape of the cavity and then sets. The part is removed while it is hot. This method is rarely used for TP resins because the mold must be cooled to permit removal of the part.

Transfer molding (Figure 15.5) is a modified version of the compression method in which the plastic can be introduced into a closed mold, and flash (seepage between the mold halves at the parting line; see Figure 15.4*b*) can thus be avoided. Another advantage of transfer molding is that inserts can be better positioned. The stage A polymer is melted in a separate chamber and is

TABLE 15.1 Characteristics of Production Processes for Plastic Parts

| | Geometrical Limits | | | | Dimensional Tolerance, in./in. | Surface Finish[b] | Undercuts[c] | Minimum Thickness, in. |
| | Overall[a] | Two Dimensions | Three Dimensions | | | | | |
			Simple	Complex				
Blow molding	M		X		±0.01	1–2	Yes	0.01
Calendering	E	X			—	1–3	No	0.01
Casting	M			X	±0.001	2	Yes	0.01
Compression molding	M			X	±0.001	1–2	N.R.	0.01
Extrusion	M	X			±0.005	1–2	No	0.001
Injection molding	M			X	±0.001	1	Yes	0.015
Rotational molding	M		X		±0.01	2–3	Yes	0.02
Slush molding	M		X		—	2–3	N.R.	0.02
Thermoforming	E	X			±0.01	1–3	No	0.002
Transfer molding	M			X	±0.001	1–2	Yes	0.01

[a]Factor governing maximum dimensions: M = mold, E = equipment, such as roll length.
[b]Surface finish: 1 = best, 5 = poor.
[c]Undercuts: Yes = requires special design, N.R. = not recommended.
Reprinted from C. A. Harper, ed., *Handbook of Plastics and Elastomers*, © 1975 by permission of McGraw-Hill Publishing Company.

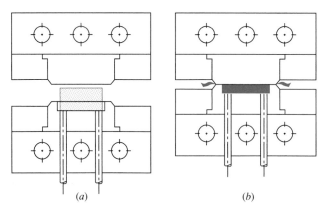

FIGURE 15.4   Compression mold. (a) Open, with preform in place. (b) Closed, with flash forced out between mold halves.

[*Modern Plastics Encyclopedia*, Copyright © 1966, reprinted by permission of McGraw-Hill Publishing Co.]

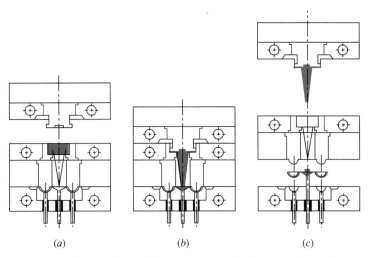

(a)   *(b)*   *(c)*

FIGURE 15.5   Typical transfer molding operation. (a) Material is fed into the pot of the transfer mold and then (b) forced under pressure when hot through an orifice and into a closed mold. (c) After the polymer has formed, the part is lifted by ejector pins. The sprue remains with the cull in the pot.

[*Modern Plastics Encyclopedia*, Copyright © 1966, reprinted by permission of McGraw-Hill Publishing Co.]

then forced by a piston into the die cavity.* This method is used for TS material, although it could also be used for TP. Injection molding is usually preferred for TP.

---

*Stage A: Lightly polymerized because of a deficiency in one of the components; soluble in organic solvents. Stage B: Further polymerized by addition of a reactant; nearly insoluble in organic solvents but still fusible under heat and pressure. Stage C: Final, infusible, crosslinked polymer shape.

In *injection molding* (Figure 15.6) the TP polymer is fed into a heated injection chamber, melted, and then forced by a ram into the mold cavity. Because the mold cavity is closed before injection, excellent accuracy is obtained.

Table 15.2 summarizes the normal processing temperatures and pressures commonly used for compression, injection, and transfer molding. The thermosetting polymers are molded while in stage A or B, depending on the particular polymer (see footnote, page 636).

The automotive industry has made extensive use of *reaction injection molding* (RIM), which is unique in that the monomers are mixed by impinging streams just prior to mold entry. The polymer forms in the mold under low molding pressures; this means that inexpensive molds may be used. Many kinds of polyurethane—ranging from elastomers to rigid foam—are formed by the RIM process into such diverse parts as bumpers, fender extensions, and spoilers. Although the RIM method is currently used predominantly for polyurethanes, any reactive monomers—such as those found in epoxy and nylon, for example—can be similarly injection-molded.

Slush molding of polymers is similar to the method of slip casting used for ceramics. A powder or liquid (unpolymerized) is poured into a heated mold. A shell of polymerized material builds up around the inside of the mold, and then the unpolymerized liquid is poured out. The part is then fully polymerized in an adjacent furnace. Cores to be used for metal castings can be made in this way (using sand filler); the advantage of these cores is that they collapse after metal solidifies around them.

In *rotational molding*, a measured quantity of TS or TP material is poured into a mold. The mold is closed, heated, and rotated on two axes until the material has hardened on the mold walls, leaving a cavity of the desired size within. Large hollow parts can be produced by this method, but production is slow.

Blow molding of polymers is similar to glass blow molding, which was described in Chapter 10. An extruded tube (parison) of hot thermoplastic is

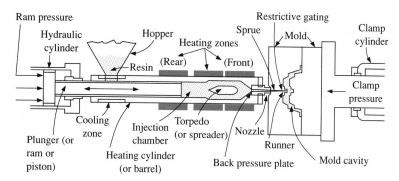

FIGURE 15.6   Schematic cross section of typical plunger (or ram or piston) injection molding machine

[*Polyolefin Injection Molding: An Operating Manual*, 1st ed., U.S. Industrial Chemicals Co. 1977. Used by permission.]

TABLE 15.2 Summary of Processing Methods for Various Polymers

Polymer	Compression Pressure,[a] psi × 10³	°C	Injection Molding Pressure,[a] psi × 10³	°C	Extrusion °C	Transfer Molding Pressure,[a] psi × 10³	°C
			Thermoplastics				
Polyethylene			5 to 22	135 to 143	80 to 94		
Polypropylene			10 to 22	204 to 288	193 to 221		
Polystyrene			10 to 24	162 to 243	90 to 107		
Polyvinylchloride	0.5 to 2	140 to 175	7 to 15	160 to 175	162 to 204		
Tetrafluorethylene[b]							
ABS[c]		162 to 190	6 to 30	218 to 260			
Polyamides			10 to 20	271 to 343	277 to 299		
Acrylics			10 to 20	160 to 260	177 to 232		
Acetals		200 to 245	15 to 25	193 to 215	188 to 204		
Cellulosics	0.5 to 5		8 to 32	215 to 254	215 to 232		
Polycarbonates			15 to 20	274 to 330	247 to 304		
			Thermosetting Polymers				
Phenolics	1.5 to 5	143 to 193	4 to 8	160 to 171		2 to 10	135 to 171
Urea–melamine	2 to 5	149 to 171				6 to 20	149 to 165
Polyesters						1 to 5	121 to 177
Epoxies						0.1 to 2	143 to 177
Silicones[d]	1 to 3	177				0.5 to 10	177

[a]Multiply psi by 6.9 × 10⁻³ to obtain MPa or by 7.03 × 10⁻⁴ to obtain kg/mm²
[b]Requires preforming of particles and fusion at 370°C
[c]Acrylonitrile–butadiene–styrene
[d]Glass-reinforced

placed between two mold halves, the mold is closed, and the parison is blown against the mold walls. The bottle-like shape is then removed.

As Table 15.1 shows, the four processes discussed above permit formation of complex three-dimensional designs, whereas the following three processes are limited processes with only two-dimensional freedom.

Calendering is essentially a rolling operation in which a putty-like mass of thermoplastic material is passed through heated rolls, which work it into a sheet of uniform thickness. The process is also used to apply plastic coatings to other materials.

Extrusion (Figure 15.7) is similar to metal extrusion except that a TS or TP plastic is forced through a nozzle with the desired cross-sectional configuration. The method can also be used to apply coating to wire or to produce complex profiles in a transverse direction.

Thermoforming involves using vacuum, air pressure, or mechanical energy to force a heated thermoplastic sheet into the shape of a mold. After cooling, the plastic part is removed from the mold and trimmed. Components range from blister or bubble packaging for food and housewares to fender liners on automobiles.

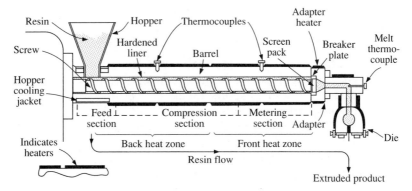

FIGURE 15.7 Schematic cross section of a typical extruder

[*Polyolefin Injection Molding: An Operating Manual*, 1st ed., U.S. Industrial Chemicals Co. 1977. Used by permission.]

EXAMPLE 15.3 *[ES]*

We are to produce a 6/6 nylon component by the RIM process. How many grams of water will be produced per gram of the two monomers if the reaction occurs as shown here?

Answer It appears from the sketch that 1 mol of the diacid + 1 mol of the diamine gives 2 mol of water. However, the last monomers do not contribute water; instead, x mol of diacid + x mol of diamine give $(2x - 1)$ mol of water. Because x would be large with a reasonable degree of polymerization, the ratio of 2:1 (water:total monomer) will be used.

$$\text{MW(diamine)} = 2 \times 14.01 + 6 \times 12.01 + 16 \times 1 = 116.08$$

$$\text{MW(diacid)} = 4 \times 16 + 6 \times 12.01 + 10 \times 1 = 146.06$$

$$\text{MW(water)} = 2(16 + 2) = 36$$

$$\frac{\text{g}(H_2O)}{\text{g(diacid + diamine)}} = \frac{36}{116.08 + 146.06} = 0.14$$

That is, approximately 14% is lost as water. In practice, not all of the water is lost, because some retention increases the impact resistance of the nylon.

15.5 Polymer Processing Compared to Metal Processing

In Chapter 6 we developed the unit operations for processing metals, and we will now apply some of these concepts to polymers. Many of the operations were covered in Chapters 13 and 14, because they are so intimately related to the physical characteristics of the polymers.

For example, thermoplastic shapes can be produced by *casting*, which is analogous to die casting of metals. Chemically reactive thermosets, such as epoxies and polyesters, can be mixed and poured to produce cast shapes without the application of heat. Because the polymerization process produces heat, it may even be necessary to cool the casting to prevent the thermal cracking of some thermosets.

Extrusion, blow molding, calendering, and thermoforming are all methods of hot *deformation* of thermoplastics. Compression and transfer molding are analogous techniques applied to the thermosets.

A number of the polymers can be produced in powder form, so processing them is similar to the *particulate processing* of metals. Examples include thermosets in stages A and B and styrofoam beads that are steam-cured to make insulating containers and packaging.

Joining of polymers by other polymer adhesives is certainly a common unit operation. In materials such as polyethylene, the lack of surface attractive forces prevents wetting by the common adhesives, and fusion welding is more appropriate.

When we *machine* polymers, we must not always assume that their lower hardness means no difficulties will arise in machining. For example, their higher coefficient of thermal expansion and poor thermal conductivity can make close dimensional control difficult. Thermoplastics can actually melt, gum up cutting tools, and result in a poor surface finish. Abrasive fillers in polymers can also lower the tool life more than expected.

We might not expect polymers to be given a *heat treatment*. However, crystallinity can be controlled by thermal processing, and we can modify residual stress by means of a stress relief anneal (relaxation).

Finishing and surface treatment are also important to polymers, as we observe in the plating, painting, and texturing of polymer surfaces. Metal coats can also provide a conducting surface.

In summary, the general processing operations applied to metals can also be used for polymers. However, the bonding in thermoplastics, thermosets, and elastomers is sufficiently different from that in metals for modifications to be required in equipment and processing sequence.

Summary

In the first part of the chapter, we discussed some of the problems that arise in the production of the polymer itself. Then we analyzed the processing of the plastic into components. It was pointed out that the two processes are now being combined in some high-production operations.

In the formulation of an addition polymer, two principal problems are rate of polymerization and temperature increase due to the reaction. The rate of reaction was related to rates of decomposition of the initiator, reaction of the initiator with the monomer, reaction of the free-radical monomer with neutral monomers, and reaction of termination. The potential temperature rise was shown to be catastrophic in a bulk reaction, a hazard that underscores the need for controlled cooling.

The processes used to produce polymer components include variations of all those used for metals—and some additional ones. Casting, injection, compression, slush molding, and transfer molding are used for complex shapes, whereas rolling, calendering, blow molding, extrusion, and thermoforming are available for simpler shapes. In Chapter 16 we will consider the processing of composite components.

Definitions

Blow molding A process in which a gob of plastic is formed and then blown into a mold of the desired shape.

Calendering Rolling a heated mass of thermoplastic.

Combination Termination of two growing chains by their joining at positions of free radicals.

Compression molding A process in which a metered amount of plastic is placed in a heated die and compressed to the desired dimensions. The process is important for thermosetting resins.

Conversion fraction The fraction polymerized $(1 - [M]/[M_0])$, where $[M]$ = total concentration of the polymer and $[M_0]$ = original concentration of the monomer.

Disproportionation Termination of two growing chains by the formation of two different end groups when their free radicals come together.

Extrusion The forcing of liquid or plastic resin through a die to obtain the required shape, such as a rod.

Free radical Electrically neutral molecule that is still reactive because of the presence of an unpaired electron.

Injection molding The injection of a resin into a die, followed by solidification or polymerization and ejection.

Reaction injection molding (RIM) Technique whereby the monomers are mixed by impinging streams just prior to mold entry during the injection molding process. The method is particularly useful for multicomponent thermosetting resins.

Slush molding Slip casting of an unpolymerized plastic shell, followed by polymerization.

Transfer molding A process in which material is fed into the pot of a transfer mold, heated, and then forced under pressure into an adjoining cavity of the same mold.

Problems

15.1 *[ES/EJ]* Refer to Problem 13.17 for the definitions of number-average and weight-average molecular weight. Explain how your interest in $\overline{M}_n$ or $\overline{M}_w$ might differ if you were a polymer user (designer) or polymer processor. (Sections 15.1 through 15.2)

15.2 *[ES/EJ]* The melt index is the rate at which a polymer is extruded under prescribed conditions through a die of specified length and diameter (ASTM-D-1238). The units are grams of polymer extruded in 10 minutes. (Sections 15.1 through 15.2)
a. How does the melt index vary with the molecular weight of the polymer?
b. Explain why $\overline{M}_w$ would be more important as a measure of the melt index than would $\overline{M}_n$ (See Problem 13.17 for definitions).

15.3 *[ES/EJ]* In Example 15.1, calculate new percentages of conversion after 10 hr when each of the following conditions is modified individually. (Sections 15.1 through 15.2)
a. The rate constant for propagation, k_p, is increased by a factor of 2.
b. The rate constant for decomposition, k_d, is increased by a factor of 2.
c. What is the practical significance of the changes in conversion that you calculated in parts a and b?

15.4 *[ES/EJ]* Using the data given in Example 15.1, calculate the amount of initiator that would be required to obtain 95% conversion in 10 hr. Explain why the assumption of a constant concentration of initiator is not correct, especially when high conversions are to be calculated. (Sections 15.1 through 15.2)

15.5 *[ES/EJ]* At the end of Example 15.1, it is mentioned that the percentage conversion would be greater at higher temperatures. (Sections 15.1 through 15.2)
a. What parameters in the calculation change as the temperature is changed?
b. How would you expect these parameters to change with temperature? For example, would doubling the temperature double the conversion?

15.6 *[EJ]* Explain why the adiabatic temperature increase in the bulk polymerization calculation in Example 15.2 would not be obtained in practice. (Sections 15.1 through 15.2)

15.7 *[ES]* Polymerization reactions are exothermic, as pointed out in the text. How would this affect your calculations in Example 15.1? (Sections 15.1 through 15.2)

15.8 *[EJ]* Which processing method would you use to make each of the following? (Sections 15.3 through 15.5)
a. Saran Wrap sheet
b. Polypropylene rope
c. Polyethylene squeeze bottle
d. Dish of melamine
e. Nylon fishing leader (0.007-in. diameter, clear)

15.9 *[ES/EJ]* A thin plastic cup from a vending machine is heated in an oven (140°C) until it collapses and assumes its original shape as a flat blank. It is noted from this experiment that the shape of the blank is elliptical. Why? [*Hint:* The blanks are cut from rolled sheet.] (Sections 15.3 through 15.5)

15.10 *[ES/EJ]* Plastic trees, shrubs, and flowers have gained some consumer acceptance. (Sections 15.3 through 15.5)
a. Indicate whether these items are made from thermoplastic or thermoset material, and speculate about the method of production.
b. What other considerations are important if such a material is to be used for a Christmas tree, which will be exposed to cold winter temperatures? [*Hint:* Recall the characteristic properties of polymers.]

15.11 *[E]* A thermoplastic polymer is injection-molded into a complex shape that has thin and thick sections. (Sections 15.3 through 15.5)
a. Which sections have the highest crystallinity?
b. Which sections are the strongest?
c. Explain the advantages of a thermal treatment after injection molding.

15.12 *[ES/E]* Figure 15.3 shows that when the conversion fraction increases, there is a wider range in chain size and more short than long chains. What effect does this have on the ease of polymer processing and on the final polymer strength? (Sections 15.4 through 15.6)

15.13 *[E]* Table 15.2 indicates that the pressures used for compression and injection molding of the thermoplastics vary. What characteristics of the polymer determine what pressure is necessary? (Sections 15.3 through 15.5)

15.14 *[ES]* A certain type of rubber has a true specific gravity of 1.40. This type of rubber is used in the manufacture of a foamed rubber that weighs 0.015 lb/in^3 (415.6 kg/m^3) when dry, and 0.025 lb/in^3 (692.6 kg/m^3) when saturated with water. (Chapter 15 and Section 6.13)
a. What is the apparent porosity of the foamed rubber?
b. What is the true porosity?

15.15 *[ES]* A particular cellular polymer is being considered for use as a sponge. The dry sponge has a bulk density of 30 lb/ft^3 (481 kg/m^3). The specific gravity is 1.34. How many pounds (kilograms) of water could a sponge 6 in. × 6 in. × 3 in. (0.152 m × 0.152 m × 0.076 m) hold if all the pores were open and available to be filled with water? (Chapter 15 and Section 6.13)

15.16 *[ES]* Some polymers are potentially toxic if they are burned and the fumes are allowed to become airborne. How much HCl would be obtained per pound of combusted PVC? (Chapter 15 and knowledge of chemistry)

PART V

Composite Materials

Of all materials, the composites are the most exciting to study. We encounter them in automobiles, tennis racquets, fishing rods, boats, golf clubs, windsurfers and aircraft. Although fiberglass products have been around for a long time, new fibers such as graphite, boron and aramid (Kevlar) have led to new products. In addition to these components, composites are important in cutting tools and heat-resistant ceramic–metal structures. Finally, we must not neglect traditional composites such as wood and concrete.

The essence of a composite is that we can take two or more very different materials and put them together to form a product with some properties that are ten times greater than those of the original materials. In addition, we can choose fibers, plate-like particles, or spheres for dispersions in the matrix. For example, high-strength graphite fibers are only 8 μm (0.0003 in.) in diameter but are commonly dispersed in an epoxy polymer matrix.

These possibilities for engineering design of a new material from a spectrum of structures are accompanied by new responsibilities. In no other group of materials is the significance of directionality (anisotropy) of properties as important. For example, if we fabricate a rod with glass fibers oriented along the longitudinal axis, we will find that the strength and modulus of elasticity are dominated by the properties of the glass. On the other hand, if we measure the same properties across the rod, we will find that the modulus and strength of the polymer (much lower values) are governing. Therefore, we have included a rather detailed section on effects of fiber orientation.

After discussing the synthetic composites in Chapter 16, we take up the traditional high-tonnage composites concrete, asphalt, and wood in Chapter 17. The use of polymers in improving wood products and in developing by-products from wood scrap is of particular interest. The bonding of aggregates with an organic, as in asphalt, or with Portland cement, as in concrete, may seem mundane, yet fundamental materials engineering principles must be followed if we are to have successful structures.

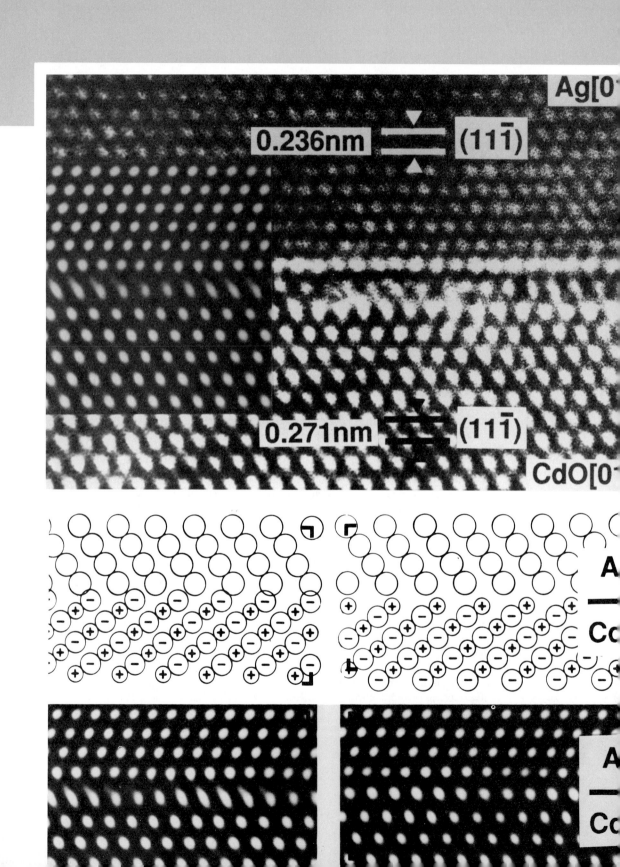

CHAPTER 16

Synthetic Composite Materials

Co-authored with Brian Flinn
Research Fellow
Materials Department
College of Engineering
University of California — Santa Barbara

One of the most advanced techniques for investigating the structure of a composite is a combination of high-resolution microscopy and computer simulation.

The most important region to observe in a composite is the interface between the matrix and the fiber. In this high-resolution micrograph (right-hand side), the positions of the atoms on (111) planes of cadmium oxide and silver are registered as spheres. The overlay box at the left is a computer-simulated calculation of the atom positions, which confirms the observations drawn from the micrograph.

It is concluded that this is an incoherent interface because no extensive strain field develops, which would be shown by a change in atom positions and by dislocations. This is in accord with the large difference in lattice misfit of 14% (0.236 vs 0.271 nm.)

Also there are two simulated models of the bonding at the interface: Ag-O-Cd upper and lower left and Ag-Cd lower right.

The Ag-O-Cd is obtained as shown by the match between the actual micrograph and the insert at the upper left.

16.1 Overview

Of all materials the composites are the most exciting to study. You have encountered them in skis, tennis racquets, golf clubs, wind surfers, boats, automobiles, and aircraft. Although fiberglass products have been around a long time, new advances in the use of graphite, boron, and Kevlar fiber have been made. The real challenge to the engineer is that by putting two or more materials together, it is possible to have a product with some properties *ten* times those of the individual components.

Rather than rushing into a discussion of glamorous applications, let us first investigate these materials from a fundamental point of view that will help you understand new materials of the future as well as current ones. We will follow this general outline:

1. Definition of a composite
2. Combinations and examples of ceramic–metal–polymer composites
3. Dispersion strengthening and particle reinforcement
4. Properties of major fiber composites
   a. Polymer matrix + ceramic
      Polymer matrix + polymer
   b. Ceramic matrix + ceramic
   c. Metal matrix + ceramic
5. Processing of composites, including preparation of fibers and processing of components

16.2 Definition of a Composite

Let us first define *composite* in its broadest sense and then narrow the definition down to the materials we will discuss in this and the next chapter. A composite consists of two or more phases that are usually processed separately and then bonded, resulting in properties that are different from those of either of the original materials tested alone. This gives us a wide range of materials—varying from mixtures of phases that can be seen only with the electron microscope to coarse *macrocomposites* such as concrete:

Examples of Synthetic Microcomposites	Examples of Macrocomposites
Fiberglass	Wood
Cemented carbides	Concrete
Graphite–epoxy	
Kevlar–epoxy	

A distinguishing feature of the *synthetic composites*, which we will discuss in this chapter, is that the composite structure is built up by selecting the nature, amounts, shapes, and sizes of the different materials and then bonding them together in a controlled orientation.

We want to emphasize that this is a very powerful tool for attaining practically any combination *of solid phases.* By contrast, we could not produce a fiberglass structure by simply melting glass and polymer ingredients in a single melt.

We will discuss the synthetic composites first because of the unlimited combinations of structures that can be achieved. Then, in the next chapter, we will take up wood, a natural composite, and the macrocomposite concrete.

16.3 Combinations and Examples of Ceramic–Metal–Polymer Composites

We are all inclined to make the mistake of visualizing as a composite only one favorite, such as fiberglass (or, more generally, FRP, fiber-reinforced plastic) when the topic is mentioned. Therefore, let us begin our basic discussion by summarizing all the possible combinations of ceramic, metal, and polymer structures in composites. We start with a simplified chart (Figure 16.1) in which we show three families with ceramic, metal, and polymer matrices, respectively. Then in each case we add a second phase from each group, for a total of nine combinations. In addition to those structures, there are three-or-more-phase "hybrids." An example is a graphite–polymer tennis racquet frame that is reinforced in highly stressed regions with boron fibers. Another variation is "interlocking matrices" in which both phases are continuous throughout the structure.

Before proceeding with a more detailed discussion of fiber reinforcement, we shall briefly consider the general types of synthetic composites, which can be divided into three categories.

1. *Dispersion-strengthened.* Small particles are dispersed in a matrix; the <u>matrix</u> is the major load-bearing constituent.

Matrix:	Ceramic	Metal	Polymer
Example:	Cer – *Cer* Al_2O_3 – SiC	Met – *Cer* Tungsten carbide in cobalt	Poly – *Cer* Fiberglass, Graphite, Boron in polymer
Example:	Cer – *Met* Al – Al_2O_3	Met – *Met* Powder metallurgy	Poly – *Met* Brake linings
Example:	Cer – *Poly* (not applicable)	Met – *Poly* (not applicable)	Poly – *Poly* Epoxy – Kevlar

FIGURE 16.1   Combinations of composite materials

2. *Particle-reinforced.* Larger particles are incorporated in a matrix, and the load is <u>shared</u> by the matrix and the particles.
3. *Fiber-reinforced.* Fibers are incorporated in a matrix. The <u>fiber</u> is the primary load-bearing component.

16.4 Composites Strengthened by Dispersion and Particle Reinforcement

Consider a metal matrix with a fine distribution of secondary particles. Because deformation in the matrix is accompanied by slip and dislocation movement, the degree of strengthening achieved is proportional to the ability of the particles to impede the dislocation movement. It follows that a finer dispersion of particles results in greater strengthening. The objective is to have the particles small enough and spaced closely enough so that dislocation movements cannot easily occur between them.

It can be shown that in *dispersion strengthening* by using particle diameters less than 0.1 micron (1 micron = 10^{-6} meter) and volume concentrations of 1–15%, dislocation movement can be effectively impeded. The strengthened matrix becomes the main load-bearing constituent.

An example of a dispersion-strengthened alloy is SAP, sintered aluminum powder. If we make a composite of fine Al_2O_3 particles in an aluminum matrix (by compacting and sintering the powders), we can significantly increase the high-temperature properties of aluminum alloys because the composite does not overage. Another example, the increased thermal resistance of iron–Al_2O_3 composites, is shown in Figure 16.2.

If we have a greater percentage of dispersion and larger particles, we obtain *particle reinforcement*, wherein the load is shared by both matrix and

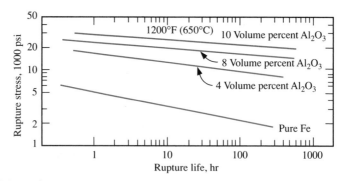

FIGURE 16.2   The increase in elevated temperature (1200°F, 650°C) stress-rupture properties obtained by dispersion strengthening iron with Al_2O_3 particles. (To obtain MPa, multiply psi by 6.9×10^{-3}.)

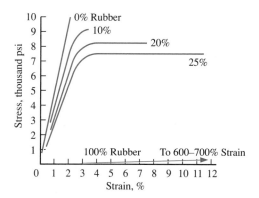

FIGURE 16.3　The influence of rubber particles on the strength of polystyrene. The stress–strain curve was obtained at high rates of strain (133 in./in. per min or m/m per min). (To obtain MPa, multiply stress by 6.9×10^{-3}.)

[From S. L. Rosen, *Fundamental Principles of Polymeric Materials*. Copyright © 1982 by John Wiley & Sons, Inc. Reprinted by permission.]

particles. In general, in a particle-reinforced composite, particle diameters are greater than 1 micron and volume concentrations are greater than 25%.

Possibly the best examples of particle-reinforced composites are the cemented carbides discussed in Chapter 11. Here the stress is supported by both the matrix and the particles, and the volume fraction of the particles is large (see Figure 11.1). Similarly, most commercial ceramics, ranging from bricks to grinding wheels, and many filled polymers are particle-reinforced composites. Figure 16.3 shows effects of rubber particle fillers on stress–strain curves of normally brittle polystyrene. The high strain rate is intended to approximate the strain imposed by impact loading in service.

16.5　Fiber Reinforcement—Polymer–Ceramic

Because composites with a polymer matrix and ceramic fiber, such as fiberglass, are at the present time the most important class, we will use them as examples. The principles, however, can be applied to the other composites.

From a simple mechanical point of view, a composite consists of particles (fibers or grains) of a given strength, modulus E, and fracture toughness K_{IC} suspended in a matrix that has a different set of properties. The critical question is "Will the composite have properties that represent the average of the two sets, weighted by the volume fractions of each?" This principle is called the rule of mixtures (ROM), and although it applies to some properties (such as density), the exceptions are more important than the rule. Sometimes, for instance, we can develop fracture toughness many times the ROM estimate: For example, in fiberglass the toughness is about ten times that of the matrix and a hundred times that of the fiber! Let us analyze, in order, the modulus of elasticity, tensile strength, and fracture toughness achieved in fiber composites.

16.6 Modulus of Elasticity in Fiber Composites

There are two important classes of fiber-reinforced composites (Figure 16.4):

1. Composites in which the fibers of the strengthener are continuous
2. Composites in which the fibers of the strengthener are discontinuous and may be merely chopped-up pieces

We will discuss each case as we consider the mechanical properties.

Composites with Continuous Fibers in a Matrix

Take the simple case of a bar 0.5 in^2 (323 mm^2) in cross section in which continuous glass fibers are in a matrix of polyester: a typical fiberglass. What effect will increasing the amount of glass have on the strength and modulus of elasticity of the composite? Assume the following values:

Material	Tensile Strength,[*] psi × 10^3	Yield Strength,[*] psi × 10^3	Modulus,[*] psi	Percent Plastic Elongation
E glass fibers	250	250	10 × 10^6	0
Polyester	5	5	4 × 10^5	~0

[*]Multiply by 6.9 × 10^{-3} to obtain MN/m^2 (MPa) or by 7.03 × 10^{-4} to obtain kg/mm^2.

The glass has a stiff stress–strain curve showing little deformation, whereas the polyester shows somewhat more (Figure 16.5).

We shall consider the glass as the fiber and the polyester as the matrix surrounding the fiber. To enable us to understand the loads in the fiber and matrix, let us conduct the following analysis.

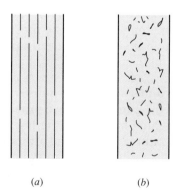

(a)   (b)

FIGURE 16.4   Variations in distribution of fiber in a composite material such as fiberglass. (a) Continuous or practically continuous fibers. (b) Chopped-up, discontinuous fibers.

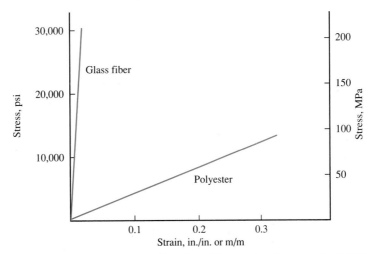

FIGURE 16.5   Stress-strain curves for glass fiber and polyester. At 0.003 strain, the stress in the fiberglass is 30,000 psi (207 MPa). At 0.003 strain, the stress in the polyester is 1200 psi (8.3 MPa).

First we apply a load P that will be carried by both the fiber and the matrix. We assume a good bond between fiber and matrix so that both stretch the same amount and the load direction is <u>parallel</u> to the fiber alignment direction. This type of loading is called *isostrain loading*, for reasons that follow.

We can write

$$P_c = P_f + P_m \qquad (16\text{-}1)$$

where P_c = total load on composite, P_f = load carried by fiber, and P_m = load carried by matrix.

We also know that the stress in the fiber, S_f, is the load divided by the cross-sectional area of the fiber:

$$S_f = \frac{P_f}{A_f} \quad \text{or} \quad P_f = S_f A_f \qquad (16\text{-}2)$$

Writing similar equations for the composite and the matrix, and substituting in Equation 16-1, we have

$$S_c A_c = S_f A_f + S_m A_m \qquad (16\text{-}3)$$

Dividing by A_c, the total cross-sectional area of the composite, yields

$$S_c = S_f \frac{A_f}{A_c} + S_m \frac{A_m}{A_c} \qquad (16\text{-}4)$$

A_f/A_c is the fraction of the total area occupied by the fiber. In this case it is also the volume fraction of the fiber, because the lengths of the fiber, composite, and matrix are all the same. Now if we let C_f denote the area (volume) fraction of fiber and C_m denote the area (volume) fraction of matrix, then $C_f + C_m = 1$, and by rewriting Equation 16-4, we obtain

$$S_c = S_f C_f + S_m C_m \qquad (16\text{-}5)$$

Now let us look at the strain produced by the load. If we call the overall strain e_c, the strain in the fiber e_f, and that in the matrix e_m, then

$$e_c = e_f = e_m$$

because we assumed a good bond, and all are strained equally under load P_c. Therefore, such a loading condition is called *isostrain loading*.

Let us divide the terms of Equation 16-5 by e_c, e_f, and e_m, separately:

$$\frac{S_c}{e_c} = \frac{S_f}{e_f} C_f + \frac{S_m}{e_m} C_m \qquad (16\text{-}6)$$

We can now substitute the overall modulus of the composite E_c for S_c/e_c, the modulus of the fiber E_f for S_f/e_f and the modulus of the matrix E_m for S_m/e_m:

$$E_c = E_f C_f + E_m C_m \qquad (16\text{-}7)$$

This enables us to determine the elastic deflection or strain.

EXAMPLE 16.1   *[ES]*

Suppose that we have a 0.5-in² (323-mm²) bar that is 60% glass and 40% polyester by volume. What are the stresses and strains in the individual components? Assume loading parallel to the fibers, or isostrain loading.

Answer $E_c = (10 \times 10^6 \text{ psi})(0.6) + (4 \times 10^5 \text{ psi})(0.4) \qquad (16\text{-}7)$

$$= 6.16 \times 10^6 \text{ psi}(0.425 \times 10^5 \text{ MPa})$$

The strain will be S_c/E_c or, because the bar is 0.5 in² (323 mm²),

$$e = \frac{P_c}{0.5 E_c} = \frac{2P_c}{E_c}$$

Next let us find the factors affecting the loads carried by the glass and the plastic.

The load carried in the fiber will be the stress in the fiber times its area, and similarly in the matrix. We can write

$$\frac{\text{Load carried in fiber}}{\text{Load carried in matrix}} = \frac{S_f A_f}{S_m A_m} \qquad (16\text{-}8)$$

or, using the relation $S = Ee$,

$$\frac{P_f}{P_m} = \frac{E_f e_f A_f}{E_m e_m A_m} \qquad (16\text{-}9)$$

But $e_f = e_m$ and $A_f/A_m = C_f/C_m$, so

$$\frac{P_f}{P_m} = \frac{E_f}{E_m} \frac{C_f}{C_m} \tag{16-10}$$

From this expression we can calculate the loads carried by the fiber and the matrix and predict the overall strength of the composite.

Let us take our 60% glass–40% polymer mixture and assume a total load of 10,000 lb (4540 kg). From Equation 16-10, the ratio of the loads will be

$$\frac{P_f}{P_m} = \frac{10 \times 10^6}{4 \times 10^5} \frac{0.6}{0.4} = 37.5$$

The load on the polymer will be

$$P_m = P_c - P_f = 10{,}000 \text{ lb} - 37.5\, P_m \text{ or } P_m = 260 \text{ lb } (118 \text{ kg})$$

$$S_m = \frac{P_m}{A_m} = \frac{260 \text{ lb}}{(0.5 \text{ in}^2)(0.4)} = 1300 \text{ psi } (9.0 \text{ MPa}) \qquad P_f = 9740 \text{ lb } (4422 \text{ kg})$$

$$S_f = \frac{P_f}{A_f} = \frac{9740}{(0.5 \text{ in}^2)(0.6)} = 32{,}500 \text{ psi } (224 \text{ MPa})$$

Checking that $e_m = e_f$, we have

$$e_m = \frac{S_m}{E_m} = \frac{1300}{4 \times 10^5} = 3.25 \times 10^{-3} \qquad e_f = \frac{S_f}{E_f} = \frac{32{,}500}{10^7} = 3.25 \times 10^{-3}$$

We have therefore accomplished our objective for composites with continuous fibers. Note that the modulus of the composite is merely the weighted sum of the moduli of the fiber and the matrix and that the load carried in the fiber is affected by the *ratio* of the moduli:

$$E_c = E_f C_f + E_m C_m \tag{16-7}$$

$$\frac{\text{Load in fiber}}{\text{Load in matrix}} = \frac{E_f}{E_m} \frac{C_f}{C_m} \tag{16-10}$$

We can show the relationship in Equation 16-10 graphically. For modulus ratios less than 100, Figure 16.6 gives the load ratios for various fiber volume fractions. The advantage of greater differences in the moduli between the matrix and the fiber (such as in Example 16.1) is that lower volume fractions of fiber are necessary to make the fiber the main load-bearing component.

Loading Perpendicular to Fiber

Assume for Figure 16.4 that the composite thickness is equivalent to the fiber diameter and that loading is perpendicular to the fiber axis. The load is supported by the series resistance of the fiber and the matrix. Therefore the

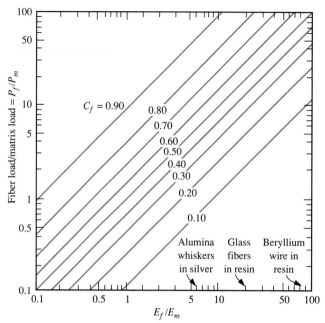

FIGURE 16.6   The interdependency of elastic moduli, load-bearing ratios, and volume fractions in fiber-reinforced composites that are to undergo isostrain loading

[Broutman/Krock, *Modern Composite Materials*, © 1967, Addison-Wesley Publishing Co., Inc., Reading, Massachusetts. Reprinted with permission of the publisher.]

stresses in the matrix, fiber, and composite are equal. We have what is termed an *isostress* condition.

$$S_c = S_m = S_f \quad \text{or} \quad S = E_c e_c \tag{16-11}$$

and

$$e_c = C_m e_m + C_f e_f \tag{16-12}$$

Therefore,

$$e_c = \frac{S}{E_m} C_m + \frac{S}{E_f} C_f \tag{16-13}$$

Substituting Equation 16-13 into Equation 16-11, we obtain

$$\frac{1}{E_c} = \frac{C_m}{E_m} + \frac{C_f}{E_f} \quad \text{or} \quad E_c = \frac{E_m E_f}{C_f E_m + C_m E_f} \tag{16-14}$$

EXAMPLE 16.2   *[ES]*

Calculate the modulus of the composite in Example 16.1, given that loading is perpendicular to the fibers.

Answer

$$E_c = \frac{E_m E_f}{C_f E_m + C_m E_f} = \frac{(4 \times 10^5)(10 \times 10^6)}{(0.6)(4 \times 10^5) + (0.4)(10 \times 10^6)}$$

$$= 9.4 \times 10^5 \text{ psi } (6.5 \times 10^3 \text{ MPa})$$

Comparison of the two loading methods, parallel and perpendicular to the fibers, in this example and the preceding one shows that parallel loading provides a composite modulus that is approximately 6.5 times higher. Figure 16.7 shows this same effect schematically. The advantage of isostrain loading, or loading along the fiber length, is apparent.

The isostrain and isostress conditions provide the limits of the design. Most experimental results fall between these limits.

Modulus with Short Fibers

These calculations are quite complex and are expressed as

$$E_c = \eta_l \eta_o E_f C_f + E_m C_m \tag{16-15}$$

where

$$\eta_l = \text{correction for fiber length}$$

$$\eta_o = \text{correction for fiber orientation}$$

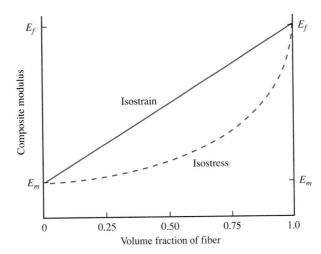

FIGURE 16.7 Schematic representation of isostrain loading vs. isostress loading of a composite material to which fiber has been added. For isostrain loading, less fiber is required to obtain the same modulus of elasticity of the composite.

As a simple example to show the effect of fiber length alone, take a case where the fibers are parallel to the uniaxial stress ($\eta_o = 1$) for a glass–nylon composite, with a fiber diameter of 16 μm and a fiber volume fraction of 0.3.

l, mm	η_l	Ratio, l/d
0.1	0.21	6.25
1.0	0.89	62.5
10.0	0.99	62.5

The data indicate that when the ratio of fiber length to fiber diameter is over 65, the modulus is 90% of that of continuous material.

Effect of Orientation and Fiber Length

Up to this point we have considered continuous fibers oriented either at 90° to the stress direction or parallel to it. We must also consider random orientation and short fibers. Akasaka derived the relation

$$\overline{E} = \frac{3}{8} E_{\|} + \frac{5}{8} E_{\perp} \tag{16-16}$$

where

$$\overline{E} = \text{average modulus}$$

$$E_{\|} = \text{parallel modulus}$$

$$E_{\perp} = \text{perpendicular modulus}$$

for randomly oriented continuous fibers in a *sheet*. For randomly oriented fibers in a bulk piece, the experimental value is less.

16.7 Tensile Strength of Ceramic Fiber Polymer Matrix

Analysis of the tensile strength of a polymer composite is more complex than analysis of the modulus. Returning to the tensile stress–strain curve (Figure 16.8), we encounter only elastic strain up to point 1. Then (from point 1 to point 2) plastic flow occurs in the matrix, although the strain is still elastic in the strong ceramic fiber.

The specimen remains uncracked until point 2, where fibers begin to fail, the interface between the fibers and the matrix fails, or both fail. The stress falls until reaching point 3 but not to zero because the ductile polymer matrix still supports load. The curve usually rises until fracture at point 4 because of the alignment of polymer molecules and the development of a microcrack structure (we will discuss this later).

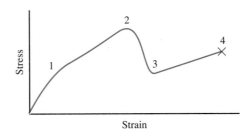

FIGURE 16.8 Schematic engineering stress–strain curve for a composite with fibers aligned in the tensile direction

For the purpose of designing on the basis of tensile strength, the most important stress is at point 2. This stress is a function of the strength of the fibers and the yield strength (flow stress of the polymer) for longitudinally aligned fibers.

$$S_2 = C_f S_f^{max} + C_m S_m^y \tag{16-17}$$

where

$$C_f = \text{volume fraction of fiber}$$

$$S_f^{max} = \text{tensile strength of fiber}$$

$$C_m = \text{volume fraction of matrix}$$

$$S_m^y = \text{yield strength of matrix}$$

There will be a small change in S_m^y from point 1 to point 2, but it is minor compared to the role of the fiber in determining point 2.

EXAMPLE 16.3 *[ES]*

Use the following data to estimate the tensile strength of a composite made of graphite and polyester.

	Vol%	Tensile strength, MPa	Yield strength, MPa
Graphite	50	2500	2500
Polyester	50	35	35

Answer

$$S_2 = C_f S_f^{max} + C_m S_m^y \tag{16-17}$$

$$= 0.5(2500) + 0.5(35)$$

$$= 1250 + 17.5$$

$$= 1267.5 \text{ MPa } (183,700 \text{ psi})$$

This result is within the range of experimental values for this combination. Obviously the fibers are the principal determining factor. If we test *across* the fibers, however, the strength of the polymer will be the limiting factor, just as in the case of the modulus. For this reason, when stresses are multidirectional or unpredictable, randomly oriented fibers or crossed laminated sheets should be used.

Effect of Fiber Length

Although continuous fibers are used in many cases (such as in winding tubing for aircraft struts or fishing rods), it is simpler in other cases to inject a suspension of short fibers in liquid polymer into a mold. Often the flow can be controlled so that the fibers are parallel to the direction of maximum stress. In this case, it can be shown from applied mechanics that if the critical fiber length l_c is greater than

$$l_c = \left(\frac{d}{2}\right)\left(\frac{\tau_{\text{fiber}}}{\tau_{\text{matrix}}}\right) \tag{16-18}$$

where τ is the shear strength (about 0.5 yield strength) and d is the fiber diameter, the optimal strength is obtained. Longer lengths do not add to the load-bearing capability.

If the fibers are randomly oriented, the tensile strength decreases in the same way that we discussed when we considered modulus of elasticity.

16.8 Fracture Toughness of Fiber-Reinforced Polymer Matrix

The fracture toughness of the polymer matrix improves tremendously when reinforcing fibers are added. For example:

Material	Polyester	Polyethylene	Epoxy	Polyester[*] 50% glass	Typical Aluminum alloys
K_{IC}, MPa$\sqrt{m}$:	0.6	1.0/6.0	0.6	42/60	28
K_{IC}, (ksi$\sqrt{\text{in.}}$)	(0.55)	(0.9/5.5)	(0.55)	(38/55)	(25)

We discussed the role of the fiber in the section on tensile testing, but it requires further analysis here. In considering the tensile test results, we observed that after an initial period wherein both matrix and fiber behave elastically, there occurs a period of plastic deformation for the matrix and elastic deformation for the fiber. Finally the fibers break. It is this step that is very important in determining K_{lc}. If the fibers pull out — that is, *debond* — from

---

[*]Fiber at most favorable orientation to the crack

the matrix before breaking, this step absorbs considerable energy that, when added to the energy required for fracture, gives a higher value of K_{IC}.

An illustration of *debonding* for contrasting specimens, one with practically no bond and one with a satisfactory bond, is shown in Figure 16.9. Table 16.1 gives the properties for this combination.

In comparing the fracture toughness of the composites with that of metals, we must remember that the latter were obtained at the most favor-

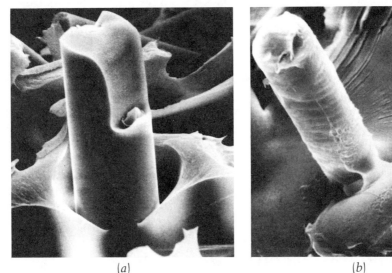

(a)   (b)

FIGURE 16.9 (a) Tensile fracture (as seen with scanning electron microscope) of an epoxy glass–polycarbonate composite molded at 190°C (374°F). Note the void around the fibers; 9100×. (b) The same material as in part (a), but molded at 275°C (527°F) followed by annealing at 245°C (473°F) for 3 hr; 8600×.

[Photographs courtesy J. L. Kardos, F. S. Cheng, and T. L. Tolbert, Washington University/Monsanto Association, St. Louis, Missouri]

TABLE 16.1 Physical Properties of Type E Glass-Polycarbonate Composite

Material	Tensile Strength* psi × 10³	Percent Elongation	Modulus of Elasticity,* psi × 10³
Molded at 190°C			
Resin alone	9.0	5.1	3.25
Resin and glass	7.6	1.3	8.19
Molded at 275°C plus 3 hr at 245°C			
Resin alone	8.9	4.9	3.11
Resin and glass	11.2	2.3	9.28

*Multiply psi by 6.9 × 10⁻³ to obtain MN/m² (MPa) or by 7.03 × 10⁻⁴ to obtain kg/mm².

able fiber orientation to the crack. If this orientation cannot be accomplished in a component, a random orientation of fibers should be used, along with a reduction in design stress, as we observed in the solution to problems involving modulus and strength.

16.9 Ceramic–Ceramic Composites

Although at first glance it might seem that there would be little advantage in making a composite of two brittle materials, interesting progress has been made in this field, and the product is important for high-temperature applications. In our discussion of ceramic materials, we found that higher fracture toughness could be attained through microcrack formation and crack deflection. If a strong second phase is introduced (such as silicon carbide whiskers in alumina), an additional mechanism of "crack stopping" — that is, energy absorption — is accomplished by *crack bridging* (Figure 16.10). In this case the strong fibers bridge the crack and act to prevent it from opening and progressing further. Another mechanism of energy absorption is pullout, or debonding, as shown in Figure 16.10. For maximum energy absorption, the optimal combination of debonding and crack bridging is needed. Ruhle *et al.* have shown that the fracture toughness of unreinforced alumina is raised from 3 MPa$\sqrt{m}$ (2.7 ksi$\sqrt{\text{in.}}$) to 7 MPa$\sqrt{m}$ (6.4 ksi$\sqrt{\text{in.}}$) by a volume fraction of 0.2 silicon carbide whiskers.

16.10 Metal–Ceramic Composites

In this field we find cases in which the metal is clearly the matrix, others in which the ceramic is the matrix, and cases in which both metal and ceramic are continuous. We will consider all cases in this section.

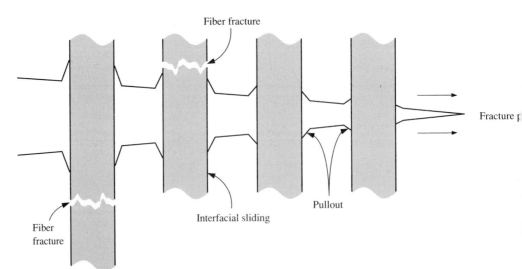

FIGURE 16.10 Bridging and debonding mechanisms for increasing fracture toughness

Let us discuss first one of the oldest metal–ceramic composites, "cemented carbide," and then a new alumina–aluminum material. Tungsten carbide cutting tools are produced by mixing tungsten carbide powder with 2–15% cobalt powder and then pressing to the desired shape in a die. (This is similar to the powder metallurgy method discussed in Chapter 6.) The compact is then sintered above the melting point of the cobalt, and a structure consisting of carbide in a tough, ductile cobalt matrix is obtained (Figure 11.1). A number of variations, including the use of complex tungsten–titanium carbides and different matrices, are employed.

This material is quite different from the fiber-strengthened materials we have discussed, because the goal is to provide wear resistance in machining tools and other wear-resistant components. The objective is to hold the hard carbides in a ductile matrix (cobalt metal). Cobalt is used rather than other metals, because it wets the carbides during sintering and has a high enough melting point to resist softening during use.

Ceramic–Metal Composites: Alumina–Aluminum and Others

After we understand the role of fibers as a second ceramic phase in crack bridging, it is natural to ask, "Would a *ductile* fiber increase the fracture toughness by undergoing plastic deformation during crack bridging?" A recently developed material called Lanxide™ consists of alumina with aluminum reinforcement (Figure 16.11). The photomicrographs show that the crack is bridged by aluminum particles called ligaments, which undergo extensive plastic deformation as well as pullout. The extent of the pullout and the amount of elongation have been measured via stereo techniques. In Figure 16.12 a cross section of these measurements is superimposed on the fracture surface. By a statistical analysis of the debonding length and the plastic stretch of the ligaments, a satisfactory comparison between theoretical and experimental data was obtained. The long-range objective is to determine what characteristics of the metal and of the debonding yield maximum toughness.[*]

Additional development work has been done on aluminum-alumina structures containing silicon carbide fibers. Typical mechanical properties are flexural strength 700 MPa (100 ksi), K_{IC} 27 MPa $\sqrt{m}$ (24.5 ksi $\sqrt{in.}$) with 15% fiber content. The structure of the fracture is shown in Figure 16.13.

In another development of considerable interest, a zirconium carbide matrix reinforced with hexagonal platelets of zirconium diboride and metallic zirconium has been produced (Figure 16.14) with K_{IC} values of 11–23 MPa $\sqrt{m}$ (10–21 ksi $\sqrt{in.}$), the higher values corresponding to greater amounts of metallic zirconium. This provides a ceramic composite in the fracture-toughness range of many metals, with potentially high wear resistance.

---

[*]B. Flinn, M. Rühle, and A. G. Evans, "Toughening in Composites of Al_2O_3 reinforced with Al", *Trans. Am. Cer. Soc.*, 1988.

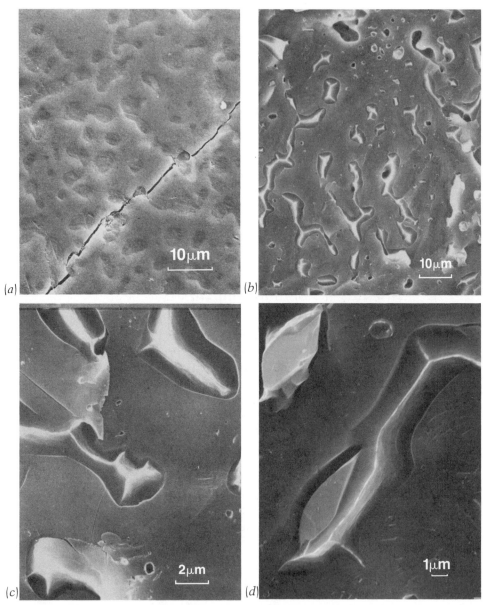

FIGURE 16.11 Crack propagation in an Al_2O_3-Al composite: (*a*) Crack in polished specimen. Note intact Al particles in crack wake. Bar = 10 μm. (*b*) Fracture of composite. Note Al particles necked to peaks and ridges. (*c*) Higher magnification of (*b*). (*d*) Higher magnification of (*b*). Note necking of Al and pullout.

[B. D. Flinn, "Fractographic Study of an Al_2O_3/Al Composite," *Proc. of 46th Annual Meeting of Electron Microscopy Society of America*, 1988, p. 745.]

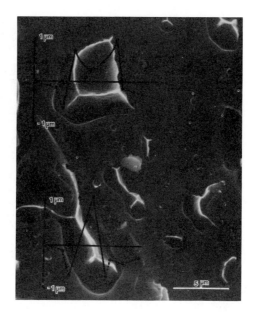

FIGURE 16.12 A stereo projection of a ligament cross section on a fracture surface in an aluminum-reinforced alumina composite

[B. D. Flinn, M. Rühle and A. G. Evans, "Toughening in Composites of Al$_2$O$_3$ reinforced with Al," American Ceramic Society 1988.]

FIGURE 16.13 Tensile surface of a flexurally tested bar of an Al$_2$O$_3$-Al Nicalon® (SiC) composite with uniaxially aligned fibers

[© Lanxide Corporation]

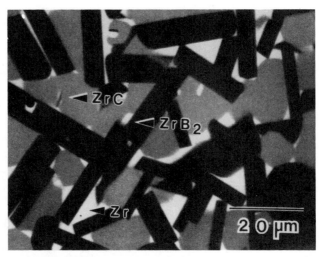

FIGURE 16.14   The microstructure of ZrC that has been reinforced with ZrB_2 platelets and zirconium. An excess of zirconium leads to higher fracture toughness.

[© Lanxide Corporation]

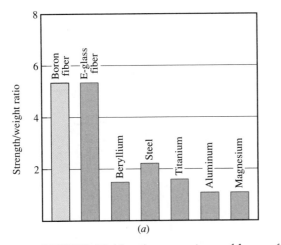
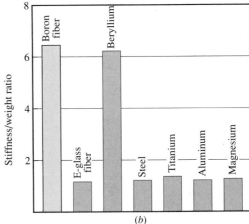

FIGURE 16.15   A comparison of boron fibers to other materials with respect to (*a*) relative strength-to-weight ratios (tensile strength/density) and (*b*) relative stiffness-to-weight ratios (elastic modulus/density)

[*Composites*, Engineered Materials Handbook, Vol. 1, ASM International, 1987, p. 852]

Aluminum with boron fibers has been used in structural parts of the space shuttle. The strength of axially reinforced aluminum boron composite compared to various metals is shown in Figure 16.15. Silicon carbide fibers are replacing boron in many applications because of lower cost and lower reactivity with different matrix materials. Reactions such as boron with liquid aluminum can generate brittle interface compounds.

PROCESSING OF COMPOSITES IN MATERIALS

16.11 Processing of Composites in General

It is not surprising that the processing of composites is closely related to the methods used to process matrix material alone. In the simplest case, for example, chopped fiberglass fibers are added to the resin and injection-molded at relatively low temperatures. At the other end of the scale, ceramic–ceramic composites are made by pressing and sintering at high temperatures. However, some very important specialized techniques greatly affect properties. We will investigate these carefully, giving less detail for modifications of methods that have already been discussed.

Fiber-reinforced plastics (FRP) are by far the largest group of composites. We will examine first the preparation of fibers and then the production of components by filament winding and by lamination or molding.

16.12 Preparation of Fibers

As we discussed different materials, we made brief reference to the production of fibers. We will summarize here the principal fibers, methods of production, and properties (Table 16.2).

Boron fibers are produced by the reduction of boron trichloride by hydrogen on a heated tungsten wire. The wire is drawn through a chamber in which the BCl_3 is decomposed, and the rate of travel determines the thickness of the coating. The fiber is then coated to promote bonding in the particular application.

TABLE 16.2 Properties of Fibers Used in Fiber Reinforced Plastics

Fiber	Structure	Tensile Strength* psi × 10³	Tensile Modulus* 10⁶ psi	Specific Gravity
Boron	Polycrystalline	500	58	2.6
Silicon carbide	Polycrystalline	225	28	3.0
Kevlar 49 (Aramid)	Molecules oriented to fiber axis	400	19	1.45
E glass	Amorphous	250	10	2.55
High-modulus graphite	Crystals oriented to fiber axis	300	55	1.94
High-tensile-strength graphite	Crystals oriented to fiber axis	400	38	1.76

None of the fibers deforms plastically, but high elastic strains are obtained (1.8% for Kevlar, 3% for glass). The range of fiber diameter is 390–980 μin. (10–25 μm)
*Multiply psi by 6.9×10^{-3} to obtain MN/m² (MPa) or by 7.03×10^{-4} to obtain kg/mm².

Silicon carbide fibers are also made by vapor deposition, on a carbon fiber base.

Graphite fibers are made from a number of starting materials. In the case of PAN (polyacrylonitrile), the fibers are stretched to align the axis of the carbon chain parallel to the fiber axis and then destructively distilled to eliminate all but the carbon atoms. Heating is continued, and the surface becomes harder and less porous. A variety of *coupling agents* are used to condition the surface for bonding with the polymer matrix. The high-modulus variety has a harder surface and less porosity than the high-strength type.

Glass fibers are made from three grades, E, S, and C.

The glasses are all silica-based and have the following nominal compositions:

E (electrical) — 15 Al_2O_3, 17 CaO, 5 MgO, 8 B_2O_3
S (high-strength) — 25 Al_2O_3, 10 MgO
C (chemical) — 4 Al_2O_3, 13 CaO, 3 MgO, 8 Na_2O, 5 B_2O_3

Over 90% of the glass fibers made are E grade. The more expensive S grade is used for the highest strength, and the C grade for chemical resistance. Liquid glass is forced through small orifices (spinnerets) to produce fibers that are coated for bonding and stored on reels.

Kevlar (poly-phenylene terepthalamide) is produced by extruding an acid solution of the polymer through spinnerets at 50–100°C (122–212°F) into air. As we noted earlier, the strength depends on aligned, rod-like molecules with a carbon chain stiffened by oxygen and nitrogen.

All fibers are coated with a variety of compounds to facilitate bonding to the polymer matrix and to prevent abrasion during processing. In many cases the fibers are woven into *cloth*, or chopped fibers are bonded with resin into *mats*. Either may be formed into *prepregs*. (This is an abbreviation of "preimpregnation"; see Example 14.6.) The glass mat is stamped as a flat shape and then placed in a heated die and formed to a three-dimensional shape. (The thin plastic coating on the glass fibers serves as a bond.)

16.13 Processing of Components

All the processes used for forming polymer components can be used for fiber-reinforced composites without modification, *if* short-chopped fibers are used as filler. However, the highest performance is obtained by new methods or by modification of the traditional methods. The techniques for optimizing the use of fiber fall into two groups:

1. Filament winding
2. Making laminates and molded parts

The major resins are epoxy, polyester, and other thermosets. We will describe filament winding here and laminates briefly in Section 16.14 and then discuss molded parts at greater length.

Filament Winding

Filament winding is a method used to produce tube shapes ranging from fishing rods to large pipes and tanks, as well as for such exotic applications as rocket housings. As shown in Figure 16.16, the *roving*, which is a strand of continuous fibers, is fed into a liquid resin and then wound onto a mandrel. A simple helical design is shown in this figure, but a number of elaborate winding systems have been developed, including a polar system for producing closed-end pressure tanks. The objective in changing the winding configuration is to have the axis of the fiber located in the direction of highest stress. For other conditions, such as biaxial tension, it is necessary to wind in several directions in different layers. After winding, the component is cured in the temperature range of 250–275°F (121–135°C).

16.14 Laminates

Instead of using chopped or continuous fibers, it is possible to weave the material into a mat or cloth. Even though the fibers are at 90° angles to each other because of the weaving process, using several layers with different orientations results in resistance to stresses in all directions. These multilayer products are called *laminates*. The binders are usually based on thermosetting polymers, as discussed in Section 14.17.

16.15 Molded Parts

We will begin our discussion of molded parts with the simplest processes (such as hand lay-up, which provides only one smooth surface) and then take up the die processes.

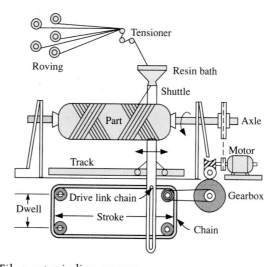

FIGURE 16.16   Filament winding process

[G. Lubin, ed., *Handbook of Composites*, Van Nostrand, 1982, p. 364. Used by permission.]

Hand Lay-Up Method

Although this method is labor-intensive, it is widely used for prototypes and short production runs. Refer to the cross section of a boat hull shown Figure 16.17 as the steps in this method are described.

The first step in any open-mold operation is the application of a release material. Without it, the part would bond permanently to the mold. Silicone spray, several layers of carnuba wax, or a film of cellophane can be used.

The second step is to apply a gel coat to provide a smooth surface finish of the desired color and to act as a barrier to ultraviolet light. In general, this should be sprayed to a thickness of 0.020 in (0.5 mm).

The third step involves cutting chopped-strand mat or continuous-fiber cloth to size and applying a mixture of resin and catalyst. A thermoset that sets at room temperature in several hours is used. The polymer is spread with a hand roller, as shown. Care should be exercised to ensure that no air bubbles are entrained between the mat and the gel coat. Additional layers of mat or woven material are added to achieve the desired thickness.

Spray Gun Method

This is really an automated lay-up procedure in which a mixture of chopped fibers and resin is sprayed onto the mold to the desired thickness. Here the strength of a given thickness is lower, because it is not possible to use the continuous fibers of cloth except as inserts. It is still necessary to roll the internal surface to remove air bubbles.

Bag Molding

The objectives of bag molding are to force out gas bubbles and excess resin and to obtain better contact between layers of the materials. After the fiber-

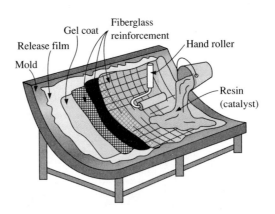

FIGURE 16.17 Hand lay-up process

[G. Lubin, ed., *Handbook of Composites*, Van Nostrand, 1982, p. 346. Used by permission.]

glass–resin mixture is sprayed into an open mold, a layer of silicone rubber (the bag) is placed over the surface of the part and clamped down at the edges. Vacuum line connections are cast into the silicone bag so that a vacuum develops between the bag and the inner surface of the component. Atmospheric pressure or increased pressure in an autoclave forces the bag against the polymer surface, causing compaction.

Closed Die Molding

Unlike the open-mold processes just discussed, the closed-mold methods provide excellent surfaces all over the component. They use greater compaction pressure and more accurate placement of prepreg inserts.

Before we explain the different closed die molding processes, we need to describe two new materials: *SMC*, sheet molding compound, is a thermoset composite of fibers—unsaturated polyester resin generally cross-linked with styrene. It is at an intermediate stage in polymerization so that it can be hot-formed to shape. *BMC*, bulk molding compound, is similar to SMC but is in the bulk plastic form of bars or "logs" that can be compressed or injected.

Thermoset matched die molding is the equivalent of the compression molding discussed earlier; BMC or SMC may be used. A measured amount of compound is compressed between two heated dies and held until polymerization is complete. The applications in the automobile of components made by this process are impressive. SMC bumpers, grilles, rear fender extensions, heater housings, and truck hoods are some of the many uses. In the appliance industry, dishwashers and air conditioners have many molded composite components.

Injection molding is limited to chopped fibers, but even so, the number of injection-molded components is increasing. It is especially important to design the mold in such a way that the fibers that are oriented in the direction of the flow are placed in the direction of maximum stress.

Pultrusion is essentially an extrusion operation in which a great variety of shapes, including hollow tubing and I-beam sections, can be obtained. Continuous fibers can be incorporated (longitudinally) by means of suitable feed controls, or chopped fibers can be used.

16.16 Processing of Other Composites

Metal–ceramic composites are produced by a wide variety of processes. In a solid-state fabrication process for boron fibers in aluminum, the fibers are placed between layers of aluminum foil and encapsulated, as shown in Figure 16.18.

Lanxide℠ aluminum-alumina is made by directed growth of alumina from an aluminum bath. The surface of the metal is oxidized and columnar alumina crystals grow upward inside a barrier, which controls the shape. At the same time metallic aluminum is pulled upward from the bath by capillary action between the alumina crystals forming the fine-grained composite.

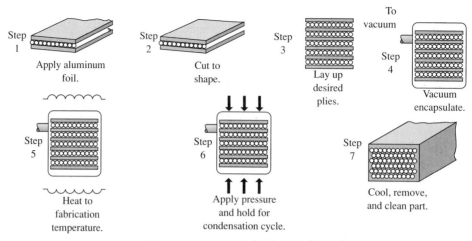

FIGURE 16.18 Typical fabrication process for boron fiber-aluminum matrix composites from a preform

[Reprinted with permission from *Engineered Materials Handbook*, Vol. I, *Composited*, ASM International, Materials Park, OH 1987.]

In a variation of the process, a mat of fibers, such as SiC or Al_2O_3, is compressed (preformed) and is placed above the melt and infiltrated by the Al_2O_3-Al composite (Figure 16.19). Shapes weighing up to 18 kg (39.6 lb) can be produced from commercial grade materials by this method.

To produce the zirconium boride composite discussed earlier, a shaped preform of boron carbide is placed in a retainer and infiltrated with zirconium. The materials react to produce zirconium carbide and zirconium boride, plus a network of metallic zirconium (Primex™).

This technique is of interest because the density of the composite is apparently less than that of the reactants due to the absence of shrinkage cavi-

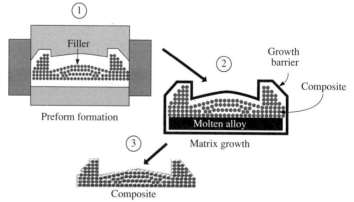

FIGURE 16.19 Simplified process schematic, illustrating (1) filler preform formation, in this case by uniaxial pressing of filler powders, (2) growth of the oxidation reaction product matrix from the molten metal bath into the preform, as limited by the growth barrier on the surfaces, and (3) recovery of the composite part

ties. In the typical case, when molten metal is added to a fiber there is a strong probability of shrinkage cavities caused by liquid-to-solid transformation, which is not present in the Primex™ product. In addition, the generation of new structures by reaction provides excellent bonding at the interface and the high K_{IC} values described earlier.

Other processes include squeeze molding, in which a suspension of the ceramic in a liquid metal is "squeezed" in a mold of the desired contour.

Ceramic–ceramic composites can be prepared by mixing the different powders in a slurry and consolidating the mixture by vacuum filtration and sintering. When fibers are used, the composite is hot-pressed in a die to give a preferred orientation.

EXAMPLE 16.4 *[EJ]*

The emphasis in our discussion of composites fell on the load-carrying capacity and elastic modulus for fibers in polymer resins. What other mechanical properties might be important for the following applications: (a) automotive leaf spring, (b) structural beam, (c) portable cooler shell?

Answer   Adhesion between the fibers and resin is important to all three cases, as is the strength and modulus.

A. Automotive leaf spring—Cyclic loading of a leaf spring is likely. Thus we would like to have data on the fatigue strength. These data are necessary over a temperature range and in the presence of a corrosive atmosphere, such as exhaust fumes and road salt.
B. Structural beam—Because the component life is long, the creep strength becomes important; we do not want the beam to undergo a permanent set. The polymer resin may undergo viscoelastic flow and may also modify its properties by aging through exposure to UV radiation or by crosslinking with oxygen, etc.
C. Portable cooler shell—Impact resistance should be considered, especially at low temperatures.

Summary

Synthetic composites are manufactured from mixtures of polymers, ceramics, and metals. By far the most popular is the dispersion of glass fiber in polymer. We analyzed the mechanical properties of this composite and found that when the fibers are aligned in the direction of uniaxial stress, the modulus of elasticity and strength are dominated by the fiber. When the fibers are aligned transverse to the load direction, however, the properties are dominated by the polymer. Accordingly, the unidirectional fiber arrangement can be used when there is only one large principal stress; otherwise a random fiber alignment is needed. Chopped fibers above a certain length-to-diameter ratio can be used in place of continuous fibers without severe loss in properties. There is a great increase in K_{IC}, fracture toughness, of the composite over the properties of the individual phases.

Ceramic–ceramic and metal–ceramic composites are excellent candidates for high-temperature service if adequate toughness can be developed. As in the case of polymer composites, the fracture toughness is much greater than that of the individual phases. This is due to the mechanism of crack bridging by the fiber phase and to the dissipation of energy by pullout. In the case of metals, added toughness is conferred by deformation of the ligaments.

The processing of composites is important in attaining optimal properties, and, for materials with a polymer matrix, most of the processes discussed earlier can be used. In addition, valuable processes for prototypes, such as hand lay-up, are available. It is important to control the location and distribution of fibers in the final product.

Definitions

Crack bridging Phenomenon whereby, when a fiber composite fractures, the fibers do not fracture but rather bridge the crack to inhibit propagation.

Debonding Pullout or sliding at the interface between a fiber and the matrix.

Dispersion strengthening A method of strengthening a material by the even distribution of particles that are small (less than 0.1 micron) in a matrix at 1–15 vol. %. The matrix is the main load-bearing component.

Fiber reinforcement A method of strengthening a material by the inclusion of fibers in a matrix; in this case the fiber is the main load-bearing component. Volume fraction requirements depend on the modulus difference between the matrix and the fiber.

Filament winding The production of a composite in tube shapes by passing the fiber through a liquid resin and winding on a mandrel.

Isostrain loading Loading of a fiber composite parallel to the fiber alignment.

Isostress loading Loading of a fiber composite perpendicular to the fiber alignment.

Macrocomposites Natural and combinations of natural and synthetic materials that have properties characteristic of composites. Examples include wood and concrete.

Particle reinforcement A method of strengthening a material by the inclusion of particles in a matrix. The load is shared by the matrix and the particles. Particles are greater than 1 micron in size, and volume concentrations are greater than 25%. (Compare *dispersion strengthening*.)

Pultrusion Impregnation of fibers by pulling through a liquid resin and then through a heated die of fixed cross section, where curing occurs.

Pyrolytic graphite Graphite fibers produced by heating a polymer in the absence of oxygen.

Synthetic composite Two or more phases that are synthetically produced and then bonded to produce a component with enhanced properties that could not be obtained in either phase by itself.

Problems

16.1 *[ES]* In dispersion-strengthened composites the following general relationship holds:

$$D_p = (2d^2/3V_p)(1 - V_p)$$

where d = particle diameter in microns, V_p = volume concentration of particles (volume fraction), and D_p = interparticle separation in microns. The normal limits are $d = 0.01$ micron to 0.1 micron and $V_p = 0.01$ to 0.15. Within these constraints, what is the range of interparticle separation necessary to inhibit dislocation movement through the matrix? (Sections 16.1 through 16.6)

16.2 *[ES/EJ]* From the data given in Figure 16.2, determine the rupture stress to give a 100-hr rupture life for the different Al_2O_3 levels. Plot the stress vs. volume percent of Al_2O_3, and from this, project the advantage of using quantities of Al_2O_3 larger than 10 vol. %. (Sections 16.1 through 16.6)

16.3 *[ES]* From the data given in Figure 16.3, determine the modulus of elasticity and plot it vs. the percent rubber. On the same plot, using another scale, plot elongation vs. percent rubber, assuming that the end point of the stress–strain curve gives the elongation. From these data, cite the advantages and disadvantages of adding rubber particles to polystyrene. (Sections 16.1 through 16.6)

16.4 *[ES/EJ]* It has been suggested that a fiber composite be produced of alumina fibers in an epoxy matrix or in a copper matrix and be loaded in an isostrain condition. It is stated that with the epoxy matrix, less fiber is necessary to have the load carried by the fiber than is necessary with the copper matrix. Is this statement correct or incorrect? Justify your answer. Modulus of elasticity values are given in Problem 16.5. (Sections 16.1 through 16.6)

16.5 *[ES]* From Problem 16.4 we have the following material properties. Determine the modulus of the composite when it is loaded in both the isostrain and the isostress condition for each matrix material. Which combination of materials gives the least difference between the two methods of loading? (Sections 16.1 through 16.6)

Modulus of Elasticity

Al_2O_3 $(C_f = 0.4)$	50×10^6 psi $(3.45 \times 10^5$ MPa$)$
Epoxy $(C_m = 0.6)$	8×10^5 psi $(5.52 \times 10^3$ MPa$)$
Copper $(C_m = 0.6)$	20×10^6 psi $(1.38 \times 10^5$ MPa$)$

16.6 *[ES]* A composite is composed of glass fibers with a modulus of 8×10^6 psi $(5.52 \times 10^4$ MPa$)$ and a polymer matrix with a modulus of 5×10^5 psi $(0.345 \times 10^4$ MPa$)$. (Sections 16.1 through 16.6)

a. Determine the composite modulus for isostrain loading if the volume fraction of glass fibers is 0.25.

b. With the modulus you calculated from part *a*, determine the volume fraction of glass fibers for isostress loading.

16.7 *[ES]* In the equation $E = \eta_1 \eta_0 E_f C_f + E_m C_m$ for short fibers, $\eta_0 = 1$ for isostrain loading and $\eta_0 = 0$ for isostress loading. Calculate the composite moduli for isostrain and isostress loading for short fibers ($\eta_1 = 0.90$) and for continuous long fibers. Explain any differences in your calculated results. (Sections 16.1 through 16.6)

Data: Glass fibers $E_f = 10 \times 10^6$
          Nylon matrix $E_m = 4.25 \times 10^5$
          $C_f = 0.30$

16.8 *[ES/EJ]* The text states that randomly oriented continuous fibers have an average modulus that can be calculated for a sheet material. In a bulk material, however, the experimental value is less than that given by the equation in the text. Explain why this might occur. (*Hint:* What is the significance of the constants 3/8 and 5/8 in the equation?) (Sections 16.1 through 16.6)

16.9 *[ES/EJ]* Referring to Figure 16.20, explain why long and short glass fibers give the same flexural modulus (an effective modulus of elasticity determined by bending a beam) for a given fiber fraction in 6/6 nylon, whereas long fibers give a significantly higher tensile strength. (Sections 16.7 through 16.10)

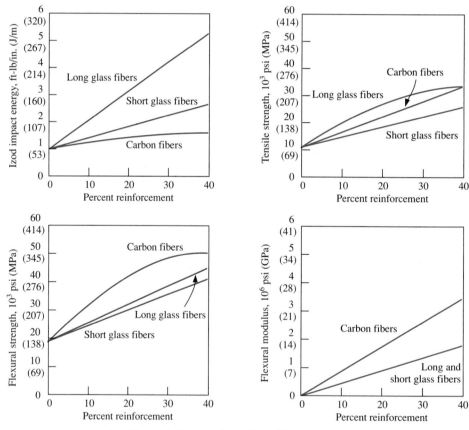

FIGURE 16.20   The effect of glass and graphite fibers on the mechanical properties of nylon 6/6

[*Metal Progress*, American Society for Metals, Metals Park, Ohio, Vol. 116, No. 6, Nov. 1979, p. 39. Used by permission.]

16.10 *[ES/EJ]* We need to know the yield strength of the polymer in order to calculate the strength in a ceramic-fiber–polymer-matrix composite. Why might it be difficult to find the yield strength value in tables? Explain how your answer would differ for a thermoset or thermoplastic matrix. (Sections 16.7 through 16.10 and Chapter 13)

16.11 *[ES]* Repeat Example 16.3 for an epoxy matrix, tensile strength and yield strength equal to 70 MPa, with the same graphite fiber fraction. Construct a plot for the graphite–polyester and graphite–epoxy composite for volume fractions of fiber between 0 and 1.0. From these results, explain whether one matrix has any strength advantage over the other. (Sections 16.7 through 16.10)

16.12 *[ES/EJ]* Indicate whether each of the following statements is true of a ceramic-fiber–ceramic-matrix composite. Give reasons for your answer. (Sections 16.7 through 16.10)
   a. The ceramic fiber should be inherently stronger than the matrix.
   b. The fracture strength at the fiber–matrix interface should be high so the composite will exhibit high fracture toughness.

16.13 *[EJ]* The text refers to examples such as aluminum fibers in an alumina matrix and alumina fibers in an aluminum matrix. Explain why each combination may be appropriate for special applications. (Sections 16.7 through 16.10)

16.14 *[EJ]* Why is so-called Type E glass, rather than just any general type of glass fibers used in fiberglass structures? (Section 16.11 through 16.16)

16.15 *[ES/EJ]* The resin used for the matrix in fiber composites is important to product performance. Assuming a constant percentage of fiber content, compare a thermoplastic to a thermoset matrix for each of the following properties. (Sections 16.11 through 16.16 and Chapter 13)
   a. Fabrication method
   b. Modulus
   c. Creep
   d. Heat resistance (thermal stability)
   e. Hardness
   f. Tensile strength

16.16 *[EJ]* Why have carbon-reinforced polymer composites found wider application in aircraft and sports equipment than in automobiles? (Sections 16.11 through 16.16)

16.17 *[EJ]* In a high-strength composite, why is it better to mold in holes than to machine them? (Sections 16.11 through 16.16)

16.18 *[EJ]* For a certain high-production-rate composite weighing 10 lb (4.5 kg), there is a choice between a thermoset and a thermoplastic matrix. Give some advantages and disadvantages of each. (Sections 16.11 through 16.16)

16.19 *[EJ]* A screw-feed mechanism is generally used for injection-molding a fiber-reinforced polymer shape. Explain how this screw can somewhat reduce the strength of the composite and why the screw pressure is maintained for a period of time after the mold is filled. (Sections 16.11 through 16.16)

16.20 *[ES/EJ]* Table 16.2 shows that the selected fibers are very strong. Why do we not use the same material in the bulk form rather than as fibers in a composite? (Sections 16.11 through 16.16)

CHAPTER 17

Traditional Composites: Concrete, Asphalt, and Wood

In studying the new high-strength composites, we are prone to neglect the traditional composites wood and concrete. As one example, we take for granted subway structures such as the one shown, which are covered after construction, making the intricate detail invisible.

17.1 Overview

It seems like a great jump in thinking to go from the fiber dispersions we have just studied to coarser composites such as concrete and wood. However, there are a number of common features. For example, we are concerned with bonding at the interfaces between phases, porosity, and directional alignment of constituents in all of the composite materials.

Concrete and wood, although traditional materials, compose the major part of all construction materials, and substantial advances are being made in their properties and uses. The need for research is great, as evidenced by the continuing history of cracking and failure of concrete structures and the decreasing availability of structural timber.

We will see that the key to an understanding of concrete lies in the proper proportioning of constituents so that the gravel and sand phases are coated with cement to develop the optimal hydraulic bond. Different cements are available and should be matched to the service requirements. The role of reinforcing rods and prestressing is of great importance.

Asphalt is simply an organically bonded aggregate. We are already familiar with the creep of roads during the summer and the development of potholes at low temperatures.

In the case of wood, it is important to realize that it is essentially a fiber-reinforced polymer and that its properties are intimately related to the microstructure. The distinction in structure between "hard" and "soft" wood, the role of mineral deposition such as silica from the sap in the wood cells, and the differences between sapwood and heartwood are important. Finally, wood is our most inhomogeneous engineering material; the properties of a large beam of lumber may differ by an order of magnitude from those of a carefully selected test piece. Despite these problems and because of its strength-to-weight ratio, wood is a major construction material. In addition, as satisfactory timber for large beams becomes less available, the use of even stronger beams that are built up by gluing together short, sound segments has increased.

17.2 Concrete in General

Concrete is a major construction material. The annual tonnage of concrete produced in the United States is greater than that of all metals combined. It offers the designer exceptional flexibility in design and form, because the material can be poured in place at room temperature, and even under water. This flexibility has its hazards, however. A group of steel girders can be tested before they are installed, but the hydraulic bond in concrete cannot be tested in advance; it depends on the skill of the contractor.

Other sharp differences between metals and concrete are also important, such as the low tensile strength, high compressive strength, and lack of ductility of concrete. In view of these differences, it is essential to study carefully not only the role of the constituents of concrete, but also the ambient conditions during curing and during service.

Concrete is a mixture of a *paste* (Portland cement, water, and sometimes entrained air) with a sand and gravel aggregate. Variations of our present-day concrete were used by the early Egyptians and improved by the Greeks and Romans. However, Portland cement as we know it was not developed until the mid-1800s. In recent years, there has been considerable research on concrete, leading to the development of many modern construction techniques.

Although concrete is widely used, it has a few limitations that must be recognized if we are to guarantee maximum service performance. The principal limitations are low tensile strength, thermal movements, shrinkage, creep under load, and permeability.

However, before we see how we can minimize these disadvantages, we must consider the contributions of the individual components in the concrete.

17.3   Components of Concrete

As we noted above, concrete is a mixture of a paste and an aggregate. We shall now consider the properties of each of the ingredients found in a concrete mixture: (1) cement, (2) water, (3) air, (4) aggregate, and (5) special additions.

Portland Cement

In Chapter 10 we described the minerals present in *Portland cement:*

C_3S	Tricalcium silicate, $3CaO \cdot SiO_2$
C_2S	Dicalcium silicate, $2CaO \cdot SiO_2$
C_3A	Tricalcium aluminate, $3CaO \cdot Al_2O_3$
C_4AF	Tetracalcium aluminoferrite, $4CaO \cdot Al_2O_3 \cdot Fe_2O_3$

The American Society for Testing and Materials (ASTM) recognizes five main grades of Portland cement. These are used for different purposes, as shown in Table 17.1. The percentage compositions in Table 17.1 do not add up to 100% because other impurities are present, notably MgO and compounds of sulfur.

Cement hardens by a hydration reaction with the formation of a gel and crystals. (Recall from Chapter 10 that a gel is a solid network with trapped liquid.) It is vital to understand that water is an important part of the final structure, and cement should *not* be allowed to dry during setting. This will stop the setting reaction! The compounds form at different rates, which is the reason for the several compositions. The characteristics of the four major components and their proportions in different cements are as follows:

C_3S Becomes jellylike in a few hours with considerable generation of heat. It is important to the early strength, which is developed in periods of less than 14 days.

TABLE 17.1   Typical Portland Cement Compositions

ASTM Grade	Use	Typical Percentage Composition			
		C_3S	C_2S	C_3A	C_4AF
I	General purpose or normal	50	24	11	8
II	Some sulfate protection,* less heat generation than Type I	42	33	5	13
III	High early strength	60	13	9	8
IV	Low heat generation	26	50	5	12
V	High sulfate* resisting	40	40	4	9

*Certain soils contain sulfates that react with concrete, causing deterioration.
Adapted from data of Portland Cement Association, Skokie, IL.

C_2S Formed by a slow hydration reaction with low heat generation. It is responsible for the development of long-term strength, or durability.

C_3A Hydrates rapidly with high heat generation. It is responsible for initial stiffening but offers the least contribution to long-term strength.

C_4AF Appears to have little effect on cement performance. It provides little strength and is added to decrease the temperature of the cement during mixing and setting.

The fineness of the cement, as well as the chemistry, is important to the rate of hydration. Finer cement tends to react more quickly (the total surface area increases as particles become finer). Therefore Type III, high early strength, is usually finer grained to further decrease the setting time.

Table 17.2 indicates the relative strengths of concrete that can be developed using the different types of Portland cement.

Under the same conditions of concrete mix, temperature, and moisture, all grades of Portland cement should achieve the same level of compressive strength after three months.

TABLE 17.2   Compressive Strength of Concrete for Different Portland Cements

ASTM Grade	Fraction of Compressive Strength Developed at Indicated Time, Based on Unity for Type I			
	1 day	7 days	28 days	3 months
I	1.0	1.0	1.0	1.0
II	0.75	0.85	0.90	1.0
III	1.90	1.20	1.10	1.0
IV	0.55	0.55	0.75	1.0
V	0.65	0.75	0.85	1.0

Adapted from data of Portland Cement Association, Skokie, IL.

TABLE 17.3   Tolerable Limits of Components in Concrete Mixing Water

Permissible Concentration (ppm)	*Components*
Less than 50	Sanitary sewage
50–500	Bicarbonates of Ca and Mg; salts of Mn, Sn, Zn, Cu, Pb; sugar
500–1000	Carbonates and bicarbonates of K and P
1000–5000	Silt; industrial waste
5000–20,000	NaCl; Na_2SO_4; acid water
20,000–40,000	$MgSO_4$; $MgCl_2$; iron salts; sea water, as salt content

Others: Oil is tolerable in small amounts; algae reduce strength; alkaline water less than 0.5% by weight of cement; $CaCl_2$ up to 2% by weight of cement can be added to accelerate hardening and strength gain.

Adapted from data of Portland Cement Association, Skokie, IL.

Mixing Water

Although any drinkable water can be used for making concrete, water unsuitable for drinking may also be used. Table 17.3 gives levels of water constituents that, when exceeded, may affect both setting time and strength. When there is some doubt about water quality, tests should be conducted on concrete mixes.

Air Entrainment

The primary purpose of air entrainment is to improve the workability of concrete and increase its resistance to freeze–thaw cycles. Although a certain amount of air is entrapped in all concrete, air entrainment is the deliberate development of air bubbles by mixing an agent such as a resin into the concrete. The amount of resin is usually low (less than 0.1% by weight of the cement). The bubbles are small in size, normally 0.001 to 0.003 in. (0.025 to 0.075 mm), and may range from 3% to 9% by volume of the concrete mix. The important point is that the bubbles not be interconnected and be well distributed. See Figure 17.1

Aggregates

The aggregate is very important to the serviceability of the concrete, because it normally makes up 60–75% of the total volume. See Figure 17.2. The aggregate is generally characterized as fine [less than $\frac{1}{4}$ in. (6.35 mm), usually sand] and coarse [greater than $\frac{1}{4}$ in., (6.35 mm), usually gravel]. For optimal strength and workability, the correct aggregate proportions must be used and the characteristics of the aggregate need to be known and controlled. The aggregate variables are size, shape, porosity, specific gravity, moisture absorption,

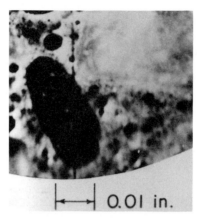

FIGURE 17.1   Enlarged and polished cross section of air-entrained concrete

[*Design and Control of Concrete Mixtures*, 11th ed., 1968, Portland Cement Assoc., Skokie, Ill.]

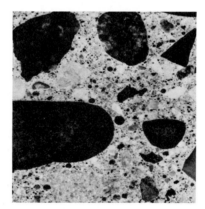

FIGURE 17.2   Polished macrosection of normal concrete, showing aggregate and distribution of paste

[*Design and Control of Concrete Mixtures*, 11th ed., 1968, Portland Cement Assoc., Skokie, Ill.]

resistance to freeze–thaw, strength, resistance to abrasion, and chemical stability. We have already discussed a number of these variables in relation to the packing of solids. The complete discussion can be found in Section 6.13.

As might be expected, the amount of cement and water required to coat the aggregate depends on the size distribution of the aggregate. Therefore a *fineness modulus* * has been developed to characterize the aggregate. The cumulative percentage of aggregate retained by each of a fixed sequence of sieves

---

*The term *modulus* is not related to the stress/strain ratio, but is a synonym for "parameter."

TABLE 17.4   Standard Sieves for Classifying Aggregate

Fine Aggregate (Sand)			*Coarse Aggregate (Gravel)*		
Sieve	*Opening, in.*	*(mm)*	*Sieve*	*Opening, in.*	*(mm)*
No. 4	0.187	(4.75)	6 in.	6.00	(152.4)
No. 8	0.094	(2.39)	3 in.	3.00	(76.2)
No. 16	0.047	(1.19)	$1\frac{1}{2}$ in.	1.50	(38.1)
No. 30	0.024	(0.6)	$\frac{3}{4}$ in.	0.750	(19.0)
No. 50	0.012	(0.3)	$\frac{3}{8}$ in.	0.375	(9.5)
No. 100	0.006	(0.15)	No. 4	0.187	(4.75)

is determined. The cumulative percentage divided by 100 gives the fineness modulus. Table 17.4 gives the sequence of sieves and their openings.

It is customary to apply the fineness modulus only to the fine aggregate and to use the following guidelines for the *maximum size* of the coarse aggregate:

1. One-fifth the dimension of nonreinforced members
2. Three-fourths the clear dimension between reinforcing bars or nets
3. One-third the depth of nonreinforced slabs on the ground

EXAMPLE 17.1 *[ES]*

A 500-g sample of fine aggregate has weights retained on each sieve as given here. Determine the fineness modulus.

Sieve No.	*Grams Retained*
4	11.3
8	74.5
16	100.0
30	80.3
50	120.1
100	98.7
Pan	14.3
	499.2

Answer   The fines passing through the no. 100 sieve may be ignored. An initial 500-g sample usually does not yield 500 g in the final analysis because of weighing errors and inevitable losses.

Sieve No.	Grams Retained	Percent Retained	Cumulative Percent Retained
4	11.3	2.2	2.2
8	74.5	14.9	17.1
16	100.0	20.0	37.1
30	80.3	16.1	53.2
50	120.1	24.1	77.3
100	98.7	19.8	97.1
Pan	14.3	2.9	Total 284.0

$$\text{Fineness modulus} = \frac{284}{100} = 2.84$$

The percent retained shows a double peak, so this sand is not well graded.

Other Additions (Admixtures)

Other agents in addition to Portland cement, aggregate, water, and entrained air may be added to concrete. A description of several *admixture agents* and their purpose follows. The list is not complete; however, it gives examples of common admixture additions.

1. *Accelerators* decrease the set time necessary at low temperatures. Calcium chloride is the most common.
2. *Retarders* increase the set time necessary in very hot weather. They are similar to water-reducing agents.
3. *Water reducers* (plasticizers) provide good workability at lower water-to-cement ratios. Lignosulfonate (a by-product of wood pulp) is an example.
4. *Pozzolans* react with lime, $Ca(OH)_2$ that is released during setting. They have no cementitious value and retard the set time. Pulverized ash from burned coal is a common pozzolan.
5. *Superplasticizers* increase the workability or flowability of the concrete mix. They make possible lower water-to-cement ratios, giving higher strength while maintaining workability. Several organic sulfonated condensates are used.

17.4 Properties of Concrete

We are now ready to consider the mechanical and physical properties of concrete. We want to maximize the strength at minimum cost. Furthermore, the strength of concrete is normally higher at lower water-to-cement ratios, yet we must arrive at a reasonable compromise to attain satisfactory workability.

Like all ceramic materials, concrete has a higher compressive strength than tensile strength. In service, therefore, the primary loading is in compression. We can increase the resistance to tensile loads by reinforcing the concrete with steel or by prestressing, which is discussed in a later section.

We shall first discuss the typical compressive strength of concrete and then examine such other properties as workability, moisture effects, temperature effects, shrinkage, creep properties, and abrasion resistance.

Compressive Strength

The hydration reaction of Portland cement paste is time-dependent. Figure 17.3 shows typical time–compressive-strength curves for air-entrainment and non-air-entrainment concrete mixtures. Low water-to-cement ratios increase the compressive strength significantly. Note that compression test coupons are normally 6 in. in diameter × 12 in. high (152.4 × 304.8 mm). However, cores or trepanned specimens can also be cut out of actual structures.

Because air entrainment appears to lower the compressive strength at a given water-to-cement ratio, its use may seem questionable. However, air entrainment makes concrete more durable, especially under freeze–thaw conditions, because the voids are discontinuous (Figure 17.1). Furthermore, air entrainment makes possible good workability at lower water-to-cement ratios. Hence we may be able to overcome the problems caused by air entrainment and obtain equivalent compressive strengths and equivalent workability (as determined by the slump test, to be discussed next) in two concretes by using two different water-to-cement ratios.

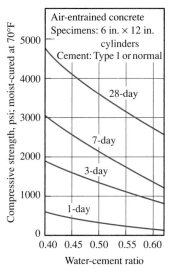 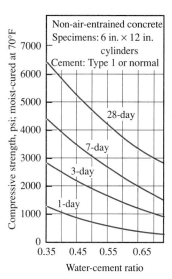

FIGURE 17.3 Typical compressive strengths for air-entrained and non-air-entrained concretes as related to curing time

[*Design and Control of Concrete Mixtures*, 11th ed., 1968, Portland Cement Association, Skokie, Ill. Reprinted by permission.]

Workability

A slump test is conducted using a truncated cone of 8 in. (203 mm) lower diameter, of 4 in. (102 mm) upper diameter, and 12 in. (305 mm) high. The slump test is used as an index of concrete workability in the field. Figure 17.4 shows a typical slump of 3 in. (76 mm). In practice a sheet steel mold is filled in one-third increments, and a rod is pushed up and down 25 times after each increment. The mold is then removed and the slump measured.

Typical slump ranges are from 1 to 2 in. (25.4 to 50.8 mm) for pavement and heavy construction and from 3 to 5 in. (76 to 127 mm) for columns, beams, and walls. These values assume the use of high-frequency vibrators to aid compaction. Without such aids, higher slumps may be necessary to achieve adequate workability. Unfortunately, the slump test does not measure the *work* required for compaction, and different mixes that have the same slump can require different amounts of work.

The workability must not be so great as to cause segregation or bleeding of the concrete. *Bleeding* is the movement of water to the surface. This leads to higher localized water-to-cement ratios and lower surface strength and durability.

Moisture

The reduction or removal of surface moisture slows down or stops the hydration reaction. Figure 17.5 shows that interruption of moist curing after a given time interval by exposure to dry air ultimately stops the curing. Interestingly, if moist-air curing is restored, the strength again increases. For example, after 28 days a dry-air-cured concrete achieves only 55% of the compressive

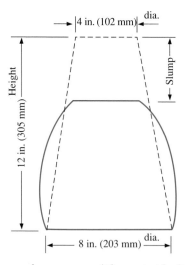

FIGURE 17.4   The slump test for concrete. (Slump is idealized here; it is usually not this symmetric.)

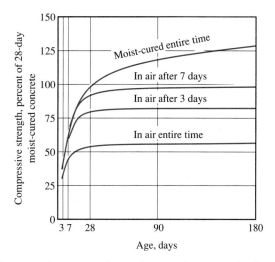

FIGURE 17.5 Absence of water in the air lowers the strength of concrete. Maximum strength depends on the length of time that moisture is present.

[*Design and Control of Concrete Mixtures*, 11th ed., 1968, Portland Cement Association, Skokie, Ill. Reprinted by permission.]

strength of a moist-cured concrete, whereas if the sample is moist-cured for the first 3 days and then dry-air-cured, the compressive strength is 80%. If after 28 days we expose both the dry-air and 3-day samples to moist air, the strength again increases.

Temperature

As with many chemical reactions, the hydration reaction in cement releases heat, and the rate of hydration is higher at higher temperatures. Therefore the correct type of cement paste, the water-to-cement ratio, and the treatment to achieve optimal strength differ depending on the ambient temperature. We shall treat this further when we discuss special concretes for high- and low-temperature setting.

Temperature plays a very important role in the control of shrinkage. When shrinkage is severe, it can result in the concrete cracking before it has reached optimal strength.

Shrinkage

Shrinkage can occur in two stages. In the first stage, it occurs while the concrete is still in a plastic state. This stage is water-, temperature-, and time-dependent. Water loss to the forms and by evaporation, along with the take-up of water of hydration, results in a net decrease in volume. Meanwhile the temperature rise often masks the shrinkage because of the thermal expansion. An extreme case can result in *plastic cracking*. To minimize this, we would like the rate of water loss to be equivalent to the natural bleeding of water to the surface.

Figure 17.6 shows the drying shrinkage that may be anticipated for concrete stored in air. It is evident that larger water-to-cement ratios result in higher shrinkage. This is another reason for minimizing that ratio.

The second stage of shrinkage occurs after initial hardening of the paste. It is due to further hydration and cooling of the mass. This usually causes little trouble, but in some cases a concrete mass may not harden uniformly because of nonuniformity of moisture in the surroundings, such as above and below ground. If this happens, complex stresses may be set up and the concrete mass may crack even after a year or more of curing.

Creep

Creep was discussed in Chapter 3, and stress relaxation and viscous flow in Chapters 13 and 10 respectively. Creep is not always detrimental; it may relieve the stresses imposed by drying shrinkage. In general, the creep rate is lower for the following conditions:

1. At higher concrete strengths (therefore at lower water-to-cement ratios and for longer curing times)
2. At lower volume percentages of cement paste
3. With larger aggregate

Figure 17.7 shows several of these variables for concrete cylinders loaded to 600 psi (4.14 MPa) after a 7-day cure.

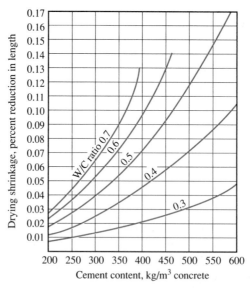

FIGURE 17.6   Average values for shrinkage of concrete stored in air, on the basis of the cement content and the water-to-cement ratio

[L. J. Murdock and K. H. Brook, *Concrete Materials and Practice*, E. J. Arnold, London. Reprinted by permission.]

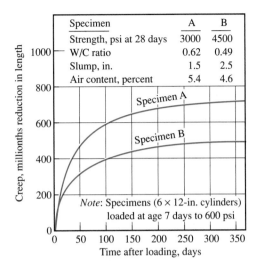

FIGURE 17.7 When stressed at 600 psi (4.14 MPa) after a 7-day age, concrete with higher compressive strength shows a lower creep rate.

[*Design and Control of Concrete Mixtures*, 11th ed., 1968, Portland Cement Association, Skokie, Ill. Reprinted by permission.]

Abrasion Resistance and Durability

Abrasion becomes very important in roads, concrete floors, and dam spillways. As might be expected, a stronger concrete has better wear resistance, as shown in Fig. 17.8. However, other agents also affect the durability of concrete. For example, some soils are high in sulfate. For applications in these soils, a Portland cement lower in tricalcium aluminate (Type V) is recommended. Sea water, salt on roads, and alternating freezing and thawing all shorten the life of concrete. Even atmospheric pollution takes its toll in destruction of the hydrated bond. This has led to international concern over the potential loss of concrete buildings and statues left by early civilizations. The problem is most acute in urban areas.

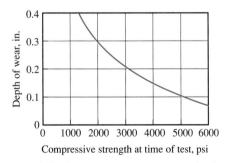

FIGURE 17.8 Concretes with higher compressive strength show a greater resistance to wear.

[*Design and Control of Concrete Mixtures*, 11th ed., 1968, Portland Cement Association, Skokie, Ill. Reprinted by permission.]

EXAMPLE 17.2 *[EJ]*

A contractor is replacing the concrete approach apron to your garage.

A. Why does the contractor not attach the apron mechanically to the existing garage floor?
B. In the summertime, why does the contractor pour the concrete late in the afternoon?
C. Why does the contractor tell you not to drive on it for at least a week?

Answers

A. Because the fresh concrete will shrink to a smaller slab and to prevent cracking, it should not be attached to the older floor. A bituminous expansion joint between the two pieces is often used.
B. In the hot sun, the moisture might be removed too rapidly. The formation of the gel and early crystals can take place overnight without excessive heat and moisture evaporation.
C. It takes a week for the concrete to develop over 50% of its potential compressive strength (see Figure 17.3).

17.5 Special Concretes

We shall now briefly identify some of the concretes used for special applications.

Air-Entrained Concrete

We have already discussed air entrainment. However, it is important to differentiate between entrainment and entrapment. All concretes contain entrapped air that cannot be removed because of the viscosity of the paste. In general, the use of finer aggregate results in more entrapped air. Entrained air is in the form of very fine bubbles retained by the addition of an organic chemical (a neutralized vinsol resin*). The air content recommended by the Portland Cement Association for severe exposure conditions is as follows:

Maximum Size Coarse Aggregate, in. (mm)	Air Content, percent by volume
$1\frac{1}{2}$ (38.1); 2 (50.8); $2\frac{1}{2}$ (63.5)	5 ± 1
$\frac{3}{4}$ (19.1); 1 (25.4)	6 ± 1
$\frac{3}{8}$ (9.5); $\frac{1}{2}$ (12.7)	$7\frac{1}{2} \pm 1$

---

*Recall that a sol is a fine suspension of solids in a liquid, as defined in Chapter 10. Sols exist in both the ceramic and organic material families. Vinsol refers to a family of complex organic materials.

Lightweight Concrete

The *lightweight concretes* may be classified in two distinct categories.

1. *Lightweight structural concrete.* 28-day compressive strength in excess of 2500 psi (17.25 MPa) and an air-dry density of less than 115 lb/ft³ (1843 kg/m³). The light weight is obtained by using heat-expanded lightweight aggregates such as shale, clay, slate, blast furnace slag, or fly ash.
2. *Lightweight insulating concrete.* 28-day compressive strength of 100 to 1000 psi (0.69 to 6.9 MPa) and an air-dry density of 15 to 90 lb/ft³ (240 to 1443 kg/m³). The low densities are achieved by the use of various porous aggregates and/or by incorporation of a cement paste into a cellular matrix with air voids. The advantages in terms of our present and future concerns with energy conservation are shown by the thermal conductivity (Figure 17.9).

Heavyweight Concrete

We can obtain protection from x rays, γ rays, and neutrons by using high-density aggregates to give a density of up to 400 lb/ft³ (6412 kg/m³). Metal punchings or shot may be used for a portion of the aggregate, although oxides of iron, titanium, and barium are more common.

Hot-Weather Concrete

Some parts of the country have very hot, dry weather all year round, and special procedures are necessary when concrete is being poured and set. More

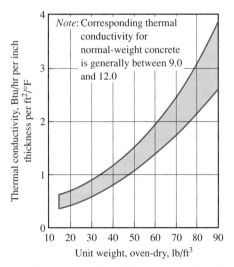

FIGURE 17.9 The thermal conductivity of lightweight (insulating) concrete is lower for low unit weights (lb/ft³).

[*Design and Control of Concrete Mixtures*, 11th ed., 1968, Portland Cement Association, Skokie, Ill. Reprinted by permission.]

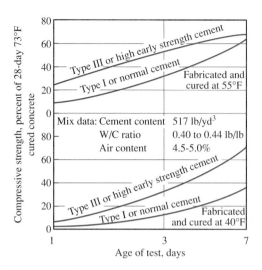

FIGURE 17.10   When cured at low temperatures, Type III cement gives higher early compressive strengths.

[*Design and Control of Concrete Mixtures*, 11th ed., 1968, Portland Cement Association, Skokie, Ill. Reprinted by permission.]

water is required to compensate for evaporation, and the concrete has a greater tendency to crack. Therefore a low-heat-generation cement such as Type II or IV is preferred (see Table 17.1).

Cold-Weather Concrete

At low temperatures the cure rate is reduced and high-early-strength cements can be used, as shown in Figure 17.10. A higher-temperature concrete mix can be poured and will provide initial setting, even when the ambient temperature is below freezing. The important point is to develop strength quickly, using the internal heat of the initial mix plus the heat of hydration. The concrete can often be protected by heated plastic bubble covers or by the use of steam-accelerated curing. The addition of calcium chloride (up to 2% maximum) accelerates curing.

17.6   Reinforced and Prestressed Concrete

The tensile strength of concrete is approximately one-tenth of its compressive strength; therefore, design is primarily in compression. (Older civilizations' recognition of this led to the development of the arch.) However, there are many applications in which tensile stresses are developed. Examples include bending of beams and loading of tall columns, where potential buckling gives rise to tensile stresses. Therefore it is common practice to apply steel reinforcement at the portion of the beam in tension, as shown in Figure 17.11.

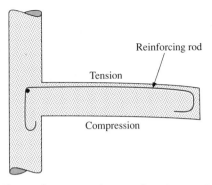

FIGURE 17.11 A cantilevered concrete beam, showing a reinforcing rod on the side to be in tension. (Deflection has been accentuated.)

Steel is used rather than other metals for the obvious reasons of its higher yield strength and elastic modulus. Recall that a large ratio E_f/E_m for fiber-reinforced polymers (Chapter 16) allowed more load to be carried by the fiber for a given volume fraction. A similar argument holds for steel bearing the load in reinforced concrete. Not so obvious is the fact that steel has a coefficient of expansion approximately the same as that of concrete. Also, concrete is alkaline, so it causes minimal corrosion to steel.

Reinforced concrete may crack as a result of tensile stresses built up during curing and shrinkage. In most cases, if there is good bonding between the cement and the reinforcement, the width of the cracks can be minimized. One major cause of cracking is lack of concrete cover over the reinforcement. Under mild weather and protected conditions, a covering of $\frac{1}{2}$ to $\frac{3}{4}$ in. (12.7 to 19.1 mm) may be adequate. However, under severe conditions and exposure to salt used for de-icing, 2 to $2\frac{1}{2}$ in. (50.8 to 63.5 mm) is required.

Road building is a familiar application of reinforced concrete. In road building, when a steel wire mesh is used for reinforcement, a single layer of reinforcement is placed at the centerline. Because of frost upheaval, movement of the roadbed, and weight of traffic, tension may develop at both the upper and lower surfaces, and a single layer of reinforcement placed at the centerline is justified. If two layers are used (above and below the centerline), the concrete must be thicker to satisfy the cover requirement.

Although air entrainment produces discontinuous pores and hence less penetration by road salt, there is considerable difficulty with road breakup even when air-entrained concrete is used. Using calcium chloride as an accelerator for cold-weather curing increases corrosion of the steel.

The use of polymers to patch and lengthen the life of concrete roads, bridges, and parking structures has gained some popularity. For example, parking garage decks have been patched successfully by cleaning the concrete down to the reinforcing rods and applying two-part epoxy primers and sand-filled epoxy resins.

The corrosion of reinforcing rods in bridge decks has led to considerable research into ways of coating and protecting the surfaces. One method requires drying of the concrete at 500–600°F (260–315°C) with heaters, pressure impregnation with a monomer such as methylmethacrylate, and finally polymerization by hot water or steam. Depths of impregnation of up to 4 in. (101.6 mm) have been obtained. However, the cost is high: $11.00/ft² (1989 prices).

The poor tensile strength and higher compressive strength of concrete suggest that we can make use of residual compression that must be overcome before the component can fail in tension. This result, found in *prestressed concrete*, is accomplished in two ways.

Pre-tensioning is placing steel rods or wire in tension, pouring the concrete around the rods or wire, and removing the tension after curing the concrete. When the tension on the wires is removed, they pull the concrete into compression. The technique requires (1) a good bond between steel and concrete and (2) end-tapered or anchored configurations for the steel to prevent stress relief.

Post-tensioning requires no bond between the concrete and the steel. In its simplest application, steel wire is placed within a tube and concrete is cast around it. After the concrete is cured, the steel is placed in tension and anchored at its ends to the concrete. When the rod end tension is released, the anchored length of the steel places the concrete in compression.

The amount of residual compression that we can make use of is limited by the compressive strength of the concrete and the attainable elastic strain in the steel. Ideally, the amount of tensile elastic strain in the steel is equivalent to the compressive elastic strain in the concrete. However, geometric placement, surface adhesion, and ambient conditions inhibit the achievement of this ideal condition. Furthermore, achieving the ideal may cause cracking of the concrete because the stress is localized near the steel insert.

EXAMPLE 17.3 *[EJ]*

The residual stress in a steel rod in prestressed concrete is found to decrease with time. What characteristics of the concrete suggest that this is normal?

Answer The principal reason for the decrease in residual stress is continued curing of the concrete and accompanying shrinkage, which relieves the strain. An associated phenomenon is creep of the concrete. Both may occur over a considerable length of time, because total curing may take years, depending on the thickness of the concrete. The similar coefficients of expansion for steel and concrete suggest no thermal effects. However, stresses from thermal gradients and mechanical fatigue may result in poorer bonding between the two components and hence some loss of residual stress.

17.7 Proportioning of Concrete Mixtures

For small amounts of concrete, the simplest—but not the most precise—method of determining the appropriate ratio of cement to aggregate is to use empirical volume ratios. The following cement-to-fine-aggregate-to-coarse-aggregate ratios can give satisfactory results.

Structure	Ratio
Reinforced concrete	1/2/4
Large concrete mass	1/3/6
Pavement and sidewalks	1/2/3

A more accurate approach is to design for a specific strength requirement in the following way.

1. Establish the conditions of exposure and geometry. Table 17.5 provides an initial guideline for maximum water-to-cement ratios.
2. Determine the minimum strength requirement after a specific cure time. Figure 17.12 provides average compressive strengths for two types of Portland cement after a 7-day and a 28-day cure.
3. Determine the optimal workability (slump requirements) and water-to-cement ratios, and establish a trial batch given the fineness modulus and the maximum coarse aggregate size. Table 17.6 gives typical trial batches.
4. On the basis of measured slump, workability, and air content, modify the mix to obtain the required characteristics.

The technique can best be demonstrated by working out an example. The calculations will be carried out in units common to the concrete industry (lb/

TABLE 17.5 Maximum Permissible Water-to-Cement Ratios

	Severe Application (Air-Entrained Only)			Mild Application (Seldom Below Freezing)		
Structure	In Air	In Fresh Water	In Sea Water	In Air	In Fresh Water	In Sea Water
Thin sections	0.49	0.44	0.40	0.53	0.49	0.40
Bridge decks	0.44	0.44	0.40	0.49	0.49	0.44
Moderate sections	0.53	0.49	0.44	*	0.53	0.44
Heavy sections	0.58	0.49	0.44	*	0.53	0.44
Concrete slabs in ground contact	0.53	—	—	0.53	—	—
Pavements	0.49	—	—	0.53	—	—

*Ratio selection is based on workability but should not be less than 470 lb cement/cubic yard.
Adapted from *Design and Control of Concrete Mixtures*, 11th ed., Portland Cement Association, Skokie, IL
Reprinted by permission.

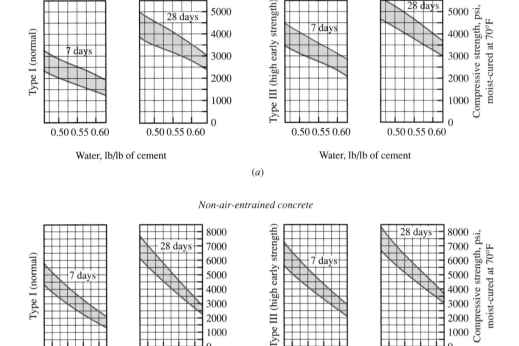

Air-entrained concrete: Air within recomended limits and 2 in. max. aggregate size

(a)

Non-air-entrained concrete

(b)

FIGURE 17.12 Development of average compressive strength in concrete for varying water-to-cement ratios, cure times, and types of cement

[*Design and Control of Concrete Mixtures*, 11th ed., 1968, Portland Cement Association, Skokie, Ill.]

yd³, sacks of cement, etc.). We shall give the conversion to metric units after the calculation.

The following conversion factors and properties are necessary to the calculations:

1. Water: density, 62.4 lb/ft³ = 1685 lb/yd³; weight, 8.33 lb/gal
2. Portland cement: apparent density, 3.15 × 62.4 lb/ft³ = 196.6 lb/ft³; 1 sack = 94 lb.
3. Aggregate: specific gravity range, 2.4 to 2.9; average true specific gravity of sand, 2.65; average true specific gravity of gravel, 2.60
4. Air: volume in cubic feet per cubic yard of concrete is approximately 0.27 times the air content in percent

TABLE 17.6 Suggested Trial Mixes for Concrete of Medium Consistency (3 to 4 in. slump*)

Water-to-Cement Ratio, lb per lb	Maximum Size of Aggregate, in.	Air Content, Percent	Water, lb per yd³ of Concrete	Cement, lb per yd³ of Concrete	With Fine Sand Fineness Modulus = 2.50			With Coarse Sand Fineness Modulus = 2.90		
					Fine Aggregate, Percent of Total Aggregate	Fine Aggregate, lb per yd³ of Concrete	Coarse Aggregate, lb per yd³ of Concrete	Fine Aggregate, Percent of Total Aggregate	Fine Aggregate, lb per yd³ of Concrete	Coarse Aggregate, lb per yd³ of Concrete
Air-Entrained Concrete										
0.40	$\frac{3}{8}$	7.5	340	850	50	1250	1260	54	1360	1150
	$\frac{1}{2}$	7.5	325	815	41	1060	1520	46	1180	1400
	$\frac{3}{4}$	6	300	750	35	970	1800	39	1090	1680
	1	6	285	715	32	900	1940	36	1010	1830
	$1\frac{1}{2}$	5	265	665	29	870	2110	33	990	1990
0.50	$\frac{3}{8}$	7.5	340	680	53	1400	1260	57	1510	1150
	$\frac{1}{2}$	7.5	325	650	44	1200	1520	49	1320	1400
	$\frac{3}{4}$	6	300	600	38	1100	1800	42	1220	1680
	1	6	285	570	34	1020	1940	38	1130	1830
	$1\frac{1}{2}$	5	265	530	32	980	2110	36	1100	1990
0.60	$\frac{3}{8}$	7.5	340	565	54	1490	1260	58	1600	1150
	$\frac{1}{2}$	7.5	325	540	46	1290	1520	50	1410	1400
	$\frac{3}{4}$	6	300	500	40	1180	1800	44	1300	1680
	1	6	285	475	36	1100	1940	40	1210	1830
	$1\frac{1}{2}$	5	265	440	33	1060	2110	37	1180	1990
0.70	$\frac{3}{8}$	7.5	340	485	55	1560	1260	59	1670	1150
	$\frac{1}{2}$	7.5	325	465	47	1360	1520	51	1480	1400
	$\frac{3}{4}$	6	300	430	41	1240	1800	45	1360	1680
	1	6	285	405	37	1160	1940	41	1270	1830
	$1\frac{1}{2}$	5	265	380	34	1110	2110	38	1230	1990

*Increase or decrease water per cubic yard by 3% for each increase or decrease of 1 in. in slump, then calculate quantities by absolute volume method. For manufactured fine aggregate, increase percentage of fine aggregate by 3 and water by 15 lb per cubic yard of concrete. For less workable concrete, as in pavements, decrease percentage of fine aggregate by 3 and water by 8 lb per cubic yard of concrete.
From *Design and Control of Concrete Mixtures*, 11th ed., Portland Cement Association, Skokie, IL. Reprinted by permission.

699

TABLE 17.6 (Continued)

Water-to-Cement Ratio, lb per lb	Maximum Size of Aggregate, in.	Air Content, Percent	Water, lb per yd³ of Concrete	Cement, lb per yd³ of Concrete	With Fine Sand Fineness Modulus = 2.50			With Coarse Sand Fineness Modulus = 2.90		
					Fine Aggregate, Percent of Total Aggregate	Fine Aggregate, lb per yd³ of Concrete	Coarse Aggregate, lb per yd³ of Concrete	Fine Aggregate, Percent of Total Aggregate	Fine Aggregate, lb per yd³ of Concrete	Coarse Aggregate, lb per yd³ of Concrete
				Non-Air-Entrained Concrete						
0.40	$\frac{3}{8}$	3	385	965	50	1240	1260	54	1350	1150
	$\frac{1}{2}$	2.5	365	915	42	1100	1520	47	1220	1400
	$\frac{3}{4}$	2	340	850	35	960	1800	39	1080	1680
	1	1.5	325	815	32	910	1940	36	1020	1830
	$1\frac{1}{2}$	1	300	750	29	880	2110	33	1000	1990
0.50	$\frac{3}{8}$	3	385	770	53	1400	1260	57	1510	1150
	$\frac{1}{2}$	2.5	365	730	45	1250	1520	49	1370	1400
	$\frac{3}{4}$	2	340	680	38	1100	1800	42	1220	1680
	1	1.5	325	650	35	1050	1940	39	1160	1830
	$1\frac{1}{2}$	1	300	600	32	1010	2110	36	1130	1990
0.60	$\frac{3}{8}$	3	385	640	55	1510	1260	58	1620	1150
	$\frac{1}{2}$	2.5	365	610	47	1350	1520	51	1470	1400
	$\frac{3}{4}$	2	340	565	40	1200	1800	44	1320	1680
	1	1.5	325	540	37	1140	1940	41	1250	1830
	$1\frac{1}{2}$	1	300	500	34	1090	2110	38	1210	1990
0.70	$\frac{3}{8}$	3	385	550	56	1590	1260	60	1700	1150
	$\frac{1}{2}$	2.5	365	520	48	1430	1520	53	1550	1400
	$\frac{3}{4}$	2	340	485	41	1270	1800	45	1390	1680
	1	1.5	325	465	38	1210	1940	42	1320	1830
	$1\frac{1}{2}$	1	300	430	35	1150	2110	39	1270	1990

EXAMPLE 17.4 *[ES/EJ]*

We wish to make a concrete slab that will be in contact with the ground and will be subjected to mild service where air entrainment is not necessary. A slump of 3–4 in. is recommended, with a coarse aggregate size of $\frac{3}{4}$ in. and a fineness modulus for the sand of 2.50. The coarse aggregate contains 2% moisture, and the fine aggregate 5% moisture. The proposed 28-day cure strength using non-air-entrained Type I Portland cement is 4000 psi.

Answer

Step 1. From Table 17.5, the recommended *maximum* water-to-cement ratio is 0.53.

Step 2. A 4000-psi concrete must be given a factor of safety, normally 15%.

Therefore $4000 \times 0.15 + 4000 = 4600$ psi concrete required. In Figure 17.12, a water-to-cement ratio of 0.56 places us in the middle of the band. However, because 0.53 is the maximum recommended ratio, we must use this number in our calculations and accept the consequence that the strength will be greater than requested.

Step 3. Table 17.6 gives us trial batches. Interpolation is necessary for differences in slump, fineness modulus, and water-to-cement ratio. The trial batch is as follows:

Water		340 lb/yd³
Cement	340/0.53 =	640 lb/yd³
Fine aggregate		1130 lb/yd³
Coarse aggregate		1800 lb/yd³
	Total =	3910 lb/yd³ (2323 kg/m³)

All of the water must not be added as liquid, because the aggregate contains water in its open pores.

Water in sand	$1130 \times 0.05 =$	55 lb
Water in gravel	$1800 \times 0.02 =$	35 lb
	Total =	90 lb (40.9 kg)

Water to be added as liquid = $340 - 90 = 250$ lb/yd³ (149 kg/m³)

Had we neglected this, the true water-to-cement ratio would have been $(340 + 90)/640 = 0.67$, which would have resulted in lower strength.

We may also calculate the volume contribution of each component as follows (assuming average densities for the aggregates):

$$\text{Water} \qquad \frac{340 \text{ lb}}{1685 \text{ lb/yd}^3} = 0.20 \text{ yd}^3$$

$$\text{Cement} \quad \frac{640 \text{ lb}}{196.6 \text{ lb/ft}^3 \times 27 \text{ ft}^3/\text{yd}^3} = 0.12 \text{ yd}^3$$

$$\begin{array}{l} \text{Fine} \\ \text{aggregate} \end{array} \quad \frac{(1130 - 55) \text{ lb}}{2.65 \times 1685 \text{ lb/yd}^3} = 0.24 \text{ yd}^3$$

$$\begin{array}{l} \text{Coarse} \\ \text{aggregate} \end{array} \quad \frac{(1800 - 35) \text{ lb}}{2.60 \times 1685 \text{ lb/yd}^3} = 0.40 \text{ yd}^3$$

$$\text{Total} = 0.96 \text{ yd}^3 \ (1.26 \text{ m}^3)$$

We want the total to be 1.00 yd^3, so the remaining 0.04 yd^3 is made up of 0.02 yd^3 anticipated entrapped air (Table 17.6) and 0.02 yd^3 of closed porosity in the aggregate.

Step 4. Let us assume that our trial mixture gave a slump that was too great. A decrease in slump of 1 in. requires 3% less water (see the footnote to Table 17.6). It is important to maintain the same water-to-cement ratio to retain the strength requirement; therefore the dry solids are increased correspondingly.

$$\text{Initial dry solids} = 640 + (1130 - 55) + (1800 - 35) = 3480 \text{ lb/yd}^3$$

$$\text{New water content} = 340 - 0.03 \times 340 = 330 \text{ lb/yd}^3$$

As a first approximation, the decrease in water should be inversely proportional to the increase in dry solids.

$$\text{Dry solids} = 3480 \times 340/330 = 3585 \text{ lb/yd}^3$$

$$\text{Portland cement} = 330/0.53 = 625 \text{ lb/yd}^3$$

$$\text{Initial dry aggregate} = 3480 - 640 = 2840 \text{ lb/yd}^3$$

$$\text{Final dry aggregate} = 3585 - 625 = 2960 \text{ lb/yd}^3$$

That is, there is a 120 lb/yd^3 increase.

Final dry aggregate:

$$\text{Fine} = (1130 - 55) + \frac{(1130 - 55)}{2840} \times 120 = 1120 \text{ lb/yd}^3$$

$$\text{Coarse} = (1800 - 35) + \frac{(1800 - 35)}{2840} \times 120 = 1840 \text{ lb/yd}^3$$

Final wet aggregate:

$$\text{Fine} = 1120/0.95 \approx 1180 \text{ lb/yd}^3 \ (702 \text{ kg/m}^3)$$

$$\text{Coarse} = 1840/0.98 \approx 1880 \text{ lb/yd}^3 \ (1119 \text{ kg/m}^3)$$

Finally,

$$\text{Liquid water addition} = 330 - (1180)(0.05) - (1880)(0.02)$$

$$\approx 235 \text{ lb/yd}^3 \ (140 \text{ kg/m}^3)$$

17.8 Asphalt

Some of our previous discussion of concrete concerned its use for road building, so it is appropriate that we also consider asphalt. *Asphalt* is a bitumen. Although it occurs naturally, it is most often obtained as a by-product of petroleum refining. In road building, approximately 6% asphalt is used to bond together an aggregate. Furthermore, asphalt is a thermoplastic material. Therefore it is applied hot and made to flow into place with the aid of heavy rollers.

Although it does not require a cure time as concrete does, the thermoplastic characteristics of asphalt cause difficulty with roads. For example, in high summer temperatures asphalt roads undergo viscous flow, or creep. In winter they become brittle (below the glass transition temperature), and stress gradients can lead to fracture, causing the familiar pothole.

The aggregrate sizing in asphalt roads is as important as it is with concrete. Various attempts have been made to use fillers that enhance the properties of asphalt. The addition of ground glass, particularly at intersections, increases traction and gives better reflectivity at night. And the addition of rubber particles gives better resiliency, especially at low temperatures (this effect is similar to that shown in Figure 16.3 for additions of rubber to polystyrene).

This concludes our discussion of composites that are combinations of synthetic and natural materials. In the last section we consider wood, a completely natural composite material.

17.9 Wood in General

Wood is a beautiful, varied, complex, and widely distributed construction material. The annual tonnage of wood used in the United States is greater than the combined tonnage of steel and concrete—more than 300,000,000 tons. More than 60 native woods are in common use, and 30 kinds are imported. Many composites, such as plywood, particle board, and paper, are also of great importance.

We have postponed our discussion of wood to this point because it is the most complex of the composite polymeric materials and because its properties are difficult to describe. We cannot use the simple approach we used for the metals and ceramics, starting with unit cells as modules, stacking these neatly together to form grains, and then showing how the properties depend on slip and rupture of these assemblies. We shall see instead that wood is a composite honeycomb structure made up of different biological cells, and that the cell walls are made up of complex arrays of cellulose fibers. These are reinforced with a matrix of polymers such as lignin and other organic compounds plus varying amounts of inorganic crystals, which are hard enough in certain cases to thwart even the teeth of the toredo worm.

A thorough understanding of its complex structure and properties will enable the engineer to select wood for design in a variety of circumstances, from siding for a house to massive beams for an industrial building. We shall present our discussion in three parts.

1. *Macrostructure.* In contrast to other materials, the macrostructure is the most important feature of wood. Many properties of a wood vary by a factor of 20, depending on the location in the log and the direction of testing.
2. *Microstructure.* An examination of the microstructure gives details essential for understanding certain features of the macrostructure and the general properties.
3. *Properties.* The background of the first two sections, makes it possible to understand the great directionality in the properties of wood and to allow for it in design.

The section concludes with a discussion of some typical uses of wood and wood products.

17.10 Wood Macrostructure

The major point we must keep in mind in analyzing the structure of wood, in contrast to that of a billet of steel or an extruded blank of plastic, is that the complex honeycomb structure of a tree was produced not in order to make homogeneous, isotropic lumber, but in response to a basic force to grow and survive. Therefore, if we are to use wood in construction, we must not become impatient with its anisotropy, but must instead strive to understand how this variation in properties with direction developed in response to growth conditions.

Let us begin by examining the cross section of a typical tree, as shown in Figure 17.13. The most important feature lies at the intersection between the bark and the wood. This is a very thin layer called the *cambium*, which is the source of both wood and bark cells. Each year the cambium grows new wood cells on its inside surface and new bark cells on its outside surface. Immediately after its formation by cell division, a wood cell begins to enlarge in both diameter and length. During this period the cell has a very thin, pliable primary wall. Once the cell attains full size, the wall is thickened by addition of a secondary wall.

The new wood cells add to the circular* band of *sapwood*, which is generally light in color, in contrast to the *heartwood*, which is darker. The sapwood is made up of living cells that carry fluids and of some older, dead cells.

The *heartwood* is composed entirely of dead cells. Its darker color is due to greater deposits of tar-like materials and minerals. It is denser, stronger, and more decay-resistant than the sapwood.

---

*The growth occurs not only in a radial direction but also vertically. As a result, the tree trunk has a slender conical shape; it is not a perfect cylinder.

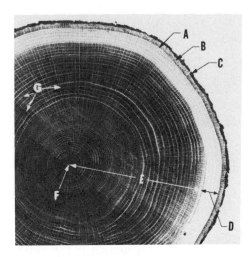

FIGURE 17.13 Cross section of a typical tree. A = cambium; B = inner bark; C = outer bark; D = sapwood; E = heartwood; F = pith; G = wood rays.

[U.S. Dept. of Agriculture Handbook No. 72 (revised August 1974), p. 2-2; courtesy Forest Products Laboratory, Forest Service, USDA]

Yearly growth rings are usually present. They result from differences in thickness of the cell walls formed in different seasons. These cells are seen clearly in the microstructure. In some tropical regions where there is little change in weather, the growth rings are poorly defined.

Wood rays are horizontal radial canals that connect the various layers from the center to the bark; their function is the storage and transfer of food.

Pith is the original soft tissue around which the first wood growth takes place. Because the tree grows vertically as well as horizontally, the pith core is found throughout the length.

Softwoods and Hardwoods

Woods are divided into classes in two ways. First, the *softwoods*, especially in the United States, are conifers — "evergreens" such as pine and spruce with needle-like leaves and exposed seeds. By contrast, the *hardwoods* lose their leaves, have true flowers, and have seeds that are covered in a fruit such as a nut. The designation of woods as either softwoods or hardwoods is as important to wood specifications as the division into ferrous and nonferrous alloys is to the metals industry. Hardwoods are *usually* harder and stronger than softwoods, but there are some exceptions. For example, softwoods such as Douglas fir and long-leaf pine are harder than the hardwoods bass and aspen. In the next section, which deals with microstructure, we shall see that there is a structural difference between the two types of wood.

Before we leave the topic of macrostructure, we should orient ourselves regarding directions in the tree, so that the correlation with microstructure will be clear. The three axes used are shown in Figure 17.14. *L* designates the

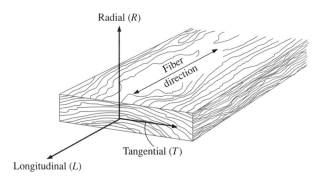

FIGURE 17.14   Axes used to specify directions in wood

[U.S. Dept. of Agriculture Handbook No. 72 (revised August 1974) p. 4-2; courtesy Forest Products Laboratory, Forest Service, USDA]

longitudinal axis in a log, which is the vertical axis of the tree, R designates radial direction, and T designates the tangential direction (normal to the radial direction). We shall see that the loss in moisture that results from drying is direction-dependent. Equally important, when wood is exposed to a humid atmosphere, the expansion is anisotropic.

17.11   Wood Microstructure

By viewing the microstructure of wood, we can understand the makeup of many of the features we saw in the macrostructure, such as growth rings and the role of the cells. In addition, we can understand the differences between softwood and hardwood and the variation in properties in different directions. Just as with other materials, the scanning electron microscope is useful at intermediate magnifications (around 300×) for analyzing the overall features and at high magnification (about 1000×) for viewing cell structure.

We shall begin with observations of softwood structure. Then we shall cover the differences we encounter in the hardwoods, and finally we shall examine the cell itself.

Microstructure of Softwood

The most important and striking feature of softwood structure (Figure 17.15*a* and *b*) is that it is composed of long, tube-like cells running in the longitudinal direction with some other cells normal to this direction (radial). A new layer of longitudinal cells (called *tracheids*) is grown each year from the cambium. The early-season (spring) cells (5) are larger and have thinner walls than the late wood cells (6). A typical cell is 4 mm long and 0.04 mm wide. When a log is sawn in cross section, this difference between early and late cells is seen as a growth ring (4). The radial rays we pointed out in the macrostructure are made up of cells (*parenchyma*) oriented in a radial direction. The rays can be

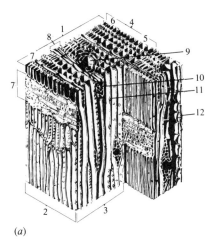
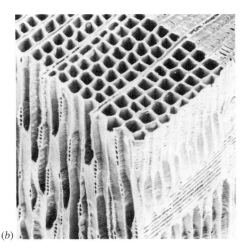

(a)   (b)

FIGURE 17.15 (a) Softwood structure (schematic). 1 = cross-sectional face; 2 = radial face; 3 = tangential face; 4 = annual ring; 5 = early wood; 6 = late wood; 7 = wood ray; 8 = fusiform ray; 9 = vertical resin duct; 10 = horizontal resin duct; 11 = bordered pit; 12 = simple pit. (b) Douglas fir; scanning electron micrograph, 300×.

[a: Classroom Demonstrations of Wood Properties, U.S. Dept. of Agriculture, Forest Products Laboratory, PA900, p. 2; b: From slide of U.S. Forest Products Laboratory, Madison, Wis.]

uniform wood rays (7) or tapered fusiform rays (8). These store food. Sometimes the ray also contains a resin duct (8 and 10). Resin ducts may also be vertical (9). As would be expected, resin ducts are found in only some softwoods. Pit-like features are visible in the cell walls (11 and 12). These are actually orifices that the cells can close like valves to control the flow of fluids through the cells. In softwoods, sap flows through the longitudinal cells (tracheids).

Microstructure of Hardwood

The microstructure of hardwood (Figure 17.16) is similar in many ways to that of softwood. Again, the majority of the cells have their long dimensions parallel to the longitudinal axis of the tree, and early wood and late wood are present, giving growth rings. These longitudinal cells are called *fibers.* The wood rays are also similar, and the cells they are composed of are called *parenchyma.* Resin ducts are not present.

The most important difference between softwoods and hardwoods is that hardwoods contain large longitudinal tubes called *vessels* (8), which transport fluids. At the ends of these vessels are perforation plates. Variations in these structures are used to identify hardwoods. Average vessel length is 0.5 mm and average width is 0.07 mm. The other important feature of the vessels is their size in a given year's growth. In the illustration shown (yellow poplar), the diameter of the vessels is the same in both early wood and late wood. This is called *diffuse porous structure* (Figure 17.16b). By contrast, in white

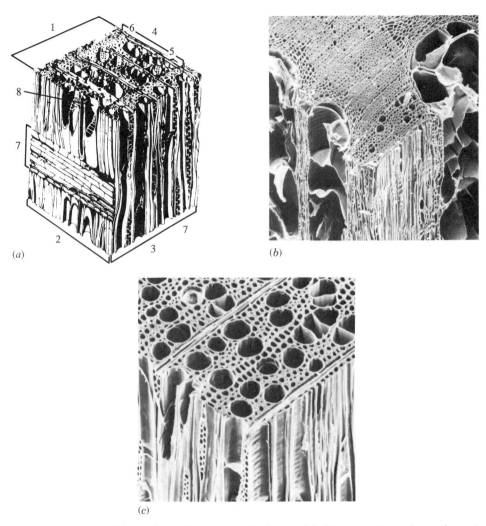

FIGURE 17.16 (*a*) Hardwood structure (schematic). For meaning of numbers 1 through 7, see legend for Figure 17.15. Here 8 = a vessel; 9 = perforation plate. (*b*) Diffuse porous structure in hardwood (yellow poplar). Scanning electron micrograph, 300×. (*c*) Ring porous structure in hardwood (white oak). Scanning electron micrograph, 300×.

[*a*: Classroom Demonstrations of Wood Properties, U.S. Dept. of Agriculture, Forest Products Laboratory, PA900, p. 4; *b*: From slide of U.S. Forest Products Laboratory, Madison, Wis.; *c*: From slide of U.S. Forest Products Laboratory, Madison, Wis.]

oak, the wide vessels appear in rings. This is called *ring porous structure* (Figure 17.16*c*). The vessels can be seen with a hand magnifying glass, and the bands contribute an attractive feature to the wood. There is an important variation in the amounts of different types of cells in hardwoods. For example, large amounts of vessels lower the apparent specific gravity:

Species	Percent Vessels	Percent Fiber (long cells)	Percent Parenchyma (food cells)	Specific Gravity
Basswood	56	36	8	0.32
Hickory	6	67	22	0.64

Cell Structure

Let us now use higher magnification to observe the details of cell structure. Despite the many variations in macrostructure and microstructure, the *basic* biological cells of wood have similar features. The differences are in the *amounts* of the constituents used to make up the structure.

We can compare the cell structure roughly to the structure you create when you build up different layers of glass fiber and resin in constructing the hull of a boat. There are two principal divisions in the cell, the primary wall and the secondary wall, as shown in Figure 17.17. Let us consider first the framework of cellulose fibers. (In Chapter 13 we illustrated the structure of cellulose as a strong linear polymer.) In the primary wall the cellulose fibers are in a loose, irregular, flexible network. This is built up first, followed by the three layers of the secondary wall: S_1, a crisscross network; S_2, a parallel, spiral-type network; and finally S_3, another irregular network. S_2 makes up 70–90% of the cell wall. The closer the fibers of S_2 approach the longitudinal direction, the stronger the cell is in this direction.

Whereas the strength in the longitudinal direction is strongly influenced by the strength of the cellulose, the transverse strength is related to the *lignin* and *hemicellulose* deposits, which form a matrix between the cellulose

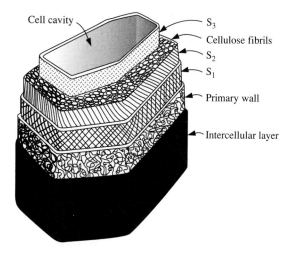

FIGURE 17.17   Cross section of a woody cell

[*Classroom Demonstrations of Wood Properties*, U.S. Dept. of Agriculture, Forest Products Laboratory PA900, p. 6]

fibers. Lignin is composed of phenol–propane network structures similar to the thermosetting network discussed earlier. Hemicellulose is made up of shorter, branched cellulose molecules with a degree of polymerization of 150 to 200 (compared to 5000 to 10,000 for the cellulose of the network).

In addition to these three constituents, there are two other important components of wood cells. There is a group of oil-like hydrocarbons, called *extractives* because they can be dissolved from the wood chemically, in contrast to the polymeric structures. The extractives can act as powerful deterrents to bacterial action and weathering, as in the case of cedar and cypress. The other component in wood cells is mineral such as silica, which can add hardness and resistance to borers by the formation of a hard mineral network.

17.12 Properties of Wood

In describing the mechanical properties of wood, we must include some additional data not required for the metals. For example, we would not expect pieces of 0.2% carbon steel from different sources to vary appreciably in specific gravity or to shrink when exposed to dry atmospheres. However, two completely dry pieces of wood of the same species may differ greatly in specific gravity, for example, if growth conditions produced different amounts of early wood cells (specific gravity 0.2) and late wood cells (specific gravity 0.80). The dimensions of a given piece also vary with moisture content. There is very little dimensional change in the longitudinal direction when wood is dried because of the continuous cellulose chain network. Radial shrinkage or expansion accounts for about 40% of the volume change, and tangential expansion for about 60%. Actual shrinkage in passing from the green to the oven-dry condition varies from 2% to 8% in the radial direction and from 4% to 11% in the tangential direction.

For these reasons, wood should be tested at a stabilized, known moisture content, and the specific gravity should be determined. The specific gravity is defined as the density of the wood divided by the density of water (a known constant). The density of wood varies with moisture content because water adds to both the volume and the mass of the wood. The density is therefore defined in a special way as the moisture-free mass (0% water) divided by the volume at a particular moisture content. We would expect the volume of a kiln-dried piece of wood to be less than the volume of a green piece because of drying shrinkage. Therefore the density of the kiln-dried wood is greater, as is its specific gravity.

Orthotropic Properties

From the previous discussions of wood structure, we expect to find different properties along the three axes: *L*, longitudinal, *R*, radial, and *T*, tangential. Wood technologists give this anisotropy a special name, *orthotropy*, to indicate that there is aniso*tropy* along the three *orthog*onal axes (axes normal to each other).

As an example, we can test specimens cut in different directions, and from the straight-line portion of the stress–strain curve we can determine the modulus of elasticity E. The variation in values with direction is striking. Example: Douglas fir: $E_L = 13,400$, $E_T = 670$, $E_R = 911$, where L, T, and R are directions and all values are in megapascals. The high value of E in the longitudinal direction is due to the continuous cellulose network, whereas the lower values in the other directions are caused by the voids and lack of continuity. The relatively higher value in the radial direction is produced by the rod-like cell structures of the rays.

A simple model will help us understand these variations better. We may consider the tree as made up of a stack of soda straws glued together. (The diameter-to-length ratio of a soda straw represents the long tube-like shapes of the tracheids and the vessel cells.) To describe mechanical properties completely, we would need to make measurements in all three orthogonal directions. However, because of practical limitations and the fluctuation in any property, the measurements are generally made "across the grain"—that is, normal to the longitudinal direction—and also parallel to the grain.

Table 17.7 gives data on typical mechanical properties of some common softwoods and hardwoods. The first line of each set refers to the green condition, and the second line refers to wood with 12% water. All specimens are selected clear grain. We shall deal with the role of defects later.

The most common test of woods is bending, because this is the typical use of structural lumber and the inhomogeneous nature of wood makes it impossible to cut a "typical" tensile specimen. The modulus of rupture is not really a modulus but reflects the maximum load-carrying capacity of a beam. (See Example 17.6.) The compression and shear properties are important in the design of load-bearing areas. Hardness is tested by measuring the load required to press a steel ball to a specified depth; therefore, the higher the load needed to indent, the greater the hardness.

We can draw several general conclusions from the data given in Table 17.7. The typical hardwoods listed are indeed harder and stronger than the softwoods. There are great differences within each group, however; the modulus of rupture of Douglas fir is 40–50% greater than that of white pine or cedar. This explains why Douglas fir is used for structural timber. However, cedar and redwood are preferred for siding because of the extractives, which give better weathering characteristics and resistance to decay. Eastern white pine is extensively used when an easy-to-work material is needed.

There are large differences among the hardwoods as well. Hickory is the hardest of the woods listed; hence its use for baseball bats. Oak, birch, and ash show both high strength and hardness, which explains their use for furniture and flooring. Cherry and maple are lower in these properties than the other hardwoods. However, they are also much used in furniture because of their beauty and workability.

TABLE 17.7 Typical Mechanical Properties of Woods Grown in the United States

Species	Specific Gravity	Static Bending Modulus of Rupture lb/in²*	Static Bending Modulus of Elasticity, 10⁶ lb/in²*	Compression Parallel to Grain; Maximum Crushing Strength, lb/in²*	Compression Perpendicular to Grain; Fiber Stress at Prop. Limit lb/in²*	Shear Parallel to Grain; Maximum Shearing Strength, lb/in²*	Side Hardness Load Perpendicular to Grain lb_f†
Softwoods							
Eastern white pine	0.34 Gr‡	4,900	0.99	2440	220	680	290
	0.35 KD	8,600	1.24	4800	440	900	380
Douglas fir (Coast)	0.45 Gr	7,700	1.56	3780	380	900	500
	0.48 KD	12,400	1.95	7240	800	1130	710
Western red cedar	0.31 Gr	5,200	0.94	2770	240	770	260
	0.32 KD	7,500	1.11	4560	560	990	350
Redwood (young growth)	0.34 Gr	5,900	0.96	3110	270	890	350
	0.35 KD	7,900	1.10	5220	520	1110	420
Hardwoods							
White ash	0.55 Gr	9,600	1.44	3990	670	1380	960
	0.60 KD	15,400	1.74	7410	1160	1950	1320
Yellow birch	0.55 Gr	8,300	1.50	3380	430	1110	780
	0.62 KD	16,600	2.01	8180	970	1880	1260
Black cherry	0.47 Gr	8,000	1.31	3540	360	1130	660
	0.50 KD	12,300	1.49	7110	690	1700	950
Hickory (pecan)	0.60 Gr	9,800	1.37	3990	780	1480	1310
	0.66 KD	13,700	1.73	7850	1720	2080	1820
Maple (big leaf)	0.44 Gr	7,400	1.10	3240	450	1110	620
	0.48 KD	10,700	1.45	5950	750	1730	850
Oak (white)	0.66 Gr	8,300	1.25	3560	670	1250	1060
	0.68 KD	15,200	1.78	7440	1070	2000	1360

*Multiply by 6.90×10^{-3} to obtain MPa.

†Load (lb$_f$) to make standard-diameter impression; multiply by 4.44 to obtain newtons.

‡Gr = green state; KD = kiln-dried to 12% moisture

The thermal expansion of wood is another property that is direction-dependent. Dry wood has a thermal expansion parallel to the grain of $3 \times 10^{-6}/°C$ to $5 \times 10^{-6}/°C$ ($1.7 \times 10^{-6}/°F$ to $2.8 \times 10^{-6}/°F$), for an average value parallel to the grain of $4 \times 10^{-6}/°C$ ($2.2 \times 10^{-6}/°F$), which is less than most metals. However, across the grain the value is 5 to 15 times greater: $20 \times 10^{-6}/°C$ to $60 \times 10^{-6}/°C$ (11.1 to $33.3 \times 10^{-6}/°F$). Although this may seem high, the dimensional changes due to moisture content are of even greater significance.

Finally, we must consider briefly the effect of the method of sawing the log on the properties of the lumber. Figure 17.18 shows two common methods, plain sawing and quarter sawing. In plain sawing all the boards are cut parallel to the same plane, whereas in quarter sawing the log is rotated 90° each time a board is cut, to produce the grain shown. Plain-sawed lumber is lower in cost, but quarter-sawed lumber shrinks and swells less in width and wears more evenly. On the other hand, plain-sawed lumber shrinks and swells less in thickness and shows more conspicuous figure patterns due to annual rings.

Effects of Moisture

Table 17.7 shows that the properties of wood vary significantly with moisture. The *fiber saturation point (FSP)* is defined as the moisture level above which the properties show little change. In all types of wood this value is 26–32% moisture; 30% is often used as an average value. Green wood has moisture levels in excess of the FSP. When the moisture content of wood is below the FSP, such as when it is kiln-dried, there is either pickup or loss of moisture, depending on the relative humidity.

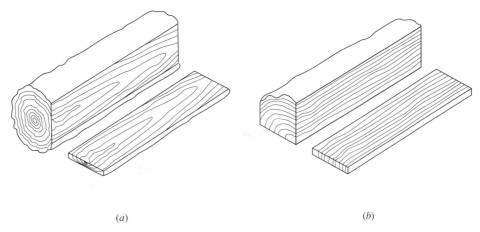

(a)   (b)

FIGURE 17.18   Boards cut from a log: (*a*) plain-sawed, (*b*) quarter-sawed

[U.S. Dept. of Agriculture Handbook No. 72 (revised August 1974), p. 3-1; courtesy Forest Products Laboratory, Forest Service, USDA]

This shrinkage or swelling due to moisture change is maximum in the tangential direction, 50–65% as much in the radial direction, and 1.5–5% as much in the longitudinal direction. Note that where dimensional change is least, the modulus of elasticity is greatest, which reflects the cellular structure of wood. Warpage is therefore dependent on moisture changes, orientation of the grain, and stresses set up during the moisture modification.

Typical values for shrinkage from the green to the oven-dried state are given in Table 17.8.

The equilibrium moisture content in wood is found to vary only slightly with species. Representative moisture levels as a function of ambient temperature and relative humidity are given in Table 17.9.

The example that follows shows how these data might be used.

TABLE 17.8   Shrinkage Values of Selected Woods from Green to Oven-Dry Condition

	Shrinkage (percent)		
Species	*Radial*	*Tangential*	*Volumetric*
Softwoods			
Eastern white pine	2.1	6.1	8.2
Douglas fir (coast)	4.8	7.6	12.4
Western red cedar	2.4	5.0	6.8
Redwood (young growth)	2.2	4.9	7.0
Hardwoods			
Yellow birch	7.3	9.5	16.8
Black cherry	3.7	7.1	11.5
Hickory (pecan)	4.9	8.9	13.6
Maple (big leaf)	3.7	7.1	11.6
Oak (white)	5.6	10.5	16.3

Note: Longitudinal shrinkage less than 0.25% from green to oven-dry state.

TABLE 17.9   Moisture Content of Wood (Percent) in Equilibrium with Surroundings

	Relative Humidity (percent)			
Temperature, °F	*20*	*40*	*60*	*80*
50	4.6	7.9	11.2	16.4
70	4.5	7.7	11.0	16.0
90	4.3	7.4	10.5	15.4
110	4.0	7.0	10.0	14.7

EXAMPLE 17.5 *[ES/EJ]*

One wall of a room is to be paneled with maple boards that are $3\frac{1}{2}$ in. wide. The boards have been stored indoors in the wintertime at 70°F and 20% relative humidity so that they have attained an equilibrium moisture content. Approximately what gap should be left between boards to account for swelling in the summer when the temperature is 90°F with 80% relative humidity. Why might the gap not have to be this large?

Answer   From Table 17.9,

$$\text{Initial moisture level} = 4.5\% \ (70°F \text{ and } 20\% \text{ RH})$$

$$\text{Final moisture level} = 15.4\% \ (90°F \text{ and } 80\% \text{ RH})$$

The average fiber saturation point (FSP) is 30% moisture (see text for discussion).

$$\text{Moisture change as fraction of FSP} = (15.4 - 4.5)/30$$

(estimate of dimensional change due to moisture change)

Assume the worst conditions for the maple boards; that is, assume they are plain-sawed boards. Therefore the shrinkage will be tangential and 7.1% (Table 17.8). Thus the *maximum* estimated shrinkage is

$$\left(\frac{15.4 - 4.5}{30}\right) \times 7.1 = 2.6\%$$

or 3.5 in. × 0.026 in. = 0.091 in. [approximately a $\frac{3}{32}$-in. (2.4-mm) gap is required].

   In reality, a gap this large would not be necessary, because

1. It is likely that not all of the boards will be plain-sawed; thus using an average of the radial and tangential shrinkage would be more appropriate.
2. The attachment (nails and adhesive) would restrain the swelling. However, the boards can split or the adhesive can fail with excessive stress.
3. The high temperature and humidity may not last long enough to establish an equilibrium wood moisture content.
4. The boards would probably be coated, which would further inhibit moisture pickup and swelling.

17.13 Role of Defects in Wood Products

It is very important to realize that the data we have covered so far were obtained on clear-grained specimens. In large pieces of lumber there may be a number of defects present, such as knots, checks, and so forth.

If we compare the properties of no. 2 lumber with those of clear-grained wood of the same species, we obtain the following data:

	Modulus of Rupture, MPa	Modulus of Elasticity, GPa	Compressive Strength, MPa
Wood (clear)	97.5	13.0	53.5
No. 2 lumber	19.4	11.0	31.9

The property showing the greatest decrease as a result of defects is modulus of rupture, one of the commonest design values. The modulus of elasticity is not greatly affected by imperfections, but the compressive strength is severely reduced.

Restructuring to Reduce the Role of Defects

The simplest method of restructuring is to glue sections of wood together to produce *laminated beams*. The position of knots can be controlled, and much larger lengths and cross sections can be produced. Most of the adhesives used in laminating are thermosetting, and they range from those that can be set at room temperature to those that require hot pressing.

Plywood is produced by machining sheets from logs and gluing the plies with the grain at 90° to each other. In this way the material is nearly isotropic and swelling is equalized.

Small pieces and shavings that were previously discarded can be used for *particle board* and *fiberboard*. Particle board can be glued with thermosetting resin, and the larger chips can be placed in the surface layers for better appearance and wear. In the case of fiberboard, the chips are converted to a pulp. In the wet process, the pulp is squeezed and dried. It derives its strength from mechanical interlocking of the fibers and bonding by lignin. In dry bonding, a resin is used as a glue.

Most newspaper is produced entirely from wood fibers, although paper such as the kind used in this textbook contains large quantities of cotton cellulose fibers, known as rag content. The process of making paper from wood fibers requires extraction of the lignin from the fibers and the addition of fillers such as clay and sizing agents such as alum, followed by washing, screening, heated rolling, and drying to produce the final product. Fibers are the major component, so the paper can have varying degrees of anisotropy, depending on the processing method. Furthermore, paper also exhibits defects that can modify the mechanical properties.

EXAMPLE 17.6 *[ES/EJ]*

The modulus of rupture is really not a modulus in the same sense as the modulus of elasticity; rather it is a measure of strength for a brittle material that may rupture under a bending load. (See a text on strength of materials for a more thorough discussion.)

For a simple-end-supported beam with a concentrated load W at its center and an unsupported length L, the following relationship is used:

$$S_f = \frac{(WL/4)c}{I}$$

where

S_f = modulus of rupture

$c = h/2$ (distance between neutral axis and outer fibers)

I = moment of inertia

$= bh^3/12$ for our example (b = beam width; h = beam height)

What concentrated load would a 2 × 8 beam of Douglas fir support if it were loaded at the center of an 8-ft span? (The actual dimensions of a 2 × 8 beam are L = 96 in., b = 1.50 in., and h = 7.25 in.) Why would the load generally be less than this?

Answer

$$W = \frac{(S_f)(I)(4)}{cL} = \frac{(S_f)\left(\dfrac{bh^3}{12}\right)(4)}{(h/2)(L)} = \frac{2S_f bh^2}{3L}$$

Because S_f = 12,400 psi (Table 17.7),

$$W = \frac{2 \times 12{,}400 \times 1.50 \times (7.25)^2}{3 \times 96} = 6790 \text{ lb}$$

The actual design load would not be this high for the following reasons:

1. The values in Table 17.7 are for clear-grained lumber, and defects could decrease the load-carrying capacity by 80% (see text).
2. As wood is loaded over a long period of time, its load-carrying capacity decreases. This is essentially a creep phenomenon as in polymers. For a 10-year load the capacity is 62.5% of the laboratory test values given in Table 17.7.
3. The wood is likely to dry out when it is part of a structure, and the moisture will be less than the 12% given in Table 17.7. The strength will therefore increase as the wood dries.
4. A safety factor should be applied to the beam design.

(These points and data used for the design of wood structural members are discussed in *Wood as a Structural Material*, Educational Modules for Materials Science and Engineering, The Pennsylvania State University, 1980.)

Summary

The amounts of aggregate and cement paste determine the properties of concrete. The objective is to obtain a concrete mass at minimum cost, and therefore the aggregate makes up a large volume fraction. Concretes with higher compressive strengths are obtained at lower water-to-cement ratios. However, one limitation on compressive strength may well be the concrete workability as measured by a slump test.

Because concrete, as a ceramic composite, has low tensile strength, the use of steel reinforcing rods in areas of high tension significantly increases the serviceability. Similarly, the concrete may be placed in residual compression at positions of tensile loading through the use of prestressing with steel reinforcement.

Asphalt uses a bitumen binder for the aggregate rather than the Portland cement found in concrete. The thermoplastic bitumen sets more rapidly than concrete, which has a somewhat slow hydration reaction that is highly moisture- and temperature-sensitive. However, asphalt also has the limitations of most thermoplastics, such as excessive creep at high temperatures and brittleness at low temperatures.

Wood can be considered a composite of cellulose fibers bonded by a matrix of polymers, primarily lignin. The macrostructure is made up of sapwood, which carries nutrients, and heartwood, which is composed of dead cells.

The two primary groups of woods are hardwoods and softwoods. The softwoods have needles and exposed seeds, whereas hardwoods lose their leaves, have flowers, and bear fruit or nuts. The microstructures of these two types of wood are similar, except that softwood has resin ducts and hardwood has large longitudinal fluid-transport vessels.

The mechanical properties of wood depend on the type, the amount of moisture, and the directional orientation of the specimen. The most common test is in bending, and the hardwoods show somewhat higher values. The presence of defects such as knots significantly decreases the strength of wood.

Definitions

Aggregate Natural filler in concrete, normally sand and gravel.

Air entrainment Addition of resin to concrete to give discontinuous small voids.

Asphalt Aggregate in a bitumen (thermoplastic) matrix.

Cambium Interface layer between the bark and sapwood of a tree.

Cement paste Portland cement, water, and entrained air.

Concrete A mixture of cement paste and an aggregate.

Extractives Oil-like hydrocarbons in wood that can be removed by chemical solution in appropriate solvents.

Fiber saturation point (FSP) A moisture level in wood above which the mechanical properties do not change (the average is 30% moisture for all wood species).

Fineness modulus   Characterization of an aggregate on the basis of a sieve analysis.

Hardwood   Trees that lose their leaves, bear fruit, and have true flowers.

Heartwood   Dead wood cells near the center of a tree.

Heavyweight concrete   Concrete made using heavyweight (high-specific-gravity) aggregates.

Lightweight concrete   Concrete made using lightweight (expanded) aggregates.

Lignin   The phenol–propane polymer that forms a matrix between the cellulose fibers in wood cells.

Orthotropy   Anisotropy or directional properties in wood.

Parenchyma   Radial wood cells.

Portland cement   Mixtures of silicates and aluminates that harden by a hydration reaction.

Post-tensioning   Pulling steel in tension after the concrete that is cast around it has cured, then anchoring the steel to the concrete before removing the load. When the load on the steel is removed, the concrete develops residual compression.

Prestressed concrete   Concrete into which residual compression has been introduced by pre-tensioning or post-tensioning steel rods.

Pre-tensioning   Pouring fresh concrete around steel rods or wires that are already in tension. The concrete develops residual compression when the concrete cures and anchors the steel when the external tension force on the steel is removed.

Reinforced concrete   Concrete with steel added to portions in tension.

Sapwood   Live wood cells near the surface of a tree that carry nutrients.

Slump test   One method of measuring workability; the amount of collapse of a standard-size truncated cone of fresh concrete.

Softwood   Trees that have needles and exposed seeds.

Tracheids   Longitudinal wood cells in softwoods.

Vessels   Longitudinal wood cells in hardwoods.

Wood   Natural composite of cellulose fibers in a polymer matrix (primarily lignin).

Wood rays   Horizontal radial rays connecting the tree center to the bark.

Workability   Ease of pouring and placing fresh concrete.

Problems

17.1   [E] Why is straw sometimes used to cover fresh concrete, especially when the weather is cold? (Sections 17.1 through 17.6)

17.2   [ES] A 600-g sample of fine aggregate has the sieve analysis shown below. Determine the fineness modulus. Also plot the percent retained vs. the sieve opening size and determine the nature of the size distribution (Sections 17.1 through 17.6)

Sieve No.	4	8	16	30	50	100	Pan
Grams Retained	13.8	78.4	134.1	156.2	107.7	85.6	23.8

17.3 *[EJ]* The following questions refer to steel-reinforced concrete. (Sections 17.1 through 17.6)

    a. Why are maximum aggregate sizes specified for a steel net or mesh like that used for a road?

    b. Why is the amount of concrete cover important?

    c. Why is aluminum reinforcement not used?

17.4 *[EJ]* In prestressed concrete, some ductility (elongation) in the steel is desirable. From the general knowledge of a stress–strain diagram, what would be the advantage of some plastic deformation in the steel? Why would a steel rod never be stressed beyond the tensile strength? (Sections 17.1 through 17.6)

17.5 *[EJ]* Justify the three criteria given in the text for the selection of the maximum size for coarse aggregate. (Sections 17.1 through 17.6)

17.6 *[ES]* A cement mix has a density of 650 lb/yd^3 (388 kg/m^3). If the water-to-cement ratio is 0.50, how much shrinkage should be expected in a garage floor measuring 22 ft × 24 ft (6.71 m × 7.32 m)? That is, what would be the new dimensions after curing? (Sections 17.1 through 17.6 and Section 16.6)

17.7 *[ES/EJ]* If a steel-reinforced concrete component behaves under isostrain loading as a composite does, what steel fraction is required in order that the steel will bear twice the load of the concrete? Give a range of values, assuming that the modulus of the concrete is the same as that of the coarse aggregate, which has a range from 1×10^6 psi to 10×10^6 psi (0.69×10^4 MPa to 6.9×10^4 MPa). Indicate whether these values seem reasonable. (Sections 17.1 through 17.6)

17.8 *[EJ]* Table 17.3 gives a maximum amount of calcium chloride that can be present in the mixing water for concrete. Indicate the effect of large additions of $CaCl_2$ on the reinforcement, cold-weather setting, and formation of the gel structure. (Sections 17.1 through 17.6)

17.9 *[ES/EJ]* What is the effect of a higher water-to-cement ratio on each of the following? (Sections 17.1 through 17.6)

    a. Creep

    b. Compressive strength

    c. Slump or workability

17.10 *[EJ]* The construction of an in-ground swimming pool may comprise the following steps:

(1) Lay reinforcing wire mesh in a prepared excavation.

(2) Blow an aggregate–cement-paste mixture to a thickness of approximately 1 ft.

(3) Finally, apply a fine white aggregate–cement-paste coating.

Indicate why each step is used and why careful control of the acidity of the swimming water is required. (Sections 17.1 through 17.6)

17.11 *[ES/EJ]* Figure 17.6 gives data on drying shrinkage for concrete. Consider the final mix used in Example 17.4, water-to-cement ratio = 0.53 and Portland cement = 625 lb/yd^3 (371 kg/m^3). (Sections 17.7 and 17.8)

    a. Determine the drying shrinkage.

    b. If the concrete was restrained at the edge after pouring, determine the stress set up by the shrinkage. The modulus of elasticity may be taken as 6×10^6 psi (4.14×10^4 MPa).

c. If the tensile strength is one-tenth of the compressive strength, will the concrete crack from the stress in part b?

d. Suggest why the tensile stress will not be so high as calculated.

17.12 *[ES]* Determine the proportions for a concrete that has to meet the following specifications for a bridge deck. (Sections 17.7 and 17.8)

Mild service: 7-day compressive strength of 4000 psi using type I Portland cement; non-air-entrained

Slump: 3 to 4 in.

Coarse aggregate: 1 in. maximum; 1.5% contained water

Fine aggregate: Fineness modulus 2.75; 3.5% contained water

17.13 *[ES/EJ]* Give several reasons why Table 17.5 indicates that lower water-to-cement ratios are required for applications in fresh water and still lower ones for setting in salt water. (Sections 17.7 and 17.8)

17.14 *[ES/EJ]* Refer to Figure 17.12, assuming water-to-cement ratios of 0.40 and 0.65 for a non-air-entrained type I Portland cement with a 28-day cure time. (Sections 17.7 and 17.8)

a. Why is a band of compressive strengths given?

b. Why is the band width not the same at the two water-to-cement ratios?

17.15 *[ES/EJ]* Figure 17.12 gives the compressive strength of concrete after 7-day and 28-day curing times. Explain why strengths are not given for other times. (Sections 17.7 and 17.8)

17.16 *[ES/EJ]* Explain why concrete is generally not recycled, whereas asphalt can be reused. (Sections 17.7 and 17.8)

17.17 *[EJ]* Why might the following additives be included in asphalt? (Sections 17.7 and 17.8)

a. Shredded pieces of rubber from tires

b. Small broken pieces of glass bottles

17.18 *[EJ]* Recalling that asphalt contains a polymer binder, explain why shipping hot asphalt paving in an uncovered truck makes laying more difficult. Assume that the asphalt does not cool and that the difficulty does not arise when the asphalt is covered or kept under a nitrogen blanket. (Sections 17.7 and 17.8 and Chapter 13)

17.19 *[EJ]* Which woods possess unique characteristics that render them more resistant than others to weathering and decay? Where in the structure are the chemicals that prevent weathering? (Sections 17.9 through 17.11)

17.20 *[ES/EJ]* Why might a hardwood tree growing in the south not be so strong as one of the same species growing in the north where the climate is colder? (Sections 17.9 through 17.11)

17.21 *[ES/EJ]* In cutting down a dead poplar tree with a chain saw, one notes that sparks fly. Closer inspection shows what appears to be small rocks in the cut. From the standpoint of microstructure and macrostructure, how did these hard, rock-like particles form? Determine whether the same particles were present in the live tree. (Sections 17.9 through 17.11)

17.22 *[ES/EJ]* Observation of a wood fire shows that wood maintains its integrity even while burning. The same is not true of many polymers. (Sections 17.9 through 17.11)

a. Why do different species of wood burn at different rates?

b. Why is wood a good thermal insulator?

17.23 *[ES/EJ]* Figure 17.19 has been used to explain why wooden structures can maintain their integrity during a fire. Give reasons why this schematic diagram can be misleading. (Sections 17.9 through 17.11)

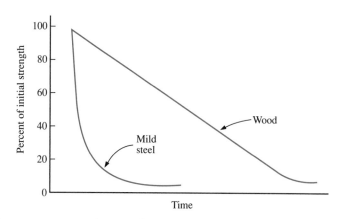

FIGURE 17.19   Retention of strength in wood and mild steel during a fire

[G. R. Moore and D. E. Kline, *Properties and Processing of Polymers for Engineers,* Prentice-Hall, Englewood Cliffs, N.J. 1984, p. 79]

17.24 *[ES]* A 1-ft^3 piece of oak weighs 32 lb dry and 45 lb when saturated with water. (Sections 17.9 through 17.11, Section 6.13, Chapter 13)

   a. What is the bulk density?

   b. What is the volume of the open pores?

   c. What is the apparent density?

   d. The oak is to be impregnated with a thermosetting polymer whose specific gravity is 1.40. As a means of quality control, the 1-ft^3 block of impregnated oak is to be weighed. Calculate the new weight, given that all the open porosity is to be filled with polymer.

17.25 *[EJ]* Why does plywood normally have an uneven number of wood plies? (Sections 17.12 and 17.13)

17.26 *[EJ]* Explain the differences in percent expansion due to absorption of moisture that would be expected in the *L, T,* and *R* directions in a piece of oak, compared with that which would be expected in a well-cemented piece of plywood. (Sections 17.12 and 17.13)

17.27 *[ES/EJ]* In refinishing a large fine wood surface, such as a table top, why is it recommended that one place the same number of coats of varnish on both the top and the bottom? (Sections 17.12 and 17.13)

17.28 *[ES/EJ]* Oak flooring, in strips $2\frac{1}{4}$-in. (57.15-mm) wide, is laid in the summer and has an equilibrium moisture content at 80°F and 70% relative humidity. (Sections 17.12 and 17.13)

   a. Using the average of tangential and radial shrinkage for white oak (flooring is usually red oak, however), estimate the new strip width in the winter at conditions of 70°F and 40% relative humidity.

   b. Why are wider strips generally not used?

17.29 *[ES]* (Sections 17.12 and 17.13)

   a. In Example 17.6, recalculate the maximum load, using a nominal 4 in. × 4 in. ($3\frac{1}{2}$ in. × $3\frac{1}{2}$ in.) eastern white pine beam with an 8-ft span.

   b. Determine a more realistic load by accounting for defects that decrease the load capability by 50% and by including an additional safety factor of 3.

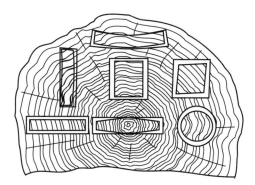

FIGURE 17.20　Potential distortion of lumber cut from a log

[R. J. Hoyle, Jr., "Lumber: Grades, Sizes, Species," in *Wood as a Structural Material*, Educational Modules for Materials Science and Engineering, The Pennsylvania State University, 1980, Fig. 6, p. 158. Reprinted by permission.]

17.30　*[ES/EJ]* Figure 17.20 shows the characteristics of sawn sections having different orientations in a log. Indicate why each variety of deformation occurs and which ones might show the greatest twist or warp. (Sections 17.12 and 17.13)

17.31　*[ES/EJ]* Indicate whether the following statements are correct or incorrect and justify each answer. (Sections 17.12 and 17.13)

a. Used lumber is stronger than new lumber.

b. The thermal expansion of wood must be considered in design.

c. The density of wood changes with time.

d. Paper generally does not tear in a straight line unless it is creased.

PART **VI** SPECIAL TOPICS

Effects of Environment on Materials

In these concluding chapters we will consider the effects of corrosion and oxidation and then take up failure analysis. Following this we will examine the performance of materials in electrical, optical, and magnetic applications. Finally we will consider the general problem of material selection. We have postponed these topics, which are all concerned with the effects of different environments, until we had discussed the structures of all the principal materials.

Although these topics come late in the course, any one of them may be the principal factor in material selection. More components fail by corrosion and oxidation than by simple mechanical forces in tension and compression. Failure analysis is necessary to understand some of the deeper aspects of fracture toughness and combined corrosion and mechanical factors, such as corrosion fatigue. The special requirements for materials to be used in electronic devices using silicon or other chips are the basis of the computer industry. New developments in optical-fiber manufacture have revolutionized communications. Magnets wound with superconducting wires have led to new diagnostic devices for medicine and power generation. Finally we summarize the factors needed for an optimal selection of materials for a component including possible processing methods related to such topics as the geometry of the part, and cost accounting.

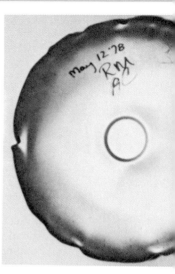
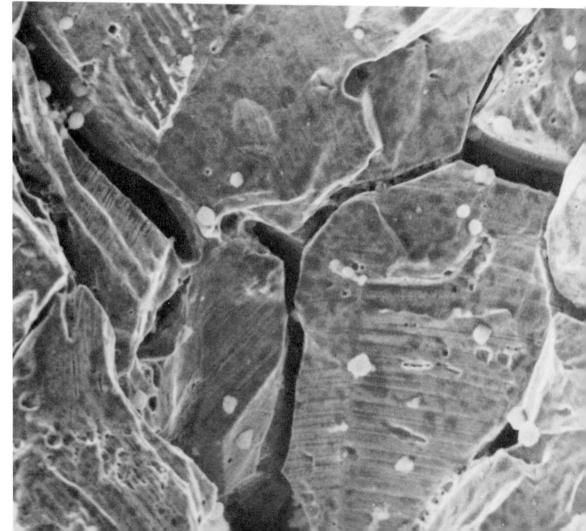

CHAPTER 18

Effects of Environment: Corrosion, Oxidation, Radiation

This 18% Cr, 8% Ni stainless steel beer keg exploded at low pressure while it was being cleaned. Although the pressure was below 25 psi, the fragments seriously injured the operator.

Microexamination showed that the heating of the stainless steel structure next to the weld caused the precipitation of carbides rich in chromium at the grain boundaries. This removed chromium from the matrix, making the grain boundaries susceptible to corrosion. The scanning electron micrograph at 1000× shows clearly the lack of continuity at the grain boundaries. At the time of the accident only a thin layer of unattacked metal remained, and this ruptured when air pressure was applied to force out the cleaning solution.

18.1 Overview

So far we have been concerned principally with the relationship of structure to stress, or, to put it in a practical fashion, how to pick a material to resist the *stresses* encountered in service. Suppose, however, that you designed a kitchen sink on the basis of mechanical stresses only and selected an easily formed, low-cost carbon steel. You would not be very popular with your family when the surface of the sink turned brown and then began to pit. It is just as important to keep corrosion in mind when designing an auto fender, to consider oxidation for aircraft turbine blades, and to take into account radiation damage to high polymers by ultraviolet light and to nuclear reactor parts by neutrons.

The junkyards of the country are filled with cars, appliances, and machines that were mechanically sound but failed in service because of reaction of the material with the environment. The annual cost of these failures is tens of billions of dollars, not to mention the suffering and loss of life caused by accidents.

One of the principal reasons for these failures due to reaction with the service environment is the relatively complex nature of the reactions involved. To illustrate the lack of understanding even among engineers, consider the phenomenon *called* "static fatigue." Many cases have been encountered where ceramic or polymer components cracked catastrophically at room temperature while in a static or rest position. For example, many a glass coffee pot has disintegrated while at rest because of previous mistreatment, such as heating without liquid. The term, *static fatigue*, however is a pathetic misnomer; the failure is due to stress and corrosion. The part usually contains residual stresses, and the water vapor or other corrodent in the atmosphere results in the progress of a small crack to a critical length, followed by rapid fracture. Obviously this is not a fatigue failure.

We have brought up this simple illustration to show that we urgently need a basic understanding of environmental problems if we are to find solutions. We can achieve a broad grasp of the field by investigating three basic concepts: corrosion, oxidation, and radiation.

Corrosion

Despite all the complex jargon (such as sacrificial anode and cathodic control), whether a metal corrodes depends on the simple electrochemical cell set up by the environment. For example, a chunk of iron rusts because iron atoms are induced by the environment to leave the component as iron ions. It may seem late in the course to go back to basic chemistry and write an equation:

$$Fe^0 \longrightarrow Fe^{2+} + 2e^- \tag{18-1}$$

But the whole issue of whether iron going into solution (rusting) continues or halts depends on how we control the iron ions and the electrons. For example, we can halt this reaction in a pipeline by hooking on a piece of magnesium to the pipe and developing a preferred reaction (chemically).

$$Mg^0 \longrightarrow Mg^{2+} + 2e^- \qquad (18\text{-}2)$$

This reaction develops enough potential in the pipe so that it does not corrode; the magnesium corrodes instead.

Therefore in the section on corrosion, we will examine the driving forces for metal solution and how to control them.

Oxidation

Here again, both the problem and its solution depend on atom movements. We know that aluminum is a good deal more active chemically than iron but that it does not disappear due to oxidation. We shall find that the answer to this anomaly is that the aluminum oxide forms a crystal structure with a stable surface film, that stops the process, whereas the iron oxide film is porous. Furthermore, the problem of oxidation at high temperature prevents the use of many otherwise excellent heat-resistant alloys.

Radiation

Once again we must understand atom–atom, photon–atom, and neutron–atom interactions in order to comprehend and prevent large-scale failure. There are two principal concerns:

(1) Damage by ultraviolet radiation principally involves polymers. We have all seen a set of shiny, plastic-covered lawn chairs crack and shrivel under the effects of ultraviolet radiation from the sun! (2) When we consider damage in nuclear reactors, we are concerned with the effect of the bombardment of construction materials in a reactor. Although this bombardment is on an atomic scale, large-scale changes in strength and ductility occur.

18.2 Corrosion of Metals

In discussing the corrosion of metals, we shall cover four areas:

1. *Chemical principles:* the driving force for corrosion, anode and cathode reactions in a corrosion cell, the solution tendency of metals and alloys, effects of concentration, inhibitors, and passivity
2. *Corrosion phenomena* based on chemical principles: galvanic action, selective leaching, hydrogen embrittlement, oxygen corrosion cells, pit and crevice corrosion, and combined mechanical-corrosive effects — that is, stress corrosion, corrosion fatigue, corrosion erosion, and cavitation
3. *Corrosion environments:* the reaction of metals to different atmospheres, fresh and salt water, and chemicals
4. *Corrosion in gas* at elevated temperatures, scaling, and growth

In each discussion we will take up the basis for prevention of corrosion as well.

CHEMICAL PRINCIPLES

18.3 Does the Metal React?

In analyzing *corrosion*, the first question is: Does the metal react with its environment? If so, what is the nature of the corrosion product? A gold wire placed in distilled water simply does not react. Therefore there is no problem. An aluminum wire reacts, but the corrosion product, aluminum oxide, is so adherent that after a layer forms there is no further reaction. An iron wire reacts more slowly than the aluminum wire at first, but the reaction continues because the product, rust, is nonprotective. Therefore, in each case we must study the reaction involved and the types of products generated.

18.4 Anode and Cathode Reactions (Half-Cell Reactions)

Whether the corrosion is spectacularly fast (such as zinc dissolving in hydrochloric acid) or quiet (such as rust insidiously forming on the back side of an automobile rocker panel), the basic types of reactions are the same.

1. There is an *anode* reaction, in which metal goes into solution as an ion; that is, it corrodes. For example, in the reactions just mentioned,

$$Zn \longrightarrow Zn^{2+} + 2e \qquad Fe \longrightarrow Fe^{2+} + 2e \qquad (18\text{-}3)$$

where e = electron where the negative charge will not be shown in our following discussion; or, in general,

$$M \longrightarrow M^{n+} + ne \qquad (18\text{-}4)$$

where M = metal.

2. The electrons flow through the metal part until they reach a point where they can be used up (*cathode* reaction). Again we use the examples given above.

In the case of zinc in acid, the electrons combine with hydrogen ions at the surface of the metal. Atomic hydrogen is formed. Most of it combines to form molecular hydrogen, which bubbles off, but some dissolves in the metal. This is important in cases of hydrogen attack, which will be discussed later:

$$2H^+ + 2e \longrightarrow 2H \longrightarrow H_2 \text{ (gas)} \qquad (18\text{-}5)$$

In the case of iron the solution surrounding the part is neutral, and we have a reaction in which oxygen and water use up the electrons from the anode reaction to form hydroxyl ions:

$$O_2 + 2H_2O + 4e \longrightarrow 4OH^- \qquad (18\text{-}6)$$
(Dissolved)

Iron does not corrode in pure water in the absence of dissolved oxygen.

TABLE 18.1   Possible Cathode Reactions in Different Galvanic Cells

Cathode Reaction	Example
$2H^+ + 2e \longrightarrow 2H^0$	Acid solutions: see text.
$O_2 + 2H_2O + 4e \longrightarrow 4OH^-$	Neutral and alkaline solutions.
$O_2 + 4H^+ + 4e \longrightarrow 2H_2O$	Using both O_2 and H^+ in acid solutions.
$M^{3+} + e \longrightarrow M^{2+}$	This is encountered when ferric ions are reduced to ferrous ions.
$M^{2+} + 2e \longrightarrow M^0$	When iron is placed in a copper–salt solution, the electrons from solution of the iron reduce copper ions to metallic copper.

For corrosion to progress, it is essential to have both anode and cathode reactions; otherwise a charge builds up, stopping corrosion. The anode reaction is generally the simple case of metal going into solution, but a variety of cathode reactions are encountered, depending on the conditions (Table 18.1). Note that in all the cathode reactions, electrons are *absorbed*.

18.5 Cell Potentials: General

Now that we have established that corrosion is the result of an electrochemical cell, we need to investigate the problem quantitatively in order to develop methods for prevention. We will begin with a discussion of the measurement of the voltages of corrosion cells under standard conditions and then take up the effects of changing the corrosive solution. We will first discuss how the voltages produced at the anode (anode half-cell) and at the cathode (cathode half-cell) are measured and used in a pragmatic way. Then we will consider briefly the valuable concept of the cell voltage from a thermodynamic point of view. This provides a logical background for introducing the Nernst equation, which is used to evaluate the effects on the cell voltage of change in concentrations of ions in the corroding solutions. When we discuss corrosion prevention by adding inhibitors, the same equation applies. (A valuable by-product of this discussion is a better understanding of the many new electrical batteries.)

18.6 Half-Cell Potentials

To evaluate the tendency of each metal to corrode we need a reference standard "half-cell" to connect to the metal half-cell. For example we might observe that a given zinc-copper cell develops 1.1 volts, but we want to evaluate separately the tendency of the zinc half of the cell to lose electrons (oxidation) and of the copper half-cell side to gain electrons (reduction) by reducing Cu^{2+} ions to Cu^0.

To do this we set up a standard half-cell of hydrogen as shown in Figure 18.1. It consists of a platinum electrode immersed in a solution with a

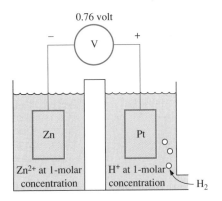

0.76 volt

FIGURE 18.1   Diagram of a half-cell for measurement of potential

one-molar concentration* of hydrogen ions produced with sulfuric acid. Hydrogen is bubbled constantly around the platinum where it is adsorbed, giving essentially an electrode of hydrogen. The platinum does not take part in the reactions. The standard half-cell is connected to a half-cell of metal in a one-molar solution of its ions*. A semipermeable membrane is used to complete the electrical circuit between the half-cells. If the metal is anodic to hydrogen we have oxidation of the metal and reduction of hydrogen.

$$M^0 - ne \longrightarrow M^{n+} \tag{18-7}$$

$$2H^+ + 2e \longrightarrow H_2 \tag{18-8}$$

or if the metal is cathodic to hydrogen we have

$$M^{n+} + ne \longrightarrow M^0 \tag{18-9}$$

$$H_2 - 2e \longrightarrow 2H^+ \tag{18-10}$$

Let us now calculate the voltage in a zinc-copper cell under standard conditions. Referring to Table 18.2, we see that zinc is anodic relative to copper and zinc will go into solution and copper ions will plate out.

We therefore write the reactions

$$Cu^{2+} + 2e = Cu^0 [+0.337 \text{ v. (sign change for reverse reaction)}]$$
$$\underline{Zn^0 - 2e = Zn^{2+} (+0.763 \text{ v.})}$$
$$Cu^{2+} + 2Zn^0 = Zn^{2+} + Cu^0 (+1.100 \text{ v.})$$

Note that because we need to write the equation for copper in the direction opposite that given in the table, we need to transfer the voltage with a

---

*Strictly speaking we say that the *activity* of the ions is one molar. If the concentration is one molar, there is an interaction between ions that reduces the *effective* concentration to a lower value. To reach unit activity the concentration must be slightly higher than one molar as determined experimentally.

TABLE 18.2　Standard Oxidation Potentials for Corrosion Reactions*

Corrosion Reaction	Potential, E_0, Volts vs. Normal Hydrogen Electrode[†]
$Au \longrightarrow Au^{3+} + 3e$	−1.498
$2H_2O \longrightarrow O_2 + 4H^+ + 4e$	−1.229
$Pt \longrightarrow Pt^{2+} + 2e$	−1.200
$Pd \longrightarrow Pd^{2+} + 2e$	−0.987
$Ag \longrightarrow Ag^+ + e$	−0.799
$2Hg \longrightarrow Hg_2^{2+} + 2e$	−0.788
$Fe^{2+} \longrightarrow Fe^{3+} + e$	−0.771
$4(OH)^- \longrightarrow O_2 + 2H_2O + 4e$	−0.401
$Cu \longrightarrow Cu^{2+} + 2e$	−0.337
$Sn^{2+} \longrightarrow Sn^{4+} + 2e$	−0.150
$H_2 \longrightarrow 2H^+ + 2e$	0.000
$Pb \longrightarrow Pb^{2+} + 2e$	+0.126
$Sn \longrightarrow Sn^{2+} + 2e$	+0.136
$Ni \longrightarrow Ni^{2+} + 2e$	+0.250
$Co \longrightarrow Co^{2+} + 2e$	+0.277
$Cd \longrightarrow Cd^{2+} + 2e$	+0.403
$Fe \longrightarrow Fe^{2+} + 2e$	+0.440
$Cr \longrightarrow Cr^{3+} + 3e$	+0.744
$Zn \longrightarrow Zn^{2+} + 2e$	+0.763
$Al \longrightarrow Al^{3+} + 3e$	+1.662
$Mg \longrightarrow Mg^{2+} + 2e$	+2.363
$Na \longrightarrow Na^+ + e$	+2.714
$K \longrightarrow K^+ + e$	+2.925

*Measured at 25°C. Reactions are written as anode half-cells. Arrows are reversed for cathode half-cells.
[†]In some chemistry texts the signs of the values in this table are reversed; for example, the half-cell potential of zinc is given as −0.763 volt. The present convention is adopted so that when the potential E_0 is positive the reaction proceeds spontaneously as written.

change in sign, as in standard algebra. The contribution of the copper half cell as written is +0.337 v. and of the zinc +0.763 v. for a total of +1.100 v.

The concept of cell voltage is important not only for understanding corrosion but also in electroplating and battery design. These all are related to basic thermodynamics. To understand the larger picture let us examine the nature of the voltage of a cell and the capacity to do work, for example in running an electric motor.

Let us consider the zinc-copper cell that we have already discussed and operate it under certain idealized conditions. Let the resistance of the cell be negligibly small so that when we connect a motor in the external circuit:

$$I = \frac{E}{R} \tag{18-11}$$

I = current

E = cell voltage

R = resistance of motor

We can also write another relation:

$$I = \frac{Ne}{t} \tag{18-12}$$

N = number of electrons passing a
given point in time t

e = charge on an electron $(1.602 \times 10^{-19}$ coulombs$)$

Then equating the Equations 18-11 and 18-12 and solving for t

$$t = \frac{NeR}{E} \tag{18-13}$$

The electrical work done in time t will equal the power EI times t. This power can also be written I^2R. Therefore

$$W = I^2Rt = \left(\frac{E^2}{R^2}\right)(R)\left(\frac{NeR}{E}\right)$$

$$= eNE \tag{18-14}$$

Because it is more convenient in calculations to talk about moles of electrons, we define n as the number of moles of electrons involved in a cell operation doing work:

$$n = \frac{N}{N_0}$$

N = number of electrons

N_0 = Avogadros number $(6.023 \times 10^{23}$ electrons/mol$)$

Then substituting

$$W = (nN_0)eE = n(N_0e)E \tag{18-15}$$

The term N_0e is a constant:

$$N_0e = F = \left(6.023 \times 10^{23} \frac{\text{electrons}}{\text{mol}}\right)\left(1.602 \times 10^{-19} \frac{\text{coulombs}}{\text{electron}}\right)$$

or

$$F = 96{,}500 \frac{\text{coulombs}}{\text{mol}} \text{ of electrons (called Faraday's constant) "}F\text{"}$$

$$W = nFE \tag{18-16}$$

Now let us turn to the free energy G which we discussed briefly in Chapter 5 in connection with solidification.

The free energy is equal to the maximum work which can be obtained from the cell. It is important to note that by thermodynamic conventions when the sign of the change in free energy is minus, the reaction will proceed spontaneously. Therefore the sign of W in the zinc-copper cell must be minus and we write

$$\Delta G = (-)nFE \tag{18-17}$$

Therefore if the sign of E is *positive* the cell will operate spontaneously if we have defined the anode and cathode correctly.

We have discussed an idealized case and now should consider two important modifications: (1) the effect of different ion concentrations; and (2) the effects of changes in the solution around the electrodes as the cell operates.

Effect of Changes in Ion Concentration

We began our discussion with a standard condition with one-molar ion concentration. We can visualize that the anode reaction $Zn - 2e \rightarrow Zn^{2+}$ will be less active if the concentration of Zn ions in the liquid is higher than one molar. (Just as the second spoonful of sugar dissolves more slowly than the first.) Conversely, at the cathode the *higher* the concentration of ions the more active will be the reduction reaction.

To calculate quantitatively the effects of ion concentration, Nernst adapted a basic thermodynamic equation to this particular case. To show the basis for the Nernst Equation, we examine the basic equation briefly without using it for calculations. If we have a chemical reaction between substances A and B to produce products C and D:

$$xA + yB = wC + zD \tag{18-18}$$

The change in free energy as we change the concentration of A, B, C, and D is:

$$\Delta G = \Delta G^0 + RT ln \frac{(C)^w (D)^z}{(A)^x (B)^y} \tag{18-19}$$

ΔG^0 = free energy (reactants − products) under *standard* conditions

R = gas constant 1.987 cal/mol K = 8.314 J/mol · K where
    1 J = 1 volt · coulomb

T = temperature, K

$(C)^W$ = concentration of C raised to the power w; similarly $(D)^z$, $(A)^x$, $(B)^y$

x, y, w, z = number of moles of each species participating in the reaction

Nernst proposed an analogous relation for a galvanic cell: using the quantity $-nEF$ to replace the free energy term ΔG and $-nE^0F$ for ΔG^0 from Equation 18-17.

$$-nEF = -nE^0F + RT ln Q \tag{18-20}$$

where

F = Faraday constant of 96,500 coulombs

n = number of electrons transferred in the reaction

T = absolute temperature, K

R = gas constant

E^0 = standard voltage

Q = concentration ratio = $\dfrac{(\text{concentration of anion})^y}{(\text{concentration of cation})^z}$

The power superscripts y and z are used to express the number of moles of electrons transferred in the reaction.

For example, in a cell involving oxidation of zinc and reduction of silver we have two silver ions for each zinc ion and

$$2Ag^+ + Zn^0 \longrightarrow Zn^{2+} + 2Ag^0 \qquad (18\text{-}21)$$

$$Q = \frac{(\text{conc } Zn^{2+})^1}{(\text{conc } Ag^+)^2}$$

We can simplify Equation 18-20 by inserting values for F, R and T at room temperature (298^0K):

$$E = E^0 - \frac{RT}{nF} lnQ = E^0 - \frac{2.303\ RT}{nF} \log_{10} Q$$

$$E = E^0 - \frac{2.303(8.314\ \text{volt} \cdot \text{coulombs/mol} \cdot \text{K})\,(298\ \text{K})}{n(96,500\ \text{coulombs/mol})} \log_{10} Q$$

or $\qquad E = E^0 - \dfrac{0.059}{n} \log_{10} Q \qquad\qquad (18\text{-}22)$

EXAMPLE 18.1 *[ES]*

A. What is the cell potential in a zinc-silver cell with a 1-molar concentration of zinc and silver ions?
B. What is the new cell voltage with a Zn^{2+} concentration of 0.1 M and a Ag^+ ion concentration of 0.01 M?

Answers

A. Zinc is more anodic than the silver (Table 18.2).

$Zn^0 = Zn^{2+} + 2e^-$ $\qquad\qquad\qquad$ E^0 = +0.763 v. (written as anode)

$2Ag^+ + 2e^- = 2Ag^0$ $\qquad\qquad$ E^0 = +0.799 v. (written as cathode)

Summing the reactions to remove the electrons:

$$Zn^0 + 2Ag^+ = Zn^{2+} + 2Ag^0 \qquad E^0 = +0.763 + 0.799 = +1.562 \text{ v.}$$

B. $E = E^0 - \dfrac{0.059}{n} \log \dfrac{[Zn^{2+}]}{[Ag^+]^2}$

$\qquad = 1.562 - \dfrac{0.059}{2} \log \dfrac{[0.1]}{[0.01]^2}$

$\qquad = 1.562 - 0.0295(3)$

$\qquad = 1.474 \text{ v.}$

The voltage is less than for the standard cell because the severe reduction in Ag^+ ions lowers the potential of the reduction reaction $Ag^+ + e = Ag^0$ more than the reduction in Zn^{2+} ions improves the solution of Zn^0. Note the value of n is taken as 2 because two moles of electrons are transferred.

18.7 Cell Potentials in Different Solutions

In practice, even more serious corrections must be made to the driving voltage of a given corrosion couple, because the metals are rarely in solutions of their own salts. For example, Table 18.3 gives the order of *half-cell potentials* in salt water for a group of commercial alloys. Note that aluminum and zinc have changed places compared with the standard sequence.

We will now review the anode, the cathode, and their electrode potentials, all of which can be difficult to identify.

Consider a piece of aluminum and a piece of iron placed in salt water, but *not connected* (Figure 18.2a). Both metals tend to corrode, because local anodes and cathodes arise within each material from inhomogeneities such as impurities and grain boundaries (to be discussed in Section 18.13). Upon *connection* of the metals, the aluminum becomes anodic to the iron (Figure 18.2b). This can be determined from Table 18.3 by comparing 1100 aluminum and mild steel. We would find accelerated corrosion of the aluminum and little or no corrosion of the iron.

The aluminum is therefore the anode, and it delivers electrons to the external circuit and hence to the iron. However, if the aluminum is to continue to deliver electrons to the cathode, the electrons must be "used up." This is accomplished at the iron–electrolyte surface by the cathode reaction shown in Table 18.1: $O_2 + 2H_2O + 4e \rightarrow 4OH^-$.

With these concepts in mind, we can now go on to look at the corrosion rate.

TABLE 18.3   Half-Cell Potentials of Various Alloys in Salt Water[a]

Metal or Alloy[b]	Potential,[c] Volts, 0.1 N Calomel Scale
Magnesium	+1.73
Zinc	+1.10
7072, Alclad 3303, Alclad 6061, Alclad 7075	+0.96
520.0-T4	+0.92
5056, 7079-T6, 5456, 5083, 514.0, 518.0	+0.87
5154, 5254, 5454	+0.86
5052, 5652, 5086, 1099	+0.85
3004, 1185, 1060, 1260, 5050	+0.84
1100, 3003, 6053, 6061-T6, 6062-T6, 6063, 6363, Alclad 2014, Alclad 2024	+0.83
413.0, cadmium	+0.82
7075-T6, 356.0-T6, 360.0	+0.81
2024-T81, 6061-T4, 6062-T4	+0.80
355.0-T6	+0.79
2014-T6, 850.0-T5	+0.78
308.0	+0.77
380.0F, 319.0F	+0.75
296.0-T6	+0.72
2014-T4, 2017-T4, 2024-T3 and T4	+0.68 to +0.70[d]
Mild steel	+0.58
Lead	+0.55
Tin	+0.49
Copper	+0.20
Bismuth	+0.18
Stainless steel (series 300, type 430)	+0.09
Silver	+0.08
Nickel	+0.07
Chromium	+0.4 to −0.18

(left margin: ↑ More anodic)

[a]53 g NaCl + 3 g H_2O_2 per liter, 25°C
[b]The potential of all tempers is the same unless temper is designated.
[c]Data are from Alcoa Research Laboratories.
[d]The potential varies with quenching rate.
Reprinted by permission of Alcoa Technical Center.

18.8   Corrosion Rates

We have introduced the concept of solution potential, and we find with further study that equilibrium and this electromotive force (emf) are closely associated. That is, the emf indicates the *tendency to corrode* as well as identifying the anode and cathode, but it does not directly govern the corrosion *rate*. For example, a battery is merely a case of controlled corrosion. If we were to ask for a 1.5-volt battery in a store, we could choose from dozens of sizes, all of which would provide 1.5 volts; the lives in ampere-hours of the batteries would be decidedly different, however. Thus the corrosion rates would be

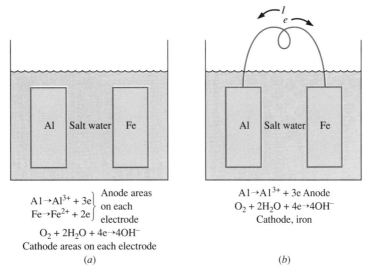

$Al \rightarrow Al^{3+} + 3e$ Anode areas
$Fe \rightarrow Fe^{2+} + 2e$ on each
electrode

$O_2 + 2H_2O + 4e \rightarrow 4OH^-$
Cathode areas on each electrode

(a)

$Al \rightarrow Al^{3+} + 3e$ Anode
$O_2 + 2H_2O + 4e \rightarrow 4OH^-$
Cathode, iron

(b)

FIGURE 18.2 (a) Electrode reactions that take place at aluminum and iron electrodes that are not connected. (b) When electrodes are connected, the iron does not corrode and the aluminum becomes the anode.

different because of the different battery sizes and the different materials used in the batteries. In effect, the corrosion rate depends on Faraday's law:

$$\text{Weight of metal dissolving (g)} = kIt \qquad (18\text{-}23)$$

where

$$k = \frac{\text{at. wt of metal}}{\text{no. of electrons transferred} \times 96{,}500 \text{ A-sec}}$$

I = current (A)

t = time (sec)

In corrosion we are usually interested in the loss in weight from a given area per unit of time. Dividing both sides of the foregoing equation by the area gives us the corrosion rate as equal to a constant times the current density [current (A)/area (cm²)]. In other words, the rate of weight loss from a given area is proportional to the *current density* (the current flowing during corrosion, divided by the area of the cathode).

When we determine the current density for an aluminum–iron corrosion cell, we find that the initial current is relatively high (4 mA/cm²) and that the current gradually falls off until it reaches 1 mA/cm² after 6 min (Figure 18.3). Also, the emf changes during this period. Such a decrease in corrosion rate with time is caused by *polarization* and is very important to corrosion control.

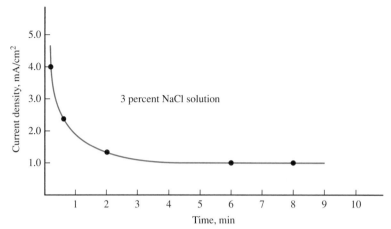

FIGURE 18.3   Decrease in corrosion current in an aluminum–iron galvanic cell (25°C, 77°F).

An example of polarization is the copper–zinc cell shown in Figure 18.4a. The zinc acts as an anode, and its emf tends to become more cathodic as the current or current density increases, as shown in Figure 18.4b. Meanwhile, the copper is cathodic and tends to become more anodic, as shown by the emf–current-density relationship. The two curves intersect, giving us a corrosion emf of less than 1.10 volts and a corresponding current density that determines the corrosion rate. Therefore, if we want to lower the corrosion rate of a component, we must find a way to decrease the current density of the cell. This involves the concept of polarization. Polarization is really composed of

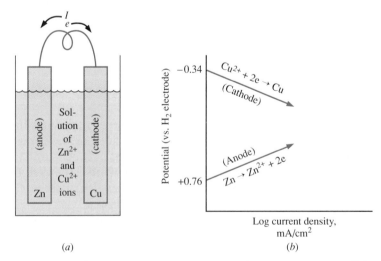

FIGURE 18.4   (a) Polarization in a copper-zinc galvanic corrosion cell (25°C, 77°F). (b) Potential vs. log of current density. Note that the cathode is the upper line, following the usual method of plotting these diagrams. The same convention is followed in Figures 18.6 and 18.8.

two effects: *activation polarization* and *concentration polarization*. The latter is easier to describe first, because it follows directly from our discussion of the effect of concentration on cell potential.

Let us again consider the copper–zinc corrosion cell in operation. As zinc goes into solution, electrons are delivered to the copper. In order for the electrons to be absorbed, positive ions are needed at the copper cathode. In the cell of Figure 18.4 these are copper ions. After current has passed for a short time, there is a scarcity of positive ions at the cathode as a result of the time needed for diffusion of ions within the solution. This is shown schematically in Figure 18.5. The greater the requirement for positive ions, as with very high initial corrosion rates, the sooner an ion-depleted layer appears at the cathode. The net effect, of course, is to decrease the corrosion rate.

The cathode voltage becomes more positive at lower ion concentrations, as shown in Figure 18.4. Therefore, depletion of ions at the cathode changes the cathode polarization, as shown in Figure 18.6, and lowers the corrosion rate.

Similarly, a high concentration of Zn^{2+} ions at the anode interface will inhibit more zinc from going into solution and cause concentration polarization at the anode. This concentration polarization will make the zinc less positive, as shown in Figure 18.6. However, the effect is negligible compared to the polarization of the cathode.

Finally, it must be emphasized that concentration polarization is related to diffusion. Conditions that change diffusion rates therefore affect concentration polarization. As an example, stirring the liquid reduces the concentration gradient of positive ions. We can expect similar effects leading to lower concentration polarization and increased corrosion rates with an increase in temperature or an increase in the concentration of the ions in solution at the cathode.

Now let us consider activation polarization, the common variety of polarization when the reaction rate is low. A good illustration of this effect is the evolution of hydrogen at the cathode, as shown in Figure 18.7. The hydrogen is first formed as atomic hydrogen, then two atoms form molecular hydrogen.

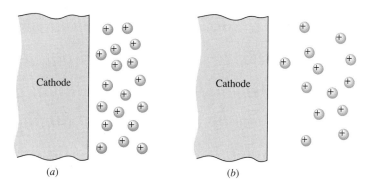

(a) (b)

FIGURE 18.5 Concentration polarization (ion depletion) due to a high rate of reaction at the cathode. (a) Time = 0. (b) Time > 0.

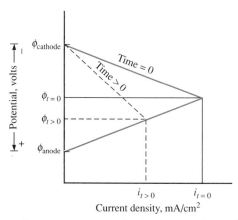

FIGURE 18.6   Effect of concentration polarization at the cathode in lowering the rate of corrosion (decrease in current density in a cathode-controlled polarization)

Next the molecules must form a bubble, and finally the bubble rises. A slow-down in any step can retard the entire sequence. The term *activation polarization* is used to give the idea of the "activation energy" needed to overcome the resistance of the slowest step. Activation polarization also occurs at the anode. This is the barrier the metal atom or ion encounters in leaving the metal specimen and entering the solution. This value is larger for the transition metals iron, cobalt, nickel, and chromium than it is for silver, copper, and zinc.

Let us summarize these effects. The concentration polarization is more of a chemical barrier, whereas the activation polarization is more of a physical or electrical barrier. Both reduce the corrosion rate.

The polarization effects at the anode and cathode are rarely equal. When

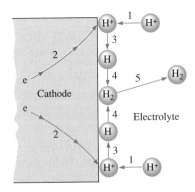

FIGURE 18-7   Steps in the formation of hydrogen gas at the cathode. (Any of the steps can be rate-limiting because of activation polarization.) Step 1: migration of hydrogen ion to interface. Step 2: motion of electrons. Step 3: formation of atomic hydrogen. Step 4: formation of H_2 gas. Step 5: detachment of gas bubble from surface of cathode.

the polarization results in a greater decrease in cathode potential, the reaction is said to be *cathodically controlled,* as shown in Figure 18.6. The term *anodic control* is used similarly. We must again emphasize that current density is our major concern, as it dictates the corrosion rate.

EXAMPLE 18.2 *[ES]*

Write the anode and cathode reactions (half-cell reactions) for the following conditions.

A. Copper and zinc in contact and immersed in sea water
B. As in part *A* with the addition of HCl
C. As in part *A* with the addition of copper ions
D. Copper immersed in fresh water
E. Iron immersed in fresh water
F. Cadmium-plated steel scratched and immersed in sea water

Answers

A. Anode: $Zn \rightarrow Zn^{2+} + 2e$
   Cathode (copper): $O_2 + 2H_2O + 4e \rightarrow 4OH^-$ (see Table 18.1)
B. Anode: $Zn \rightarrow Zn^{2+} + 2e$
   Cathode (copper): $O_2 + 4H^+ + 4e \rightarrow 2H_2O$
   (Because H^+ ions can help use up electrons generated by corrosion of zinc, the corrosion rate is higher than in part *a.*)
C. Anode: $Zn \rightarrow Zn^{2+} + 2e$
   Cathode: $Cu^{2+} + 2e \rightarrow Cu$
D. There is little or no corrosion in this case because copper is noble. Compare the emf for a solution of copper with that for the reaction $O_2 + 2H_2O + 4e \rightarrow 4OH^-$ (see Table 18.2).
E. Anode: $Fe \rightarrow Fe^{2+} + 2e$
   Cathode: $O_2 + 2H_2O + 4e \rightarrow 4OH^-$
F. Anode: $Cd \rightarrow Cd^{2+} + 2e$
   Cathode: $O_2 + 2H_2O + 4e \rightarrow 4OH^-$
   [*Note:* Cadmium is cathodic to iron in the standard oxidation–reduction potentials (Table 18.2) but anodic to iron in salt water.]

18.9 Inhibitors

We have given a detailed account of polarization because one of the most important methods for controlling corrosion is the use of *inhibitors*. The action of these chemicals is related to polarization. The corrosion rate can be reduced drastically by inhibiting the reaction at *either* the anode or the cathode. In many cases the role of the inhibitor is to form an impervious, insulating film of a compound on either the cathode or the anode. A common case of inhibitor use is the use of chromate salts in automobile radiators. The iron ions liberated at the anode surface combine with the chromate to form an insoluble coating. Another case is the use of gelatin, which is adsorbed and limits the ion reaction with the electrode. Most oils, waxes, and greases behave similarly to gelatin and provide only temporary protection of exposed metal surfaces.

18.10 Passivity

A logical extension of inhibition is the development of a passive film on the surface of a metal. A classic example is the corrosion of iron in nitric acid. When a piece of iron is placed first in concentrated nitric acid and then in dilute nitric acid, no appreciable corrosion occurs. A thin adherent *passive* film of iron oxide is produced in the concentrated acid. If the sample is then scratched, the iron corrodes rapidly in the dilute acid because of the rupture of the film. If the sample is not immersed first in concentrated acid, both the initial and the continuing corrosion are rapid in the dilute acid.

The importance of the experiment is that *self-repairing passive films* are formed when more than 10–13% chromium is present in the iron. In general, iron, chromium, nickel, titanium, and aluminum alloys can form passive layers.

We can also explain such *passivation* with the aid of a polarization curve obtained by measuring cell current for a material as a function of voltage. A material that passivates exhibits the anode polarization curve shown in Fig-

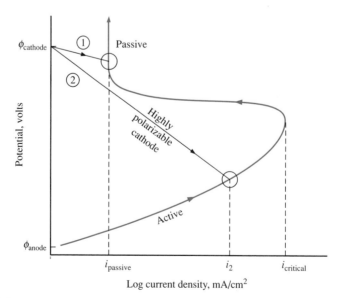

FIGURE 18.8 Polarization curve of a passive metal. High cathode polarization results in a high corrosion rate i_2; low cathode polarization maintains a passive anode and a lower corrosion rate $(i_{passive})$.

ure 18.8. As the anode is driven to higher current densities (generally by exposure to an oxidizing acid), it reaches a maximum value called $i_{critical}$. Upon reaching this value, the anode forms a protective layer, and its corrosion decreases to a value called $i_{passive}$. It is important to realize that if the initial corrosion rate is not great enough, $i_{critical}$ will not be exceeded, and this can cause some peculiar effects. For example, stainless steel corrodes more readily in a weak oxidizing acid ($i < i_{critical}$) than in a concentrated oxidizing acid (i falls to $i_{passive}$) after initially reaching $i_{critical}$.

The polarization of the cathode can determine whether the anode reaches the passive condition. In Figure 18.8, for instance, two possible *cathode* polarization curves have been superimposed on the anode polarization curve. In the case of cathode 1, the intersection with the anode is in the passive region. Cathode 2 polarizes at a higher rate (such as through concentration polarization), and it intersects the anode polarization curve within the active region, giving a higher corrosion rate. Therefore, care must be exercised when coupling passive metals to highly polarizable cathodes. Otherwise passivation may be destroyed and a high corrosion rate obtained.

CORROSION PHENOMENA

18.11 Types of Corrosion

The basic principles we have just discussed can explain practically any case of corrosion. However, there are a number of corrosion situations or phenomena that are encountered often, are given special names, and deserve careful analysis. These common cases also serve as good illustrations of how to apply the basic principles. The phenomena to be covered are

1. Corrosion units
2. Galvanic corrosion, macroscopic and microscopic cases
3. Selective leaching (dezincification, etc.)
4. Hydrogen damage
5. Oxygen-concentration cells, water-line attack
6. Pit and crevice corrosion
7. Combined mechanical-corrosive effects
   a. Stress corrosion
   b. Corrosion fatigue
   c. Liquid-velocity effects (corrosion erosion and cavitation)

18.12 Corrosion Units

In the standard corrosion tests, which we will discuss later, samples of a metal are measured and weighed before and after immersion for a given period of time. The following units are commonly used:

mdd: milligrams lost per square decimeter per day
ipy: inches corroded per year
mpy: mils corroded per year (1 mil = 0.001 in.)

The last two units are preferable because they enable us to visualize easily the long-range effect. These values apply to uniform corrosion. They may be used in design only when nonuniform corrosion, such as pit corrosion, is not present.

18.13 Galvanic Corrosion

In a general way all corrosion depends on galvanic action, but the term *galvanic corrosion* means specifically a type of corrosion that occurs because two materials of different solution potential are in contact. We shall consider galvanic effects on the microscopic scale after we discuss the macroscopic case.

Macroscopic Cases of Galvanic Corrosion

Perhaps the best illustration of the phenomenon of galvanic corrosion on a macroscopic scale was given by an inspired sculptor who produced a figure for a central plaza in New York City. He used a bronze body, an aluminum crown, an iron sword, and a stainless steel base. The result was a fine collection of galvanic cells, and the original form did not last long.

Many less spectacular cases of galvanic action exist. Often the more reactive member of the couple is deliberately used as a *sacrificial anode*. For the common case of galvanized steel or wire, the steel is coated with zinc either by being dipped into molten zinc or by electroplating. The effects shown in Table 18.4 were obtained in an experiment in which zinc and steel samples of equal size were tested separately for the same time period in the attached or coupled condition. Note that when uncoupled, both zinc and steel corrode in the solutions, but when coupled, the zinc protects the steel, corroding at an accelerated rate as a sacrificial anode.

Another common couple is found in the "tin" can, which is composed of steel covered with a thin layer of tin. When a cut section is allowed to cor-

TABLE 18.4   Weight Loss of Zinc and Steel in the
Uncoupled and Coupled Conditions

| | Weighted Change of Each Sample, g | | | |
| | Uncoupled | | Coupled | |
Solution	Zinc	Steel	Zinc	Steel
0.05 molar Na_2SO_4	−0.17	−0.15	−0.48	+0.01
0.05 molar NaCl	−0.15	−0.15	−0.44	+0.01
0.005 molar NaCl	−0.06	−0.10	−0.13	+0.02

rode, it is usually the steel rather than the tin that corrodes. This confirms a prediction that could be made from Table 18.2, because iron has a greater solution potential. This situation is found as long as the electrolyte is oxidizing. However, oxygen is usually deficient inside a tin can to prevent food spoilage. It is believed that under these conditions, tin and iron exchange galvanic series positions, and the tin becomes anodic as a result of the formation of complexes. On a scratched surface the tin becomes sacrificial and protects the steel. The tin is nontoxic when it goes into solution, and the corrosion rate is quite low.

The question might also be asked: Why aren't cans made of galvanized steel? The answer is that although the zinc would protect the steel, the zinc ions would be present in the food at an undesirable level. In the manufacture of tin cans, great care is taken to avoid contact between the steel and the food (through the use of soldered joints and lacquering). Therefore, the corrosion rate is only that of tin or lacquer, which is very low when the coating is not ruptured.

As another example of galvanic corrosion, rivets of a material different from the basic structure are often used, especially as an expedient in repair. When copper rivets are used in steel sheet, there is a *large anode area* (the steel), and the galvanic action is not serious. However, when steel rivets are used in a copper sheet, all the metal loss is concentrated in a small anodic region, and there is a large area for cathode reactions to absorb electrons. The corrosion is catastrophic. Figure 18.9a shows the effect of the ratio of anode area to cathode area on the corrosion rate (current) of just such an iron–copper system. Therefore, galvanic coupling is to be avoided, especially in systems exposed to aqueous solutions. Figure 18.9b shows another example, in which a brass pipe and a cast-iron fitting were used as part of a steam condensate trap. Because the iron is anodic, it has corroded almost completely away.

Microscopic Cases of Galvanic Corrosion

Suppose that we have a sample of an alloy with several phases present. Won't these act like small galvanic couples and corrode? As a matter of fact, this effect is the reason we can distinguish different phases in the microstructure. Usually there is not much to be seen in the as-polished condition; it is only after etching that the details of structure are visible. There are many commercial cases in which this behavior is important. We shall now discuss some examples.

Weld Decay of 18% Cr, 8% Ni Stainless Steel Type 304 stainless steel contains 18% Cr, 8% Ni, and 0.08% C maximum. It is usually delivered in the single-phase austenitic structure obtained by rapid cooling from elevated temperatures. The $M_{23}C_6$* phase can precipitate if the steel is reheated in

---

*$M_{23}C_6$ is an iron–chromium carbide that can contain up to 30 wt % iron. With 30 wt % Fe, the chromium content is in excess of 60 wt % for the indicated stoichiometry.

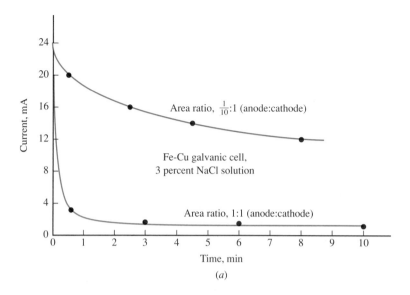

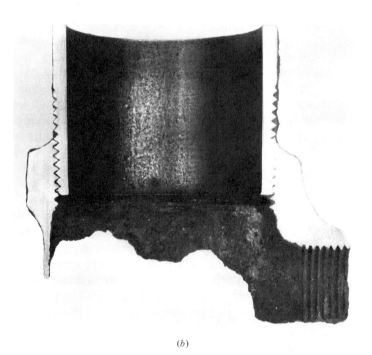

FIGURE 18.9   (*a*) Rate of corrosion (current) related to ratio of anode area to cathode area (25°C, 77°F). (Upper curve shows higher corrosion with large cathode and small anode.) (*b*) Condensate trap of iron fitting in brass pipe (cross section), with iron fitting badly corroded.

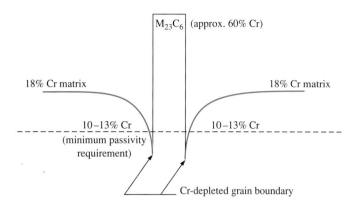

FIGURE 18.10   The lowering of the amount of chromium in a stainless steel below the critical amount required for passivity. Corrosion along the grain boundary results when chromium carbides precipitate.

the two-phase field, however. During welding the weld zone is heated to the liquid state and cools rapidly enough to avoid carbide precipitation. However, there is a region adjacent to the weld that has been heated just enough to precipitate $M_{23}C_6$. This usually takes place at grain boundaries. Because of the high chromium content of the carbide, the nearby regions are impoverished of chromium as the carbide is formed. The chromium level falls below 10% in these regions near a grain boundary. Hence the low-chromium regions are not passive ($<$10–13% Cr), whereas the remainder of the matrix is passive, as shown in Figure 18.10. The result is galvanic action called *weld decay* between the grain boundary region and the higher-chromium regions within the grain. It should be emphasized that it is not the $M_{23}C_6$ particles that are corroded, but the low-chromium–iron matrix. The $M_{23}C_6$ particles can be recovered after corrosion. Under the electron microscope they appear as plate-like crystals. This type of corrosion can be avoided by using a solution heat treatment at 2000°F (1093°C) after welding.

We can also minimize this type of grain boundary corrosion by using 18% Cr, 8% Ni steels to which a stronger carbide former than chromium (for example, titanium or niobium) is added or by specifying a very low-carbon level, as indicated in Table 18.5.

TABLE 18.5   Chemical Compositions of Several Stainless Steels

	Chemical Analysis, percent			
Stainless Steel	*C*	*Cr*	*Ni*	*Other*
Type 304	0.08 max.	18	8	
321	0.08 max.	18	8	Ti = 5 × percent C
347	0.08 max.	18	8	Nb = 10 × percent C
304L	0.03 max.	18	8	

The use of the special stainless steels is not always a safeguard against galvanic action, because chromium carbide may still form in preference to other carbides in a certain temperature range. The special case called "knife-line attack" can be encountered in type 347. All carbides are dissolved in the weld, and niobium carbide then precipitates in the range 2250–1450°F (1232–788°C) upon cooling. However, if the material at the edge of the weld is cooled too rapidly in the range for niobium carbide formation but more slowly from 1450–950°F (788–510°C), complex carbides containing chromium can precipitate, again forming an active–passive condition. The weld corrodes in a narrow region where this time-temperature relationship is present, which gives rise to the name "knife line."

Plain-Carbon Steel   Cases of galvanic action are not limited to the high-alloy steels. An interesting case in plain-carbon steel piping is called *ring-worm corrosion.* This selective attack takes place near the end of the pipe that has been *especially heated* to forge the flange portion. This treatment results in a spheroidized iron carbide structure with a different solution potential than the untreated balance of the pipe. A circular-shaped area of attack occurs near the junction of the two regions.

Aluminum Alloys   There is a marked difference between the behavior of pure aluminum and some of the age-hardened aluminum alloys, particularly those containing copper, in corrosive liquids such as salt water. In a 4% copper alloy, for example, the potential after solution heat treatment is uniform (+0.69 volt). After aging, however, there is a change in potential near the higher copper $CuAl_2$ particles that precipitate. The potential of the matrix near the grain boundaries when the precipitation is heavy is +0.78 volt, whereas the potential of the balance of the structure remains at +0.69 volt. In this situation the alloy is subject to *intergranular corrosion.* For protection a pure aluminum coating to produce "Alclad" is used.

Single-Phase Alloys   Even in single-phase alloys, galvanic effects develop. A common source of potential difference is segregation during solidification (Chapter 5). Figure 18.11 shows an as-cast cupronickel alloy. By reheating the alloy to below the solidus, allowing diffusion to take place, and hot rolling, the material can be homogenized.

In a homogeneous alloy the material at the grain boundaries is at a high solution potential because the concentration of dislocations leads to higher strain energy. After cold working, the material tends to corrode more than annealed material. Figure 18.12 shows an interesting illustration of the use of this effect in criminal investigation. A series of identification numbers was stamped on an engine part. Next the identification was ground away. However, upon etching of the ground surface, the numbers reappeared. The reason is that the metal is cold-worked below the visible bottom of the impression. It therefore still etches differentially after the impression is ground off. To avoid this action in the corrosion of cold-worked parts, at least partial annealing is necessary (recovery).

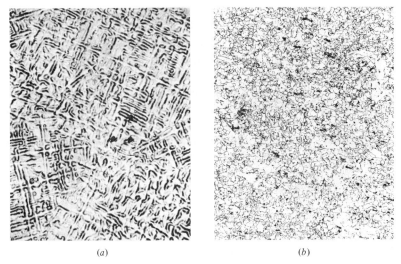

FIGURE 18.11 (*a*) 70% Cu, 30% Ni alloy as-cast, dendritic structure; 100×, chromate etch. (*b*) Same as (*a*), but after rolling and homogenization at 1700°F (927°C); 100×, chromate etch.

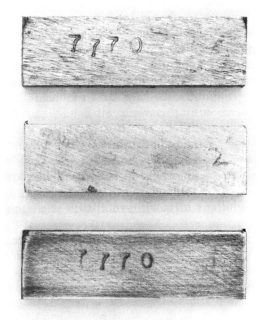

FIGURE 18.12 Process by which one can reveal stamped numbers after they have been ground away. Top, numbers are stamped on a bar of metal. Middle, numbers are ground away. Bottom, after 10 min in boiling 50% sulfuric acid, the numbers reappear because the severely cold-worked region beneath the stamped number etches more rapidly. This illustrates the higher solution potential of cold-worked metal.

18.14 Selective Leaching (Dezincification, etc.)

Selective leaching is a term that covers a number of classifications, such as dezincification and dealuminization. The case of brass illustrates this type of failure.

After exposure to fresh or salt water, brass may develop copper-colored layers or plugs of material (Figure 18.13a and b). We find that these are in fact spongy low-strength regions of copper from which the zinc has leached out. The usually accepted mechanism is that the brass dissolves slightly, then the copper ions are displaced by more zinc going into solution, and the copper plates out. This is obviously a dangerous phenomenon when spongy plugs of copper form in a pipe under pressure. To avoid the problem, the zinc content of the pipe should be lowered. The most sensitive alloys contain 40% zinc and have a second phase (β) that aggravates the problem. In severe cases, in which the 30% zinc alloys are troublesome, the addition of tin and arsenic (1% Sn, 0.04% As) or reduction of the zinc content to below 20% alleviates the problem.

Another example of selective leaching can be seen in gray cast-iron pipe in which the iron matrix may be slowly leached away, leaving the insoluble graphite behind. This is called *graphitic corrosion* or, improperly, *graphitization*. (It is the matrix and not the graphite that corrodes!) It is usually a very slow effect, and pipe showing this condition has been in service for centuries. On the other hand, it can occur in a few years under extreme soil conditions, such as where cinders have been used for backfill. Figure 18.13c shows the microscopic appearance of this type of corrosion.

Selective leaching of aluminum from copper alloys and of cobalt from a cobalt–tungsten–chromium alloy have also been encountered.

18.15 Hydrogen Damage

In discussing the basic movements of atoms in corrosion, we said that hydrogen discharged at the cathode could form bubbles *or* dissolve in the metal and diffuse through it. There are cases involving either hydrogen evolution from applied currents, as in electroplating, or hydrogen evolution from corrosion in acid solution. In *hydrogen blistering* the atomic hydrogen (H) diffuses through the metal and, finding a void, diffuses into it. It then forms molecular hydrogen (H_2) and exerts high pressure, which tends to spread out the void. The pressure of molecular hydrogen in equilibrium with atomic hydrogen is more than 10^5 atm $(1.033 \times 10^6 \text{ kg/m}^2)$. In *hydrogen embrittlement* the dissolved interstitial atoms lead to low ductility and low impact strength, perhaps because of interaction with microscopic cracks. These effects can be minimized by baking the metal part for a long time to remove hydrogen by diffusion and by avoiding couples that produce it.

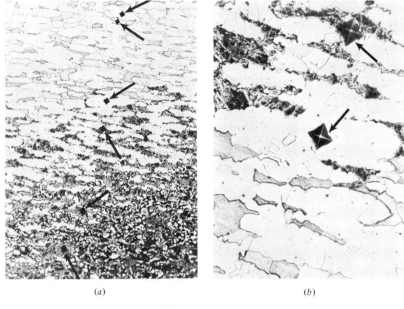

(a)

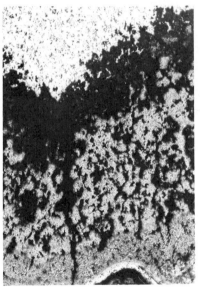

(c)

FIGURE 18.13 (a) 60% Cu, 40% Zn brass. The light portions are FCC α phase, HV 100. The gray portions are BCC β phase, HV 130. The dark portions are spongy copper, HV 52; 100×, chromate etch. (b) 60% Cu, 40% Zn brass. The light portion is FCC α. Gray portion is BCC β phase. Spongy copper is dark. 500×, chromate etch. Note that corrosion takes place in the β regions (which have higher zinc content and are more active). The selective leaching is called dezincification. (c) Gray cast-iron pipe showing graphitic corrosion. The original diameter of pipe included all the black and gray corrosion product in the lower two-thirds of the photomicrograph. Only a spongy mass of graphite and some silicate remain. The white area at the top is unattacked gray iron, in which graphite flakes can be seen. 50×, unetched.

18.16 Oxygen-Concentration Cells; Water-Line Corrosion

In contrast to the galvanic effects just discussed, corrosion can appear to occur without any starting potential. For example, a drop of water corrodes a polished iron surface, and a homogenized steel tank corrodes at the water line. To explain these cases, let us refer to the following equations shown in the emf series in Table 18.2:

$$O_2 + 2H_2O + 4e \longrightarrow 4OH^- \qquad +0.401 \text{ volt} \quad \text{(note reversed sign)}$$

$$Fe \longrightarrow Fe^{2+} + 2e \qquad +0.440 \text{ volt}$$

We see, therefore, that in an *oxygen-concentration cell* involving solution of iron at the anode and discharge of electrons by the formation of OH⁻ at the cathode, a potential of 0.841 volt exists under standard conditions.

In the apparently innocuous case of the drop of water, we can predict the following sequence of events: Some iron dissolves in the water, and oxygen and water react to absorb the electrons. The oxygen in the water can be replaced rapidly only at the edge of the drop. This region of high oxygen concentration develops as the cathode (Figure 18.14). Iron dissolves inside the drop, but more goes into solution closer to the edge regions than at the center. This is because there is a higher resistance to electron travel from the center; that is, formation of anodes and cathodes is favored when they are close together. A ring of rust is formed at the edge of the drop because the iron ions migrate faster than the hydroxyl ions. The ferrous hydroxide is oxidized to hydrated ferric hydroxide (rust) in a secondary reaction.

The case of water-line corrosion is similar. We find that a tank that is kept only partly filled with water corrodes more rapidly than a completely filled one. Corrosion takes place at the water line because of the gradient in oxygen concentration that develops between the solution at the water surface and that at some depth. The greatest attack is *just below* the water line, where there is a pronounced dropoff in oxygen concentration as well as a short electron path to the water line, where hydroxyl ions are formed.

There are, of course, other examples wherein oxygen affects the corrosion rate. Home hot-water-heating systems are usually "closed systems." The initial oxygen inside the system is very rapidly used up, and the corrosion rate becomes very low because of the low amount of oxygen. On the other hand, if

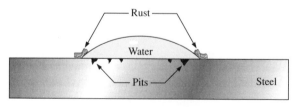

FIGURE 18.14   Corrosion of steel by a drop of water

an "open system" is used (fresh water is continuously added), the high amount of dissolved oxygen results in a continued high corrosion rate. Another example is the stains on the bottom of cooking utensils from "cooked-on" food, which are usually the result of concentration cells. Finally, the liquid under barnacles on a ship's hull develops a different concentration of ions from normal salt water. Hence a concentration corrosion cell forms.

EXAMPLE 18.3 *[EJ]*

Allowing water to dry on stainless steel usually results in "water spots." How do these spots occur?

Answer This is an oxygen-concentration-cell effect that can be increased by the presence of other ions. When a large area of water, say several inches in diameter, is observed during evaporation, it will be evident that the spot occurs in the very last portion to evaporate. This may be explained by the higher concentration of oxygen and other ions in the last liquid. (Dissolved salts and gases remain in the liquid. Because the volume of liquid becomes smaller, these concentrations become greater.)

18.17　Pit and Crevice Corrosion

In many applications it is *pit corrosion* and *crevice corrosion*, rather than the overall corrosion rate, that dictate the choice of materials. There is little to be said for owning a gasoline storage tank that is 99.9% intact but has numerous pits that have penetrated to the outside. Such an unhappy situation can be avoided by understanding the causes of pitting and realizing that certain well-known combinations of materials and environments are prone to this phenomenon.

　　Until recently the formation of a pit was considered merely a special case of an oxygen-concentration cell. However, this explanation did not account for the important role of ions such as chlorides. The most recent concept is that a pit begins at a surface discontinuity such as an inclusion or grinding mark. An oxygen-concentration cell develops between the discontinuity and the surrounding material. The chloride ions are involved as follows: Within the incipient pit, positive metal ions dissolve and accumulate. These attract chloride ions. The metal chloride concentration begins to build up in the pit. If the metal chloride is iron chloride, for example, this hydrolyzes to give HCl as follows:

$$M^+Cl^- + H_2O \longrightarrow MOH + H^+Cl^- \tag{18-24}$$

The combination of chloride and hydrogen ions then accelerates the attack. In proof of this mechanism, it has been found that the fluid within crevices exposed to an overall neutral dilute sodium chloride solution contains 3 to 10 times as much chloride ion as the bulk solution and has a pH of 4 rather than 7.

Crevice corrosion has the same mechanism as pit corrosion; the crevice serves as a ready-made pit in which the oxygen concentration is low.

To combat pitting, the most important point is to avoid combinations of materials and environments known to be susceptible. For example, many materials with passive surfaces exhibit pitting because of the large difference in potential between the passive and active regions of the pit. Furthermore, chloride ions are known to destroy passive layers locally. Among the stainless materials, the following sequence, from the most to the least susceptible, is found for salt water:

Type 304 stainless steel (18% Cr, 8% Ni)
Type 316 stainless steel (18% Cr, 8% Ni, 2% Mo)
Titanium

It is also interesting that titanium forms a very stable passive layer under oxidizing conditions. Therefore, it has excellent corrosion resistance and is extensively used in the chemical industry.

Finally, resistance to pit corrosion is not easily evaluated, because such corrosion is often the initial form of uniform corrosion. However, average pit depths can be measured. Corrosion products tend to cover the pit so the pit area must be carefully cleaned prior to measurement. (Interestingly, the pit base will appear to be clean because this is where corrosion is taking place.) Large specimens should be used; there is a higher statistical probability that a pit of a specific depth will occur in larger areas. It is important that results for average pit depth not be confused with average weight loss in uniform corrosion. It is also observed that a pit exudes its electrolyte and corrosion products and breaks down the passivity in the direction of the gravitational influence. This is why pits often appear in a line whose orientation is determined by gravity.

18.18 Combined Mechanical-Corrosive Effects

In many cases a component fails because of the combined effect of mechanical or hydraulic factors and corrosion. Such cases are of three types: stress corrosion, corrosion fatigue, and liquid-velocity effects (corrosion erosion and cavitation).

The typical test for *stress corrosion* is to take a sample of metal in the form of a small beam, apply a permanent bending load, and place the beam in a corrosive liquid. Failure occurs far more rapidly than it would in an unstressed part. Another testing method is to take a deformed sample such as a stamping and observe the onset of cracking. Classic examples are the *season cracking* of drawn-brass cartridge cases in the tropics and the *caustic embrittlement* of steel in boilers. In both of these cases we have a highly deformed structure with high residual stress and a hostile environment. In the case of the brass cartridge cases, the cracking is associated with the presence of ammonia from decaying organic matter plus the high humidity of the tropics. In the boiler tubes the elastic strain is produced by the cold rolling of the ends, and the caustic is present in the liquid.

Stress corrosion is often accompanied by an intergranular fracture in which the material shows no ductility, even though tensile test results may indicate a high plastic elongation capability. Although many examples are due to residual stresses from processing, the stresses can also result from the service conditions. Thermally induced stresses resulting from different coefficients of expansion have caused stress-corrosion failure in some bimetal plumbing fixtures.

Because of the importance of stress-corrosion cracking, we shall treat it more extensively in Chapter 19.

EXAMPLE 18.4 *[EJ]*

Where does corrosion normally occur first in a car bumper?

Answer   The portions that have been cold-worked the most are usually the first to corrode. Just as on a microscopic scale, grain boundaries are *anodic* to the surrounding grain, the macroscopic area of high residual stress is anodic to the nonworked portion. Of course, a corrosive medium such as salt spray from roads is necessary to cause the corrosion. Chromium (cathodic to steel) does not offer any galvanic protection for the steel. Once ruptured, it makes the problem worse because of the large cathode-to-anode-area relationship.

If the chromium plate cracks or breaks, giving the corrosive solution access to the steel, the corrosion rate is higher than with steel alone. The entire chromium surface serves as a large cathode.

In Chapter 2 we discussed fatigue failures, or cyclic stress failures. We pointed out that such failures are very sensitive to surface conditions. Pits from corrosion cause a surface stress concentration that easily propagates as a crack under a cyclic stress; hence the name *corrosion fatigue.*

Normally an increase in liquid velocity is accompanied by an increase in corrosion rate. Therefore a spinning disk in water has a higher corrosion rate at its periphery. (High velocities of the water decrease concentration polarization.) A normal manifestation of liquid-velocity effects is a gouging out, or *erosion.* However, not all erosion results from high liquid velocities. A dripping action has also been known to "wear a hole" in materials such as copper sewer pipe, as shown in Figure 18.15. *Cavitation,* on the other hand, results from bubbles of vapor popping against a surface such as a ship propeller. Because of the propeller geometry and relatively constant propeller speed, these bubbles break at the same localized positions. Cavitation then also appears to wear holes in the surface where the bubbles constantly break.

FIGURE 18.15   Copper sewer pipe with holes eroded by dripping waste fluids. Copper has excellent resistance to corrosion, but here waste fluids that dripped and became concentrated upon standing finally penetrated the pipe.

CORROSIVE ENVIRONMENTS

18.19   General Concepts

The most important corrosion problems occur in three types of environments:

1. The atmosphere
2. Water, fresh and salt
3. Chemicals: acids, alkalis, salts

We shall discuss the types of reactions encountered in each case.

18.20   Atmospheric Corrosion

Although atmospheric corrosion is not spectacular, the cost of it is. The annual loss in the United States is more than $2 billion. After much experimentation, corrosion engineers have found three distinctly different corrosion rates in the atmospheres of industrial, marine, and rural environments.

The problems in the industrial environment arise from SO_2 in the atmosphere, which leads to H_2SO_4 and H_2SO_3. Salt and other contaminants from roads also lead to accelerated corrosion rates.

In marine atmospheres the chief problem is salt spray.

In rural atmospheres rain and dust cause the principal problems. By contrast, in the Atacama Desert in Chile, automobiles wrecked 50 years ago appear just as they did at the time of the accident.

Recently a change in atmosphere has increased the life of galvanized eave troughs. In the past, in highly industrialized areas it was not uncommon to have to replace galvanized eave troughs in as few as 5 years. Recent efforts to reduce air pollution have resulted in an extension of their life in these areas to more than 20 years, which compares favorably with their life span in a rural atmosphere.

In selecting a material for atmospheric exposure, the first decision is whether a shiny metallic surface such as that given by stainless steel is necessary. If not, there are two principal alternatives:

1. Use an alloy that forms a protective coating, such as a steel with small amounts of copper and nickel, which develops a brown, but adherent, surface. The extreme case is the use of copper alloys that develop an attractive green patina. Styles change, and although shiny metal surfaces used to be in fashion, muted surfaces have recently found favor.
2. Apply paint or plastic coatings.

18.21 Water

The attack from fresh water varies widely, depending on the dissolved salts and gases. The principal contaminants are chloride ions, sulfur compounds, iron compounds, and calcium salts. There is little difference in the effects of water on plain and low-alloy steels. Cast iron and ductile iron are widely used for water pipe. At critical junctions such as valves, the mating surfaces are usually specified as copper alloys. In general, in cases of dezincification or dealuminization, alloys with more than 80% copper are used. Monel, aluminum, some stainless steels, and cupronickel are also employed, depending on the application.

Sea water attacks ordinary steel and cast iron fairly rapidly, and protection by painting or a sacrificial anode is used. For example, ocean-going vessels have zinc sacrificial anodes bolted at intervals to the hull. In salt water, pitting is encountered in stainless steel, and brass with less than 80% copper may dezincify. Titanium has excellent resistance to salt water.

18.22 Chemical Corrosion

The petroleum and chemical industries have the most severe chemical corrosion problems. In the petroleum industry, salt water, sulfide, organic acids, and other contaminants accelerate corrosion. Stainless steel, Stellite (a cobalt-base alloy), and Monel are used. There are a few specific cases of the successful combating of chemical corrosion:

Corrosive Chemical	Resistant Material
Nitric acid	Stainless steels
Hot oxidizing solutions	Titanium
Caustic solutions	Nickel alloys
Concentrated sulfuric acid	Steel
Dilute sulfuric acid	Lead
Pure distilled water	Tin

To show the effects of concentration and temperature, Figure 18.16*a* lists materials used to resist sulfuric acid. When sulfur trioxide (SO_3) is dissolved in 100% H_2SO_4, the mixture is treated as equivalent to H_2SO_4 of concentrations in excess of 100%, as Figure 18.16*b* shows. Note that the temperature–concentration graph in part *b* is divided into 10 zones, corresponding to those in the list, of increasing severity of attack. As the attack becomes more severe, the number of materials that can be used to give a corrosion rate of less than 0.020 in./yr dwindles rapidly until only gold, glass, and platinum are left. The behavior of type 316 stainless steel is interesting. It can resist higher acid concentrations better than it can resist intermediate concentrations. This is related to passivation, whereby a high concentration of oxidizing acid is required to form a *stable* passive layer. Passivity is not achieved at low concentrations, so the corrosion rate is higher.

Materials in Designated Zones Having Reported Corrosion Rates Less than 20 MPY

	Zone 1
10% aluminum bronze (air-free)	Gold
Glass	Platinum
Lead	Silver
Copper (air-free)	Zirconium
Monel (air-free)	Tungsten
Rubber (up to 170°F)	Molybdenum
Impervious graphite	Type 316 stainless steel (up to 10% H_2SO_4,
Tantalum	aerated)

	Zone 2
Glass	Tantalum
Silicon iron	Gold
Lead	Platinum
Copper (air-free)	Silver
Monel (air-free)	Zirconium
Rubber (up to 170°F)	Tungsten
10% aluminum bronze (air-free)	Molybdenum
Austenitic ductile iron	Type 316 stainless steel (up to 25% H_2SO_4
Impervious graphite	at 75°F, aerated)

FIGURE 18.16   (*a*) Materials used to resist sulfuric acid

	Zone 3
Glass	Tantalum
Silicon iron	Gold
Lead	Platinum
Monel (air-free)	Zirconium
Impervious graphite	Molybdenum

	Zone 4
Steel	Impervious graphite (up to 96% H_2SO_4)
Glass	Tantalum
Silicon iron	Gold
Lead (up to 96% H_2SO_4)	Platinum
Austenitic ductile iron	Zirconium
Type 316 stainless steel (above 80% H_2SO_4)	Hastalloy B (Ni base plus Mo, Fe, and Co)

	Zone 5
Glass	Tantalum
Silicon iron	Gold
Lead (up to 175°F and 96% H_2SO_4)	Platinum
Impervious graphite (up to 175°F and 96% H_2SO_4)	Hastalloy B

	Zone 6
Glass	Gold
Silicon iron	Platinum
Tantalum	Hastalloy B

	Zone 7
Glass	Gold
Silicon iron	Platinum
Tantalum	

	Zone 8
Glass	Gold
Steel	Platinum
18% Cr, 8% Ni stainless steel	

	Zone 9
Glass	Gold
18% Cr, 8% Ni stainless steel	Platinum

	Zone 10
Glass	Platinum
Gold	

FIGURE 18.16   (*a*) (Continued)

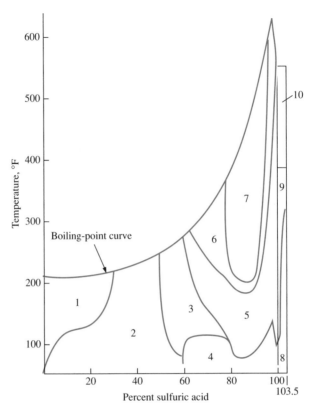

FIGURE 18.16(b)   Resistance to corrosion of materials exposed to sulfuric acid
[M. B. Fontana and N. D. Green, *Corrosion Engineering*, McGraw-Hill, New York, 1967; used with permission.]

CORROSION OF CERAMICS AND PLASTICS

18.23 General Concepts

Ceramics and plastics, often in conjunction with metals, offer alternative solutions to corrosion problems. The main drawback of ceramics is their lack of
ductility. For plastics, the main drawback is the effects of low and elevated
temperature and in some cases water absorption and flammability. On the
other hand, none of the effects associated with metal corrosion, such as pit-

ting and galvanic action, is encountered when plastics or ceramics are used. In many cases plastic parts can be used to insulate metal parts from corrosive interaction, as long as care is taken to ensure that oxygen-concentration cells do not form under the plastic. We shall consider the resistance of ceramics and plastics to the same group of corrosive environments that attack metals: atmosphere, water, and chemicals.

An important general rule in predicting the performance of an organic material in a solvent is that *like dissolves like*. Structures that contain a polar group such as OH, C≡N, or COOH are attacked or swelled by polar solvents such as acetone, alcohol, and water. (Remember that a polar group is one in which there is nonsymmetry of atoms or electrons, which results in an attractive force between molecules.) These polar groups resist dissolution by the balanced nonpolar solvents, such as carbon tetrachloride, gasoline, and benzene.

On the other hand, polymers with nonpolar groups such as methyl (CH_3) and phenyl (C_6H_5) are resistant to polar solvents, but they are swollen or dissolved by nonpolar solvents.

Furthermore, straight-chain (aliphatic) polymers tend to dissolve in straight-chain solvents such as ethyl alcohol, whereas those with benzene rings (aromatic polymers) tend to dissolve in aromatic solvents such as benzene. As molecular weight increases, solubility decreases. Also, with greater crystallinity and consequently denser packing and stronger intermolecular forces, solubility decreases.

18.24 Atmosphere

Ceramics are only slowly affected by the atmosphere, as is evident from the many structures made of brick and cement that have stood for centuries. The principal dangers in weathering are the effects of water entering cracks or joints and expanding upon freezing. Salt in water aggravates this problem.

Air pollution has caused concern in many European cities with rich collections of ceramic structures and statuary. It is believed that acids in the atmosphere have caused more rapid degradation of these ceramics in the past twenty-five years than in all the previous centuries. You may recall that specific types of sulfate-resistant Portland cements (Chapter 17) have been recommended for use under conditions of exposure to sulfuric acid. Unfortunately, such materials were not available to earlier civilizations.

Plastics are affected only slowly by the atmosphere, but particularly by sunlight. Table 18.6 shows the wide differences in this effect on various plastics. Polymers are susceptible to crosslinking with oxygen and loss of the plasticizer, especially if they are externally plasticized. This leads to stress cracks, or crazing, often observed in polymer coatings (paints and varnishes). The effect is accelerated by UV radiation from sunlight. Engineers

TABLE 18.6 Corrosion Resistance of Plastics*

Material	Acids		Alkalis		Organic Solvents	Water Absorption, percent/24 hr	Oxygen and Ozone	High Vacuum	Ionizing Radiation	Temperature Resistance, °F	
	Weak	Strong	Weak	Strong						High	Low
Thermoplastics											
Fluorocarbons	Inert	Inert	Inert	Inert	Inert ·	0.0	Inert		P	550	G, 275
Polymethylmethacrylate	R	A-O	R	A	A	0.2	R	Decomp.	P	180	G, 70
Nylon	G	A	R	R	R	1.5	SA		F	300	G
Polyether (chlorinated)	R	A-O	R	R	G	0.01	R			280	G
Polyethylene (low-density)	R	A-O	R	R	G	0.15	A	F	F	140	G, 80
Polyethylene (high-density)	R	A-O	R	R	G	0.1	A	F	G	160	G, 100
Polypropylene	R	A-O	R	R	R	<0.01	A	F	G	300	P
Polystyrene	R	A-O	R	R	A	0.04	SA	P	G	160	P
Rigid polyvinylchloride	R	R	R	R	A	0.10	R		P	150	P
Vinyls (chloride)	R	R	R	R	A	0.45	R	P	P	160	P
Thermosetting plastics											
Epoxy (cast)	R	SA	R	R	G	0.1	SA		G	400	L
Phenolics	SA	A	SA	A	SA	0.6			G	400	L
Polyesters	SA	A	A	A	SA	0.2	A		G	350	L
Silicones	SA	SA	SA	SA	A	0.15	R		F	550	L
Ureas	A	A	A	A	R	0.6	A		P	170	L

*Abbreviations: R = resistant, A = attacked, SA = slight attack, A-O = attacked by oxidizing acids, G = good, F = fair, P = poor, L = little change, decomp. = decomposes

From M. B. Fontana and N. D. Green, *Corrosion Engineering*, McGraw-Hill, New York, 1967. Used with permission.

and polymer chemists have developed plastic components with excellent resistance to atmospheric corrosion. Though these have many legitimate applications, environmentalists are concerned because unnecessary and inappropriate use of polymers creates "plastic trash," which is not readily biodegradable and will remain for many years.

18.25　Water

Nonporous ceramics are widely used for containers and piping. Glass-lined and enameled-steel tanks have been used for many years with great success.

Plastics are also generally resistant to water; they make excellent protective coatings (see Chapter 14). However, there is a small percentage of water absorption in all cases except for Teflon, polyethylene, and polypropylene. This absorption can aggravate problems such as creep in fibers that are used for clothing; thus our clothes get wrinkled in summer weather.

18.26　Chemical Corrosion

There are wide differences in the resistance of ceramics to chemicals. Among the glasses, pure silica and borosilicate are very resistant, but the soda-lime glasses are slowly attacked by alkalis. Basic refractories such as magnesia are attacked by acids. Organic solvents have no effect on the typical ceramic.

Plastics also show a great deal of variation in their resistance to chemicals. Most are resistant to weak acids and alkalis. However, strong acids decompose cellulose acetate, and some oxidizing acids decompose melamines and phenol-formaldehyde. Strong alkalis and organic solvents also attack certain plastics. The most resistant plastics are Teflon, polyethylene, and vinyl.

Sometimes the corrosive media come from unexpected sources. As an example, we often have considerable difficulty trying to maintain eyeglass frames. A combination of body fluids, exposure to airborne hydrocarbons, and stress often results in failure in plastic eyeglass frames in 12 to 24 months, as shown in Figure 18.17.

18.27　Summary of Steps Used to Prevent Corrosion

In summary, the methods commonly used to inhibit corrosion are

1. Careful selection of materials, to prevent the formation of galvanic cells
2. Alteration of environment—temperature, velocity, ion concentration—and the use of inhibitors
3. Design changes, such as removing pockets that may hold corrosive fluids
4. Cathodic protection, which can be applied to almost any metal
5. Anodic protection, which can be used only for materials that passivate
6. Application of coatings—metallic, ceramic, or organic—including metallic coatings as sacrificial anodes and organic coatings such as paints

FIGURE 18.17 Plastic eyeglass frames that failed because of applied stress and corrosive media. Although thermoplastics generally have good corrosion resistance, the combination of stress and body fluids caused this failure at normal temperature.

Finally, as we pointed out at the beginning of the chapter, corrosion is seldom totally prevented. Economics plays a most important role, and a compromise — optimizing the combination of costs and required life span — is usually the best solution.

EXAMPLE 18.5 *[EJ]*

For the cases below, indicate the type of corrosion cells and state how the corrosion might be minimized.

A. Bolts and nails stored in a basement corrode.
B. A painted trash barrel corrodes on the inside.
C. A so-called mag wheel for an automobile corrodes between the aluminum hub and the steel rim.
D. Paint blisters form on an automobile fender.

Answers

A. The corrosion cells are localized galvanic cells accentuated by residual stress in the bolts and nails that resulted from cold forming and machining. The basement suggests dampness, and storage in a drier atmosphere would reduce corrosion. Bolts and nails are also available in zinc- or cadmium-plated forms that have increased corrosion resistance. In hardware stores, prepackaging in plastic containers excludes high-humidity atmospheres.

B. This is another example of localized galvanic action. One should exclude fluids from the barrel by covering it to keep out rain and by placing small holes in the bottom to drain off fluids. A corrosion-inhibiting paint would also decrease the corrosion rate.

C. The design of the mag wheel ignores a fundamental principle in corrosion control: Do not join dissimilar materials. The result of doing so in this case is a galvanic cell made worse by the environmental factor of the salt used for deicing roads. The remedy is to use only one material, to use a coating over the materials that will maintain their integrity over many years, or to electrically insulate the two materials.

D. Although the corrosion under the paint on the fender may have been initiated by a scratch or stone bruise, its continuation is the result of a concentration cell. The obvious solution is sanding and repainting. To prevent recurrence, we must completely remove the rust and make sure the paint has no pinhole porosity.

CORROSION IN GAS

18.28　General Concepts

A topic such as oxidation may seem out of place in the same chapter with corrosion. However, as we delve into the actual mechanisms, it will be evident that oxidation is closely related to other forms of corrosion, because in both processes ions and electrons must be transferred. We shall consider first the types of *scales*, or oxides, encountered, then how they form, next the rate of oxidation, and finally the scale-resistant alloys and special cases such as catastrophic oxidation and internal oxidation.

18.29　Types of Scales Formed

When a metal is exposed to air, for example, it is important to determine first whether the scale that forms occupies a smaller or a larger volume than the metal it came from. For instance, if we oxidize the outer layer of a piece of magnesium, the volume of the scale formed is less than the volume of metal from which it was formed. Magnesium oxidizes rapidly because cracks appear in the thin scale, giving ready access to the metal beneath. In the case of iron, however, the volume of the scale is greater than that of the parent metal. Therefore, the scale is protective if it is made adherent by small additions of copper.

EXAMPLE 18.6　*[ES]*

What is the ratio of oxide volume to metal volume for the oxidation of magnesium? (The specific gravity of Mg is 1.74, and that of MgO is 3.58.)

Answer   Assume that 100 g of Mg is oxidized to MgO:

$$Mg + \tfrac{1}{2}O_2 \longrightarrow MgO$$

$$\text{Volume of Mg} = \frac{100 \text{ g}}{\text{density of Mg}} = \frac{100 \text{ g}}{1.74 \text{ g/cm}^3} = 57.5 \text{ cm}^3$$

Letting x = the amount of MgO produced, we have

$$\frac{100 \text{ g}}{\text{mol. wt. Mg}} = \frac{x \text{ g}}{\text{mol. wt. MgO}}$$

$$x = \frac{100 \times 40.32 \text{ g}}{24.32} = 167 \text{ g}$$

$$\text{Volume of MgO} = \frac{167 \text{ g}}{\text{density of MgO}} = \frac{167 \text{ g}}{3.58 \text{ g/cm}^3} = 46.6 \text{ cm}^3$$

or 57.5 cm³ Mg produces 46.6 cm³ MgO. The ratio, then, is 46.6/57.5 = 0.810.

The ratio of the volume of the scale to the volume of the parent metal is called the *Pilling-Bedworth ratio.* It may be calculated from the formula

$$\text{P-B ratio} = \frac{Wd}{Dw} \tag{18-25}$$

where W = molecular weight of the oxide, d = density of the metal, D = density of the oxide, and w = molecular weight of the metal, because more than 1 mol may be required to achieve the stoichiometry (see Example 18.7).

When this ratio is much less than 1, the scale is nonprotective because it will crack. On the other hand, if the ratio is much greater than 1, the scale may crack off because the volume difference (P-B ratios greater than 2.3) is so great. This ratio, therefore, serves as a rough screening test to determine good or poor oxidation resistance. However, we must realize that other factors such as coefficient of expansion, melting point, vapor pressure, and high-temperature plasticity are also important.

The P-B ratio is not just an experimental observation but is based on the concept that ratios less than 1 result in tensile strains in the oxide scale. As pointed out in Chapter 10, oxides and ceramics in general do not possess high tensile strengths. Therefore scale cracking can occur. However, their compressive strengths are also not infinite, and large P-B ratios (greater than 2.3) can result in spalling off of the protective scale.

18.30 Mechanism of Scale Formation

Once a relatively adherent scale forms, how does it grow? There are two possibilities: The oxygen or oxygen ion can diffuse through the scale to the metal, *or* the metal or metal ion can diffuse through the scale to the surface and react with oxygen. This is a rather important distinction; in the first case the scale grows at the metal–oxide interface, whereas in the second it grows at the oxide–air interface. Wagner, who carried out early experiments on corrosion phenomena, suspected that because metal ions are generally smaller than oxygen ions, the growth occurs at the outer surface, and he performed in 1933 the experiment illustrated in Figure 18.18. Instead of oxidation with air, he chose the reaction of silver with liquid sulfur in a cell. Two weighed cakes of silver sulfide were placed on a specimen of silver, and then liquid sulfur was placed over the cakes. At a given temperature two reactions were possible. Ag^+ ions could travel through the sulfide and react at the upper surface with sulfur, or S^{2-} ions could migrate to the metal–sulfide interface. After the cell was disassembled, Wagner found that the upper cake had gained much more weight than the lower one, indicating that the metal ion migration was most important. The usual oxidation mechanism, therefore, is the migration of metal ions and electrons to the outer surface to react with oxygen. The mobility of electrons is usually an order of magnitude higher than that of the ions, so this mobility is usually not a limiting factor except under very special circumstances.

Oxide Defect Structures

As we will discuss later in the section on semiconductors, many oxide structures do not follow the exact chemical formula but may have an excess of one ion or the other. This leads to vacancies at certain lattice points which develop in order for the oxide to attain electroneutrality. The presence of vacancies increases the diffusion rate. However, if an element that reduces the vacancies is added, the oxidation rate is lowered. A rather startling case is that the addition of a small amount of lithium, an easily oxidized element, lowers the oxidation rate of nickel by occupying vacancies in the nickel oxide lattice. Anion and cation defects in oxides, discussed in Chapters 10 and 11, also increase diffusion rates.

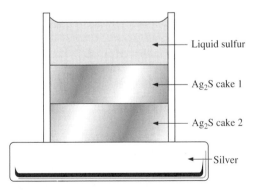

FIGURE 18.18   Diffusion experiment to show whether metal ions (Ag^+) or nonmetal anions (S^{2-}) diffuse more rapidly. The results can be compared to the formation of oxide scale in metal systems.

18.31 Oxidation Rates

When a scale is adherent, build-up usually follows the *parabolic law.* If the rate of oxidation is proportional to the thickness of the scale, we obtain the relation

$$W^2 = k_p t + C \qquad (18\text{-}26)$$

where W = weight of scale, t = time of exposure, and k_p and C = constants for the particular reaction.

If the scale cracks, as occurs at P-B ratios less than 1 or greater than 2.3, the rate of oxidation is *linear:*

$$W = kt \qquad (18\text{-}27)$$

In cases of fairly thick scale containing voids or when a limiting thickness is approached (such as aluminum in air), an empirical *logarithmic* relation is obeyed:

$$W = k \log(Ct + A) \qquad (18\text{-}28)$$

where k, C, and A are constants.

18.32 Scale-Resistant Materials

Scale-resistant materials are most needed at temperatures in the range of 800–2000°F (427–1093°C). Iron, nickel, and cobalt oxidize appreciably above 1000°F (538°C). The addition of chromium, silicon, and aluminum to iron leads to the formation of protective, adherent scale. Although these elements oxidize more readily than iron, they form new structures with iron, such as the spinel types discussed in Chapter 10. Even gray cast iron with 2% silicon has reasonable oxidation resistance.

18.33 Special Cases

The importance of forming the proper scale is illustrated especially well by the oxidation of pure molybdenum at elevated temperatures. Despite the excellent creep resistance, the volatility of MoO_3 leads to reduction of section thickness at a rapid rate. Similar problems are encountered with tungsten and niobium.

Certain alloys also oxidize very rapidly under special conditions. This phenomenon is called *catastrophic oxidation.* The most insidious type of catastrophic oxidation is the effect of small amounts of vanadium pentoxide or lead oxide in the gas phase. These oxides can combine with the normal metal scale to provide a low-melting-point phase in the scale. This results in a liquid that provides rapid transport of oxygen to the metal, and attack is accelerated.

Another interesting case is called *internal oxidation.* In this case the scale is more permeable to oxygen than to the metal ion. The oxygen dissolves in the metal, migrates, and forms oxide particles in the metal itself.

This causes some hardening, but the embrittling effect is usually undesirable in high-temperature alloys.

EXAMPLE 18.7 *[ES/E]*

Determine by calculation whether aluminum is more resistant to oxygen or to chlorine gas. The density of Al_2O_3 is 3.8 g/cm^3 and that of $AlCl_3$ is 2.44 g/cm^3.

$$2Al + \tfrac{3}{2}O_2 \longrightarrow Al_2O_3 \qquad Al + \tfrac{3}{2}Cl_2 \longrightarrow AlCl_3$$

Answer   The P-B ratio is equal to Wd/Dw.   (18-25)

$$\text{P-B ratio of } Al_2O_3 = \frac{(2 \times 26.98 + 3 \times 16.0)(2.699)}{(3.8)(2 \times 26.98)} = 1.34$$

We therefore expect aluminum in oxygen to show a parabolic rate of scale formation.

$$\text{P-B ratio of } AlCl_3 = \frac{(26.98 + 3 \times 35.46)(2.699)}{(2.44)(26.98)} = 5.47$$

Although compressive stresses result because the P-B ratio is greater than 1.0, they are too high to maintain without fracture (the P-B ratio is greater than 2.3). Hence we expect a linear rate of scale formation with chlorine.

Therefore, from these P-B ratio calculations, we know that aluminum is more resistant to an oxygen atmosphere than to a chlorine atmosphere.

We might also anticipate, in the absence of reaction between oxygen and chlorine or the formation of complexes with aluminum, that pre-oxidation of aluminum might form a scale that would be adherent and protective against chlorine and the formation of $AlCl_3$. In practice this might not offer complete protection, because some aluminum ions might migrate through the Al_2O_3 scale. However, the chlorination rate would be lower than it would be if the Al_2O_3 scale were not present.

RADIATION DAMAGE

18.34   General Concepts

Radiation damage to engineering materials can be due either to electromagnetic waves or to particles. The *electromagnetic* spectrum ranges from radio waves several meters in wavelength to γ rays with a wavelength of a fraction of a nanometer. Because the energy of the radiation is inversely proportional to the wavelength, the γ rays possess significantly greater energy than the radio waves. (A more extensive discussion of the physics of electromagnetic radiation is given in Chapter 20.)

A second variety of radiation is due to particles, most notably the neutrons as encountered in a nuclear reactor environment.

Both electromagnetic and particle radiation have the potential to modify the structure, and hence the properties and integrity, of the engineering materials. As we might expect, the amount of radiation damage depends on the energy of the radiation, the radiation density, and the inherent bond strength of the material. Therefore, weakly bonded thermoplastic polymers are more susceptible to radiation damage than strongly bonded metals.

In the following sections, we will consider a few examples of how the individual classes of materials react to a radiation environment.

18.35 Electromagnetic Radiation Damage

Although a number of the principles are deferred to Chapter 20 because they also apply to electrical and optical properties, a few general observations can be made here.

Electromagnetic radiation can be absorbed by a material or transmitted through it. Examples of transmission include that of visible light through glass and that of x rays or γ rays through materials for radiographic examination. Such transmission results in essentially no radiation damage to materials.

However, absorption or partial absorption can disrupt the electron or atom structure. As an example, excitation of orbital electrons by electromagnetic radiation (absorption) induces an instability that results in the release of energy when the electrons return to their unexcited state. When the released energy is within the visible spectrum in wavelength, we may note fluorescence and phosphorescence. Analogously, the removal of inner-shell electrons and their replacement by electrons from adjacent orbits produce characteristic x rays, as explained in Chapter 1.

Now let us take up the specific case of radiation damage of polymers by ultraviolet light. Outdoor exposure is more hostile to polymers than to metals and ceramics, because the energy of photons of *ultraviolet light* is sufficient to rupture the bonds within the molecules. This rupture causes changes in molecular weight, formation of cross links, and reaction with oxygen. These structural changes lead to gross physical changes such as chalking, cracking, surface embrittlement, and loss of tensile and impact strength.

Let us examine the relationship between the energy of the photons in different portions of the spectrum of sunlight and the bonds in polymers. The range of dissociation energies between atoms in polymer molecules is 70–100 Kcal/mol. Ultraviolet light in the range 2900–3200 Å has an energy range of 89–98 Kcal/mol, but it accounts for only 0.5% of the sunlight, even in tropical regions. Light in the range of 3000–3600 Å (79–89 Kcal/mol) accounts for up to 2.5% of the sunlight. Fortunately, then, only a small percentage of sunlight is active for bond rupture. However, there is concern that the amount of ultraviolet light reaching the earth will change if the ozone layer continues to be modified.

EXAMPLE 18.8 *[ES]*

Show that, as illustrated in Table 13.2, a photon of ultraviolet radiation of wavelength 2530 Å has enough energy to break an O—H bond of energy 110 Kcal/g · mol.

Answer   From the cornerstone of modern physics, we know that

$$E = h \times \nu \tag{18-29}$$

where h = Planck's constant and ν = frequency. This can be simplified because

$$\nu = \frac{c}{\lambda} \left(\frac{\text{velocity of light}}{\text{wavelength}} \right)$$

Therefore

$$E = h\nu = \frac{hc}{\lambda} \tag{18-30}$$

$$= \frac{(6.62 \times 10^{-34} \text{ J} \cdot \text{sec})(3 \times 10^{10} \text{ cm/sec})}{\lambda} = \frac{1.986 \times 10^{-23} \text{ J} \cdot \text{cm}}{\lambda}$$

Converting our units to electron volts and angstroms yields

$$E = \frac{1.986 \times 10^{-23} \text{ J} \cdot \text{cm}}{\lambda(1.6 \times 10^{-19} \text{ J/eV})} \left(\frac{1}{10^{-8} \text{ cm/Å}} \right) = \frac{1.24 \times 10^{4} \text{ eV} \cdot \text{Å}}{\lambda}$$

Then

$$E = \frac{1.24 \times 10^{4}}{2530 \text{ eV} \cdot \text{Å}} = 4.90 \frac{\text{eV}}{\text{photon}} \quad \text{or} \quad \frac{\text{eV}}{\text{atomic bond}^{\star}}$$

$$4.90 \frac{\text{eV}}{\text{atomic bond}^{\star}} \times \left(1.610^{-22} \frac{\text{KJ}}{\text{eV}} \right) \left(6.02 \times 10^{23} \frac{\text{atomic bonds}}{\text{mol}} \right) \left(\frac{1 \text{ Kcal}}{4.186 \text{ KJ}} \right)$$

$$= 113 \frac{\text{Kcal}}{\text{mol}}$$

Polymers can be divided into two groups: those that contain chromopores — atom groupings in the molecule that react with sunlight — and those that do not. The first group comprises benzene rings such as polyesters, polycarbonates, and polysulfones; the second is made up of aliphatic polymers such as

---

*We assume we have 1 mol of OH, or 6.02×10^{23} atomic bonds to break.

polyethylene and polyamide. The distinction is not sharp, however, because the second group contains traces of chromopores and foreign chemicals that also absorb sunlight. Degradation may occur more severely in the surface layers of the sunlight-absorbing polymer and is usually limited to the first 100–200 μm (4 to 8 thousandths of an inch). Two methods have been used to protect polymers from sunlight: incorporation of material into the polymer, as in fillers, and application of coatings.

Additions to the polymer are of two types: ultraviolet-absorbing pigments (such as carbon black or titanium dioxide) and materials that quench those molecular fragments that are in an excited state as a result of photon bombardment by combining with the excited molecules. *Coatings* have the advantage that they protect the polymer surface from any degradation and can add to appearance. A number of precautions are necessary, however. The coating must not result in harmful solution of the component that leads to crazing. Also, the application of a brittle coating to a ductile interior can lead to exterior crack initiation, followed by progression of the crack through to the interior. And when a baking cycle is needed for the coating, the effect on the base polymer must be evaluated.

18.36   Radiation Damage by Particles

Here we will limit our discussion to the effect of neutron bombardment of materials. In all cases irradiation with neutrons is damaging; however, the flux density and time of radiation play an important role.

When fast neutrons collide with atoms in metals, atom displacement from normal lattice positions can occur, resulting in vacancies that can then collect as dislocations. The effect on mechanical properties is similar to that

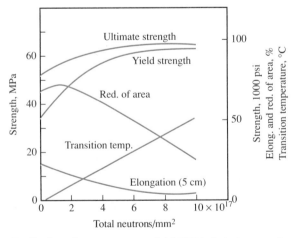

FIGURE 18.19   Radiation damage to steel (ASTM A-212-B Carbon–Silicon Steel) (Adapted from C. O. Smith, ORSORT, Oak Ridge, Tenn.)

of work hardening, as shown in Figure 18.19. The electrical and thermal conductivities are also decreased, as with cold working. Annealing will restore the properties, and as might be expected, irradiation at higher temperatures is less damaging because there is simultaneous annealing. In the event that the atomic nucleus should capture a neutron, the nucleus becomes unstable. It emits α, β, or γ radiation* and can become a radioactive isotope.

Ceramic materials exhibit both ionic and covalent bonding, and their resistance to radiation damage depends on the predominant type of bond. When a ceramic is highly ionically bonded, ion displacement and the creation of vacancies can occur much as they do in metals. This effect can be reversed by annealing, which is desirable because ion defects can lead to lower strengths, as pointed out in Chapters 10 and 11. When bonds in a ceramic are covalent, they can be irreversibly broken by radiation and result in complete modification in structure. Furthermore, high-temperature neutron radiation causes greater damage because the thermal excitation makes the covalent bond even weaker. Like electromagnetic radiation, then, neutron damage can be most severe in the polymers because of their covalent and van der Waals bonds. If the neutron flux is low and of short duration, there can be crosslinking and branching similar to those caused by exposure to x rays or γ rays. On the other hand, larger total neutron exposure of polymers can result in degradation.

The relative effects of neutron flux on engineering materials are summarized in Table 18.7.

Summary

The environment can cause failure of a component that is competently designed to resist service stresses. It causes such failure by three mechanisms: corrosion, oxidation, and radiation.

Corrosion takes place because the surface of a component goes into solution or undergoes another reaction with its environment. In the case of metals, each instance of corrosion can be studied as an electrochemical cell in which the metal goes into solution — that is, corrodes at the anode. For the process to continue, the electrons left in the solid at the point of solution must migrate through the structure and be consumed at the cathode. By far the most common reaction, that seen in the rusting of iron, is the reaction with water and dissolved oxygen to produce hydroxyl ions: $2H_2O + O_2 + 4e \rightarrow 4OH^-$. Rust is in fact ferric hydroxide, which may also be considered a hydrated ferric oxide.

A starting point in estimating the tendency of a metal or alloy to corrode is its half-cell potential, the voltage it develops when connected to a standard electrode. However, although this shows that elements such as aluminum and magnesium have a high tendency to corrode, it does not tell us anything

---

*α radiation: He^{2+} ions
β radiation: highly excited electrons

TABLE 18.7   Effects of Radiation on Various Materials*

10^{14}	Germanium transistor — loss of amplification
	Glass — coloring
10^{15}	Polytetrafluoroethylene — loss of tensile strength
	Polymethyl methacrylate and cellulosics — loss of tensile strength
	Water and least stable organic liquids — gassing
10^{16}	Natural and butyl rubber — loss of elasticity
	Organic liquids — gassing of the most stable ones
	Butyl rubber — large change, softening
10^{17}	Polyethylene — loss of tensile strength
	Mineral-filled phenolic polymer — loss of tensile strength
	Natural rubber — large change, hardening
10^{18}	Hydrocarbon oils — increase of viscosity
	Metals — most show appreciable increase in yield strength
	Carbon steel — reduction of notch-impact strength
10^{19}	Polystyrene — loss of tensile strength
	Ceramics — reduced thermal conductivity, density, crystallinity
	All plastics — unusable as structural materials
10^{20}	Carbon steels — severe loss of ductility, doubled yield strength
	Carbon steels — increased fracture-transition temperature
	Stainless-steels — yield strength tripled
10^{21}	Aluminum alloys — reduced but not greatly impaired ductility
	Stainless steels — reduced but not greatly impaired ductility

Integrated fast neutron flux, n · cm/cm³ (or nvt)

*Indicated exposure levels are approximate. Indicated changes are at least 10 percent. From C. O. Smith, *Nuclear Reactor Materials*, ORSORT, Oak Ridge, TN.

about the nature of the corrosion product, which is important. The test does-reveal something about the corrosion rate, which is determined by the current, or electron flow. If a passive or unreactive film forms, corrosion stops. A great many specialized terms have been coined in the study of corrosion, and practically all involve applications of simple electrochemical cell theory.

Ceramics and polymers corrode by dissolving, not by galvanic action. Ceramics are usually quite resistant to water, and special acid-resistant cements are available. Polymers vary in performance and, in general, dissolve in organic liquids of similar structure (like dissolves like). Most resist water and dilute chemicals, and some (such as Teflon) have outstanding resistance.

Oxidation of metals in gases is a specialized case of corrosion that is governed by a preference of the metal to form either a protective or a nonprotective scale. When the scale is in tension, it easily ruptures because of the inherently low tensile strength of the oxides.

Radiation can also modify the properties of materials via interaction between the energy source and the atomic structure. With electromagnetic energy such as ultraviolet radiation, polymeric materials can degrade because of the rupture of covalent bonds. High-energy neutrons can similarly cause damage to polymers but may increase the strength and decrease the ductility of metals. The effect in metals is analogous to cold working and is reversible by annealing.

Definitions

Activation polarization   Polarization that causes a decrease in cell potential because of rate-limiting intermediate steps necessary to complete the reaction.

Anode   The electrode of a cell at which metal goes into solution; the negative pole in the external circuit.

Anodic control   Polarization at the anode that results in a lower corrosion rate. (*See also* Passivation.)

Cathode   The electrode of a cell at which metal ions are plated out or negative ions are created, as in the reactions $2H_2O + O_2 + 4e \rightarrow 4OH^-$ and $Cu^2 + 2e \rightarrow Cu^0$.

Cathodic control   Polarization at the cathode that decreases the corrosion rate.

Cathodic protection   Plating or attachment of a more active metal that dissolves, preventing solution of the component to be protected.

Cavitation   Stress corrosion induced by the local collapse of vapor bubbles against a surface such as a propeller in sea water.

Cell potential   The electromotive force (emf) developed by a cell.

Concentration polarization   Polarization that causes a decrease in the emf of a cell because of the build-up of ions around the electrodes.

Corrosion   The solution or harmful reaction of the structure of a material with its environment.

Corrosion fatigue   Fatigue failure aggravated by corrosion.

Current density   The current flowing during corrosion divided by the area of the cathode. Large cathodes (small anodes) cause rapid corrosion rates.

Erosion   The wearing away of a surface because of the combination of mechanical and corrosive effects.

Galvanic corrosion   Corrosion that is accelerated by the presence of electrically connected dissimilar metals.

Graphitic corrosion   Selective corrosion of gray cast iron in which only a structure of graphite and some oxides remain.

Half-cell potential of an element   The emf developed when an element is coupled with a hydrogen half-cell.

Hydrogen blistering   The development of blisters as a result of the diffusion of hydrogen into voids.

Hydrogen half-cell   A reference half-cell involving hydrogen adsorbed on a platinum wire in a standard solution of hydrogen ions.

Inhibitor   A chemical added to produce a film or coating that slows down the corrosion reaction.

Intergranular corrosion   Preferential corrosion at grain boundaries resulting from the formation of an anode because of precipitation of a second phase at the grain boundaries.

Oxygen-concentration cell   A galvanic cell caused by differences in oxygen concentrations. An example is the water line in a partially filled tank.

Passivation   The formation of a film of reaction product that inhibits further reaction.

Pilling–Bedworth ratio   The ratio of volume of scale to volume of parent metal. Values much different from 1 may lead to high scaling rates.

Pit corrosion   Corrosion resulting from differences in oxygen and ion concentrations at the base of a pit compared with these concentrations at the surface.

Radiation damage   Modification of the properties of a material by exposure to electromagnetic radiation or to particles such as neutrons.

Sacrificial anode   An active metal that, when attached to the object to be protected, makes the object cathodic and therefore causes solution to take place only on the active metal.

Scale   A reaction product formed on a surface that is attacked by gas or liquid.

Season cracking   The acceleration of corrosion as a result of high residual stresses. *Caustic embrittlement* is a similar effect in an alkaline environment.

Selective leaching   Preferential solution of one element in an alloy. For example, in brass the zinc dissolves and a spongy copper deposit remains (dezincification).

Ultraviolet radiation   Electromagnetic radiation that is of modestly shorter wavelength and hence higher energy than that found in the visible spectrum.

Weld decay   Corrosion at or adjacent to a weld as a result of galvanic action resulting from differences in structures produced by the welding.

Problems

Let us be perfectly candid about the problems that follow. We could have put together a group of electrochemical calculations that would be a great exercise in calculator manipulation. But this might give the erroneous impression that it is possible to calculate such things as the corrosion rate of a car fender in the spring mush of Michigan roads. M. Pourbaix has done some excellent work in the application of thermodynamics to corrosion, but this cannot yet be applied directly to the average complex situation. Quantitative calculations of potentials necessary for protection of pipes have been made, however.

Therefore, in this problem section, we shall illustrate the application of the general principles set forth in the sections on chemical principles of metallic corrosion, corrosion phenomena, and gas corrosion.

18.1   *[ES/EJ]* Write the ion–electron equations for the anode and cathode reactions (half-cell reactions) in the following cases. (If you believe corrosion will not occur, write "no reaction.") (Sections 18.1 through 18.7)

a. An opened (punctured) tin can at the bottom of a fresh-water lake. (A tin can is a steel can with a tin coating. See Section 18.13)

b. An opened (punctured) tin can in the ocean. Why would the rate of corrosion be more rapid here than in part a?

c. A stainless steel (type 302) piece of trim on a car that is exposed to intermittent splashing with salt solutions

d. A car fender (type 1010 steel) covered on the inside with a coat of asphalt in which there are a few holes

    e. A copper heating coil brazed to a steel pipe with a 70% Cu, 30% Zn braze. The junction is exposed to dripping from a hot-water tank.

    f. The effect of attaching a "copper ground" from a "live" electric stove to a steel water pipe that sweats in the summer

    g. A punctured Alclad aircraft wing (pure aluminum covering alloy 2014) exposed to salt water

18.2 *[ES/EJ]* Large amounts of copper are obtained from copper mine water by immersing iron scrap in the solution and later collecting fine copper powder (cement copper). Write the half-cell reactions and discuss the economics of this technique compared with other possible methods, such as electrolysis, evaporation, and reduction of the salt. (Sections 18.1 through 18.7)

18.3 *[ES/EJ]* Why does gold exist in nature as a pure element, whereas iron is not generally found this way? (Sections 18.1 through 18.7)

18.4 *[ES]* If a mercury–sodium alloy is used, the sodium can be effectively oxidized in a battery without detrimental side reactions. (Sections 18.1 through 18.7)

    a. For a battery containing an electrolyte with a sodium ion concentration of 1 g-mol/liter (1 molar) and a standard hydrogen electrode, what should be the battery voltage?

    b. If the sodium ion concentration in the electrolyte were only 0.1 g-mol/liter, what would be the battery potential?

    c. If the reference hydrogen electrode in part a is replaced with a silver electrode, what should be the potential?

    d. In the following table, indicate the correct answers for the case described in part *a* by filling in the blanks correctly.

Electrode	Anode	Cathode	Positive	Negative
Sodium				
Hydrogen				

18.5 *[ES]* Chrome plate is in reality three plates: Steel is coated with copper, then coated with nickel, and finally given an outer coating of chromium. (Sections 18.1 through 18.7)

    a. List the four metals in order from most anodic to most cathodic in salt water. In Table 18.3 the two emf values for chromium are for the active and passive conditions.)

    b. A scratch cuts through the chromium plate just enough to expose only the nickel plate. Write the half-cell reactions at the anode and cathode.

18.6 *[ES]* The greater the emf of a cell (anode emf minus cathode emf), the greater the tendency for corrosion to occur. For the following corrosion cells, write the anode and cathode reactions in salt water and the initial cell emf. (Sections 18.1 through 18.7)

    a. Mild steel and zinc

    b. 1100 aluminum and copper

18.7 *[ES]* Rust can be thought of as a hydroxide of iron, $Fe(OH)_3$. Indicate how this can occur in water by showing appropriate half-cell reactions. Remember that iron can have two valence states. (Sections 18.1 through 18.7)

18.8 *[ES/EJ]* The discussion of Figure 18.2 indicates that the materials are placed in salt water. Why is salt water used? Would the results have been any different if tap water had been substituted? (Sections 18.1 through 18.7)

18.9 *[EJ]* A battery is nothing more than a controlled corrosion cell. What is the purpose of a battery charger, or alternator, in your car? (Sections 18.8 through 18.12)

18.10   *[ES]* A more common way to present the data given in Table 18.4 is in units of mpy, or mils of uniform corrosion per year (1 mil = 0.001 in.). What is the area (in^2) of the sheet samples used in the test if a 0.15-g loss in steel is equivalent to 5 mpy and the test is run for 48 hr? Assume that steel has a density of 7.8 g/cm^3 and that weight loss from the steel edges can be ignored. (Sections 18.8 through 18.12)

18.11   *[ES/EJ]* Stainless steel takes its name from its corrosion resistance. However, explain the following observations about the material.(Sections 18.8 through 18.12)
a. It corrodes more rapidly in dilute nitric acid than in concentrated acid.
b. It corrodes more rapidly when the electrolyte contains chloride ion.

18.12   *[ES/EJ]* Explain how to interpret the data in Figure 18.3 using Faraday's law. [*Hint:* Faraday's law assumes constant current.] (Sections 18.8 through 18.12)

18.13   *[EJ]* Indicate whether the corrosion rate of a piece of iron placed in tap water is increased or decreased by doing the following. (Sections 18.8 through 18.12)
a. Adding NaCl to the water
b. Imposing electron flow *into* the iron (by means of a battery)
c. Placing nickel in contact with the iron
d. Adding chromate ion to the water
e. Freezing the water

18.14   *[ES/EJ]* Indicate whether the following statements are correct or incorrect and justify each answer. (Sections 18.8 through 18.12)
a. A high concentration of oxidizing acid may render a metal more corrosion-resistant.
b. When we see an etched microstructure under the microscope, the grain boundaries were anodic during the etching process.
c. Oxygen dissolved in water has no effect on the corrosion rate of iron that is exposed to the water.
d. Aluminum rivets in a steel structure should have longer life against corrosion than steel rivets in an aluminum structure.

18.15   *[EJ]* Antique brass pots and kettles may not be great "finds;" they often leak because of a corrosion phenomenon. What would the corrosion be called if it had the following appearance? (Sections 18.13 through 18.18)
a. Fine cracks that appear to follow grain boundaries
b. Copper rather than the normal yellow brass color

18.16   *[ES/EJ]* How would Figure 18.9 appear if current density rather than current were used? (Sections 18.13 through 18.18)

18.17   *[ES/EJ]* (Sections 18.13 through 18.18)
a. In recent years automobile manufacturers have used galvanized steel sheet in body parts to combat corrosion. Considering the principal corrosive to be a dilute NaCl solution, write the ion–electron equations for the corrosion taking place before this changeover (in ordinary steel) and after (in galvanized steel).
b. A sign manufacturer makes small signs of 18% Cr, 8% Ni, 0.08% C stainless steel by welding letters to a plate of the same material, using a weld rod of the same material. Corrosion occurs $\frac{1}{4}$ in. from the weld.
   1. Write the ion–electron equations.
   2. Why (in 10 words) is the stainless steel not "stainless"?
   3. What could be done to prevent the corrosion *without* changing the composition of the parts and without painting?

18.18   *[ES]* Galvanizing is a coating of essentially pure zinc on steel, and a common designation for it is 0.50 oz Zn/ft^2 of steel. What is the thickness of the zinc coating? (Sections 18.13 through 18.18)

18.19   *[ES/EJ]* Why does the surface of a corroding automobile fender feel raised when you run your hand across it? (Sections 18.13 through 18.18)

18.20   *[ES/EJ]* Indicate whether the following statements are correct or incorrect and justify each answer. (Sections 18.13 through 18.18)
   a. Corrosion pits are generally of the same depth.
   b. If you do not change the oil in your automobile, some components will be more susceptible to fatigue failure.
   c. The head of a nail is more likely to corrode than the shank.

18.21   *[EJ]* Compare the merits of using tin and zinc (*a*) as a protective coating for cans for food and (*b*) for outdoor fencing. (Sections 18.13 through 18.18)

18.22   *[ES]* As noted in the text, $M_{23}C_6$ is the grain boundary carbide that can precipitate in stainless steels. The maximum amount of iron present in this carbide corresponds to the stoichiometry represented by the formula $Fe_7Cr_{16}C_6$. Calculate the weight percentages of iron, chromium, and carbon present in the carbide. (Sections 18.13 through 18.18)

18.23   *[ES/EJ]* Why might the corrosion rate in an automotive cooling system be higher if the water is changed several times during the year? Assume that no antifreeze or corrosion inhibitors are added. (Sections 18.19 through 18.27)

18.24   *[ES/EJ]* (Sections 18.19 through 18.27)
   a. What material would you recommend using to contain 70% sulfuric acid between 100°F and 200°F (38°C and 93°C)?
   b. What supplementary data would it be useful to have before making a final decision?

18.25   *[EJ]* Medicine has advanced so that more and more synthetic components are considered as replacements for natural bone and tissue. What are some of the requirements of such a material if it is to resist corrosion? (Sections 18.19 through 18.27)

18.26   *[EJ]* Under what conditions might you expect concrete to corrode? (Recall the type of bonding in concrete.) (Sections 18.19 through 18.27)

18.27   *[EJ]* Explain why the clouding of inexpensive drinking glassware after washing in an automatic dishwasher might be a corrosion phenomenon. (Sections 18.19 through 18.27)

18.28   *[ES/EJ]* Use Table 18.6 to answer each question. (Sections 18.19 through 18.27)
   a. Why is polyethylene more resistant to organic solvents than polyvinylchloride is?
   b. Why is water absorption included in the table?
   c. What is the significance of ionizing radiation?

18.29   *[EJ]* A nail is immersed completely in oxygenated water. At which locations will corrosion occur? Write the anode and cathode reactions. (Sections 18.19 through 18.27)

18.30   *[EJ]* Home and industrial water systems often use different piping materials. Local codes often suggest using "dielectric unions," which are commonly made of nonmetallic materials. What is the advantage of using them? (Sections 18.19 through 18.27)

18.31   *[ES]* A heat-treating furnace is to operate at 1800°F (982°C). Parts are to be placed on molybdenum trays. There is some question about the oxidation resistance of molybdenum. Show by calculation whether molybdenum will stand up under these conditions. (Sections 18.28 through 18.36)

$$Mo + \tfrac{3}{2}O_2 \longrightarrow MoO_3$$

Densities (g/cm³): Mo = 10.22; MoO_3 = 4.50; O_2 = 1.43×10^{-3}

18.32   *[ES/EJ]* Magnesium parts are heat-treated in an atmosphere containing 1% SO_2, which forms a sulfate coating. Why is this atmosphere preferable to air? (The density of $MgSO_4$ is 2.66 g/cm³, and that of MgO is 3.58 g/cm³.) (Sections 18.28 through 18.36)

18.33 *[ES]* Cesium metal is known to form an oxide, Cs_2O. From the following data, show by calculation whether cesium can be expected to exhibit a linear or a parabolic oxidation law (that is, to give a porous or nonporous oxide scale). (Sections 18.28 through 18.36)

	Cs	O_2	Cs_2O
Crystal structure	BCC	Gas	—
Atomic radius, Å	2.62	0.62	$Cs^+ = 1.65 \quad O^{2-} = 1.32$
Molecular weight	132.9	32	281.8
Specific gravity	1.87	Gas	4.36
Melting point	28°C	−218.4°C	400°C

18.34 *[ES]* Show by calculation whether the nitrification of titanium would be expected to be linear or parabolic (related to oxidation). The reaction is

$$2Ti + N_2 \longrightarrow 2TiN$$

The specific gravity of titanium is 4.50, and that of TiN is 5.43. (Sections 18.28 through 18.36)

18.35 *[ES/EJ]* Outline a laboratory experiment in which the objective is to determine the oxidation resistance of copper. Be specific about the measurements that are to be taken. (Sections 18.28 through 18.36)

18.36 *[ES/EJ]* Explain why the addition of a small amount of a given element to one alloy may increase its oxidation resistance but adding the same element may decrease the oxidation resistance of another alloy. (Sections 18.28 through 18.36)

18.37 *[ES]* Show by calculation whether a scale of P_2O_5 formed on the surface of phosphorus will crack or remain intact. The specific gravity of P_2O_5 is 2.39. (Sections 18.28 through 18.36)

18.38 *[EJ]* There are essentially two major types of biodegradable polymers, those that decay in the local environment (corrode) with the possible aid of microorganisms and those that degrade because of UV radiation. (Sections 18.28 through 18.36)
a. Indicate how the UV-biodegradable type might function and its response in a landfill.
b. Why is human tissue sensitive to UV radiation and how do sun-block coatings provide protection?

18.39 *[EJ]* Table 18.7 indicates that gassing can occur in several polymeric materials when they are exposed to a neutron flux. Explain the mechanism and identify the possible gasses produced. (Sections 18.28 through 18.36)

18.40 *[ES/EJ]* Cite some advantages and some disadvantages of exposing a plain-carbon steel to a neutron flux. (Sections 18.28 through 18.36 and Chapter 8)

18.41 *[ES/EJ]* Correctly name each of the following corrosion phenomena and suggest a remedy (galvanic action, concentration cell, stress corrosion, etc.). (Summary of Chapter 18)
a. An aluminum rivet falls out of a steel component.
b. A steel shovel corrodes at areas left covered with mud.
c. A buried cast-iron water pipe shows deep spongy areas that are local and that penetrate the pipe in several places.
d. A piece of stainless steel in a corrosive medium fractures along the grain boundaries.
e. An impeller on a water pump develops pits, but only in a few small areas.

18.42 *[ES/EJ]* (Summary of Chapter 18)
a. For each of the corrosion-related situations shown in Figure 18.20:
1. Draw and label with arrows touching the cathode and anode region(s), respectively.
2. Write balanced ion–electron equations, giving the reactions taking place at the anode and cathode.

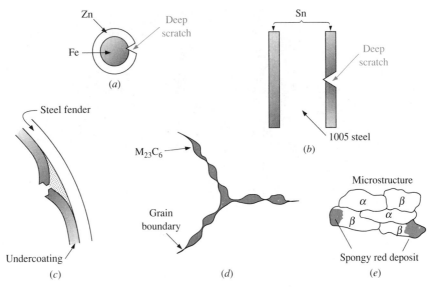

FIGURE 18.20 (*a*) Cross section of galvanized-iron fence wire in moist air. (*b*) Cross section of tin can in the rain. (*c*) Hole in undercoating of fender of car driven in the rain. (*d*) Microstructure of a weld in type 304 stainless steel (18% Cr, 8% Ni, 0.08% C, bal. Fe) immersed in sea water and showing particles of carbide precipitate at grain boundaries. (*e*) 60% Cu, 40% Zn brass propeller corroding in sea water.

3. Label the corrosion phenomenon, using the one of the following terms that describes it most closely. *Do not use any given term more than once.*

Galvanic corrosion	Graphite corrosion
Intergranular corrosion	Selective leaching
Caustic embrittlement	Sacrificial anode
Oxygen-concentration cell	

b. How could corrosion be avoided in situation *d*? Select the *one best* remedy from the following list. (Indicate your choice by number.)

   1. Heat the metal to 750°F (399°C), hold for 5 hr, then water-quench.
   2. Use a similar alloy containing 1% aluminum.
   3. Heat the metal to 2000°F (1093°C), hold for 5 hr, then water-quench.
   4. Attach a platinum plate.
   5. Increase the copper content of the metal to 80%.
   6. Increase the zinc content of the metal to 80%.
   7. Use a similar alloy containing 0.4% niobium.

c. How could corrosion be avoided in situation *e*? Select the *one best* remedy from the list in part *b*. (Indicate your choice by number.)

CHAPTER 19

Analysis and Prevention of Failure

Co-authored with J. W. Jones
Associate Professor, Materials Science and Engineering
University of Michigan, Ann Arbor

The catastrophic failure of certain steel ships, often while they are moored quietly at dockside, has led to extensive investigation of the fracture toughness of materials. A tensile specimen cut from the steel plate of these ships invariably exhibits good strength and ductility. The brittle fracture of the ship results from a combination of constraints and lowered temperature, as we will discuss in the text.

19.1 Overview

For many years engineers have been pressed to develop strong, lightweight alloys for aircraft. Recently, with the need to conserve energy, people have been seeking similar lightweight and high-strength alloys for cars, railroad equipment, and moving parts of machinery. As long as designers worked at ambient temperatures with the lower-strength, high-ductility materials, they could generally avoid failures by designing with stresses below the yield strength. However, when they used the same methods with the new high-strength, low-ductility materials, there were many catastrophic failures. In addition, the fractures were brittle and did not exhibit even the lower levels of ductility of the tensile test bar. New design criteria have been developed for the safe use of these high-strength alloys on the basis of the concept of fracture toughness and equations developed from fracture mechanics.

In the second part of the chapter we will discuss fracture analysis.

19.2 What Is Fracture Toughness?

A simple experiment illustrates the physical significance of *fracture toughness*, the behavior of material with a crack present. If we try to bend and break a glass rod, we find that it takes considerable force. However, if we place a small notch on the surface of the glass rod, the force needed to cause fracture is greatly reduced. When we repeat the experiment on a copper rod, we find that the small notch has no effect on the force required to bend the rod. In fact, the notched copper rod can be bent into a U-shape without fracturing. Figure 19.1 shows specimens on which such tests were performed.

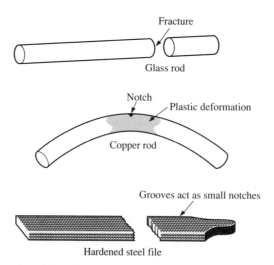

FIGURE 19.1 Results of bending tests of glass and copper rods, of equal diameter, each containing a notch of the same depth. (Note that a file makes a poor lever because—even though it is very high in strength—it is low in fracture toughness.)

Note that, in the case of the glass rod, no detectable local plastic deformation accompanied fracture. However, considerable plastic deformation occurred in the copper rod, and yet no fracture occurred. From the results of this simple experiment, we may conclude that glass has a very low fracture toughness and that copper has a much higher fracture toughness. We may generalize these observations and say that if a material fractures without gross local yielding, it is brittle, whereas if considerable plastic deformation occurs before fracture, the material is ductile.

Copper and glass represent two extremes in fracture behavior: extremely ductile and extremely brittle. Common sense tells us to avoid extremely brittle materials in structural applications and to use ductile materials instead. Unfortunately, the more ductile materials are not strong enough for many applications. For example, high-strength alloys must be used where minimizing the weight of the component is critical. As we have mentioned, the aircraft and automotive industries are prime examples of areas where the strength-to-weight ratio of materials is important. Unfortunately, as the strength of alloys increases, their ductility and fracture toughness generally decrease and their susceptibility to brittle fracture increases.

We can understand why some materials have low fracture toughness and some have high fracture toughness if we consider two important aspects of fracture:

1. The response of materials to high local stresses (which can be many times greater than the average stress of the cross section)
2. The role of notches, holes, cracks, and other defects in producing very high local stresses in a part

19.3 Fracture Energy

To fracture a material, work must be performed. This work supplies the energy needed to create the fracture surfaces and to plastically deform the material if local yielding occurs prior to fracture. This "energy-balance" approach to fracture can be summarized as follows:

$$\begin{pmatrix} \text{Energy input (work)} \\ \text{to produce fracture} \end{pmatrix} \geq \begin{pmatrix} \text{surface energy } (\gamma_s) \\ \text{of fracture} \\ \text{surfaces} \end{pmatrix} + \begin{pmatrix} \text{energy of plastic} \\ \text{deformation } (\gamma_p) \end{pmatrix}$$

Here γ_s is the *surface energy*[*] per unit surface area (in.-lb/in^2 or J/m^2), and γ_p is the energy of plastic deformation per unit volume (in.-lb/in^3 or J/m^3).

One measure of the work required for fracture is the area under the true stress–true strain curve. To illustrate this, let us reconsider the fractures of

---

[*] γ_s may be thought of as the energy required to break the atomic bonds and create the new surfaces.

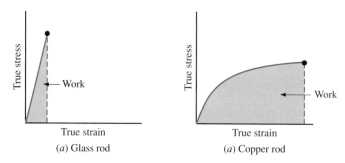

FIGURE 19.2   True stress–true strain curves for glass and copper rods

the glass and copper rods. If we performed a tensile test on both types of rods, the result would be curves such as those shown in Figure 19.2a and b.

For glass, no plastic deformation occurs prior to fracture, and the energy required for fracture is quite small. In fact, a portion of the elastic deformation (the stored elastic energy term) is recovered during fracture.* The total energy consumed, which is considerably less than the area under the true stress–true strain curve, is equal to the increase in surface energy (γ_s × fracture surface area); see Figure 19.3.

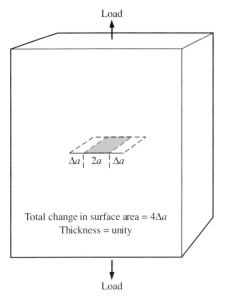

FIGURE 19.3   Extension of a center crack of width 2a in a large sheet. Extension of the crack by Δa in both directions produces an increase in surface area of $4\Delta a$ (two new surfaces, top and bottom, are produced on each end).

---

*In springback of the fractured parts

For copper, considerable plastic deformation occurs before fracture. The area under the true stress–true strain curve is quite large, and the energy required for fracture is correspondingly large (γ_s × fracture surface area + γ_p × volume of plastic deformation). Generally in metals, $\gamma_p \approx 10^4 \gamma_s$, which explains why the fracture toughness of copper is much greater than the fracture toughness of glass.

19.4 Stress Concentration

If no flaws are present, all important structural alloys deform, at least locally, before failure. Yet every year catastrophic failures occur in components in which the operating stress was well below the yield stress. Often such failures occur because the stresses in particular regions of the components have been amplified by the presence of notches, holes, cracks, and other geometrical discontinuities. The following example illustrates the local increase in stress at a defect.

If we drill a hole of radius r through a bar, we reduce the cross section of the bar (Figure 19.4). We calculate the nominal stress in the bar at the hole as

$$S_{nom} = \frac{P}{(w - 2r)t} \tag{19-1}$$

where S_{nom} = nominal stress, P = load, w = width, and t = thickness. (The normal area of the bar would be wt, and we reduce this by $2rt$ to take care of the area of the cross section lost by the hole.)

However, defects such as this have led to the disastrous failure of many highly stressed components, especially in the presence of cyclic stresses, because the stress measured at the sides of the hole is not the nominal stress but is much higher. This is because the stresses that would normally be supported by the material where the hole was drilled are *concentrated* at the edge of the hole (Figure 19.4). Furthermore, the smaller the ratio of the radius of the hole to the width of the bar, the *higher* the *stress-concentration factor* K_σ, which reaches a value of about 2.8 at a radius-to-width ratio of 0.05. Measurements beyond this point are difficult to make.

EXAMPLE 19.1 *[ES]*

Given P = 20,000 lbf, r = 0.1 in., w = 1 in., and t = 1 in., calculate the maximum stress in a bar similar to that shown in Figure 19.4.

Answer

$$S_{nom} = \frac{20,000}{(1 - 0.2)1} = 25,000 \text{ psi } (172.5 \text{ MPa})$$

but because $K_\sigma = 2.7$ $(r/w = 0.1)$,

Stress at edges of hole $= 2.7 \times 25,000 = 67,500$ psi (466 MPa)

(This is why so many fatigue failures start at holes and other "stress raisers" such as fillets and notches.)

When a hole or a notch approaches the geometry of a sharp crack, we can no longer readily measure the stress-concentration factor K_σ. (Note that the curve in Figure 19.4(c) is not continued for low values of r/w.) For very sharp cracks, we must resort to the methods of fracture mechanics described in the following section.

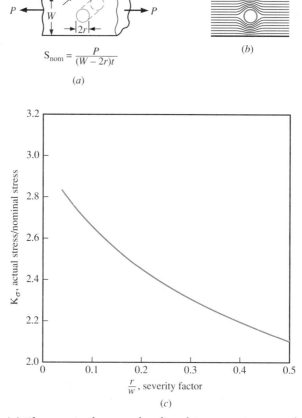

FIGURE 19.4 (a) The nominal stress developed in a tension member containing a round transverse hole. (b) The concentration of stresses in the vicinity of the hole. (c) The stress-concentration factor, K_σ, caused by a transverse hole in the tension plate member.

[Reprinted from C. Lipson, G. C. Noll, and L. S. Clock, *Stress and Strength of Manufactured Parts*, © 1950, by permission of McGraw-Hill Publishing Company.]

19.5 Need for Fracture Mechanics

We have discussed how stress concentration can lead to very high stresses locally, and how we may expect materials to respond to this region of locally high stresses. It is the response of the material to these high stresses that in large measure determines the fracture resistance or fracture toughness of the material. Let us again refer to our example of the fracture of copper and glass rods. If we assume that both rods have the same diameter and that the notches are identical, we must conclude that the stress concentration is the same in each rod. For equal applied loads, the stresses at the tip of the notch are the same in both materials. Yet only in the glass rod did the notch cause catastrophic brittle fracture. We may draw two important conclusions from these observations:

1. The ability of a particular *flaw*, or stress concentrator, to cause catastrophic failure depends on the *fracture toughness*, a material property.
2. The *stress concentration* depends on the *geometry* of the flaw and the *geometry* of the component, but not on the properties of the material.

Therefore, to predict the fracture strength of a component, we must know both the severity of the stress concentration *and* the fracture toughness of the material. This theme will recur throughout our discussions of the analysis of fracture and failure.

With this general background, let us investigate the quantitative evaluation of fracture toughness. We shall begin with the simplest case, glass, in which no plastic deformation is involved, and then advance to the metallic materials in which plastic deformation is dominant.

19.6 The Griffith Analysis: Fracture Mechanics of Glass

In the 1920s, A. A. Griffith conducted a series of experiments to determine the fracture strength of glass specimens that contained small flaws. From his experiments, he concluded that the stress required to cause failure decreases as the size of the flaw increases, even after correction is made for the reduced cross section of good material. He also showed that very thin fibers of glass have much higher tensile strength than coarse fibers because the thin fibers are essentially flaw-free. Griffith then developed an expression relating the fracture stresses to the flaw size:

$$\sigma_f = \sqrt{\frac{2E\gamma_s}{\pi a}} \qquad (19\text{-}2)$$

where σ_f = *fracture stress* (psi or MPa), a = one-half of the crack length (this was chosen to simplify the derivation) (in. or m), E = Young's modulus of elasticity (psi or MPa), and γ_s = energy required to extend the crack by a unit area (in.-lb/in^2 or J/m^2).

For glass, γ_s is simply equal to the surface-energy term we discussed earlier. For a given glass, E and γ_s are constants and Equation 19-2 can be simplified:

$$\sigma_f = \frac{C}{\sqrt{\pi a}} \qquad (19\text{-}3)$$

Note that the constant C, which has units of psi$\sqrt{\text{in.}}$ (MPa$\sqrt{\text{m}}$), is proportional to the energy required for fracture.

EXAMPLE 19.2 *[ES]*

Fused silica has a surface energy of $\sim 17.1 \times 10^{-5}$ in.-lb/in^2 (4.32 J/m^2) and an elastic modulus of 10×10^6 psi (69,000 MPa). A large plate of this material is to withstand a nominal stress of 5000 psi (35 MPa). What is the largest flaw that can be tolerated without fracture occurring?

Answer   Rearrange the Griffith equation, 19-2, as follows:

$$a = \frac{2E\gamma_s}{\pi \sigma_f^{\,2}}$$

Substitute the appropriate values and solve for a.

$$a = \frac{2 \times (10 \times 10^6 \text{ psi})(17.1 \times 10^{-5} \text{ in.-lb/in}^2)}{3.14(5000 \text{ psi})^2}$$

$$= 4.4 \times 10^{-5} \text{ in. } (1.1 \times 10^{-3} \text{ mm})$$

Because a is half the crack length, the actual crack length is 8.8×10^{-5} in. $(2.2 \times 10^{-3}$ mm).

   Note: In this example the critical crack length is only 88 microinches (2.2 microns) in length. Such a small flaw is invisible to the naked eye yet is common in commercial glass. This explains why the actual strength of glass is far below the theoretical strength calculated from the breaking of atom–atom bonds.

19.7 Application of Fracture Mechanics to Metals

With slight modification, the Griffith analysis can be applied to the fracture of metals. As we mentioned earlier, the difference in the fracture of metals is the occurrence of plastic deformation at the tip of the propagating crack. The fracture toughness is proportional to the energy consumed in the plastic deformation. Unfortunately, it is hard to measure accurately the energy required for this plastic deformation.

We use a parameter called the *stress-intensity factor*, or K_I, to determine the fracture toughness of most materials. The stress-intensity factor, as the name suggests, is a measure of the concentration of stresses at the tip of a sharp crack. It is similar to the stress-concentration factor K_σ, but the two are not equivalent. For a given flawed material, catastrophic failure occurs when the stress-intensity factor reaches a critical value, denoted as K_{IC}. This critical value is called the *fracture toughness* of the material.

An expression relating the fracture stress to the fracture toughness and the flaw size is

$$\sigma_f = \frac{K_{IC}}{Y\sqrt{\pi a}} \qquad (19\text{-}4)$$

where K_{IC} = fracture toughness (psi$\sqrt{\text{in.}}$ or MPa$\sqrt{\text{m}}$), σ_f = nominal stress at fracture (psi or MPa), a = crack length (or one-half of the crack length, depending on geometry) (in. or m), and Y = a dimensionless correction factor that accounts for the geometry of the component containing the flaw. Note that when $Y = 1$, K_{IC} is equivalent to the constant C in Equation 19-3. The actual derivation of Equation 19-4 is rather complex.*

Confusion sometimes arises as to how the crack length is measured. For an *edge crack*, the crack length is a. For a *center crack*, the crack length is $2a$. Similarly in Equation 19-4, for an edge crack the a value is the crack length, and for a center crack the a value is the crack length divided by 2.

The relationship between stress intensity K_I and fracture toughness K_{IC} is similar to the relationship between stress and tensile strength. The stress intensity K_I represents the level of "stress" at the tip of a crack in a test specimen or component containing a crack (stress-dependent), and the fracture toughness K_{IC} is the *highest value* of stress intensity that the specimen can withstand at this crack sharpness without fracturing (material-dependent). The units of stress intensity and fracture toughness, psi$\sqrt{\text{in.}}$ (MPa$\sqrt{\text{m}}$), may seem strange. They can best be thought of as a combination of the units of stress (psi) and crack length (in.). When we note the general expression for stress intensity,

$$K_I = S\sqrt{\pi a}\, Y \qquad (19\text{-}5)$$

where S is the nominal applied stress and a is the crack length, we see that, for a given crack length, the stress intensity is zero when $S = 0$, and increases linearly with applied stress and the square root of the crack length. As we shall see later, a knowledge of the stress intensities developed at flaws during service is also important in predicting failure due to fatigue and stress corrosion cracking. We may think of the stress intensity, K_I, as any value, whereas

---

*It is discussed further in John Knott, *Fundamentals of Fracture Mechanics*, Halsted Press, New York, 1973.

K_{IC} is a particular value; this is very much like the relationship between any stress and a particular stress (such as the one corresponding to tensile strength).

EXAMPLE 19.3 *[ES]*

Consider a plate containing a crack of length $2a$ extending through the thickness. Assume that the length of the crack is negligible compared with the width of the plate. For a fracture toughness of 25 ksi$\sqrt{}$in. (27.5 MPa$\sqrt{}$m) and a yield strength of 65 ksi (448.5 MPa), calculate the fracture stress σ_f and the ratio $\sigma_f/S_{\mathrm{Y.S.}}$. Assume $Y = 1$.

Answer

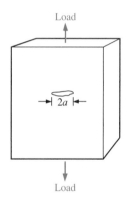

a		$\sigma_f = \dfrac{K_{\mathrm{IC}}}{\sqrt{\pi a}}$		$\sigma_f/S_{\mathrm{Y.S.}}$
in.	*mm*	*ksi*	*MPa*	
0.05	1.27	63.0	435	0.97
0.1	2.54	44.6	308	0.69
0.15	3.81	36.4	251	0.56
0.20	5.08	31.5	217	0.48
0.25	6.35	28.2	195	0.43
0.30	7.62	25.8	178	0.40
0.35	8.89	23.8	164	0.37
0.40	10.16	22.3	154	0.34

Note: With extremely small cracks (less than 1.27 mm), the general yield strength is reached before catastrophic failure occurs. However, with longer cracks and/or a low-toughness material, catastrophic failure occurs long before the yield strength is reached.

19.8 The Role of Specimen Thickness in Fracture Toughness

In presenting Equation 19-4, we introduced a geometric correction factor Y. This factor takes into account the width and thickness of a specimen, the structural behavior, and the simple stress-concentration effects. The factor Y includes all of these geometric effects except thickness, which is important because a material may show ductile behavior in a thin sheet but may fracture in brittle fashion in a thick plate. We shall see later that this ductile-to-brittle transition takes place only at very thick sections for low-strength, very ductile materials but that it occurs at thinner sections for high-strength, low ductility alloys. To explain this effect, we need to define conditions of plane stress and plane strain.

19.9 Plane Stress and Plane Strain

Let us consider the conditions leading to plane stress and plane strain in a given material.

Plane Stress

Consider a notch in a thin plate loaded in simple tension in the y direction. Because there is a volume of material missing in the notch, the tensile stress normally borne by this volume is transferred to the notch region. Figure 19.5 shows the higher level of stress in the y direction near the notch, which finally falls off to the average value away from the notch. This is a typical stress-concentration effect.

There is, however, a *local stress* in the x direction even though the specimen is stressed only in the y direction overall. To understand this, consider the material in the region of the notch as small tensile specimens (Figure 19.5b). If these specimens were separate and free to deform, the diameter of each would contract as the specimen was stretched. (The ratio of the strain in the x direction to the strain produced by the tension in the y direction is called *Poisson's ratio*; for steel it is about 0.3.) However, in the real material the small tensile specimens cannot contract away from each other because this would leave voids in the material. A local stress, σ_x, results, which prevents the contraction in the x direction. This stress is zero at the notch because there is no material in the empty space to react with. It reaches a maximum at a short distance from the notch and then falls off (Figure 19.5c).

Note particularly that because the specimen is thin (only one bar thick in our example), a similar condition does not develop across the thickness that we call the z direction in this case. We therefore have finite stresses only in the x and y directions, and these vectors are in a plane. This condition of

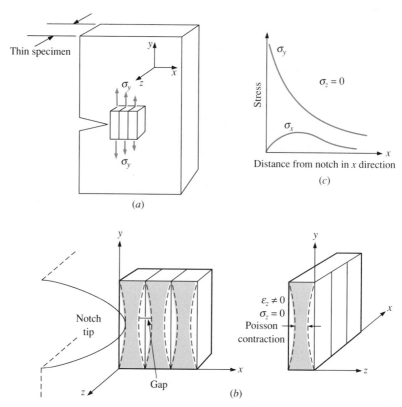

FIGURE 19.5 Plane stress in a thin sheet. (*a*) The region near the notch can be thought of as several small tensile bars. (*b*) If each bar were free to contract in the *x* direction, gaps would form between bars. Tensile stresses develop in the *x* direction to keep such gaps from forming. (*c*) In the *z* direction (thickness), the specimen is only one bar thick. Each bar is free to contract in this direction, so $\sigma_z = 0$ whereas $\epsilon_z \neq 0$.

plane stress exists for thin plates. Also, because the thin specimen can contract in the *z* direction all across the thickness, *the strain in the z direction is not zero.*

As a further explanation, we should point out that this stress arises as a result of a *strain gradient.* If we again assume small tensile specimens, in the *x* direction (Figure 19.5*b*), the gap between our small ligaments becomes large near the crack tip, reaches a maximum, and then becomes smaller. In fact, if we move far enough away from the crack tip, there is no tendency for a gap to form. However, the gap represents a strain. The width of the gap changes, so there is a strain gradient and hence a stress. This stress reaches a maximum and then falls, because the strain gradient does the same (Figure 19.5*c*).

In an analogous fashion (Figure 19.5*b*), the ligament or bar in the *z* direction has a width equal to the sheet thickness. A gap does not exist, there is no stress because there is no strain gradient, and we do not have plane strain as discussed below. In other words, ϵ_z is constant and uniform across the thickness.

Plane Strain

Let us now extend the same reasoning to a <u>thick</u> plate with a notch. We see that the tensile cylinders are restrained in the z direction as well, a stress develops in the z direction, and we no longer have a condition of plane stress. However, the material at the center of the notch (thickness) is no longer free to neck inward from the sides, as it is in the thin specimen, because it is restrained by the mass of material at the sides in the z direction. Therefore the *strain* in the z direction is approximately zero. The caption for Figure 19.6 explains the development of a stress in the z direction. The strain is finite in the x and y directions, so these vectors are in a plane. Figure 19.6 sums up this condition which is called *plane strain*.

In summary, for this material we have a condition of plane stress in thin specimens and plane strain in thick specimens. The condition of plane strain or plane stress influences the development of plastic deformation at the notch tip. A large plastic zone develops under plane stress, and consequently

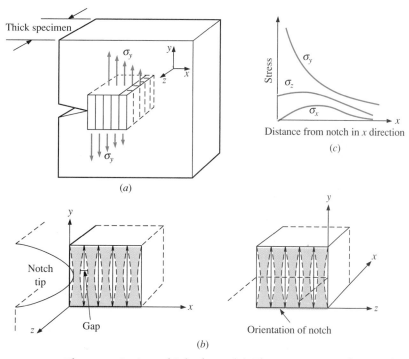

FIGURE 19.6 Plane strain in a thick sheet. (*a*) The region near the notch can be thought of as small tensile bars, as in Figure 19.5. However, the sheet is several bars thick. (*b*) As with plane stress, contraction of the bars in the x direction is prevented by the development of stresses σ_x. In the z direction, stresses now develop because the greater number of bars (thickness) prevents the bars from contracting. (*c*) Schematic of stresses developed in plane strain.

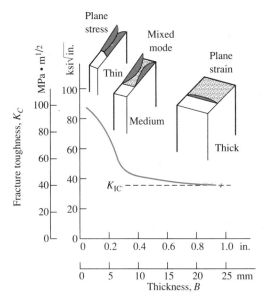

FIGURE 19.7 Variation in fracture toughness with thickness. The appearance of the fracture varies with thickness.

[*Welding Handbook*, Vol 1, 7th ed., American Welding Society, 1976. Used by permission.]

thin specimens have a higher value of fracture toughness. As shown in Figure 19.7, the minimum fracture toughness at greater thicknesses is called the *plane-strain fracture toughness* and is denoted as K_{IC}. In thinner sections, the *plane-stress fracture toughness* is denoted as K_{C}.[*] For plane-stress conditions, the thickness must be reported with fracture toughness, because K_{C} depends heavily on thickness. We shall see later that the transition from plane-stress to plane-strain fracture toughness depends on the yield stress and varies for different materials.

Years of testing have shown that plane-strain conditions generally prevail when

$$\text{Thickness} \geq 2.5 \left(\frac{K_{\text{IC}}}{S_{\text{Y.S.}}} \right)^2 \tag{19-6}$$

This relation is incorporated in the ASTM standards for fracture toughness testing. If a material of unknown fracture toughness is tested and the resulting fracture toughness does not satisfy the requirements of Equation 19-6, the thickness of the test sample must be increased and the test performed again.

---

[*]This is also called K_{Q}.

EXAMPLE 19.4 *[ES]*

What are the minimum thicknesses of specimens from which valid plane-strain fracture toughness values can be determined for two structural alloys, one with $S_{Y.S.} = 70$ ksi and $K_{IC} = 150$ ksi$\sqrt{\text{in.}}$ and the other with $S_{Y.S.} = 77$ ksi and $K_{IC} = 27$ ksi$\sqrt{\text{in.}}$?

Answer For the high-toughness alloy, Equation 19-6 becomes

$$\text{Minimum thickness} = 2.5\left(\frac{K_{IC}}{S_{Y.S.}}\right)^2 = 2.5\left(\frac{150 \text{ ksi}\sqrt{\text{in.}}}{70 \text{ ksi}}\right)^2 = 11.5 \text{ in. (292 mm)}$$

For the low-toughness alloy, Equation 19-6 becomes

$$\text{Minimum thickness} = 2.5\left(\frac{K_{IC}}{S_{Y.S.}}\right)^2 = 2.5\left(\frac{27 \text{ ksi}\sqrt{\text{in.}}}{77 \text{ ksi}}\right)^2 = 0.3 \text{ in. (7.6 mm)}$$

Note: Except in thick-walled pressure vessels, structural alloys are not commonly used in thicknesses approaching 12 in. The use of K_{IC} for thinner sections would be overly conservative. It would be more appropriate to use K_C for the thickness. However, for a high-strength alloy with relatively low toughness, thicknesses of 0.3 in. and greater are common, and K_{IC} would be the appropriate parameter to use.

19.10 Fracture Mechanics and Design

The application of fracture mechanics to the design of fracture-resistant structures and to the prediction of catastrophic failure in existing structures depends on the determination of three important parameters:

1. Fracture toughness, K_{IC} or K_C. This parameter is determined from pre-cracked test coupons using procedures set forth by the American Society for Testing and Materials (ASTM E399).
2. Existing crack length, a. This can be determined from rigorous inspection, proof testing, or (if inspection or testing is not possible) a conservative estimate.
3. Operating stress, S. The distribution of nominal stress can be determined by stress analysis or experimental measurements. This is usually, although not always, a design variable.

The magnitude of any two of these parameters determines the third, through their interdependence as given by Equation 19-4. Whether plane stress or plane strain is used, the correction factor Y is necessary. For example, if the design stress is fixed and the choice of materials depends on properties other than toughness and is also fixed, then the size of the critical flaw is automatically established. Proper inspection must be made to ensure that cracks of this size or larger do not exist prior to service, and periodic inspection must also be made to ensure that such flaws do not develop during service. Actually, as an additional safety factor, design should be based on a crack length of only a fraction of the critical size.

19.11 Fracture Toughness vs. Yield Strength

In a manner similar to the variation of ductility with yield strengh, the fracture toughness of metals and alloys generally decreases as the yield strength increases. Thus, when we use very high-strength materials to minimize the size or weight of components, the size of flaws that can be tolerated becomes smaller. Figure 19.8 illustrates the variation of fracture toughness with yield strength.

19.12 Delayed Failure: Fatigue and Stress Corrosion Cracking

Even if we design for stress levels in accordance with the relationships we have discussed, catastrophic failure may still occur after the component has been in service for some time. Cyclic stresses or corrosive conditions are the cause of failure in many cases. In both these situations, small flaws are initiated and grow until the critical crack size is reached; then rapid failure transpires. These delayed failures are more dangerous than early failures because consumers have false confidence in a component that has been in service for months or years.

19.13 Initiation of Fatigue Cracks

In Chapter 2 we discussed fatigue in terms of the *S-N* curve, which is the overall summary of the cycles for failure at different stresses for a given mate-

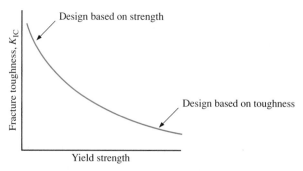

FIGURE 19.8   Variation of fracture toughness with yield strength

rial. The new insight into the relation between flaws and fracture toughness has made it possible to carry out a more detailed analysis of fatigue failure.

We may begin by dividing the region of the *S-N* curve prior to failure into two zones that have been found by experimental observations with the scanning electron microscope: initiation of cracks and growth of cracks (Figure 19.9). Cracks may initiate from essentially flaw-free regions or from existing defects such as inclusions. This variation in initiation is one of the factors leading to the scatter in the data for failure.

In the case of a defect-free region, cracks initiate from minute stress concentrators caused by localized plastic deformation on particular slip bands. Figure 19.10*a* shows an example of such localized plastic flow. After repeated cycling, a small crack develops at the intersections of these localized deformation bands with the surface of the specimen. Figure 19.10*b* shows the early stages of such crack development. Even minute surface scratches may eliminate or greatly shorten the crack-initiation phase of fatigue failure. This explains the strong dependence of total fatigue life on surface finish.

The presence of internal defects also shortens the time required for initiation of cracks. Figure 19.11 shows fatigue cracks arising from a nonmetallic inclusion and from a gas pore in a nickel-base alloy that was tested in fatigue.

In the early stages of crack propagation, the direction of propagation depends on orientation of the grain. Cracks proceed on specific crystallographic planes, so the direction may change abruptly when a grain boundary is crossed, as in Figure 19.12. This is called stage I crack growth. As the crack lengthens, the direction of growth becomes independent of grain orientation and develops normal to the applied stress. This phase of growth, called stage II, involves most of the propagation life. It will be considered in the remainder of our discussion of fatigue. Figure 19.13 is a schematic illustrating the various phenomena of the initiation and growth of fatigue cracks.

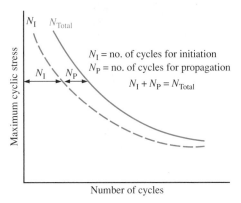

FIGURE 19.9   Schematic of *S-N* curve, showing that the total fatigue life consists of two phases: crack initiation and crack propagation. Note that, at high stresses, $N_I < N_P$ and that at low stresses, $N_P < N_I$.

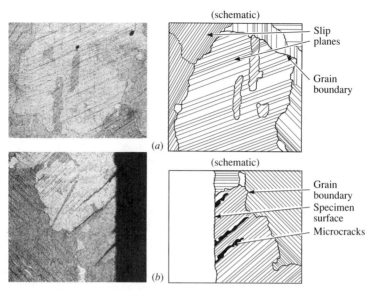

FIGURE 19.10   (a) Regions of intense slip that developed during fatigue of a nickel alloy. (b) Cracks forming at the intersection of surface with slip bands; 200×.

[Courtesy of J. M. Hyzak, Sandia National Laboratory]

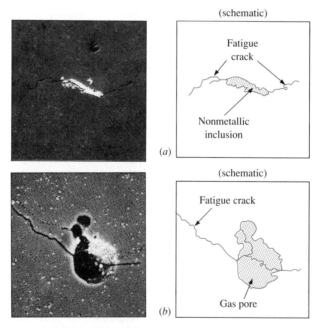

FIGURE 19.11   (a) Fatigue crack initiating from a nonmetallic inclusion in a nickel alloy. (b) Fatigue crack initiating from a gas pore in a nickel alloy; 1000×.

[Courtesy of J. M. Hyzak, Sandia National Laboratory]

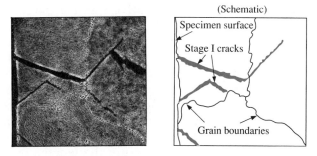

FIGURE 19.12 Propagation of a stage I fatigue crack; 200×

[Courtesy of J. M. Hyzak, Sandia National Laboratory]

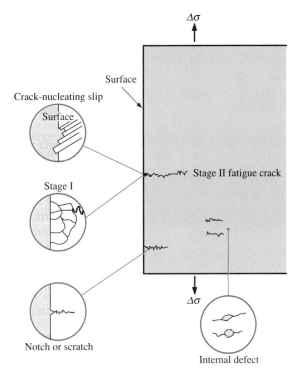

FIGURE 19.13 Schematic of the various ways in which fatigue cracks initiate and grow

19.14 Propagation of Fatigue Cracks

It is often advantageous to assume that small flaws do exist in a component. This bypasses the need for information on the early stages of initiation and growth of cracks. If we assume that small flaws do exist, we are able to estimate fatigue life on the basis of the measured rates of crack propagation. Such an approach may be conservative if in fact no flaws exist, but it eliminates the danger of premature failure occurring if flaws are present initially and are not detected. As we shall see, this method also has the advantage that fracture mechanics can be used in a quantitative manner in predicting the life of the component.

Let us consider the growth of fatigue cracks in two identical specimens subjected to identical fatigue loading. However, let us assume that the initial length of the crack in specimen I, or a_I, is greater than the initial length of the crack in specimen II, or a_{II}. During the course of fatigue tests, we may stop the test periodically and measure the lengths of the cracks. If we plot the lengths of the cracks versus the number of fatigue cycles accumulated at the time of each measurement, we get the curves shown in Figure 19.14.

Also note several important features of these curves.

1. When the length of the crack is small, the growth rate of the crack, $\Delta a / \Delta N$, is also small.
2. As the length of the crack increases, the growth rate of the crack also increases.
3. Under identical cyclic stressing, larger initial cracks propagate to failure in fewer cycles.
4. If the cyclic stressing is the same and if the geometries of the specimen and the crack are the same, the *length of the crack at failure* will be the same, regardless of the starting length of the crack and the number of cycles to failure.

For many years it was assumed that some function of the stress controlled the rate of propagation of cracks. It is now generally accepted that the rate of

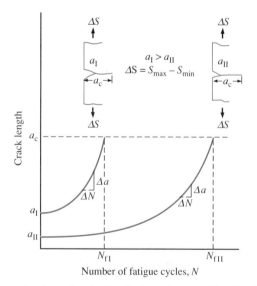

FIGURE 19.14 Growth of cracks during fatigue for two identical specimens that are under identical cyclic loading but have cracks that were initially of different lengths

propagation of cracks is a function of the stress-intensity factor, K_I. For many engineering alloys, the rate of propagation of cracks, treated as a differential da/dN, can be expressed as a function of the *stress-intensity range*, ΔK_I, that the crack experiences during the stress cycle.

$$\frac{da}{dN} = C\Delta K_I^{\,m} \tag{19-7}$$

Here C and m are constants that depend only on the material. We can calculate ΔK_I as follows: At the maximum stress of the fatigue cycle, S_{max}, we set up the equation for K_I using the formula

$$K_{I\,max} = S_{max}\sqrt{\pi a}\,Y \qquad \text{(See Eq. 19–4)}$$

and at the minimum stress,

$$K_{I\,min} = S_{min}\sqrt{\pi a}\,Y$$

(Note that we are considering K_I, any stress intensity, and not the fracture toughness, K_{IC}.) Then the range of stress intensity is

$$\Delta K_I = K_{I\,max} - K_{I\,min} = (S_{max} - S_{min})\sqrt{\pi a}\,Y \tag{19-8}$$

Note that ΔK_I is not a constant but rather varies as the crack length a changes in response to increasing stress cycles. Y, π, S_{max}, and S_{min} are constant.

Figure 19.15 shows a schematic representation of rate of crack growth vs. stress-intensity range. There are three regions on this curve that generally appear in actual data. In region I, at low values of ΔK_I, the rate of growth increases rapidly with a small increase in ΔK_I. In region II, at intermediate values of ΔK_I, Equation 19-7 is followed. The curve is linear when plotted on log–log scales and has a slope of m. In region III, where ΔK_I approaches K_{IC}, da/dN increases sharply with increasing ΔK_I. Many components spend the

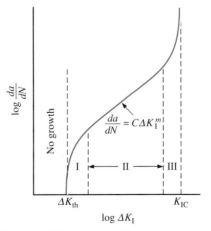

FIGURE 19.15 Growth rate of fatigue cracks as a function of the stress-intensity range, ΔK_I. (Note that this is a reproduction of Figure 2.30.)

majority of their service life in region II, and Equation 19-7 can be used to predict fatigue life.

In region I a limiting value of ΔK_I is reached below which no measurable growth of cracks takes place. This limiting value ΔK_{th} is called the *threshold stress-intensity range*. Its significance is similar to that of the fatigue limit determined from an *S-N* curve. If the combination of starting size of the crack and applied cyclic stress is such that the stress-intensity range is below ΔK_{th}, then no growth of cracks should occur and failure of the component by fatigue will be not a problem. This assumes, of course, that no larger flaws are introduced during service.

In Example 19.6 we consider how we can use data on crack growth to predict the safe lifetime of a component operating under cyclic stress. First, however, let us see how ΔK_I is determined.

EXAMPLE 19.5 *[ES]*

A metal strip (4 in. wide and 0.2 in. thick) is subjected to a cyclic load ranging from 6000 to 43,000 lb. There is a crack in the center of the strip that extends through the thickness. For $a = 0.1$ and 0.4 in., calculate ΔK_I. (Recall that a is one-half of the crack length for a center crack.)

Answer   For this geometry, $K_I = S\sqrt{\pi a}\,Y$, and in this example we assume that $Y = 1$.

$$\text{Compute stresses:} \quad S_{max} = \frac{43,000 \text{ lb}}{0.2 \times 4} = 53,750 \text{ psi (371 MPa)}$$

$$S_{min} = \frac{6000 \text{ lb}}{0.2 \times 4} = 7500 \text{ psi (51.8 MPa)}$$

Compute stress intensities for $a = 0.1$ in.

$$K_{I\,max} = S_{max}\sqrt{\pi a} = 53,750\sqrt{\pi(0.1)} \approx 30.1 \text{ ksi}\sqrt{\text{in.}}\,(33.1 \text{ MPa}\sqrt{\text{m}})$$

$$K_{I\,min} = S_{min}\sqrt{\pi a} = 7500\sqrt{\pi(0.1)} = 4.2 \text{ ksi}\sqrt{\text{in.}}\,(4.6 \text{ MPa}\sqrt{\text{m}})$$

$$\Delta K_I = 30.1 \text{ ksi}\sqrt{\text{in.}} - 4.2 \text{ ksi}\sqrt{\text{in.}} = 25.9 \text{ ksi}\sqrt{\text{in.}}\,(28.5 \text{ MPa}\sqrt{\text{m}})$$

Repeat the calculations for $a = 0.4$ in.

$$K_{I\,max} = 53,750\sqrt{\pi(0.4)} = 60.3 \text{ ksi}\sqrt{\text{in.}}\,(66.3 \text{ MPa}\sqrt{\text{m}})$$

$$K_{I\,min} = 7500\sqrt{\pi(0.4)} = 8.4 \text{ ksi}\sqrt{\text{in.}}\,(9.2 \text{ MPa}\sqrt{\text{m}})$$

$$\Delta K_I = 60.3 \text{ ksi}\sqrt{\text{in.}} - 8.4 \text{ ksi}\sqrt{\text{in.}} = 51.9 \text{ ksi}\sqrt{\text{in.}}\,(57.1 \text{ MPa}\sqrt{\text{m}})$$

Note: For a constant load range, the stress-intensity range increases as the crack length increases. Thus the crack extension per cycle increases as the crack grows.

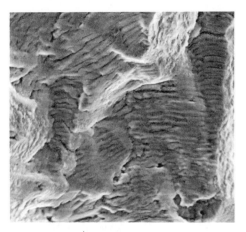

↑ Growth direction

FIGURE 19.16   Fatigue striations observed on the fatigue fracture surface of an iron–nickel–chromium alloy; 1200×

For many materials, distinct markings on the fracture surface indicate the occurrence of stage II crack growth. These markings, shown in Figure 19.16, are called *fatigue striations*. Generally the distance between striations represents the crack growth per stress cycle. We shall discuss this in greater detail later.

EXAMPLE 19.6   *[ES]*

A large panel with a central crack through the thickness, with $a = 0.10$ in., is cyclically stressed between 14,000 psi and zero stress. The panel is 30 in. wide and 0.5 in. thick and is fabricated from a material with a fracture toughness of 25,000 psi$\sqrt{\text{in.}}$. Growth of fatigue cracks follows the law [Equation 19-7]:

$$\frac{da}{dN} = C\Delta K_I^{m}$$

with $C = 1.8 \times 10^{-18}$ in./(cycle · psi$\sqrt{\text{in.}}$) and $m = 3.0$. Estimate the number of fatigue cycles it takes to cause failure of the panel.

Answer   Calculate the length of crack at which failure occurs. For the "center-cracked" panel,

$$K_I = S\sqrt{\pi a}\, Y(a/w)$$

Because $a \ll w$, we shall set $Y(a/w) = 1$. [$Y(a/w)$ means that Y is normally a function of w. See Example 2.6.]

$$a_c = \left(\frac{K_{IC}}{\sigma_f}\right)^2 \frac{1}{\pi} = \left(\frac{25{,}000 \text{ psi}\sqrt{\text{in.}}}{14{,}000 \text{ psi}}\right)^2 \frac{1}{\pi} = 1.0 \text{ in. } (25.4 \text{ mm})$$

Note: Keep in mind that the values for the crack lengths used here are one-half the total length in the panel. At failure we would see a fatigue crack of 2 in. total length.

We now estimate the number of cycles it takes to extend the crack by a small amount, then repeat this operation until the crack reaches the critical length, where we sum the cycles to failure.

We choose an increment of growth of 0.05 in. For extension from 0.10 in. to 0.15 in., we use the average crack length of 0.125 in. to compute an average growth rate over this interval. From this we compute the number of cycles required. A sample calculation follows.

$$a_0 = 0.10 \text{ in.}, \qquad a_1 = 0.15 \text{ in.} \qquad a_{av} = 0.125 \text{ in.}$$

$$\Delta K_I = \Delta S \sqrt{\pi a_{av}} = 14{,}000 \text{ psi} \sqrt{\pi(0.125)}$$

$$= 8770 \text{ psi} \sqrt{\text{in.}} \, (9.65 \text{ MPa} \sqrt{\text{m}})$$

Next, compute the growth rate for this value of ΔK_I.

$$\frac{da}{dN} = C \Delta K_I{}^m = (1.8 \times 10^{-18})(8770)^3$$

$$= 1.21 \times 10^{-6} \text{ in./cycle} \, (30.7 \times 10^{-6} \text{ mm/cycle})$$

Compute the number of cycles needed to extend the crack by $\Delta a = 0.05$ in.

$$N = \frac{\Delta a}{da/dN} = \frac{0.05 \text{ in.}}{1.21 \times 10^{-6} \text{ in./cycle}} \qquad \text{(First interval) } N = 41{,}100$$

Repeat the foregoing procedure for each interval of growth. The results are as follows:

a (in.)	a_{av} (in.)	N (cycles)	ΣN (cycles)
0.10	0	0	0
0.15	0.125	41,100	41,100
0.20	0.175	24,500	65,600
0.25	0.225	16,900	82,500
0.30	0.275	12,600	95,100
0.35	0.325	9,800	104,900
0.40	0.375	7,900	112,800
0.45	0.425	6,500	119,300
0.50	0.475	5,500	124,800
0.55	0.525	4,800	129,600
0.60	0.575	4,200	133,800
0.65	0.625	3,700	137,500
0.70	0.675	3,300	140,800
0.75	0.725	3,000	143,800
0.80	0.775	2,700	146,500
0.85	0.825	2,400	148,900
0.90	0.875	2,200	151,100
0.95	0.925	2,000	153,100
1.0 (a_c)	0.975	1,800	154,900

Note: Because $Y = 1$, we may use calculus to solve this problem in closed form.

$$\frac{da}{dN} = C\Delta K_I^m \quad \text{or} \quad \frac{da}{C\Delta K_I^m} = dN$$

Substituting in Equation 19-8,

$$N = \int_{a_0}^{a_f} \frac{1}{C} \Delta S^{-3} (\pi a)^{-3/2} \, da$$

$$= \frac{1}{C(\Delta S)^3 \pi^{3/2}} \int_{a_0}^{a_f} a^{-3/2} \, da$$

$$= \left(\frac{-2}{C(\Delta S)^3 \pi^{3/2}} \right) \left(a^{-1/2} \Big|_{a_0}^{a_f} \right)$$

$$= \frac{2}{C(\Delta S)^3 \pi^{3/2}} \left(\frac{1}{\sqrt{a_0}} - \frac{1}{\sqrt{a_f}} \right)$$

$$= \frac{2}{(1.8 \times 10^{-18})(14{,}000)^3 \pi^{3/2}} \left(\frac{1}{\sqrt{0.1}} - \frac{1}{\sqrt{1}} \right)$$

$$= 1.57 \times 10^5 \text{ cycles}$$

The actual and approximated answers differ. However, when we use smaller increments of crack growth in our calculations, the answers agree more closely. For a more thorough description of fatigue, consult *Deformation and Fracture Mechanics of Engineering Materials*, by R. W. Hertzberg, published by John Wiley, New York, 1976; also *Fracture and Fatigue Control in Structures*, by S. T. Rolfe and J. M. Barson, published by Prentice-Hall, Englewood Cliffs, N. J., 1977.

19.15 Stress Corrosion Cracking

As we saw in Chapter 18, the environments in which most engineering materials are used are seldom inert. In many cases environmental attack can limit the useful lifetime of components. Corrosion is the most common result of environmental attack. But there are more subtle interactions that may limit service life even if general corrosion problems have been eliminated. One such particularly insidious interaction is stress corrosion cracking (SCC).

Stress corrosion cracking, as the name suggests, is the advance of a crack in a material subjected to stress in the presence of a hostile gaseous or liquid environment. SCC may be particularly difficult to detect for the following reasons:

1. Environments that are only mildly corrosive to the material may cause severe SCC.

2. The required concentration of the harmful component in the environment may be extremely small and its presence difficult to detect.
3. The attack may be highly localized as one or a number of small cracks that propagate undetected to failure.
4. Residual stresses in components are often great enough to cause stress corrosion cracking even in the absence of applied stresses.

As with fatigue cracking, there are two approaches to the measurement of lifetimes in components experiencing SCC. One approach involves determining the time required to cause failure of smooth, uncracked specimens subjected to stresses in the environment in question. Figure 19.17 shows that, as the applied stress level decreases, the time to failure from SCC increases. If the stress is low enough, time to failure becomes excessively long and an apparent threshold stress below which SCC does not occur can be defined. In such a test, the time to failure necessarily involves the time required to initiate the crack by localized chemical attack and the time required to propagate the crack to failure. As in fatigue, when small cracks are already present, the time to failure may be much shorter than that predicted from tests on smooth specimens.

If we measure the rate of growth of stress corrosion cracks for a particular alloy–environment combination, we find, just as in fatigue, that the rate is determined by the stress-intensity factor, K_I. Figure 19.18 illustrates schematically how the growth rate of stress corrosion cracks varies with K_I. Note that, because loading is static rather than cyclic, K_I is used rather than ΔK_I. This figure has three regions of interest. In region I the rate of crack growth increases with increasing K_I. In region II the rate of crack advance is independent of the intensity of the stress. In region III, further increases in K_I cause a rapid increase in rate of growth of the crack.

As in fatigue, a threshold value of stress intensity exists. Below this threshold value—designated K_{ISCC}—no growth of cracks from stress corro-

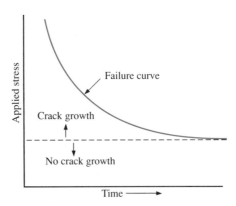

FIGURE 19.17   Variation of time to failure due to stress corrosion cracking as a function of applied stress

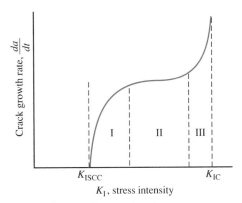

FIGURE 19.18   Variation of rate of growth of stress corrosion crack with applied stress intensity, K_I

sion occurs. Values of K_{ISCC}, which can be determined from laboratory testing, are an indication of the relative susceptibility of a material to SCC. High values of K_{ISCC} are desired.

Figure 19.19 shows an example of cracking due to the combination of corrosive environment and high residual stresses. The cup was made by deep-drawing stainless steel. After drawing, the cup, which contained no cracks, was placed on a laboratory shelf. When it was inspected the following day, the cracks shown in Figure 19.19 were observed. Apparently stress corrosion

FIGURE 19.19   A deep-drawn cup that cracked while sitting overnight on a shelf. Atmospheric contaminants, combined with high residual tensile stress, can cause failure by stress corrosion cracking.

[Photo courtesy D. Meuleman]

TABLE 19.1   Relative Susceptibility of Copper Alloys to Ammoniacal SCC

Very low susceptibility	Cupronickels, tough pitch copper, silicon bronze
Low susceptibility	Phosphorized copper
Intermediate susceptibility	Brasses containing less than 20% Zn, such as red brass, commercial bronze, aluminum bronze, nickel silver, phosphor bronze, and gilding metal
High susceptibility	Brass containing over 20% Zn, with or without small amounts of Pb, Sn, Mn, or Al (such as leaded brass, naval brass, admiralty brass, manganese bronze, or aluminum brass). The higher the zinc content, the higher the susceptibility.

From B. F. Brown, *Stress Corrosion Cracking Control Measures,* National Bureau of Standards Monograph No. 156, 1977.

resulted from the laboratory air or from contaminants in the lubricating oil. In many cases, drawn parts must be annealed immediately after drawing to remove residual stresses and to avoid the start of stress corrosion cracking by immediate reaction with the atmosphere.

As an illustration of resistance to stress corrosion, Table 19.1 shows the considerable variation in crack susceptibility among copper alloys exposed to an ammonia-bearing atmosphere.

19.16   Fracture of Ceramics and Polymers

Because of the low fracture toughness of ceramics and polymers, the study of failure by brittle fracture is very important.

The origin of the crack and the crack pattern often disclose the loading conditions that caused the crack (Figure 19.20). In another common cause, thermal shock, a single crack is often obtained. Figure 19.21 shows a champagne glass that cracked suddenly when the ice-cold liquid was poured to the level of the crack. The glass layers chilled by the liquid contracted to a smaller diameter, imposing a severe shear stress relative to the unchilled upper layers.

Because of the low fracture toughness of these materials, even a small flaw is very important. Examination of the origin often discloses an inclusion, a void, or even a region of large grains (Figure 19.22). Around the fracture origin a "fracture mirror" (a smooth region) is found, surrounded by an irregular region of "mist" and "hackle." In the mirror region, a single crack develops and branches in the outer regions.

Another feature that is particularly important in polymer fracture is the phenomenon of *crazing* (Figure 19.23, page S-91). A test for this effect is to place a light beneath the lower surface of a beam during a bend test in which the upper surface is in tension. The transparent section clouds and whitens— that is, crazes—at a stress well below the stress that would be necessary to

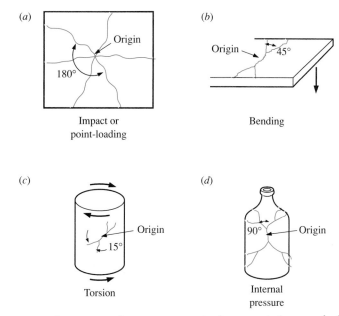

FIGURE 19.20 Information that one can gain by examining crack direction and crack branching

[Reprinted from D. W. Richerson, *Modern Ceramic Engineering*, 1982, p. 326. Courtesy of Marcel Dekker, Inc.]

FIGURE 19.21 Photograph of a champagne glass with a circumferential fracture due to thermal stress

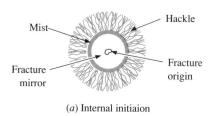

(a) Internal initiaion

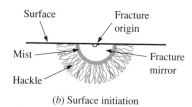

(b) Surface initiation

FIGURE 19.22   Schematic showing the typical fracture features that surround the fracture origin: (a) internal initiation; (b) surface initiation

[Reprinted from D. W. Richerson, *Modern Ceramic Engineering,* 1982, p. 329. Courtesy of Marcel Dekker, Inc.]

produce gross fracture. Examination with the electron microscope discloses that ligaments of the polymer bridge microcracks, just as the ductile phase bridges a crack in the brittle phase of a composite. It is of further interest that various chemicals, and even water, can reduce the stress required for crazing. This phenomenon is important in the design of polymer windows and as a warning prior to gross cracking. The clouding or whitening is caused by the cavities produced by the craze.

EXPERIMENTAL METHODS FOR EXAMINING FAILED PARTS

19.17   General Approach, Nondestructive Methods

It is essential to divide experimental methods into two groups: nondestructive and destructive. In many cases the investigator is limited to nondestructive methods, although it can be reasoned that the careful cutting of a specimen for microexamination and for exhibit is not really destructive, compared with grinding up a sample for chemical analysis.

The budding metallurgist is often too eager to cut up the part and etch a microspecimen, as is indicated by the following case. Our laboratory was notified that a failed gear was being sent in for examination. After waiting several months, we checked with a colleague who was in charge of the engineering design laboratory next door to see whether the part had gone astray.

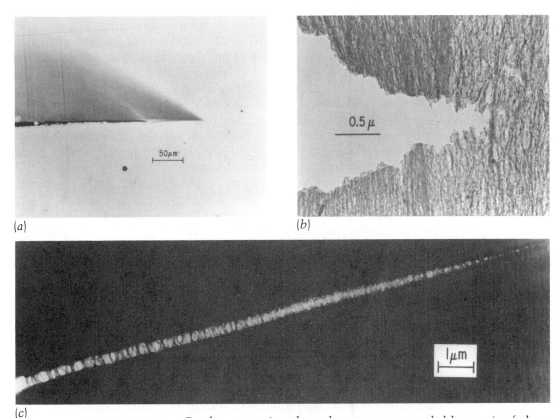

(a)

(b)

(c)

FIGURE 19.23   Crack propagation through a craze surrounded by a pair of shear bands (an epsilon crack) in polycarbonate

[(a) R. P. Kambour, A. S. Holik and S. Miller, *Polymer Science*, Vol. 16, 91, 1978. Courtesy R. P. Kambour, General Electric Corporate Research and Development, Schenectady, N.Y. (b) R. P. Kambour and A. S. Holik, *Polymer Science*, part A-2, Vol. 13, 93, 1969. Courtesy R. P. Kambour. (c) R. P. Kambour and R. R. Russell, *Polymer Science*, Vol. 12, 237, 1971. Courtesy R. P. Kambour]

We received a pretty blunt answer: "Sure, the gear came in a month ago. We checked the dimensions against the blueprint and found some bad conflicts with the mating gear. It's a good thing it didn't go to you or you would have cut it up right away and we never would have found the trouble!"

This story illustrates that, before proceeding with the actual examination, we need to obtain all relevant background material, such as drawings and material specifications, conditions of service, history of similar parts (failures and successes), performance data, and design calculations. Also, all possible nondestructive examinations should be completed. An equally important precaution is to repress the urge that possesses laypeople and engineers alike — to force the two pieces of the fracture together and exclaim, "They fit!" Imagine the effect on the scanning electron photomicrographs shown throughout this text of gashes looking like Alpine landslides caused by mashing fractures together. Some investigators have even interpreted such gashes as fatigue striations!

The first step in nondestructive examination is careful viewing in a good light with the unaided eye, then with a 10× hand lens or low-power binocular microscope. This examination is often adequate to distinguish between different types of fracture, such as between a ductile tensile failure and fatigue failure. This is also an essential step in beginning to locate the point of initiation of the fracture and in selecting specimens for fractographic examination with the scanning electron microscope.

*Scanning electron microscope (SEM) examination** is of great value for investigating the fracture in detail. If the part cannot be placed in the vacuum chamber of the SEM or if a specimen cannot be cut, another method called the *replica technique* may be used. We place a few drops of acetone (generally noncorrosive) on the surface and press a strip of cellulose acetate plastic against the surface. After a few seconds, the film is stripped off, bearing an accurate replica of the fracture. For examination in the *transmission electron microscope* (TEM), a thin layer of metal such as gold is vapor-deposited on the replica. This does not alter the detail, and excellent photomicrographs are obtained.

Other nondestructive techniques include magnaflux examination, use of penetrating dye, electrical tests, and radiography. For *magnaflux inspection*, the component must be ferromagnetic. The part is magnetized and a colored magnetic powder is shaken on the surface. High concentrations of powder are attracted to cracks, just as iron filings are attracted to the gap in a horseshoe magnet. The part should be magnetized in several directions to provide favorable conditions for revealing all cracks. This method is called *dry magnafluxing.* Some people use a more sensitive method that employs a fluorescent magnetic powder contained in a low-viscosity oil. In this case, when the part is viewed under ultraviolet light, the fluorescent powder delineates the defects.

Dye-penetrant inspection is used for nonmagnetic materials. First the part is sprayed with a red-colored oil that penetrates defects. Next the oil is washed from the surface, leaving a residue in the cracks. Finally, a white coating is sprayed over the surface. The oil in the cracks gradually oozes through the white surface, delineating the defects.

Electrical and sonic methods encompass a variety of techniques. One technique is *ultrasonic testing.* Sound waves are transmitted into the part from a vibrating crystal pressed against the outer surface. These sound waves are then received and displayed on an oscilloscope. We can determine discontinuities and variations in structure from the oscilloscope trace or evaluate them quantitatively and display them as numerical values. Other techniques involve evaluating hysteresis and eddy-current losses as well as acoustic emission from the part.

---

*The principles of the SEM and TEM are described in standard texts on physics.

Radiography by x ray, cobalt 60, or other high-energy sources is well known from its uses in medical diagnosis; an extensive description is not needed here. We should point out that excellent ASTM standard radiographs are available for specification of quality.

19.18   Analyzing the Appearance of the Fracture

Before we assign responsibility for failure, we must first determine how the component failed. For example, components may fail because the stresses imposed during service exceeded the ultimate strength of the material. If this is the case, we say that the part failed because of a tensile overload. If fatigue is responsible, we say that the component failed as a result of fatigue. The fracture surfaces produced by a tensile overload appear quite different from those produced by fatigue failure. In fact, it is often possible to classify the type of failure by macroscopic observation of the fracture surfaces themselves.* In this section we shall briefly discuss features of the more common types of failure, including:

1. Tensile overload
2. Propagation of brittle cracks
3. Fatigue
4. Stress corrosion

Tensile Overload

Tensile overload results when the applied stress reaches the ultimate tensile strength of the material. The appearance of the fracture depends strongly on the material's ductility. Figure 19.24a shows the general types of fractures for the ductile and brittle cases. The appearance of the fracture is generally dull. In low-strength, high-ductility materials, evidence of plastic deformation can be found.

Propagation of Brittle Cracks

We may think of brittle cracks as initiating at a pre-existing defect and rapidly propagating through the component to cause failure. In many materials this rapid propagation leaves characteristic marks on the surfaces of the fracture. We can use these marks to locate the origin of fracture. Figure 19.24b shows two examples.

---

* Although there are some general characteristics of fracture surfaces that we shall summarize in this section, you should be aware that in many cases a combination of fracture types may be present. Also the characteristics of one type of failure may differ from material to material, as evidenced by our earlier discussion of fracture in polymers and ceramics.

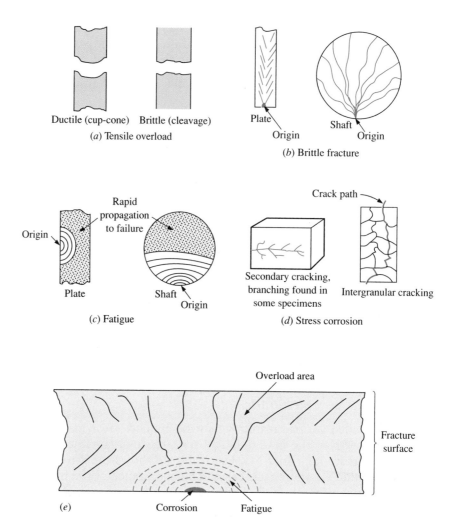

FIGURE 19.24 Features common to various types of fractures: (*a*) tensile overload fracture; (*b*) brittle fracture; (*c*) fatigue fracture; (*d*) stress corrosion cracking; (*e*) a combination of types of fracture that results in failure of a plate

Fatigue

Often we can readily note the presence of fatigue by observing the surface of the fracture. Fatigue cracks generally propagate normal to the stress axis. No shear lips are present. (A *shear lip* is a slanting ridge at the edge of a fracture.) Many times a series of concentric rings called clamshell marks, or beach marks, are visible on the surface (see Figure 2.27). These marks represent abrupt changes in fatigue loading during service and help pinpoint the origin of the fatigue crack. Figure 19.24c shows two cases of fatigue failure. In the typical case, a region of fast fracture is present where the final fracture occurred. This region is usually a bright crystalline region.

Stress Corrosion

Familiarity with each material under study is important in assessing stress corrosion failure. Some materials may show branching or multiple cracking. The fracture path may be transgranular or intergranular (Figure 19.24d).

In some instances a number of different events combine to produce the final failure. Such a case is shown schematically in Figure 19.24e. Here a fatigue crack initiated from a corrosion pit. The crack propagated, and final failure resulted from a rapid propagation of the crack. Each area shows different features. The analyst must rely on experience and a knowledge of the properties of the failed material to sort out the multiple types of degradation. We suggest consulting *The Metals Handbook*, Vols. 9, 10, 11 and 12, 9th ed., 1985, 1986, 1986, 1987, for a detailed coverage of the macroscopic and microscopic aspects of fracture surfaces.

In ceramics and polymers we encounter a phenomenon called *static fatigue,* which is related to stress corrosion cracking. In this case a component under stress, such as a tight screw or fitting, or a component under residual tensile stress fails after a period of time. One of the important contributors is the environment; many times a humid environment leads to attack at the surface, and then the crack progresses aided by corrosion effects. In general, these cases should be considered stress corrosion cracking, because no cyclical loading (as in fatigue testing) is present.

EXAMPLE 19.7 *[EJ]*

Why is a fatigue failure sometimes referred to as a catastrophic brittle failure?

Answer   The catastrophe is that a crack can grow to the critical length and not be observed. Final failure can be in a single stress cycle because of an overload on the remaining area. Upon visual examination, the fracture may appear brittle and the striations (crack opening with each cycle) may be too close to observe.

Microscopic examination, especially with an electron microscope, can show the striations (Figure 19.16), or beach marks, and the final failure area may show some ductility by microscopic cup-cones. These cup-cones are from the production of microvoids and their coalescence during failure. Because the ductility scale is so small, the visible fracture appears brittle all the way through.

19.19   Destructive Methods of Analysis

Although not literally destructive, these methods do involve sectioning and some etching of the component. Indeed, a well-prepared, etched cross section is often the best permanent exhibit of all. We shall not discuss preparation of specimens in detail (see *ASM Metals Handbook*, Vol. 9, 9th ed., 1985), but we shall indicate some important precautions.

Macroexamination

The purpose of macroexamination is to disclose two types of structure: (1) flaws and (2) macroscopic inhomogeneity. In many cases it is helpful to cut sections of the specimen in two perpendicular directions and then to polish and etch the sections. This brings to light larger inclusions, such as slag, sand, or entrapped metal oxides (also called *dross*) and shrinkage cavities (see Figs. 6.10 and 6.32). Also, if the structure near the surface is different as a result of conditions of casting or heat treatment, this is evident from differences in etching. Finally, the etch discloses the direction of hot or cold working, which affects the properties.

Extensive tables of etchants are given in the *ASM Metals Handbook*, Vol. 9, 9th ed., 1985, but we shall give a few favorites for most materials here. (In all cases, consult a standard chemistry safety manual before proceeding, because the reagents and fumes are dangerous.)

Material	*Method*
Steel or cast iron	Add 1 part conc. HCl to 1 part water. Heat in Pyrex or other suitable container to 160°F (71°C). Immerse sample for 10–40 min to desired amount of etch. Use well-ventilated hood. Handle sample with tongs and rubber gloves and rinse carefully with water.
Aluminum alloys	Dissolve 10 g sodium hydroxide in 90 ml water, use at 150°F (66°C), 5–15 min. Remove etching products with concentrated nitric acid, then wash with water.
Magnesium alloys	Mix ethylene glycol (75 ml), distilled water (24 ml), and concentrated nitric acid (1 ml). Etch sample with mixture at room temperature; wash with water.
Copper alloys	Etch sample with concentrated nitric acid, under hood. Remove acid with cold water.

In all cases, avoid creating spurious indications by excessive heating of the cut surface during cutting or pronounced cold working by machining. For example, it is extremely difficult to section a hard material such as white cast iron or hardened steel with an abrasive cutoff wheel without altering the structure near the cut surface. Use the softest cutoff wheel, with a copious flow of coolant, followed by extensive wet grinding after sectioning.

Microexamination

The specimen should be cut for *microscopic examination* only after careful macroexamination has indicated critical areas. Observe the same precautions in cutting microspecimens as in cutting macrospecimens. Polishing and etching procedures are given in detail in the *ASM Metals Handbook*, Vol. 9, 9th ed., 1985, and in earlier editions. A few useful etches for microspecimens of different alloys are recommended and listed below.

Material	Method
Iron and steel (less than 8% alloy)	Use 2% nitric acid in ethanol.
Stainless and high-alloy steels	Use a mixture of nitric acid (10 ml), hydrochloric acid (25 ml), and glycerol (25 ml).
Aluminum alloys	Use a mixture of conc. hydrofluoric acid (0.5 ml) and water 99.5 ml). Hydrofluoric acid is particularly dangerous. It penetrates the skin even in low concentration.
Magnesium alloys	Use a mixture of ethylene glycol (60 ml), glacial acetic acid (20 ml), conc. nitric acid (1 ml), and distilled water (19 ml).
Copper alloys	1. Use ammonium hydroxide (5 ml), 3% hydrogen peroxide (5 ml), and water (5 ml). Use fresh for brass. 2. Use a mixture of 2 g of $K_2Cr_2O_7$, 1.5 g of NaCl, 8 ml of conc. H_2SO_4 and 100 ml of H_2O (for more resistant alloys, such as silicon brass).

When you conduct a metallographic examination of the structure, you should examine the entire polished surface, especially the areas adjacent to the fracture, at low magnification (50–100×) before proceeding to higher magnification. You can miss the cause of failure if you do not follow this procedure. In many cases the cause of failure has been improperly assessed because a non-representative field was selected for study and photographing at high power.

Mechanical Testing

Hardness tests may be made on carefully prepared surfaces of the component or on microspecimens. Microhardness tests, as illustrated in Chapter 7 and elsewhere, are particularly valuable in determining the properties of thin layers or of individual phases. Tensile, compressive, torsion, and impact test blanks may be cut carefully from selected areas to check the mechanical properties of the material in the component. If the results are slightly at variance with the specified values, the manufacturer is not necessarily at fault. Only if the specifications require minimum values *in specimens cut from specified regions of components* can the cut specimen be used in place of special test bars as an index of responsibility.

Summary

To provide a basis for understanding the analysis and prevention of failure, the discussion is divided into two parts.

1. *Fracture toughness and fracture mechanics,* plus additional material on fatigue, and stress corrosion cracking. To avoid catastrophic failures in high-strength, low-ductility alloys, one must consider fracture toughness as well as yield strength. In cases in which the plane-strain fracture toughness is the criterion, the important formula is

$$\sigma_f = \frac{K_{IC}}{Y\sqrt{\pi a}}$$

where σ_f is the fracture stress, Y is a factor calculated from the crack length and the geometry of the part, K_{IC} is the plane-strain fracture toughness, and a is the crack length.

Fatigue fracture depends on the initiation of a crack, its growth to a critical length for the geometry of the part, and the stress involved. Parts can be designed to avoid fatigue failure using the same concepts of fracture toughness as for static stresses. Stress corrosion cracking also depends on the concept of development of a crack under the combined effects of stress and corrosion.

2. *Examination of failed parts.* Once we have obtained all data relevant to the failure, we should subject the part first to nondestructive testing. This includes macroexamination, with photography and, where indicated, radiography, dye penetrant, magnaflux, and ultrasonic inspection. Small pieces may be viewed directly in the scanning electron microscope. If metallography is permitted, microspecimens should be cut from relevant areas, polished, etched, and examined. Hardness and mechanical tests may be performed to determine how the properties of the material compare with specifications.

Definitions

Dye-penetrant inspection A nondestructive method of examining a failed part, in which a penetrating oil is sprayed on the part to penetrate cracks, followed by a background spray that delineates oil-saturated cracks.

Fracture stress, σ_f The nominal stress at fracture. When computing the nominal stress, one generally ignores the presence of a flaw.

K_{ISCC} The value of stress intensity above which cracks grow because of stress corrosion. Below this value, crack growth due to stress corrosion does not occur.

Macroscopic examination Examination with the unaided eye or low magnification (usually up to $50\times$).

Magnaflux inspection Use of a magnetic field in conjunction with magnetic powder to find surface flaws or cracks.

Microscopic examination Examination at greater magnification than $50\times$, usually using a light microscope with a polished and etched section.

Nondestructive testing (NDT)   The examination of a part by methods such as x ray or magnaflux that do not change the part's surface or interior, so that it can be used afterwards in service.

Nonmetallic inclusions   Brittle oxides, sulfides, or silicates that result from the entrapment of foreign matter in liquid metal during pouring or from reactions within the liquid metal.

Plane-strain fracture toughness, K_{IC}   The minimum stress intensity required to cause catastrophic failure. K_{IC} can be measured only above a particular thickness, which depends on the material, and is constant for all thicknesses above this value.

Plane-stress fracture toughness, K_C   The value of stress intensity required to cause catastrophic failure in components with thicknesses below that required for plane strain. K_C varies with thickness.

Scanning electron microscope (SEM) examination   Use of an electron beam with scanning equipment that magnifies 50–50,000×. A fractured surface can be examined without sectioning and polishing.

Static fatigue   Failure that occurs in ceramics and polymers with time as a result of a stress and a corrosive medium.

Stress-concentration factor, K_σ   The ratio of the stress near a notch or hole to the nominal stress.

Stress-intensity factor, K_I   A measure of the magnitude of the stresses near the tip of a sharp crack. K_I is a function of nominal stress, crack length, and component geometry.

Stress-intensity range, ΔK_I   The range of cyclic stressing that occurs between K_{max} and K_{min}; $\Delta K_I = K_{max} - K_{min}$.

Surface energy, γ_s   The increase in energy of a system per unit area of new surface created by a fracture. Related to the number of atomic or molecular bonds broken to create the new surface.

Threshold stress-intensity range, ΔK_{th}   The value of ΔK_I below which crack propagation due to fatigue does not occur.

Ultrasonic testing   A nondestructive method of examining a failed part, in which ultrasonic impulses, coupled with a crystal receiver, are used to monitor reflections and speed of sound waves in the part. Reflections by internal surfaces indicate defects.

Problems

19.1   *[E]* An air-operated hoist had a hook 1 in. in diameter at its end. The shank portion of the hook was too long, so the engineer shortened the shank by flame-cutting out a portion and rewelding the end sections. Although the shank of the hook was 1 in. in diameter, the diameter at the weld was excessive ($1\frac{3}{4}$ in.). Subsequently the rewelded shank failed in service at the junction of the weld and the shank. It was found that grinding back the weld eliminated the service failures (see the sketch at the top of the next page). Explain why removal of the excess material seemed to strengthen the component. (Sections 19.1 through 19.7)

19.2   *[ES/E]* Sketch true stress–true strain curves similar to those in Figure 19.2 for a ther-

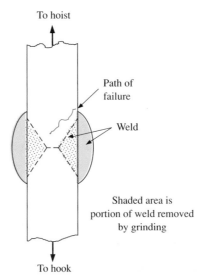

To hoist

Path of
failure

Weld

Shaded area is
portion of weld removed
by grinding

To hook

moplastic polymer above and below its glass transition temperature. Also include the energy required for fracture. (Sections 19.1 through 19.7)

19.3 *[ES]* What is the maximum load that can be applied to a bar 1 in. thick by 1 in. wide (25.4 mm × 25.4 mm) with a hole of $\frac{1}{4}$-in. (6.35-mm) diameter as shown in Figure 19.4 if the stress at the edge of the hole must not exceed 50,000 psi (345 MPa)? (Sections 19.1 through 19.7)

19.4 *[ES]* Answer the following questions for a given material. (Sections 19.1 through 19.7)
   a. What happens to the fracture stress in Equations 19-2, 19-3, and 19-4 when the flaw size increases?
   b. What happens to the stress in Equation 19-5 for a given flaw size when the stress intensity or sharpness of the crack increases?

19.5 *[EJ]* Why are wire mesh or plastic interleaves put in glass windows? (Sections 19.1 through 19.7)

19.6 *[ES]* A transparent thermoplastic polymer below its glass transition temperature has a surface energy γ_s of 0.2 in.-lb/in^2 (50.5 J/m^2) and an elastic modulus of 4×10^5 psi (2.76×10^3 MPa). After injection molding, a center flaw 0.0354 in. (0.90 mm) long can be seen. (Sections 19.1 through 19.7)
   a. Calculate the fracture stress.
   b. What flaw size would be required for the fracture stress to equal the tensile strength of 8000 psi (55.2 MPa)?

19.7 *[ES/EJ]* Indicate why it should make little difference in our discussion of fracture mechanics whether we use true or engineering stress and strain values in calculations. (Sections 19.1 through 19.7)

19.8 *[ES]* In your own words, differentiate among the following terms: stress-concentration factor, stress-intensity factor, and fracture toughness. (Sections 19.1 through 19.7)

19.9 *[ES/EJ]* (Sections 19.8 through 19.11)
   a. When a material exhibits high ductility or low yield strength in a tensile test, what happens to the specimen size requirement for plane-strain fracture toughness (K_{IC})?
   b. How might fracture toughness values for specimen thicknesses less than those required for plane strain be used in design?
   c. Schematically plot fracture toughness versus specimen thickness.

19.10 *[ES/EJ]* The fracture toughness of a section of an alloy 0.6 in. (1.5 cm) thick is 25,000 psi$\sqrt{\text{in.}}$ (27.5 MPa$\sqrt{\text{m}}$). The yield strength of this alloy is 35,000 psi (241.5 MPa). We wish to use this fracture toughness value in designing a component from the same material, but with a thickness of 1.6 in. (0.041 m). Is this approach valid? (Sections 19.8 through 19.11)

19.11 *[ES]* A steel that exhibits the relationship between fracture toughness and yield strength given in the accompanying graph is to be used for a pipeline. The wall stress is to be as high as possible without either yielding or fracturing—that is, minimum wall thickness is desired. Inspection techniques ensure that there will be no cracks or defects longer than 2 mm ($a < 1$ mm). What level of yield strength should be specified? Assume a geometric correction factor of $Y = 1.0$. (Sections 19.8 through 19.11)

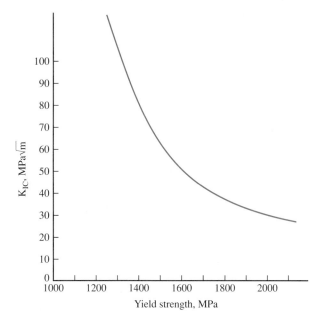

19.12 *[ES]* The stress distribution in plane stress near the tip of a crack is as follows. (Sections 19.8 through 19.11)

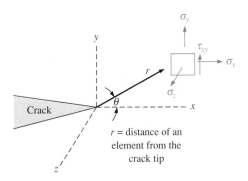

a. For a stress intensity of 20,000 psi$\sqrt{\text{in.}}$ (22 MPa$\sqrt{\text{m}}$), show the variation of σ_y along the projected path of the crack (x direction, $\theta = 0$) from $r = 0.05$ to $r = 1.0$ in. (1.3–25.4 mm).

b. Repeat part *a* for $K_I = 40,000$ psi$\sqrt{\text{in.}}$ (44 MPa$\sqrt{\text{m}}$).

c. Given that the yield stress is 35,000 psi (241.5 MPa), approximate the size of the plastic deformation field at the crack tip for both K_I values.

19.13 [ES] Assume that the tension support member described in Example 2.6 must support a load of 320,000 lb (145,280 kg). Because of inspection difficulties, the smallest edge flaw that can be detected with confidence is 0.15 in. (3.8 mm) long. Which material— 4340V given a 500°F (260°C) temper or 4340V given an 800°F (427°C) temper—should you choose to minimize the weight of the support member? Assume that the width remains constant at 4 in. (Sections 19.8 through 19.11 and Section 2.16)

19.14 [ES] A container is made of steel with a yield strength of 85,000 psi (586.5 MPa) and a plane strain fracture toughness of 100,000 psi$\sqrt{\text{in.}}$ (93.5 MPa$\sqrt{\text{m}}$). (Sections 19.8 through 19.11)

a. At what container thickness would plane-strain fracture toughness be a concern?

b. Calculate how much the fracture stress would increase or decrease if the detectable minimum crack length were doubled. Assume a constant geometric correction factor.

19.15 [ES/EJ] Consider two metal alloys exposed to the same stress: K_{IC} for alloy A = 75,000 psi$\sqrt{\text{in.}}$ (82.5 MPa$\sqrt{\text{m}}$), K_{IC} for alloy B = 50,000 psi$\sqrt{\text{in.}}$ (55 MPa$\sqrt{\text{m}}$). (Sections 19.8 through 19.11)

a. Which material can tolerate the largest flaw before fracture?

b. Calculate the ratio of flaw sizes, a_A/a_B. Assume the geometric correction factor $Y = 1.0$.

c. Which alloy is likely to have the highest yield strength?

19.16 [ES/EJ] Considerable scatter is generally found in *S-N* fatigue data. What are the possible causes of this scatter? [*Hint:* Review the section on initiation of cracks.] (Sections 19.12 through 19.15)

19.17 [EJ] A thin section of a fan blade has failed in service. You believe that the mode of failure is fatigue. Close examination of the fracture surfaces does not reveal the characteristic clamshell marks (as shown in Figure 19.24), although a portion does show a shear lip and rapid propagation of cracks. In fact, a large portion of the surface of the fracture appears worn and shiny. Why might this still represent a fatigue failure? (Sections 19.12 through 19.15)

19.18 [ES/EJ] Stress corrosion cracking (SCC) occurs in stainless steels and is often intergranular (along the grain boundaries). As a result, a common conclusion is that SCC is *always* intergranular. Explain why intergranular SCC may occur in stainless steel and under what circumstances it may also be transgranular. (Sections 19.12 through 19.15)

19.19 [ES/EJ] Indicate whether the following statements are correct or incorrect and justify each answer. (Sections 19.12 through 19.15)

a. In cyclic stressing under a constant load range, the crack growth rate is constant.

b. The machining method used on a component has no effect on fatigue life.

c. Fatigue striations are visible to the naked eye.

19.20 [ES/EJ] Evaluate the statement of a design engineer who says that a few thousand cycles at a high load should cause no damage because normal operating loads are much lower where the anticipated life is greater than 10^7 cycles. (Sections 19.12 through 19.15)

19.21 [ES/EJ] Might a rubber mount for an automotive engine fail by fatigue? How might you tell from the fracture appearance? (Sections 19.12 through 19.15)

19.22 *[ES]* The fatigue life predicted in Example 19.6 assumes stage II fatigue. What happens to this prediction if the ΔK_{I} values become very low or very high? (Sections 19.12 through 19.15)

19.23 *[ES]* In Example 19.6, values for m and C have been provided. Outline a procedure to determine these constants experimentally. (Sections 19.12 through 19.15)

19.24 *[ES]* Refer to Example 19.6. (Sections 19.12 through 19.15)
   a. For the initial crack length given, calculate the cycles to failure if the fracture toughness of the material is doubled.
   b. Using the original toughness given in Example 19.6, calculate the cycles to failure if the initial size of the flaw is decreased to $a = 0.05$ in. (1.3 mm).
   c. Explain the reasons for the differences between your answers in parts *a* and *b*.

19.25 *[ES]* The K_{ISCC} for a particular alloy is 15,000 psi$\sqrt{\text{in.}}$ (16.5 MPa$\sqrt{\text{m}}$). What is the largest flaw that will not be subject to growth by stress corrosion, given that this flaw is located in the central portion of a large plate that is subjected to a static stress of 25,000 psi (172.5 MPa)? (Assume $Y = 1$.) (Sections 19.12 through 19.15)

19.26 *[ES/EJ]* Cold drawn brass cylinders such as cartridge cases have fractured on prolonged exposure to tropical climates. The occurrence is more pronounced during the rainy season, so the name *season cracking* has been coined. What would be a more accurate term, what causes the cracking, and what is a possible remedy? (Sections 19.12 through 19.15)

19.27 *[ES/EJ]* Explain whether the point of origin in the fracture of ceramics shown in Figure 19.20 would be on the tension side or the compression side. How might the appearance of the fracture be different if the ceramic contained a residual stress? (Chapters 10 and 19)

19.28 *[EJ]* In a soda-lime glass, the long-term tensile stress to obtain cracking is less in a moist than in a dry atmosphere. Explain this observation and indicate the role of a residual stress. (Chapters 10, 18, and 19)

19.29 *[ES/EJ]* Explain the following observations for crazing in polymers. (Chapter 13 and 19)
   a. A crazed polymer can still support a load.
   b. Crazing while under load can increase with time.
   c. Actual polymer failure by craze growth depends on whether the loading is above or below a critical stress-intensity factor, K_C.

19.30 *[EJ]* Why is it so important to determine the point of origin in a failed component? How do we differentiate between defects in material and defects in design (geometry)? (Sections 19.16 through 19.19)

19.31 *[EJ]* Hardness values, especially of cast irons, are often used to predict the tensile strength. Explain why this might be a useful technique for quality control. What might the limitations be? (Chapters 9 and 19)

19.32 *[EJ]* A serrated paneling nail is hardened so that it will penetrate modern wood-composite panels. It is noted that a slightly decarburized surface minimizes the possibility that the nail will snap off to become a projectile during hammering. Explain why decarburization might be an advantage for the serrated paneling nail but a disadvantage for a wear design. (Chapters 8 and 19)

CHAPTER 20

Electrical and Optical* Properties of Materials

Co-authored with Paul A. Flinn
Senior Research Scientist
Intel Corporation
Santa Clara, California

This illustration shows a linear integrated circuit used in a power amplifier in a radio. The grayish background is a 0.1-mm-wide "chip" of silicon alloyed with a small amount of material to make it an *n*-type material, as described in the text. By a combination of diffusion, etching, insulation with SiO_2, and further diffusion, a system of transistors, capacitors, and resistors is built up into a complete circuit on the silicon chip. There are eight transistors grouped in two columns of four on the left-hand side of the photograph.

This is a medium-scale integrated circuit. In the past decade, large-scale integrated circuits with up to ten layers have become very important as memory units in computers.

---

*The contributions of John R. Flinn, staff engineer, NEC, to the section on optical properties are greatly appreciated.

20.1 Overview

The field of electronic materials for computers, tapes, television components, and lasers is exhibiting tremendous growth. It is also a very exciting field because there are few limits on the cost of materials — for example, gold is used widely for contacts, and rare metals such as indium and gallium are commonly employed, although in small quantities.

Materials engineers are often reluctant to enter this field because the emphasis is on the relationship of electronic rather than mechanical properties to structure. It is worthwhile, therefore, to outline in a general way the subjects to be covered in this chapter and explain how they are related to, as well as how they differ from, subjects we have covered before. The whole thrust of the chapter is to show how the principles already covered can be employed to modify electronic materials. Perhaps it may be reassuring to point out that it was a metallurgist (W. G. Pfann) who developed the zone refining method for producing the high-purity silicon on which the semiconductor industry depends. The method is based on a simple but ingenious application of phase diagrams (Chapter 4).

If the materials engineer is to contribute to this field, he or she must recognize that, instead of focusing on mechanical properties, which depend to a larger extent on atomic movements, it is necessary to concentrate on the movement of small charged particles such as electrons inside the relatively immense atomic framework. We have to set aside the solid-ball models of the atoms that served us well in discussing mechanical properties, and realize that there is a great deal of empty space between the positively charged nuclei at the centers of the "solid" spheres. To develop this new perspective, imagine a bar of copper with positively charged nuclei in an FCC array and valence electrons randomly distributed as an electron gas. Now let us apply a potential to the ends of the bar, causing the negative electrons to drift toward the positive terminal. It can be shown that the average electron travels several hundred atom diameters before colliding with an atom or another electron. This fact certainly suggests that refinement of the hard-sphere model is needed.

One other point to adjust to is the small charge values and the large quantities of the charged particles. The charge per electron is 1.60×10^{-19} coul. Therefore, we have 6.25×10^{18} electrons per coulomb. A current of 1 A is 1 coul/sec, so a current of 1 pA (picoampere, or 10^{-12} A, the smallest we can measure) represents a flow of 6.25×10^6 electrons per second. From this concept of the extremely small size of the electron, we can conceive how a photon of light can deliver sufficient energy to an electron to enable it to leave its atomic orbit.

Despite the importance of these fine-scale phenomena, we will begin our discussion with familiar engineering properties, such as the resistivity and conductivity of materials. Then we will show how these large-scale properties are related to electron movement.

To introduce the very important field of semiconductors, we will consider the movement of charge carriers in insulators, as well as in metals, developing the concept of the electron hole. Then we will investigate the carriers in pure (intrinsic) semiconductors and in "doped" (extrinsic) semiconductors.

Next the carrier movement in *p-n* junctions and other devices will be taken up. To complete the discussion of electrical materials, we will take up dielectrics, mechanical–electrical coupling, and thermoelectric effects.

Finally, turning to optical materials, we will first consider such classic properties as laser operation and then discuss fiber optics.

ELECTRICAL CONDUCTIVITY

20.2 Conduction and Carriers

Let us first consider conductivity from the large-scale engineering point of view and then see how we can develop the same relations from a knowledge of elementary particles. If we have a wire to which we apply a potential V, the current I that flows depends on the circuit resistance R, as given by the well-known Ohm's law: $I = V/R$ (Equation 18-11 where here we have substituted the symbol V, voltage, for the electromotive force, E). The resistance depends on the nature of the wire itself; for example, a copper wire has a lower resistance than an iron wire of the same size (length and cross section). We use the term *resistivity* (ρ) to characterize the inherent ability of the wire to affect current flow, and we multiply this by l/A to give the *resistance:*

$$R = \rho \frac{l}{A} \quad \text{or} \quad \rho = R\frac{A}{l} = \text{ohm}\frac{\text{m}^2}{\text{m}} = \text{ohm-m} \tag{20-1}$$

where R = resistance, ρ = resistivity, l = length of wire, and A = cross-sectional area of wire.

EXAMPLE 20.1 *[ES]*

A student wants to build a dc heating coil rated at 110 volts and 660 watts. She has some Chromel wire 0.1 in. (2.54 mm) in diameter with a resistivity of 1.079 μohm-m. What length of wire should she use?

Answer

$$\text{Power} = VI \quad I = \frac{660}{110} = 6 \text{ A} \quad R = \frac{V}{I} = \frac{110}{6} = 18.3 \text{ ohm}$$

$$R = \rho\frac{l}{A} \quad \text{or} \quad l = \frac{RA}{\rho}$$

$$l = \frac{18.3 \text{ ohm} \times (0.1 \text{ in.} \times 0.0254 \text{ m/in.})^2 \times (\pi/4)}{1.079 \times 10^{-6} \text{ ohm-m}} = 86 \text{ m}$$

It is more positive and simpler to think of the material as conducting rather than resisting the passage of current, so from now on we will use the well-known parameter *conductivity* instead of resistivity. This is simply the reciprocal of resistivity:

$$\sigma = \frac{1}{\rho} = (\text{ohm-m})^{-1} = \frac{\text{mho}}{\text{m}} \tag{20-2}$$

Now let us examine the structural factors that go into conductivity. Given a cube of material 1 m on the edge, the conductivity between opposite faces depends directly on three factors:

1. The number of charge carriers, n (carrier/m³)
2. The charge per carrier, q (coulomb/carrier)
3. The *mobility* of each carrier, μ (m/sec)/(volt/m)

The conductivity depends on the product of all three of these factors:

$$\sigma = nq\mu \tag{20-3}$$

Checking the units, we have

$$\sigma = \frac{\text{carrier}}{\text{m}^3} \times \frac{\text{coulomb}}{\text{carrier}} \times \frac{\text{m}}{\text{sec}} \times \frac{\text{m}}{\text{volt}} \tag{20-4}$$

Because coulomb = A-sec and volt = A-ohm,

$$\sigma = \frac{1}{\text{ohm-m}} = (\text{ohm-m})^{-1} = \frac{\text{mho}}{\text{m}}$$

as in the large-scale example. This is important because we shall be studying how large-scale electrical properties are due to the movement of the elementary carriers.

20.3 Types of Carriers

There are four different types of *charge carriers* that give the phenomenon of "current flow."

Type 1: *The electron* (charge = 1.6×10^{-19} coul). An ampere is a coulomb per second. Therefore, a movement of 6.25×10^{18} electrons is the motion of a coulomb of charge. If it occurs across a cell each second, we have an ampere of current flowing.

Type 2: *The electron hole* (charge = 1.6×10^{-19} coul). In Chapter 10, we discussed briefly the concept of an electron hole in the $(Fe^{2+}, Fe^{3+})O^{2-}$ lattice. Associated with each Fe^{3+} ion is an electron hole, a place to which a traveling electron is attracted because the electrical field of O^{2-} ions around the Fe^{3+} is adjusted for an Fe^{2+}. We can visualize the motion of three electrons, each moving one unit distance to the left (Figure 20.1). As a result of the movement of the first electron, the Fe^{2+} marked (A) becomes Fe^{3+} in step 1,

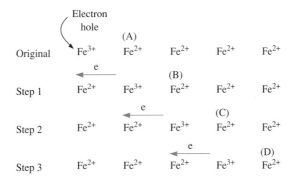

FIGURE 20.1   Schematic representation of countercurrent flow of electrons and vacancies

as does the Fe^{2+} marked (B) in step 2, and so forth. Because we have electron motion, we have conductivity. However, instead of keeping track of small movements of many electrons, it is simpler to focus on the movement of the hole. Instead of saying that three electrons (negative charges) moved three unit distances to the left, we say that one positive charge moved the same total distance to the right, from (A) to (D). Thus the adjacent *electron hole* is a charge carrier with the same magnitude of charge as an electron but with an opposite sign.

Types 3 and 4:   *Positive and negative ions* (charge = $n \times 1.6 \times 10^{-19}$ coul, where n = valence). The current is the net charge transferred per second, so when a positive ion such as Ca^{2+} moves from left to right, the electrical effect is the same as if two electrons moved from right to left. In this way ion movement can contribute to conductivity. Of course, the movement of an O^{2-} ion from right to left would be equivalent (in terms of charge) to the movement of two electrons to the left. In a perfect crystal the movement of ions would be difficult, but real crystals have vacancies that make ion movement possible. It is important to distinguish between electron hole movement that takes place by *electron* jumps from ion to ion and ion movement that takes place by *ion* jumps from one lattice position to another. Naturally, we expect ions to be much less mobile than electrons because of their larger size.

20.4 Conductivity in Metals

Let us now begin to discuss why different materials differ vastly in conductivity (Table 20.1).

For the metals we already have a partial picture of the metallic bond, in which the valence electrons are contributed to an electron cloud by the atom as it becomes an ion in the unit cell. We therefore expect to find a high mobility of these charge carriers. We have to refine this picture a little to explain why a metal with two electrons, such as magnesium (2.5×10^7 mho/m), does

TABLE 20.1   Electrical Conductivity of Selected
                        Engineering Materials

Material	Conductivity, mho/m
Silver (commercial purity)	6.30×10^7
Copper (high conductivity)	5.85×10^7
Aluminum (commercial purity)	3.50×10^7
Ingot iron (commercial purity)	1.07×10^7
Stainless steel (type 301)	0.14×10^7
Graphite	1×10^5
Window glass	2×10^{-5}
Lucite	10^{-12} to 10^{-14}
Borosilicate glass	10^{-10} to 10^{-15}
Mica	10^{-11} to 10^{-15}
Polyethylene	10^{-15} to 10^{-17}

not have better conductivity than copper and silver with one valence electron
and why silicon with four electrons has low conductivity (4.3×10^{-4} mho/m).
To explain this and other concepts, we have to use the "band model," which
runs contrary to classical thinking. Let us study the model and its explanation of conductivity.

First let us review the structure of sodium and the energy levels of its
electrons (Figure 20.2a). The electrons closer to the nucleus in the inner
shells, such as $1s^2$, $2s^2$, and $2p^6$, need not concern us here. It is the $3s^1$ electron,
which is farthest from the nucleus, that leaves the atom to form the electron

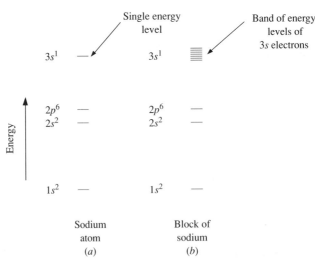

FIGURE 20.2   Development of a band of energy levels for (a) a single sodium atom,
and (b) a block of sodium. The inner electrons ($1s, 2s, 2p$) are sufficiently shielded so
that they form no bands.

gas. It is important to examine this electron gas more closely. When we discussed the structure of an atom, we mentioned the *Pauli exclusion principle*, which holds that only two electrons in an atom can have the same energy level (and these have opposite spins). The electrons in a group of atoms in a metal block have the same tendency. The electrons that form the electron gas in sodium, for example, must have energies slightly different from one another. Therefore, instead of having the sharp energy level of the single 3s electron in an isolated sodium atom, the electrons in a block of sodium exhibit a band of energy levels; hence the name *band model* (Figure 20.2b). It is postulated further that the number of energy levels in a band is equal to the number of electrons that can occupy the energy level times the number of atoms present in the block. This leads to the model of the conduction band of sodium having a number of "states" in the 3s band equal to twice the number of 3s electrons. (Recall that two electrons are possible at the 3s level due to opposite spins, yet only one 3s electron is present in each sodium atom.)

The energy levels or states at the top of the band are higher in energy. Also, because there is only one 3s electron per atom of sodium and the number of states equals twice the number of atoms, half of the states are unoccupied. The higher states have higher energy, so we would expect that at absolute zero only the lower half of the states would be occupied, and this is indeed the case (Figure 20.3a). As we warm up the metal, some electrons have higher energy and leave lower states unoccupied (Figure 20.3b). It is useful for later discussion to define the *Fermi level* (E_F), the energy level where the probability of occupancy of a state by a conduction electron is 0.5. In other words, at the Fermi level half the states are occupied, and there are as many occupied states above this level as there are unoccupied states below it.

Note that the number of available states at different energy levels is a complex function, as shown by the dotted areas in Figure 20.3. Recognizing this is important in analyzing the action of a thermocouple.

Returning now to sodium, we explain its conductivity by saying that the

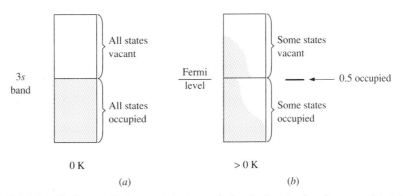

FIGURE 20.3　Enlarged (schematic) view of the 3s band of sodium at (a) 0 K and (b) higher temperatures

electrons in the 3s band can move readily to empty states in that band. Therefore, when an electrical potential is applied, electrons are accelerated and act as charge carriers. The important point is that in a monovalent metal the energy band is half filled, and the electrons in the upper part of the band can be energized and accelerated easily because there are open higher-level states nearby.

How do we explain the conductivity of magnesium, which has two valence electrons? Using the Pauli exclusion principle, we could calculate that all the spots in the 3s band, for example, would be filled because we have two valence electrons per atom. Because there are no empty adjacent spots, the element would be an insulator. However, the energy levels of the 3p band overlap the 3s band, and there is a continuous series of possible states (Figure 20.4). Electrons can be accelerated into the upper levels of the combined band and serve as conductors.

20.5 Applications

Let us review the conductivity of a few alloys (Figure 20.5). We see that the monovalent elements silver, copper, and gold are best. Pure aluminum also has good conductivity because there are many unfilled p levels above its higher-energy electrons. (That is, the p level is only half filled for a trivalent metal. See Figure 20.4.) Iron and the other transition metals have lower con-

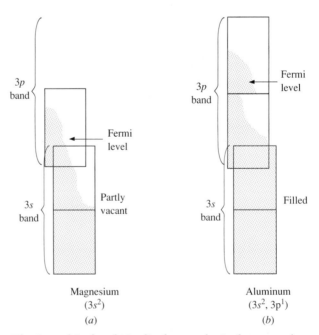

FIGURE 20.4   The 3s and 3p bands in divalent and trivalent metals

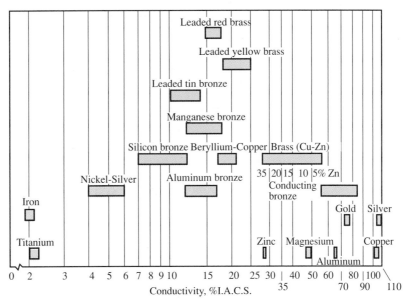

FIGURE 20.5   Conductivity of common metals and alloys in terms of percentage of the international annealed copper standard (which is not of optimal purity)

ductivities because of the complex energy levels in the region where *s* and *p* and *d* bands overlap.

The wide range of conductivity in any family of alloys deserves careful analysis. The conductivity of all elements decreases as a second element is added in solid solution. Figure 20.6 shows data for copper. The more dissimilar the added element is to copper, the greater the change in resistivity. As an example, compare the effect of phosphorus to the effect of silver. When two phases are present, the conductivity is determined by the volume fraction of each.

Because of the severe increase in resistivity that accompanies alloying, it is often useful to improve the mechanical properties and yet maintain the conductivity of a part by cold-working pure metal rather than by adding an alloy. Cold working increases the resistivity only a few percent because there are large blocks of pure, undistorted metal available. On the other hand, even a small percentage of another element in solid solution produces irregularities in the lattice every few atoms. This has a pronounced effect in impeding electron motion, thereby raising the resistivity.

As an example of conductor selection, let us examine the competition between copper and aluminum. The conductivity of copper is higher, but the *conductance* (reciprocal of resistance) of a wire of the same weight per foot is lower than it is for aluminum. Therefore aluminum is used for high-tension wires. However, the distance between towers is an important cost factor, so a steel core is used to strengthen the aluminum cable. Once the power is close to the consumer, it is more convenient to use less bulky copper wires, which

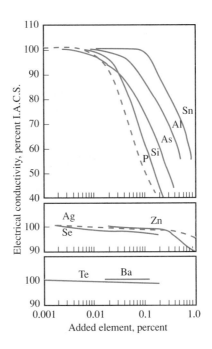

FIGURE 20.6   The effect of various elements on the conductivity of annealed copper

have the added advantage of being easier to solder. For high frequencies, the current flows only in the outer regions near the surface. Therefore copper tubing is also used extensively in this application.

Resistors are of great importance in circuitry and in heating. In industrial furnaces, resistivity is not as important as oxidation resistance. For this reason the heat-resistant nickel–chromium alloys such as 80% Ni, 20% Cr (nichrome) are widely used in such furnaces. In laboratory furnaces, for temperatures over 1800°F (982°C) silicon carbide resistors and even platinum are employed.

20.6 Conducting Glasses

Although glass is normally an insulator, there are a number of applications in which slight conductivity is useful. For example, in high-voltage glass devices such as x-ray tubes, it is desirable to avoid a gradual build-up of charge. To avoid a build-up, a glass containing lead oxide is heated in hydrogen, producing a superficial layer of metallic lead. This is then grounded. In another application a transparent layer of tin oxide, SnO_2, is deposited on a glass surface. When a voltage is applied, a small current can flow against the high resistance, and the heating prevents fogging. It is possible to develop bulk conductivity in glass by adding substances such as iron oxide. This leads to conduction by electron holes, as described earlier and in Chapter 10.

20.7 Superconductivity

The interest in superconductors has undergone several highs and lows of historical technical enthusiasm for essentially two reasons. First, if we could produce a material with very low electrical resistivity, the increased efficiency in power transmission and in computer switching speeds would be advantageous. Second, and possibly not so obvious, a high-current-density conductor showing little resistance loss would exhibit a large magnetic field that would be difficult to achieve with conventional magnetic materials.*

Although either one of these reasons would seem to warrant a high degree of technical interest, the history of the discovery of superconductivity suggests some of the engineering difficulties that needed to be solved. In 1911, H. Kamerlingh Onnes was experimenting with liquefaction of helium around 4 K and as a byproduct discovered that mercury had zero electrical resistivity below this temperature — in other words, it became a superconductor. It was found that if a potential was applied to a continuous mercury loop to develop a current and then the potential was shunted off, the current would flow indefinitely.

Other pure metals such as lead and tin also showed a *critical temperature*, θ_c, below which their resistivity was zero. In each case, however, superconductivity was obtained only at very low temperatures approaching absolute zero. Furthermore, it was found that the magnetic field due to current-density limitations could be built up only to a dissappointingly small value before superconductivity vanished. This latter point will be discussed more fully in the section on superconducting magnets in Chapter 21.

These difficulties were somewhat resolved in the late 1950s by workers in the United States (Matthias and Kunzler), who found that certain alloys of niobium such as niobium–tin and niobium–titanium could be made superconducting at higher critical temperatures with higher critical magnetic field strengths. For example, Nb_3Sn exhibited a θ_c of 18 K (below which it was superconducting). However, to maintain components below this temperature still required the evaporation of liquid helium. Furthermore, the inherent brittleness of intermetallic compounds made fabrication difficult. (See Chapter 22 for further discussion of fabrication methods.)

It was then proposed that if superconductors could be found that have a critical temperature above the boiling point of liquid nitrogen, 77 K, a number of engineering applications would become viable. (Nitrogen is abundant and readily liquefied so that it may be used as a coolant.) Unfortunately, however, some physicists had calculated that all materials had a theoretical critical temperature threshold of 30 K and that it would be impossible to find materials with a θ_c even close to the desired 77 K of evaporating liquid nitrogen.

---

*Note that a current loop conductor creates a magnetic field around it that is analogous to the field in a bar magnet.

Even though research budgets for superconductor research were significantly decreased in the face of these theoretical limitations, a few research groups remained active. Then in 1986 Bednorz and Müller showed that an unlikely material based upon copper oxide, $(La, Ba)_2CuO_4$, exhibited a θ_c of 35 K—above the theoretical threshold! A period of intense research following this discovery led to compounds with a θ_c of 125 K, and even higher critical temperatures have been reported.

The potential for the many engineering applications of these new superconductors is extraordinary, and an international race is under way to discover still higher θ_c materials and to develop methods of fabricating them.

The phenomenon of superconductivity in metal superconductors was explained by Bardeen, Cooper, and Schreiffer, using the concept of electron pairing. They reasoned that below θ_c, conduction electrons of the same energy but opposite spin form pairs. The atomic or ionic vibrations are also less below θ_c, as is the probability of collision by electrons. In the paired condition and below θ_c, the electrons are exempt from collision and scattering and therefore exhibit perfect conductivity.

The reason for the high θ_c of the superconductor oxides is unknown at the present time, but a number of structural features have been investigated, showing the basic structure to be perovskite, as discussed in Chapter 1. In Figure 20.7 we see the basis for these complex structures, such as $YBa_2Cu_3O_6$ and $YBa_2Cu_3O_7$, which are variations of the Cu–O sheet structure. Although the sheet has a repeating simple building block, as outlined in Figure 20.7, distortions are created by the yttrium and barium in the central position at a ratio of 1 to 2. (These compounds are called the 1, 2, 3 type because of the ratios of Y, Ba, and Cu.)

The crystallographic structures are quite complex and are related to the ionic character and the bond angles. However, the following observations have been proposed:

1. An optimal number of oxygen defects is required. The O_7 configuration appears to show more desirable superconducting properties (a higher θ_c) than the O_6.

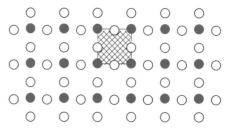

FIGURE 20.7   The Cu–O sheet common to all copper oxide superconductors. Open circles are oxygen; filled circles are copper. A repeating structure is cross-hatched. This is a "perovskite" structure.

[A. W. Sleight, "Chemistry of High-Temperature Superconductors," *Science*, Dec. 16, 1988, Vol. 242, p. 1520–1521. Copyright 1988 by the AAAS. Used by permission.]

2. Because of the oxygen defects and multiple valences of copper, assuming that its valence is simply 1 or 2 is not appropriate. It can be shown that the valence charge varies in small increments from 1 to 3, depending on the attraction of adjacent atoms for electrons in the $3d$ and $4s$ bands.
3. The compounds undergo an orthorhombic-to-tetragonal transformation. As the transformation temperature decreases, θ_c increases.
4. Substitution of elements such as thallium (T1), lead, and bismuth still maintains the copper oxide sheet structure. As in the $1, 2, 3$ type, higher oxygen and a greater number of Cu–O sheets lead to a higher θ_c.

A number of investigations are directed toward the explanation of superconductivity in these new oxide materials and the subsequent rational development of higher θ_c values. The development of higher critical temperatures to date is summarized in Figure 20.8, which reflects the sudden surge in development after 1980.

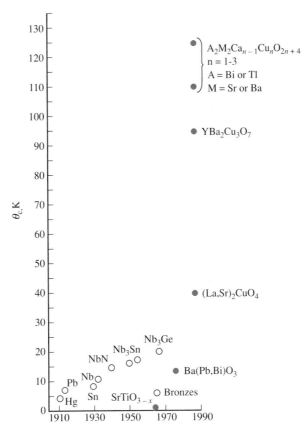

FIGURE 20.8 The critical temperature for superconduction as a function of the date. The open circles are metallic in nature, while the closed circles are ceramic.

[A. W. Sleight, "Chemistry of High-Temperature Superconductors," *Science*, Dec. 16, 1988, Vol. 242, p. 1520–1521. Copyright 1988 by the AAAS. Used by permission.]

EXAMPLE 20.2 *[ES/EJ]*

Explain the relative effect of cold working, alloying, and temperature on the resistivity of a metal.

Answer   We recognize that the resistivity is equal to the sum

$$\rho_{\text{microstructure}} + \rho_{\text{thermal}} + \rho_{\text{cold work}} \qquad (20\text{-}5)$$

$\rho_{\text{microstructure}}$ is likely to have the largest effect as long as foreign elements are in solution. A coarsely distributed second phase with little solid solubility in the matrix may not significantly increase the resistivity.

$\rho_{\text{cold work}}$ has only a slight effect on resistivity because some grains in cold-worked materials have little slip or distortion to inhibit electron flow.

ρ_{thermal} has an intermediate effect, because thermal vibration of the atoms and electrons increases the probability of electron collisions.

Therefore anything that distorts the lattice or grains and causes increased atom vibration increases the resistivity or correspondingly decreases the conductivity.

20.8   Semiconductors in General

We come now to a fascinating group of materials. Although they are lower in electrical conductivity than the metals, they are essential in a number of newly developed devices, from wrist calculators to tiny remote communication units in distant space satellites. The transistor, the solar battery, and the integrated miniature circuit in TV games, computers and auto engine controls all depend on the properties of semiconductors.

The heart of all these appliances is not complex circuitry but materials that can marshal and direct electron motion in a tiny space with a precision never before attained. The key to these devices is a very small crystal of a semiconductor, such as silicon, with controlled amounts of impurities in solid solution.

To understand the operation and vast potential of these materials, we shall first observe how the band structures of the semiconducting elements differ from those of the metals. Then we shall see the effect of adding different types of impurities (called *dopants*) to create *n*- and *p*-type semiconductors. Finally, we shall take up a few applications of these materials.

20.9   Semiconductors and Insulators

To understand the semiconductors, such as silicon and germanium, it is helpful to begin with an insulator, diamond (C), which is in the same column in the periodic table. All three elements, along with tin, have the diamond cubic structure. Whereas the electronic structure of the *carbon atom* is $1s^2, 2s^2, 2p^2$, in the *diamond crystal* the second-quantum-level electrons all form covalent bonds and have similar energy levels. This process, called *hybridization*, was discussed in Chapter 1. In other words, at the temperature and pressure ranges at which diamond is formed, this type of bonding is the equilibrium state. This deviation should not be too disturbing when we recall other minor shifts between quantum levels in other cases, such as the unfilled $4s$ shell in copper, which allows the formation of a stronger $3d$ group (Table 1.2). This type of covalent bond also forms in silicon and germanium, and a modification of the band graph used for the metals is necessary. Instead of showing the s electrons in one band, overlapped by the p electrons as for aluminum (Figure 20.4), the four (s plus p) electrons are considered equal and form a valence band. Above this lies a band for electrons at a higher energy level, called the conduction band. In the case of diamond (and silicon and germanium), the valence band is filled. The reason why diamond is an *insulator* is that the energy level of its conduction band is considerably above that of its valence band and only a very few electrons attain entry to the conduction band. (The distribution of energy levels among electrons, as among atoms, follows a statistical curve and some high-energy electrons are always present.) Models of the energy bands for metals, semiconductors, and insulators are shown in Figure 20.9, where E_g is the *energy gap*.

In the case of silicon, the energy gap between bands is smaller, and electrons that attain enough energy to leave their positions of covalent bonding occupy positions in the conduction band. An important consequence of an electron leaving the valence band is the creation of an electron hole, often called simply a *hole*, in the valence band, as discussed earlier (Figure 20.10).

It is important to quantify the value of the energy gap, using the electron volt (eV) as a convenient energy unit. When an electron falls through a potential of 1 volt, its energy is increased by 1 electron volt. The increase in kinetic energy is the product of the charge q and the voltage V, or

$$qV = (1.60 \times 10^{-19} \text{ coul/electron}) (1 \text{ V}) = 1.60 \times 10^{-19} \text{ J} = 1 \text{ eV} \qquad (20\text{-}6)$$

It should be noted that the electron volt is a general energy term and can be used in any type of problem, not necessarily only those involving electrons.

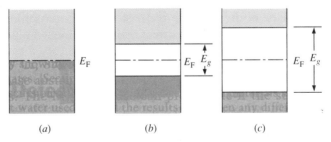

(a)   (b)   (c)

FIGURE 20.9 Models of energy bands of (*a*) a metal, (*b*) a semiconductor, and (*c*) an insulator. E_F is the Fermi level (0.5 of possible states are occupied).

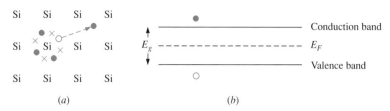

FIGURE 20.10   (a) Formation of a conduction electron and an electron hole in pure silicon (an intrinsic semiconductor). Through the acquisition of thermal energy, an electron attains enough energy to leave its valence position. (b) The process shown in (a) has led to the presence of an electron in the conduction band and a hole in the valence band. The Fermi level is halfway between the bands (50% occupation).

The difference in gap energy between diamond and the semiconductors is shown below.

Material	Gap Energy, E_g (20°C) eV
Diamond	5.30
Silicon	1.06
Germanium	0.67

Tin is yet another element with four electrons in its outer shell. However, the energy gap is only 0.06 eV, which is small enough to make the element a conductor. Note that the energy gap decreases as the atomic number increases.

In addition to the Group IV elements, semiconductors have been produced by combining elements that have three outer-shell electrons with elements that have five outer-shell electrons to give an average of four, as, for example, in AlP and GaAs. These are called III-V compounds. The band structure differs from that of Group IV elements, but a range of energy gaps is obtained for different combinations as in pure elements.

There is an important detail in the band structure of III-V compounds such as gallium arsenide compared to silicon, which is important in light emission and electron movement. The energy gap is measured as the shortest distance between an electron in the conduction band and a hole in the valence band. It is important to realize that these bands exist in space, not in the plane of the paper. In the case of gallium arsenide it is possible for an electron to reach a position directly above an electron hole. On falling to the electron hole the energy of recombination is released as a photon of light.* This is called a direct-band structure. In the case of silicon, the electrons in the conduction band fall to the valence band by an angular path. This results in dif-

---

*Photons will be discussed more fully in Section 20.20. II–VI compounds such CdS follow the same principles.

ferent momentum vectors and momentum must be conserved. This cannot be done by momentum transfer to photons. Consequently, instead of light emission, heat is evolved as phonons or mechanical strains develop. This is called indirect-band structure. For this reason the Group III-V semiconductors are needed for light-emitting devices and in optical communications systems. To meet this need, an extensive processing development to deposit thin films of III–V semiconductors on silicon chips as bases has been funded.

20.10 Extrinsic vs. Intrinsic Semiconductors

Up to now we have discussed only semiconductors that have an average of four electrons per atom, or *intrinsic semiconductors*. If we vary this balance by *doping* with impurities, we obtain *extrinsic semiconductors* of two types, *n* and *p*. To understand these important materials, let us consider the effect of adding impurities, or dopants, to an intrinsic semiconductor such as silicon. If we add a small quantity of phosphorus, its atoms form a substitutional solid solution with silicon, giving the structure shown in Figure 20.11.

We have denoted the valence electrons contributed by the four silicon atoms adjacent to a phosphorus atom by the symbol ×. However, phosphorus has five outer-shell electrons, symbolized by •, so we have an extra electron. The phosphorus, therefore, is an electron donor, and the structure is called an *n*-type extrinsic semiconductor (*n* refers to the *negative* extra electron). The electrons from the phosphorus are not in exactly the same positions as those that would be produced by an electron in the valence band, but they lie just below it in energy.[*] This is because the electrical field of the phosphorus is

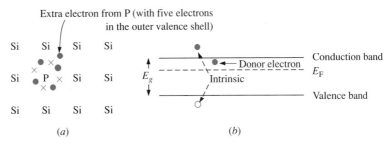

FIGURE 20.11 (a) Structure of silicon with a phosphorus atom as dopant, giving rise to an extra electron, thus making an *n*-type semiconductor (extrinsic). (b) The electron from the phosphorus shown in (a) is found just below the conduction band. The Fermi level E_F is higher than in Figure 20.10 because of the addition of an electron above the old Fermi level and no increase in the electron holes below.

---

[*]However, at room temperature, there is enough excitation by thermal energy to raise the electron to the conduction band.

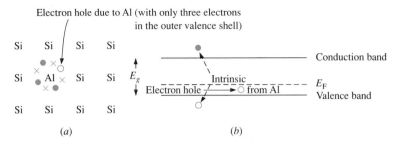

FIGURE 20.12   (*a*) Structure of silicon with an aluminum atom added as a dopant, giving rise to an electron hole, thus making a *p*-type semiconductor (extrinsic). (*b*) The electron hole produced by the aluminum shown in (*a*) is found just above the valence band. The Fermi level E_F is lower than in Figure 20.10 because of the addition of a hole below the old Fermi level and the lack of any increase in the number of electrons above.

not identical to that of a silicon atom. The Fermi level E_F is close to the conduction band. We have added electrons to the structure without adding holes, so E_F shifts upward. By doping silicon in this way, we can increase its conductivity 10,000 times!

Now let us consider the effect of adding an element with less than enough electrons to satisfy the covalent bonds of silicon. Aluminum is such an element (Figure 20.12). Again the silicon electrons are shown by $\times$ and the aluminum electrons by $\bullet$. We see that there is an electron missing. Aluminum is an acceptor element, and the structure is called a *p*-type semiconductor (the electron holes are *positive*). The band structure is also shown in Figure 20.12, where the holes are just adjacent to the valence band. In this case the conduction is mainly from electron holes, and the 50% occupied level (E_F) is closer to the valence band, because we have added holes without adding electrons.

EXAMPLE 20.3 *[ES]*

Calculate the conductivity of the *intrinsic* semiconductor germanium from the following characteristics (300 K):

$$\text{Carrier density (where holes = electrons)} = 2.4 \times 10^{19} \text{ carrier/m}^3$$

$$\text{Electron mobility} = 0.39 \text{ m}^2/\text{volt-sec}$$

$$\text{Hole mobility} = 0.19 \text{ m}^2/\text{volt-sec}$$

$$\text{Charge/electron} = \text{charge/hole}$$

$$= 1.6 \times 10^{-19} \text{ coul/carrier}$$

Answer

$$\sigma = nq\mu \quad \text{(from Equation 20-3)}$$

$$= nq_n\mu_n + pq_p\mu_p \quad (20\text{-}7)$$

where n = number of electrons/m^3, q_n = charge/electron, μ_n = mobility of electron carrier, p = number of holes/m^3, q_p = charge/hole, and μ_p = mobility of hole carrier. Because $n = p$ and $q_n = q_p$ in an intrinsic semiconductor, we have

$$\sigma = nq(\mu_n + \mu_p)$$

$$= (2.4 \times 10^{19} \text{ carrier/m}^3) \times (1.6 \times 10^{-19} \text{ coul/carrier})$$

$$\times [(0.39 + 0.19) \text{ m}^2/\text{volt-sec}]$$

$$= 2.23 \frac{\text{coul}}{\text{volt-sec-m}} = 2.23 \frac{\text{A}}{\text{volt-m}}$$

$$= 2.23 \frac{1}{\text{ohm-m}} = 2.23 \frac{\text{mho}}{\text{m}}$$

Another quantitative relationship that holds for both intrinsic and extrinsic semiconductors is

$$n \cdot p = \text{a constant for a given element and temperature} \quad (20\text{-}8)$$

For example, the following values are obtained for n-type germanium at 300 K $(n \cdot p = 6.2 \times 10^{38} \text{ carrier}^2/\text{m}^6)$.

	Parts Impurity per Part Germanium		
Parts of Impurity	10^{-10}	10^{-8}	10^{-5}
Donor conc., N_d/m^3	4.4×10^{18}	4.4×10^{20}	4.4×10^{23}
Free electron conc., n_n/m^3	2.5×10^{19}	4.4×10^{20}	4.4×10^{23}
Hole conc., p_n/m^3	2.5×10^{19}	1.4×10^{18}	1.4×10^{15}
Resistivity, ohm-m	47×10^{-2}	3.4×10^{-2}	1.0×10^{-5}

Note that whereas $n \cdot p = $ constant, a very high value of n leads to greatly reduced resistivity because $(n + p) \neq$ constant. Similarly for p-type materials, a high value for holes reduces resistivity. The data in the table also show that for very small amounts of impurity (10^{-10}), the intrinsic electrons and holes are the dominant carriers. However, when a concentration of 10^{-8} is reached, the donor electrons dominate and the number of holes declines to keep $n \cdot p$ constant.

It should be added that the dominant type of carrier is called the *majority carrier* and the less concentrated type the *minority carrier*, such as the holes in Example 20.3. However, the movement of minority carriers is often of great importance, as in the solar cell to be described in Sec. 20.12. Finally, it is important to note that doping involves much smaller quantities of "alloy" than in conventional materials, so it is necessary to begin with high-purity silicon to avoid the effects of tramp elements.

20.11 *p-n* Junctions and Rectification

The *p-n junction*, the interface between *p*-type and *n*-type materials, is the critical region of a number of electronic devices, including rectifiers, transistors, solar cells, semiconductor lasers, and integrated circuits. For this reason we will attempt to explain the carrier movements in this region simply and carefully and then look at a few devices. The many complicated and intricate uses are best left to a course in electrical engineering, but the problems in materials can be illustrated here because they involve questions of crystal growth, impurities, diffusion, etc.

Although the typical *p-n* junction is made by beginning with *p*-type silicon and diffusing in a donor to make an *n*-type material to a specific thickness, let us take a simpler case for illustration, in which we bring a block of *p*-type material into intimate contact with *n*-type material (Figure 20.13). No external voltage is applied. The following effects develop:

1. Let us focus only on the dopant atoms (acceptors and donors) because these are responsible for the phenomena. Note that the *p*-type atoms are electrically neutral but are ready to donate a positive charge (a hole) to the surroundings. *After* this is done, the remaining atom is *negative*. On the other hand, the *n*-type atoms are ready to donate an electron and ultimately become positive.
2. When contact is made, the electrons and holes from the atoms in the interface diffuse and neutralize (there is no external potential yet). This results in a *carrier-depleted zone* about 0.5 μm (5000 Å) wide at the junction, a distance of about the wavelength of visible light.
3. No further carrier diffusion takes place across the junction. This important phenomenon is contrary to our normal concept of diffusion of *atoms* across an interface. In this case, however, the development of planes of charged positive and negative atoms acts as an energy barrier to holes and electrons alike. That is, the holes are repelled by the positive barrier on the *n* side of the junction, and the electrons are repelled by the negative barrier on the *p* side.
4. When an external potential is applied so that the positive connection is on the *p* side and the negative connection is on the *n* side (called *forward bias*), a *current flows*. The positive potential neutralizes the negative side

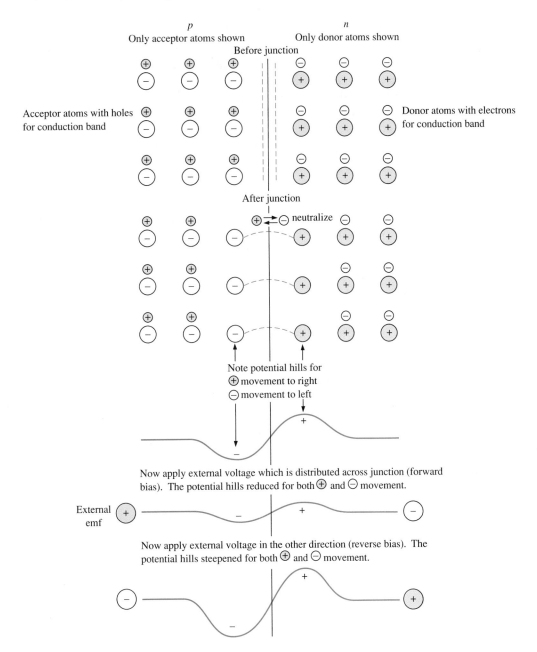

FIGURE 20.13 Relationship of a *p-n* junction to an applied external voltage

of the barrier, and the negative potential neutralizes the positive side of the barrier. (Recall that a combination of hole movement in one direction and electron movement in the other provides a current.)

5. When an external potential is applied so that the negative connection is on the *p* side and the positive connection is on the *n* side (called *reverse bias*), practically no current flows. (The negative potential adds to the potential hill on the negative side of the barrier, and the positive potential adds to the potential on the positive side.)[*]

Figure 20.14 shows the net result of these effects. When a small voltage is applied in the forward direction, a large current is obtained, whereas very little current results with a normal range of voltages in the reverse direction. Therefore, if we impose the output of an alternating current transformer, it is rectified to a series of dc pulses.

Now let us analyze these effects quantitatively.

The current *I* across the *p-n* junction is given by the equation[†]

$$I = I_s \left[\exp\left(\frac{qV}{kT}\right) - 1 \right] \tag{20-9}$$

where V = applied voltage, I_s = maximum current obtained with reverse bias (a small value), q = charge per carrier (1.602×10^{-19} coul), k = Boltzmann's constant (1.381×10^{-23} J-K^{-1}), and T = absolute temperature (K).

The maximum or limiting current I_s obtained with reverse bias may be expressed as follows:

$$I_s = Aqn_i{}^2 \left(\frac{D_e}{L_e N_d} + \frac{D_p}{L_p N_s} \right) \tag{20-10}$$

where A = area of junction, n_i = number of intrinsic carriers (a function of material and temperature, as discussed in Example 20.6), D_e = diffusion coefficient

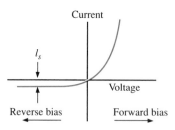

FIGURE 20.14 Amplification of current obtained by applying a current with a forward bias. I_s is maximum current obtained with reverse bias.

---

[*]A very slight current called the limiting current does occur, because a very small number of holes are present on the *n* side among the predominating number of electrons. Also, a small number of electrons are driven from the *p* side.

[†]See D. A. Fraser, *The Physics of Semiconductor Devices*, Clarendon Press, Oxford, 1977.

for electrons, D_p = diffusion coefficient for holes, L_e and L_p = constants (diffusion lengths related to the distance that carriers travel before recombination), and N_d and N_s = concentrations of donor and acceptor elements, respectively.

EXAMPLE 20.4 *[ES]*

Calculate the ratio of the currents with forward and reverse bias, respectively, with imposed voltages of +1 volt and −1 volt at room temperature (300 K).

Answer   The current in the forward direction, I_F:

$$I_F = I_s\left[\exp\left(\frac{qV}{kT}\right) - 1\right] \tag{20-9}$$

$$I_F = I_s\left[\exp\left(\frac{(1.602 \times 10^{-19} \text{ coul}) (1 \text{ volt})}{(1.381 \times 10^{-23} \text{ J-K}^{-1}) (300 \text{ K})}\right) - 1\right]$$

$$= I_s\left[\left(6.21 \times 10^{16} \frac{\text{coul-volt}}{\text{A-volt-sec}^{-1}}\right) - 1\right]$$

$$= I_s(6.21 \times 10^{16} - 1)$$

The current in the reverse direction, I_R:

$$I_R = I_s\left[\exp\left(\frac{qV}{kT}\right) - 1\right] \tag{20-9}$$

$$I_R = I_s\left[\exp\left(\frac{(1.602 \times 10^{-19} \text{ coul}) (-1 \text{ volt})}{(1.381 \times 10^{-23} \text{ J-K}^{-1}) (300 \text{ K})}\right) - 1\right]$$

$$= I_s(1.61 \times 10^{-17} - 1)$$

Then $$\frac{I_F}{I_R} = \frac{I_s(6.21 \times 10^{16} - 1)}{I_s(1.61 \times 10^{-17} - 1)} \approx \frac{6.21 \times 10^{16}}{-1} \approx -6.21 \times 10^{16}$$

Note: With a negative reverse bias current, the forward bias current is positive.

EXAMPLE 20.5 *[ES]*

Calculate the change in I_s caused by doubling the concentration of the dopants in the p and n sides of a semiconductor.

Answer

$$I_{s,\text{orig}} = \text{const}\left(\frac{D_e}{L_e N_d} + \frac{D_p}{L_p N_s}\right) \tag{20-10}$$

$$I_{s,\text{new}} = \text{const}\left(\frac{D_e}{2L_e N_d} + \frac{D_p}{2L_p N_s}\right) = \tfrac{1}{2} I_{s,\text{orig}}$$

In other words, although the conductivity of both the *n* and *p* legs is increased, the limiting current is decreased and rectification is improved. However, in silicon *p-n* junctions the reverse current is already very low—approximately 0.001 μA. This is why a *p-n* junction may also function as a capacitor.

20.12 Solar Cells

The solar cell (Figure 20.15) is another interesting use of *p-n* junctions. In the preparation of a *solar cell*, an *n*-type layer is formed on the surface of a relatively thick *p*-type silicon substrate by diffusion of an element such as phosphorus. Contact strips are imprinted on the *n* and *p* layers. The *n* layer is exposed to sunlight or other radiation. Electrons receiving the radiation are raised to higher energy levels and move to the conduction band, and so holes are formed. This is called *electron–hole-pair production*. It raises the Fermi level of the *n* layer. It is equivalent to attaching an electric generator across the diode.

Because $n \cdot p$ must remain constant, the *p* region is negative at the *n-p* junction. The flow of holes to the *p* region forward-biases it, giving a current flow in the forward direction. An open-circuit voltage develops. This is essentially an equilibrium among the formation of hole pairs, the light current, and the migration of carriers across the *p-n* junction.

The equation for the migration across the interface, as before, is

$$I = I_s \left[\exp\left(\frac{qV}{kT}\right) - 1 \right] \tag{20-9}$$

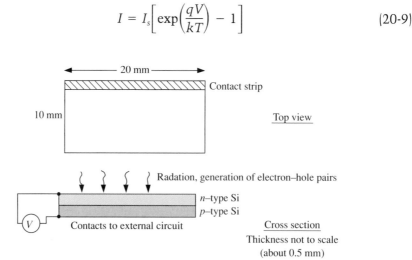

FIGURE 20.15 Schematic representation of a solar cell

The V is the open-circuit voltage of the junction. At equilibrium, $I = I_L$, where I_L is the current produced by the radiation, and

$$I_L = I_s \left[\exp\left(\frac{qV}{kT}\right) - 1 \right] \tag{20-11}$$

Solving for V, we obtain

$$V = \frac{kT}{q} \ln \left(\frac{I_L}{I_s} + 1 \right) \tag{20-12}$$

The laser diode is an interesting reversal of the solar cell. In this case, a current is passed in a reverse direction (reverse bias) to produce intense coherent light. A condition for emission is that when the minority carrier diffuses away from the *p-n* junction and combines with an electron, the electron falls from a higher-energy conduction band state to a valence band state, giving off a photon. Only a few semiconductors meet this condition; they include gallium arsenide.

20.13 Amplifier Circuits

To illustrate the use of *p-n* junctions in more complex circuits, let us consider two different devices in amplifier circuits, the bimodal transistor and the MOSFET. The principle of the amplifier circuit is that the magnitude of a strong current passing between a point A at ground potential and a point B at higher potential is generally greatly modified by a relatively weak signal voltage. The signal is said to be amplified because the strong current varies by a greater amount than the pulses of the signal. The current also has the strength to drive a radio speaker, for example.

When a *transistor* is composed of two *p-n* junctions it is called bimodal. It always has an emitter, a base, and a collector and is either an *n-p-n* or a *p-n-p* variety. Consider the *p-n-p* type of transistor shown in Figure 20.16. At the emitter a small voltage V_e is applied in a forward bias—that is, electron holes or current flows into it. A larger voltage V_c in a reverse bias is present at the collector. For the transistor to operate, there should be hole generation at the emitter-base junction, little recombination in the base, and hole extraction (usually by an applied load) at the collector.

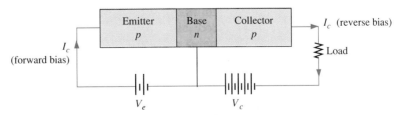

FIGURE 20.16   Schematic representation of a *p-n-p* transistor

Of course, it is necessary to prevent the holes from recombining in the *n*-type base, which is accomplished by making the base thin and having a relatively long recombination time (a characteristic of the material). Hence a small change in the voltage at the emitter causes a large change in the voltage at the collector.

More important, the collector current changes exponentially with the emitter voltage according to the relationship

$$I_c = I_0 e^{V_e/B} \tag{20-13}$$

where I_c is the current at the collector and I_0 and B are constants that depend on temperature. Therefore a weak current can be amplified rapidly by very small variations in emitter voltage. The high-frequency applications of this principle are numerous and extend from radios to computers.

In the *MOSFET* (*m*etal-*o*xide-*s*emiconductor *f*ield-*e*ffect *t*ransistor) (Figure 20.17), *p*-type silicon is altered to *n*-type in small patches. A thin silicon dioxide layer is deposited, followed by an aluminum conducting layer called a *gate*. When we apply a potential between the two *n* spots, no appreciable current flows because one spot is reverse-biased and we can obtain only I_s. When we now apply a signal flowing through the aluminum gate, the field affects the *p* layer below. If the signal is in the "right" direction, the *field effect* distorts the layer of *p* material close to the cap, converting it to an *n*-type conduction band. This permits current to pass easily. On the other hand, if the signal is in the opposite direction, only a small limiting current can flow. The signal is therefore amplified.

The CMOS (Complimentary Metal-Oxide Silicon) paired transistor is an essential part of modern VLSI (very large-scale integrated circuits) (see Figure 22.2). If MOS transistors only were used in a switching operation half of the circuits would be open and the other half closed. This latter group would be passing current from line voltage in the chip to ground. An intolerable amount of heating would develop from thousands of transistor circuits. In the CMOS a pair of opposed transistors in series stores the required information without current flow. For example, a left-hand transistor can register ON and a paired right-hand OFF and vice versa for the opposite information. Because the transistors are in series, one is always off and no current flows. As mentioned later, under processing, special materials and layer-growth techniques are required.

20.14   The Hall Effect

We have spoken about the motion of holes and electrons as carriers without explaining how we know which major carriers are present. We accomplish this by using the *Hall effect*. We make even more important use of the Hall effect when we turn the principle around to measure magnetic fields in the analog multiplier.

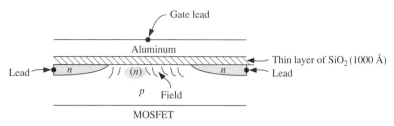

FIGURE 20.17   Schematic representation of a MOSFET (metal-oxide-semiconductor field-effect transistor)

Recall that when a wire carrying a current is placed in a magnetic field, there is a force on the wire—or, more accurately, on the carriers in the wire. This is of course the principle of the electric motor. Now consider the case shown in Figure 20.18, where a current is flowing in a block of metal because of the motion of carriers. We want to find out whether the carriers are holes or electrons. When we superimpose a magnetic field B, it exerts a force on electrons or holes that acts at right angles to both the direction of current flow and the magnetic field—that is, toward the vertical side walls of the block. If the majority of the carriers are electrons, the motion of the side current is toward one wall, whereas if they are holes, the motion is toward the opposite wall. These differences can be transmitted, as voltages of opposite sign, to the voltage-measuring probes on the side walls. The magnitude of the voltage measured is also an indication of the number of carriers.

In actual experimentation we measure the voltage developed by a given magnetic field. The Hall constant is derived from the data. It is large for semiconductors and varies inversely with the number of carriers. The Hall voltage is a function of the applied magnetic field and of the normal current flowing through the conductor.

We may also use the Hall effect to measure the intensity of a magnetic field if we use a conductor with known carriers.

20.15 Effects of Temperature on Electrical Conductivity

One of the characteristics of a metal is the decrease in its conductivity with increasing temperature. We discussed the reasons for this briefly in the section on superconductivity. By contrast, the conductivity of semiconductors and insulators increases with temperature. When an intrinsic semiconductor is heated, more electrons are pumped up to energies where they can enter the conduction band, leaving electron holes. This effect is disadvantageous in a circuit that depends on a *p-n* junction, because the number of natural carriers can obscure the effects of the carriers added by the dopant.

On the other hand, because of the sensitivity of the semiconductor to temperature, the change in resistance can be used to indicate temperature

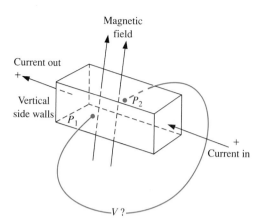

FIGURE 20.18   A block of metal containing a flow of current and exposed to a magnetic field exhibits Hall voltage, *V*. This is called the Hall effect.

accurately from 1–723 K (−272–450°C). Semiconductors used in this way are called *thermistors* (Figure 20.19).

EXAMPLE 20.6 *[ES]*

The general formula for the number of electrons and holes in an intrinsic semiconductor is

$$n \cdot p = AT^3 e^{-11,600 E_g /T} \tag{20-14}$$

where $n \cdot p$ = number of carriers^2/m^6, A = constant, T = absolute temperature in K, E_g = energy gap (eV), and e = 2.718.

We wish to construct a thermistor from silicon. What is the change in the number of carriers per cubic centimeter in going from 27°C to 47°C? For silicon we are given that $n = p = 1.5 \times 10^{16}$ carriers/m^3 at 300 K (27°C). Also, $E_g = 1.06$ eV and remains constant over the temperature range in question.

Answer   First we must arrive at the general equation, which requires calculation of *A* from the known conditions.

$$2.25 \times 10^{32} \text{ carriers}^2/\text{m}^6 = A(300 \text{ K})^3 e^{(-11,600 \times 1.06)/300} \qquad A = 5.26 \times 10^{42}$$

At 47°C,

$$n \cdot p = n^2 = 5.26 \times 10^{42}(320)^3 e^{(-11,600 \times 1.06)/320} = 3.54 \times 10^{33}$$

$$n = 5.95 \times 10^{16} \quad \text{and} \quad \frac{5.95 \times 10^{16}}{1.5 \times 10^{16}} = 3.97$$

Therefore, for the 20°C change, the number of carriers increases 3.97 times, which of course increases the conductivity by this factor plus a factor for the corresponding increase in mobility of both holes and electrons.

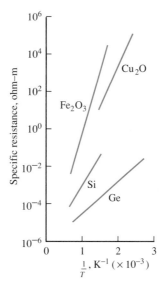

FIGURE 20.19   Variation in resistivity of semiconductors with temperature

The conductivity of ionic materials also increases with temperature. This is shown quantitatively by the effect on mobility:

$$\mu = \frac{qD}{kT} \tag{20-15}$$

where μ = mobility (m²/volt-sec), q = charge on carrier (coulomb/carrier), D = diffusion coefficient (m²/sec), T = absolute temperature (K), and k = Boltzmann's constant (1.38×10^{-16} erg/K).

Because of their large size, the ions in a solid have a mobility that is orders of magnitude lower than that of electrons or holes. Ions move by jumps, using vacancies. In liquid salts the conductivity is higher because ion movement in the liquid state is easier and more vacancies exist.

OTHER ELECTRICAL PROPERTIES

20.16 Dielectric Properties

To the designer of electrical equipment, there are five characteristics of insulating materials that are as important as mechanical properties are to the designer of mechanical components.

1. Dielectric strength
2. Dielectric constant
3. Power factor and dissipation factor
4. Arc resistance
5. Resistivity (already discussed)

Dielectric Strength

To test the dielectric strength we take a plate of material, place an electrode on either side, and gradually increase the voltage until electrical breakdown occurs and a surge of current flows through the section. The dielectric strength is expressed as

$$\frac{\text{Breakdown voltage}}{\text{Thickness}} = V/\text{mil} \ (0.001 \text{ in.}) \qquad (20\text{-}16)$$

This is not a constant, because a higher voltage/mil is required for breakdown in thin sections (several mils) than in thicker sections. Therefore, the thickness is specified in the ASTM procedure.

Dielectric Constant

The relative permittivity (*dielectric constant*) ϵ_r is the quantity used to evaluate the charge-storing capacity of a dielectric in a capacitor:

$$\epsilon_r = \frac{\epsilon}{\epsilon_0} \qquad (20\text{-}17)$$

where ϵ is the permittivity of the dielectric and ϵ_0 is that of a vacuum. Typical values of the dielectric constant are given in Table 20.2.

In absolute values, for a parallel-plate capacitor, the *capacitance C* is

$$C = \frac{0.224\epsilon_r A}{10^6 d}(n - 1) \qquad (20\text{-}18)$$

where C = microfarads (μF), A = area in square inches of one plate, d = distance between plates in inches, n = number of plates, and ϵ_r = dielectric constant.

The value of the dielectric constant depends on the ability of the material to react and orient itself to the field. The greater the reaction, the greater the energy stored, and hence the higher the dielectric constant. The behavior of the dielectric can be made up of the following effects:

1. *Electronic polarization.* This is present in all dielectrics. The positions of outer-shell electrons around atoms are affected by the field. Electronic polarization takes place very rapidly.
2. *Ionic polarization.* Ions of opposite sign move elastically because of the effect of the field. This is also rapid and takes place only in ionic solids.
3. *Orientation of molecules.* When asymmetric (polar) molecules are present, their orientation is changed by the field.
4. *Space charge.* This is the development of charge at the interface of phases.

Of the foregoing effects, the orientation of molecules contributes most heavily to the differences in the dielectric constant. As an example, a greater charge can be built up on the plates of a capacitor when a material with a

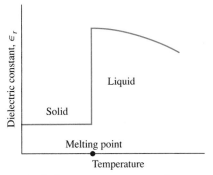

FIGURE 20.20   Variation in dielectric constant with temperature (polarizable crystalline polymer)

strong *dipole* is placed between the plates. In other words, the material with larger dipoles has a higher dielectric constant. Similar reasoning suggests that liquids have higher dielectric constants than solids, because polarization or dipole orientation is easier. This effect is shown in Figure 20.20. After the change due to the melting, the gradual fall-off with increasing temperature of the liquid reflects a higher atomic or molecular mobility, which decreases polarization.

Polymers have the same general characteristics. The amorphous polymers are not so tightly bonded as the crystalline polymers. Therefore, amorphous polymer dipoles are more easily aligned, giving a higher dielectric constant. The crystalline polymers exhibit a large increase in the dielectric constant at their melting point, as shown in Figure 20.20.

The effect of frequency is different for the various materials. The small physical movements encountered with electronic and ionic polarization suggest that these changes take place over a broad range of frequencies. This is not the case, however, for dipole motion of molecules. Generally, the dielectric constant decreases as the frequency increases, because it becomes difficult for the dipole to shift at high frequencies.

Figure 20.21 shows that the dipoles can rotate when the polarity of the capacitor changes. But it takes time to shift from (*a*) to (*c*), as represented in

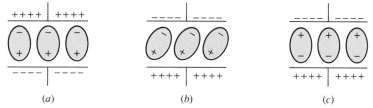

FIGURE 20.21   Dipole motion caused by a change in polarity. When the polarity of the capacitor reverses, as shown here in going from (*a*) to (*b*), it takes finite time to make the complete shift to (*c*).

TABLE 20.2 Dielectric Strength and Dielectric Constant for a Number of Engineering Materials

*Dielectric Strength of Nonmetallics, * volt/mil*

Material	High	Low	Material	High	Low	Material	High	Low
Micas, natural and synthetic	2,000	1,000	Nylon, glass-filled	500	400	Silicones (molded)	400	250
Polymethylstyrene	1,950	890	Nylons 6 and 11	500	420	Ureas	400	300
Polyvinylchloride	1,400	24	Polyesters (cast), allyl type	500	330	Melamines, shock-resistant	370	130
Acetal copolymer	1,200		TFE fluorocarbons	500	400	Phenolics (molded), very high shock resistance	370	200
Polyvinyl formal	1,000	860	Polyethylenes	480		Alkyds	350	300
Plastic laminates, high pressure	1,000	70	Polycarbonate, filled	475		Polycrystalline glass	350	250
Polypropylene	800	520	Nylons 6/6 and 6/10	470	385	Phenolics, heat-resistant	350	100
Plastic laminates, low pressure	800		Epoxies (molded)	468	334	Melamines, GP†	330	310
Phenolics (cast), GP†	800	100	Cellulose propionate	450	300	ABS resins, extra high impact	312	
Modified polystyrenes	650	300	Diallyl phthalate	450	275	Beryllia	300	250
Polyallomer	650		Phenolics (cast), GP†	450	300	Alumina ceramics	300	200
Cellulose acetate	600	250	Melamines, electrical	430	350	Standard electrical ceramics	300	55
Cellulose nitrate	600	300	Phenolics (molded), GP†	425	200	Zircon	290	60
CFE fluorocarbons	600	530	Polystyrenes, glass-filled	425	340	Steatite	280	145
Hard rubber	600	344	ABS resins, high impact	416	350	Forsterite	250	
Mica, glass-bonded	600	270	Cellulose acetate butyrate	400	250	Phenolics (cast), GP† transparent	250	75
Polyesters (cast), rigid	570	340	Chlorinated polyether	400		Cordierite	230	140
Epoxies (cast)	550	350	Melamines, cellulose, electrical grades	400	350	Polyethylene foam, flexible	220	
Acrylics	530	400	Phenolics (cast), mechanical and chemical grades	400	350			
Polystyrenes, GP†	>500		Polycarbonate	400				
Acetal	500		Polyesters (cast), nonrigid	400	220			
Ethyl cellulose	500	350	Polyvinyl butyral	400				

Dielectric Constant of Nonmetallics[‡]

Material	High	Low	Material	High	Low	Material	High	Low
Mica, glass-bonded	40.0	6.9	Alkyds, GP[†]	5.0	4.8	Acrylics	3.2	2.7
Phenolics (cast)	11.0	4.0	Rubber phenolics	5.0		Polyvinyl formal	3.0	
Alumina ceramics	9.6	8.2	Vinylidene chloride	5.0	3.0	Polycarbonate	2.96	
Lead silicate glass	9.5	6.6	Hard rubber	4.95	2.90	Chlorinated polyether	2.92	
Zircon	9.2	5.3	Polyesters (cast), allyl type	4.8	3.3	Methylstyrene-acrylonitrile	2.81	
Polyvinylchloride	9.1	2.3	Alkyds, electrical and impact	4.8	4.2	Epoxies, resilient	2.8	2.6
Micas, natural and synthetic	8.7	5.4	Diallyl phthalate	4.5	3.3	Polystyrenes, GP[†]	2.65	2.45
Phenolics (molded)	8.0	4.0	Nylons 6 and 11	4.5	3.5	Polymethylstyrene	2.48	
Soda-lime glass	7.4	7.2	Epoxies (cast)	4.4	2.6	CFE fluorocarbons	2.37	2.30
Melamines	7.2	4.7	Epoxies, GP[†]	4.4	3.4	Polyethylenes	2.3	
Beryllia	7.0	6.4	Boron nitride	4.2		Propylene-ethylene polyallomer	2.29	
Cellulose acetate	7.0	3.2	ABS resins	4.1	2.8	Polypropylene	2.1	2.0
Standard electrical ceramics	7.0	5.4	Epoxies, heat-resistant	4.0	3.5	TFE fluorocarbons	2.0	
Ureas	6.9	6.4	Modified polystyrenes	4.0	2.5	Prefoamed epoxy, rigid	1.55	1.19
Plastic laminates, high pressure	6.8	3.3	Polyesters (cast), rigid	4.0	2.8	Polyethylene foam, flexible	1.49	
Forsterite	6.5	6.2	Nylon, glass-filled	3.9	3.4	Urethane rubber foamed-in-place, rigid	1.40	1.05
Steatite	6.5	5.5	Phenoxy	3.8	3.7	Silicone foams, rigid	1.26	1.23
Aluminum silicate glass	6.3		Silica glass	3.8		Polystyrene foamed-in-place, rigid	1.19	
Cellulose acetate butyrate	6.2	3.2	Acetal	3.7		Prefoamed cellulose acetate, rigid	1.12	1.10
Cordierite	6.2	4.0	Cellulose propionate	3.6	3.4	Prefoamed polystyrene, rigid	<1.07	
Polyesters (cast), nonrigid	6.1	3.7	Ethyl cellulose	3.6	2.8			
Plastic laminates, low pressure	5.6	3.4	Nylons 6/6 and 6/10	3.6	3.4			
Polycrystalline glass	5.6		Polycarbonate, filled	3.50				
Borosilicate glass	5.1	4.0	Polystyrenes, glass-filled	3.41	2.74			
Silicones (molded)	5.1	3.6	Modified polystyrenes, extra high impact	3.3	1.9			
			Polyvinyl butyral	3.3				

*Values represent high and low sides of a range of *typical* values. To obtain volt/cm, multiply by 393.7.
[†]GP = general-purpose grades.
[‡]Values represent high and low sides of a range of *typical* values at 10[6] cycles.
From "Materials Selector Guide," *Materials and Methods*, Reinhold Publishing Corp., New York, 1973. Reprinted by permission.

Figure 20.21, and as the complexity of the molecule increases, it takes a longer time to shift the dipole.

If we impose a frequency that causes a slight shift each time, we can have dielectric heating because of the energy loss in each cycle. This phenomenon has been used in industry in such applications as the setting of glues in furniture manufacture, and now it has also found use in the home. Most of our foods are made up of organic molecules with dipoles (such as proteins and water), so it is possible to rapidly "cook them from within" by exposing them to suitable frequencies delivered by a microwave oven.

Power Factor and Dissipation Factor

As a result of molecular and electron motion in a changing field, we obtain microwave heating, which can be useful in cooking, as we have mentioned, but is often troublesome in power transmission and communication. It is important to measure this loss of power, which is converted into heat and can produce breakdown of insulation. The power factor P.F. is defined as

$$\text{P.F.} = \frac{\text{watts lost}}{\text{(effective) volts} \times \text{amps}} \tag{20-19}$$

The dissipation factor is defined from the vector diagram of the circuit containing the dielectric material (tan δ), as shown in Figure 20.22. For insulating materials, the dissipation factor is the same as the power factor. (This should not be confused with the power factor cos θ, which is commonly used to express relative values of capacitance and resistance.)

Arc and Track Resistance

Arc resistance is a measure of the breakdown of the surface of an insulator caused by an arc that tends to form a conducting path. A number of tests of arc resistance have been devised. The commonest is ASTM D 495-61, which is a high-voltage, low-current test under clean-surface conditions. The result is expressed in seconds to obtain "tracking." Other tests take into account the presence of contaminants such as fog and dust on the surface.

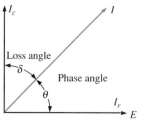

FIGURE 20.22   Vector diagram showing loss angle δ and angle θ between the current I and the voltage E. With pure resistance, I has the same angle as E (I_r), whereas with pure capacitance, I is at an angle of 90° to E (I_c).

In general, polymers with aromatic rings, such as phenolics and styrenes, track easily, whereas long straight-chain polymers and mineral fillers are more resistant. The presence of fibers generally reduces resistance if the test is run under wet conditions.

Excellent resistance to surface arcing is displayed by a ceramic spark plug, which is glazed to prevent surface adsorption of contaminants and has ribs to increase the discharge path.

20.17 Barium Titanate–Type Dielectrics (Ferroelectrics)

As manufacturers tried to reach objectives such as pocket-size television sets, they demanded materials with higher dielectric constants to reduce the size of the capacitors. Because the mineral rutile, TiO_2, was known to have a dielectric constant ϵ_r of about 10, the search led to the titanates. Finally, it was found that barium titanate showed values of ϵ_r over 1000, several orders of magnitude better than any known material. In addition, it was noticed that a permanent charge was developed. This is called *ferroelectric behavior,* and materials that exhibit it are called *ferroelectrics.*

It is interesting to examine how these effects result from the structure of $BaTiO_3$. In Chapter 1 we discussed the fact that the Ti^{4+} ion is larger than the octahedral site formed by the six oxygen ions (radius $Ti^{4+} = 0.64$ Å vs. 0.625 Å for the site) (Figure 20.23). As a result, the titanium ion is located to one side of the center. Thus in each unit cell, one side of the center is positive and the other negative, or, in other words, a dipole develops.

When an electric potential is applied across the capacitor plates, the Ti^{4+} ions are attracted to the negative side. This leads to a high charge storage in the plates and therefore to a higher dielectric constant.

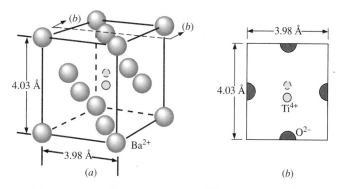

FIGURE 20.23   (a) Structure of barium titanate. (b) The source of the dipole is due to two possible positions for the Ti^{4+} ion.

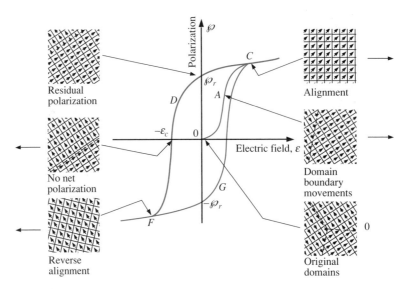

FIGURE 20.24   Ferroelectric hysteresis. The spontaneous polarization can be reversed; however, energy is consumed with each cycle.

Let us now examine the hysteresis effect (Figure 20.24). By placing a solution of magnetized powder over an etched sample of $BaTiO_3$, we can show that groups of unit cells within a grain of the ceramic are organized into *domains* even before the electrical field is applied. In the unit cells of a given domain the Ti^{4+} ions are oriented in the same direction. When the field is applied, domains that have Ti^{4+} ions situated in the *right* direction (toward the negative plate) grow at the expense of those with the *wrong* orientation by means of a change in orientation of the adjacent unit cells. This domain growth finally slows down, leading to a smaller increase in charge (polarization) as applied fields are increased. If the electric field is cut to zero, a charge remains in the material because of the domain alignment. This is called the *residual polarization* $\mathscr{P}_r$. To remove this, one must reverse the electric field from the original condition to $-\mathscr{E}_c$.

There is an important effect of temperature on this ferroelectric behavior of barium titanate. At 120°C (248°F) the structure changes from tetragonal to cubic, and the effect disappears. This temperature is called a *Curie temperature*, in honor of Pierre Curie. The general term is used to specify the temperature at which any change in electrical or magnetic properties takes place. The change may or may not be accompanied by a phase change.

To obtain ferroelectric materials for use at higher temperatures, researchers investigated other compounds with similar structures and found lead titanate, with a Curie temperature of 480°C (896°F), and lead meta-niobate, with a Curie temperature of 570°C (1058°F). The ferroelectric material most commonly used at present is $Pb(Zr, Ti)O_3$, which is known as PZT.

EXAMPLE 20.7 *[ES/EJ]*

Show that lead titanate, $PbTiO_3$, can be expected to have a structure equivalent to that of $BaTiO_3$.

Answer

	Ba^{2+}	Ti^{4+}	O^{2-}	
Ionic radius, Å:	1.43	0.64	1.32	(From Table 1.6)
	Pb^{2+}	Ti^{4+}	O^{2-}	
Ionic radius, Å:	1.32	0.64	1.32	

In general, in the perovskite structure (Figure 20.23) the ions in the unit cell corners have an average valence of 2 and radii close to that of Ba^{2+}. To obtain extensive solid solution, the divalent radii should be ±15% or 1.21–1.64 Å if we assume a relationship for ions that is analogous to the Hume-Rothery rule for metal atom substitution. Ions in the $\frac{1}{2}, \frac{1}{2}, \frac{1}{2}$ sites must average 4+ (3+ in conjunction with Nb^{5+} in niobates, for example) and must be of about the radius of Ti^{4+}. See the problems for further data.

20.18 Interrelated Electrical–Mechanical Effects (Electromechanical Coupling)

Every radio amateur has heard of the piezoelectric effect of a quartz crystal. When pressure is applied to a quartz crystal, the ends become charged, and conversely, when an electric field is applied, the crystal changes length. When an alternating voltage is applied, the crystal oscillates, giving out a sound wave of constant frequency.

Figure 20.25 shows the reason for the generation of a sound wave. Let us assume that we have a single crystal with the dipoles of the unit cells aligned

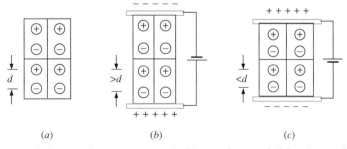

| (a) | (b) | (c) |

FIGURE 20.25 (a) Piezoelectric material. (b) An electric field induces dimensional expansion. (c) Reverse polarity causes a corresponding contraction. The procedure can be inverted by applying a pressure and obtaining a change in the voltage.

[Lawrence Van Vlack, *Elements of Materials Science*, © 1964, Addison-Wesley Publishing Co., Inc., Reading, Massachusetts. Reprinted with permission of the publisher.]

as shown. When we apply an electric potential as shown, the crystal lengthens because the ions are attracted to the pole plates. If we use an ac voltage, the crystal alternately expands and contracts, sending out a wave into the surrounding medium, whether air or water. On the other hand, if we supply the mechanical motion, we receive an electric charge by the reverse procedure. This phenomenon is called *piezoelectricity*, from the Greek word meaning "to press."

With ferroelectric materials we can obtain effects similar to, but stronger than, the piezoelectric effect. We must distinguish between piezoelectric and ferroelectric effects. In piezoelectricity, such as in quartz, a single crystal is required if the phenomenon is to be used in a device. However, ferroelectric materials contain domains that grow in response to the applied field. This has two important consequences. First, the charge storage is greater in ferroelectric materials because of the greater structural change. Second, a single crystal is not required. The ability to use a polycrystalline material is naturally of great commercial advantage.

20.19 Thermocouples; Thermoelectric Power

When we produce a temperature difference in a rod, there are more high-energy electrons in the hotter region (Figure 20.26). High-energy electrons flow toward the cold end, producing a difference in charge. If we now attach a voltmeter with contact wires of the same material, no voltage difference registers at the meter. If the wires are of a different material than the rod, however, a different voltage is induced and the net voltage appears on the meter.

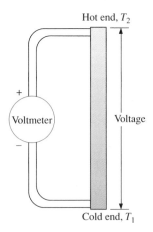

FIGURE 20.26   Development of a voltage in a thermocouple

[R. M. Rose, L. A. Shepard, and J. Wulff, *The Structure and Properties of Materials*, Vol. 4: Electronic Properties, John Wiley, New York, 1965. By permission of John Wiley & Sons, Inc.]

This voltage is the *Seebeck potential S*, which is used in *thermocouples.*
Some examples of thermocouples follow.

Thermocouple	Maximum Useful Temperature, °C (°F)	
Copper vs. constantan (60% Cu, 40% Ni)	315	(599)
Iron vs. constantan	950	(1742)
Chromel (90% Ni, 10% Cr) vs. Alumel		
(94% Ni, 2% Al, 3% Mn, 1% Si)	1200	(2192)
Platinum vs. platinum with 10–13% rhodium	1500	(2732)
Tungsten vs. tungsten–rhenium,		
also tungsten vs. molybdenum	>1500	(>2732)

The basic reason for the potential difference of a thermocouple goes back
to the concept of the Fermi level. In different metals the Fermi level changes
at different rates with increased temperature. Thus in a pair of metals, the
element with the greater temperature coefficient of change has more energetic electrons, which flow into the element with the lower temperature coefficient of change.

When we survey the metals used for thermocouples, we see that many
are transition metals. This is because the distribution of electrons at different
energy levels shows the greatest difference with temperature in metals with
unfilled 3*d* and 4*f* shells.

EXAMPLE 20.8 *[EJ]*

A number of thermocouple pairs are given in the foregoing table. What criteria have probably been used to select these useful candidates?

Answer

1. The maximum useful temperature is an obvious criterion, as is the cost.
   Wires with a higher melting point generally cost more to refine and
   fabricate.
2. The atmosphere in which the thermocouple will be used must be considered. For example, platinum is useful in neutral or oxidizing atmospheres
   but is contaminated by reducing atmospheres.
3. The output voltage of the thermocouple per degree of temperature increase
   should be high in order to minimize error and the cost of the instrumentation needed to read the voltage.
4. The increase in voltage per degree of temperature increase should be close
   to linear over the thermocouple's useful temperature range. This simplifies
   instrumentation and gives a consistent degree of precision over the temperature range.

FIGURE 20.27   Schematic drawing showing both particle and wave characteristics of a photon

OPTICAL PROPERTIES

It is appropriate to discuss optical properties here because of the tremendous growth of electro-optic materials involved in fiber-optics and related devices. We begin with a definition of classic optics and optical devices and conclude with a discussion of fiber-optics.

20.20 Emission

There are two main types of emitted light: light with a continuous spectrum, such as "white" light from a tungsten light bulb, and light confined to definite wavelengths, such as that from a sodium-vapor lamp. We shall concentrate on the emission of light of definite wavelengths because of its great importance in new devices.

Light emitted from any source can be thought of as traveling in discrete units, or little energy packets, called *photons* (Figure 20.27). Furthermore, we know that light has a definite wavelength that is related to the energy of the photon. We shall need to use both these aspects of light in explaining its interaction with materials, and especially with the electrons of materials.

Visible light can originate from radiation of nonvisible wavelengths. Let us review a broad portion of the visible spectrum. On one side of the visible spectrum we find the infrared and longer-wavelength photons, and on the other, the ultraviolet and x-ray wavelengths (Figure 20.28).

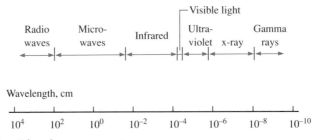

FIGURE 20.28   The electromagnetic spectrum

From physics we have the relation between the energy and the wavelength of a photon:

$$E = \frac{hc}{\lambda} \tag{20-20}$$

where E = energy, h = Planck's constant = 6.62×10^{-27} erg-sec, c = velocity of light = 3×10^{10} cm/sec, and λ = wavelength in centimeters.

When we express this relation in simple terms of electron volts, which we used earlier, with λ in units of cm we have

$$E = \frac{1.24 \times 10^{-4}}{\lambda} \, eV \tag{20-21}$$

EXAMPLE 20.9 *[ES]*

Calculate the energy of photons of infrared light (λ = 11,500 Å), yellow light (λ = 5800 Å), and ultraviolet light (λ = 2000 Å).

Answer

Infrared: $E = \dfrac{1.24 \times 10^{-4}}{11,500 \times 10^{-8}} = 1.08$ eV

Yellow: $E = \dfrac{1.24 \times 10^{-4}}{5800 \times 10^{-8}} = 2.14$ eV

Ultraviolet: $E = \dfrac{1.24 \times 10^{-4}}{2000 \times 10^{-8}} = 6.20$ eV

This interrelation of energy and wavelength is important to both the emission and the absorption of light. We shall consider emission first.

We are all familiar with the emission of yellow light that occurs when we throw salt into a fire. Let us examine the reason for this phenomenon. We recall from Chapter 1 that the electrons in an atom may occupy different energy levels. For example, Figure 20.29 shows the energy levels above the 3s level in sodium. When an electron in the 3s is raised to the 3p level and then falls back to the 3s level, a photon of $\lambda \cong 5890$ Å, or yellow light, is emitted. The steps involved in the energy absorption (raising of the electron to an upper level) and the fall need not be the same. For example, we could supply the energy to raise the electron to the 6s level, but it could jump down to the 3p level and then to the 3s. (This is somewhat like the movement of the ball in a pinball machine. The ball is sent to the top in one stroke, but it

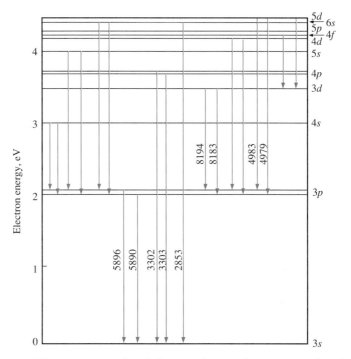

FIGURE 20.29   Electron energy-level diagram for a sodium atom. The photon wavelengths associated with some of the transitions of electrons from excited states are given in units of 10^{-8} cm.

can come down by a series of jumps. Some of these may give a blink of light, others just noise or heat.)

At this point we should also consider the companion phenomenon of the emission of shorter wavelengths (x rays) when an electron falls to a still lower level. If we bombard sodium atoms (or ions) with sufficiently energetic electrons, we can knock out an electron from the first (K) shell. This creates a hole into which an L-shell electron can fall. For example, the transition $2p \rightarrow 1s$ for sodium gives a photon of wavelength 11.91 Å, an x-ray wavelength. Just as the color of visible light emitted from a given jump depends on the particular levels involved, so the wavelength of the x ray depends on the specific case. For instance, because of the greater charge on the nucleus it takes much more energy to knock out an electron from the inner ring (K shell) of an atom of high atomic number, and we expect to find a shorter-wavelength x ray emitted when an L-shell electron falls into the K shell. Typical atoms and the wavelengths emitted illustrating this effect for the K_α x ray are manganese ($\lambda = 2.10175$ Å), iron ($\lambda = 1.93597$ Å), and tungsten ($\lambda = 0.208992$ Å), as we indicated in Chapter 1.

The scanning electron microprobe that we mentioned in Chapter 1 is also based on the emission of these characteristic x rays. A scanning electron beam is swept across a given microstructure. Each tiny region of the structure produces x rays with wavelengths that depend on the elements present. For example, if we want an indication of the amount of silicon in a given region, we put in a filter that allows only rays of the wavelength characteristic of silicon to pass through. Then the picture screen showing the microstructure is irradiated by these x rays and shows bright spots only at regions of high silicon concentration (see Figure 1.37). Similarly, we can find regions of high aluminum concentration by allowing only the x rays that are characteristic of aluminum to pass through in another picture. In another procedure we can scan a given region and record the relative intensity of all of the characteristic x rays as shown in the color section.

We should discuss one more phenomenon, called *luminescence*, which is quite important in the new light-emitting materials. A familiar sight in natural history museums is a case of fluorescent and phosphorescent minerals. After exposure to ultraviolet light the specimens glow different colors, and some continue to glow after the ultraviolet light source is turned off. This conversion of ultraviolet light to visible light is given the general name *luminescence*. If the glow ceases when the ultraviolet stimulus is removed, it is called *fluorescence*. If it persists for over 10^{-8} sec after the ultraviolet light is turned off, the light is called *phosphorescence*. In either case we have to explain why light of a visible wavelength results from exposure to ultraviolet light or bombardment by electrons.

A typical luminescent material consists of a solid, such as zinc sulfide, that contains a small amount (1 part per million) of an impurity, such as copper. This impurity is the key to luminescence. Again the process is a little like a sophisticated pinball machine, in which an electron is raised to a high level and may fall into traps from which it is later released (Figure 20.30). There are essentially three steps in this process.

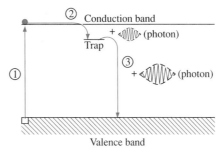

FIGURE 20.30   Development of luminescence. Step 1: The electron is excited by ultraviolet radiation into the conduction band. Step 2: The electron falls into the trap, emitting an infrared photon (the trap is an intermediate energy level that is caused by an impurity). Step 3: The electron falls into a lower level, emitting a photon of visible light. *Note:* The electron hole is also attracted to the trap and also emits light.

1. A photon or electron is shot at a solid, such as zinc sulfide, and its energy is absorbed. This enables an electron of the solid to rise to the conduction band, which is separated from the valence band by a definite energy gap. A hole is created in the valence band at the same time.
2. The electron wanders through the structure until if falls into a trap, or luminescent center. These traps are produced by the impurity — in our example, the copper ions. The "energy level" traps lie between the more elevated conduction band of the solid and its valence band.
3. After a time the electron acquires enough energy to leave the trap and fall to the valence band, giving up a photon of definite wavelength (in the visible spectrum) that is related to the ion producing the trap — in this case, copper.

Phosphorescence occurs when materials have deep traps. In such materials it takes more time for the trapped electrons to receive enough energy to escape. As a consequence, they are liberated over a longer period by continued electron movement. It is possible to produce approximately "white" light by using a mixture of impurities, or luminescent centers, such as silver (which gives a peak at 4300 Å), copper (at 5100 Å), and manganese (at 5900 Å).

Applications

These variables in color and in duration of luminescence are of great practical importance. In the dots of a color television picture tube, for example, we want phosphors that give pure primary colors and do not have long decay times. The development of materials with these characteristics provides an example of the extremes to which manufacturers will go for color quality. It was found that europium, a rare earth, provided good red-color centers. As a result, a new mining and chemical-separation industry developed to supply europium, and the other rare earths that are found with europium are now available as low-cost by-products. Another application of the effects of luminescence is in electroluminescent panels, in which the initial radiation step is eliminated. Here the application of a voltage across a strip provides directly the energy required for electrons of the material to move to the phosphor centers, and the jumps that occur later give off light. The color of the panel is a function of the active ions chosen.

Lasers

Light amplification by *stimulated* emission of *radiation* (*laser*) is an even more sophisticated effect of luminescence. In the ruby laser, for example, a single crystal rod of alumina, Al_2O_3, contains a small quantity of Cr^{3+} ions as an impurity (Figure 20.31). The rod is ground with flat surfaces at each end. One end is heavily silvered so that it is opaque, and the other is lightly silvered for partial reflection. Shining on the rod is a tube containing xenon gas, which emits constant-wavelength light (5600 Å) when energized. The se-

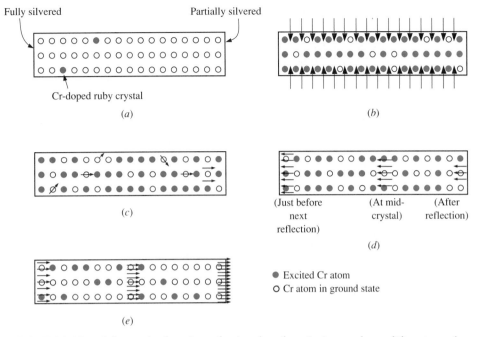

FIGURE 20.31 Schematic drawing of stimulated emission and amplification of a light beam in the ruby laser. (*a*) At equilibrium. (*b*) Excitation by xenon light flash. (*c*) A few spontaneously emitted photons start the stimulated emission chain reaction. (*d*) Reflected back, the photons continue to stimulate emission as they travel. (*e*) Increasing in power, the beam is finally emitted.

[R. M. Rose, L. A. Shepard, and J. Wulff, *The Structure and Properties of Materials*, Vol. 4: Electronic Properties, John Wiley, New York, 1966. By permission of John Wiley & Sons, Inc.]

quence is as follows: At first only a few electrons in the chromium ions are above the ground state, but after the xenon lamp is flashed, the electrons are energized. These electrons either fall to the ground state or take up a higher-energy metastable state for an average of 3 msec. A high percentage are pumped up to this metastable state. Next a few electrons fall to the ground state, emitting photons of a characteristic wavelength. A percentage of these emitted photons are reflected from the ends and stimulate emission throughout the structure in a short period, 6 msec.* A pulse of coherent wavelength emission thus travels through the lightly silvered end. (By *coherent* we mean that the photons are in phase with one another.) It is a beam with very low divergence because photons emitted through the sides of the rod are lost to the surroundings.

Many uses have been developed for lasers, including surgical and metallurgical heating, surveying, and precision measurement. Various materials,

---

*There is a critical relationship between the length of the rod and the wavelength of the emission, so that a "standing wave" of photons is built up within the rod.

including certain gases, can be used in the laser. In a semiconductor laser with a *p-n* junction, an electrical impulse energizes the carriers, so that as they fall through an energy gap, a stimulated emission of photons of longer wavelength (6500–8400 Å) is obtained. Also, lasers with *continuous* emission of light (rather than pulsed emission) have been developed.

Emission from Solid Masses

As distinguished from the emission of light by separated ions or electrons, another important form of light emission is the radiation from heated solids such as a tungsten wire filament in a conventional light bulb. In this form of emission we obtain white light, or a continuous spectrum, because in the solid tungsten metal we have a *band* of electron energies and a band of sites into which excited electrons can fall. Another case of emission involves the "boiling off" of electrons from the incandescent filament of a radio tube to be accelerated through space by the potential between the electrodes.

20.21 Absorption

It is important to distinguish between color produced by *light emission* and color produced by *absorption*. We see a yellow color *emitted* when salt is thrown in a fire because of the characteristic wavelength that results when the heat causes electron jumps in the sodium atom. However, if we look through a yellow glass filter at a white light, we see yellow light transmitted because the filter has absorbed the other colors and has merely allowed the already existing photons of yellow wavelength to pass through. The interaction in the second case, therefore, is between the electrons of the material and the photons of nonyellow wavelengths. The range of colors produced this way is called the *absorption spectrum*.

For example, glass is colored only by ions of the transition elements. Copper is considered a transition element in the Cu^{2+} state because it has an unfilled $3d$ shell. Typical colors produced by the ions of transition elements are Cr^{2+}, blue; Cr^{3+}, green; Cu^{2+}, blue-green; Cu^+, colorless; Mn^{2+}, orange; Fe^{2+}, blue-green; and U^{6+}, yellow (unfilled $5f$ shell). These colors result from the interaction of the inner-shell electrons of the transition elements with the photons of the various wavelengths of white light. The characteristic color that emerges from each element is the unabsorbed part of the spectrum. Because the ions do not act independently of the glass matrix, the color is not a sharp wavelength, as it is in the emission spectrum; the ions in the vapor have no such interaction with a matrix. The rare earths also produce colored glasses because of the electron interaction in the $4f$ shell.

Another example of absorption is the *photoelectric effect*, in which the absorption of a photon results in the emission of an electron.

When a photon of light of energy $h\nu$ [ν = frequency = c/λ (Equation 20-20)] strikes an electron in a block of metal and the electron is sufficiently energized to leave the block, we say that the energy of the electron is made up of two components:

1. The energy required to leave the block $(e\phi)$
2. The kinetic energy (K.E.) of the electron outside the block

Therefore, $\qquad\qquad\qquad hv = e\phi + \text{K.E.}$ $\qquad\qquad\qquad$ (20-22)

The value $e\phi$ is a constant for the material at a given temperature. It is called the *work function*. Multiplying the work function by the charge on the electron gives the energy barrier that is surmounted as the electron leaves the material.

We would expect the alkali metals to make good emitters because of the ease with which their electrons can be stripped. The yield of electrons varies with the wavelength of the light and the element. As the atomic number increases from sodium to cesium, the maximum yield occurs at higher wavelengths. In other words, it takes a photon of higher energy to knock out an outer-shell electron in sodium than to do so in cesium, because the outer electrons in cesium are further from the nucleus and are more loosely held.

EXAMPLE 20.10 *[ES]*

The work function $e\phi$ for tungsten is 4.52 eV. What is the maximum wavelength of light that will give photoemission of electrons from this element?

Answer The energy of the photons will just equal the value they require to escape; in other words, kinetic energy is equal to zero in Equation 20-22. Therefore,

$$hv = e\phi = 4.52 \text{ eV}$$

$$(6.62 \times 10^{-27} \text{ erg-sec})\left(10^{-7}\frac{\text{J}}{\text{erg}}\right)v = 4.52 \text{ eV}\left(1.6 \times 10^{-19}\frac{\text{J}}{\text{eV}}\right)$$

or $\qquad\qquad\qquad v = 1.09 \times 10^{15} \text{ sec}^{-1}$

and $\qquad\qquad \lambda = \dfrac{c}{v} = \dfrac{(3 \times 10^{10} \text{ cm/sec})(10^8 \text{ Å/cm})}{1.09 \times 10^{15} \text{ sec}^{-1}} = 2750 \text{ Å}$

20.22 Reflection

Reflection can be an apparently simple phenomenon or a rather complex one. In the simple case we merely shine a beam of white light at a surface and receive a reflected beam of almost the same intensity and "spectrum"; the angle of incidence equals the angle of reflection. In the more complex case, we find a red color in the beam reflected from copper, a yellow color from gold, and similar but less noticeable variations from other metals. In all cases the photons interact with the electrons of the material, because a photon can be considered an electromagnetic wave. However, the interaction is not the same for all wavelengths of light and depends on the energies of the different electron levels of the material. In metallic reflection, significant interaction can occur between the photons and the outer-shell electrons of the metal. Thus in the case of copper, the photons corresponding to blue wavelengths are absorbed by $3d$ electrons, leaving a reddish color to be reflected.

20.23 Transmission

Now let us consider why metals are opaque in all but very thin sections and why ionic and covalent solids are transparent to visible light. At the same time we can discuss the subject of why materials we consider opaque, such as silicon and germanium, are transparent to infrared.

Metals

Recall that we can think of the valence electrons of a *metal* as having a band of energies with many unfilled energy states at the upper energy levels. When a photon of light strikes an electron in a metal, the electron can be raised to an unoccupied higher level, and the radiation is absorbed rather than transmitted. Metals generally exhibit *opacity*; we can transmit visible light in metals only in very special cases, such as in a very thin sheet of gold, in which the probability of collision of the photons during transmission is reduced.

Ionic and Covalent Solids

In the case of ionic and covalent solids, the electrons are bound to the atoms. It takes a high-energy photon to break an electron bond and accelerate the electron to the conduction band. (Recall that there is a gap between the valence and conduction bands in these materials.) For this reason photons of the lower-energy visible and infrared wavelengths pass through ionic solids with ease; these solids exhibit *transparency*. However, very-high-energy radiation such as ultraviolet is absorbed because the energy of these photons can be used to accelerate the electrons to the conduction band. If atoms of an impurity are present, absorption can occur. Such is the case with chromium ions in ruby, which absorb the photons of blue and green wavelengths. The remaining photons (or other wavelengths) in white light pass through, giving an overall red spectrum.

Semiconductors

Semiconductors provide a very interesting special case of transmission. Relatively little energy is required to raise the carriers to the conduction band and therefore cause the semiconductor to absorb radiation. It happens that the energy of the photons in infrared radiation is not sufficient (infrared radiation has a long wavelength and, therefore, low energy), so that semiconductors are transparent to infrared radiation and can be used as windows or lenses for efficient transmission. However, the energy of visible light is sufficient to raise the carriers to the conduction band; thus semiconductors appear opaque in visible light.

EXAMPLE 20.11 *[ES]*

Explain the transmission effect of white light on each of the following materials.

Material	Energy Gap, eV	Color
Diamond	5.6	Colorless
Sulfur	2.2	Yellow
Silicon	1.1	Opaque

Answer Referring to Example 20.9, we see that none of the photons of the visible spectrum is energetic enough to react with the electrons of diamond and raise them to the conduction band. Therefore, the light is transmitted unchanged. In sulfur the green and blue wavelengths are absorbed, leaving yellow (and red). In silicon all the visible light is absorbed, because only 1.1 eV is required for photon–electron interaction, and this is available from the light. Another interesting example is the use of HgTe with an energy gap of only 0.2 eV as a filter in taking far infrared photographs of the earth.

The use of light to activate electrons of a semiconductor to the conduction band is called *photoconductivity*. It is possible to make a semiconductor with an energy gap that corresponds to the energy of photons of light of a certain wavelength. If light of a wavelength longer than the required wavelength is used, the photons are not energetic enough to activate electrons to the conduction band. If the wavelength is much shorter than required, the light is so heavily absorbed that only the surface is affected; because most of the crystal is inactive, the overall photoconductivity is small. At just the critical wavelength the response is high, as shown in Figure 20.32. This technique is applied to the detection of missiles or aircraft, both rich infrared emitters, from hundreds of miles away. Cadmium sulfide is useful as a photographic lightmeter because of its sensitivity within the visible spectrum (energy gap 2.42 eV).

Color Centers

The characteristics of transmission of light of a given material may be altered by impurities. This is especially evident in the colors developed in transparent crystals such as ruby and sapphire, and it is due to the development of

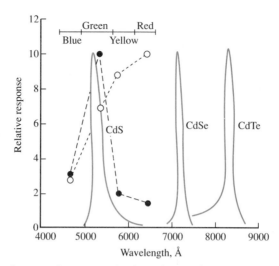

FIGURE 20.32 Photoconductivity of CdS, CdSe, and CdTe, showing a maximum response at wavelengths corresponding to the width of the forbidden energy gap. The forbidden energy gap is the area between the valence and conduction bands where electrons cannot reside. The data circles show the response of commercial CdS and CdSe photocells to colored light bulbs.

[Reprinted from J. J. Brophy, *Semiconductor Devices,* © 1964, by permission of McGraw-Hill Publishing Company.]

color centers. You can find the most striking evidence of this effect by heating a transparent uncolored crystal of sodium chloride in sodium vapor and observing the development of a yellow color in the material. The mechanism is as follows: Sodium atoms condense on the surface of the crystal, and chloride ions migrate from the inner regions to combine with the sodium. This leaves ionic vacancies within the crystal. To balance the charges in the crystal, these vacancies attract electrons. These electrons are not held as rigidly as the normal valence electrons and can therefore be raised to the conduction band by lower-energy photons, such as those in the visible spectrum. Thus, when white light passes through the crystal, certain wavelengths are absorbed and the remaining beam is colored. Ions of impurities can cause the same effect by providing electrons that are easily raised to the conduction band. Therefore photons of certain wavelengths are selectively absorbed.

A number of important new glasses depend on the interaction of light with impurities in the glass. For years people noticed that the windowpanes in old buildings became purple with age. It was found that this was due to the reaction of ultraviolet light with trace amounts of manganese in the glass. The Mn^{2+} ion is colorless, but reaction with a photon causes it to lose another electron and become Mn^{3+}, which absorbs most of the visible spectrum except violet, thereby giving the glass a purple hue. In this case the electron from the Mn^{2+} ion is probably trapped by iron impurities, so the process is not reversible. However, the popular photochromic eyeglasses and windows contain silver ions and work by a reversible process. When the light is bright, the photons reduce Ag^+ ions to metallic silver, causing a darkening. When the light source is removed, the process reverses and colorless Ag^+ ions are regenerated.

20.24 Refraction

Everyone has seen that a stick immersed in water appears to be bent, starting at the water–air interface. It is useful to remember that the velocity of light is lower in water. The index of refraction n is defined as

$$\eta = \frac{c}{v} \tag{20-23}$$

where c = velocity of light in a vacuum and v = velocity of light in the material.

Liquids or solids with dense atomic packing or elements of high atomic number generally have high indices of refraction (lower velocities of light). In some crystals the index of refraction is not the same in all directions, just as there can be differences in magnetic or mechanical properties in different directions. Crystals of lower symmetry, such as calcite and quartz, split the incident beam into polarized beams. This is called *birefringence*, or *double refraction.* Not only is this characteristic useful in identifying minerals, but it is what makes it possible to produce Polaroid filters. In this case crystals of iodoquinone are encased in plastic. These filters absorb the glare component of the light that is polarized.

20.25 Advanced Optical Materials in Communication

Fiber-optics have become so important and depend so heavily on exact control of structure that they deserve special attention. Some of the striking advantages of fiber-optic cable vs. conventional copper wire are

	Fiber-Optic	*Copper*
Distance between repeater stations, km	200	1–2
Cable size for equivalent transmission, in.	3/8	3
Cable weight per kilometer, lb	132	16,000
Electrostatic or electromagnetic interference	none	present
Error rate	lower	higher

Microwave towers are also used in transmission but are being replaced because of interference and error rate. It is necessary to obtain a right of way for cable, but many ingenious schemes are used, such as incorporating the thin cable in power transmission cable or inside pipe lines.

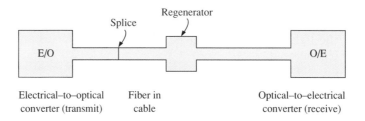

(channel banks and multiplexing are not shown)

FIGURE 20.33   Schematic representation of a fiber-optic communications system

The principal components of a fiber-optic communications circuit are shown in Figure 20.33. In the operation of a system, information is generated in a digital system as bits — that is, either 1 or 0 signals. This electrical input is fed into an electrical-to-optical (E/O) converter, which reproduces the bits as light pulses. A crude illustration is to consider a solar cell being operated in reverse. In other words, an intermittent current is provided to produce flashes of light. Actually a laser diode or a light-emitting diode is used.

The pulses of light are then fed into the core of a fiber-optic cable. Here the important point is to provide cladding or a gradient in refractive index in the glass so that no light is lost through the sidewalls. In modern designs this is always done so that all losses are internal, as discussed shortly. If the distance between stations is over 100 km, repeater units are used to reinforce the signal. At the other end, the optical pulses are reconverted to electrical output and decoded. The system can transmit over 500 megabits per second.

We will begin with a discussion of the structure and properties of optical fibers, the heart of the system, and then will describe the transmitter and receiver. The multiplex system for transmitting more than 8000 different conversations over a pair of fibers is standard equipment in switching systems.

The low loss of signal per mile — attenuation — in a silica fiber is principally due either to cladding or to a composition gradient that changes the index of refraction so that no light is lost through the sidewalls. The relationship is given by Snell's law (Figure 20.34a). A ray striking the sidewall at a steep angle is partly reflected and partly refracted. As the angle changes, a *critical angle* θ_c is reached at which the refracted ray travels only parallel to the longitudinal axis. Beyond this angle, all light is retained in the fiber (except for minor losses such as scattering by foreign particles). The numerical aperture of a fiber is $\eta_1 \cos \theta_c$, Equation 20-25, as shown in Figure 20.34b.

The *acceptance angle* is a measure of the amount of incident light that is transmitted by the fiber. It is equal to twice the half-angle shown for the numerical aperture (Figure 20.34c, Equation 20-27).

The scientific definition of attenuation or, more strictly, attenuation coefficient is the rate of diminution of optical power with respect to distance along a fiber. It is defined by the equation

$$P_Z = P_0 \times 10^{-(\alpha Z/10)} \tag{20-28}$$

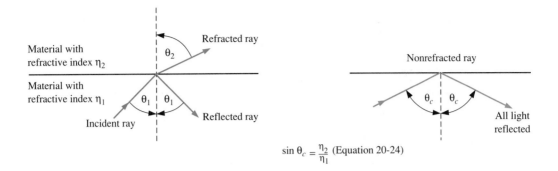

$$\sin \theta_c = \frac{\eta_2}{\eta_1} \text{ (Equation 20-24)}$$

(*a*) Snell's Law

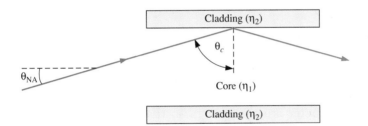

Critical angle $= \theta_c = \sin^{-1} \frac{\eta_2}{\eta_1}$

NA (numerical aperture) $= \sin \theta_{NA}$

$= \eta_1 \cos \theta_c$ (Equation 20-25)

$= \sqrt{\eta_1{}^2 - \eta_2{}^2}$ (Equation 20-26)

$\eta_2 < \eta_1$

(*b*) Numerical aperture

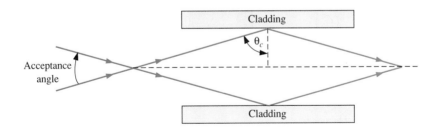

One half the acceptance angle = the numerical aperture (Equation 20-27)

(*c*) Acceptance angle

FIGURE 20.34   Laws governing signal loss in fiber-optic materials: (*a*) Snell's law; (*b*) numerical aperture; (*c*) acceptance angle

where

P_z = power at distance Z along the fiber

P_0 = power at $Z = 0$

α = attenuation coefficient, in units of decibels per kilometer, (dB/km), Z is in kilometers

Rearranging terms in Equation 20-28 yields

$$\alpha Z = -10 \log_{10}(P_Z/P_0) \qquad (20\text{-}29)$$

The *fiber bandwidth* is the lowest modulation frequency at which the ratio of output to input optical power decreases to half of the zero-frequency value.

Optical Fiber Types

There are three different types of optical fibers: (1) multimode step index, (2) multimode graded index, and (3) single-mode step index (Figure 20.35). *Index* here refers to index of refraction.

The *multimode step index* has a core of constant index surrounded by a cladding of lower index to avoid transmission of light into the cladding from the core. The core diameter is large enough so that rays can have different path lengths. The shortest is the "fundamental mode," which passes through the centerline without reflection. The disadvantage of this multimode distribution is that a sharp pulse of input is spread out in time at the output.

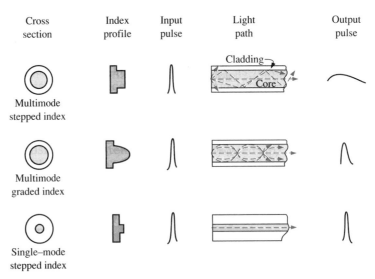

FIGURE 20.35   Characteristics of the three basic types of optical fibers

[E. R. Berlekamp, R. E. Peile, and S. P. Pope, "The Application of Error Control to Communications," *IEEE Communications Magazine*, Vol. 25, No. 4 (April 1987) p. 34. © 1987 IEEE. Used by permission.]

In the *graded-index fiber* the composition of the glass changes gradually to a lower index at the outer diameter, and the output is sharpened somewhat.

In the *single-mode fiber* the diameter is less than the wavelength of the light. For example, a core of diameter 10 μm surrounded by a cladding of 125 μm would be designated as a 10/125 single-mode fiber. The single-mode fiber gives the sharpest output; because there is only one transmission distance for the light rays, the bits of light can be spaced closer than for other methods without overlapping. Until recently, single-mode fiber was more expensive to produce, but large production volume has made its cost the same as that of the types it is replacing.

Sources of Transmission Loss in Optical Fibers (Light Guides)

The reason for the recent interest in optical fibers is suggested by Figure 20.36, which portrays the improvement (reduction) in loss over historical time.

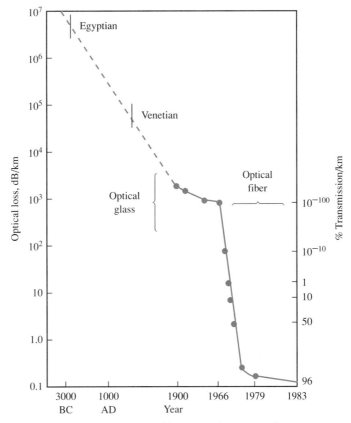

FIGURE 20.36 Historical reduction of loss as a function of time

[E. R. Berlekamp, R. E. Peile, and S. P. Pope, "The Application of Error Control to Communications," *IEEE Communications Magazine*, Vol. 25, No. 4 (April 1987) p. 24. © 1987 IEEE. Used by permission.]

Only in the late sixties were methods developed for producing fiber with losses below 1 dB/km. These techniques involve the vapor deposition of silica on a mandrel and subsequent drawing operations; they will be discussed in Chapter 22.

At this point, we will examine the three aspects of fiber structure that contribute to attenuation: absorption, scattering, and waveguide use (Table 20.3).

Intrinsic absorption losses are related to the interaction of the photons of the light with the electronic and molecular structure. For a given glass this represents a minimum loss that cannot be reduced. The loss varies with the frequency of the transmitted light.

Extrinsic absorption losses are related to deliberate additions of foreign elements and to impurities (Figure 20.37). For silica there is a good transmission window around 1 μm, but the frequencies corresponding to OH^- peaks must be avoided. Minimum absorption occurs at 1.57 μm, where losses as low as 0.16 dB/km have been observed. Small amounts (1 part per billion) of transition elements such as iron and copper must be avoided.

The fibers used for the multimode graded index have a higher loss because of the materials added to change the index of refraction. Lower additions are needed for single-mode fiber.

TABLE 20.3   Sources of Transmission Loss in Lightguides

1. • Intrinsic	—UV electronic transitions
	—IR molecular vibrations
• Impurity	—Transition metals
	—Rare earths
	—Interstitials
	—Matrix impurities
	—OH^- vibration
	—H_2 vibration
• Defects	—Vacancies
	—Radiation-induced
	—Thermally induced
	—H_2-induced
2. Scattering*	
• Rayleigh	—Minute density and concentration fluctuations
• Bulk imperfections	—Bubbles, inhomogeneities, cracks
• Waveguide imperfections	—Core, clad interfacial irregularities
• Brillouin, Raman	—Spontaneous
3. Waveguide*	
• Macrobending	—Curvature-induced
• Microbending	—Perturbation-induced
• Design	—Radiative
• Stimulated Raman, Brillouin	—Depends on power density

E. R. Berlekamp, R. E. Peile, and S. P. Pope, "The Application of Error Control to Communications," *IEEE Communications Magazine*, Vol. 25, No. 4 (April 1987), p. 36. © 1987 IEEE. Used by permission.

*For a complete description of some of these losses, reference to a textbook on optical physics may be necessary.

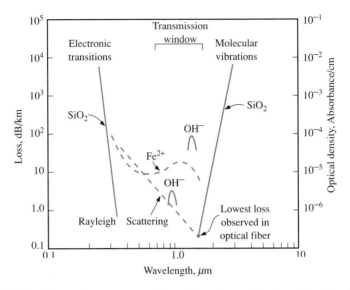

FIGURE 20.37   Schematic representation of loss in mechanisms in silica-based light-guides as a function of wavelength

[E. R. Berlekamp, R. E. Peile, and S. P. Pope, "The Application of Error Control to Communications," *IEEE Communications Magazine*, Vol. 25, No. 4 (April 1987), p. 37. © 1987 IEEE. Used by permission.]

Other glasses and materials (which are difficult to fabricate) are shown in Figure 20.38. ZBLA is a zirconium–barium–lanthanum–aluminum–fluoride glass under development in which actual losses as low as 0.7–0.9 dB/km at the 2.3-μm wavelength have been obtained. If the other scattering mechanisms listed in Table 20.3 can be avoided, very low attenuation can be expected.

The other causes of loss, such as bubbles and cracks, are related to processing or installation (examples include bending beyond a critical radius or propagation of elastic strain) and will be discussed in the section on processing. The packaging of several fibers in a single cable is shown in Figure 20.39.

Emitters

The two principal types of *optical emitters* are laser diodes and LEDs (light-emitting diodes). The coupling of an LED with a fiber (Figure 20.40) shows that the light output pattern of 60–100 degrees is considerably larger than the acceptance cone of a fiber (about 23° for multimode with a numerical aperature, NA = 0.2 or 12° for single mode with an NA of 0.1.) Therefore the coupling efficiency is fairly low. The LED also has a fairly wide spectrum, which leads to added dispersion in passing through a fiber.

Laser diodes operate similarly to an LED until a threshold current is reached, and then output rises sharply (Figure 20.41). The spectral width is much narrower than that of an LED. This permits closer spacing of bits. The use of thermistors for temperature control to maintain the laser diode at an optimal temperature of operation is shown in Figure 20.42.

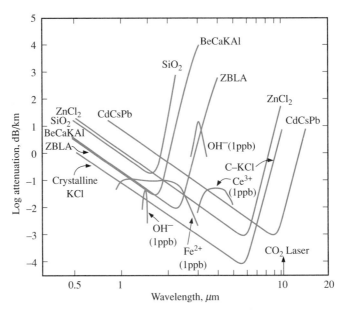

FIGURE 20.38 Theoretical losses of lightguide material vs. wavelength. BeCaKAl are fluoride glasses made with those cations. ZBLA represents a zirconium–barium–lanthanum–aluminum–fluoride glass. CdCsPb is a chloride glass made from those cations.

[E. R. Berlekamp, R. E. Peile, and S. P. Pope, "The Application of Error Control to Communications," *IEEE Communications Magazine*, Vol. 25, No. 4 (April 1987), p. 37. © 1987 IEEE. Used by permission.]

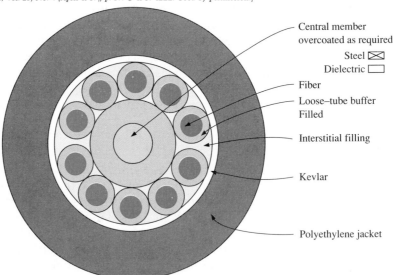

FIGURE 20.39 A fiber-optic cable made up of 10 fibers. The loose-tube design allows for relative motion between fibers.

[*Principles and Applications of Fiber-Optic Communication Systems*, Educational Course package by LighTech, Inc., p. 7–36. Kenneth H. Lewis and Joe R. Bass, authors, of LighTech, Inc., 2116 Arapaho, Suite 515, Richardson, TX 75081. Reprinted by permission.]

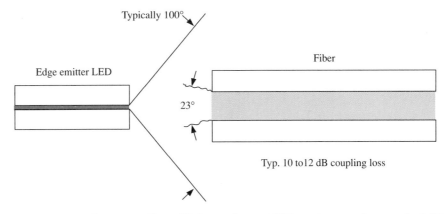

FIGURE 20.40 Low coupling efficiency for an LED emitter and an optical fiber. Even butting directly allows the loss of light rays that exit the LED at angles greater than the fiber acceptance angle.

[*Principles and Applications of Fiber-Optic Communication Systems*, Educational Course package by LighTech, Inc., p. 7-36. Kenneth H. Lewis and Joe R. Bass, authors, of LighTech, Inc., 2116 Arapaho, Suite 515, Richardson, TX 75081. Reprinted by permission.]

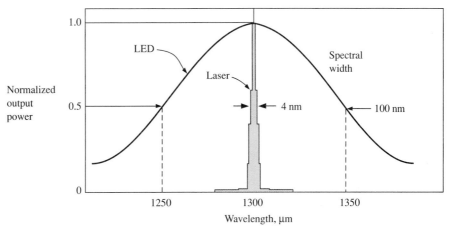

FIGURE 20.41 Comparison of the spectral width of a laser to that of an LED for emitters in fiber-optics

[*Principles and Applications of Fiber-Optic Communication Systems*, Educational Course package by LighTech, Inc., p. 7-44. Kenneth H. Lewis and Joe R. Bass, authors, of LighTech, Inc., 2116 Arapaho, Suite 515, Richardson, TX 75081. Reprinted by permission.]

Optical Detectors

To convert the light bits to electrical currents, PIN (positive, intrinsic, negative) diodes or avalanche photo diodes are used. In both cases photons entering the *optical detector* collide with hole–electron pairs. The energy from the photon separates the pair, providing two carriers. With proper DC bias, increased electron flow takes place, as in the simple transistor described earlier (Figure 20.43).

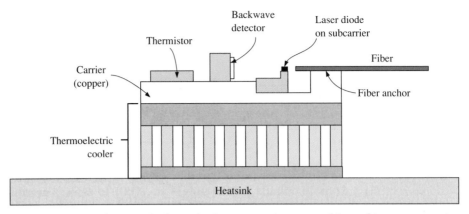

FIGURE 20.42   The use of a laser diode as part of an assembly making up an emitter in a fiber-optic system

[*Principles and Applications of Fiber-Optic Communication Systems,* Educational Course package by LighTech, Inc., p. 7-48. Kenneth H. Lewis and Joe R. Bass, authors, of LighTech, Inc., 2116 Arapaho, Suite 515, Richardson, TX 75081. Reprinted by permission.]

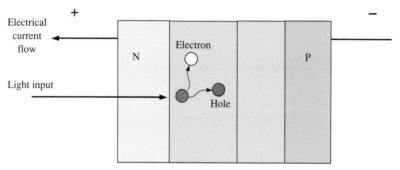

FIGURE 20.43   Schematic representation of the conversion of light pulses (photons) to electron flow via a solid-state optical detector

[*Principles and Applications of Fiber-Optic Communication Systems,* Educational Course package by LighTech, Inc., p. 7-57. Kenneth H. Lewis and Joe R. Bass, authors, of LighTech, Inc., 2116 Arapaho, Suite 515, Richardson, TX 75081. Reprinted by permission.]

Summary

The electrical properties of materials depend mainly on the number and mobility of the charge carriers in the structure: electrons, electron holes, and ions.

The flow of current is related to the conductivity, σ = (number of carriers/m³) × (charge/carrier) × mobility. The number of carriers per cubic meter is relatively great in a metal because the electrons donated by the atoms exist in a band of energies that has many vacant states to which the electrons can be accelerated and thereby carry current. By contrast, in an insulator the valence electrons are bound in a filled energy band. To raise an electron to an energy level at which it can conduct (the conduction band) requires considerable energy. The intrinsic semiconductors occupy an intermediate position between conductors and insulators. The number and type of carriers in a semiconductor can be increased by "doping" it with elements that are electron donors or acceptors. By setting up junctions of these "extrinsic" semiconductors, we can form *p-n* junctions. Certain metals and alloys exhibit superconductivity below critical temperatures near the boiling point of helium (4.2 K). However, new ceramic materials have been developed that show superconduction above the boiling point of nitrogen (77 K), with accompanying growth in boosting engineering applications *considerably*.

Dielectrics with greatly improved dielectric constants can be made from ferroelectric materials such as barium titanate. In these materials there is a permanent dipole or charge imbalance in the unit cell. These dipoles result in the formation of electrical domains or regions of similar dipole alignment. This configuration leads to a high capacitance when the material is used in a capacitor. Electromechanical coupling from these materials or materials with fixed dipoles leads to valuable devices for conversion of electrical impulses to sound. Thermocouples are important for measuring temperature. They operate because of the development of an emf by pairs of metals subjected to two different temperatures.

The optical properties of major importance are emission, absorption, reflection, transmission, and refraction. The basis for understanding each of these phenomena is the relationship between electrons and photons, which may be thought of as little energy packets with definite energy and wavelength.

Emission of a photon occurs when an electron falls from a higher to a lower energy state, or quantum level, in an atom. For example, sodium atoms emit yellow light because the jump of the predominant electron is accompanied by the emission of a photon of $\lambda \cong 5890$ Å. In fluorescent and phosphorescent materials, electrons absorb photons of higher energy, such as x rays, and then fall back through a series of electron traps formed by atoms of impurities. The fall from an intermediate trap to the normal level is accompanied by the emission of visible light.

Absorption is the reverse of emission. A photon strikes an electron that can be raised to an existing higher energy state by absorbing energy. Metals are opaque because there is such a selection of empty energy states that are continuous with valence bands that photons of practically any wavelength can be absorbed.

Reflection is the re-emission of light from a surface. If there is no selective absorption (absorption of photons of certain energy levels by electrons), the light is re-emitted with the color unchanged. In some cases, such as copper and gold, photons in the blue and green portions of the spectrum are absorbed.

Transmission is closely related to absorption. A diamond, for instance, is transparent to visible light because photons of these wavelengths do not have enough energy to raise the electrons through the energy gap to the conduction band. Therefore, the light passes through unaffected. Color centers may cause a material to transmit colored light when it is exposed to white light. These centers result from impurities or lattice defects that form regions in the structure with electrons that can absorb certain photons of the visible spectrum. The remaining transmitted light is colored.

Refraction is due to the fact that light travels slower in some materials than in others (and always more slowly than in a vacuum). The index of refraction increases with the density of the material.

Fiber-optic lightguides are being used extensively for communications because of long distances (about 200 km) between repeaters, small diameter, lightweight cables, and accuracy. The attainment of low attenuation losses depends on the wavelength of the light beam, design of the cross section (multimode vs. single-mode), mechanical defects and the particular glass dopants used to alter the refractive index.

Definitions

Absorption   The preferential capture of light of certain wavelengths.

Band model   A model of the structure of atoms in which the outer electrons of the atoms are thought of as combined in a band of energy levels. If there are empty levels within a band, the material is a good conductor, because upper-level electrons can easily be raised to a higher level of energy and therefore move to conduct current.

Birefringence   Splitting of a single light ray into two waves by an optically anisotropic material such as quartz.

Capacitance   The charge-storing ability of a capacitor. Capacitance is a function of the geometry of the device and the dielectric constant of the material between the plates. The unit of capacitance is the farad (F).

Charge carrier   A tiny particle with an electric charge, specifically electrons, holes and ions.

Color centers   Areas in crystals showing preferential absorption and transmission of photons of certain (colored) wavelengths because certain ions (transition elements) have inner-shell electrons that interact with photons of certain wavelengths and do not interact with others.

Conductance   The inverse of resistance. Note the similar relationship between resistivity and conductivity.

Conductivity, σ   A measure of the ease of passage of an electric current. $\sigma = 1/\text{resistivity}$.

Critical temperature, θ_c   The temperature in some pure metals, alloys, and ceramics below which superconductivity (zero resistivity) occurs.

Dielectric constant, ϵ_r   A quantity that indicates the charge-storing capacity of a dielectric in a capacitor. The ratio of the capacitance obtained using the material between the plates of a capacitor to that obtained when the material is replaced by a vacuum.

Dipole An atomic or ionic grouping such as a molecule or unit cell in which the positive and negative centers of charge do not coincide. The *dipole moment* is the product of the charge and the distance between the charges.

Electromechanical coupling The behavior exhibited when a material develops a charge at its ends when compressed or, conversely, changes dimensions when an electric potential is applied.

Electron A negative charge carrier with a charge of 1.6×10^{-19} coulomb.

Electron hole A positive charge carrier with a charge of 1.6×10^{-19} coulomb.

Electronic polarization The movement of the electrons in a dielectric toward the positively charged plate.

Extrinsic semiconductor A semiconductor material that has been doped with an *n*-type element (such as phosphorus) that donates electrons that have energies close to the conduction band, or with a *p*-type element (such as aluminum) that provides electron holes close to the valence-band level.

Fermi level, E_F The energy level at which the probability of occupancy of an energy state is 0.5. In other words, at this level the possible states that an electron or electron hole can occupy are half-filled.

Ferroelectrics Materials such as barium titanate in which the permanent dipoles are aligned in domains.

Fiber-optic A conductor of optical pulses that can be used for high-speed transmission of data.

Fluorescence The emission of light in the visible spectrum by a material exposed to ultraviolet light.

Forward bias Electron flow into the *n*-material of a *p-n* semiconductor.

Gap energy, E_g The energy required to move an electron from the valence band to the conduction band in an atom; usually expressed in electron volts.

Hall effect The development of a voltage by applying a magnetic field perpendicular to the direction in which a current is flowing. Electron carriers are forced to one side and electron holes to the other, giving a voltage that is perpendicular to both the field and the current.

Insulator A material with a filled valence band and a large gap in energy between the valence band and the conduction band.

Intrinsic semiconductor A semiconductor material that is essentially pure and for which conductivity is a function of the temperature and the energy gap of the material.

Ionic polarization The displacement of ions in the material toward oppositely charged plates.

Laser *L*ight *a*mplification by *s*timulated *e*mission of *r*adiation.

Light emission The emission of photons of wavelengths in the visible spectrum, caused by electron jumps from higher to lower energy levels.

Luminescence The emission of light from a material after exposure to ultraviolet stimulation. (*See* fluorescence and phosphorescence.)

Metal   A material with an unfilled valence band, leading to a very high concentration of carriers.

Mobility   A measure of the ease of motion of the carrier.

MOSFET   *Metal-oxide-semiconductor field-effect transistor.*

Opacity   The state in which photons interact with electrons and raise electrons to unfilled energy states at higher levels. Because no electrons are given off, no light is transmitted. Opacity occurs, for example, in metals.

Optical detector   A diode that converts photons of light bits to electric currents in a fiber-optic system.

Optical emitter   A laser diode or light-emitting diode that is used to convert electrical signals to optical signals in a fiber-optic system.

p-n junction   The boundary between n- and p-type materials in electronic devices.

Pauli exclusion principle   The principle that states that no more than two electrons in an atom or block of material can occupy the same energy level.

Phonon   A quantum of elastic energy, usually acoustic or vibrational, used to explain thermal properties of materials.

Phosphorescence   The continued emission of light in the visible spectrum, by a material exposed to ultraviolet light, after the source of the ultraviolet light has been removed.

Photoelectric effect   The absorption of photons of definite wavelengths, leading to emission of electrons, which causes a photoelectric current.

Photon   A quantum of electromagnetic energy used to explain the optical properties of materials.

Piezoelectric material   A material that shows electromechanical coupling. All ferroelectric materials are piezoelectric, but only materials with electrical domain structures are ferroelectric.

Rectification   The conversion of alternating to direct current.

Resistance $R = V/I$, where V is the voltage and I the current. The resistance of a conductor to the passage of a current increases with length and decreases with increasing cross section. Note that this applies to direct current, not to high-frequency alternating current, in which case most of the current flows in the surface layers.

Resistivity   The inherent ability of a material to resist current flow. $\rho = R(A/l)$, where A is the area of a conductor, l is its length, and R is its resistance.

Resistor   An electrical component designed to provide a desired resistance. Examples include a tiny component in a radio and the heating element in a toaster.

Reverse bias   Electron flow into the p-material of a p-n semiconductor.

Semiconductor   A material with a filled valence band and a small energy gap between the valence band and the conduction band.

Solar cell   A type of semiconductor in which current is made to flow by forward biasing of an n-p junction by exposure of the n-material to radiant energy.

Superconductivity   The phenomenon of zero resistivity that occurs in some pure metals, alloys, and ceramics and is a function of temperature.

Thermistor   A device that uses the change in resistivity with temperature to measure temperature.

Thermocouple   A pair of materials that, when joined, develop an emf when one junction is at a different temperature from the other.

Transistor   A three-element, two-junction, semiconductor device (*n-p-n* or *p-n-p*) composed of an emitter, a base, and a collector. Through forward bias of the emitter and reverse bias of the collector, the transistor becomes a current-amplification device.

Transparency   The state in which photons of transmitted light do not interact with electrons of the structure. The photon energy is insufficient to liberate bound electrons. An example of a transparent material is diamond.

Problems

20.1   *[ES]* Suppose that in Example 20.1 the student considers the length of the wire too awkward and decides to use a 10-ft (3.05-m) length. What diameter of Chromel wire should she buy? Suppose that she decides to use only a 1-in. (0.0254-m) length. What diameter should she use? What practical difficulties might she encounter in using such short wires? (Sections 20.1 through 20.7)

20.2   *[EJ]* Many analogies can be drawn between electrical and thermal properties of materials. Estimate the percent difference in thermal conductivity between copper and stainless steel. Indicate how the two materials are used together in cookware. (Sections 20.1 through 20.7)

20.3   *[ES]* The electrical carriers in a pure metal are essentially the valence electrons. (Sections 20.1 through 20.7)
   a. Calculate the carrier density per cubic meter for pure aluminum.
   b. Estimate the carrier mobility. [*Hint:* Determine the electrical conductivity first.]

20.4   *[EJ]* From the calculations of Problem 20.1, explain why the manufacurer's catalog of Chromel wire gives recommendations for diameters for different currents. Why is the allowable current lower for wires embedded in refractory than for wires in air when the thermal conductivity of air is lower than that of refractory? (Sections 20.1 through 20.7)

20.5   *[ES]* The resistance of a 1-ft (0.305-m) length of metallic conductor was increased 20% by cold working the wire and decreasing its cross-sectional area by 10%. By what percentage was the mobility of the electrons changed per foot of length? The calculation of a new length is not required. (Sections 20.1 through 20.7)

20.6   *[ES]* A well-insulated wire-wound small-tube furnace requires 900 W to achieve a temperature of 2000°F (1093°C). To prevent the wire from melting, the maximum current draw is limited to 6 A. (Sections 20.1 through 20.7)
   a. Calculate the voltage requirement and the wire resistance.
   b. The diameter of the wire is 0.050 in. (0.0013 m) and a piece 1-foot (0.305-m) long has a resistance of 0.205 ohm. Calculate the resistivity of the wire and the required length of wire to consume the 900 W at 6 A.
   c. The outside diameter of the tube furnace is 2.00 in. (0.051 m). How many turns of wire are required for the length in part *b*?

20.7 *[ES/EJ]* Indicate whether the following statements are correct or incorrect and justify each answer. (Sections 20.1 through 20.7)

a. The electrical conductivity of a solid solution decreases linearly with increased amount of solute.

b. Minimum electrical resistivity of a pure element is obtained at absolute zero.

c. Metal tubing can be used as a current carrier, especially at low frequencies.

20.8 *[ES/EJ]* Explain why higher temperatures and the addition of agents to produce ion defects are advantageous for increasing the conductivity of glass. (Sections 20.1 through 20.7)

20.9 *[ES/EJ]* We need a solid wire that is 1000 ft long to carry an electric current of 25 A and a mechanical load of 500 lb. The materials of choice are aluminum and copper. There is to be a maximum power loss of 25 W over the wire length, and the allowable stress is one-third of the yield strength. Aluminum and copper cost $0.95 and $1.25 per pound, respectively. Calculate the cross-sectional areas that will be necessary for the two materials to meet the stress and electrical requirements. Then select one of them on the basis of minumum cost per foot of length. (Sections 20.1 through 20.7 and Chapter 7)

20.10 *[ES/EJ]* In the discussion of ceramic superconductors, considerable importance is attached to a change in crystal structure, multiple valences for copper, and oxygen defects. How might these contribute to the superconducting characteristics? (Sections 20.1 through 20.7)

20.11 *[EJ]* Give several reasons why metallic superconductors might still be preferred over ceramic superconductors despite the higher θ_c for the ceramics. (Sections 20.1 through 20.7)

20.12 *[ES]* An intrinsic semiconductor is produced from a compound containing Group III and Group V elements. The density of the electrons is 2×10^{15} electron/cm^3, with an electron mobility of 0.30 m^2/volt · sec and a hole mobility of 0.10 m^2/volt · sec. (Sections 20.8 through 20.15)

a. Calculate the conductivity due to the combined movement of holes and electrons.

b. Suppose that antimony is added to produce an extrinsic semiconductor with a conductivity of 200 mho/m. Calculate the number of antimony atoms per cubic meter of semiconductor. Assume the same hole and electron mobilities.

20.13 *[ES/EJ]* Explain what would happen to the number of intrinsic carriers in the extrinsic semiconductor schematically shown in Fig. 20.11 if the temperature were increased or decreased. (Sections 20.8 through 20.15)

20.14 *[ES]* Using the data for germanium in the table in Section 20.10, determine the weight percent increase in *n* carriers (electrons) in going from 10^{-8} to 10^{-5} parts of impurity. Assume the dopant to be phosphorus. The specific gravity of germanium is 5.323. (Sections 20.8 through 20.15)

20.15 *[ES/EJ]* (Sections 20.8 through 20.15)

a. Explain why the electrons and electron holes do not recombine across the junction interface of a diode when there is no imposed voltage.

b. Why can diodes of different physical sizes be purchased?

20.16 *[ES]* An intrinsic silicon semiconductor is made extrinsic by having 1 electron per 10,000 electron holes (the hole density is increased by 99.99%). Assume that the mobility of the electrons is approximately twice that of the electron holes. Using the data for electron mobility and carrier density from Problem 20.17, calculate the resistivity for the doped material and the percent change in resistivity compared with that of pure silicon. (Sections 20.8 through 20.15)

20.17 [ES] The following table summarizes the characteristics of the two intrinsic semiconductors silicon and germanium at 300 K.

Characteristic	Silicon	Germanium
Electrical resistivity, μohm · m	2.3×10^9	4.5×10^5
Energy gap, eV	1.06	0.67
Density of carriers, (electron carriers plus hole carriers)/m^3	1.5×10^{16}	2.4×10^{19}
Lattice drift mobility, m^2/volt · sec		
Electrons	0.135	0.390
Electron holes	0.048	0.190

The conductivity of germanium was calculated in Example 20.3. Calculate the conductivity of silicon, and check the way it corresponds to the resistivity given in the table. (Sections 20.8 through 20.15)

20.18 [ES] A p-type semiconductor is produced by adding aluminum to silicon. The resistivity is 5×10^{-4} ohm · m, with an extrinsic carrier mobility of 0.1625 m^2/volt · sec. Calculate the following. (Sections 20.8 through 20.15)
a. The density of the carriers
b. The carriers per atom of added aluminum
c. The carriers per kilogram of silicon (the density of silicon is 2.33×10^3 kg/m^3)
d. The kilograms of aluminum per kilogram of silicon

20.19 [ES/EJ] Assume that we have an intrinsic silicon semiconductor wire 1 cm long and 0.01 cm^2 in cross-sectional area. Using the data in Example 20.6, calculate the temperature-measuring capability of the semiconductor, given that resistance can be measured to the nearest 0.001 ohm. Compare this result to a thermocouple that has an accuracy of 0.1°C within the same temperature range. (Sections 20.8 through 20.15)

20.20 [ES] Say we were to use germanium rather than silicon, as proposed in Example 20.6. Show by calculations whether the germanium thermistor would be more or less sensitive within this temperature range (27–47°C). (Sections 20.8 through 20.15)

20.21 [ES/EJ] One of your associates says that the following equation relates the electrical resistance of a conductor to its temperature. (Sections 20.8 through 20.15)

$$R = \frac{lkT}{nq^2DA} \tag{20-30}$$

a. List the fundamental equations necessary to the derivation of the relationship.
b. What limitations, if any, should be included with the relationship?

20.22 [ES] The electrical conductivity of a compound AX is controlled by the mobility of A$^+$ ions (4.8×10^{-14} m^2/volt · sec). Suppose that AX has the NaCl structure with radii of A$^+$ = 0.88 Å and X$^-$ = 1.61 Å. Calculate the following. (Sections 20.8 through 20.15)
a. The volume of the AX unit cell
b. The electrical conductivity of AX

20.23 [ES/EJ] (Sections 20.8 through 20.15)
a. Repeat Example 20.4 for a temperature of 50°C.
b. What is the significance of these results for the operation of diodes at elevated temperatures?

20.24 *[ES/EJ]* (Sections 20.8 through 20.15)
   a. In Equation 20-10,

$$I_s = Aqn_i^2\left(\frac{D_e}{L_e N_d} + \frac{D_p}{L_p N_s}\right)$$

   determine the units for each parameter and show the final units for I_s.
   b. How is the mobility related to the diffusion coefficients in this equation? [*Hint:* A unit analysis is helpful.]

20.25 *[ES]* In a solar cell, what ratio of I_L/I_s is required to produce 1 volt at 100°C? (Sections 20.8 through 20.15)

20.26 *[ES/EJ]* Suggest an experiment that would provide the values for I_0 and B in the equation $I_c = I_0 e^{V_e/B}$, Equation 20-13, which applies to a transistor. [*Hint:* The equation can result in a linear plot by suitable manipulation. Understanding of how a transistor functions will dictate what must be controlled and measured during the experiment.] (Sections 20.8 through 20.15)

20.27 *[ES/EJ]* The loss angle increases if a material lags more in polarization when in a reversing field, and an energy loss in the form of heat usually occurs. Predict how the dielectric constant and an increased frequency will affect this energy loss. (Sections 20.16 through 20.19)

20.28 *[ES]* Figure 20.23 shows the structure of barium titanate as tetragonal. Above 120°C (248°F) the structure is cubic with O^{2-} at the facial centers and Ba^{2+} at the cube corners. (Sections 20.16 through 20.19)
   a. Calculate a_0 for the cube if no Ti^{4+} is present at the center.
   b. What happens to your calculated value of a_0 when Ti^{4+} is present at the cube center?
   c. Why is 120°C (248°F) called a Curie temperature, and why is the tetragonal structure a piezoelectric material?

20.29 *[EJ]* Given a pair of wires to be used for a thermocouple application, the first requirement is their calibration. Outline an experimental procedure to establish an emf-vs.-temperature curve for the thermocouple pair. (Sections 20.16 through 20.19)

20.30 *[ES]* Show that the following complex materials meet the criteria for the perovskite structure and therefore might react in a way similar to barium titanate. (Sections 20.16 through 20.19)
   a. $KLaTi_2O_6$
   b. Sr_2CrTaO_6
   c. $BaKNbTiO_6$
   d. $Pb(Zr, Ti)O_3$

20.31 *[EJ]* How do the characteristics of the barium titanate structure lead to its use in the following devices? (Sections 20.16 through 20.19)
   a. Ultrasonic applications such as emulsification of liquids, mixing of powders and paints, and homogenization of milk
   b. Microphones and phonograph pickups
   c. Accelerometers
   d. Strain gages
   e. Sonar devices

20.32 *[ES]* Indicate which material of each pair has the higher dielectric constant and explain why. (Sections 20.16 through 20.19)
   a. Polyvinylchloride (PVC) and polytetrafluoroethylene (PTFE)
   b. PVC at 25°C and 100°C
   c. PVC at 10^2 Hz (hertz) and PVC at 10^{10} Hz

20.33 *[ES/EJ]* Many capacitors are not parallel plates but rather continuously rolled cylindrical sandwiches of two metal foils separated by a dielectric. (Sections 20.16 through 20.19)

a. How should the parallel-plate equation given in the text be modified to account for the wrapped geometry?

b. Why do defects in the foil or the dielectric have an important effect on the capacitor function?

c. Why do capacitors have both a capacitance and a voltage rating?

20.34 *[ES/EJ]* One type of a room water vaporizer uses a transducer to generate the fine water particles, which are then distributed by a fan. How does the transducer accomplish this, and why might the directions caution the user about employing the vaporizer where electronic equipment might also be in service? (Sections 20.16 through 20.19)

20.35 *[EJ]* Why might a circuit board be made of a filled thermosetting polymer? (Chapter 13 and Sections 20.16 through 20.19)

20.36 *[EJ]* Why are surface track and arc resistance measured in units of seconds to obtain tracking? (Sections 20.16 through 20.19)

20.37 *[ES/EJ]* The text states that white light is emitted from a heated tungsten filament. However, color slide film for outdoor photography exhibits a reddish hue if it is exposed to illumination by tungsten light. Why does this occur and how may it be corrected? (Sections 20.20 through 20.24)

20.38 *[ES]* An absorption-vs.-wavelength curve for a material shows a strong maximum at a wavelength equal to 1500 Å, but it shows little absorption at longer wavelengths. (Sections 20.20 through 20.24)

a. Calculate the gap energy.

b. Would the material be transparent, colored, or opaque in transmitting white light?

c. Would you expect the material to be metallic or nonmetallic?

20.39 *[ES]* Calculate whether barium, with a work function of 2.50 eV, would be more suitable than the tungsten in Example 20.10 for use as a photocell with visible light. (Sections 20.20 through 20.24)

20.40 *[ES/EJ]* To express the transmission of visible light, the following relationship is often used:

$$I = I_0 e^{-\alpha x} \qquad (20\text{-}31)$$

where I = intensity of transmitted light, I_0 = intensity of incident light, α = absorption coefficient, and x = thickness of the material. The value of α is a material constant dependent on the wavelength of the photons. Qualitatively, how does α, and hence I, vary with the atomic number of the transmitting medium and the wavelength of the radiation? (Sections 20.20 through 20.24)

20.41 *[ES/EJ]* A further refinement in the description of the transmission of light uses the relationship

$$I/I_0 = (1 - R)^2 e^{-\alpha x} \quad \text{(see Problem 20.40)} \qquad (20\text{-}32)$$

where

$$R = \left(\frac{\eta - 1}{\eta + 1}\right)^2 = \text{reflectivity} \qquad \text{and} \qquad \eta = \text{index of refraction} \qquad (20\text{-}33)$$

The absorption coefficient and the index of refraction are functions of the wavelength. (Sections 20.20 through 20.24)

a. With reflectivity expressed as a percentage, explain the "brilliance" of lead crystal glass ($\eta = 2.0$) compared to window glass ($\eta = 1.5$).

b. Explain what happens to the transmission of visible light (wavelength of approximately 400–800 nm) when a window is tinted blue by the addition of a suitable ion.

c. What is the principle used to produce the gold-colored windows seen in many office buildings? [*Hint:* The transmission appears different looking into and out of such a window.]

20.42 *[ES/EJ]* Gamma rays have higher energy than x rays. Why is lead shielding used more often than iron shielding? When one is radiographing a metal component to find internal imperfections, why does a pore appear darker than the solid material on the developed plate? [In Problem 20.40, for high-energy radiation, $\alpha = (\mu/\rho)\rho$, where (μ/ρ) is the mass absorption coefficient and ρ is the density.] (Sections 20.20 through 20.24)

20.43 *[ES/EJ]* Why are some lasers pulsed but others can be operated continuously? (Sections 20.20 through 20.24)

20.44 *[ES/EJ]* From the text discussion of emission, explain the classifications of fluorescent, neon, and xenon lamps. (Sections 20.20 through 20.24)

20.45 *[ES]* GaAs and GaP can be used as light-emitting diodes that give off red and green light, respectively, when recombination occurs at the junction.

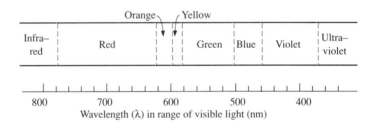

From the spectrum given above, calculate the range of values for E (in electron volts) to obtain the red and green colors. (Sections 20.20 through 20.24)

20.46 *[ES]* Glass is more easily fabricated into laser rods than are single crystals of ruby. Early attempts to make lasers of glass with Cr^{3+} ions as an impurity failed because the Cr^{3+} ions interacted with the ions of the glass rather than acting independently, as they do when substituted for Al^{3+} ions in alumina. However, satisfactory glass lasers have been made using neodymium in silicate glass. Look up the electronic structure of neodymium and explain what characteristic will lead to less interaction between the adjacent ions of the glass compared with Cr^{3+}. [*Hint:* Determine which electrons are the color centers in the neodymium.] (Sections 20.20 through 20.24 and Chapter 10)

20.47 *[ES]* Explain why we are interested in the numerical aperture of an optical fiber. (Section 20.25)

20.48 *[ES]* An optical glass fiber has an index of refraction of 1.4434 and an acceptance angle of 22°. Calculate the index of refraction of the cladding and, from this, determine the critical angle. (Section 20.25)

20.49 *[ES]* Figure 20.35 shows three types of optical fibers. Explain why the input and output pulse profiles are different for the three fiber types as they are described schematically. (Section 20.25)

20.50   *[ES/EJ]* In Figure 20.37, determine where the transmission window is relative to the visible spectrum. What conclusions might be reached with respect to losses within the visible spectrum? (See also Problem 20.45) (Section 20.25)

20.51   *[ES/EJ]* Figure 20.40 indicates a typical coupling loss for an LED–optical-fiber junction. Estimate the equivalent loss in kilometers for a present-day optical fiber and suggest what this calculation might imply. (Section 20.25)

20.52   *[ES/EJ]* The text mentions that repeater units are used in fiber-optic systems whenever the distance between stations exceeds 100 km. Estimate the power loss over this distance as a ratio of output to input power. (Section 20.25)

20.53   *[EJ]* Indicate how viewing a remote or hazardous location can be accomplished by using optical-fiber systems. Consider the number of fibers and how a light source may also be incorporated. (Section 20.25)

20.54   *[ES/EJ]* From the rather brief description of optical-fiber systems given in the text, what characteristics do you think would be preferred for the fibers? That is, what would you consider as variables to control? (Section 20.25)

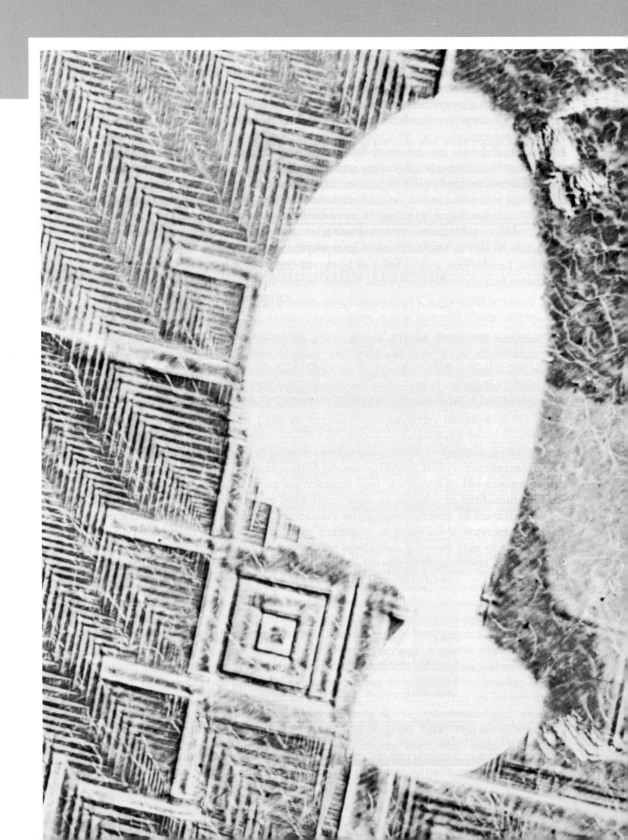

CHAPTER 21

Magnetic Properties of Materials

This is an x-ray topograph pattern from an iron–3% silicon single crystal. The interaction of x rays with the magnetic domains in the iron caused the features shown. The "magnetic fish" shows no structure because the orientation of this region does not satisfy the Bragg law, discussed in Chapter 1, for diffraction. The square region to the right of the tail of the "fish" is a ferromagnetic domain. The chevron markings show other domains of different orientation.

In this chapter we shall take up the relationship of magnetic properties to structure in two groups of materials: the metallic and the ceramic. We shall see that the importance of each group has increased recently to meet the demands of the electronics industry.

21.1 Overview

It is not generally recognized that the development of materials with new high levels of magnetic properties is responsible for the proliferation of many new devices, such as the tape recorder, ferrite magnets in television sets, memory cores in computers, permanent magnets in motor controls, and particle accelerators used in basic research. We shall see how these new developments involve two important groups of materials:

1. The metallic materials in which the advances have taken place in recent years, from the conventional iron-based alloys to Alnico and, more recently, samarium–cobalt and $Nd_2Fe_{14}B$ magnets.
2. The ceramic materials called *ferrites*, which are spinels and related structures. The growth of this field has been comparable to that of the semiconductors in the electrical materials.

To understand the specification of magnetic materials and how their structure is related to their magnetic properties, we must first review a few definitions of the most important properties. These are relatively simple and are all related to the magnetic flux that is produced in a material as a result of a magnetic field. These properties, such as permeability and remanent induction, are given quite simply by the graph of the flux density B as a function of the field strength H, called a *hysteresis loop* or a *magnetic hysteresigraph*.

Following this review we shall take up important metallic magnetic materials, in which we encounter magnetic domains and domain movement, just as we found domains in ferroelectric materials. We shall see that the structure required for soft, or low-hysteresis, magnets is quite different from that required for the hard, or permanent, magnets. Finally, we shall discuss the magnetic properties available in the ceramic magnets, or ferrites, and their relationship to structure.

21.2 The Magnetic Circuit and Important Magnetic Properties

First recall that, in an electric circuit (Figure 21.1), if we apply a voltage V, the current I that flows in the conductor is related to the conductivity σ of the material. Mathematically,

$$V = IR \text{ (Ohm's law)} \quad \text{and} \quad R = \frac{\rho l}{A} = \frac{l}{A\sigma} \text{ (Equation 20-1)}$$

so

$$V = \left(\frac{I}{A}\right)\left(\frac{l}{\sigma}\right) \tag{21-1}$$

Therefore,

$$\sigma = \left(\frac{I}{A}\right)\left(\frac{1}{V/l}\right) \tag{21-2}$$

FIGURE 21.1   Analogy between (a) electric circuit and (b) magnetic circuit. (Assume unit length and unit cross-sectional area.)

where I/A is the current density with units of A/m^2 and V/l is a voltage gradient with units of volt/m. Then

$$\sigma \text{ (electrical conductivity)} = \frac{\text{current density}}{\text{voltage gradient}} \qquad (21\text{-}3)$$

In an analogous way (Figure 21.1), if we apply a magnetic field across a gap, the flux in the gap is proportional to the *permeability* μ of the material in the gap. If we represent the field strength or gradient by H and the flux density by B, then

$$\mu = \frac{B}{H} = \frac{\text{flux density}}{\text{field strength or gradient}} \qquad (21\text{-}4)$$

The usual meter-kilogram-second (mks) units for the quantities in this equation are derived in physics as

$$H = 1 \text{ A/m} \quad B = 1 \text{ weber/m}^2 = 1 \text{ tesla} = 1 \text{ volt-sec/m}^2$$

For a vacuum,

$$\mu_0 = 4\pi \times 10^{-7} \text{ henry/m}$$

where 1 henry = 1 ohm-sec. Often centimeter-gram-second (cgs) units are used, and the following conversions apply:

H: 1 A/m $= 4\pi \times 10^{-3}$ oersted

B: 1 tesla $= 10^4$ gauss

μ_0: $4\pi \times 10^{-7}$ henry/m = 1 gauss/oersted

The important difference between the electric and magnetic circuits is that the conductivity in an electric circuit is constant (independent of the values of V and I), whereas the permeability μ of a magnetic material changes with the applied field strength H. The ratio B/H is not constant. Also, the flux can persist after the field is removed. This is called *remanent induction*, as in a permanent magnet. Clearly, we need to examine these important effects in a magnetic circuit and to relate them to the structure of the material if we are to understand and construct magnetic devices.

Let us first discuss the *B-H* curve, which is as important in expressing magnetic properties as the stress–strain curve is for understanding mechanical properties. We place a sample in a gap in which we provide a controlled magnetic field strength *H* (as, for example, by changing the current in an electromagnet with its poles at the sides of the gap), and we measure the magnetic flux density *B* and plot the *B-H* curve (Figure 21.2). In a typical virgin curve, the value of *B* rises rapidly and then practically levels off at B_s, which is called the *saturation induction*. (The curve obtained the first time the material is magnetized — the virgin curve — is the dashed curve that appears in the figure.) As the field strength is reduced to zero, there is still a magnetic flux density B_r, called the *remanent induction* or *residual induction*. To lower the flux to zero, we need a definite field strength H_c in the opposite direction to the original, and this is called the *coercive magnetic force* (or the *coercive field*). Because the graph never retraces the virgin curve but traces out a loop, it is called a *hysteresis loop* and is an index of the energy lost in a complete cycle of magnetization.

The shape of the *B-H* curve varies greatly with different magnetic materials, and various types of curves have different uses. To illustrate this, consider the *B-H* curves for a transformer core (soft magnet), a computer memory unit (square-loop magnet), and a permanent magnet (hard magnet) shown in Figure 21.3. In the curve for a transformer core the hysteresis loop encloses only a small area (Figure 21.3*a*). This represents work done per cycle. It is related to power loss and is kept as small as possible. By contrast, in a memory unit we wish to magnetize easily to a sizable value of *B* and then retain most of this flux density when the power is off, so that it can be used as "memory" to activate the controls when called on. For this we want a "square loop," so that even with a reversed field we still obtain a strong flux for "readout" (Figure 21.3*b*). In the permanent magnet the "power" of the magnet is related

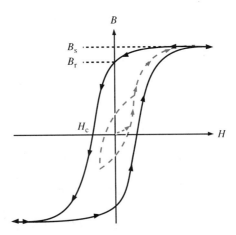

FIGURE 21.2 Hysteresis loop for a ferromagnetic material, showing B (flux density) versus H (field strength)

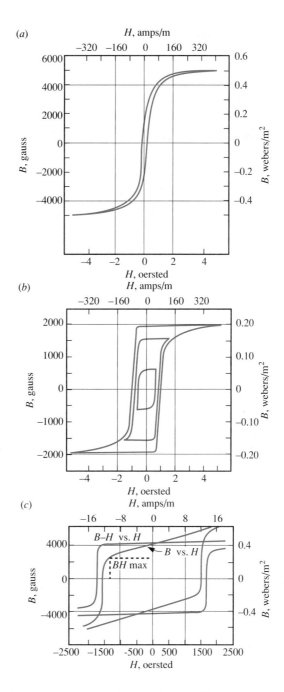

FIGURE 21.3 (*a*) Hysteresis curve of a soft ferrite. (*b*) Hysteresis curves of a square-loop ferrite. (*c*) Hysteresis curve of a permanent-magnet ferrite.

to the *BH* product.* We note that in this graph the *x* axis, the field strength, is many times more compressed than in the other two graphs, which reflects the large value of *H* needed to produce a permanent magnet with a large B_r. Note this large value of remanent induction (Figure 21.3c).

21.3 Magnetic Permeability

As stated earlier, the permeability of a vacuum is $\mu_0 = 4\pi \times 10^{-7}$ henry/m. If we insert a material into the same space, we obtain a new value for μ because the same applied field produces a different flux density *B*. Instead of discussing the absolute value, it is useful to compare the relative value with that of the vacuum. We define the *relative permeability* as

$$\mu_r = \frac{\mu}{\mu_0} \qquad (21\text{-}5)$$

For a vacuum, $\mu = \mu_0$ and $\mu_r = 1.0$. After determining μ_r for a wide variety of materials, we find that they fall into three general classes:

1. If μ_r is less than 1 by about 0.00005, the material is called *diamagnetic*.
2. If μ_r is slightly greater than 1 (up to 1.01), the material is called *paramagnetic*.
3. If μ_r is much greater than 1 (up to 10^6), the material is called *ferromagnetic* or, in the case of ceramic magnets, *ferrimagnetic*.

All three of these phenomena are related to the structure of the material. *Diamagnetism*, which is negligible from an engineering standpoint, can be explained as follows: When a magnetic field is applied to a material, the motions of all the electrons change. This produces a local field around each electron. By a law of physics known as Lenz's law, this field is in the opposite direction to the applied field and gives a small decrease in the total magnetism. *Paramagnetism* is also a small effect noted in all atoms and molecules having an odd number of electrons. Here the electrons line up with the applied field, reinforcing the field and giving a value of permeability slightly greater than 1.

Ferromagnetism is of the greatest engineering importance because a high permeability leads to a great flux density *B* for a given magnetizing force *H* compared with a vacuum or normal materials.† Without this high flux density and the ability to change or reverse it (as in soft magnetic materials such as electromagnets) or maintain it after the field is removed (as in hard or permanent magnets), most electrical devices such as transformers, motors, tape recorders, and computers would not work.

---

*The maximum *BH* product in the second quadrant of the *B-H* curve is used.

†Superconducting magnets, wherein diamagnetism is very important, will be treated in a later section.

We encounter ferromagnetism in only a few elements: those with partially filled 3*d* shells, such as iron, cobalt, and nickel, and those with partially filled 4*f* shells, such as gadolinium. Although only these four elements exhibit ferromagnetism at room temperature, alloys or ceramic compounds of the other elements with partially filled 3*d* or 4*f* shells, such as manganese, have important magnetic properties.

Now let us examine the structural factors responsible for these magnetic properties. We shall see that, as with mechanical properties, both the atomic structure and the microstructure govern the magnetic properties. We shall first discuss the effects of atomic structure on permeability and remanent induction in general terms; then we shall take up metallic and ceramic magnets.

21.4 Atomic Structure of Ferromagnetic Metals

Recall from Chapter 1 that, in addition to its other quantum numbers, each electron revolving about the nucleus has the quantum number m_s, which describes the way an electron is spinning on its axis. When $m_s = +\frac{1}{2}$, we think of the electron as spinning in one direction, and when $m_s = -\frac{1}{2}$, in the other. A spinning charge generates a magnetic field, and we can say by definition that a quantum number of $+\frac{1}{2}$ means that the axis of the field is up and $-\frac{1}{2}$ means it is down. Furthermore, we find that in an atom of manganese, for example, the five 3*d* electrons all have their spins in the same direction rather than alternating in direction. This is an example of *Hund's rule*, which states that as electrons are added as we progress to higher elements in the periodic table, there is a strong tendency to attain the maximum number of spins in one direction, such as five for the 3*d* level, before electrons of opposite spin are added. This leads to a magnetic moment for the atom as a whole. Each unpaired electron (an electron not balanced by one of opposite spin) contributes a magnetic moment that is given the special name of the *Bohr magneton*. This has the value of 9.27×10^{-24} A-m^2 in mks units or 0.927 erg/gauss in cgs units. The isolated manganese atom, therefore, has a magnetic moment of 5 Bohr magnetons. It is important to establish the magnetic moments of some other nearby elements (Figure 21.4).

We note that iron has one less Bohr magneton than manganese because there are only four unpaired electrons, and that the moments for cobalt and nickel similarly reflect lower numbers of unpaired electrons. The magnetic moment of copper is zero, because the tendency to attain a second completed group of electrons of the same spin is stronger than that to add a second 4*s* electron. (The 4*s* electrons do not contribute to the magnetic moment.)

On the basis of Figure 21.4, we would predict that a block of manganese metal would have the greatest magnetic moment because its atoms have the highest moment. This is not the case, however, because when the atoms are placed together in a block, complex alignment called *exchange interaction* occurs. In the case of *metallic* manganese, the atoms line up such that the directions of the magnetic fields of the individual atoms are antiparallel and the

Magnetic moment	Element	Number of electrons	Electronic structure 3d shell					4s electrons
1	Sc	21	↑					2
2	Ti	22	↑	↑				2
3	V	23	↑	↑	↑			2
5	Cr	24	↑	↑	↑	↑	↑	1
5	Mn	25	↑	↑	↑	↑	↑	2
4	Fe	26	↑↓	↑	↑	↑	↑	2
3	Co	27	↑↓	↑↓	↑	↑	↑	2
2	Ni	28	↑↓	↑↓	↑↓	↑	↑	2
0	Cu	29	↑↓	↑↓	↑↓	↑↓	↑↓	1

FIGURE 21.4   Magnetic moments (un-ionized) of the transition elements

fields cancel one another. This is called *antiferromagnetism*. (We shall find later that this is not the case for manganese in ceramic magnets, where the ions are isolated.) However, in the case of iron, cobalt, and nickel, a fraction of the atoms do line up to produce ferromagnetism.

To get a feel for the relationship between the atomic-scale property of magnetic moment and the engineering-scale measurement of the gross effect called saturation induction, which we discussed earlier, let us make a calculation to compare theory with experiment, just as we related atomic radius and density in Chapter 1.

First we write again, for a vacuum with a given field strength H,

$$B = \mu_0 H \qquad (21\text{-}6)$$

If we place a ferromagnetic material in the circuit, B rises greatly, and we can satisfy the equation by changing μ_0 to a new μ to show the new permeability. However, it is simpler to express the increase in B over the original value by adding a term $\mu_0 M$, where M is called the *magnetization*.

$$B = \mu_0 H + \mu_0 M \qquad (21\text{-}7)$$

The term $\mu_0 M$, therefore, is the increase in field strength, which is the same as the increase in magnetic moment caused by the new material.

As a final step, for ferromagnetic materials, we can usually omit $\mu_0 H$, because it is so small compared with $\mu_0 M$. Let us now consider an example.

EXAMPLE 21.1   *[ES]*

Calculate the saturation induction B_s that you would expect from iron in webers per square meter (teslas).

Answer $B = \mu_0 H + \mu_0 M$, Equation 21-7, but because $\mu_0 M$ is always many times greater than $\mu_0 H$ for a ferromagnetic material,

$$B \simeq \mu_0 M$$

The magnetic moment of an iron atom is 4 Bohr magnetons. Let us assume that when $B = B_s$, all magnetic moments are aligned. Converting to mks units, we have

$$\frac{\text{Magnetic moment}}{\text{Atom}} = \frac{4 \text{ Bohr magneton}}{\text{Fe atom}} \left(9.27 \times 10^{-24} \frac{\text{A-m}^2}{\text{Bohr magneton}}\right)$$

The density of iron is 7.87 g/cm^3, so there are n atoms per cubic meter:

$$n = \frac{(7.87 \times 10^6 \text{ g/m}^3)(0.602 \times 10^{24} \text{ atom/at. wt.})}{55.85 \text{ g/at. wt.}}$$

$$= 8.5 \times 10^{28} \text{ Fe atom/m}^3$$

The magnetization is the magnetic moment per atom times n, or

$$M = \frac{4 \text{ Bohr magneton}}{\text{Fe atom}} \left(9.27 \times 10^{-24} \frac{\text{A-m}^2}{\text{Bohr magneton}}\right)$$

$$\times \left(8.5 \times 10^{28} \frac{\text{Fe atom}}{\text{m}^3}\right)$$

$$= 3.15 \times 10^6 \frac{\text{A}}{\text{m}}$$

$$B_s = \left(4\pi \times 10^{-7} \frac{\text{volt-sec}}{\text{A-m}}\right) \left(3.15 \times 10^6 \frac{\text{A}}{\text{m}}\right) = 3.96 \frac{\text{volt-sec}}{\text{m}^2} \text{(3.96 tesla)}$$

We find experimentally that the actual B_s is less; it is 2.1 tesla. This is because the actual interaction between iron atoms in the metallic state does not lead to the sum of the magnetic moments. In other words, not all the magnetic moments are aligned. Only when the atoms are isolated and ionized, as in a ceramic, is the full magnetic moment realized.

At this point a question arises: If the atoms (and therefore their electrons) are lined up in a block of iron, why is the block not spontaneously a magnet without the necessity of applying a magnetic field? This leads us to the topics of magnetic domains and Bloch walls.

21.5 Magnetic Domains

For many years physicists speculated that a block of iron contained many tiny magnets that would line up under the influence of a magnetic field. In 1912

Pierre Weiss went further and postulated that there were *domains* (regions) where the atoms were aligned to produce a magnetic field in one direction and that these were balanced by domains of opposite alignment, so that no magnetic flux was present outside the specimen. Finally, in 1931, Francis Bitter decided to look for these domains under the microscope. He sprinkled magnetic powder on a polished sample and found something that appeared like grain boundaries within the individual grains of the material. These were indeed the domains of Weiss, and the structure is shown in Figure 21.5.

In the unmagnetized sample the domains are unaligned and assume balanced orientation, with a resultant magnetic moment of zero. However, when a magnetic field is applied, those domains with magnetic directions parallel to the field grow at the expense of those opposed. If the microstructure is such that the new domain structure is retained, we have a permanent magnet; if, on the contrary, the material returns to the original state, we have a soft magnet.

As one domain grows at the expense of another (Figure 21.6), a change in alignment of the individual atomic magnets takes place. The domain boundary is called a *Bloch wall*. It is highly important because it helps us to understand the reasons for the magnitude of the magnetic field required to magnetize a material. First, the ease of magnetization varies with direction in a crystal, just as the ease of straining elastically varies in a single crystal, as shown by the modulus of elasticity (Chapter 2). Figure 21.7 gives typical magnetization data for iron and nickel. Note that the preferred orientation for iron is [100] (higher induced magnetization for low magnetic field strengths). We would expect to encounter a domain polarized in the [001] direction on one side of a Bloch wall and another in the [00$\bar{1}$] direction on the other side. In the wall itself (about 1000 Å wide) we find intermediate polarizations where the direction of magnetization changes through the directions that are difficult to magnetize (Figure 21.8). Some magnetic domains are shown clearly in Figure 21.9. These were revealed by a technique that uses the interaction of the domains with polarized light. Figure 21.9 shows the effect of a magnetic probe in growing domains.

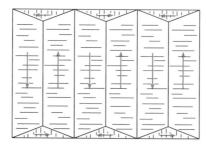

FIGURE 21.5   Domains shown by a magnetic powder etch in a single crystal of iron

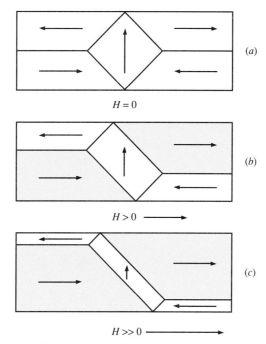

FIGURE 21.6 Two-dimensional representation of the growth of domains. (*a*) No magnetic field (orientation of domains shown by arrows). (*b*) Application of a field causes growth of those domains with correct orientation. (*c*) Applying fields with still greater strengths causes further selective growth of correctly oriented domains at the expense of those with incorrect orientation.

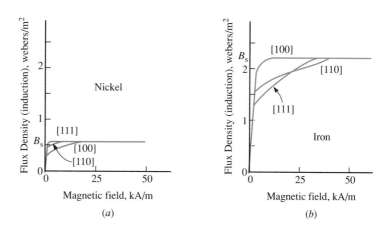

FIGURE 21.7 Magnetic anisotropy. (a) Nickel, (b) iron.

[After S. Kaya et al. *Sci. Reports* (Tohoku Imp. Univ.). (1)**15**: 721 (1926); (1)**17**: 1, 157 (1928)]

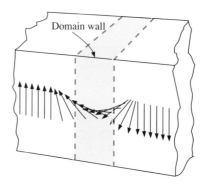

FIGURE 21.8 Domain wall, showing gradual change in direction of the field between adjacent domains

FIGURE 21.9 Magnetic domains in a thin (0.1-mm) single crystal of a magnetic ceramic, samarium terbium orthoferrite (perovskite structure). A beam of polarized light passes through the crystal and then through an analyzer. The domains show on the screen because of the effect of their magnetic field on the light beam (Faraday effect). The magnetic needle (white finger) has caused growth of those domains in the crystal (black areas) that have the same magnetic direction as the domains in the needle.

[Courtesy of P. C. Michaelis, Bell Telephone Laboratories]

EXAMPLE 21.2 *[EJ]*

A physics experiment suggests that one can make a magnetic compass using a glass of water, a small cork, a sewing needle, and a horseshoe magnet. How can this be accomplished?

Answer   The needle is ferrous and hence ferromagnetic. By rubbing the magnet along the length of the needle in *one direction*, we may partially align the domains in the needle. Placing the needle on the small cork floating on the water surface allows the needle to align with the earth's magnetic field. Because the domain alignment in the needle is not permanent, the effect is lost with time, depending on chemistry, the heat treatment of the needle, and the ambient temperature.

21.6 Effect of Temperature on Magnetization

Every ferromagnetic material loses its magnetic properties in some temperature range upon heating, and the ferromagnetism finally disappers at a temperature called the *Curie temperature*. For example, the Curie temperature for iron is 770°C (1418°F). The effect of temperature on ferromagnetism is illustrated by the graph in Figure 21.10. At the Curie temperature, the thermal oscillations of the atoms overcome the orientation that is due to exchange interaction, and a random grouping of the atomic magnets results. Domains are re-formed on cooling, so the application of a magnetic field from the Curie temperature to room temperature can be helpful in producing a permanent magnet.

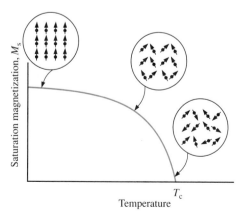

FIGURE 21.10   Effect of temperature on the saturation magnetization M_s of a ferromagnetic material below the Curie temperature T_C

Let us now, for review, consider the characteristics of the *B-H* curve in terms of domain theory.

21.7 Magnetic Saturation

If we start with a small field strength H and then increase it, the flux density B rises rapidly and attains saturation induction (Figure 21.2). After this point the flux increases only slightly with H. This phenomenon can be readily explained by the effect of increasing field strength on domains, as we noted earlier. To begin with, as the field is increased, the domains in line with the field grow and the permeability is high. As the flux approaches saturation, the atomic magnets in the domains are practically all aligned and contributing to the high flux density. Beyond this point of saturation induction, any further increase in field strength produces only the small increases in flux density that would be obtained without the core of magnetic material.

21.8 Remanent Induction

When the field strength H that was applied originally is decreased to zero, a magnetic flux B_r persists in a magnetic material (Figure 21.2). In a permanent magnet B_r has a high value, whereas in a soft magnetic material its value is low. The reverse field H_c that is needed to demagnetize the bar and produce zero flux is called the *coercive field*. Figure 21.3 shows the extreme differences in remanent induction between soft and hard magnets.

To explain this phenomenon structurally, we must observe the effects of the metal structure on the retention of the domain structure that is present at high imposed field strengths. In a bar of pure iron, a soft magnetic material, the domains revert to the original balance when the field is removed. However, in a magnetic material with obstacles to uniform domain movement, such as inclusions, or in highly cold-worked material or material with strained regions, a considerable portion of the magnetism is retained (B_r). For example, a quenched or cold-worked bar of high-carbon steel makes a good permanent magnet. Methods of observing these effects include isolating domains by using powders suspended in an insulator so that each particle is a domain and, as in Alnico magnets, the development of a fine precipitate in which the domains are isolated from each other.

21.9 Metallic Magnetic Materials

Let us now compare the magnetic properties of some typical engineering materials with their structures. There are two principal groups: soft magnetic materials and hard or permanent magnetic materials.

Soft Magnetic Materials

In *soft magnetic materials,* such as those used in a transformer core, the most sought-after properties are minimum hysteresis loss (area of the *B-H* loop) and maximum saturation magnetization. The attainment of these properties depends on the ease of domain movement, and for this reason cold work and the presence of second phases must be avoided. In the silicon–iron alloys, the grain orientation in the sheet is also controlled. The ease of magnetization varies with direction in a crystal; it is easiest in the ⟨100⟩ directions. The sheet for transformers is prepared with the (100) plane parallel to the plane of the sheet and the [001] direction parallel to the rolling direction. This is the so-called cubic texture. The advantage of this is that the greater the ease of magnetization, the smaller the hysteresis loop and, consequently, the lower the power loss in the transformer core.

In Chapter 1 we introduced the glassy metals. These materials are being considered for iron-based transformer cores because they possess both high electrical resistivity, which dampens eddy currents, and high magnetic permeability. This combination of properties gives low I^2R loss (heating) and a small hysteresis loop (small magnetic energy loss) over a wide frequency range.

We obtain the special properties of "permalloy" (very low hysteresis) by controlled cooling, which prevents the ordered structure that occurs in these iron–nickel alloys. This type of alloy is especially sensitive to work hardening, and for this reason the shape or wire is formed first and then annealed.

Hard Magnetic Materials

As we would expect, the characteristics we want in *hard* (or permanent) *magnetic materials* are quite the opposite of those we want in soft materials. When the field is removed, we want high values of remanent magnetization B_r and of coercive field H_c. It has been found, furthermore, that the lifting force of a magnet is related to the area of the largest rectangle that can be drawn in the second quadrant of the hysteresis curve, called the *BH* product. This energy product is taken as an index of the "power" of a magnet (see Figure 21.3c).

The key to retaining magnetism is preventing the domains from relapsing into a balanced orientation, giving no external magnetic moment. There are two principal methods of accomplishing this:

1. Use of a strained metal structure
2. Division of the magnetic part of the structure into small volumes that reorient with difficulty

In the first method high-carbon or alloy steels are quenched to produce martensite, which has a high internal stress. The second method uses alloys such as Alnico (Al, Ni, Co, and Fe) and Cunife (Cu, Ni, and Fe). The key to the Alnico alloys is that a single-phase BCC structure, stable at 1300°C (2372°F), decomposes into two different BCC phases, α and α', at 800°C (1472°F). The α' phase, rich in iron and cobalt, has a higher magnetization and precipitates as fine particles. It takes a substantial field to magnetize these particles because each is essentially a single domain and must be rotated. However, once this is done, B_r is high and the coercive force to demagnetize is also high. Many interesting techniques to improve the characteristics of hard magnets are used, such as controlled solidification to develop preferred orientation and heat treatment in a magnetic field.

More recently, samarium–cobalt and platinum–cobalt permanent magnets have been developed with very high maximum-energy BH products, as shown in Table 21.1.

In the platinum–cobalt magnets a phase transformation (order to disorder) results in isolated domains, as in Alnico. This material is machinable but expensive (prices fluctuate with the platinum commodity price).

The samarium–cobalt magnets and magnets involving rare earths such as praseodymium and lanthanum are prepared as intermetallic compounds such as Co_5Sm (37% by weight samarium). Then the material is ground to fine powder, so that each particle is domain size. These tiny grains (of hexagonal structure) exhibit great differences in ease of magnetization, depending on the direction in the crystal — easy along the c axis and difficult in the a direction. The powder particles are aligned by a strong magnetic field in the die and are then pressed together and sintered. Care is taken to avoid grain growth during sintering. One use of Co_5Sm magnets is in step motors in electronic wristwatches. The name of these motors arises from the fact that each time the motor receives a pulse from the vibration of a quartz crystal, it moves the hands a step. The present cost of the samarium–cobalt magnet is high, but significant decreases are expected. A new iron base material of even higher BH product, $Nd_2Fe_{14}B$, has been developed.

TABLE 21.1   Maximum BH Products for Several Magnetic Materials

Material	Maximum BH Product, A-weber/m^3
Samarium-cobalt	120,000
Platinum-cobalt	70,000
Alnico	36,000
Ferroxdur*	12,000
Iron-cobalt sinter	7,000
Carbon steel	1,450

*A synthetic ceramic magnet ($BaFe_{12}O_{19}$) of a type described in the following section.

21.10 Ceramic Magnetic Materials

For many years a search was conducted for magnetic materials with high electrical resistance. For example, we know that a transformer heats up because the alternating electrical field generates a stray current in the core, which leads to heating (eddy currents). In high-frequency equipment this heating is very severe, so a high-resistance magnetic material was sought. It was recognized that the mineral magnetite, Fe_3O_4, was a natural high-resistance ceramic magnet, so the search led to synthetic ceramic magnets of similar structure, which are generally called *ferrites* although many do not contain iron. (It is important not to confuse the use of the word *ferrite* when it is applied to these magnetic oxides with the same name for α iron!)

As a starting point, we should look at the unit cell structure of magnetite itself. If we write the formula $Fe^{2+}Fe_2^{3+}O_4^{2-}$, we see that it is analogous to that of the spinel $Mg^{2+}Al_2^{3+}O_4^{2-}$, which was discussed in Chapter 10 as a refractory, and we find that magnetite has a similar unit cell. We have waited until this point to describe the unit cell (Figure 21.11), because it is most important here to understand the magnetic properties. To explain the important ion positions, it is necessary to take a multiple of 8 times the chemical formula written in general terms, with the divalent and trivalent metal ions written as M^{2+} and M^{3+}: $[M^{2+}M_2^{3+}O_4^{2-}]_8$.

The 32 O^{2-} ions are arranged to form eight FCC cells. We can determine from inspection that there are twice as many tetrahedral sites as FCC atoms in a unit cell, whereas the number of octahedral sites is equal to the number of FCC atoms (four per unit cell). Figure 21.11 shows examples of these sites. In the model there are 64 tetrahedral and 32 octahedral sites, but only a por-

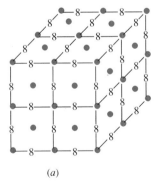 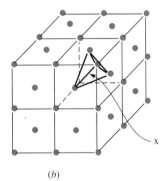

(a)   (b)

FIGURE 21.11 Octahedral and tetrahedral sites in a spinel. (See the table at the top of the next page.) Oxygen atoms are indicated by a •; (a) possible octahedral sites are indicated by the number 8; and (b) location of a tetrahedral site is indicated by an x.

tion of them are occupied in spinel. In the ferrites the iron ions are distributed as follows in the so-called *inverse spinel* structures:

Iron Ions	Tetrahedral Sites *	Octahedral Sites *
16 Fe^{3+}	8	8
8 Fe^{2+}	0	8

*A *site* is a position that can be occupied by an atom. A tetrahedral site is at the center of a tetrahedron formed by other atoms. Six atoms form an octahedron and surround an octahedral site.

The reason this is called *inverse* is that in *normal spinel*, $MgAl_2O_4$ for example, the M^{2+} ions are in the tetrahedral sites and the M^{3+} ions are all in the octahedral sites.

Now, just as in ferromagnetic iron, the magnetism of a ceramic magnet is due to the magnetic moments of the ions. These are calculated in the same way as the magnetic moments of the atoms, but they are not the same if $3d$ electrons are removed in ionizing. For example, Fe^{2+} is formed by removing the two $4s$ electrons, so that it has the same moment as Fe^0. However, in forming Fe^{3+} we develop a magnetic moment of 5 Bohr magnetons because one electron of opposite spin is removed. Table 21.2 gives magnetic moments for ions commonly found in spinels.

The eight Fe^{3+} ions in the tetrahedral sites are always opposite in magnetic alignment to the eight Fe^{3+} ions in the octahedral sites, so no net magnetic moment results from the Fe^{3+} ions. This is called *antiferromagnetism*, as in metallic manganese. However, the moments of the eight Fe^{2+} ions that also occupy octahedral sites are all aligned in the same direction in the unit cell and are responsible for the net magnetic moment — the ferrimagnetic condition (Figure 21.12). Let us proceed to check this arrangement with a calculation.

TABLE 21.2   Magnetic Moments for Ions Found in Spinels

Ion	Total Electrons	3d Electrons	Bohr Magnetons
Fe^{3+}	23	$3d^5$	5
Mn^{2+}	23	$3d^5$	5
Fe^{2+}	24	$3d^6$	4
Co^{2+}	25	$3d^7$	3
Ni^{2+}	26	$3d^8$	2
Cu^{2+}	27	$3d^9$	1
Zn^{2+}	28	$3d^{10}$	0

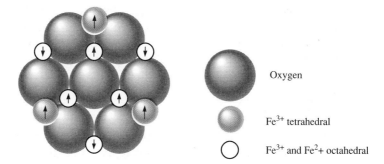

FIGURE 21.12 Spin directions in Fe_3O_4. Fe^{3+} ions in tetrahedral sites have opposite spin directions from Fe^{3+} ions in octahedral sites. The net magnetization is due to the Fe^{2+} ions in the octahedral sites.

EXAMPLE 21.3 *[ES]*

Calculate the saturation induction of $[NiFe_2O_4]_8$. The unit cell is cubic, $a_0 = 8.34$ Å.

Answer The magnetic moment is due entirely to nickel ions, because the spins for the 16 Fe^{3+} ions cancel each other out. Therefore, we have

$8 \times 2 = 16$ Bohr magneton/unit cell contributed by the Ni^{2+} ions

$$M = \left(\frac{16 \text{ Bohr magneton/unit cell}}{(8.34 \times 10^{-10})^3 \text{m}^3/\text{unit cell}}\right)\left(9.27 \times 10^{-24}\frac{\text{A-m}^2}{\text{Bohr magneton}}\right)$$

$$= 2.56 \times 10^5 \frac{\text{A}}{\text{m}}$$

$$B_s = \mu_0 M \qquad \text{(neglecting the } \mu_0 H \text{ term in Equation 21-7)}$$

Assuming that all magnetic moments are aligned at B_s,

$$B_s = \left(4\pi \times 10^{-7}\frac{\text{volt-sec}}{\text{A-m}}\right)\left(2.56 \times 10^5 \frac{\text{A}}{\text{m}}\right)$$

$$= 0.32\frac{\text{volt-sec}}{\text{m}^2} \quad (0.32 \text{ tesla})$$

The experimental value 0.37 tesla is even greater than the calculated value because, as a result of cation vacancies in the crystal structure, some Fe^{3+} is present with Ni^{2+} and contributes to a higher magnetic moment per atom (see Table 21.2).

However, note that although the saturation induction is greater than the theoretical value, it is less than that for pure iron as found in Example 21.1. There are two reasons for this: the lower magnetic moment per atom and the lower density of the atoms in the ceramic that contribute to the magnetic moment.

Returning to the development of ceramic magnets in general, we find that in most cases the Fe^{2+} ions are replaced partly or completely with other ions. Because the ions in these positions are responsible for the magnetization, a partial replacement with an ion with lower moment reduces the overall magnetization. Another technique is to add nonmagnetic ions that will replace some of the Fe^{3+} ions and force them into the Fe^{2+} sites.

For *soft* magnets for electronic and radio communications the following substitutions for Fe^{2+} ions are used: $Mn^{2+} + Zn^{2+}$, $Ni^{2+} + Zn^{2+}$, $Ni^{2+} + Cu^{2+} + Zn^{2+}$, all of which give a hysteresis loop similar to that shown in Figure 21.3(). The Mn^{2+}, Ni^{2+}, and Cu^{2+} ions have magnetic moments, but the role of the Zn^{2+} ions is interesting in that they have no magnetic moment themselves but raise the overall moment in the following way. We mentioned that in the normal spinel structure the divalent ions occupy the tetrahedral sites. The Zn^{2+} ions continue to follow this tendency when they are added to the inverse spinel. They thereby force some Fe^{3+} from their "unproductive" tetrahedral sites to the octahedral sites, where they align to increase the magnetic moment. Just as we added alloying elements in solid solution to change mechanical properties, in this case we are adding ions in solid solution to modify magnetic properties.

To provide magnets with square loops (Figure 21.3b), the substitutional ions are $Mn^{2+} + Mg^{2+}$ and Co^{2+}. These magnets are important for computer operations.

The field of ceramic *permanent* or *hard* magnets is especially exciting. Already there are a variety of uses for these materials, such as the magnetic inserts in refrigerator doors and in small motors. Formerly an electromagnet was needed in these motors, but the use of small, powerful permanent magnets to provide a field greatly simplifies the design. Such motors operate windshield wipers, heater blowers, air conditioners, and window raisers in modern automobiles. The 30-billion-eV accelerator at the Brookhaven National Laboratory uses over 7 tons of ferrites. As an illustration of possible future use, Figure 21.13 shows a drawing of an elevated magnetic roadway in which the car would be floated by magnetic repulsion.

The crystal structure of permanent ceramic magnets is slightly different from the spinel structure (see the problems), but because the same principles are followed as for the spinel structure, the materials are still called ferrites. Just as the ferrites we have discussed have the structure of a naturally occurring mineral (spinel), so these hard ferrites have the hexagonal structure of a

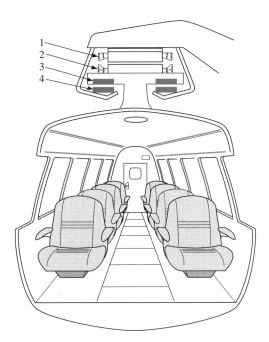

FIGURE 21.13 An artist's conception of a magnetic roadway. Electric drive: (1) Portion of linear motor attached to roadway and serving the same function as the stator of a conventional electric motor. (2) Portion of linear motor attached to the car, functioning as a rotor. Magnetic suspension: (3) Ceramic (ferrite) permanent magnets attached to the car. (4) Ferrite magnets attached to the roadway.

[Courtesy of Westinghouse Corporation]

mineral called *magnetoplumbite*, $Pb(Fe, Mn)_{12}O_{19}$. The source of the magnetism is similar: the alignment of ions such as Fe^{2+} and Mn^{2+} with magnetic moments and the formation of domains. The synthetic ceramic magnets have simpler formulas, such as $PbFe_{12}O_{19}$ or $BaFe_{12}O_{19}$. One of the ways of obtaining a strong initial alignment of domains is to heat the material above its Curie temperature and then cool it under the influence of an externally imposed magnetic field. This is analogous to polarizing a ferroelectric material by cooling it under an electric field.

In addition to the soft and hard types of ceramic magnets, there are two subtypes of great importance.

The square-loop materials retain their full remanent magnetization as the field is reversed until a sharp cutoff is reached (Figure 21.3*b*). Therefore, in using the magnet as an information-giving device, we can get full strength without weakening until the cutoff is reached. If we want to clear out the information, we can do this cleanly and the device is ready to remagnetize. For these magnets, manganese–magnesium–iron oxides and cobalt–iron oxides are used.

For microwave communication it is important to have low losses per cycle because of the high frequencies. In iron-core transformers we can tolerate the hysteresis losses at 60 Hz, but at 10^6 Hz they would be prohibitive. Here the high resistance and very narrow loops of certain ferrites are important. Typical oxides with these properties are of the (Al, Ni, Zn)O type.

Another family of magnetic ceramics is the iron garnet group. The general formula is $3M_2O_3 \cdot 5Fe_2O_3$. The unit cell of these ceramics is more complex than that of spinel. Their chief advantage is that elements with high magnetic moment, such as samarium and yttrium, can be accommodated as M in the formula.

EXAMPLE 21.4 *[ES/EJ]*

Explain the following observations, using the concept of magnetic domains.

A. A tie bar of soft iron is placed on a horseshoe magnet when it is not being used.
B. An induction furnace operating at 3000 Hz is used to melt steel.

Answers

A. A more random orientation of domains is the lowest-energy configuration. Therefore the residual magnetism decreases as a random orientation develops with time. The use of the tie bar maintains the domain alignment between the magnet ends and thus maintains the residual magnetism.
B. The heating of steel in an alternating magnetic field *is not* completely explained by magnetic domains. It is true that one circuit around a hysteresis loop causes an energy loss that occasions heating. However, the amount of energy loss is insufficient for melting. A second and more important source of energy loss is eddy currents. In an alternating field a voltage, and hence a current (eddy current), is induced in the core (the steel in our example). These I^2R losses can be large enough to cause melting. In induction hardening, we use a higher frequency with lower depth of penetration, because melting is not desired. The total heat in the core then depends on frequency, time, heat transfer to the surroundings, and conductivity.

21.11 Superconducting Magnets

When an external magnetic field is applied to a body, the induced flux density increases significantly in ferromagnetic and ferrimagnetic materials. In paramagnetic materials the increase in flux density is slight; in diamagnetics there is a slight decrease. Therefore, the magnetic permeability has values close to, but modestly above or below, a value of 1.0 in these latter materials (Figure 21.14).

In a superconductor we have an approximately perfect diamagnetic material, as shown schematically in Figure 21.15. However, the flux exclusion is present only up to a certain critical value of field strength called H_0. Once this field strength is exceeded, the superconductivity is destroyed. We therefore have a relationship between field strength H_0 and critical temperature θ_c for each superconducting material, as shown in Figure 21.16. Correspondingly, as the temperature is decreased below θ_c we can have higher field strengths.

As pointed out in Chapter 20, we could utilize superconductors for their low resistivity and they would then be current carriers. However, any current

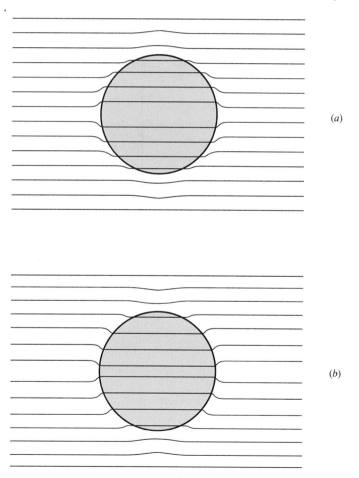

(a)

(b)

FIGURE 21.14   Idealized lines of magnetic flux in (a) diamagnetic materials and (b) paramagnetic materials

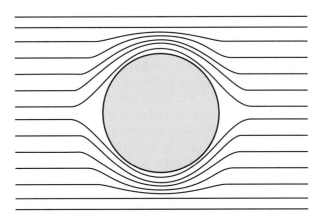

FIGURE 21.15 Nonpenetration of a magnetic field in a perfect diamagnetic material, as found in a superconductor

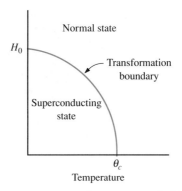

FIGURE 21.16 Effect of temperature and a magnetic field on a superconductor

[Z. D. Jastrzebski, *The Nature and Properties of Engineering Materials*, 2nd ed., John Wiley, New York, 1977, p. 430]

loop produces a magnetic field around it, and superconductivity can be destroyed once a critical current density is obtained. Normal resistance plus nonideal diamagnetism would return.

Three factors are of vital importance in a superconductor: the *critical temperature* θ_c, the *critical field strength* B_c *, and the *critical current density* J_c. These three factors are interrelated, as shown for a niobium–titanium alloy in Figure 21.17. It is important to note that we cannot operate close to θ_c if we want large values of B_c and J_c, which are certainly needed in a magnet. We must remain within the $B_c J_c \theta_c$ envelope to retain superconductivity.

---

*Even though B has been used for flux density and H for field strength in bar magnets, it is not uncommon to use B with units of teslas and to refer to it as a field strength in superconductors because of the end use. B_c is often referred to as B_{2c}, the upper field strength, because a small value of field strength, B_{1c}, is an initial value to be passed before building up to B_{2c}.

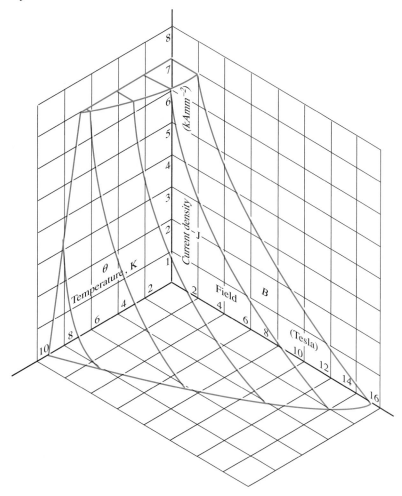

FIGURE 21.17   Critical-current surface for a commercial superconducting alloy of niobium–titanium. These data are based on recent measurements at 4.2 K, together with earlier measurements at variable temperature by Hampshire, R., Sutton, J., and Taylor, M. T. (1969).

[M. N. Wilson, *Superconducting Magnets*, Clarendon Press, Oxford, England, 1983, p. 2. Reprinted by permission.]

Therefore, we still operate the niobium–titanium alloy near the boiling point of helium (4.2 K).

We can now observe why the alloys of niobium were a significant advance over the pure metals—apart, that is, from the modest gain in θ_c. The critical field strength of 1/20 tesla for pure metals was increased to approximately 15 teslas in the niobium alloys. These alloys were then called type II superconductors to distinguish them from the type I pure metals. Methods of fabricating them will be treated in Chapter 22.

The critical density J_c is also a function of microstructure. These relations are quite complex, and as an approximation we may begin by visualizing magnetic lines of force (also called fluxoids) as occupying separate positions produced by a low current. As we increase the current, these positions tend to crowd together and overlap—and we lose superconductivity. However, higher values of J_c can be reached by pinning the fluxoids, just as higher tensile strengths are obtained by pinning dislocations. As a matter of fact, cold working to produce dislocations gives a material in which the fluxoids *are* pinned!

A fine second-phase precipitate is even more effective in attaining a higher J_c because of the pinning effect. In advanced treatments, electron microscope investigations have shown that combinations of cold working and precipitation hardening can produce structures that allow current densities of 3000 A/mm^2 at 5 teslas. This is a current of over 6000 A in a wire the diameter of a pencil lead without resistance loss! At the present time, the exact treatments are closely guarded secrets.

It is significant that the attainment of higher critical temperatures is only a part of the puzzle to be solved for superconductors. Data such as those given in Figure 21.17 will have to be developed for the new superconducting oxides before the engineering applications can be completely identified.

For example, the diamagnetic characteristics of a superconductor indicate that it would be strongly repelled by a magnet. This feature can be used in a magnetic levitation railway similar to that shown in Figure 21.13, except that a superconducting magnet would replace ferrite. Although some designers have proposed superconductors operating at liquid helium temperatures (~4 K), superconducting oxides operating at critical temperatures above the boiling point of nitrogen (77 K) would offer great economic benefits. This assumes, of course, that suitable current densities and field strengths can also be attained.

Summary

In the electrical materials the most important concept is the effect of structure on the ease of movement of charge carriers. In an analogous way, in magnetic materials the most important concept is the effect of structure on the value of magnetic flux. There are two important principles to be established immediately.

First, the value of magnetic flux density B obtained by the application of a magnetic field of strength H is a function of the permeability μ, but above an elevated value of H there is very little increase in B. This is the saturation magnetization B_s.

Second, the flux density does not fall to zero when the field is removed but only drops to the residual magnetization B_r, which is low in a soft magnetic material and high in a permanent or hard magnetic material.

In reviewing the features of hard and soft metallic magnetic materials, it is established that the key feature is the magnetic domain, a region of structure in which the magnetic moments of the atoms are all aligned in one direction. When a field is applied, the domains with moments aligned parallel to the field grow at the expense of others. In a soft magnetic material, domain wall movement is made easy by simple grain structure, avoidance of cold work, and other techniques. Conversely, in hard magnetic materials the desired structure consists of tiny (300-Å) particles, each with only one domain, aligned to provide a strong magnetic field.

In ceramic magnets the attainment of the inverse spinel structure produces soft magnets of the desired characteristics. For hard magnets fine domain-sized particles of the magnetoplumbite structure are used.

The superconducting magnets offer great potential for numerous engineering applications. However, their limitations include a low critical temperature, as well as critical field strengths and current densities that must not be exceeded if the superconductivity characteristic of zero resistivity is to be maintained.

Definitions

Antiferromagnetic   Material with ferromagnetic atoms that are aligned with magnetic moments in opposing directions, leading to no net magnetic moment.

Bloch wall   The boundary between two domains. On either side of the wall, the magnetic moments are oriented at 90° or 180° to each other.

Bohr magneton   The magnetic moment produced in a ferromagnetic atom by one unpaired electron. 1 Bohr magneton = 9.27×10^{-24} A/m^2 (0.927 erg/gauss).

Coercive field, H_c   The field needed to reduce the flux density to zero after initial application of a magnetic field to obtain B_s.

Critical current density, J_c   The maximum current density that a superconductor can be exposed to and still retain its characteristic of zero resistivity.

Critical field strength, B_c   The limiting field strength in a superconducting magnet; exceeding it destroys the magnet's superconducting quality.

Curie temperature   Temperature at which ferromagnetism disappears upon heating.

Diamagnetism   Very small negative devation from $\mu_r = 1$.

Domain   A region of a metal or ceramic structure in which the magnetic moments due to the individual atoms are aligned in the same direction.

Exchange interaction   The relationship between electron spins in neighboring atoms. When the spins are aligned, the material is ferromagnetic; when the spins are opposite, antiferromagnetism results.

Ferrimagnetic   A ceramic material with some ferromagnetic atoms that are aligned in opposing directions in the unit cell. However, ferromagnetic atoms at other unit cell positions are aligned, and the total unit cell exhibits a net magnetic moment.

Ferrite   A magnetic inverse spinel. Do not confuse this term with the same word used earlier to mean α iron.

Ferromagnetism Large values of μ_r produced by alignment of certain atoms (such as iron, cobalt, and nickel) into domains.

Hard magnetic material Material that is magnetized and demagnetized with difficulty and that has a large area within the *B-H* loop.

Inverse spinel Material with the same general formula as a normal spinel but with the trivalent ions occupying tetrahedral and octahedral sites and the divalent ions occupying only octahedral sites.

Magnetization, *M* The magnetization times the permeability of a vacuum gives the increase in flux that is due to the insertion of a given material into a field of strength *H*. Because μ_0 is a constant $(4\pi \times 10^{-4}$ volt-sec/A-m), the magnetization is a measure of the increased flux density produced by the material.

Normal spinel A ceramic of the general formula $M^{2+}M_2^{3+}O_4$. There are divalent ions in tetrahedral sites and trivalent ions in octahedral sites.

Paramagnetism Very small positive deviation from $\mu_r = 1$.

Permeability, μ The ratio B/H, where B is the flux density, developed when a magnetic field of strength H is applied. This is analogous to the concept that the conductivity σ is equal to the ratio I/V, where I is the current produced by a voltage V in a wire (for unit length and area of the conductor).

Relative permeability, μ_r The ratio of permeability of a material to the permeability of a vacuum; $\mu_r = \mu/\mu_0$.

Remanent induction, B_r The flux density remaining when the magnetic field is removed.

Saturation induction, B_s The maximum magnetic flux obtainable in a material. For a ferromagnetic material, B_s may be taken as $\mu_0 M$.

Soft magnetic material An easily magnetized and demagnetized material, with a small area within the *B-H* loop.

Problems

21.1 *[ES]* Indicate which are the material constants and which the dependent and independent variables in the discussion in Section 21.2 of the parameters σ, I, V, μ, H, and B. (Sections 21.1 through 21.5)

21.2 *[ES/EJ]* Explain how the following characteristics of a ferromagnetic material could affect its *B-H* curve. (Sections 21.1 through 21.5)
   a. Grain size
   b. Inclusions
   c. Residual stresses
   d. Heat treatment (specify)

21.3 *[ES/EJ]* Explain the values of relative permeability for each of the following materials from the given description of its magnetic activity. (Sections 21.1 through 21.5)
   a. Paramagnetic — weakly attracted by a magnet
   b. Diamagnetic — weakly repelled by a magnet
   c. Ferromagnetic — strongly attracted by a magnet

21.4 *[ES/EJ]* Figure 21.3 includes some phenomena not discussed in the text. (Sections 21.1 through 21.5)

a. Why are there three hysteresis curves in Figure 21.3(b), and what do they represent?

b. Why is *BH* plotted against *H* as a second curve in Figure 21.3(c)?

21.5 *[ES]* Calculate the saturation magnetization for pure cobalt, and compare your answer to the handbook value of 1.87 tesla. (Sections 21.1 through 21.5)

21.6 *[ES]* Estimate the saturization magnetization of gadolinium (atomic weight = 157.26 g/mol, density = 7.8 g/cm^3), assuming that the magnetic moment per atom is equal to the number of unpaired 4*f* electrons and that the atom follows Hund's rule. In this case Hund's rule states that the first seven 4*f* electrons will have the same spin direction and that the next seven will have the opposite spin direction. (Sections 21.1 through 21.5)

21.7 *[ES]* Problem 21.27 shows several hysteresis loops. Estimate quantitatively the "power" of ferromagnetic alloys B and C. (Sections 21.1 through 21.5)

21.8 *[ES]* Another way of describing magnetic characteristics is in terms of *magnetic susceptibility*, χ. The relationship of this value to the magnetic permeability is as follows:

$$\mu = \frac{B}{H} = 1 + 4\pi\frac{M}{H} = 1 + 4\pi\chi \tag{21-8}$$

Give reasons why magnetic susceptibility values are tabulated more often than magnetic permeability. (Sections 21.1 through 21.5)

21.9 *[ES/EJ]* What is meant by the phrase *pinning domain boundaries*? How can this be used in a practical sense? (Sections 21.1 through 21.5)

21.10 *[ES]* Explain the relationship between a normal magnetic hysteresis loop and the data given in Figure 21.7. (Sections 21.1 through 21.5)

21.11 *[ES]* Why does a glassy metal have a high magnetic permeability and a small hysteresis loop compared to the same metal in polycrystalline form? (Sections 21.6 through 21.11)

21.12 *[ES]* Calculate the saturation induction B_s of $[Li_{0.5}Fe_{2.5}O_4]_8$, assuming that $a_0 = 8.37$ Å in an inverse spinel. (Sections 21.6 through 21.11)

21.13 *[ES/EJ]* Theoretically, metallic or ceramic magnets can be used up to their respective Curie temperatures. Explain why this is seldom done in practice. (Sections 21.6 through 21.11 and Section 1.26)

21.14 *[ES]* Show how the net magnetic moment of $[(Zn_{0.5}Ni_{0.5})Fe_2O_4]_8$ is greater than that of $[NiFe_2O_4]_8$. [*Hint:* The zinc ions displace Fe^{3+} ions from tetrahedral sites.] (Sections 21.6 through 21.11)

21.15 *[ES]* (Sections 21.6 through 21.11 and Section 10.6)

a. Assume an FCC unit cell for element A. Tell whether each of the following sites is classified as *octahedral*, *tetrahedral*, or *neither* when an atom B is added: (1) $\frac{1}{2}, \frac{1}{2}, \frac{1}{2}$; (2) $\frac{1}{2}, 1, 1$; (3) $\frac{3}{4}, \frac{3}{4}, \frac{1}{4}$; (4) 1, 1, 1.

b. Consider an FCC lattice of an atom A *with no B present*, extending through the space. (1) What is the coordination number of the atom at 1, 1, 1? (2) List the coordinates of all its nearest neighbors in the unit cell. (3) If B atoms are added to occupy all nearby tetrahedral sites, what will the coordination number of atom A become?

c. What is the significance of tetrahedral and octahedral sites in an inverse spinel structure?

21.16 *[ES]* In the magnetoplumbite structure of hard magnets, the general formula is $AO \cdot 6B_2O_3$, where A is barium, lead, or strontium, O is oxygen, and B is aluminum, chromium, gadolinium, or iron. (Sections 21.6 through 21.11)
a. The valences of the A and B metals are constant but different. What are they?
b. Using these valences, find the percent variation in ionic radius for the A and B metals.

21.17 *[ES]* It has been suggested that we synthesize a ferrite to achieve higher saturation induction than is obtained in $[NiFe_2O_4]_8$, which we found to be 0.32 tesla in Example 21.3. In the new synthesized ferrite, manganese ions would be substituted for half of the nickel ions to produce $[Ni_{0.5}Mn_{0.5}Fe_2O_4]_8$. The size of the unit cell would increase to 9.03 Å because manganese is larger than nickel. (Sections 21.6 through 21.11)
a. What is the magnetic moment contribution from each ion?
b. What is the magnetism M?
c. What is the saturation magnetization B_s?

21.18 *[ES/EJ]* (Sections 21.6 through 21.11)
a. What is the effect of defects on the net magnetic moment obtained in a ceramic magnet?
b. In $[Fe^{2+}Fe_2^{3+}O_4]_8$, why do the Fe^{2+} ions preferentially occupy the octahedral rather than the tetrahedral sites?

21.19 *[ES/EJ]* You wish to obtain maximum residual magnetization B_r in a component. It has been suggested that a compaction and sintering process be used. Explain why adhering to each of the following guidelines may enhance the maximization of B_r. (Sections 21.6 through 21.11)
a. Powder size should be equal to domain size.
b. Compaction, sintering, and cooling should be done in the presence of a magnetic field.
c. Incomplete sintering or the presence of inclusions at the interface between particles may be an advantage.

21.20 *[ES]* Why are some alloys severely cold-worked and annealed in order to maximize their magnetic properties? (Sections 21.6 through 21.11)

21.21 *[ES]* We can have a perfect superconductor just below θ_c. Why, then, would we still like to operate a superconducting magnet well below θ_c? (Sections 21.6 through 21.11)

21.22 *[EJ]* Cold working and precipitation hardening have been identified as important to the attainment of high current densities in superconducting magnets. What other metallurgical features might influence the critical current density? (Sections 21.6 through 21.11)

21.23 *[EJ]* Metallic magnets generally have higher saturation magnetization values than ceramic magnets do. Why, then, are ceramic magnets so widely used? (Summary of Chapter 21)

21.24 *[ES/EJ]* Why must magnets for power and high-frequency applications have high magnetic permeabilities and low coercive field requirements? (Summary of Chapter 21)

21.25 *[ES/EJ]* Induction heating is due to so-called I^2R loss in which induced eddy currents result in energy loss and internal heating. (Summary of Chapter 21)
a. Why is copper more difficult to heat by induction than iron is?
b. Why are loose powders difficult to heat by induction?
c. Why are laminated metal sheets used for power transformer cores?

21.26 *[ES/EJ]* An important characteristic in a power transformer core is the power loss, which leads to heating and lowered efficiency. The power loss is made up of stray currents induced in the core and to the hysteresis loss, which is related to the area inside the *B-H* loop for a complete cycle. (Summary of Chapter 21)

a. Describe the type of metallic material you would select for a transformer, and explain how it would be fabricated (laminations, preferred orientation, etc.).

b. Describe the type of ceramic material you would select and the ideal *B-H* curve you would specify.

21.27   *[ES/EJ]* The figure shows magnetization curves for three different ferromagnetic alloys. On the basis of these curves, answer the following questions. (Summary of Chapter 21)

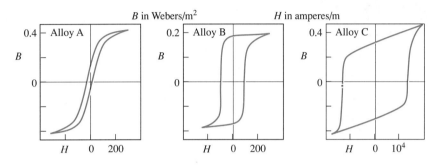

a. Which alloy has the highest coercive field?
b. Which alloy has the highest hysteresis loss?
c. Which alloy has the highest residual magnetization?
d. Which alloy would be preferred for a permanent magnet?
e. Which alloy would be best for a computer memory?
f. Which alloy would be best for a transformer core?
g. Which alloy probably has the highest fraction of domains oriented in parallel when $H = 0$ (after first having been magnetized to saturation)?
h. Which alloy probably has the highest mechanical hardness?
i. Are ferrite magnets generally preferred to metallic magnets in applications involving high-frequency or low-frequency electric fields?
j. What is the temperature called above which ferromagnetic materials lose their ferromagnetic properties?

21.28   *[ES]* The terms *ferromagnetic, antiferromagnetic,* and *ferrimagnetic* are used to describe the different types of alignment of electron spins in iron, manganese, and magnetite (Fe_3O_4, spinel), respectively. For each case, sketch a group of atoms illustrating the difference in spin alignment. In the ferrimagnetic structure, mark which group of ions is antiferromagnetic and which group is aligned to produce magnetization. (Summary of Chapter 21)

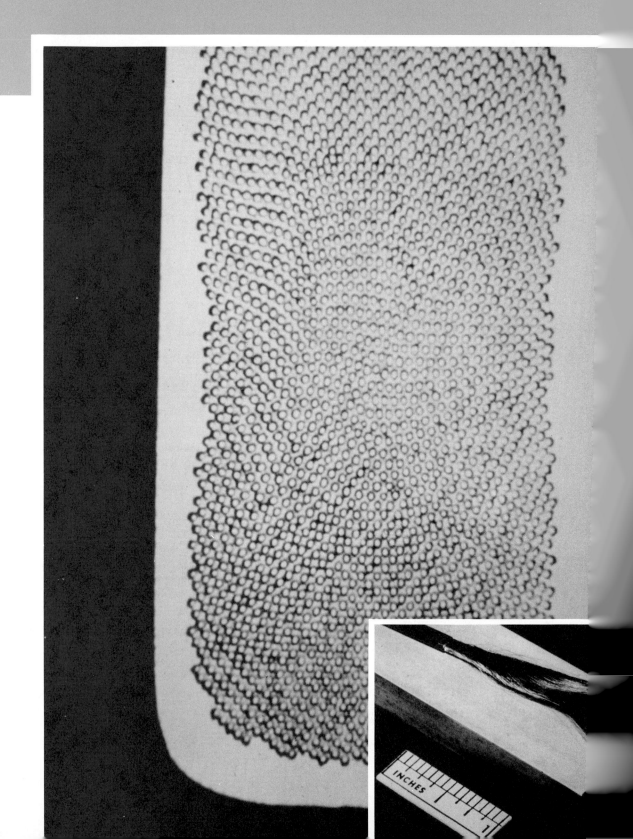

CHAPTER 22

Processing of Electrical, Magnetic, and Optical Materials and Components

Co-authored with Paul A. Flinn
Senior Research Scientist
Intel Corporation
Santa Clara, California

The processing of electronic materials employs many of the steps we have studied earlier, but often with some interesting innovations based on fundamental concepts.

The manufacture of wire incorporating superconducting filaments of niobium titanium in a ductile copper matrix is an intriguing case. First a billet of niobium titanium is made and then placed in an extrusion can of pure copper. To promote satisfactory bonding, all metal surfaces are carefully cleaned and the can is evacuated and sealed by electron-beam welding. Hot extrusion is used to produce hexagonal shapes of copper containing the niobium titanium. These are then cleaned, stacked, and placed in another copper can to be reextruded to produce a multifiber wire. At appropriate steps the wire is given a precipitation heat treatment to produce an optimum current density and a twist to decouple the filaments electrically.

22.1 Overview

Although we can apply many of the concepts that govern the conventional processing of metals, ceramics, and polymers to the processing of electrical and optical materials, there are also advanced processes that are essential to the very existence of these industries. As one example, we reviewed in Chapter 20 how minute amounts of dopants are required in the production of *p* and *n* silicon; special procedures are needed to introduce these, beginning with a single crystal of high-purity silicon. As a second example: After producing a chip it is necessary to encapsulate it to protect it from the atmosphere. One of the encapsulants is ceramic. It took the suppliers of this ceramic some time to realize that the specification of less than 1 part thorium per *billion* in a glass ceramic was "for real." (Ordinary beach sand has 100 times this amount.) The reason for the specification is that the emanations from the radioactive decomposition of thorium can change the memory of a chip and lead to error.

Another difference in the processing of electrical and optical materials is that we are dealing with *small devices* that control electron movements, magnetic fields, and photons rather than with large components that transmit stress. As a result, we have much more latitude in the use of expensive materials: gold, rare-earth elements, and their compounds such as yttrium–aluminum–garnet.

Finally, we can use exotic processing methods that would be prohibitively expensive for larger components. These include chemical vapor deposition (CVD), ion implantation, and molecular beam epitaxy. On the other hand, we have to lay down accurate line deposits of aluminum only 1 μm (10^{-6} m) wide!

At this point you may be a little awed at such an array of new materials and processes. Let us reassure you with the observation that many have entered this field without even the background you have acquired in this short course — that is, without understanding such basics as diffusion and phase diagrams. For example, many serious failures were observed in service from ruptures at the junction between aluminum conductors in the circuit and the gold lead wires. A purple deposit was noted, and the phenomenon was called "the *purple plague.*" Understanding the problem and finding a solution depended on a simple application of the relevant phase diagram and diffusion concepts. In another case, an aluminum–silicon alloy was used as a base for a device that had to withstand a high level of earthquake stress. It was supposed that because the material was over 90% aluminum, it had to be ductile! (See Chapter 7.)

How should we explore this new field of microdimensions, exotic processes, and unlimited materials? Because processes and materials are changing so rapidly, it seems best first to examine the basic architecture of a device and then to see what processes and materials are used to produce it. For example, by far the greatest development effort in the electrical field is directed toward VLSI (very large-scale integrated circuit) chips, and this application will provide us with an interesting case for studying thin-film techniques. In the new field of ceramic superconductors, special techniques are also needed. The preparation of small magnets is vital in the production of magnetic devices. And turning to the optical field, we will study the fabrication of optical fibers.

22.2 Processing of an MOS Transistor

Let us review first the single *metal-oxide-semiconductor*(MOS). After this we will be in a position to understand the significance of some of the new, advanced procedures (Figure 22.1).

1. *Preparation of silicon-chip base.* First we must prepare high-purity silicon, to eliminate traces of *n*- and *p*-type elements which might affect the conductivity. We do this by a process called *zone melting,* whose principle depends on the phase equilibria discussed in Chapter 4. In most cases, if we have a given impurity *B* dissolved in solid solution in the silicon, the solubility of *B* is higher in liquid silicon than in the solid. We can express this by the partition coefficient discussed earlier. Now let us consider a mass of solid silicon contained in a long refractory boat. If we melt one end by placing an induction coil around and across the boat, then move the boat slowly so that the liquid zone sweeps from one end to the other, the impurity concentrates in the liquid and is swept to the far end of the boat. If we make a number of passes in this manner, we can reduce the impurities in the main body of the boat to one part per billion. We discard the end portion containing the impurities.

   A single crystal of silicon is required to avoid the effects of grain boundaries and change in grain orientation. To get a single crystal, we remelt the silicon in a crucible under a protective gas atmosphere to avoid oxidation. If we want *p*-type silicon, we add a small amount of an acceptor element to the melt. A small crystal of silicon of the desired orientation is placed at the end of a rod that is lowered to touch the melt, which is close to the freezing point. As the silicon freezes onto the nucleus, the crystal is pulled upward and rotated, and as material continues to freeze on the rod, the crystal is withdrawn. The long, rodlike crystal, about 70 mm in diameter, is then carefully sectioned into 0.5-mm slices with a diamond-tipped saw wheel, and the slices are polished. These form the silicon wafers.

   The key to the economics of the process from this point onward is to produce many identical circuits on adjacent squares of the chips *before* the wafer is broken up into "dice" that must be handled separately. Although the prices of silicon and of processing have risen moderately in recent years, the price of the circuits has fallen drastically. This is because the size of the dice has been constantly decreased, allowing many dice to be processed for the same overall cost.

2. *Growing a film of silicon dioxide.* A film of SiO_2 is then grown on the surface of the wafer by exposing the disk to an atmosphere of pure oxygen at 1000–1200°C (1832–2192°F). For example, a layer of oxide 0.1 mm thick

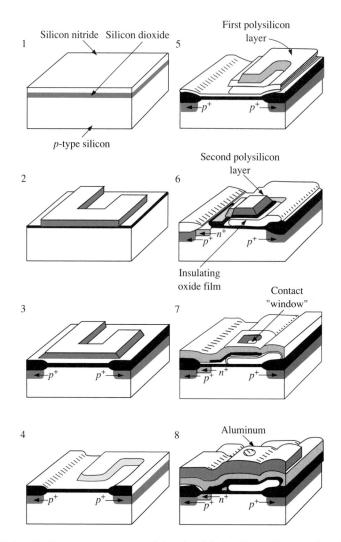

FIGURE 22.1   Various steps required in the fabrication of a two-level n-channel polysilicon gate metal-oxide semiconductor (MOS) circuit element. (1) Selective oxidation of silicon with the aid of vapor-deposited silicon nitride film. (2) Selective removal of a layer of silicon nitride by photolithography (first masking). (3) A p-type dopant is added, followed by oxidation of silicon; silicon nitride acts as a mask. (4) Silicon nitride is removed by chemical etching, leaving the unattached silicon and silicon dioxide. (5) The first polysilicon layer is deposited (second masking). (6) Silicon is oxidized to produce an insulating layer; then a second polysilicon layer is deposited (third masking). (7) Hydrofluoric acid etch attacks silicon dioxide to expose some regions to diffusion of n-type dopant. More silicon dioxide is deposited and contact windows are opened in this fourth mask. (8) Deposition of aluminum provides the fifth mask.

[From page 122 of "The Fabrication of Microelectronic Circuits," by William G. Oldham. Copyright © September 1977 by Scientific American, Inc. All rights reserved.]

can be produced at 1050°C (1922°F) in 1 hour. The ease of production and the insulating qualities of this SiO_2 layer have resulted in the dominance of silicon as a semiconductor material over germanium or compound materials such as gallium arsenide. The silica can also be used as a dielectric in a capacitor formed between layers. Several hundred wafers can be processed simultaneously with computer control of the operation.

3. *Etching the circuit design.* We now wish to etch away selectively the oxide to expose the silicon for a selected circuit design. To do this, we want to deposit a protective film in the desired pattern. A special organic lacquer called a *photoresist* is dropped on the wafer. When the wafer is spun, a uniform film of this lacquer is formed and dried. The organic film has the special property that, if it is exposed to ultraviolet light, the polymer cross-links and is insoluble in organic solvents, whereas the original material is soluble. Therefore light is passed through a carefully produced mask that defines the desired circuit on the wafer surface. After exposure, the unexposed lacquer is washed off, leaving a bare silica surface in the desired circuit pattern. The wafer is then immersed in a solution of hydrofluoric acid that cuts through the silica layer to the silicon. The photoresist protection layer covering the SiO_2 is removed by another chemical treatment.

We can now produce *p-n* junctions at exposed areas in the silicon chip by diffusion. A boat of silica wafers is placed at 1100°C (2012°F) in an atmosphere containing phosphorus. A layer of phosphorus-rich material builds up in the surface of the wafers, and the reaction is governed by relations similar to those studied in the carburization of steel. For example, a layer 1 μm deep is attained in 1 hour. A second heat treatment in a neutral atmosphere, called a *drive-in treatment*, is used to develop a deeper layer and reduce the concentration gradient. Ion bombardment may also be used to implant dopant atoms. This has the advantage that it can be performed *through* a thin layer of silica.

When circuits are to be formed above the lower layers, thin films may be evaporated in predetermined patterns, using the same photoresist techniques just discussed. Polycrystalline silicon is laid down by deposition of chemical vapor by heating silane, SiH_4, which decomposes into silicon and hydrogen. The polycrystalline silicon is used as a resistor, not as a *p-n* junction. In these steps silicon nitride is used in place of silicon oxide to produce a thinner, flatter film.

After production, the individual circuits on each wafer are tested with a computer-controlled probe. Unsatisfactory circuits are automatically marked with a tiny point dot and later discarded.

The wafers are then notched and broken into dice and assembled into packages. This is a crucial step to protect against corrosion. Carefully formulated ceramic materials are generally used to package the wafers.

The transistor may be even more complex, as the CMOS (complimentary *metal-oxide semiconductor*) transistor illustrates (Figure 22.2), but the important concept is that the desired layers and combinations are built up by way

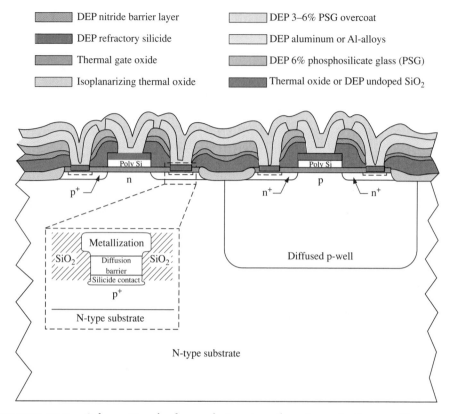

FIGURE 22.2   Schematic of a layered structure forming a bulk complimentary metal-oxide semiconductor (CMOS) device. The inset shows a cross-sectional view of a shallow silicide contact on a shallow junction, in the case of a bipolar configuration. Eight of the ten layers used to fabricate this structure are deposited by PVD or CVD techniques.

[Reprinted with permission from *Journal of Metals*, Vol. 38, No. 2, Feb. 1986, p. 56, a publication of The Metallurgical Society, Warrendale, Pennsylvania.]

of a series of operations involving *thin films:* (1) doping, (2) masking, (3) etching, (4) deposition of a thin film, and (5) encapsulation.

As the number of transistors on a chip increases and the dimensions of all the parts of the circuit become smaller, new processes are needed for accurate miniaturization. Many of the processes can be used for several operations. And in some cases the beam can be focused so that masking is eliminated. Let us review physical vapor deposition (PVD), chemical vapor deposition (CVD), and their modifications.

22.3   Physical Vapor Deposition (PVD)

Physical vapor deposition is the transport of a material in finely divided form from a target to the chip surface without chemical change in the material. Let us review the simple principle of *evaporation* first. When we evacuate a glass chamber and then apply sufficient voltage to a thin gold wire inside, the wire melts and evaporates, and the walls become coated with a film of gold. (The same procedure is used to aluminize mirrors.) For higher-melting materials, electron-beam evaporation, in which a beam of electrons strikes and evaporates the target, is used.

Sputtering is an improvement to PVD in which a target of almost any material is bombarded in a vacuum chamber with gas ions accelerated by high voltage. Particles of atomic dimensions are ejected from the target, traverse the chamber, and are deposited as a thin film on the chip surface. The advantage of sputtering is that the target may be kept cool, thereby avoiding the fractionation of different elements of different vapor pressure. A number of modifications of the process have been developed, including the use of magnetic fields to focus the bombarding ions.

An important new method is *ion beam sputtering* (Figure 22.3), in which an ion beam is directed at the target from an ion gun. The energy and flux density of the beam are independently controlled. Furthermore, compounds such as Si_3N_4 can be formed by introducing a nitrogen atmosphere with a silicon target.

22.4 Chemical Vapor Deposition (CVD)

In *chemical vapor deposition*, gases such as AsH_3, B_2H_6, and PH_3 are passed over the chip in a controlled-atmosphere furnace at 800–1100°C (1472–2012°F) to change its surface analysis. The arsenic, boron, or phosphorus

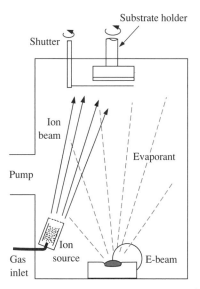

FIGURE 22.3   Typical ion-beam-assisted evaporation as used in a thin-film deposition

[Reprinted with permission from *Journal of Metals*, Vol. 38, No. 2, Feb. 1986, p. 56, a publication of The Metallurgical Society, Warrendale, Pennsylvania.]

atoms precipitate at the surface and diffuse inward. In this way a region of the silicon that was previously p-type can be made n-type or a complete film or mask may be formed. Three types of CVD reactors are shown in Figure 22.4. Modifications developed recently include photochemical CVD (PH CVD), in which ultraviolet light is used to catalyze the reactions.

It is sometimes difficult to decide whether to use PVD or CVD methods. Technicians have had a good deal of experience with the physical methods, but the chemical methods offer a greater throughput for high-volume operations.

EXAMPLE 22.1 *[ES/EJ]*

Figures 22.1 and 22.2 and the descriptions of them contain notation that can be confusing. What is the difference between diffusion barriers and dielectrics and between n^+ and n in planar semiconductors?

Answer   A dielectric is an electric insulator that separates charged bodies. As such, it can be used as a capacitor whose temperature range of operation is usually close to room temperature. A diffusion barrier may also be a dielectric, but its primary purpose is to prevent migration of atoms during the high-temperature processing of the electronic components. The objective is to prevent contamination that would result in unpredictable electrical characteristics.

Certainly, an n-type semiconductor means excess electrons are available. During processing of integrated circuits, however, the conduction characteristics are locally modified by masking and diffusion of other elements containing excess electrons. These new local doped areas then become n^+. We could differentiate similarly between p and p^+.

22.5 Defects in Chips

A number of unusual defects can develop that are related to basic concepts. We will discuss a few that are related to the structure.

Diffusion problems are important because very short diffusion distances are involved. We have already spoken of the "purple plague" caused by diffusion into gold to form a brittle compound. Another problem is that after p-type silicon is doped with the desired shallow layer of n-type conductor, if the chip is heated to high temperatures in subsequent operations, the n-type conductors diffuse into the lower regions of the silicon. Electromigration is another problem: Aluminum atoms (ionized) are driven along the grain boundaries by an electric field because they are less tightly bound than the atoms at definite lattice positions within the grains. This causes grain boundaries aligned in the direction of flow to erode faster than those aligned across the flow and leads to a "bamboo" structure.

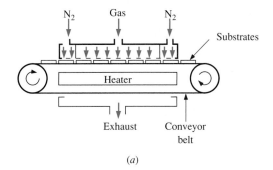

(a)

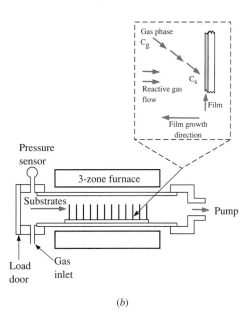

(b)

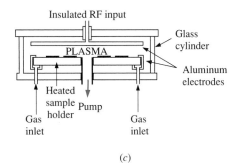

(c)

FIGURE 22.4   Schematic diagrams of CVD reactors: (a) continuous, atmospheric-pressure reactor; (b) hot wall, low-pressure reactor (shows a schematic representation of standard CVD process); (c) parallel-plate plasma-enhanced reactor

Production Problems and Limitations in Chip Production

Now that we have in mind the production sequences and minute dimensions of a chip, it is interesting to examine the necessary variations in conventional methods.

The production of the starting material, usually silicon (but also gallium arsenide), may involve either the Czochralski (CZ), floating zone (FZ), or Bridgman method (Figure 5.8).

One of the major factors in choice of process is the number of dislocations encountered. By the floating zone method dislocation-free crystals can be made. A few dislocations are encountered in CZ crystals and a higher number in Bridgman crystals.

Most crystals for small chips are made by the CZ method where the dislocation density can be tolerated. If, for example, a given slice of silicon contains 5 dislocations and there are 100 chips, the 5 imperfect chips can be discarded during inspection.

In all cases the orientation of the silicon crystal is controlled by using a seed crystal of the desired orientation. The usual case is (111) in the plane of the chip but in special cases (100) orientation is used. Some of the reasons are complex, but one is to facilitate epitaxial growth (as discussed shortly).

Crystal purity is of paramount importance. This is illustrated in Chapter 20, in which the effect of small amounts of dopants was discussed. To minimize contamination from the atmosphere, processing is conducted in Class 10 rooms, which have a specification of less than 10 particles above 1 mm diameter per *cubic meter.*

The reason most silicon crystals for small chips are made by the CZ method is because of intrinsic gettering. It is relatively easy to incorporate a very small controlled amount of oxygen (27 ppm) during production. This oxygen reacts with silicon, forming fine SiO_2 particles, which develop elastic strain fields. In turn, these attract, "getter," the trace amounts of harmful foreign atoms in the crystal, and essentially eliminate their effects on electrical properties.

For large rectifiers, however, where no dislocations can be tolerated in a 5-in. (127 mm)-diameter slice, the CZ material cannot be used and the FZ technique is needed. Let us digress to point out how a seemingly small detail like this can have tremendous industrial significance. Formerly a number of producers of FZ silicon in the United States supplied crystals for the principal domestic manufacturer of rectifiers. Then a foreign source eliminated the U.S. FZ producers by price cutting. The foreign source then refused to supply the U.S. rectifier manufacturer, who then was forced to close his plants.

Now let us turn to the Bridgman method, which is rarely used for silicon production but is important for gallium arsenide. GaAs production is more difficult than silicon production because in addition to the usual requirements of purity it is necessary to obtain equal amounts of Ga and As. While silicon melting and crystallization can be performed at atmospheric pressure, the loss of As is variable in making GaAs and a sealed chamber is needed. This is more easily provided with the Bridgman technique. We should add that the major use of GaAs is as a thin (1 mm) layer grown epitaxially on silicon.

Epitaxial growth is an application of a phenomenon known to mineralogists for many years. It occurs when the composition of the solution or vapor surrounding a mineral crystal changes and a new mineral is deposited on the old crystal maintaining the same structure and orientation. To make this possible the natural space lattice of the new mineral must be close to that of the original mineral. It should be emphasized that the result is still a single crystal without a grain boundary.

Two common methods for epitaxial growth are liquid-phase epitaxy, in which the liquid that is the source of the new structure is flushed over the base material, and molecular-beam epitaxy, where the new material is projected toward the matrix as a charged beam of molecular-sized particles. The silicon base is heavily doped to serve as a good conductor to either ground or to line voltage. The complex circuitry is then built up in the GaAs. We recall from Chapter 20 that the reasons for the use of GaAs are light emission or greater switching speed. It is possible to provide regions of GaAs at selected locations in the circuit where fast switching is essential.

Another important application of epitaxy is in the manufacture of CMOS transistors (Figure 22.2). When the manufacture is limited to silicon, an instability can develop called "latching," which can reverse the on and off switching of the two sides with disastrous results. It is found that by using a base of silicon and building up the circuit in an epitaxially deposited layer of GaAs the problem can be circumvented.

Miniaturization is another fascinating production problem because the number of circuits per chip has grown to over 10,000. If the chip size remains the same, the dimensions of the transistors and even the conductor lines must decrease. At the present time the dimensions are of the order of 1 μm. This means that diffusion distances are short and that interface boundaries and carefully graduated differences in composition may be erased with exposure to sufficient combinations of temperature and time. We must bear in mind, for example, that in the later stages of production, dielectric layers on silicon nitride are applied at temperatures over 400°C (752°F), which can cause substantial diffusion between existing layers. Accordingly, in recent practice it is customary to develop a "thermal budget" of time and temperature which must not be exceeded for a given chip design. As distances have become shorter, expedients have been developed to reduce the exposure of the chip to the temperatures required for the formation of coatings. One method is to deliver the required temperature to the *additive* by a plasma circuit. However the process must be carefully controlled to obtain satisfactory bonding with the substrate.

22.6 Residual Stresses and Stress Relief

We have mentioned many cases in which a film of a different material is applied to a substrate or over another film at elevated temperature. From our earlier discussions, we can expect that residual stresses would develop in two cases:

1. A temperature difference exists between film and substrate at the time of application.
2. The coefficients of expansion are different.

These two cases can lead not only to residual stresses but also to cracking and "punch through," which is the passage of current through an insulating layer.

At first glance you might reason that because we are dealing with thin films, the stresses should be low. To convince yourself that this is not the case, consider a bimetal such as that in Example 3.1. If we change the thickness of both sides of the bimetal proportionately, there will be no change in the residual stress!

The problem of measuring residual stress for different processing techniques and material combinations is both an important and challenging one. An elegant experimental technique by which to measure the residual stress in actual films deposited on silicon or other combinations has been developed.[*] This method depends on the fact that when a film such as aluminum is deposited on a circular silicon wafer, the wafer is bent (curved) inward or outward depending on the residual stresses. For example, if we deposit the aluminum at an elevated temperature and neglect any temperature gradient, we would expect the disc to curve inward toward the aluminum layer because of the greater coefficient of expansion (actually contraction in this case) of the aluminum.

From mechanics, we can calculate the film stress S_f as follows:

$$S_f = \frac{E_s \cdot t_s^2}{6 t_f R} \tag{22-1}$$

where

E_s = modulus of elasticity of substrate

t_s = thickness of substrate

t_f = thickness of film

R = radius of curvature

---

[*]P. A. Flinn, D. S. Gardner, and W. D. Nix, "Measurement and Interpretation of Stress in Aluminum-Based Metallization as a Function of Thermal History," *IEEE Transactions on Electron Devices*, Vol. ED-34, No. 3 (March 1987), p. 689.

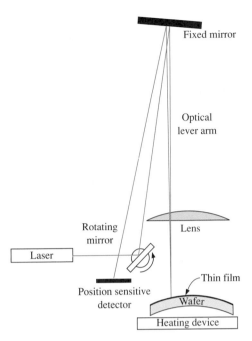

FIGURE 22.5 Diagram of stress-measuring apparatus. In order to construct a more compact apparatus, the optical path has been folded by the addition of a fixed mirror.

[P.A. Flinn, D.S. Gardner, and W.D. Nix, "Measurement and Interpretation of Stress in Aluminum-Based Metallization as a Function of Thermal History," *IEEE Transactions on Electron Devices*, Vol. ED-34, No. 3 (March 1987). © 1987 IEEE. Used by permission.]

The radius of curvature is measured accurately by a laser beam and a position-sensitive detector (Figure 22.5). To appreciate the precision of the equipment, consider that when it is used to measure the bending of the wafer under the force of gravity, it gives the difference in curvature $(1/R)$ as -0.0063 m^{-1}.

Figure 22.6 shows the application of the equipment to a typical case of an aluminum–1%-silicon film with a 1 μm film on a wafer 100 mm in diameter. There is a residual stress of 50 MPa (7245 psi) tension as a result of the sputtering. Because of the lower coefficient of expansion of the substrate, the stress falls upon uniform heating and becomes compressive to 100°C (212°F). As the temperature is raised the strain becomes increasingly plastic, and although heating to 400°C (752°F) produces about 1% strain because of differences in expansion, the elastic strain is relieved. Therefore the residual stress decreases. As the sample is cooled, tensile stress builds up rapidly for about 20°C (36°F), then plastic strain as well as elastic strain develop until the stress of about 280 MPa (40,580 psi) is reached at room temperature.

On the second cycle the heating curve begins at a higher value of tensile stress, but because of plastic deformation at higher temperatures, we reach the same value at 450°C (842°F). The second cooling curve is a duplicate of the first, as would be expected.

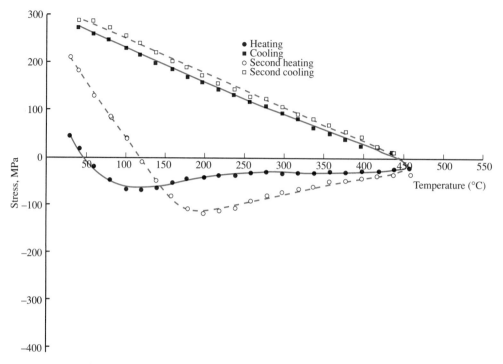

FIGURE 22.6 Stress as a function of temperature for a sputtered film of Al–1%-Si deposited over thermal oxide on a silicon wafer with no substrate heating

[P. A. Flinn, D. S. Gardner, and W. D. Nix, "Measurement and Interpretation of Stress in Aluminum-Based Metallization as a Function of Thermal History," *IEEE Transactions on Electron Devices*, Vol. ED-34, No. 3 (March 1987). © 1987 IEEE. Used by permission.]

This technique makes it possible to evaluate the stress build-up in different thin films and with different processes, such as PVD vs. CVD. And other interesting effects, such as precipitation in aluminum films containing copper (to reduce electromigration), can be investigated.

22.7 Ceramic Superconductor Processing

Because of the great research activity under way and the many potential engineering uses of ceramic superconductors, it is important to discuss the preparation and processing of these materials, even though they are still in development. The media indicate that a high school student in possession of a few chemicals and equipped with a microwave oven can produce them. However, to prepare *in useful shapes* material that can conduct high current density is another problem entirely.

The best-known 1, 2, 3 superconductors of the $[(RE)Ba_2Cu_3O_{7-x}]$ $(RE = $ rare earth) type have been prepared by powder methods, as described earlier, and even as thin films and single crystals. The typical brittleness of ceramic oxides prevents the processing used for ductile materials, and encapsulating

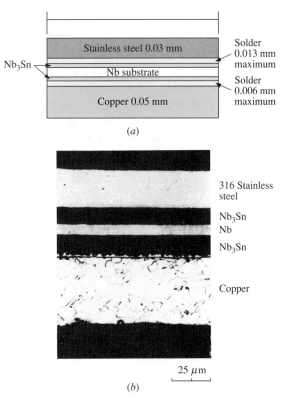

FIGURE 22.7 Cross section of a superconducting tape. (*a*) The tape design in diagrammatic form; Nb₃Sn is made at the surfaces of an Nb inner substrate. Cladding layers of stainless steel and Cu are then soldered to the Nb₃Sn surface. (*b*) Cross section of an actual tape. The cross section has been etched to permit differentiation of the components.

[E.B. Forsyth, Brookhaven National Laboratory and *Science*, vol. 242, page 397, Oct. 21, 1988. c 1988 by the AAAS.]

techniques (Figure 22.7) are also needed for the niobium materials, which are essentially brittle intermetallic compounds.

For ceramic superconductors, after pressing or extruding with an organic binder or casting on a tape substrate, the product is sintered in the 900–975°C (1652–1787°F) range because melting occurs at 980°C (1796°F). An oxygen flow is used to oxidize the organic binders and to develop a high oxygen content close to that of $Ba_2YCu_3O_7$. Wires can be produced as a composite containing the superconductor and a ductile metal matrix.

A major problem with superconductor specimens involves the maximum current density that can be attained. Above a critical current density J_c, the superconductivity breaks down. In polycrystalline specimens, J_c is notably lower than in a single crystal because of the low J_c at grain boundaries.

The unique challenges of processing superconducting metals warrant separate treatment, which appears in the following section.

22.8 Processing of Magnets with Superconducting Wires

Although low temperatures must be maintained in the regions of the equipment that contain the superconducting wires, other portions of the equipment can be at room temperature. For example, motors and generators have been built in which only the stationary-field section contains superconducting wires that require cooling. In another very significant use, medical equipment for *nuclear magnetic resonance* (NMR) scanning of body structures provides greater detail than x rays and (unlike x rays) has no harmful effects.

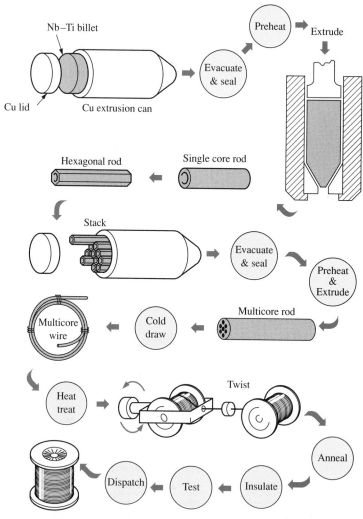

FIGURE 22.8   Production of filamentary niobium titanium–copper composite wire. (Based on sketches by A. Woolcock.)

[H. N. Wilson, *Superconducting Magnets*, Clarendon Press, Oxford, England 1983, p. 290. Reprinted by permission.]

Let us follow the processing, from the production of wire and its forming into coils to the cryogenic engineering and the construction of the final machine.

Niobium–tin has a higher critical temperature θ_c (at which the alloy reverts to normal resistivity), but niobium–titanium alloys with about 46.5 to 50 wt.% Ti are more ductile and can be fabricated into fine wires (Figure 22.8).

Ingots (100–250 mm in diameter) are cast from melts made under controlled atmospheres with less than 1000 ppm of O, C, and N. However, small amounts of tantalum may be present from the ore and, because of similarity to niobium, have no deleterious effect. The ingot is machined and sealed in a can of high-purity copper after careful cleaning. After preheating to 550°C (1022°F), the billet is extruded to 1/20 of the original section. The extrusion is then cold-drawn to a hexagonal rod. After cleaning, the single wire rods are stacked together in another copper can and extruded again.

Precipitation heat treatments and twisting to decouple the filaments are applied at the desired stages in drawing. The wires are finally extruded to a hexagonal cross section in order to obtain maximum density of packing.

The use of the wires in magnets for NMR chambers is shown schematically in Figure 22.9. The operation of this device is an interesting application of the principles introduced in Chapters 20 and 21. In paramagnetic elements such as hydrogen, fluorine-19 and sodium-23, the atoms align with a magnetic field, and hydrogen is used most commonly. The superconducting magnets are used to produce this alignment. Then if an alternating magnetic field (such as a radio frequency) is superimposed, the alignment shifts with the release of photons at the locations of hydrogen atoms. Because the concentration of hydrogen atoms varies with location in the body, the variation in emission can be used to map out the portion of the body structure that is of interest (Figure 22.10). A pulsed combination transmitter/receiver is used to scan the body.

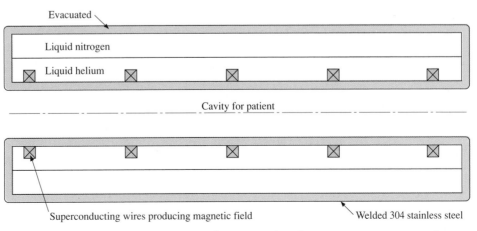

FIGURE 22.9   Schematic section of an NMR (nuclear magnetic resonance) unit. (Vents and the supplies of liquid helium and liquid nitrogen are not shown.)

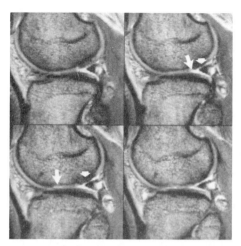

FIGURE 22.10 NMR images of a human knee as thin, continuous slices. The long arrows point to cartilage degeneration, the short arrows to cartilage loss and irregularity.
[Siemens Medical Systems, Iselin, NJ]

The advantage the superconducting magnets confer is an extremely uniform magnetic field—a maximum variation of 10 parts per million over a 10-cm diameter area at a field strength controlled between 0.5 and 2 teslas. One of the reasons for this uniformity is the constant current flowing in the magnet coils. An initial voltage is applied to build up the desired current, and then the potential is switched off—that is, shunted out—and a constant current flows. The other advantage of using the special magnets is that they are small and can be located closer to the patient while providing a substantial magnetic field.

22.9 Processing of Ceramic Magnetic Materials

The methods for producing ceramic magnets exploit many of the same principles as the manufacture of engineering ceramics. For example, the desired structure (such as a spinel or garnet) is developed by careful chemical production of the oxides, mixing, and sintering. In addition, control of grain size is important in attaining desired domain size. Other techniques include controlled cooling from high temperatures and quenching.

To illustrate the effects of heat treatment, magnesium ferrite transforms at high temperatures to the normal spinel structure as the thermally excited Mg^{2+} ions occupy the tetrahedral sites. If the sample is quenched, this structure is retained and the saturation magnetic moment is 2.23 Bohr magnetons (BM). With slow cooling, the magnesium reverts to the octahedral sites, giving a saturation magnetic moment of 1.28 BM. The lower value is caused by displacement of iron ions from sites where they produce a magnetic moment, as we noted in Chapter 21.

The production of magnetic bubble materials involves elegant processing of magnetic garnets. A thin (5-μm) film of a magnetic garnet is deposited on a nonmagnetic crystal substrate. The crystal structure of the film is a continuation (matching) of the base crystal (*epitaxial* deposition, as discussed in Sections 22.5 on electrical materials and 22.12 on optical materials). However, the slight difference in thermal expansion results, upon cooling, in a stress that produces magnetization perpendicular to the plane of the film. This leads to the magnetic domains and the magnetic bubble shown in Chapter 21.

In the hexagonal ferrites, which are widely used as permanent magnets, the compounds are produced by sintering the proper oxides at 1300°C. To obtain magnetization in a preferred direction, a magnetic field is applied during pressing and sintering. Particles rotate so that the direction of easy magnetization is parallel to the field. At the same time, the grain size is controlled; a change in grain diameter from 10 μm to 1 μm increases the coercive force from 100 to 2000 oersteds. This is due to the fact that at the smaller grain size, the grain boundaries lie within each domain and impede movement (growth to align with a new imposed field).

22.10 Processing of Ceramics for Other Electrical Uses

Although many minerals occur in nature, they are generally synthesized when high purity is needed as in electrical materials. Let us examine the production of barium titanate (perovskite structure).

Proper amounts of barium carbonate and titania are mixed and heated (calcined). There are two steps to the reaction:

1. $2BaCO_3 + TiO_2 \rightarrow Ba_2TiO_4 + 2CO_2$
2. $Ba_2TiO_4 + TiO_2 \rightarrow 2BaTiO_3$

The second step requires a higher temperature because the reactivity of the Ba_2TiO_4 is lower. If a small excess of TiO_2 is added, it acts as an inhibitor to crystal growth. The finer-grained material has a lower dielectric loss. When a lead containing perovskite is desired, special precautions are needed because of the volatility of lead oxide. The components are embedded in precalcined material to reduce the loss due to volatilization.

22.11 Processing of Optical Materials — Optical Fibers

Glass is the principal optical material, and we have already discussed the processing of glass in general. It is worthwhile here, however, to investigate the special care taken in the processing of optical fibers. Very low losses have been obtained in these fibers as a result of (1) careful control of the analysis of the fiber and (2) proper application of cladding. Both objectives are met via one of several processes for *vapor deposition* in fiber-optics.

As an example, let us consider the Corning outside vapor deposition process (Figure 22.11). A carrier gas (oxygen) is saturated with the desired reactants ($SiCl_4$ and $GeCl_2$). Passing the vapors through a burner forms SiO_2 and GeO_2 soot, which deposits on a rotating mandrel. The core of a cylinder (called a blank), which is many times the desired final diameter of the fiber, is built up first. Then the reactants are changed to produce the desired cladding—either pure SiO_2 or fluorosilicate with a lower refractive index. Other processes for blank manufacture are also shown in the figure. In any case, the blank is transferred to the heating and drawing apparatus shown. The preform is heated to 1150–1250°C (2102–2282°F), at which temperature the glass is soft enough to be drawn into a fiber. After fiber of the desired diameter is pulled through the die, it is coated with an organic polymer to protect it against nicks during further processing and installation. It is important to remember that if the glass temperature is too high during coating, degradation of the polymer will occur.

The technology of the process has advanced to the point where it pro-

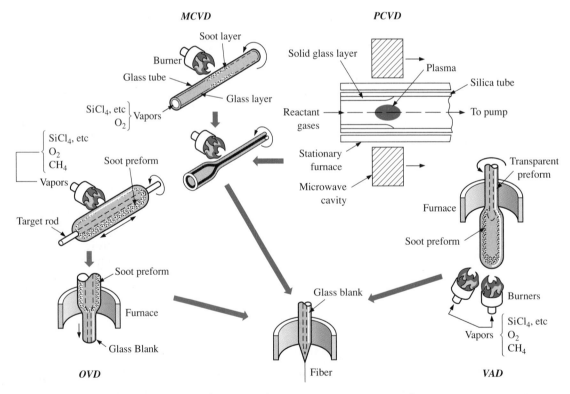

FIGURE 22.11   Schematic of OVD, VAD, MCVD, and PCVD processes: OVD— outside vapor deposition; VAD—vapor axial deposition; MCVD—modified chemical vapor deposition; PCVD—plasma chemical vapor deposition

[*Corguide*, Corning Glassworks Technical Report, "Optical Waveguide Fiber Fabrication and Drawing," *G. Kar*, TR 42/August 1984. Reprinted by permission of Corning, Inc.]

vides drawing and coating speeds of 1–10 m/sec. Lengths of 15–250 km are drawn from a single blank, maintaining satisfactory cladding!

The fiber is then encased in a cable (see Figure 20.39) that has two functions: It acts as a moisture barrier to prevent access of OH$^-$ ions to the fiber, and it provides mechanical protection.

A central member of steel- or glass-reinforced plastic is used to protect against impact loading and to allow flexing of the cable. A buffer jacket is used for added protection against outside stresses. The strength member is the primary load-bearing member; it is made of metal, glass yarn, or synthetic yarn. An outer jacket made of an extruded polymer such as PVC, PE, or PTFE protects the entire core.

Splicing techniques have been highly developed. They involve exposing the fibers, cleaving in a fixture with a diamond tool to produce fresh surfaces, and joining by an epoxy cement with the same refractive index or fusing with an electric arc.

The effect of stress on the optical characteristics of the fibers is pronounced. For this reason, the fibers are coiled in a helix in the buffer tube. When the cable length changes with temperature, the fiber merely coils or uncoils in the loose material in the buffer tube without developing elastic strain.

22.12   Special Problems in Emitter Materials

We should consider briefly the production of light-emitting diodes and laser diodes for electro-optics — that is, the changing of an electrical signal to an optical one. We have already pointed out that there is a "window" of desirable wavelengths for minimum attenuation in the silica fiber. The problem is to get the emitter to provide optical bits in this wavelength band.

It turns out that gallium arsenide is an excellent material but quite nasty to fabricate because of the toxic nature of the chemicals. It is simpler, though still demanding, to grow indium phosphide crystals (Figure 22.12) as a substrate and then deposit a thin film of InGaAsP by a process called *epitaxial growth*. The InP and GaAs have closely matching crystal structures. Therefore, when we take a crystal of InP and expose the surface to GaAs vapors, the crystal structure of the underlying layers is reproduced. This epitaxial growth of GaAs can be developed either by exposing the InP to a melt doped with GaAs or by molecular beam epitaxy. In the latter case, the desired composition is deposited by directing a beam of charged particles of the desired composition at the substrate by an electrical field. As an example, Matsushita Electric has developed an optical transmitter chip consisting of a driving circuit and a laser of layers of InP and InGaAsP. It emits light at 1.3 μm, an ideal wavelength (Figure 22.13). The companion receiver consists of layers of InGaAs and InP.

22.13   Unit Processing Operations

You will recall from Chapters 6 (metals), 12 (ceramics) and 15 (polymers) that we developed processing as a series of unit operations. Because the electrical, magnetic, and optical materials are composed of all of these classes of materials,

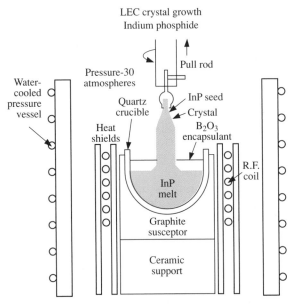

FIGURE 22.12   Schematic of a laboratory-size LEC growth system developed at AT&T Bell Laboratories

[Reprinted with permission from *Journal of Metals*, Vol. 36, No. 8, Aug. 1984, p. 56, a publication of The Metallurgical Society, Warrendale, Pennsylvania.]

the general processing principles still apply, although the specific terminology may be confusing.

The unit operations we discussed are casting, deformation, particulate processing, joining, machining, thermal treatment, and surface treatment. You will recognize that each of these figured in some form in the processing we have discussed in this chapter. More important, the role of processing defects assumes new dimensions in component performance. For example, in microelectronics a 1 μm defect can penetrate a deposit layer completely; a defect of the same size would be insignificant in an I-beam.

The key to quality, then, is identification and testing for defects and the ability to control the processing at microscopic levels. This indicates a heavy dependence on computer-controlled systems.

EXAMPLE 22.2   *[ES/E]*

Review the unit processing operations that were incorporated into the brief description of optical fibers in Section 22.11.

Answer   Hot *deformation* and *surface treatment* are certainly apparent in the cladding and drawing sequence. The adhesion between the cladding and fiber, raising the possibility that flaws will occur, suggests problems characteristic of any *joining* process. Some of these flaws may be healed during the drawing operation.

Most of the operations take place at an elevated temperature, so the variables involved in *thermal treatment* are important. For example, the drawability and interdiffusion of the glasses are temperature-sensitive, and it may be necessary to control the atmosphere. The cooling rate is also important, because residual stress influences the optical performance.

Summary

The closely controlled chemistry required in the processing of electrical, magnetic, and optical materials imposes new constraints on traditional processing methods. This was demonstrated by the processing sequence in the

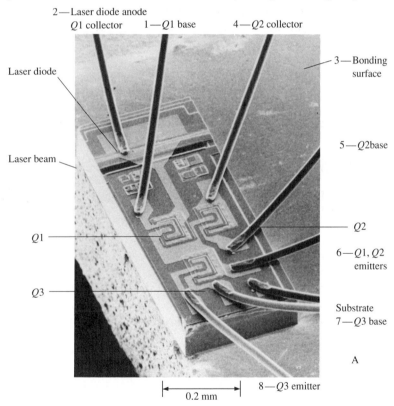

FIGURE 22.13   An indium phosphide integrated transmitter chip, shown in a scanning electron micrograph, measures 910 by 350 μm; its laser at the rear of the InP chip emits a light beam to the left; the three transistors that drive and modulate the laser are in the forward part of the chip (Q_1, Q_2, Q_3).

[J. Shibata and T. Kajiwara, "Optics and Electronics Living Together," *IEEE Spectrum*, p. 35, February 1989. © 1989 IEEE]

manufacture of MOS and CMOS microelectronic devices. Thin films are particularly important; techniques such as evaporation, sputtering, and chemical vapor deposition are required.

Even when extreme precautions are taken, defects can occur. Difficulties in diffusion control are apparent. Not so apparent is the bowing of components as a result of the high-temperature application of films of different chemistry and the development of residual stress upon cooling.

Superconductor processing presents its own set of unique problems. Some metal superconductors, such as niobium–titanium alloys, are widely used because they are relatively easy to fabricate. Fine wire in superconducting magnets is used in medical equipment for scanning tissue (nuclear magnetic resonance).

In ceramic superconductors such as $YBa_2Cu_3O_7$, control of anion vacancies (O^{2-}) during sintering is important. In conventional ceramic magnets, the entire particulate processing may be carried out in a magnetic field in order to maintain maximum domain alignment.

In the processing of optical fibers, special vapor deposition methods are employed in the cladding operation. Jacketing with polymers is necessary to provide mechanical strength for handling. Emitters in particular require special attention in the growing of diodes that are layered and in the production of single crystals of materials that are difficult to handle, such as GaAs and InP.

Definitions

Chemical vapor deposition (CVD)   Changing the surface chemistry of an electronic device by exposing it to metallic hydrides.

Epitaxial growth   Growth of a new layer of different chemistry on a crystal that has the same crystallographic orientation as the underbody (substrate).

Ion beam sputtering   Redeposition, as a film, of gas ions produced by a focused ion gun.

Metal-oxide semiconductor (MOS)   A transistor in which the gate metal is insulated from the base by a thin oxide film.

Nuclear magnetic resonance (NMR)   Release of photons from paramagnetic materials such as hydrogen when they are exposed to an alternating magnetic field generally produced by superconducting magnets.

Physical vapor deposition (PVD)   Melting or atomization of a material in a vacuum chamber and deposition of the vapor on another surface.

Purple plague   An aluminum and gold corrosion product that can occur at semiconductor junctions.

Vapor deposition in fiber optics   The methods used to clad optical fibers with a glass of lower index of refraction are descriptively classified according to the technique of application. They include modified chemical vapor deposition (MCVD), outside vapor deposition (OVD), plasma chemical vapor deposition (PCVD), and vapor axial deposition (VAD).

Problems

22.1 *[ES/EJ]* Explain why, in doping a semiconductor chip by diffusion, it might be necessary to expose the layer to more than one temperature in order to achieve the desired chip characteristics. (Sections 22.1 through 22.6)

22.2 *[ES/EJ]* The production of semiconductor devices requires modification of both films and the local surface chemistry. Give several reasons why PVD or CVD methods might be more adaptable as films for the modification of surface chemistry. (Sections 22.1 through 22.6)

22.3 *[EJ]* In the discussion of defects in microchips, the text emphasizes problems arising from difficulties in diffusion control and from residual stresses. What other defects might occur in the semiconductor devices? (Sections 22.1 through 22.6)

22.4 *[ES/EJ]* The metal-oxide semiconductor shown in Figure 22.1 is finally encapsulated in glass, although this is not shown. What would you consider to be the most important characteristics of the glass? (Sections 22.1 through 22.6 and Chapter 10)

22.5 *[ES]* Show that units of MPa would be correct in the equation for the stress in the surface film of a semiconductor, and indicate the significance of each variable in the equation (Equation 22-1). What is the physical meaning of the difference in the radius of curvature (-0.0063 m^{-1}) that results from the force of gravity? (Sections 22.1 through 22.6)

22.6 *[ES/EJ]* Cite some of the advantages and disadvantages of having films that are either very strong or very weak relative to the residual stress that is due to the processing of semiconductors. (Sections 22.1 through 22.6)

22.7 *[ES/EJ]* In making magnetic sealer sticks for refrigerators, the magnetic material is placed in a thermoplastic in the liquid state, and a magnetic field is maintained while the plastic sets. Why? (Sections 22.7 through 22.13)

22.8 *[EJ]* Outline a general method that could be used to produce a ceramic magnet shape. Include processing methods and subsequent thermal treatments if necessary. (Sections 22.7 through 22.13)

22.9 *[EJ]* Figure 22.7 indicates that solder is used to join the individual layers in the superconducting tape. What is the purpose of the solder, and what processing precautions should be undertaken to ensure a good joint? (Sections 22.7 through 22.13 and Section 6.15)

22.10 *[EJ]* What is the purpose of the copper used in the production of niobium–titanium wire for superconductors? (Sections 22.7 through 22.13)

22.11 *[ES/EJ]* In the processing of ceramic magnetic ferrites, what are the effects of material purity, particle size, and cooling rate after sintering? (Sections 22.7 through 22.13)

22.12 *[ES]* In Section 22.10, a two-step process to produce barium titanate is discussed. Calculate the weight of $BaTiO_3$ obtained from 1 kg of $BaCO_3$ with no excess TiO_2. (Sections 22.7 through 22.13)

22.13 *[EJ]* Explain how a stress can modify the operating characteristics of an optical fiber. (Sections 22.7 through 22.13)

22.14 *[EJ]* Figure 22.12 shows a schematic diagram for growing indium phosphide crystals. Explain the purpose of each of the components listed in the diagram. (Sections 22.7 through 22.13)

22.15 *[EJ]* Although machining is an important unit process, it is less significant in the manufacture of electrical, magnetic, and optical materials. Why? (Sections 22.7 through 22.13)

CHAPTER 23

Materials Selection and Design

The selection of materials for lightweight bike frames is a fascinating, competitive process. Some of the most advanced materials substituted for steel tubes are metal matrix composites with boron or silicon carbide fibers in an aluminum matrix. The aluminum composite is four times stronger and three times stiffer than aluminum alloys. However, the joints are still made of metal-steel or titanium. The long range objective is to attain a frame weight of 1 pound.

23.1   Overview

There are two broad concepts that are important in this chapter. The first is the *overall* relationship among *materials, processes,* and *service conditions.* We will discuss the role of service conditions first, along with their effect on performance, and then take up materials and processes. (In practice, you should try to keep all three in mind simultaneously as you attempt to reach an optimal solution.)

The second issue we will take up in this chapter is the *special considerations* that influence the process of selection. These topics, which are discussed in separate sections, are (1) secondary processing, (2) strength–density ratio, (3) the reading and use of specifications, (4) safety and reliability, (5) quality control, (6) computer data bases, (7) prototypes and experimentation, and (8) cost analysis for a component, including the recycling of scrap.

23.2   The Selection of Materials

The whole thrust of this text is to help prepare you to select the "best" material for a given application. An important, related goal is to help you analyze failures and suggest a remedy.

As an example, how do you find the best material for a pump? The wrong way is to reach for the nearest handbook and look up pump materials. The proper way is to take a deep breath and realize that what you really want is quiet, safe, reliable performance at minimum cost.

What does this performance really depend upon? The three main factors are *materials, processes,* and *service conditions.* These can be thought of as the three legs of a stool: not dealing properly with any one of them will lead to failure.

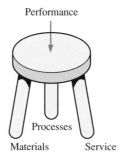

23.3   Service Conditions

Before selecting a material, process, or design for a part, we must thoroughly understand the leg representing service conditions. It is possible to obtain detailed knowledge of the service the part is supposed to deliver either by observing the actual equipment or similar equipment in service or by discussing the matter with personnel familiar with previous performance problems. At the same time, the mechanisms of potential failure—stress, corrosion, abrasion, creep, and obsolescence—or any environmental requirement should be assessed. Finally a statement of the initial cost of the component, installation and repair costs, and service life should be developed.

In cases of new equipment and new designs, a great deal of the desired data may be lacking. For example, if the pump you have been asked to design must handle a new chemical solution, corrosion data may be unavailable. Here you have two options. A part that will see limited production does not justify extensive testing, so you can consider the possible chemical reactions and use a material that is known to perform well under more severe conditions. For extensive production, however, because there is a wide cost differential among corrsion-resistant materials, you can test discs of materials in the solution under simulated service conditions and then in prototypes. (Simulated conditions must be selected with care; brief simulations may not truly represent long-term service.) In development work it is desirable to use a wide spectrum of materials. Examination of materials that fail can provide insight into the mechanism of attack and help lead to an optimal solution.

Another important reason for understanding the application is related to the location of test specimens. A separately poured or otherwise produced test specimen may have properties quite different from the material in the actual component at the location of maximum stress. Also, a residual stress condition may develop in production *or in service* and may drastically affect performance. This is an important concept, so let us proceed to examine an actual case.

A number of years ago, railroad freight car wheels of the type shown in Figure 23.1 were made of gray cast iron. In production, liquid iron was cast against a circular metal chill to produce a wear-resistant white cast-iron layer on the tread and a gray iron interior. This wheel behaved well at speeds of up to 15–20 mph, but as railroads began to operate at higher speeds, the wheels failed in large numbers. The source of failure was thermal cracking of the

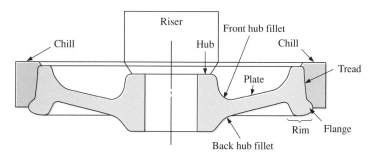

FIGURE 23.1   Cross section of a 750-lb car wheel

[Flinn, *Fundamentals of Metal Casting,* © 1963, Addison-Wesley Publishing Co., Inc., Reading, Massachusetts. Reprinted by permission of the publisher.]

brittle white iron, which was caused by the application of brake shoes on the tread surface.

After many experiments, investigators developed a new process: material combination in which steel wheels (0.8% C) were produced by pouring liquid steel into semipermanent graphite molds. The early specimens of the new wheels were simply annealed after casting and gave excellent performance compared to the old wheels. However, occasional failures under heavy braking conditions still occurred owing to catastrophic cracking (Figure 23.2). These cracks began at smaller flaws due to braking and spread across the wheel.

We found that the braking action developed high hoop tensile stress in service. What happens is that the outer layers of the rim are heated by braking to the plastic region for steel—over 1000°F (538°C). Because the steel is plastic, there is very little *elastic* strain or corresponding residual stress *while the rim is hot.* Upon cooling, however, the rim contracts and builds up hoop tension, because the underlying (cool) interior of the wheel does not contract. When the hoop tension reaches a value high enough for rapid crack propagation, catastrophic failure takes place. One remedy is to produce the wheels with residual hoop *compression in processing* by cooling the rim *before* the interior by a controlled quench from the annealing temperature. A second remedy is to use an S-shaped or parabolic cross section for the interior. Doing so provides a cushion for the radial compression that develops upon cooling after braking, and it thus results in lower hoop tension for the rim. Understanding the sources of failure made possible the development of a successful combination of material, design, and process for a product involving the safe performance of millions of freight cars.

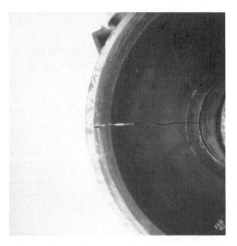

FIGURE 23.2   Radial thermal crack in a railroad car wheel due to heavy braking

[Photo courtesy G. Lazar, Griffin Wheel Co., Chicago]

23.4 Materials and Primary Processes

Throughout the text we have separated the discussions of *materials* and those of *processes* so that the interrelationship of *structure* and *physical and mechanical properties* could be described clearly. Now it is time when selecting a material for an actual component to ask you to think of processing and material at the same time. To put it bluntly, the most important property of a material is its processability. For example, one of the great advantages the metals offer is that, as a result of the slip systems in their structure they can be rolled into thin sheets.

To illustrate the relationship between properties and processability, we have prepared Table 23.1, which follows on the next four pages. Many compilations of mechanical properties exist, but this table represents a first attempt to tabulate the properties of materials and their processability in one place. We have already spent a good deal of effort in earlier chapters comparing the properties of materials, so now we will go on to concentrate on their processing.

Primary material processes include casting, deformation, and powder agglomeration. The casting process is probably the most widely applicable. In the metals we can cast any alloy in practically any shape, from a tiny 30-gram turbine blade to an ore crusher machine base weighing 150 tons. In the ceramics we can cast large glass shapes such as a 200-inch telescope lens or huge structures such as dams of concrete, which is a ceramic composite. In the case of polymers few really large castings are made, but a very high tonnage of small parts is produced by injection molding, which is essentially die casting. Here the dimensions are limited by the die size. In the composites the casting of fiber–liquid polymer mixtures by injection molding is also important.

The *deformation methods* such as rolling and forging can be used for all metals and alloys that offer sufficient ductility. Glasses are uniquely suited to a great variety of deformation methods. However, the typical oxide ceramic is brittle and must be formed by powder methods. Thermoplastics have excellent formability.

It should be pointed out that the "max" dimension given for *sheet* material is the *width* and that practically any *length* may be obtained. This is also true for other shapes of constant cross section, such as the rail, rod, plate, and honeycomb designs.

Powder methods applied to metals must take into account possible oxidation in the sintering step. Special atmospheres are needed for aluminum. The dimensions are limited by the die size. The ceramics are usually prepared by some powder method. Atmosphere control in sintering is less important with ceramics than with metals, because oxides are generally used as the starting materials.

(text continues on page S-244)

TABLE 23.1   Properties and Processes for Engineering Materials

	Properties*							
Metals	Y.S.,[a] MPa	E,[a] GPa	el., %	Hard. HV	Start Creep, °C	K_{IC} MPa $(m)^{1/2}$	Th. exp., °C $\times 10^{-6}$	Dens. Mg, m^{-3}
Aluminum alloys	100 / 630	75	2 / 40	50 / 160	>150	20 / 45	24	2.6 / 2.9
Magnesium alloys	80 / 300	41 / 45	3 / 20	50 / 90	>150	—	—	1.9
Copper alloys	60 / 960	120	1 / 50	50 / 300	>300	—	18	7.5 / 9.0
Nickel alloys	250 / 1600	200	1 / 50	75 / 350	>500	—	13	7.8 / 9.2
Titanium alloys	180 / 1320	115	6 / 30	115 / 380	>500	50 / 110	—	4.3 / 5.1
Zinc alloys	160 / 420	43 / 96	1 / 10	80 / 90	>100	—	—	5.2 / 7.2
Semiconductors (Si)	—	107	—	—	—	—	—	2.3
Carbon and Low-Alloy steels	260 / 1300	200	5 / 20	150 / 600	>500	50 / 150	14	7.8
Stainless steels								
Ferritic	600	200	20	150 / 300	>600	—	11	7.5 / 8.1
Austenitic	600	200	40	150 / 300	>700	—	17	7.8
Martensitic	400	200	2 / 10	300 / 1000	>600	—	11	7.8
Gray iron	220 / 420	80 / 140	0.05 / 2	120 / 300	>500	6 / 20	—	6.9 / 7.5
Ductile iron (nodular)	420 / 1400	140 / 170	2 / 20	120 / 350	>500	—	—	6.9 / 7.5
Traditional Ceramics								
Soda Glass	3600	69	0	>500	>500	0.7	7	2.5
Silica (fused)	69	69	0	>500	>800	—	0.05	2.6
Pottery — Whiteware	—	—	0	>500	>700	—	6	2.5
Brick Materials	—	—	0	>500	>700	—	6	2.5
Abrasives (Al_2O_3)	—	390	0	>1000	—	—	9	2.5 / 3.5
Advanced Ceramics								
Alumina–Zirconia	4000	390	0	>1000	>2000	15	8	3.9
Graphite (pyrolitic)	2000 / 2750	260 / 380	0	<10	>3000	—	1 / 2.7	2.2
Silicon carbide	2500 / 4000	450	0	>1000	>2000	3	4.7	2.2 / 2.5
Silicon nitride	8000	305	0	>1000	>1500	4 / 5	—	3.2
Diamond	50,000	1000	0	>3000	>3000	—	1.2	3.5

*Table footnotes are on page S-242.

	Casting		Deformation		Powder Methods	
	Thickness range, mm	Max. dim., mm	Thickness range, mm	Max. dim., mm[b]	Thickness range, mm	Max. dim., mm[c]
Aluminum alloys	1–70	1000	0.02–50	1000	1–25	120
Magnesium alloys	1–70	1000	0.02–50	1000	n.a.	
Copper alloys	1–140	2000	0.02–50	1000	1–25	120
Nickel alloys	2–140	2000	0.02–50	1000	1–25	120
Titanium alloys	3–25	300	0.05–20	1000	n.a.	
Zinc alloys	1–70	300	0.05–20	—	n.a.	
Semiconductors (Si)	10–200	400	—	—	n.a.	
Carbon and Low-Alloy steels	3–300	5000	0.05–300	2000	1–25	120
Stainless steels						
Ferritic	1–300	5000	0.05	—	1–25	120
Austenitic	—	—	—	—	—	—
Martensitic	1–300	5000	0.05	—	1–25	120
Gray iron	1–300	5000	n.a.		1–25	120
Ductile iron (nodular)	1–300	5000	n.a.		1–25	120
Soda Glass	1–70	5000	0.02–50	1500	n.a.	
Silica (fused)	—	—	0.04–30	200	n.a.	
Pottery—						
Whiteware	1–70	3000	1–70	3000	—	—
Brick Materials	25–100	500	25–100	500	25–100	500
Abrasives (Al$_2$O$_3$)	n.a.		n.a.		—	—
Alumina–Zirconia	n.a.		n.a.		1–25	250
Graphite (pyrolitic)	n.a.		n.a.		n.a.	
Silicon carbide	n.a.		n.a.		1–25	250
Silicon nitride	n.a.		n.a.		1–25	250
Diamond	n.a.		n.a.		n.a.	

TABLE 23.1 *(Continued)*

Polymers	Y.S.,[a] MPa	E,[a] GPa	el., %	Hard. HV	Start Creep, °C	K_{IC} MPa (m)$^{1/2}$	Th. exp., °C × 10^{-6}	Dens. Mg, m^{-3}
Polyethylene HD	$\frac{14}{28}$	0.8	$\frac{15}{100}$	<30	50	2	180	0.95
Polystyrene	49	$\frac{3.2}{3.4}$	$\frac{1}{2}$	<30	180	2	—	$\frac{1.0}{1.1}$
Nylon	80	$\frac{2}{4}$	60	<30	220	3	—	$\frac{1.1}{1.2}$
Polycarbonate	62	2.5	62	<30	275	$\frac{1}{2.5}$	—	$\frac{1.2}{1.3}$
PVC	41	2.8	$\frac{2}{30}$	<30	150	—	—	$\frac{1.3}{1.6}$
Kevlar	3000	100	2	<30	—	—	—	1.45
Melamine	48	10	0	<30	265	—	15	1.5
Epoxy	69	7	0	<30	350	0.4	72	$\frac{1.1}{1.4}$
Rubbers	30	$\frac{0.01}{1}$	5	<30	—	—	670	$\frac{0.8}{0.9}$
Composites								
CFRP (graphite)	650	$\frac{70}{200}$	1	<30	350	$\frac{30}{45}$	—	$\frac{1.4}{2.2}$
GFRP (epoxy)	$\frac{100}{300}$	$\frac{7}{45}$	1	<30	350	$\frac{20}{60}$	—	1.8
WC-Co cermet	—	$\frac{400}{530}$	1	>1000	800	15	9	$\frac{11}{12.5}$
Concrete Comp. (cement)	$\frac{20}{65}$	$\frac{45}{50}$	0	>500	—	$\frac{10}{15}$	—	$\frac{2.4}{2.5}$
Wood grain (parallel)	—	$\frac{9}{16}$	0	<30	—	$\frac{0.05}{1}$	—	$\frac{0.4}{0.8}$
Wood grain (perpendicular)	—	$\frac{0.6}{1.0}$	0	<30	—	$\frac{1}{9}$	—	$\frac{0.4}{1.8}$

[a]Divide by 6.9 MPa by 6.9 × 10^{-3} to obtain psi and GPa by 6.9 × 10^{-6} to obtain psi.
[b]This is sheet width, not length.
[c]Limited by die size.
Abbreviations:

Y.S. = Yield strength (usually 0.2% offset), $\frac{MPa}{6.9}$ = psi × 10^3

E = Young's modulus of elasticity, $\frac{GPa}{6.9}$ = psi × 10^6

% el. = % plastic elongation at rupture
HV = Vickers hardness
Start creep = Temperature (°C) at which creep becomes important. For polymers the softening temperature is used; it is above the creep temperature.
K_{IC} = Plane strain fracture toughness, MPa$\sqrt{m}$ = MNm$^{-3/2}$ (divide by 1.1 to obtain ksi (in.)$^{1/2}$
Th. exp. = Coefficient of thermal expansion per °C
n.a. = not applicable

	Processes, Sizes, and Shapes					
	Casting		Deformation		Powder Methods	
	Thickness range mm	Max. dim., mm	Thickness range, mm	Max. dim., mm[b]	Thickness range, mm	Max. dim., mm[c]
Polyethylene HD	1–100	1000	0.02–200	2500	n.a.	
Polystyrene	1–100	1000	0.02–200	2500	n.a.	
Nylon	1–100	1000	0.02–200	2500	n.a.	
Polycarbonate	1–100	1000	0.02–200	2500	n.a.	
PVC	1–100	1000	0.02–200	2500	n.a.	
Kevlar	n.a.		n.a.		n.a.	
Melamine	n.a.		n.a.		1–100	100
Epoxy	1–100	1000	n.a.		1–100	100
Rubbers	1–100	1000	n.a.		1–100	100
	(inj.)		(BMC)		(winding)	
CFRP (graphite)	1–10	100	1–10	100	1–100	1000
GFRP (epoxy)	1–10	100	1–10	100	1–100	1000
WC-Co cermet	n.a.		n.a.		1–100	100
Concrete Comp. (cement)	10–5000	>10,000	10–250	2500	n.a.	
Wood grain (parallel)	n.a.		n.a.		n.a.	
Wood grain (perpendicular)	n.a.		n.a.		n.a.	

23.5 Secondary Processes

Up to this point we have considered only the primary processes of casting, deformation, and powder methods in relation to type of material. Now let us consider the relationship of the *secondary material processes* (assembly and finishing) to material.

Welding

Welding can be used readily for metals in which the structures of the base metal before and after welding and the weld structure itself are *ductile.* The effect of the welding temperature as a heat treatment of the base metal must also be accounted for. In cases of low-ductility welds, the service requirements must be carefully assessed. In corrosive environments, the development of corrosion cells in the microstructure and between the welded materials should be investigated. In the ceramics, elegant welding of glass (glass blowing) is legendary. One way of splicing optical fibers is by welding. The joining of other ceramics is accomplished by sintering, wherein glassy phases are often used. Ceramic–metal and glass–metal seals can be formed by building up transition layers of material with graded coefficients of thermal expansion that fall between the values for the metal and the ceramic.

Thermoplastic polymers are very easily welded, as are thermosets that exhibit only a small amount of cross linking. Thermoplastic *composites* can be welded, but the strength of the weld zone should be investigated because of lack of continuity of the fiber.

If we expand our definition to joining processes in general, we should acknowledge that polymer adhesives are important in assembly processes.

Machining

One of the most challenging problems in material selection is the machining cost or (at times) whether it is even possible to machine the component to final shape. Let us consider several cases. The nameplate on an automobile does not have to possess high mechanical properties, so a material such as die-cast zinc or polymer is used to provide accurate detail and a surface that is easily chromium-plated without machining. On the other hand, an automobile crankshaft, which must transmit stress, may be cast or formed close to final shape, but machining is needed at bearing surfaces. In this case we need a strong, high-modulus material (to reduce deflection) that is readily machinable; either ductile iron or steel is used. In the preparation of cermet (cemented carbide) inserts for machining tools, great effort is invested to produce the insert close to final dimensions because of the high cost of grinding the final shape. Conventional, low cost, machining such as drilling or milling can generally be done on material below HB 350; above this value, grinding is usually required.

Although polymers and polymer composites are machinable, several precautions are needed. Threads have relatively low strength and wear resistance, and holes drilled through composites rupture and disturb the fiber structure near the hole. It is generally better to mold in a machinable or already-machined metal insert for highly stressed regions.

Thermal Treatment

Three important types of thermal treatment are stress relief, annealing, and heating and quenching to alter polyphase structures.

Stress-relief treatment can be used for metallic, ceramic, polymeric, and composite materials. It is necessary to heat to a temperature high enough to permit creep to occur, replacing elastic strain with plastic strain. The stress-relief temperatures for ceramics and for metals are therefore much higher than for the polymers and the polymer composites.

Annealing is used to remove the effects of cold work or to soften a polyphase structure. The temperature required is related to the recrystallization of cold-worked material or to structural changes such as the spheroidization of polyphase structures.

Heating and quenching operations have been discussed in detail. In general, to produce a hard, nonequilibrium structure, we begin by heating to develop a high-temperature structure that is then transformed at lower temperatures to develop some type of precipitation-hardened structure. Although these operations are used predominantly for metals, an interesting modification in ceramics is the transformation-toughening mechanism. Here, a metastable structure is developed by heat treatment that will transform and expand at the tip of an advancing crack, developing compressive stress (Chapter 11).

Finishing Operations

As we pointed out in Chapter 6, there are four major categories of finishing operations that may be necessary before a component is shipped.

1. *Mechanical surface modification*, such as sand blasting or buffing.
2. *Thermal surface modification*, which may be similar to thermal heat treatment but modifies only the surface. An example is the introduction of surface compressive stresses in tempered glass.
3. *Chemical surface modification* such as carburizing.
4. *Surface coating*, which often employs polymer coats such as painting.

Each of these kinds of operations can be applied equally well to materials other than metals, as shown in Example 23.1.

EXAMPLE 23.1 *[ES/EJ]*

Indicate the precautions we must bear in mind when performing the following processes.

A. Heat treatment of a polymer
B. Machining of a ceramic
C. Joining of a glass-fiber-reinforced polymer

Answers

A. As in metals, the microstructure can be controlled through heat treatment if we have more than one phase present. For example, the percentage of crystallinity can be controlled through heat treatment. We may also have to stress-relief-anneal a polymer shape that cooled nonuniformly during molding.
B. Although our earlier discussion suggests that machining cemented carbides is difficult, it is still possible. We could machine in the green or partially sintered state where the lower hardness is advantageous. In the advanced ceramics, machining of the final sintered shape can be disastrous because it induces surface flaws that decrease the material's strength and fracture toughness.
C. If we joined two fiber-reinforced polymers together with another polymer without fibers, we would expect to have different (though not necessarily lower) properties at the joint. We could weld or use an adhesive that would require testing before the joint strength could be evaluated. We would also like to know what service stresses the joint will undergo, because assemblies seldom require the same strength everywhere.

23.6 Strength-to-Density and Modulus-to-Density Ratios

These variables are important in attaining minimum weight for a component. Weight reduction used to be important principally in aircraft design, where it was commonly said that: "For every ounce of weight removed, the change was worth its weight in gold." The reasoning was that the aircraft could then carry an ounce more payload on each flight. Today, weight reduction has become desirable in practically every moving part because of the importance of saving energy.

Several important relations in material selection involve design, density, strength, and modulus.

As an illustration, let us explore two key considerations in design: how much load a component will carry and how much it will deflect. We will analyze three of the commonest designs in terms of *optimal strength* and *optimal stiffness*.

Let us consider the relationship between yield strength, σ_y, and density, ρ, for (1) a bar in simple tension, (2) a beam, and (3) a plate (Figure 23.3).

Mode of loading	Optimal strength	Optimal stiffness
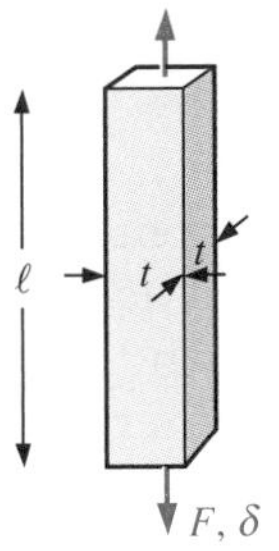 **Tie**	$\sigma_y = \dfrac{F}{t^2}$ $m = \rho \ell t^2$ $= F\ell\left(\dfrac{\rho}{\sigma_y}\right)$ **Maximize** $\dfrac{\sigma_y}{\rho}$	$\delta = \dfrac{F\ell}{Et^2}$ $m = \rho \ell t^2$ $= \left(\dfrac{\ell^2 F}{\delta}\right)\dfrac{\rho}{E}$ **Maximize** $\dfrac{E}{\rho}$
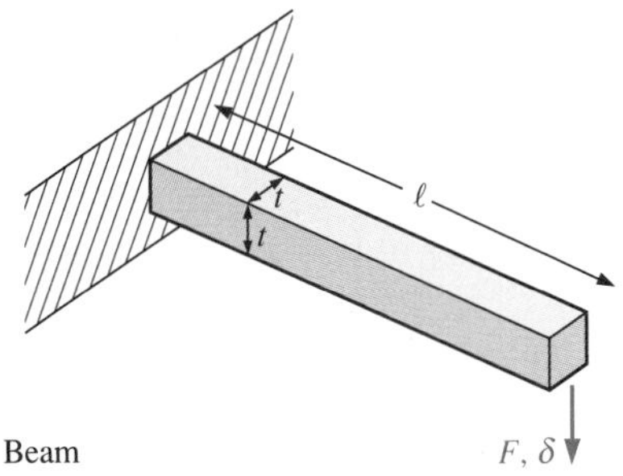 **Beam**	$\sigma_y = \dfrac{6F\ell}{t^3}$ $m = \rho \ell t^2$ $= \ell(6F\ell)^{2/3}\left(\dfrac{\rho}{\sigma_y^{2/3}}\right)$ **Maximize** $\dfrac{\sigma_y^{2/3}}{\rho}$	$\delta = \dfrac{4F\ell^3}{Et^4}$ $m = \rho \ell t^2$ $= 2\left(\dfrac{F\ell^5}{\delta}\right)^{1/2}\left(\dfrac{\rho}{E^{1/2}}\right)$ **Maximize** $\dfrac{E^{1/2}}{\rho}$
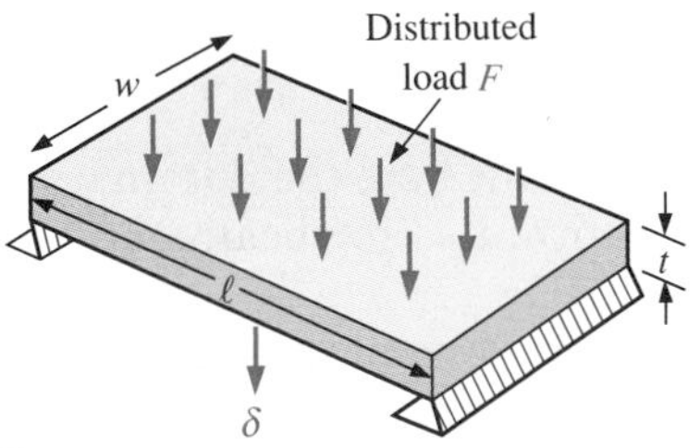 **Plate**	$\sigma_y = \dfrac{3F\ell}{4wt^2}$ $m = \rho \ell w t$ $= \left(\dfrac{3F\ell^3 w}{\delta}\right)^{1/2}\left(\dfrac{\rho}{\sigma_y^{1/2}}\right)$ **Maximize** $\dfrac{\sigma_y^{1/2}}{\rho}$	$\delta = \dfrac{5F\ell^3}{32Ewt^3}$ $m = \rho \ell w t$ $= \ell^2\left(\dfrac{5Fw^2}{32\delta}\right)^{1/3}\left(\dfrac{\rho}{E^{1/3}}\right)$ **Maximize** $\dfrac{E^{1/3}}{\rho}$

E = Young's modulus; σ_y = yield strength; ρ = density

FIGURE 23.3 Important properties required to maximize the stiffness-to-weight and strength-to-weight ratios under various loading conditions

[Reprinted with permission from *Engineering Materials*, Vol. 2, by M. F. Ashby and D. R. H. Jones. Copyright © 1980, Pergamon Press]

From Chapter 2, we recall that

$$\sigma_y = \frac{F}{t^2} \tag{23-1}$$

F = force

t^2 = area of square bar of thickness and width t

The mass $m = \rho l t^2$, where ρ = density and l = length.
Substituting for $t^2 = F/\sigma_y$ yields

$$m = Fl\left(\frac{\rho}{\sigma_y}\right) \tag{23-2}$$

Therefore, for a given geometry (component), the mass is a function of the ratio of ρ (density) to σ_y (yield strength). When we must have the lightest weight component, we select the material with the lowest ratio.

However, the ratio is different for the other designs. For the cantilever beam in Figure 23.3, for example,

$$\sigma_y = \frac{6Fl}{t^3} \qquad \text{(from mechanics)} \tag{23-3}$$

$$m = \rho l t^2 \tag{23-4}$$

$$= l\,(6Fl)^{\frac{3}{2}}\ \left(\frac{\rho}{\sigma_y^{\frac{2}{3}}}\right) \tag{23-5}$$

Therefore, for a given geometry and load, the weight of the component is directly proportional to the density divided by the yield strength to the $\frac{2}{3}$ *power*.

By a similar analysis for a plate with a uniformly distributed load, we find (Figure 23.3) that the weight is proportional to

$$\rho/\sigma_y^{\frac{1}{2}} \tag{23-6}$$

Deflection is calculated similarly, but here the dominating variable is the power of Young's modulus. (The formulas for beams and plates are taken from standard texts on mechanics.)

The analysis of other shapes can be done in like fashion. Special attention should be given to I-beam and hollow-shaft designs in which the mass is concentrated at a distance from the central axis to minimize deflection. Some of these concepts will be addressed in the problems.

EXAMPLE 23.2 *[ES/E]]*

We can refer to tables such as Table 23.1 and obtain the values of yield strength and density that we need to calculate optimal strength-to-weight ratios. Explain, however, why these calculations may provide only a first screening test in the material selection procedure.

Answer   In calculations of optimal strength, we are concerned with only two characteristics: yield strength and density. We can first assume that density values can be obtained accurately, although they may have to be calculated for a fiber composite because there is such a wide variation in the amount of fiber that might be used for a given application.

The yield strength presents another problem when we recall how it is obtained. Let us review the different classes of materials.

Metals — The yield strength is generally obtained by the 0.2% offset method in which there is a permanent set of 0.002 in./in.

Ceramics — Because of their inherent brittleness, the yield strength and fracture strength are likely to be coincident. This may also be the case in brittle metals and polymers.

Polymers — Here we may not even find a true yield strength, because many of the polymers can relax or creep at the test temperature, and tensile strengths may be the only recorded strength value.

Composites — These materials show the same sensitivity as their metallic, ceramic, or polymeric composition.

The best strength-to-weight comparisons may then be most reliable within a given material class, yet even this approach may not provide a final answer when the values for several materials are close together.

We might have to base our final selection on an *allowable stress* rather than a yield or failure stress. This would incorporate factors of reliability into the decision, such as:

1. Limits to be anticipated in test values for yield strength or tensile strength
2. The confidence in finished product quality that enables us to design closer to the tabular strength values
3. Incorporation of possible environmental degradation of properties, which might simply be loss of strength in a corrosive atmosphere

The objective is to resist the urge to plug and chug without knowing some of the limitations that the calculations are subject to.

23.7 Reading and Using Specifications

Some of the dullest reading anywhere is to be found in engineering *material specifications,* but they enable the Purchasing Department to obtain the material and shape the engineer needs. The legal obligations of the producer and consumer also depend on them. The challenge to the engineer is to examine the documents of the specification and make sure he or she understands the structure and properties of the component that will result. It is easy to become distracted by the legal phrasing and fail to realize that the specifications must be the bridge between the *engineer's* concept of structure, properties, and processing and the need for a document that *producers* can use to estimate costs and determine what quality controls will be needed.

One of the most important groups in formulating specifications is ASTM (the American Society for Testing and Materials). The full set of ASTM specifications includes over seventy volumes covering raw materials, processed materials such as bar stock, components such as cast and forged components, composites, paints, and a very important compilation of test methods. Let us analyze a specification that is more interesting than the usual type because it specifies microstructure as well as properties, chemical analysis, and mechanical testing.

Let us assume that we are interested in considering ductile iron (Chapter 8) as a material that may offer cost savings when used in a part often made of forged steel, such as a crankshaft.

When we turn to the ASTM standards we find a variety of specifications, but one general one is of special interest (Figure 23.4).

This particular standard is important because it covers different ranges of ductile iron that are available. Let us look at the different sections and comment on their meaning and on precautions to observe in writing the purchase order.

1. *Scope.* Note that this calls for substantially spheroidal graphite. We will take this up quantitatively under Section 9, "Special Requirements."
2. *Applicable Documents.* To avoid repetition of standard test methods, such as tension testing, in this specification, special volumes are devoted to test details usually preceded by the letter E, meaning to refer to ASTM Testing Procedures.
3. *Ordering Information.* This alerts the purchaser to details to be included on the order. For example, if the microstructure is to be specified, this is covered in Paragraph 3.3.
4. *Chemical Requirements.* It is interesting to note that no specific range of chemical analysis is called for unless specified by agreement. The objective here is to enable the largest possible number of producers to bid on a given contract on the basis of the *mechanical* properties of their output. Usually a given producer attempts to standardize on a few analyses. If these are outside the limits of a chemical specification, that producer does not bid.
5. *Tensile Requirements.* The available grades are summarized in Table 1 of the specification, along with the test bar design. At the present time, the pearlitic grades are generally used for crankshafts because of their combination of strength and elongation.
6. *Heat Treatment.* Here the consumer is warned that given *tensile* mechanical properties can be attained by different heat treatments but that the quenching treatment may give lower fatigue strength.

(text continues on page S-256)

ASTM Designation: A 536 – 84

AMERICAN SOCIETY FOR TESTING AND MATERIALS
1916 Race St., Philadelphia, Pa. 19103
Reprinted from the Annual Book of ASTM Standards, Copyright ASTM
If not listed in the current combined index, will appear in the next edition.

Standard Specification for

DUCTILE IRON CASTINGS[1]

This standard is issued under the fixed designation A 536; the number immediately following the designation indicates the year of original adoption or, in the case of revision, the year of last revision. A number in parentheses indicates the year of last reapproval. A superscript epsilon (ε) indicates an editorial change since the last revision or reapproval.

This specification has been approved for use by agencies of the Department of Defense to replace Military Specification MIL-I-11466B (MR) and for listing in the DoD Index of Specifications and Standards.

1. Scope

1.1 This specification covers castings made of ductile iron, also known as spheroidal or nodular iron, that is described as cast iron with the graphite substantially spheroidal in shape and essentially free of other forms of graphite, as defined in Definitions A 644.

1.2 The values stated in inch-pound units are to be regarded as the standard.

1.3 No precise quantitative relationship can be stated between the properties of the iron in various locations of the same casting or between the properties of castings and those of a test specimen cast from the same iron (see Appendix X1).

2. Applicable Documents

2.1 *ASTM Standards:*
A 370 Methods and Definitions for Mechanical Testing of Steel Products[2]
A 644 Definitions of Terms Relating to Iron Castings[2]
A 732/A 732M Specification for Steel Castings, Investment, Carbon and Low-Alloy, For General Application[2]
E 8 Methods of Tension Testing of Metallic Materials[3]
2.2 *Military Standard:*
MIL-STD-129 Marking for Shipment and Storage[4]

3. Ordering Information

3.1 Orders for material to this specification shall include the following information:
3.1.1 ASTM designation,
3.1.2 Grade of ductile iron required (see Table 1, and Sections 4 and 5),

3.1.3 Special properties, if required (see Section 9),
3.1.4 If a different number of samples are required (see Section 11),
3.1.5 Certification, if required (see Section 14), and
3.1.6 Special preparation for delivery, if required (see Section 15).

4. Chemical Requirements

4.1 It is the intent of this specification to subordinate chemical composition to mechanical properties; however, any chemical requirements may be specified by agreement between the manufacturer and the purchaser.

5. Tensile Requirements

5.1 The iron represented by the test specimens shall conform to the requirements as to tensile properties presented in Tables 1 and 2. The irons listed in Table 1 cover those in general use while those listed in Table 2 are used for special applications (such as pipes, fittings, etc.).

5.2 The yield strength shall be determined at 0.2 % offset by the offset method (see Methods E 8). Other methods may be used by mutual consent of the manufacturer and purchaser.

---

[1] This specification is under the jurisdiction of the ASTM Committee A-4 on Iron Castings and is the direct responsibility of Subcommittee A04.04 on Ductile Iron Castings.
Current edition approved June 15, 1984. Published December 1984. Originally published as A 536 – 65T. Last previous edition A 536 – 80.
[2] *Annual Book of ASTM Standards*, Vol 01.02.
[3] *Annual Book of ASTM Standards*, Vol 03.01.
[4] Available from Naval Publications and Forms Center, 5801 Tabor Ave., Philadelphia, PA 19120.

FIGURE 23.4 An ASTM standard specification

6. Heat Treatment

6.1 The 60-40-18 grade will normally require a full ferritizing anneal. The 120-90-02 and the 100-70-03 grades generally require a quench and temper or a normalize and temper, or an isothermal heat treatment. The other two grades can be met either as-cast or by heat treatment. Ductile iron, that is heat treated by quenching to martensite and tempering, may have substantially lower fatigue strength than as cast material of the same hardness.

7. Test Coupons

7.1 The separately cast test coupons from which the tension test specimens are machined shall be cast to the size and shape shown in Fig. 1 or 2. A modified keel block cast from the mold shown in Fig. 6 may be substituted for the 1-in. Y-block or the 1-in. keel block. The test coupons shall be cast in open molds made of suitable core sand having a minimum wall thickness of 1½ in. (38-mm) for the ½-in. (12.5mm) and 1-in. (25-mm) sizes and 3-in. (75-mm) for the 3-in. size. The coupons shall be left in the mold until they have cooled to a black color (approximately 900°F (482°C) or less). The size of coupon cast to represent the casting shall be at the option of the purchaser. In case no option is expressed, the manufacturer shall make the choice.

7.2 When investment castings are made to this specification, the manufacturer may use test specimens cast to size incorporated in the mold with the castings, or separately cast to size using the same type of mold and the same thermal conditions that are used to produce the castings. These test specimens shall be made to the dimensions shown in Fig. 1 of Specification A 732 or Figs. 5 and 6 of Methods and Definitions A 370.

7.3 The manufacturer may use separately cast test coupons or test specimens cut from castings when castings made to this specification are nodularized or inoculated in the mold. Separately cast test coupons shall have a chemistry that is representative of castings produced from the ladle poured and a cooling rate equivalent to that obtained with the test molds shown in Figs. 1 through 5 or Appendix X2. The size (cooling rate) of the coupon chosen to represent the casting should be decided by the purchaser. If test coupon size is not specified, the manufacturer shall make the choice. When test bars will be cut from castings, test bar location shall be agreed on by the purchaser and manufacturer and indicated on the casting drawing. The manufacturer shall maintain sufficient controls and control documentation to assure the purchaser that properties determined from test coupons or test bars are representative of castings shipped.

7.4 The test coupons shall be poured from the same ladle or heat as the castings they represent.

7.5 Test coupons shall be subjected to the same thermal treatment as the castings they represent.

8. Tension Test Specimen

8.1 The standard round tension test specimen with a 2-in. or 50-mm gage length shown in Fig. 3 shall be used, except when the ½-in. (12.7-mm) Y-block coupon is used. In this case, either of the test specimens shown in Fig. 4 shall be satisfactory.

9. Special Requirements

9.1 When specified in the contract or purchase order, castings shall meet special requirements as to hardness, chemical composition, microstructure, pressure tightness, radiographic soundness, magnetic particle inspection dimensions, and surface finish.

10. Workmanship, Finish, and Appearance

10.1 The castings shall be smooth, free of injurious defects, and shall conform substantially to the dimensions of the drawing or pattern supplied by the purchaser.

10.2 Castings shall not have chilled corners or center chill in areas to be machined.

11. Number of Tests and Retests

11.1 The number of representative coupons poured and tested shall be established by the manufacturer, unless otherwise agreed upon with the purchaser.

11.2 In the case of the Y-block, the section shall be cut from the block as shown in Fig. 5. If any tension test specimen shows obvious defects, another may be cut from the same test block or from another test block representing the same metal.

12. Responsibility for Inspection

12.1 Unless otherwise specified in the contract or purchase order, the supplier is responsible for the performance of all inspection requirements as specified herein. Except as otherwise specified in the contract or order, the supplier may use his own or any other facilities suitable for the performance of the inspection requirements specified herein, unless disapproved by the purchaser. The purchaser reserves the right to perform any of the inspections set forth in the specification where such inspections are deemed necessary to assure supplies and services conform to prescribed requirements.

13. Identification Marking

13.1 When size permits, each casting shall be identified by the part or pattern number in raised numerals. Location of marking shall be as shown on the applicable drawing.

14. Certification

14.1 When agreed upon in writing by the purchaser and the seller, a certification shall be made the basis of acceptance of the material. This shall consist of a copy of the manufacturer's test report or a statement by the seller, accompanied by a copy of the test results, that the material has been sampled, tested, and inspected in accordance with provisions of this specification. Each certification so furnished shall be signed by an authorized agent of the seller or manufacturer.

15. Preparation for Delivery

15.1 Unless otherwise specified in the contract or purchase order, cleaning, drying, preservation, and packaging of casting shall be in accordance with manufacturer's commercial practice. Packing and marking shall also be adequate to ensure acceptance and safe delivery by the carrier for the mode of transportation employed.

15.2 *Government Procurement*—When specified in the contract or purchase order marking for shipment shall be in accordance with the requirements of MIL-STD-129.

TABLE 1 Tensile Requirements

	Grade 60-40-18	Grade 65-45-12	Grade 80-55-06	Grade 100-70-03	Grade 120-90-02
Tensile strength, min, psi	60 000	65 000	80 000	100 000	120 000
Tensile strength, min, MPa	414	448	552	689	827
Yield strength, min, psi	40 000	45 000	55 000	70 000	90 000
Yield strength, min, MPa	276	310	379	483	621
Elongation in 2 in. or 50 mm, min, %	18	12	6.0	3.0	2.0

TABLE 2 Tensile Requirements for Special Applications

	Grade 60-42-10	Grade 70-50-05	Grade 80-60-03
Tensile strength, min, psi	60 000	70 000	80 000
Tensile strength, min, MPa	415	485	555
Yield strength, min, psi	42 000	50 000	60 000
Yield stength, min, MPa	290	345	415
Elongation in 2 in. or 50 mm, min, %	10	5	3

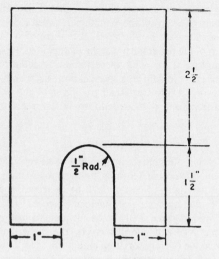

Metric Equivalents

in.	mm	in.	mm
½	12.7	1½	38.1
1	25.4	2½	63.5

NOTE—The length of the keel block shall be 6 in. (152 mm).

FIG. 1   Keel Block for Test Coupons

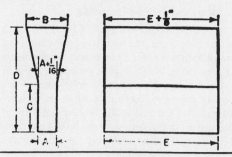

	"Y" Block Size					
Dimensions	For Castings of Thickness Less Than ½ in. (13 mm)		For Castings of Thickness ½ in. (13 mm) to 1½ in. (38 mm)		For Castings of Thickness of 1½ in. (38 mm) and Over	
	in.	mm	in.	mm	in.	mm
A	½	13	1	25	3	75
B	1⅝	40	2⅛	54	5	125
C	2	50	3	75	4	100
D	4	100	6	150	8	200
E	7	175	7	175	7	175
	approx	approx	approx	approx	approx	approx

FIG. 2   Y-Blocks for Test Coupons

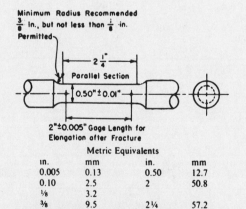

Metric Equivalents

in.	mm	in.	mm
0.005	0.13	0.50	12.7
0.10	2.5	2	50.8
⅛	3.2		
⅜	9.5	2¼	57.2

NOTE—The gage length and fillets shall be as shown but the ends may be of any shape to fit the holders of the testing machine in such a way that the load shall be axial. The reduced section shall have a gradual taper from the ends toward the center, with the ends 0.003 to 0.005 in. (0.08 to 0.13 mm) larger in diameter than the center.

FIG. 3   Standard Round Tension Test Specimen with 2-in. or 50-mm Gage Length

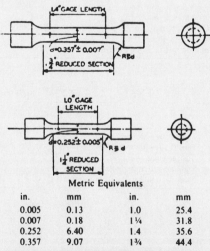

Metric Equivalents

in.	mm	in.	mm
0.005	0.13	1.0	25.4
0.007	0.18	1¼	31.8
0.252	6.40	1.4	35.6
0.357	9.07	1¾	44.4

NOTE—If desired, the length of the reduced section may be increased to accommodate an extensometer.

FIG. 4   Examples of Small-Size Specimens Proportional to Standard ½-in. (12.7-mm) Round Specimen

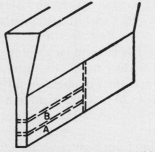

(a) ½-in. (12.7-mm) Y-block—Two blanks for 0.252-in. (6.40-mm) diameter test specimens.

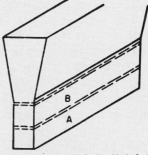

(b) 1-in. (25.4-mm) Y-Block—Two blanks for 0.50-in. (12.7-mm) diameter tension test specimens.

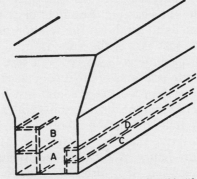

(c) 3-in. (76.2-mm) Y-Block—Two blanks for 0.50-in. (12.7-mm) diameter tension test specimens.

FIG. 5   Sectioning Procedure for Y-Blocks

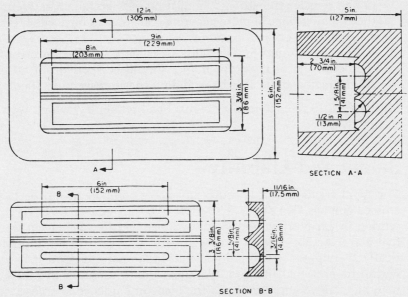

FIG. 6   Mold for Modified Keel Block

7. *Test Coupons.* The subject of test coupons is very important regardless of the material or process. In this case the specification calls for separately cast coupons from the same heat as the castings. In our experience this is inadequate to ensure performance. The spheroidal graphite of ductile iron is produced by adding a magnesium alloy to a liquid iron–carbon–silicon melt. About 0.05% Mg dissolves and then oxidizes or diffuses as a function of time. Therefore, when a series of castings is poured over, say, a 15-min period, the graphite in the last castings may not be so nearly spheroidal as in the first. We should add to the specification that test bars must be poured *after* the last casting from a given ladle. Also the designer must not assume that the properties of the casting are necessarily those of the test bar. If the designer depends on certain mechanical properties in a given location in a casting, a number of castings should be sectioned for tensile specimens to establish the relation of the strength in the critical section for the test bars of the same heat.

8. *Tension Test Specimens.* Two types of test specimens are available because two different groups of producers were engaged in the development work. Either design produces a sound test bar, but the "Y"-block series offers a range of different test sections. For example, if the actual castings are relatively light in section (about $\frac{1}{2}$ in.), the $\frac{1}{2}$-in. Y-block would be more representative than the massive keel block.

9. *Special Requirements.* Additional specifications are detailed under this heading. For example, it is often specified that the graphite shall be 80% Type I or II, as shown in Specification A 395-77 (Figure 23.5), at a given location in the test bar or test lug on the casting. Although the combination of tensile strength and elongation is a semiquantitative indication of the graphite shape, a more accurate control may be needed to ensure wear resistance or some other property.

10. *Workmanship and Finish.* This is vaguely worded. In critical castings, the definition of *injurious defect* should be spelled out by reference to ASTM radiographic standards and magnetic inspection standards.

Sections 10–15 refer to quality control and reliability, which we will discuss separately.

23.8 Safety and Liability

The rapid escalation of lawsuits involving component failure has led to increased emphasis on safety. The topic of failure analysis was given special attention in Chapter 19, but we should also consider it here in the context of material selection. It is best to begin with the realistic assumption that practically every product will fail and that the important point is to avoid failure within its projected lifetime. We know that a garden shovel will eventually rust, but can we afford to make the blade of hardened stainless steel?

The decision as to what probability of failure is tolerable often depends on the consequences. In most automobiles, certain parts are designated

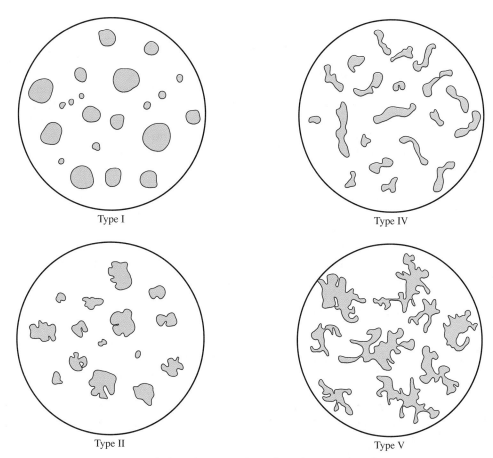

FIGURE 23.5   Suggested classification of graphite form in ductile cast iron
[From ASTM Specification A 395-77. Copyright ASTM. Reprinted with permission.]

"safety components" because a serious accident can result if they fail. Here we would find steering and braking mechanisms, for example, have either a very low failure rate or zero failure is dictated. On the other hand, auto body corrosion is expected and tolerated under severe conditions, because the cost of using stainless steel is prohibitive. Polymer composites have been substituted in some cases, although there is concern about the difficulty of recycling such materials.

23.9 Quality Control and Quality Assurance

A thorough knowledge of quality control requires a firm grounding in statistics, followed by training in the application of statistics to production problems. However, we can point out here some of the salient features of quality control in the application of materials.

The primary purpose of quality control (QC) is to make sure that a component will provide satisfactory service *with no more than a small probable percentage of failures.* The italicized portion seems a little callous and might seem the antithesis of good engineering, so let us explain by citing two examples. Very occasionally, upon opening a new can of tennis balls, we find a "dud" that has either low pressure or a defective seam. On the other hand, a defective artillery shell might explode before leaving the cannon. Obviously, the tennis ball manufacturer can set a much lower standard of inspection for defective balls than that which the shell manufacturer must impose. We note in passing that a producer *can* set the probable percentage of failures at zero. A well-known chip producer (Intel) has instituted a program called ZOD (zero outgoing defects), but even in this case failures can occur in service.

There are really two issues related to the overall performance or reliability of a component. We can best explore them by considering a simple example, such as producing a small ceramic laboratory melting crucible.

First, the service requirements will define the quality. These will be demonstrated by the published specifications, which may also be used in marketing. There is an implied warranty that any crucible we sell will meet these specifications, which might include such things as dimensions, strength, and thermal shock resistance. However, experience tells us that average specifications can be misleading and that limits or tolerances are necessary if we wish to market the crucible successfully.

A simplified sketch of our crucible is shown in Figure 23.6. We randomly select a large number of our product, measure the heights, and find that there is an equal probability that the height will be above or below the desired value of 75.0 mm. If we now plot these results as a *statistical distribution*, we obtain what is referred to as a *normal distribution* curve (Figure 23.7). It would appear that we could sell all of our crucibles within the limits of our

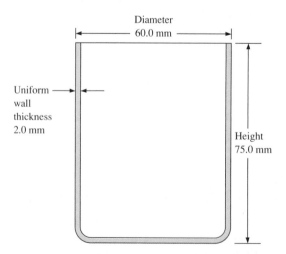

FIGURE 23.6   Sketch of the cross section of a small ceramic crucible

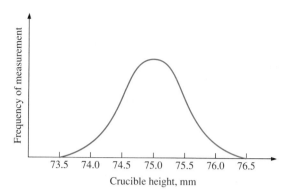

FIGURE 23.7   A normal distribution of crucible heights obtained by measuring a large number of the ceramic crucibles shown in Figure 23.6

data—that is, a specification of 75.0 mm plus or minus 1.5 mm. This might cause us some difficulty, however, because we have not determined whether we have a truly random sample and whether it is of adequate size.

It is not our purpose here to discuss sample size and selection but rather to point out that any crucible with a height not within the range of 73.5–76.5 mm should be scrapped because it does not meet our quality standard. Next week, we might find that all of our crucibles show a similar normal distribution but with a new mean and a different range. It is therefore necessary to treat quantitatively this deviation from the average.

One method is to determine the mean and then the deviation of each point from the mean. Because there will be plus and minus values, we will square each deviation, add them all up, divide by the total number of points, and then take the square root. This value can be treated mathematically as follows and is called the *standard deviation*, σ.

$$\sigma = \sqrt{\frac{1}{n} \sum_{i=1}^{i=n} (x_i - \overline{x})^2} \qquad (23\text{-}7)$$

where

$$n = \text{number of values}$$

$$x_i = \text{a particular value}$$

$$\overline{x} = \text{the arithmetic mean}$$

Most inexpensive hand-held calculators will readily solve for σ from a given set of data. What is more important here is the percentage of the values within the distribution that are included in a standard deviation.

$$\overline{X} \pm 1\sigma = 68.26\% \text{ of all values in the distribution} \qquad (23\text{-}8a)$$

$$\overline{X} \pm 2\sigma = 95.46\% \text{ of all values in the distribution} \qquad (23\text{-}8b)$$

$$\overline{X} \pm 3\sigma = 99.73\% \text{ of all values in the distribution} \qquad (23\text{-}8c)$$

It would appear, then, that our quality standard for the crucible is about $\overline{X} \pm 3\sigma$. Once the quality is defined, it must be controlled. We might be able to make a profit even if, over a period of time, 5% of our product fell outside of the range and had to be scrapped. On the other hand, our profit margin might be low or higher scrap rates might occur, and either of these outcomes calls for immediate action. (We might, for example, find ways to recycle or re-manufacture our product, although this is doubtful for crucibles.)

The second aspect of quality assurance is that we must recognize that we may pay a high penalty when we measure the crucible heights *as a final quality check only.* Many intermediate processing steps occur that could affect the crucible height. These include powder consolidation and sintering, both of which have easily measured variables. For example in sintering, the temperature, heating rates, time at temperature, and cooling rates all influence shrinkage and hence crucible height.

Each of these processing steps can also be treated statistically by developing their distribution curves and the three sigma limits. This is known as *statistical process control (SPC).* With all of the individual processes in control, our final product should be within specifications, because the same methods can be applied to requirements other than dimensions. Furthermore, we can use similar techniques to evaluate the quality of our incoming raw materials.

EXAMPLE 23.3 *[EJ]*

Assess the validity of the following statement: Our three sigma limits on a particular process X are never exceeded, so we should tighten up the limits.

Answer   We must first determine how the original three sigma limits were established. In other words, failure to draw a genuinely random sample may be part of the problem. Furthermore, we might be controlling our process X better because of modifications in processing steps prior to process X. An example would be rough-cut machining closer to tolerance such that finish grinding at station X shows less deviation from the average attributable to tool wear.

We certainly could add to the data base and thereby tighten the limits in response to the observation that the "new process" shows less deviation from the average. On the other hand, it is equally important to ask whether the tightened limits would result in a finished product that has a better service life.

A definitive answer to the question, then, is not possible. Remember that the objective is to establish a useful life for the component and to establish the QC and SPC that will be required to accomplish this goal.

Finally, it should be pointed out that many operations do not produce a normal distribution curve. Several examples of interval distributions are shown in Figure 23.8. The data are usually grouped into eight or more intervals so that proper characterization of the process can be obtained. The loca-

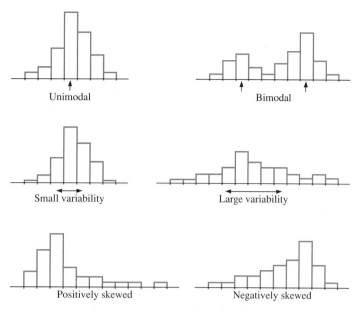

FIGURE 23.8   Several varieties of frequency distributions

tion of the intervals on the *x* axis is governed by a desire to maximize symmetry. Drawing a continuous curve through the centers of the intervals in the unimodal distribution in Figure 23.8 would result in a curve similar to that shown in Figure 23.7.

If we use examinations as an example of Figure 23.8, skewed distributions can result from very easy or very difficult exams. The variability range can be modified by the number of exam questions and the provision for partial credit on a given question. A bimodal distribution might be expected when the students have different backgrounds, as in an advanced mechanics course in which both mechanical and electrical engineers are enrolled.

In addition to helping us set specifications, probability data are very helpful in finding reasons for scrap. For example, in trying to reduce the number of rejects by a customer, a manufacturer plots graphs of number of rejects versus different variables, including month of the year and time of day when machined. The results are expressed quantitatively as correlation coefficients, which denote the probable significance of a given variable. (The methods are given in standard texts on statistics and in computer software support manuals.) In the case just described, the greater reject rate by the consumer during the winter months was caused by the expansion produced by the transformation of austenite to martensite during cold shipment (Chapter 7).

23.10   Help from the Computer

The computer can be very useful in the selection of materials. We will not discuss the use of computers in maintaining records of costs and quality control or in simple "number-crunching" tasks. Here we will emphasize the computer's contribution to assembling the necessary information for optimal material selection. For this we require data that may not be easy to find.

Once an application is defined it usually suggests certain classes of materials, although historical precedent is being broken every day. (Optical fibers for data transmission would never have been developed if we had limited our attention to metallic conductors only.) In any case, our first step is to find out what the "state of the art" is. This suggests a literature search that can best be conducted via a computer tied into one of the many literature abstract services.

It is important to maintain a broad perspective at the beginning of the search and to refine it as necessary later in the process. This is accomplished by selection of the keywords during the search. For example, *biomaterials* and *implants* might be early keywords in a search where the ultimate goal is specification of a replacement for a leg bone. Furthermore, textbooks and reference books are necessary for the background information, whereas monthly journals are more likely to contain recent developments. Remember that the objective is to find out as much as possible about the application and to determine what materials will be good candidates and what their properties are. (Steel yourself not to be totally inhibited by historical success stories.)

Applicable codes and standards should also be a part of the literature search. In some cases the computer-driven search will not be very rewarding and will have to be augmented by traditional literature abstract and title searches. This is especially true when we are interested in something that may have sources more than ten years old.

The constraints of the application must also be considered, and here the computer is again useful through what is broadly defined as *computer-aided design (CAD)*. In a simple example, we might merely change the section size of a component, recalculate the stress distribution, and see what impact the new geometry will have on mating parts. This *simulation* can be carried out as many times as necessary to obtain an optimal solution. The necessary data come from our literature search.

Remember, though, that the data base also has limitations that will carry through any CAD approach. Mechanical properties are averages and show a broader range when we include different processing. For example, nylon 6/6 may be purchased in cast or extruded forms. The extruded shapes exhibit anisotropy, and different vendors may also show variation in their tables of nominal properties.

Integration of the material characteristics, processing methods, and service environment is necessary to optimal material selection. Use of the computer can shorten the procedure by providing a data base and making simulation possible. *However, experimentation and prototyping are still required to ensure a successful product.*

Finally, the use of the computer can be expanded to include *computer-aided manufacturing (CAM)*, wherein the manufacturing process is computer-controlled. A more global approach to the use of computers in addressing all engineering problems is called *computer-aided engineering (CAE)*. Do not overlook the significance of the fact that all these acronyms include the term *aided*. Just as it is important to recognize when to use a computer, it is also important to know when it is not appropriate to use one.

23.11 Prototypes and Experimentation

When selecting a material for a given component, the engineer has two options in addressing issues of safety, production volume, and cost. The weaker choice is to decide only on the basis of existing production facilities and previously used materials.

The more satisfactory option (though it is one that can be exercised only when high-volume production is involved) is to select a group of materials, processing methods, and designs to make up pilot parts to test. The annual production of millions of engine crankshafts, for example, makes it possible to test a variety of designs, materials (ductile iron, forged steel, cast steel), and processes (forging, shell mold casting, green sand casting). And even though many producers recognize the advantages of ductile iron, some continue to produce forged shafts because doing so is cheaper, using fully depreciated equipment, than using ductile iron and having to purchase new casting equipment.

Several basic principles should be followed in prototype testing. Instead of selecting a narrow range of materials, the broadest range possible should be tested. For example, if we are going to use a ductile-iron crankshaft, we should not merely test structures with a small range of ferrite–pearlite mixtures but should test from 100% ferrite to 100% pearlite, even though the 100% ferrite might seem too soft. The dividend would be that the extreme cases would illustrate more clearly the interaction between crankshaft matrix and the bearing material and provide a better basis for setting the final specification. Also we are often surprised at the difference between the test results and our preconceived opinions!

A final word on testing: be cautious with accelerated tests or bench tests. In many cases, "simulated service" data list materials in an order of merit that is the reverse of actual service. The key to understanding anomalies of this type is to observe the difference in the response of the *structure* to the applied conditions.

23.12 Cost Analysis for a Component

The average engineer has an inadequate background in cost accounting, and the typical required economics course does not cover costs. Many good engineers have become embittered, after several years of work in developing a reliable, high-performance component such as a new gas turbine, to be told, "Sorry, it would just cost too much." In some cases the basis for the decision is inaccurate but the typical engineer is helpless to protest. In one example let us assume that a turbine has been made and tested, and the accounting team moves in and questions the engineer about the materials and processes to be used for the components. Some of the most important factors in the cost estimate are answers to the following questions:

What are the specifications for the material?
How stringent must they be for adequate safety?
The processing—what will be the reject rate?
The assembly—how do reasonable labor hours compare to those used in making the prototype?

If the input of the engineer is too conservative or the accountants are overly careful, it is easy to wind up with an estimate that is *twice* the production cost that can actually be achieved.

To appreciate the contribution of various factors to the cost (and financial success) of a product, the engineer must understand a few accounting terms and procedures.

Let us take a typical simple case to illustrate the various estimating pitfalls. In this project, 20 plants were faced with extinction because of an outdated cast-iron car wheel and the development by competitors of a more expensive but safer and longer-lasting *forged* steel car wheel (Figure 23.9). We believed as engineers that a wheel of the same pearlitic steel structure could

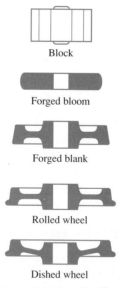

Block

Forged bloom

Forged blank

Rolled wheel

Dished wheel

FIGURE 23.9   The steps used in production of a forged railroad car wheel

[*ABC's of Car and Locomotive Wheels*, Simmons–Boardman Books, Omaha, Neb., p. 11. Reprinted with permission.]

be cast to shape at a lower cost than that of the forged wheel. We reasoned that the forged wheel had to begin as an ingot. Then slices had to be cut from the sound portions of the ingot and the centers punched out. Next, after reheating, the "cookie-shaped" rounds had to be rolled to shape with intricate sets of rolls. Finally, the tread and other portions required machining. To compete, a process was developed wherein a wheel was cast to shape in a graphite mold (Figure 23.10).

It should be emphasized that the costs we are about to give are merely estimates made some time ago. The purpose of the illustration is to show how estimates are developed and to cite different types of expense that must be considered in any operation.

Before calculating costs, however, we need to have an overall understanding of the process and of the process locations where costs are developed. To

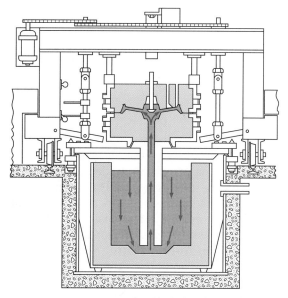

FIGURE 23.10   Pressure pouring a steel wheel in a graphite mold. Controlled pressure pouring is a patented process for the production of steel freight car and diesel locomotive wheels. The process involves forcing molten steel into graphite molds with air pressure. In steel wheel plants, a ladle of molten metal is placed in a tank. The tank is sealed hydraulically with a steel cover. This cover has a ceramic tube attached. Air forced into the sealed chamber pushes molten steel up through the tube into a graphite mold that has been automatically positioned over the pouring tube. The steel fills the mold to form the freight car or diesel locomotive wheel. The controlled rate at which steel is forced into the mold results in a product cast to close tolerances and a surface finish that eliminates nearly all surface conditioning and machining.

[Griffin Wheel Company, Chicago, Ill.]

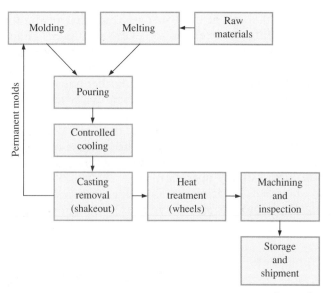

FIGURE 23.11   A flowchart linking the major operations performed in a railroad car wheel casting plant

this end, Figure 23.11 gives an overall description of the plant operation, itemized as follows.

Molding. The process begins with the assembling of semipermanent graphite molds with a cavity of the desired shape. The labor for pouring, shakeout, and mold preparation is included in this operation.

Melting. Steel of the desired analysis is melted in electric-arc furnaces, using a cold charge made up of pig iron, steel scrap and risers, and defective castings.

Pouring. The molds are clamped and poured at a controlled rate.

Shakeout. The wheels are extracted from the molds after a fixed cooling period and taken to hub cutting. The molds are cooled and conditioned for pouring again. The bore of the wheel is cut out to the desired diameter by means of an oxygen lance.

Heat Treatment. The wheels are heat-treated to attain the desired hardness and stress distribution via the methods described in Chapter 6.

Machining, Inspection, and Shipment. The wheels are cleaned. The hub bore is then machined, inspected using Magnaglo (fluorescent magnetic particles), and shipped.

Let us examine the costs summarized in Table 23.2 and discuss how they are determined in each department or category.

1. *Melting* The labor costs for charging and handling the hot metal in the electric furnaces are obtained by dividing the daily cost by the tonnage (38 tons of good wheels).

TABLE 23.2 Summary of Wheel Costs (Pilot plant, a hypothetical case) for Production of 38 Net Tons of Good Wheels per Day

	Cost per Net Ton of Good Wheels
1. *Melting* (exclusive of metal cost in good wheels)	
Labor: Furnace charging, melters and helpers, slagman, lab testing, cleanup, ladle lining	$ 52.
Materials: Electrodes, refractories, power	160.
2. *Metal*	
Scrap steel, pig iron, alloys, (5% scrap wheels not included)	150.
3. *Molding*	
Labor: Repair molds, apply wash, apply sand for risers, close molds, pour, separate mold, shakeout	44.
Materials: Graphite for molds, mold reconditioning	76.
4. *Heat treatment*	
Labor: Place wheels in pits, remove to cooling area, control furnaces	10.
Material: Fuel, lining repair	20.
5. *Cleaning, machining hubs*	
Labor: Grind and machine, inspect	40.
Material: Grinding wheels, tools	6.
6. *Maintenance*	
Labor: General maintenance	60.
Material: General supplies	12.
7. *General reserve*	
Labor: Unassigned plant labor and supervision	152.
Materials: General equipment and supplies	80.
8. *Reserve*	
Fixed: Depreciation, taxes, insurance	200.
Variable: Social Security, etc.	100.
9. *Administrative, sales, and research expenses*	70.
	$1232.

The power cost is somewhat complex in that it is made up of a demand factor and a use factor. We are all familiar with paying cents per kilowatt hour for the electricity we use in our homes. In every industrial plant, however, the power company charges a demand factor that is assessed not per kilowatt but per kilowatt-hour. A meter attached to the power line registers the kilowatts being drawn by the plant at any time. The *highest* reading for any 15-min period is the basis for the demand charge for the month.

2. *Metal* The conversion from cost per ton of charge to cost per net ton of good wheels requires explanation. Two questions arise: (1) What happens to the risers that are cut off the wheels and to scrap (defective) wheels? (2) What happens to the iron units that are lost by oxidation and in the slag?

The answer to the first question is that the simplest way to account for scrap is to visualize a *constant* amount of it in process in the plant. We simply put in new material and take out the same tonnage of good wheels (less oxidation loss). Therefore, although the actual furnace charge contains scrap, we do not place a value on it (because we regard it as staying in the plant for accounting purposes).

The answer to the second question is that the oxidation loss is a real cost and is estimated at 3% of the charge. A number of factors contribute to this oxidation loss: iron in the slag, iron oxidized during heat treatment and hub cutting, and so on. Therefore the cost of metal per ton of good wheels is 3% higher than the cost of the charge.

3. *Molding* The labor charges are labor hours times hourly rate per net ton of good wheels (NTGW).

The principal material charge is the cost of the semipermanent graphite mold. It is easier to calculate this as cost per wheel and then to convert to cost per net ton of good wheels.

To arrive at this charge, we begin with the cost of the graphite blocks needed to form the cope and drag of the mold — approximately $8000. The machining labor is added here because it is not a direct labor cost in daily production. After 100 wheels have been poured, the mold is not discarded but is remachined with a form tool in the contour of the wheel surface. The refractory costs for the stopper seat, plunger, pouring sleeve, and riser sand are included as materials.

4. *Heat treatment* The labor involved in heat treatment consists of loading and unloading wheels and checking furnace temperatures. The only significant material cost is the fuel.

5. *Cleaning, machining* The labor consists of moving wheels from the cleaning stations, chipping, grinding, machining, inspecting, and loading into cars or to inventory. Costs for materials include the cost of grinding wheels, machine tools, and the like.

6. *Maintenance* This account covers the maintenance of all equipment in the different departments. It is not subdivided unless a detailed cost analysis is needed.

7. *General reserve* This account covers labor and services that are difficult to allocate to any one department.

8. *Reserve* Depreciation expense is one of the important, yet controversial, elements of cost. To illustrate this let us consider the electric arc furnaces used for melting, and assume that a furnace costs $500,000 installed. Furthermore we know from previous experience that with good maintenance the furnace will last 15 years and will have negligible scrap value at that time. Even in this straightforward case there are a number of procedures for calculating the cost per year, but we will use the simple averaging method. The important point is that we must consider this a fixed cost, regardless of increases or decreases in production. We might argue that lower production would cause less wear and tear on the furnace. Nevertheless its

value decreases, because newer designs will make the furnace obsolete. Another subtle point is that the money for depreciation expense is put in a reserve fund, not paid out as it is for labor or materials. For example, even though the company has already paid $500,000 for the furnace, a cost of $33,333 per year must be charged as an expense for 15 years and accumulated to replace the furnace. Molding and pouring equipment, heat-treatment equipment, and the building structures must be similarly depreciated. Decisions about inflation and whether to replace or repair should also be included in the depreciation accounting.

The variable reserve fund is used to take care of such expenses as pension changes with number of employees, hours worked, and federal and state regulations governing unemployment insurance and retirement pensions.

9. *Administrative, Sales, and Research Expenses* It is vital to recognize the importance and cost of marketing, research, and general administration. Because these costs may be substantial in a large corporation, proper allocation of the burden among different plants is often the subject of considerable debate. It can be argued, for example, that a minor fraction should be assigned to a new plant because these expenses are fixed and are already being absorbed by existing plants. And by the same token, when a high percentage of effort in the research laboratory has been directed toward the development of a highly profitable new product, it is only fair to allocate a high percentage of this overhead to the new production facility.

The main point of this discussion of costs and projected costs is that the numbers are not fixed but have to be arrived at through flexible thinking and sound engineering input. The engineer must realize that items such as direct labor and materials make up only a portion of the total expense.

EXAMPLE 23.4 [EJ]

Someone in the car wheel plant has suggested that the costs for metal can be reduced 10% by purchasing in larger quantities and by changing vendors. List some concerns that the company should address before implementing this idea.

Answer Table 23.2 indicates that a 10% metal cost savings ($15) translates to more than a 1% cost saving in NTGW ($1232 total cost). There are other considerations, however.

1. Are we dissatisfied with our present vendor and are the proposed vendors reliable?
2. If we buy in larger quantities, is there sufficient plant storage space for the raw materials?
3. Does taking advantage of the proposed price necessitate a long-term commitment? If so, does the industry trend in metal prices suggest that such an arrangement will be advantageous?

4. Can the new vendors meet our specifications for incoming materials? The answer to this question may, in fact, be a key to our decision. It is important to recognize that the quality of input material may influence the quality of the final product.

For example, consider steel scrap that is very dirty with excessive rust, oil, paint, or other surface contaminants. If we do not modify our melting operation to accommodate the characteristics of the new input material, our control of the car wheel quality could suffer. And if we therefore modify the melting process, the cost savings in raw materials could be offset by increased costs for melting that could range from labor requirements to refractory costs and difficulty in pollution control.

Seemingly simple cost decisions can influence many aspects of the total operation, and the engineer can play an important role in making these decisions.

23.13 The Recycling and Reuse of Materials

You may wonder why we are introducing the topics of recycling and reuse at this point. Why have we waited so long, and what do recycling and reuse have to do with materials selection? Actually we have not ignored these topics but have interwoven them into both the text and the problems.

More importantly, we have noted that the engineering approach to design requires an optimal solution. Everything may ultimately fail, so the *lifecycle* of a component is important as are economic considerations and the anticipation of abuse and safety hazards.

However, recycling or reuse may not always figure into the lifecycle of an engineering component. We may remanufacture large pieces of equipment because they are too costly to replace, or we may generate scrap (as in the car wheel example) that is recycled because doing so is economical. The common thread is economics.

Recycling requires that we adopt a perspective broader than that of short-term economics. For example, the simple and "economical" discarding of unwanted material has resulted in a shortage in landfill space. Energy costs have also increased because of resource depletion and because of competition between those who process and market oil as a fuel and those who use it as a raw material for many polymer products.

The objective here is not to identify all of the factors that promote or inhibit recycling; these can range from politics to tax law. Recycling should, however, become a more active component in the optimal solutions that engineers are asked to come up with. Finding those solutions may not, of course, be easy.

Consider, for example, the wider use of fiber composites in automobile bodies. The payoff of using fiber composites as a substitute for steel is difficult to ignore. The corrosion resistance is better, and the strength-to-weight ratios provide a lighter vehicle with higher fuel economy. Both of these advantages offer the potential for higher consumer satisfaction. However, multiply the anticipated use over a number of years and ask what we as a society will do with millions of fenders that may be too stable for their own good. The recycling, reuse, and remanufacture of fiber-reinforced polymers may become an important environmental issue.

In short, engineers must look beyond short-term considerations and be sensitive to these secondary impacts on our quality of life.

EXAMPLE 23.5

Figure 23.12 shows six cups, each made of a different material. They will all hold fluids, but beyond that, all were developed for different purposes. For each cup, indicate the primary advantage of the material chosen, and assess its potential for recycling, remanufacture, or reuse.

Answer The wooden cup was probably developed for esthetic reasons. It exhibits a crack due to moisture and anisotropy sensitivity. In recycling, it would decay naturally or could be burned for its fuel value. Note also that wood is a renewable resource.

Both the china cup and the champagne glass (ceramics) were developed for their utility and attractiveness and should enjoy a long life until broken. When discarded, the china pieces could last for many years; the firing process is more or less irreversible. The champagne glass could be remelted and formed into a new shape.

As for the metals, the silver cup has not only esthetic value but also intrinsic value due to its silver content. It will last for a long time and its recycling is guaranteed. The steel backpacker's cup was developed for its durability and should last indefinitely unless it is abused. It could be remelted along with other steel scrap. If discarded, it will corrode in a landfill over an extended period.

The cup (polymer) is made of Styrofoam and is intended for a single use. It can be manufactured very cheaply and is an excellent insulator to keep fluids hot or cold. It does not readily degrade when discarded, so the beads may be identified for many years. If Styrofoam is burned for its fuel value, toxic fumes may be produced.

Summary

The performance of a component depends on the service conditions, the material, and the combination of design and process. Failure to take into account any one of the three can lead to failure.

FIGURE 23.12   Cups produced from several engineering materials. Top row — wood goblet (arrow points to a crack). Middle row — china tea cup and silver cup. Bottom row — champagne glass, steel backpacker's cup, and Styrofoam cup.

A number of additional special factors influence the success of a component.

Secondary processing (such as machining, welding, or heat treatment) can affect structure, properties, and residual stresses.

The proper use of *strength/density and modulus/density* ratios can optimize the choice of materials and the design of cross sections.

Understanding and using *specifications* such as those published by ASTM play a major role in obtaining the desired material and structure in a component.

Quality control gives assurance that the desired performance will be obtained from a given percentage of parts and provides a method for controlling the percentage of failures.

Prototype production and testing is a valuable tool for obtaining optimal performance, because it yields information well beyond that which routine tests of mechanical properties supply. It is more important to analyze the mode of failure of prototypes than the successes!

Cost accounting is an important tool for the engineer, who must achieve the financially responsible combination of materials and processing for a given component. Considerable imagination and courage are needed to provide proper estimates.

The computer can play an important role in the selection of materials. It can facilitate data searches, design, and manufacturing.

Finally, consideration of the life cycle of a component should, whenever possible, include recycling as an important—though often neglected—part of the decision process of correctly matching an engineering material to the required service performance.

Definitions

Most of the terms used in this chapter have been defined in earlier portions of the text. See the index for the relevant page numbers. Supplementary definitions are given below.

Computer-aided engineering (CAE)   Use of the computer to aid in the decision making and execution associated with engineering processes; includes *computer-aided design (CAD)* and *computer aided-manufacturing (CAM)*.

Material specifications   Standards designating the properties and geometry that a material must exhibit in order to meet a specific service condition.

Optimal stiffness   The relationship between elastic modulus and material density that achieves maximum stiffness at minimum weight.

Optimal strength   The relationship between strength (normally yield strength) and material density that achieves maximum strength at minimum weight.

Primary material processes   Casting, deformation, and powder agglomeration.

Recycling   The reuse or remanufacture of engineering materials.

Secondary material processes   Processes such as joining, machining, thermal treatment, and finishing.

Standard deviation   A mathematical method used to determine the deviation of points from the arithmetic mean.

Statistical distribution   A plot that relates the variation in a measured parameter to its frequency of occurrence. Such a distribution may be normal (in which case the plot is a bell curve), it may be skewed with one peak, or it may exhibit several peaks (in which case it is called multimodal).

Problems

Chapter 23 summarizes a number of principles covered more fully in earlier chapters. It may be necessary to consult portions of the text other than those sections noted within each problem.

23.1   *[ES/EJ]* Indicate whether the following statements are correct or incorrect, and justify your answer. (Sections 23.1 through 23.5)
a. Components must be uniformly sound and have the same structure throughout.
b. Ceramic materials are not widely used because of their inherent brittleness.
c. Polymers have higher strength-to-weight ratios than metals.

23.2 *[EJ]* Review the three-legged stool analogy of performance and the major factors that can lead to unsatisfactory performance. Indicate how each of the three "legs" can contribute to the following kinds of failure. (Sections 23.1 through 23.5)
a. A prototype fails under simulated service conditions.
b. A small fraction (0.1%) of a component fails after extended service.

23.3 *[EJ]* Comment on the statement that properties after long-term storage may be an important consideration in the design and process analysis of a component. (Sections 23.1 through 23.5)

23.4 *[ES/EJ]* Give reasons for the current interest in primary processing to near-net shape — that is, in minimizing secondary processing. (Sections 23.1 through 23.5)

23.5 *[ES/EJ]* Table 23.1 summarizes a great deal of information. Indicate how the information represented by each column heading might be used in the material selection process, that is, why the information is important. (Sections 23.1 through 23.5)

23.6 *[EJ]* Deformation processing in Table 23.1 refers to sheet materials. Using metals as an example, indicate what other information about deformation processing would be significant to the material specification process. (Sections 23.1 through 23.5)

23.7 *[ES]* In Section 23.6, there appears a mathematical notation wherein σ_y (true stress) rather than S_y (engineering stress) is used for the yield strength. Explain why this should make little difference in the outcome of the strength-to-weight calculations we make when comparing materials. (Sections 23.6 through 23.9)

23.8 *[ES/EJ]* Indicate why wood was an excellent material to select in early aircraft design. (Sections 23.6 through 23.9 and 17.9 through 17.13)

23.9 *[EJ]* Using nylon as an example, explain why steel inserts may also be necessary in a component such as an interior door handle of an automobile. Note that new vehicles have thin doors and that the handles do not protrude into the passenger compartment. (Sections 23.6 through 23.9)

23.10 *[ES]* How would you calculate the density of a composite material in order to determine the strength-to-weight ratio? (Sections 23.6 through 23.9)

23.11 *[ES/EJ]* Indicate how each of the following individuals might use the ASTM specification for ductile-iron castings. (Sections 23.6 through 23.9)
a. Designer
b. Quality control supervisor in the casting plant
c. Purchaser of castings for secondary processing
d. Field service engineer for the final product

23.12 *[ES/EJ]* Figure 23.5 shows several types of graphite shapes that might occur in ductile cast iron. Assume that the matrix does not change. What might happen to the mechanical properties when considering a change from type I to type V. Do types IV and V offer any advantage over gray cast iron? (Sections 23.6 through 23.9)

23.13 *[EJ]* The automobile is designed to meet certain safety requirements that add to the cost. For example, survival of a federally mandated crash test is required in new vehicles. As an engineer, consider what might happen if the vehicle were subjected to these tests after it has undergone several years of service. In what way would the geographical location of the service play a role? (Sections 23.6 through 23.9)

23.14 *[ES]* In the manufacture of a particular crankshaft, the following machining specifications are required:
Rough turning 2.508 ± 0.002 in.
Finish grind $2.5000 + 0.0000 - 0.0015$ in.
What is the meaning of these dimensions, and what would be the effect on a statistical frequency distribution? (Sections 23.6 through 23.9)

23.15 *[ES]* Given below is an ordered distribution of spot weld strengths in shear. (Sections 23.6 through 23.9)

Pounds to Cause Shear Failure*	Frequency of Occurrence
161.5	5
157.5	12
153.5	17
149.5	35
145.5	19
141.5	4
137.5	2
	Total 94

*Midinterval values

a. Determine the arithmetic mean and plot the interval distribution.
b. Determine the average deviation (AD) from the mean.

$$AD = \frac{1}{n} \sum_{i=1}^{i=n} |X_i - \overline{X}| \qquad (23\text{-}9)$$

c. Determine the standard deviation.
d. What is the significance of the two calculated values for the deviation?

23.16 *[ES/EJ]* Refer to the figure in Problem 6.34 that treats yield strength for test bars cut from sand-cast AZ91C magnesium in the T6 condition. (Sections 23.6 through 23.9)
a. Determine the arithmetic mean, the average deviation from the mean (see Problem 23.15b), and the standard deviation.
b. How would these data be incorporated into the material specifications for the alloy?

23.17 *[EJ]* A simplified model for a computer-controlled process follows. (Sections 23.10 through 23.13)

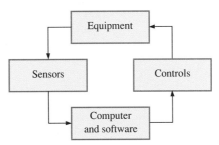

a. Indicate briefly how each of the components of the model contributes to the process.
b. What cost considerations would arise in deciding whether to retrofit an old piece of equipment or purchase new, computer-controlled equipment? Compare table-top equipment with a large die press.

23.18 *[EJ]* Indicate how computers might be used in the prototyping and experimentation discussed in the text. (Sections 23.10 through 23.13)

23.19 *[EJ]* In the discussion of the manufacture of railroad car wheels, it is observed that the bore is cut out to the desired diameter, by means of an oxygen lance, right after shakeout. Explain how this might be accomplished, including the source of heat to cut the hole. (Sections 23.10 through 23.13)

23.20   *[ES/EJ]* It has been suggested that we increase the production of our railroad car wheel plant by 10% because spreading the costs out over more wheels will reduce the final cost per wheel. Indicate why this argument is overly simplistic. (Sections 23.10 through 23.13)

23.21   *[ES/EJ]* Historically, casting and forging have been alternatives for producing a given component. List a few of the advantages and disadvantages of combining the two processes to optimize the cost and performance. (Sections 23.10 through 23.13)

23.22   *[ES/EJ]* Indicate why the cost of a hipbone replacement might be more like that of a turbine bucket blade than like that of an automotive wheel axle. (Sections 23.10 through 23.13)

23.23   *[EJ]* Discuss the potential impact of each of the following on the materials selection process. For example, what is the effect on the life cycle and the cost associated with the selection? (Sections 23.10 through 23.13)

a. Requiring deposits on all beverage containers

b. A government proposal that a $2 surcharge per tire be imposed as a recycling subsidy

c. The chemical industry's discovery of an economical way to convert Styrofoam containers into insulating boards

d. The disappearance of aluminum siding from abandoned houses

References

Without trying to be comprehensive, we list here a few references for the convenience of students and instructors. References are divided into four groups: (1) metals, (2) ceramics, (3) polymers, and (4) general. In each group, the first references are those that can be used readily by students who have completed this course, in order to find specifications for materials in their professional work. These are followed by a list of texts for study beyond the scope of this book.

Metals

American Society of Testing Materials Specifications (*see under* General)

Metals Handbook, 8th ed., American Society for Metals, Metals Park, Ohio. This is not one handbook, but eleven large volumes, beginning with volume 1, *Properties and Selection of Metals*, and advancing through volume 11 on topics such as heat treatment, casting, welding, phase diagrams, microstructures, failure analysis, and nondestructive testing. An authoritative source, written under the supervision of committees of technical experts. The 9th edition presently has 17 volumes treating many topics not covered in previous editions. A single-volume desk edition also exists (1984) and is supported by two floppy disks.

Metals Reference Book, 4th ed., C.J. Smithells, Plenum Publishing Corp., New York, 1967. Three volumes; has many tables of properties: diffusivity, thermochemical data, etc.

The Structure of Metals, 3rd ed., C.S. Barrett and T.B. Massalski, McGraw-Hill, New York, 1966. This text features: (1) a thorough discussion of crystallographic methods such as x-ray diffraction techniques, and (2) excellent discussions of metallic structures and transformation mechanisms. Much of the material is at the senior or graduate level.

Textbooks and Journals on Metallurgy

Atomic Theory for Students of Metallurgy, W. Hume-Rothery, Institute of Metals, London, 1955.

Elements of Mechanical Metallurgy, W.J.M. Tegart, Macmillan, New York, 1966.

Elements of Physical Metallurgy, 3rd ed., A.G. Guy and J.J. Hren, Addison-Wesley, Reading, Mass., 1974.

Elements of X-Ray Diffraction, B.D. Cullity, Addison-Wesley, Reading, Mass., 1956.

High Strength Materials, V. F. Zackay, Wiley, New York, 1965.

Journal of Materials Education, Pennsylvania State Univ., University Park, PA.

Mechanical Metallurgy, 2nd ed., G. E. Dieter, McGraw-Hill, New York, 1976.

The Mechanical Properties of Matter, A. H. Cottrell, Wiley, New York, 1964.

Modern Physical Metallurgy, 4th ed., R. E. Smallman, Butterworths, Boston, 1985.

Phase Diagrams in Metallurgy, F. N. Rhines, McGraw-Hill, New York, 1956.

Physical Metallurgy, E. Birchenall, McGraw-Hill, New York, 1959.

Physical Metallurgy, Bruce Chalmers, Wiley, New York, 1959.

Physical Metallurgy, 2nd ed., P. Haasen (trans. J. Mordike), Cambridge Univ. Press, New York, 1986. University Press, Cambridge, MA.

Physical Metallurgy Principles, 2nd ed., R. E. Reed-Hill, Van Nostrand Reinhold, New York, 1973.

The Physical Metallurgy of Steels, W. C. Leslie, McGraw-Hill, New York, 1981.

Science of Metals, N. H. Richman, Blaisdell, Waltham, Mass., 1967.

Structure and Properties of Alloys, 4th ed., R. M. Brick, A. W. Pense, and R. B. Gordon, McGraw-Hill, New York, 1977.

Ceramics

American Society of Testing Materials Specifications (*see under* General)

Concrete-Making Materials, S. Popovics, Hemisphere Publishing Corp., McGraw-Hill, New York, 1979.

Design and Control of Concrete Mixtures, 11th ed., Portland Cement Association, Skokie, Ill., 1968.

Phase Diagrams for Ceramics, E. M. Levin, C. R. Robbins, and H. F. McMurdie, American Ceramic Society, Columbus, Ohio, 1964.

Textbooks on Ceramics

Electronic Ceramics, E. C. Henry, Doubleday, Garden City, N.Y., 1969.

Elements of Ceramics, 2nd ed., F. H. Norton, Addison-Wesley, Reading, Mass., 1974. Excellent discussions of technique, as well as basic material

Glass-Ceramics, P. W. McMillan, Academic, London, 1964.

Glass Science, R. H. Doremus, Wiley, New York, 1973.

Introduction to Ceramics, 2nd ed., W. D. Kingery, H. K. Bowen, and D. R. Uhlmann, Wiley, New York, 1976. General coverage of ceramic structures and properties

Physical Ceramics for Engineers, L. H. Van Vlack, Addison-Wesley, Reading, Mass., 1964. Basic coverage with many illustrative problems

Polymers

American Society for Testing Materials Specifications (*see under* General)

"Materials Selector Guide," *Materials and Methods*, Reinhold, Stamford, Conn., published yearly.

Modern Plastics Encyclopedia, McGraw-Hill, New York, published yearly.

Textbooks on Polymers

Fundamental Principles of Polymeric Materials for Practicing Engineers, S. L. Rosen, Wiley, New York, 1982.

Handbook of Plastics and Elastomers, C. A. Harper, ed., McGraw-Hill, New York, 1975.

Mechanical Properties of Polymers, L. E. Nielsen, Van Nostrand Reinhold, New York, 1962.

Organic Polymers, T. Alfrey and E. F. Gurnee, Prentice-Hall, Englewood Cliffs, N.J., 1967.

Plastics, 6th ed., J. H. DuBois and F. W. John, Van Nostrand Reinhold, New York, 1981.

Principles of Polymer Engineering, N. G. McCrum, C. P. Buckley and C. B. Bucknall, Oxford Univ. Press, New York, 1988.

Properties and Structure of Polymers, A. V. Tobolsky, Wiley, New York, 1960.

Textbook of Polymer Science, 2nd ed., F. W. Billmeyer, Wiley, New York, 1971.

General

American Society for Testing Materials Specifications, American Society for Testing Materials, Philadelphia. A number of separate volumes issued triennially that cover metals, ceramics, and polymers; the most commonly used standard for specification.

Crystal Structures, R. W. G. Wykoff, Wiley, New York, 1963. Several volumes giving details of crystal structures

Materials Handbook, G. S. Brady, McGraw-Hill, New York, 1951.

"Materials Selector Guide," *Materials and Methods*, Reinhold, Stamford, Conn., published yearly.

Textbooks on Materials in General

Corrosion Control, 2nd ed., H. H. Uhlig, Wiley, New York, 1971.

Corrosion Engineering, 2nd ed., M. B. Fontana and N. D. Green, McGraw-Hill, New York, 1978.

Diffusion in Solids, P. G. Shewmon, McGraw-Hill, New York, 1963.

Electronic and Magnetic Properties of Materials, A. Nussbaum, Prentice-Hall, Englewood Cliffs, N.J., 1967.

Electronic Processes in Materials, L. V. Azaroff and J. J. Brophy, McGraw-Hill, New York, 1963.

An Introduction to Materials Science and Engineering, K. M. Ralls, T. H. Courtney, and J. Wulff, Wiley, New York, 1976.

Introduction to Materials Science for Engineers, 2nd ed., J. F. Shackelford, Macmillan, New York, 1988.

Introduction to Properties of Materials, D. Rosenthal, D. Van Nostrand, Princeton, N.J., 1964.

Introduction to Solids, S. V. Azaroff, McGraw-Hill, New York, 1960.

Materials Science and Engineering, W. D. Callister, Wiley, New York, 1985.

Materials Science in Engineering, 3rd ed., C. A. Keyser, Merrill, Columbus, Ohio, 1980.

Mechanical Behavior of Engineering Materials, J. Marin, Prentice-Hall, Englewood Cliffs, N.J., 1962.

Modern Composite Materials, L. J. Broutman and R. H. Krock, eds., Addison-Wesley, Reading, Mass., 1967.

The Nature and Properties of Engineering Materials, 2nd ed., Z. D. Jastrzebski, Wiley, New York, 1977.

Physics of Solids, 2nd ed., C. A. Wert and R. M. Thomson, McGraw-Hill, New York, 1970.

Principles of Engineering Materials, C. R. Barrett, W. D. Nix, and A. S. Tetelman, Prentice-Hall, Englewood Cliffs, N.J., 1973.

Principles of Materials Science and Engineering, W. F. Smith, McGraw-Hill, New York, 1986.

Science and Engineers of Materials, 2nd ed., D. R. Askeland, PWS Kent, Boston, 1989.

The Science of Engineering Materials, 2nd ed., C. O. Smith, Prentice-Hall, Englewood Cliffs, N.J., 1977.

The Structure and Properties of Materials, A. T. Di Benedetto, McGraw-Hill, New York, 1967.

The Structure and Properties of Materials, 4 vols., J. Wulff et al., Wiley, New York, 1965.

Index

Boldface numbers indicate pages on which terms are defined.

PHYSICAL PROPERTIES OF SELECTED ELEMENTS

Element	Symbol	Atomic Number	Atomic Weight	MP (°C)	Density (g/cm^3)	Crystal Structure	Atomic Radius (Å)	Ionic Radius (Å)	Most Common Valence
Aluminum	Al	13	26.98	660	2.699	FCC	1.43	0.57	+3
Argon	A	18	39.99	−189	1.78×10^{-3}	FCC	1.92	—	—
Barium	Ba	56	137.36	714	3.5	BCC	2.17	1.43	+2
Beryllium	Be	4	9.01	1277	1.85	HCP	1.14	0.54	+2
Boron	B	5	10.82	2030	2.34	Ortho.	0.97	0.2	+3
Bromine	Br	35	79.92	−7.2	3.12	Ortho.	1.19	1.96	−1
Cadmium	Cd	48	112.41	321	8.65	HCP	1.50	1.03	+2
Calcium	Ca	20	40.08	838	1.55	FCC	1.97	1.06	+2
Carbon[1]	C	6	12.01	3727	2.25	Hex.	0.71	<0.20	+4
Cerium	Ce	58	140.13	804	6.77	HCP	1.82	1.18	+3
Cesium	Cs	55	132.91	28.7	1.90	BCC	2.65	1.65	+1
Chlorine	Cl	17	35.46	−101	3.21×10^{-3}	Ortho.	1.07	1.81	−1
Chromium	Cr	24	52.01	1875	7.19	BCC	1.25	0.64	+3
Cobalt	Co	27	58.94	1495	8.85	HCP	1.25	0.82	+2
Copper	Cu	29	63.54	1083	8.96	FCC	1.28	0.96	+1
Fluorine	F	9	19.00	−220	1.70×10^{-3}	—	—	1.33	−1
Germanium	Ge	32	72.60	937	5.32	Dia.	1.22	0.44	+4
Gold	Au	79	197.00	1063	19.32	FCC	1.44	1.37	+1
Helium	He	2	4.00	−270	0.18×10^{-3}	HCP	1.79	—	—
Hydrogen	H	1	1.01	−259	0.09×10^{-3}	HCP	0.46	1.54	−1
Iodine	I	53	126.91	114	4.94	Ortho.	1.36	2.20	−1
Iron	Fe	26	55.85	1536	7.87	BCC	1.24	0.87	+2
Lead	Pb	82	207.21	327	11.36	FCC	1.75	1.32	+2
Lithium	Li	3	6.94	180	0.534	BCC	1.52	0.78	+1
Magnesium	Mg	12	24.32	650	1.74	HCP	1.60	0.78	+2
Manganese	Mn	25	54.94	1245	7.43	Cubic	1.12	0.91	+2
Mercury	Hg	80	200.61	−38.4	13.55	Rhomb.	1.50	1.12	+2
Molybdenum	Mo	42	95.95	2610	10.22	BCC	1.36	0.68	+4
Neon	Ne	10	20.18	−249	0.90×10^{-3}	FCC	1.60	—	—
Nickel	Ni	28	58.71	1453	8.90	FCC	1.25	0.78	+2
Niobium	Nb	41	92.91	2468	8.57	BCC	1.43	0.74	+4
Nitrogen	N	7	14.01	−210	1.25×10^{-3}	Cubic	0.71	0.1 to 0.2	+5
Oxygen	O	8	16.00	−219	1.43×10^{-3}	Ortho.	0.60	1.32	−2
Phosphorus[2]	P	15	30.98	44.3	1.83	Ortho.	1.09	0.3 to 0.4	+5
Platinum	Pt	78	195.09	1769	21.45	FCC	1.38	0.52	+2
Potassium	K	19	39.10	63.7	0.86	BCC	2.31	1.33	+1
Scandium	Sc	21	44.96	1539	2.99	FCC	1.60	0.83	+2
Silicon	Si	14	28.09	1410	2.33	Dia.	1.17	0.39	+4
Silver	Ag	47	107.88	961	10.49	FCC	1.44	1.13	+1
Sodium	Na	11	22.99	97.8	0.971	BCC	1.86	0.98	+1
Strontium	Sr	38	87.63	768	2.60	FCC	2.15	1.27	+2
Sulfur[3]	S	16	32.07	119	2.07	Ortho.	1.06	1.74	−2
Tin	Sn	50	118.70	232	7.30	Tetra.	—	0.74	+4
Titanium	Ti	22	47.90	1668	4.51	HCP	1.47	0.64	+4
Tungsten	W	74	183.86	3410	19.3	BCC	1.37	0.68	+4
Uranium	U	92	238.07	1132	19.07	Ortho.	1.38	1.05	+4
Vanadium	V	23	50.95	1900	6.1	BCC	1.32	0.61	+4
Zinc	Zn	30	65.38	419	7.13	HCP	1.33	0.83	+2
Zirconium	Zr	40	91.22	1852	6.49	HCP	1.58	0.87	+4

[1] Present as graphite—sublimes rather than melts.
[2] White phosphorus.
[3] Yellow sulfur.